CHEMICAL ENGINEERING THERMODYNAMICS

The Study of Energy, Entropy, and Equilibrium

PRENTICE-HALL INTERNATIONAL SERIES
IN THE PHYSICAL AND CHEMICAL ENGINEERING SCIENCES

NEAL R. AMUNDSON, EDITOR, *University of Minnesota*

AMUNDSON *Mathematical Methods in Chemical Engineering: Matrices and Their Application*
ARIS *Elementary Chemical Reactor Analysis*
ARIS *Introduction to the Analysis of Chemical Reactors*
ARIS *Vectors, Tensors, and the Basic Equations of Fluid Mechanics*
BALZHISER, SAMUELS, AND ELIASSEN *Chemical Engineering Thermodynamics*
BERAN AND PARRENT *Theory of Partial Coherence*
BOUDART *Kinetics of Chemical Processes*
BRIAN *Staged Cascades in Chemical Processes*
CROWE ET AL. *Chemical Plant Simulation*
DOUGLAS *Process Dynamics and Control, Volume 1, Analysis of Dynamic Systems*
DOUGLAS *Process Dynamics and Control, Volume 2, Control System Synthesis*
FREDRICKSON *Principles and Applications of Rheology*
HAPPEL AND BRENNER *Low Reynolds Number Hydrodynamics: With Special Applications to Particulate Media*
HIMMELBLAU *Basic Principles and Calculation in Chemical Engineering, 2nd Edition*
HOLLAND *Multicomponent Distillation*
HOLLAND *Unsteady State Processes with Applications in Multicomponent Distillation*
KOPPEL *Introduction to Control Theory with Applications to Process Control*
LEVICH *Physicochemical Hydrodynamics*
MEISSNER *Processes and Systems in Industrial Chemistry*
PERLMUTTER *Stability of Chemical Reactors*
PETERSEN *Chemical Reactor Analysis*
PRAUSNITZ *Molecular Thermodynamics of Fluid-Phase Equilibria*
PRAUSNITZ AND CHUEH *Computer Calculations for High-Pressure Vapor-Liquid Equilibria*
PRAUSNITZ, ECKERT, ORYE, O'CONNELL *Computer Calculations for Multicomponent Vapor-Liquid Equilibria*
WHITAKER *Introduction to Fluid Mechanics*
WILDE *Optimum Seeking Methods*
WILLIAMS *Polymer Science and Engineering*
WU AND OHMURA *The Theory of Scattering*

PRENTICE-HALL, INC.
PRENTICE-HALL INTERNATIONAL, INC., UNITED KINGDOM AND EIRE
PRENTICE-HALL OF CANADA, LTD., CANADA

CHEMICAL ENGINEERING THERMODYNAMICS

The Study of Energy, Entropy, and Equilibrium

Richard E. Balzhiser
University of Michigan

Michael R. Samuels
John D. Eliassen
University of Delaware

P T R PRENTICE HALL
Englewood Cliffs, New Jersey 07632

Prentice-Hall, Inc.
A Paramount Communications Company
Englewood Cliffs, New Jersey 07632

ISBN: 0–13–128603–X
Library of Congress Catalog Card Number: 72–167633

Printed in the United States of America
20 19 18 17 16

Prentice-Hall International (UK) Limited, *London*
Prentice-Hall of Australia Pty. Limited, *Sydney*
Prentice-Hall Canada Inc., *Toronto*
Prentice-Hall Hispanoamericana, S.A., *Mexico*
Prentice-Hall of India Private Limited, *New Delhi*
Prentice-Hall of Japan, Inc., *Tokyo*
Simon & Schuster Asia Pte. Ltd., *Singapore*
Editora Prentice-Hall do Brasil, Ltda., *Rio de Janeiro*

Preface

Most thermodynamicists, having wrestled with the complexities of their subject for years, remember rather vividly that point at which comprehension began to overtake confusion and the full power and significance of the fundamental concepts began to unfold. This onset is frequently accompanied by a compulsive desire to share with others the "unique" process by which the mysteries of the subject were finally penetrated. In part we, as authors, have been motivated by this urge and, as the reader will observe, have expended considerable effort trying to anticipate and clarify those aspects of thermodynamics that ordinarily cause difficulties for the student.

An equally important reason for initiating this project was our desire to place under a single cover an engineering treatment of mechanical and chemical thermodynamics, which draws on microscopic considerations for added clarity in studying entropy. The approach is admittedly simplified, but hopefully it will be sufficient to help the student understand entropy without requiring extensive preparatory work in statistical mechanics or quantum theory, the time for which does not exist in most undergraduate thermodynamics courses.

Emphasis is placed on the fundamental concepts: energy, entropy, and equilibrium; their interrelations; and the engineering relationships to which they give rise. Considerably more attention is given to entropy than is characteristic of most undergraduate texts; this topic is introduced early in the book as an indicator of the effectiveness with which man utilizes his energy reserves. The concept of lost work is used in developing the entropy generation term to further emphasize the significance of dissipative processes that degrade energy potentials. The customary approach of defining entropy in terms of heat and temperature is postponed until the student has grasped the more fundamental point that entropy is related to the most probable spatial and energy distribution of a system's components. Employing the insight obtained from such an introduction, an effort is then made to provide a rational explanation for the quantitative expression linking entropy to heat and temperature.

The concept of the equilibrium state as the most probable configuration naturally evolves from such an approach. More important, however, is the derivation of a macroscopic criterion for the equilibrium state, which permits us to deal readily with physical and chemical equilibria in multicomponent systems. An operational relationship is thus established by recognizing that an isolated system in equilibrium is incapable of delivering any work. This approach has particular utility in dealing with the impact of centrifugal, gravitational, and other field effects on a system's equilibrium state. For many years these latter points have been developed by Professor J. J. Martin for his students. As students, two of us were priviliged to study under him; as authors, we have drawn liberally from his teaching for which we are deeply appreciative. We hope our embellishment of this material with our own ideas justifies the publication of this thermodynamics textbook.

The student has been foremost in the minds of the authors throughout the preparation of this text. Sufficient detail and sample problems are used in the development of important principles so that we believe it is possible to rely heavily on the text, rather than lectures, for development of the principles, allowing the instructor to devote class time to problem analysis and discussion of student questions. Sample problems have been selected for their instructional value and should be studied carefully by the student to derive maximum benefit. The computer has been used in the solutions of several of these problems to demonstrate its potential for solving complex engineering and/or computational problems.

A variety of engineering oriented problems that can be assigned for homework are found at the end of each chapter. Their complexity varies from simple discussion questions to several worthy of a computer solution, with many of the problems suitable alone as a homework assignment. We have resisted the temptation to incorporate comprehensive property tabulations and plots, thereby competing with the handbooks. We have included sufficient data and information for most problems in the text, but have not felt it inappropriate to expect the student to use other reference material in the library on occasion. We believe the data included will permit the student to derive the full educational value from the text without forcing him to lug a data bank around campus.

The first eight chapters relate to the thermodynamics of single-component systems. Much of the first three chapters is devoted to developing an understanding of the thermodynamic properties, particulary entropy. In contrast to the descriptive material in the early chapters, Chapters 4 and 5 develop quantitative relationships accounting for energy and entropy changes occurring in a system. The energy and entropy balances combined with mathematical manipulations that apply to state properties form the basis for all thermodynamic analysis, whether mechanical, thermal, chemical, or whatever. Thus Chapters

4, 5 and 6 are the backbone of any 1 or 2 term course. Chapters 7 and 8 represent applications in the areas of energy conversion and fluid flow respectively. Although either can be skipped without creating a continuity problem for someone interested principally in chemical phenomena, Chapter 7 is of basic importance in that the limitations relating to the conversion of thermal to mechanical energy are treated quantitatively. No student of thermodynamics should by-pass a thorough treatment of this subject.

Chapter 9 introduces the student to multicomponent systems, building directly on the material of Chapter 6. Various schemes for treating nonideal solution behavior are discussed from both theoretical and empirical points of view. Phase separation and equilibrium between phases for nonreacting systems are introduced in Chapter 10. In Chapters 11, 12, and 13 reacting systems, both chemical and electrochemical, are considered. Chapter 14 introduces the concepts of irreversible thermodynamics. This chapter is brief and is meant to serve soley as an introduction to this field.

The qualitative concept of equilibrium permeates the development throughout. An operational definition of equilibrium between different phases of a single component system is developed in Chapter 6. This logic is employed again in Chapter 10 for multicomponent phase equilibria, and finally in Chapter 11 for systems undergoing chemical reaction. Some continuity will be sacrificed if those three developments are not treated in their original sequence.

There are a variety of packages available to the user of this text:

Package	Chapter
(A) Mechanical Thermodynamics	1–8 [14]
(B) Thermodynamic Fundamentals	1–6, 9 [7, 14]
(C) Chemical Thermodynamics	1–6, 9–11 [7, 12, 13, 14]

The above selections without the optional materials represent coherent packages which we feel can be adequately treated in one 15 week term. The entire text requires two full terms (3 hours per week) to cover adequately; where two terms are available we suggest Chapters 1–8 be covered in the first term, and Chapters 9–14 in the second.

In addition to Professor J. J. Martin, we are indebted to numerous of our colleagues for their helpful suggestions after reviewing and/or using the manuscript. The students who used the preliminary edition have been most tolerant of mistakes of all sorts and equally helpful in eliminating them. Most importantly, their enthusiasm for the text in spite of the long list of errata was most encouraging to us as authors. Professor Stanley I. Sandler of the Chemical Engineering Department (University of Delaware) provided many interesting discussions, particularly concerning the development of the material in Chapter 11. David M. Wetzel and Robert Thornton, both graduate students in chemical engineering at the University of Delaware, have provided much assistance in checking the numerical solutions of the sample and end of chapter problems.

Particular thanks must go to Mrs. Alvalea May and Mrs. Sharon Poole who have handled the bulk of the typing for the final manuscript, and to the secretarial staff of the Chemical Engineering Department at the University of Delaware who have assisted with duplication and compilation of the final manuscript.

Finally we must acknowledge the patience of our wives, whose sacrifices have been many throughout the preparation of this book. To them we dedicate this effort in sincere appreciation for the understanding and encouragement they have exhibited since the inception of our efforts some six years ago.

Richard E. Balzhiser
Michael R. Samuels
John D. Eliassen

Contents

CHEMICAL ENGINEERING THERMODYNAMICS

The Study of Energy, Entropy, and Equilibrium

Engineering, Energy, Entropy, and Equilibrium

1

Introduction

Many readers of this textbook will have encountered the subject of thermodynamics in earlier chemistry and physics courses. To these students energy, entropy, and equilibrium, the "three E's" of thermodynamics, will be familiar if not fully appreciated concepts. In this book these three concepts will be used as the basis for presenting an array of engineering thermodynamic relationships which provide valuable analytical assistance to the engineer. A thorough understanding of these concepts is important enough to justify the concerted effort made here to acquaint the student with the three E's or to refine his knowledge of them. This task begins in an introductory way in this chapter and continues in Chapter 3 in greater detail. After Chapter 3 these concepts are used to formulate useful engineering relationships such as the energy and entropy balances which are discussed in Chapters 4 and 5, respectively.

The main difference between the treatment in this book and that typical of physics or physical chemistry courses which precede or accompany the engineering thermodynamics course is the applications aspect (engineering orientation) of the relationships that are generated. For this reason it seems equally appropriate to discuss engineering, a fourth E, in the introductory pages.

1.1 Engineering

It has been said that politics is the bridge across which ideas march to meet their destiny. It seems equally appropriate to characterize engineering as that bridge which links the laboratories of the scientist with the products of industry. Indeed, as we examine more carefully the role of the engineer in society, we become increasingly aware of the sizable gap that must be spanned by the engineering profession. In a technical sense the engineer represents the interface between the scientific community and the business world. When it is economically feasible, the engineer translates the scientists' successes in the laboratories to the production operations of industry. This "when" is an important one, which requires the engineer to have both technical and economic training to successfully perform his role.

The engineer's responsibilities range from devising a process complete with the specification of operating conditions and the type and size of all equipment needed, to figuring the depreciation, cash flow, and profitability associated with a manufacturing process. He draws heavily on the principles of mathematics, science, economics, and common sense in performing such a role. However, these alone are insufficient to handle all aspects of his professional responsibilities. Equations and formulas can seldom completely replace experience and intuition in the engineering scale-up of a process. The latter commodities would be difficult to produce on a scale that would match the needs of a highly industrialized society if the engineering profession itself had not advanced from an "art" to a science in its own right.

By sharing and correlating the "experiences" of investigations through the years, an improved understanding of many physical phenomena has been realized. Nevertheless, many seemingly simple processes such as the flow of a fluid through a pipe or the flow of heat across an interface defy complete understanding. To circumvent his inability to describe completely certain processes, the engineer will commonly choose to approximate the process by a simplified model. By identifying the key parameters affecting the process and then varying them in an ordered way, the engineer is frequently able to develop an expression that relates all of the important parameters.

The flow of a fluid in a pipe serves as an excellent illustration of the power of this approach. By varying pipe diameter, pressure drop, and fluid properties while determining flow rates, it is possible to devise an empirical correlation that relates all these parameters. Thus, although we may not completely understand turbulence or the resistance to flow occurring at the wall, we are able to use such a correlation to size the pipe and pumps for a given application. Pooling of all such information soon permits refinements in the correlating expressions which can then be used with reasonable confidence for fluids and conditions beyond those specifically studied.

Such procedures have been used over and over by the engineer to organize and coordinate the experiences of many for the collective good of the profession. Postulate, experiment, correlate, refine—that is the sequence by which the engineering profession has developed the tools of the trade. The procedure is not unlike that used by Mendeleev and other chemists in devising the periodic chart of the elements. In spite of all the observations made to date with respect to elements and their chemical behavior and in spite of the fact that the system has permitted man to predict the existence of elements before they were experimentally detected, the concept of the atom, on which the periodic chart is based, could be completely erroneous.

Such is often the case when extending the frontiers of technology. The process is one of trial and error. Each observation consistent with the model tends to increase the likelihood of its aptness and hence increases its utility and reliability. Thus, although many "laws" are considered as absolute truths, they could in fact become fiction as investigators continue their explorations.

Many engineering relationships have been devised by these procedures. They have offered great utility in making reliable engineering estimates of certain process specifications. Frequently it is deemed desirable to incorporate a safety factor in finalizing a design. This is merely the engineer's way of compensating for the reliability of the procedures available to him. Similarly, efficiencies are often used to relate actual performances to a highly idealized model of a process that is not capable of exact description. These techniques are all part of an increasingly sophisticated array of procedures available to the engineer.

Included in this collection of engineering "tools" is thermodynamics, which is often referred to as an engineering science. This designation arises in part from its origin in the sciences and in part from the increasing sophistication and utility of the relationships developed. Frequent references are made to the laws of thermodynamics. In reality they are laws only in the sense that no one has yet disproved them. Their formulation followed the pattern (described earlier) that has been used by scientists and engineers to expand progressively their understanding of the physical world. Energy and entropy are in effect "models" of a conceptual sort; both have been refined and expanded since their initial introduction many years ago.

Additional changes may well occur in the future as man continues to open new frontiers of understanding. In this development an attempt will be made, while building on the concepts as presently understood, to illustrate to the student where progress in future years is likely to lead. The areas of irreversible and statistical thermodynamics are two such possibilities. Both are intimately tied to the concepts of entropy and equilibrium. Although somewhat tangential to a basic treatment of the subject for the undergraduate, both will likely play an increasingly prominent role in graduate thermodynamics courses.

1.2 Energy

The relationships between heat and temperature were poorly understood until the early 1800s. Indeed, the thermodynamic concept of heat itself had not developed until that time. Evidence of this confusion between the concepts of heat and temperature is still encountered occasionally, in expressions of the following type: "The body was brought to a high *heat* (instead of *temperature*) through contact with the flame." The proper relationship between heat and temperature was not fully recognized until the concept of energy was developed.

The observation that a body's temperature could be changed by contact with a hotter (or colder) body is almost as old as mankind itself. Since the body's temperature has changed, scientists have concluded that something must have been gained or lost by the body. Around the middle of the 1700s this something was termed "caloric" and was considered to be a massless, volumeless substance that could flow between bodies by virtue of a temperature difference. According to the caloric theory, the higher the caloric content of a body, the higher its temperature.

In the late 1700s, however, Count Rumford (a native of Woburn, Mass., who was working for the Bavarian government at the time) observed that the temperature of cannon barrels became quite hot when the barrels were subjected to the mechanical action of boring tools. Rumford questioned how this heating effect was produced, and he finally concluded that the "heating" was brought about by friction from the boring tool and that it would continue as long as the boring bits were kept in motion.

The consequences of Rumford's observations had a profound influence on the scientific thinking of his time. In a relatively small number of years James P. Joule had conclusively shown that many forms of mechanical work (that is, the effect of a force moving through a distance) could be used to produce a heating effect. Joule also observed that a given unit of mechanical work, regardless of its form, always produced the same temperature rise when applied to the same body.

In an effort to unite these apparently unrelated phenomena, the concept of energy evolved. The basis of this concept assumes that a body may contain several forms of energy. From physics we are familiar with the observation that a body may possess potential energy by virtue of its position within a force field (such as height within a gravitational field), and kinetic energy by virtue of its average bulk motion. In each of these instances the energy is determined by the position or velocity of the center of gravity of the mass, not by the microscopic particle motions or interactions that occur on an atomic scale. Although molecular motions and interactions are excluded from our conventional concepts of potential and kinetic energy, they must clearly be accounted for in any overall discussion of energy. *Internal energy* is defined in Chapter 3 to include all forms of energy possessed by matter as a consequence of random molecular motion or intermolecular forces.

We shall divide energy in transit between two bodies into two general categories: Heat is the flow of thermal energy by virtue of a temperature difference; work is the flow of mechanical energy due to driving forces other than temperature, and which can be completely converted (at least in theory) by an appropriate device to the equivalent of a force moving through a distance.

Included in the broad class of mechanical energies are kinetic, potential, electrical, and chemical energies. Thermal energy, on the other hand, is unique in that there is no known device whose sole effect is to completely convert thermal energy into a mechanical form. Therefore, we are led to the conclusion that heat, a flow of thermal energy, cannot be completely converted into work, a flow of mechanical energy. This statement expresses the essence of what is frequently called the "second law of thermodynamics" and will be considered in great detail later.

As originally envisioned, energy was a conservative property. That is, it was assumed that the energy content of the universe remained constant—energy could change its form, but the total amount always remained fixed. Findings by nuclear physicists in relatively recent times have shown this assumption to be incorrect. Their findings suggest that it is mass and energy together which are conserved. They even succeeded in relating mass to energy through the Einstein relation $E = mc^2$. For our purposes as thermodynamicists the conversion of mass to energy by nuclear reactions is of little current interest, and for applications outside the realm of nuclear processes, it is possible to think of energy itself as being conserved.

As an example of the restrictions just discussed, let us consider a roller coaster as it accelerates down a hill. Since its height is decreasing, the roller coaster is losing some of its potential energy. However, since its velocity is increasing, the kinetic energy is also increasing. If the roller coaster is frictionless, the decrease in potential energy is exactly counterbalanced by the increase in kinetic energy. On the way up the hill, this kinetic energy is reconverted into potential energy as the roller coaster returns to its original height. Thus we find that kinetic and potential energy are directly and completely interconvertible—a characteristic common to all forms of mechanical energy.

If, on the other hand, the roller coaster is not frictionless, some of its potential energy will be converted to thermal energy by the rubbing action of bearing surfaces. Consequently, less potential energy gets converted to kinetic energy and the kinetic energy at the bottom of the hill is less than the original potential energy (by an amount equal to the amount of thermal energy produced). Thus the roller coaster is unable to return to its original height (or potential energy), since the thermal energy it now possesses leads only to a higher temperature and is of no value in returning the roller coaster to its original position. Only by performing work (pushing on the cart) can you return it to its original elevation.

At this point it may be asked: Couldn't this work be supplied by converting the thermal energy generated by friction back into work? The answer to this is: Not completely. We have already indicated that it is not possible to *completely* convert thermal energy into mechanical energy. Thus, although we could theoretically recover some of the frictional losses, we could never recover all of them. Frictional effects reduce forever the amount of energy in the universe which is capable of transformations that produce work.

Indeed, it is this distinction between our ability to obtain useful work from the various forms of energy and the necessity for a quantitative measure of the usefulness of a unit of energy which leads us to our discussion of a second fundamental property of matter—entropy.

1.3 Entropy

The significance of the entropy concept can best be illustrated by an example. Consider a gas as it flows through a wind tunnel. The gas molecules possess kinetic energy, a portion of which is random and a portion of which is ordered and contributes to the bulk velocity of the gas as it moves through the duct. The ordered portion is similar to the kinetic energy of any macroscopic object and is mechanical in form. As such it is capable of being converted to work by an appropriate device such as a turbine or windmill. Extraction of this ordered kinetic energy as work by a perfectly designed turbine would reduce the overall velocity of the gas and hence its kinetic energy, but would not affect the random behavior of the collection of molecules as the gas passed through the blades.

The random contribution to the total energy of the gas is effectively superimposed on the oriented flow. It contributes nothing to the energy flow to the turbine blade, as its random character produces as many collisions which tend to prevent the turbine from rotating as those which would assist it. Thus the random, or thermal, component does not decrease through its interaction with the turbine. Theoretically one could extract all the oriented kinetic energy possessed by the gas and leave only the thermal component. Such a total conversion would require many stages and an unreasonably large device in proportion to what the latter stages of the conversion would yield. Nevertheless, this process represents (in theory) the most efficient use of the energy available and provides the thermodynamicist with a standard with which he is able to compare less ideal conversions.

Contrast this process in which the ordered kinetic energy of the gas is completely converted to work with the condition that would exist if the inlet and outlet ducts of the turbine were suddenly closed. Clearly the total energy of the gas trapped inside would remain unchanged, as it would become essentially isolated. As the flowing molecules strike the closed outlet they rebound and interact with one another such that after a short period of time all kinetic energy will be random in nature. The extent of randomness within the collection will have increased significantly, as the original thermal component will now have added to it a thermal component equal to the kinetic component of the oriented flow.

Although the total energy remains unchanged, any attempt to convert any portion of this energy to work with the turbine is now impossible. We will show in Chapter 7 how a fraction of this thermal energy may be converted to work. However, this fraction is always less than unity and, under the circumstances pictured, would actually be quite close to zero. Thus, in fact, little work can be recovered from this energy once it has been converted to the thermal form. The change that took place in this element of gas by simply trapping it inside the turbine is quite revealing. The net effect was to leave its total energy unchanged but to convert a portion of the mechanical form (kinetic) to the thermal form with an accompanying loss in the ability of the gas to convert its energy to work. (The gas will also experience an increase in pressure as a result of the flow stoppage; since this is mechanical in form, it could be used to convert a portion of the internal energy of the gases to a mechanical form at a subsequent time. However, the amount of mechanical energy gained during the pressure rise will be quite small in comparison with the kinetic energy lost, so we still experience a net loss in the usefulness of our original energy.)

Entropy will be shown to provide a measure of the effectiveness of such energy conversion processes. Although in theory all mechanical forms of energy are completely interconvertible or transferable as work, in practice these ideal conversions cannot be realized. Frictional and other dissipatory effects inevitably lead to a downgrading of the available energy resources to a thermal form from which it can *never*, even by an ideal process, be completely reconstituted to a mechanical form. Entropy is effectively a measure of the extent of randomness within a system and thus provides an accurate indication of the effectiveness of energy utilization. When the mechanical forms of energy possessed by a body are permitted to degenerate by any process whatsoever to the thermal (or random) form, the entropy of the body is increased.

If we were to watch the gas molecules of the previous example, we would find that the likelihood that the random molecular motions will ever reorient themselves without some external input is extremely small. That is, the random thermal energy will not freely revert to a mechanical form. Since the entropy of a substance is related to its randomness, the entropy will not decrease without some external interaction. However, the only manner known to man to reduce molecular randomness is to transfer the randomness to another body, thereby increasing the randomness and entropy of the second body. Thus as thermal energy is transferred from one body to another, entropy is effectively transferred. The body receiving the thermal energy experiences an increase in entropy while the body releasing the thermal energy experiences a reduction in entropy. The transfer of thermal energy (or randomness) as heat in this fashion is the *only* manner by which it is possible to reduce a body's entropy. Since, as we shall see later, this process at best results in a constant amount of randomness in the universe (and generally produces an increase), the entropy of the universe is forever increasing. Unlike energy, it is nonconservative, and therein lies much of the mystery that continues to perplex students. Hopefully the development of the concept in Chapter 3 will help clarify students' understanding of this most important and useful concept.

1.4 Equilibrium

A body at equilibrium is defined to be one in which all opposing forces, or actions, are exactly counterbalanced (subject to the restraints placed upon the system), so that the macroscopic properties of the body are not changing with time. Experience tells us that all bodies tend to approach an equilibrium condition when they are isolated from their surroundings for a sufficient period of time. For example, if a ball is placed on a surface as shown in Fig. 1-1, it tends

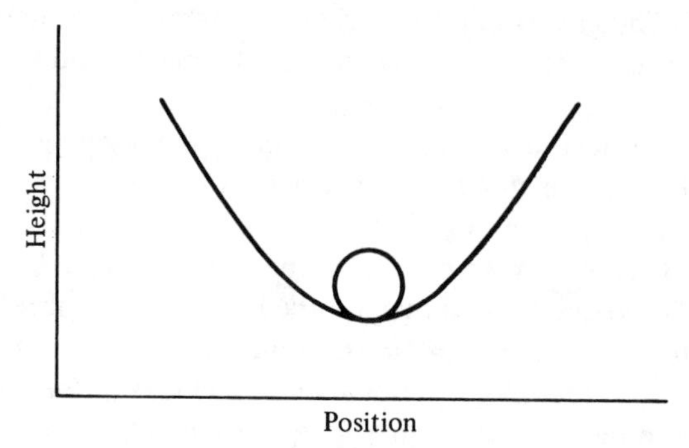

FIG. 1-1. The equilibrium condition.

to settle in the lowest portion of the surface, where the force of gravity is exactly counterbalanced by the supporting force of the surface. Thus the ball has settled in its equilibrium position—subject to the constraint that it remain in contact with the surface.

The equilibrium condition described in Fig. 1-1 is termed a *stable equilibrium* because the ball will always return to this condition after it has been moved away (or disturbed) from the equilibrium position. In addition to stable equilibrium conditions, we may have metastable and unstable equilibrium conditions, as shown in Fig. 1-2.

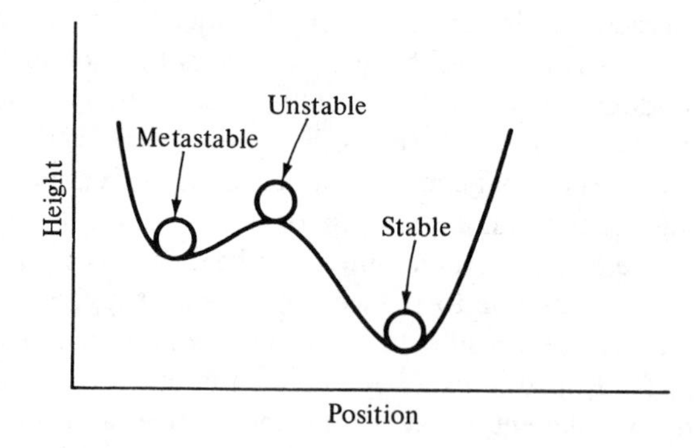

FIG. 1-2. Types of equilibrium.

A metastable equilibrium is one in which the system will return to its original state if subjected to a small disturbance, but which will settle at a different equilibrium condition if subjected to a disturbance of sufficient magnitude. For example, a mixture of hydrogen and oxygen can remain unchanged for great periods of time if not greatly disturbed. However, if the mixture is disturbed sufficiently, say by an electrical spark or a mechanical shock, then a violent reaction between the hydrogen and oxygen can be expected to occur.

An unstable equilibrium condition is one in which the system will not return to its original condition whenever it is subjected to a finite disturbance. For example, if we carefully balance a dime on its edge, we have a body that is essentially in unstable equilibrium, because a small disturbance will cause the dime to topple.

Although the three kinds of equilibrium have been illustrated in the previous paragraphs, we shall now restrict the remainder of our consideration to the discussion of *stable equilibrium*, as this is the condition to which most systems will ultimately move.

For the simple ball-and-surface illustration, we may picture equilibrium to be a static unchanging condition. However, for many of the problems we shall encounter, this static description is far too simple. For example, if we examine the equilibrium between vapor and liquid water on a molecular scale, we would find constant motion and change: Molecules from the liquid are constantly entering the vapor phase and molecules from the vapor are constantly entering the liquid phase. Equilibrium between the liquid and vapor phases occurs not when all changes cease, but when these molecular (or microscopic) changes just balance each other, so that the macroscopic (or gross) properties remain unchanged. As seen in this broader context, equilibrium is actually a dynamic process on a microscopic scale, even though we treat it as a static condition in macroscopic terms.

All spontaneous (naturally occurring) events tend toward more probable and more random molecular configurations. (If the new configuration were not more probable, the system would not tend toward it spontaneously.) Thus, if a body is isolated from its surroundings and allowed to interact with itself, all changes in body's properties must lead to more probable and random configurations. When the body finally attains its most probable configuration, it can undergo no additional change. Since the body may undergo no further change in its properties, we observe that it must then be in equilibrium with itself, or simply in *equilibrium* (subject, of course, to the constraint of isolation from its surroundings). Conversely, if an isolated body is in its equilibrium condition—so that its properties do not change with time—the body must be in its most probable and random configuration. (If it were not in the most probable configuration, it would tend to move toward that configuration.) As we have shown, the more random or probable a configuration, the higher is the entropy of the material in this arrangement. Since the equilibrium condition of an isolated body corresponds to the most probable conditions, the equilibrium conditions must correspond to the conditions of maximum entropy—subject to the constraints placed upon the body. In our discussion of entropy in Chapter 3, we shall examine the relations between entropy and equilibrium in much greater detail.

In macroscopic terms the equilibrium condition requires that all energy potentials, such as temperature and pressure (which measure the availability of energy), be uniform throughout the body. If this were not true, energy flows would occur and could theoretically be used to produce work if the flow were channeled through an ideal engine. Thus a body in equilibrium (with itself) can be described as one from which no work can be derived if any part of the body is allowed to communicate with any other part through an ideal engine. Later we shall deal with many types of problems in which the criteria for equilibrium will be extremely difficult to specify. In these cases we shall find it useful to rely on this last observation to develop other useful criteria for equilibrium.

Introduction to Thermodynamics 2

2.1 System—Definition

A thermodynamic *system* may consist of any element of space or matter specifically set aside for study, while the *surroundings* are thought of as representing the remaining portion of the universe. The system *boundary*, which may be real or imaginary, separates the system from the surroundings. A system that is not permitted to exchange mass with the surroundings is termed a *closed system*, whereas a system that exchanges mass with the surroundings is called an *open system*. A closed system that exchanges no energy with its surroundings, either in the form of heat or work, is called an *isolated system*. Examples of a closed system might include a block of steel or a fixed amount of gas confined within a cylinder; a pipe or a turbine through which mass is flowing are examples of an open system. However, if one were to identify a given mass of fluid and follow its passage through a pipe, this fixed amount of mass, surrounded by an imaginary boundary, would constitute a closed system. Two blocks of steel at different temperatures but perfectly insulated from their surroundings constitute an isolated system.

The thermodynamic system will provide a basis for analysis in subsequent chapters of this text. In the solution of any problem the first step will always involve a clear definition of the system under consideration. Such a choice is often arbitrary, because several possibilities often exist. Experience demonstrates that in many cases an open system may have inherent advantages, whereas in other cases a closed system might be preferable.

2.2 Characterization of the System

The condition in which a system exists at any particular time is termed its *state*. For a given state the system will possess a unique set of properties, such as pressure, temperature, and density. A change in the state of a system brought about by some interaction with its surroundings always results in the change of at least one of the properties used to describe the state. However, if by a series of interactions with the surroundings the system is restored to its original state, then all the properties by which that state was originally characterized must return to their original values. A true state *property*, therefore, is one whose value corresponds to a particular state and is completely independent of the sequence of steps by which that state was achieved. Many of the properties with which we are familiar, such as pressure, temperature, and density, are state properties.

A system may consist of one or more phases. A *phase* is defined as a completely homogeneous and uniform state of matter. (This definition is strictly valid only for an *equilibrium phase*. However, since all phases encountered in this text are at equilibrium, no ambiguity will result if the term phases is used alone.) Although both ice and water have a uniform composition, they do not have a uniform consistency or density and would therefore be considered as two different phases. On the other hand, two immiscible liquids would possess different compositions and thus regardless of densities would be considered as separate phases.

In describing a system, one might assume either a microscopic or macroscopic point of view. A microscopic description might consider the atoms or molecules of which the system is composed. Specification of their individual masses, positions, velocities, and interactions would be required. If the number of particles is sufficiently great, statistical procedures may then be used to predict the behavior of the total collection of molecules.

A macroscopic description does not consider the individual molecules or particles that make up the system. The system is described in terms of the cumulative interaction of the collection with the surroundings. Properties such as temperature, pressure, or density have meaning when used to characterize such a system. However, if applied to an individual molecule, such parameters are meaningless.

To contrast the two points of view just described, let us consider a container of gas as our system. From the microscopic point of view this system is composed of many molecules constantly moving about within the container. Although the molecules might all be identical in structure, they would be in continuous motion, occupying different positions and possessing different velocities as they move and interact with one another and the container. If one were to focus his attention on several of these molecules, it would seem that the system is forever undergoing change and that the condition we call equilibrium would never be achieved. However, as we consider an increasing number of molecules, we observe that although some may slow down or change their direction, the effects of such a change are offset by the movements of other molecules. We soon realize that although individual particles within the system experience continuous change, an average velocity and a uniform distribution throughout the container which does not change perceptibly with time eventually results. Thus the gas is in an equilibrium state.

Although it is difficult to think in terms of equilibrium when considering individual particles, the concept does have significance when applied to the averaged behavior of many molecules. Kinetic theory and statistical mechanics enable one to convert the averaged behavior of molecules into state properties, such as temperature, pressure, and density, ordinarily used to characterize thermodynamic systems. It is these properties, like the averaged molecular velocities or energies, which remain constant in the equilibrium state.

The development of classical thermodynamics has been based on the macroscopic point of view. It is this point of view that will provide an operational procedure for system analysis later in this text. However, in recent years man's understanding of the microscopic nature of matter has been enlarged and the principles of statistical and quantum mechanics better defined. These developments have added another dimension to the development of the subject and will provide additional insight into the fundamental concepts of energy, entropy, and equilibrium in subsequent chapters. Indeed, many courses and textbooks cover just such relationships and demonstrate the interrelationship between molecular mechanics and the properties of matter used in classical macroscopic thermodynamics.

2.3 *Processes—Interactions of a System and Its Surroundings*

Changes in the physical world are brought about by *processes*. In the thermodynamic sense a process represents a change in some part of the universe. This change may affect only a single body, as in the approach to equilibrium of an isolated system that was not initially in its equilibrium state, or a process may involve changes in both system and surroundings. The expansion of a gas as it flows through a turbine to produce work is an example of a process in which work is developed by utilizing a pressure difference to extract energy in the form of work from the gas. A thermodynamic analysis might consider either a fixed mass of gas (closed system) or the turbine (open system) as the system. In either of these cases the process involves an interaction between the system and its surroundings.

The fixed mass of gas expands against its surroundings, a part of which is represented by the turbine blades. As it impinges, the gas exerts a force that causes the shaft to rotate. The rotating shaft, if attached to an appropriate device, can do useful work in the surroundings. In this instance both the system (the gas) and the surroundings experience a change in their state as the process proceeds.

In the choice of the turbine as the system, the nature of the interaction between system and surrounding is different, although the net effect on the universe remains unchanged. The interaction of the turbine and the surroundings is represented by passage of mass into the system at a given energy level followed by an internal conversion of the kinetic energy of the gas to shaft work, and the subsequent discharge of a lower-energy gas at the turbine's exhaust. In this case the surroundings are being changed, whereas the system itself may operate indefinitely with no change in its state.

This example illustrates the relationship of a system to a process. Whereas the process may be defined specifically in terms of a specific change to be accomplished, such as the production of shaft work by expanding a gas from a high pressure to a low pressure, various systems might be specified in a thermodynamic analysis of the problem. The system chosen must in some way involve the process of interest if the analysis is to yield meaningful results. In some instances the change may occur totally within a system such that a redefinition of the system is essential to permit a more meaningful analysis.

As an illustration of this point, let us take a process in which two metal blocks are originally at different temperatures. If the two blocks are brought together, we observe that the temperatures of the two blocks approach the same value. The process involves energy transfer from the hotter block to the colder one. If one defines the system to be both blocks, the process occurs within the boundaries of the system and no net change in the energy content takes place in either the system or the surroundings. If one desires to know the amount of energy transferred, such an isolated system is a poor choice. On the other hand, if one chooses either block as the system, the process would involve a flow of energy between the system and part of its surroundings. Physically the choice of system makes no difference in the final state of the universe; that is, both blocks end up at the same temperature regardless of which system is specified.

2.4 Reversible and Irreversible Processes

One objective of thermodynamics is to describe the interactions between system and surroundings (such as heat and work) which take place as a system moves from one (equilibrium) state to another. For instance, suppose that the gas in a well-insulated cylinder (the system) expands against a piston and transfers mechanical energy to the surroundings. If the piston moves frictionlessly within the cylinder and slowly enough so that viscous losses can be neglected, the work done by the gas will be just equal to the mechanical energy received by the surroundings. Moreover, the mechanical energy received by the surroundings can be stored (by the raising of a weight, for example) and used to return both the system and the surroundings to precisely their original states.

On the other hand, if there is friction between the piston and the cylinder, a part of the work done by the gas in its expansion will be converted to thermal energy and cannot be stored in the surroundings in a mechanical form. Similarly, on the return stroke of the piston, not all the work done by the surroundings will be transferred to the gas, but some will be converted to thermal energy by the friction. Thus, if the piston–cylinder arrangement is to be returned to its initial state, the surroundings will have to supply more work to the gas during the compression than was received during the expansion. Similarly, the gas would have to transfer an equivalent amount of heat to the surroundings (in order to return to the original energy level). Thus the overall effect of the frictional expansion–compression cycle is a net transfer of mechanical energy into the gas and an equivalent net transfer of thermal energy back to the surroundings. However, as we have previously indicated, it is impossible to conceive of a device whose only effect is to completely convert thermal energy back into mechanical energy. Thus we find that there is a net change in the universe which can never be completely reversed. A similar result would have been obtained for a frictionless piston if the expansion were allowed to take place so rapidly that nonuniformities in pressure could occur within the gas in the cylinder. The dissipation of a shock wave is an extreme example of this type of nonuniformity.

In order to provide a criterion by which to distinguish between the two types of processes discussed above, we define a *reversible process* as one which occurs in such a manner that both the system and its surroundings can be returned to their original states. Any process that does not meet these stringent conditions is called an *irreversible process.*

In the previous discussion the process in which the piston moved frictionlessly and at moderate speed was a reversible process, because both system and surroundings could be returned to their original states by using the work produced in the expansion to recompress the gas. *For such a process it is a straightforward matter to completely describe the interaction of system and surroundings.* This behavior is a major virtue of reversible processes. For an irreversible process, such as the frictional expansion, it is not such a straightforward matter to describe the interactions between system and surroundings. For example, knowledge of either the amount of mechanical energy transferred from the surroundings to the piston or that transferred from the piston to the gas does not give us the other unless we also know the amount of mechanical energy dissipated in overcoming the friction between piston and cylinder. Thus for the frictional process, not only do we need to describe the changes that take place in the system, but we need to know how these changes are transferred to the surroundings before our description of the process is complete. This additional complexity of irreversible processes carries over into other types of systems as well and will be the subject of much discussion in the next several chapters.

Before leaving the topic, let us examine some of the general conclusions that can be reached concerning reversible and irreversible processes. Since it is much easier to enumerate those things which make a process irreversible than vice versa, the following is a discussion of the sources of irreversibility and their consequences.

We have already noted that mechanical friction leads to irreversibility because the mechanical energy dissipated in overcoming friction is transformed into heat. Similar irreversibilities are exhibited by a large class of processes, of which the following are examples:

1. The flow of a viscous fluid through a pipe. System: the fluid.
2. The transfer of electricity through a resistor. System: the resistor.
3. The inelastic deformation of a solid material. System: the solid.

Each process has in common the fact that a part of the work provided by the surroundings to carry out the process can be stored and is recovered when the direction of the process is reversed, but part of the work is lost to friction (molecular and electronic in these examples) and is converted to heat. Thus the system and surroundings cannot both be returned to their original states and these processes are irreversible.

Not all irreversibilities involve the degradation of work to heat immediately, although it can be shown that restoration of the system to its original state will eventually lead to such a conversion. Consider the following processes in isolated systems:

1. A sealed and insulated cylinder containing one high-pressure chamber and one vacuum chamber with the connecting valve then opened (free expansion). System: the gas.

2. A hot and a cold block of metal insulated from their surroundings but brought into thermal contact with each other. System: both blocks.

To restore the gas to its original pressure the gas must be compressed using work supplied by the surroundings. During the compression an equivalent amount of heat must be transferred back to the surroundings to restore the gas to its original energy level. Thus the surroundings undergo a net change that cannot be reversed, and the process is irreversible.

To restore the second system to its original state, the temperature of one block must be increased by transferring heat to it from a high-temperature reservoir in the surroundings. The other block must be cooled by transferring heat to a low-temperature reservoir. As we shall show later, it is not possible to transfer heat from the low-temperature reservoir to the high-temperature reservoir without supplying work. Thus, although the system can be returned to its original state, the surroundings cannot; the process is again irreversible.

The common factor in these two processes which can be generalized is that energy has been transferred from one part of the system to another and an energy potential (pressure difference or temperature difference) has been reduced without the production of work. If, in the first case, the energy potential (pressure difference) had been reduced by expanding the gas against a frictionless piston until its volume equaled that of the combined chambers, the work produced in the surroundings would have been just that necessary to recompress the gas. Although the final state of the work-producing system would not be exactly the same as for the free expansion, this process, in contrast, is reversible.

In the second case a heat engine (for example, a steam engine) might have been employed to produce work during the transfer of heat from the high-temperature block to the low-temperature block. This work could then be used to drive a heat pump (refrigerator) which could drive heat from the low-temperature block back to the high-temperature block. Under ideal conditions it is possible to return both blocks to their initial states, thereby reversing the original process.

An irreversible process always involves a degradation of an energy potential without producing the maximum amount of work or a corresponding increase in another energy potential other than temperature. The degradation can result either from frictional effects, as in the case of the piston, or from an imbalance of mechanical-energy potentials, as in the case of the free expansion. In the latter case we can generalize this observation to state that a process will be irreversible if that process (by virtue of a finite driving force) occurs at a rate that is large compared to the rate at which molecular adjustment in the system can occur. Since molecular processes occur at a finite (even if rapid) rate, a truly reversible process will always involve an infinitesimal driving force to assure that the energy transfer occurs without degradation of the driving potential. Hence they will take place at an infinitesimally low rate. Under such conditions it is always possible for a system to readjust on the molecular level to the process change such that the system moves successively from one equilibrium state to another.

In practice, we cannot afford to operate processes at an infinitely slow rate, because our output and hence profit would also be infinitesimal. Thus we intentionally sacrifice our energy potentials to accomplish immediate change. However, many process changes, while occurring at a finite rate, do not occur so rapidly that the system is unable to adjust on a molecular level (owing to the very rapid rate at which molecular processes occur). Such processes are referred to as *quasi-static* or *quasi-equilibrium* and are frequently amenable to analysis by much the same technique as reversible processes.

Although it is generally much easier to describe a system's behavior under reversible conditions rather than irreversible ones, most processes of interest are irreversible. Thus to facilitate thermodynamic analysis we often invent reversible processes that closely approximate actual process behavior and yet provide some basis with which to compare irreversible processes between the same end states.

2.5 System Analysis

In thermodynamics the behavior of a system is studied by monitoring the changes in state that it experiences during a process and the flows of mass or energy that cross the boundary. An effective accounting scheme is necessary so that changes may be carefully analyzed quantitatively just as the flow of money to and from a checking account is carefully monitored. The scheme used in keeping a record of one's bank balance is really rather simple and exactly the same as the procedure we choose to use in thermodynamics.

Let us consider a checking account as the "system." The balance of dollars might be thought of as the state of the system, where dollars represent a property of the system. Flows between system and surroundings are represented by deposits, withdrawals, or checks. Since service charges or interest can result in a change in the balance without any flow, some mechanism must be provided for such a change. In balancing such an account a simple system is used:

$$\overbrace{\text{Dollars deposited} - \text{checks written}}^{\text{Flows}}$$
$$\underbrace{+\ \text{interest} - \text{service charge}}_{\text{Generation terms}} = \text{change in amount of dollars}$$

In more general terms, such a scheme simply relates the change, or accumulation, of a quantity in a system to what enters or leaves the system plus any production (or destruction) of that quantity that occurs totally within the system.

The simple mass balance is an application of this simple scheme. Since mass is known to be conserved in all nonnuclear transformations, the generation term becomes unnecessary. For a given system

$$\text{mass}_{\text{in}} - \text{mass}_{\text{out}} = \text{change in mass} \tag{2-1}$$

or

$$\delta M_{\text{in}} - \delta M_{\text{out}} = dM \tag{2-2}$$

where the symbol δ is used to indicate the differential flow of a quantity (in this case mass) and the symbol d is used to indicate the differential change in a system's properties (in this case the system's mass).

If we are interested in the mass of a component in a system rather than the total mass, we must allow for a generation term, because a given component can be produced, or consumed, within the system by chemical reaction. Thus for a component, B, the following balance can be written:

$$(\delta B)_{\text{in}} - (\delta B)_{\text{out}} + (\delta B)_{\text{gen}} = dB \tag{2-3}$$

where we have used the term $(\delta B)_{\text{gen}}$ to represent the differential amount of B generated within the system. Thus our use of the differential operator δ is extended to include both flow and generation terms.

This same accounting scheme will be used later to analyze changes in energy and entropy experienced by a system. Although the time rate of change of these variables is not of particular concern in a thermodynamic analysis, one could incorporate the notion of rates by simply dividing each term by a time factor. The same scheme can also be used to formulate basic rate expressions for different processes in later courses.

2.6 Units

Any physical measurement must be expressed in units. For example, if the length of a football field was expressed as 100 long, it would convey no meaning.

However, if we add a unit and say the field is 100 yards long, the statement has meaning. The term "yard" is a unit of length whose value is defined in terms of a reference length which is kept at the National Bureau of Standards. There are many units which the engineer is likely to encounter. Some of these units consist of groupings of other units. In certain cases different units may be used to express the same physical property (for example, "foot" and "yard" are both units of length) and therefore are interconvertible.

The subject of units has been greatly confused by the use of different sets of units in the scientific and engineering communities. The scientific community generally uses the metric (CGS or MKS) system; the engineering community uses the engineering, or British, system. Although the engineering and metric systems appear at first glance to be basically different, we shall see that they are actually quite similar.

A vast number of physical units exist. However, the great majority of these units can be expressed in terms of the four fundamental dimensions of mass, length, time, and temperature. The dimensions for both force and energy can be expressed from these four fundamental dimensions. Most of the difficulty encountered in the use of dimensions and units arises because the relation between the dimensions (and units) of mass and force is not properly understood. This problem is compounded greatly by the different mass–force relations used in the engineering and metric systems.

Let us now examine the similarities and differences between the engineering and metric systems of units. In the metric system the unit of length is either the meter or the centimeter; in the English system the unit of length is the foot or inch. In both systems the basic unit of time is the hour, minute, or second. The metric unit of mass is the gram or kilogram; the English unit of mass is the pound (abbreviated lb_m, for pound mass). Up to this point the metric and engineering systems are very similar, except for the numerical factor for converting feet to meters or lb_m to kg. However, as we stated before, the unit of force is a derived quantity and the metric unit of force is defined in a slightly different manner from the English units of force.

The unit of force may be derived from Newton's law of acceleration, which says that the force F necessary to uniformly accelerate a body is directly proportional to the product of the mass of the body and the acceleration it undergoes. That is,

$$F \propto M \cdot a \tag{2-4}$$

or

$$F = \frac{Ma}{g_c} \tag{2-5}$$

where g_c is the universal conversion factor, whose magnitude and units depend on the units chosen for F, M, and a.

For example, in the metric system the unit of force is either the newton (MKS) or the dyne (CGS). The newton is defined as the force needed to accelerate a 1-kg mass at 1 m/sec^2; the dyne is the force necessary to accelerate a

1-g mass at 1 cm/sec². Substitution of these values into equation (2-5) gives

$$1 \text{ newton} = \frac{1 \text{ kg} \cdot 1 \text{m/sec}^2}{g_c} \tag{2-6}$$

and

$$1 \text{ dyne} = \frac{1 \text{ g} \cdot 1 \text{ cm/sec}^2}{g_c} \tag{2-6a}$$

We may now solve for g_c as

$$g_c = 1 \frac{\text{kg m}}{\text{newton sec}^2} = 1 \frac{\text{g cm}}{\text{dyne sec}^2} \tag{2-7}$$

Thus, in the metric system g_c has the value unity, and the dimensions (where a dimension is a whole class of units)

$$g_c [=] \frac{\text{mass} \cdot \text{length}}{\text{force} \cdot \text{time}^2} \tag{2-7a}$$

where [=] represents "has the dimensions of."

It would seem natural to define the units of force in the engineering system in a manner similar to that of the metric system, that is, as the force necessary to accelerate a 1-lb_m mass at 1 ft/sec². The *poundal* is, in fact, defined in just this manner. However, the poundal has never received great acceptance as a unit of force and is hardly ever seen. The pound force, lb_f, is the most frequently used unit of force in the engineering system; it is defined as the force necessary to accelerate 1 lb_m at 32.174 ft/sec². (It should be noted that 32.174 ft/sec² is the acceleration of the earth's gravity field at the equator.) Therefore, in the engineering system g_c is defined as

$$g_c = \frac{1 \text{ lb}_m \cdot 32.174 \text{ ft/sec}^2}{1 \text{ lb}_f} \tag{2-8}$$

or

$$g_c = 32.174 \frac{\text{lb}_m \text{ ft}}{\text{lb}_f \text{ sec}^2} \tag{2-8a}$$

The weight of an object can be calculated from equation (2-5) by remembering that the weight of a body is identical to the force necessary to accelerate the body at the same rate it would accelerate during free fall in a vacuum. Therefore, the weight of a body is expressed as

$$W = \frac{Mg}{g_c} \tag{2-9}$$

where g is the acceleration of gravity and W the weight.

In the engineering system of units we may express the weight of a 1 lb_m as

$$\begin{aligned} W &= \frac{1 \text{ lb}_m \cdot 32.174 \text{ ft/sec}^2}{32.174 \text{ lb}_m/\text{lb}_f \cdot \text{ft/sec}^2} \\ &= 1 \text{ lb}_f \end{aligned} \tag{2-10}$$

That is, in the engineering system of units the magnitudes of the weight and mass of a body are identical at sea level, where $g = 32.174$ ft/sec², when the weight is expressed as lb_f and the mass as lb_m, and the acceleration of gravity is 32.174 ft/sec². Thus the lb_f might alternatively have been defined as the force necessary to support a 1-lb_m mass against the forces of gravity at the equator.

The weight in newtons of a 1-kg mass is given by

$$W = \frac{1\ \text{kg}\cdot 9.8\ \text{m/sec}^2}{1\ \text{kg}\cdot\text{m/sec}^2}\cdot\text{newtons} \tag{2-11}$$

where $g = 9.8\ \text{m/sec}^2$, or

$$W = 9.8\ \text{newtons} \tag{2-11a}$$

Thus in the metric system the weight of a mass does not have the same numerical value as its mass, whereas in the engineering system it does. On the other hand, in the metric system the magnitude of $g_c = 1$; in the engineering system the magnitude of g_c = the magnitude of g at the equator.

Since most scientific fields prefer to use metric, rather than engineering, units, the omission of the term g_c causes no serious problem, and the practice of not including g_c in the equations involving the conversion of mass units to force units is almost universal. *However, in the engineering system neglect of g_c can be catastrophic, because the magnitude of g_c is not equal to unity.* Thus it is of primary importance that the proper use of g_c be fully understood.

Table 2-1 lists the more commonly used systems of units and their respective mass, length, time, and force conversion.

TABLE 2-1 Common Systems of Units

System of Units	Unit of Length	Unit of Time	Unit of Mass	Unit of Force	g_c	Definition of the Force Unit
Engineering	foot	second	lb_m	lb_f	$32.174\ \frac{lb_m\text{-ft}}{lb_f\ \text{sec}^2}$	Force needed to accelerate a 1-lb_m mass at 32.174 ft/sec²
Engineering	foot	second	lb_m	poundal	$1\ \frac{lb_m}{\text{poundal}}\ \frac{\text{ft}}{\text{sec}^2}$	Force needed to accelerate a 1-lb_m mass at 1.0 ft/sec²
Metric (CGS)	centimeter	second	gram	dyne	$1\ \frac{\text{g}}{\text{dyne}}\ \frac{\text{cm}}{\text{sec}^2}$	Force needed to accelerate a 1-g mass at 1.0 cm/sec²
Metric (MKS)	meter	second	kilogram	newton	$1\ \frac{\text{kg}}{\text{newton}}\ \frac{\text{m}}{\text{sec}^2}$	Force needed to accelerate a 1-kg mass at 1.0 m/sec²
Combined	centimeter	second	gram	g_f	$\frac{980\ \text{g cm}}{g_f\ \text{sec}^2}$	Force needed to accelerate a 1-g mass 980 cm/sec²
Combined	meter	second	kilogram	kg_f	$\frac{9.8\ \text{kg m}}{kg_f\ \text{sec}^2}$	Force needed to accelerate a 1-kg mass 9.8 m/sec²

Let us now examine the dimensions and units of some of the quantities most frequently encountered in engineering studies. Since all units that have identical dimensions must be interconvertible, conversion factors between these units must exist. Lists giving the commonly needed conversion factors and some other useful constants are presented as Tables 2-2 and 2-3.

1. *Work* has the dimensions of force · length, or, in the engineering system, ft-lb_f. We shall see in later chapters that because both work and heat are energy terms, they must have the same dimensions. In the engineering system heat is expressed in British thermal units (Btu's). The conversion factor between ft-lb_f and Btu is

$$1 \text{ Btu} = 778 \text{ ft-lb}_f$$

2. *Kinetic energy*, being a form of energy, must also have dimensions of force · length. Kinetic energy is evaluated from the formula

$$\text{KE} = \frac{1}{2}\frac{Mu^2}{g_c}$$

where M = mass, lb_m
u = velocity, ft/sec
g_c = 32.17 ft/sec^2 · lb_m/lb_f

SAMPLE PROBLEM 2-1. Calculate the kinetic energy, in Btu, of a 4000-lb_m car traveling at 60 mph.

Solution:

$$60 \text{ mph} = 88 \text{ ft/sec}$$

$$\text{KE} = \frac{1}{2}\frac{4000 \text{ lb}_m(88 \text{ ft/sec})^2}{32.17 \text{ lb}_m \text{ ft/lb}_f \text{ sec}^2}$$

$$= 4.8 \times 10^5 \text{ ft lb}_f$$

$$= 618 \text{ Btu}$$

---

3. *Potential energy* also has the dimensions of force · length and for gravitational potential energy is evaluated from the formula

$$\text{PE} = \frac{MgZ}{g_c} = WZ$$

where M = mass, lb_m
Z = height, ft
g = 32.17 ft/sec^2
g_c = 32.17 lb_m ft/(sec^2 lb_f)
W = weight, lb_f

4. *Pressure* has the dimensions of force/area. Often hydrostatic pressure is measured by the formula

$$P = \frac{\rho gh}{g_c}$$

TABLE 2-2 Energy-Conversion Factors†

	joules = 10^7 ergs	kg_f m	ft-lb	kw hr	hp hr	liter atm	kcal	Btu	g cal
joules	1	0.102	0.738	2.77×10^{-7}	3.73×10^{-7}	9.69×10^{-3}	2.39×10^{-4}	9.48×10^{-4}	0.239
kg_f m	9.80	1	7.23	2.72×10^{-6}	3.65×10^{-6}	9.68×10^{-2}	2.34×10^{-3}	9.29×10^{-3}	2.344
ft-lb	1.36	0.14	1	3.77×10^{-7}	5.05×10^{-7}	1.34×10^{-2}	3.24×10^{-4}	1.29×10^{-3}	0.324
kw hr	3.6×10^{6}	3.67×10^{5}	2.66×10^{6}	1	1.34	3.55×10^{4}	8.61×10^{2}	3.41×10^{3}	8.61×10^{5}
hp hr	2.68×10^{6}	2.74×10^{5}	1.98×10^{6}	0.746	1	2.65×10^{4}	6.42×10^{2}	2.55×10^{3}	6.42×10^{5}
liter atm	1.01×10^{2}	10.3	74.7	2.82×10^{-5}	3.77×10^{-5}	1	2.42×10^{-2}	9.60×10^{-2}	24.2
kcal	4.18×10^{3}	4.27×10^{2}	3.09×10^{3}	1.17×10^{-3}	1.56×10^{-3}	41.3	1	3.97	1×10^{3}
Btu	1.06×10^{3}	1.07×10^{2}	7.78×10^{2}	$2.93 + 10^{-4}$	3.93×10^{-4}	10.4	0.252	1	2.52×10^{2}
g cal	4.18	0.427	3.09	1.16×10^{-6}	1.56×10^{-6}	4.13×10^{-2}	1×10^{-3}	3.97×10^{-3}	1

† To find a value in units of those along the top row, take the value you have in units of the left column and *multiply* by the number in the corresponding box of that row and column. Example: Btu × 778 = ft-lb_f.

where ρ = density, lb_m/ft^3
g = acceleration of gravity
h = height of liquid, ft
g_c = universal gravity constant, 32.17 lb_m-ft/lb_f sec^2

The discussion of the fourth fundamental dimension, temperature, is taken up in Chapter 3. Its role in thermodynamics is of paramount importance, and a more detailed analysis of its true meaning is imperative.

TABLE 2-3 Miscellaneous Factors

1. Miscellaneous Units

Multiply	*by*	*to get*
ft^3	7.48	U.S. gal
ft^3	28.3	liters
U.S. gal	3.79	liters
lb_m	454	g
in.	2.54	cm

2. Pressure

$$1 \text{ atm} = 760 \text{ mm Hg} = 14.7 \text{ psia} = 0.00 \text{ psig}$$
$$= 33.9 \text{ ft } H_2O = 29.9 \text{ in. Hg}$$

3. Temperature

$$0°C = 32.0°F = 273.15°K† = 491.67°R†$$

† These can be rounded to 273°K and 492°R.

Problems

2-1 A 1-kg mass is accelerated with a force of 10 lb_f. Calculate the acceleration in ft/sec^2 and cm/sec^2.

2-2 What is the weight of a 50-g mass (in CGS units)?

2-3 CO_2 is contained in a vertical cylinder at a pressure of 30 atm by a piston with a mass of 57 lb_m. If g is 32.4 ft/sec^2 and the barometric pressure is 29.7 in. Hg, what is the area of the piston?

2-4 A spring scale is calibrated to read pounds mass at a location where the local acceleration of gravity is 32.40 ft/sec^2. (That is, 1-lb mass placed on the scale will give a scale reading of 1, a 2-lb mass will give a reading of 2, etc.) If the scale is moved to a location where g is 32.00 ft/sec^2, what will be the mass, in pounds mass, of an object which, when placed on the scale, gives a reading of 20.00? Justify your method of calculation.

2-5 A $\frac{1}{2}$-lb_m ball traveling at 60 mph is struck by a bat. If the bat imparts an impulse of 4 lb_f sec on the ball, what is the velocity in ft/sec at which the ball leaves the bat? (*Hint:* You should remember from physics that the impulse is equal to the change in momentum—both vector quantities.)

Thermodynamic Properties 3

3.1 Measurable and Conceptual

The description of a state of matter in terms specific enough to permit a unique characterization is an important first step in the development of thermodynamics. Certain parameters, because of their fundamental significance and our ability to measure them, have long been used to describe a system. Mass, composition, temperature, pressure, and volume are such properties. The mass of a substance can be determined by measuring its interaction with the earth's gravitational field, either with a simple spring scale or by comparing its mass with a known mass by means of a balance. Composition and volume can also be determined by fairly standard procedures. In the case of temperature and pressure we are familiar with instruments or devices that can provide us with an accurate measure of these properties over a large range of each parameter.

Experience has shown that these properties are not sufficient to describe completely all the transformations that matter might undergo. It was in response to such a need that the concepts of energy and entropy were introduced. Neither energy nor entropy can be measured directly on an energy or entropy meter. Values are usually expressed in relation to an arbitrary reference state, as is the case with the steam tables, where all values are calculated relative to liquid water at 32°F. Thermodynamics has been called by many "the science of energy and entropy." Such a statement seems appropriate in that much of the effort in this book will be devoted to giving the student a better understanding of each of these conceptual properties and the laws and principles that have been developed to relate them to physical changes in the world.

Our inability to measure energy and entropy directly frequently leads to apprehensions on the part of the student to accept them as he does temperature, pressure, or volume. He generally has fewer reservations about energy than entropy, because he has discussed the former in relation to heat, work, motors, and engines in the home or in earlier courses. Thus he has at least a superficial understanding of energy in a practical, if not a thermodynamic, sense. The biggest challenge for any instructor or textbook is to develop a similar appreciation for the role of entropy.

Although both energy and entropy will be discussed in greater depth later in this chapter, it seems appropriate to make several observations which help provide a better overview of all properties. Once the concepts of energy and entropy have been fully developed, it will be possible to provide a more significant definition of temperature in terms of these conceptual properties. The Fahrenheit and centigrade temperature scales to which we are accustomed will be shown to be special cases of a more general temperature scale in which the zero value has a special physical significance.

3.2 Intensive and Extensive Properties

The properties discussed in Section 3.1 can be subdivided into two categories: those which relate to the amount of matter present and those which do not. If matter existed in a uniform state and one wished to characterize it with the properties just discussed, the composition, pressure, and temperature would be the same regardless of whether one were to describe all the matter or any part of it. Properties such as these, which are independent of the amount of matter, are termed *intensive properties.*

On the other hand, the mass or volume of the matter would vary directly as the amount is changed, even though the state remains unchanged. Properties that are dependent on the extent or size of the system are termed *extensive properties.* In addition to mass and volume, energy and entropy also fall into the extensive category, as our later discussions will demonstrate.

Extensive properties can be converted to a unit-mass basis by simply dividing the extensive property by the mass of the system. Extensive properties presented on a per-unit-mass basis are termed *specific properties* and are represented by placing a bar beneath the property symbol. Thus for volume, $\underline{V} = V/M$ and has the units ft^3/lb_m. Since the specific properties are independent of the amount of matter considered, they are also intensive in nature. Thus it is possible to convert an extensive parameter to an intensive form.

3.3 Mass and Volume

The mass of a system is a measure of the amount of matter present and is directly proportional to the number of molecules in the system (the proportionality constant being the molecular weight divided by Avogadro's number). The standard unit of mass is the kilogram in the metric (MKS) system and the pound mass in the engineering system. The mass of an unknown body may be determined by comparison with standard masses on a balance, or by use of a scale. However, a scale measures weight rather than mass, so one must convert the weight of the unknown object into its mass by means of the mass–weight expression: $M = Wg_c/g$. Normally the symbol M is used to express mass.

Volume is a measure of the physical size of system and is defined as that portion of space occupied by the system. The volume of a body may be determined directly by measuring its physical dimensions or indirectly by measuring the amount of fluid (say water) that the body displaces. The units of volume are cubic meters in the metric (MKS) system and cubic feet in the engineering system. The symbol V is used to indicate volume.

3.4 Internal Energy

Energy associated with matter is referred to as internal energy and is unrelated to position or velocity (and hence to potential or kinetic energy of the total mass). Internal energy is given the symbol U and is normally expressed in calories in the metric system and Btu in the English system. In an effort to better understand internal energy, it is helpful to examine matter from a microscopic point of view. Inasmuch as internal energy is to be related to the structure of matter, it seems reasonable to assume that an absence of mass in a given system would result in an absence of what we define to be internal energy. Although such an assumption fixes a zero value for internal energy, the problem of determining the internal energy when several molecules are then added to a system is extremely perplexing. Let us now discuss the various types of energies we may associate with atoms and their interactions.

Translational Energy

In the gaseous phase, molecules are known to move about randomly. Each molecule possesses kinetic energy by virtue of its finite mass and velocity. In the liquid state, we find the molecules somewhat more restricted in their movement. Thus the contribution of translational energy is reduced. However, since clusters of molecules can still undergo short-range random motion, translational effects are not completely absent. In the solid state the lattice structure essentially eliminates any possibility of significant translational contributions to internal energy.

Rotational Energy

Consider the CO_2 molecule, which has a dumbbell configuration, as shown in Fig. 3-1. In addition to its translational energy, the molecule may also rotate about some axis through its center of gravity, as indicated in Fig. 3-1. Just as a rotating flywheel possesses rotational energy, so does a rotating molecule.

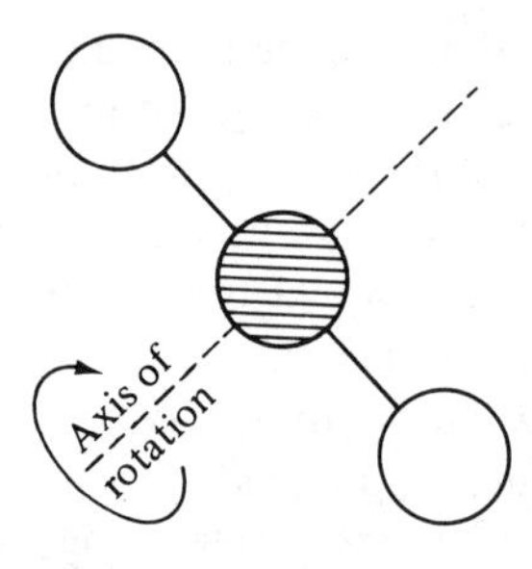

FIG. 3-1. CO_2 molecule.

Rotational energy can be exchanged with neighboring molecules or converted to translational energy by means of molecular collisions. For liquids and solids, rotational contributions are less important because molecular motion is restricted by interactions with other nearby molecules.

Vibrational Energy

Further consideration of the CO_2 molecule suggests the possibility of a vibrational motion within the molecule. Strong bonding forces, which act like a spring, hold the carbon and oxygen atoms together. When excited, the oxygen atoms can vibrate with respect to the central carbon atom as shown in Fig. 3-2.

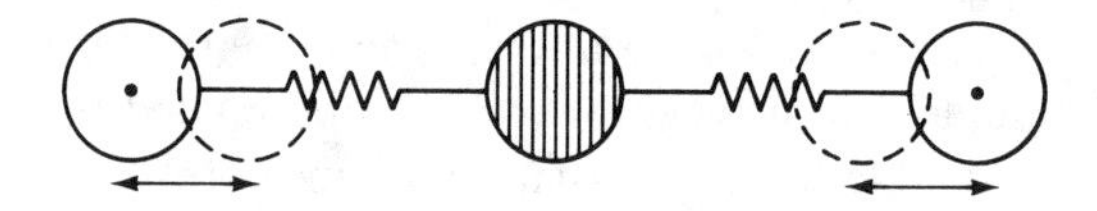

FIG. 3-2. Vibrational energies.

Since these oscillations are capable of activating other molecules through molecular collisions, they, too, must be considered in computing the total internal energy of a substance. Both vibrational and rotational contributions are associated only with polyatomic molecules in the gaseous phase.

In the solid state the lattice structure virtually eliminates the possibility of translational or rotational energy, but the vibrational contribution of atoms about their lattice points is one of the principal contributions to the internal energy of matter.

Atomic and Molecular Interactions

When molecules, or atoms, approach each other, strong intermolecular, or interatomic, forces come into play. Although these forces are repulsive at very short intermolecular spacing, they rapidly become attractive as the spacing increases. Since the energy levels associated with intermolecular forces are quite significant, these forces exert a controlling influence on the state of matter. As long as the molecules possess sufficient translational energy to overcome these attractive forces, matter remains in the gaseous state. As the energy content is reduced, the translational energy may no longer be sufficient to overcome the intermolecular forces; then a change from the gaseous to the liquid state occurs.

In the liquid state the molecules continue to translate, although not nearly as freely as in the gas phase. As the energy content is further reduced, intermolecular forces become more important, as all translation ceases and the liquid state transforms to the solid state.

Nuclear and Electron Contributions

In attempting to account for the internal energy possessed by matter, we have thus far examined only the modes of energy that atoms or molecules might possess by virtue of their relative motion or position. An attempt to further subdivide matter into the more fundamental particles poses great difficulties. Many of our analogies with macroscopic phenomena vanish and quantum mechanical principles must be introduced if these subatomic contributions are to be considered. Until recently thermodynamics did not concern itself with processes that disturbed the atomic makeup of matter. The atomic structure was thus assumed to remain constant, and internal energies have always been referred to a reference state that ignored the enormous energy content of the atom itself. Processes were always assumed to involve only changes in the molecular-energy modes previously cited.

In recent years the plasma (highly ionized gas) has become of engineering significance; therefore, the energies associated with the plasma state warrant some discussion at this point. Modern atomic theory pictures the atom as highly concentrated nuclear matter surrounded by orbiting electrons. Because of the positively charged nucleus, the electrons experience an attractive force that tends to hold them in orbit. If the internal energy of a gas is raised to a sufficiently high level, the outer electrons are energized to a point where they overcome the bonding energy holding them in orbit and move freely throughout the sea of atoms, thus causing the gas to become conductive. This ionization process is similar to the changes that occur in the phase transitions from solid to liquid to gas and thus is frequently said to lead to a fourth state of matter. The movement of electrons from higher or excited levels down to the lower or more stable levels occurs with a release of energy. Thus it can be seen that the energy possessed by electrons within an atom can also contribute to the internal energy of matter.

Finally, the advent of fission has demonstrated that the atom itself possesses a quantity of energy which dwarfs all contributions cited so far. Inasmuch as our concerns in thermodynamics will not involve nuclear reactions, no attempt is made beyond the Einstein relationship to account for this form of internal energy. Thermodynamic analyses generally are concerned with processes producing changes in the state of matter which do not affect its atomic makeup. Thus only those energy changes arising from the translational, rotational, vibrational, and intermolecular modes will be of concern to us when treating systems of fixed composition. If processes are to be studied in which matter experiences a change in its chemical composition, the energy changes associated with chemical bonding must also be considered.

Calculation of Internal Energy

In thermodynamics we are seldom interested in absolute values of internal energy. The working relationships that will be developed in later chapters call for dU or ΔU, the differential or finite internal energy change between two states. Since it is the difference that is of interest to us, we can use an arbitrary reference state as a basis for calculating internal energy in all other states. Since the reference state cancels out when we difference the internal energy between two states, the value we assign to U in the reference state is unimportant. Typically, the reference state is assigned a value of zero at a convenient temperature and pressure, often 32°F and 1 atm.

If, for example, the reference state of water was specified to be liquid water at 32°F, its internal energy could be evaluated in higher and lower energy states by evaluating the changes in the various energy modes which contribute to the internal energy. These changes can be determined experimentally through calorimetric measurements (as discussed in Chapter 4) or, for certain special molecules, theoretically using statistical mechanics and kinetic theory. The calorimetric measurements being macroscopic measurements do not allow us to determine the changes in the various energy modes but give us only the total energy change.

3.5 Entropy

The concept of entropy was first suggested by the German physicist Clausius. Entropy was conceived as a measure of the change in the ability of the universe to produce work in the future, as a result of past or presently occurring transformations or processes. Initially it was considered in a purely macroscopic sense and was related quantitatively to heat, work, and temperature. Subsequent advances in our understanding of the microscopic nature of matter and the application of statistical and quantum mechanical principles to the structure of matter have led to a microscopic interpretation of entropy which is considerably more revealing than that given by macroscopic considerations alone. The introduction to entropy that follows utilizes a microscopic treatment to establish entropy as a state property and attempts to give the student a physical appreciation of its significance. Equally important in this introduction to entropy is the interrelationship it possesses to the fundamental state properties: mass, volume, and energy.

We will begin our discussion of entropy by considering the microscopic characterization of matter. We will quickly observe that a true microscopic characterization is impossible—the amount of information involved is simply too vast. On the other hand we will observe that although microscopic characterization of individual particles is not possible, the large numbers of particles involved makes the use of statistical, or average characterizations extremely valuable. After that, the relationships between statistical characterizations and the entropy function will be examined. Finally, we will close the discussion by considering the relationships between entropy and the macroscopic concepts of heat and work.

Microscopic Characterization of Matter

The complete description of a collection of monatomic molecules in the gaseous state would require the specification of a tremendously large number of coordinate positions and particle velocities. In normal systems where the number of molecules (particles) exceeds billions of billions, a procedure for completely describing such a system is difficult to imagine, even in the age of computers. Furthermore, since the position and velocity of each particle are continuously changing, the problem becomes even more complex. The problem in completely describing a polyatomic molecular system is orders of magnitude more difficult because of the need to account for rotational and vibrational behavior as well. In addition, the Heisenberg uncertainty principle tells us that it is impossible to know exactly what is happening to any single particle at a given instant of time. However, the application of statistical procedures to an entire collection can provide much insight into the behavior of a "typical" particle. Just as insurance companies employ statistical analyses to estimate the number of deaths among a certain element of our population without knowing just who will die, the scientist may use such procedures to predict the position and velocity distribution possessed by a system without relating it to specific particles. The insurance statistician needs certain pieces of information that characterize the population before making such an analysis. Similarly, the statistical characterization of a system of many particles requires the specification of certain parameters: the number of particles, the volume in which they are free to move, and the total energy they possess.

The specification of the number of particles and the volume of the container in which they are free to move is relatively straightforward. If a simple monatomic species is studied under conditions where rotational, vibrational, nuclear, and electronic energy changes are either absent or ignored, the problem of describing the energy of the collection of particles reduces to one in which only the translational contributions are of importance. Under these conditions internal energy changes can be directly related to changes in particle velocities and a much simplified problem results. For this initial exposure to entropy such a simplification is highly desirable.

The specification of internal energy as a system parameter then serves to limit the total translational energy. Without such a limit a given number of particles in a given volume could possess any distribution of velocities (or momenta) and a unique, or most likely, distribution would be impossible to specify. Thus by limiting the internal energy of the entire collection, we have said in effect that the sum of the individual particle energies must total a given amount. Each particle could have the same energy (equal to the total internal energy, U, divided by the number of particles, N), or some presumably could have none and others correspondingly greater amounts.

Ultimately entropy will be shown to serve as an indicator of the equilibrium, or most probable, distribution among a system of particles given values of U, V, and N for the system as a whole. However, its relationship to microscopic behavior can be illustrated equally well, and with greater clarity, for the undergraduate student by considering initially just the spatial distributions of many-particle systems in space. The development that follows considers the placement of a small number of particles (molecules) in a small number of boxes (volume cells in space). The numbers are intentionally small so that the student can follow clearly each step of the development which relates entropy to the most probable configuration.

Consider first the many possible positional arrangements that exist for six particles, numbered 1 through 6, in two volume elements, A and B. Table 3-1 identifies each of the 64 distinguishable distributions that result for such a system. The number of distinguishable distributions may be predicted directly by means of equation (3-1), which is obtained from the mathematics of permutations and combinations:

$$\Omega = M^N \tag{3-1}$$

where Ω = number of distinguishable arrangements
M = number of distinguishable cells
N = number of distinguishable particles

Equation (3-1) illustrates the effect that increasing the number of particles or volume elements has on the number of distinguishable distributions realizable. For example, if we take 1 lb-mole of a gas at standard conditions of temperature and pressure, it will occupy approximately 359 ft^3. If we divide the volume into 1-in.^{3} elements, there would be 620,352 1-in.^{3} volume elements into which approximately 2.5×10^{26} molecules could distribute themselves. Application of equation (3-1) yields 620,352 raised to the 2.5×10^{26} power different ways in which such a system could distribute itself. Needless to say, that number is virtually beyond our comprehension, but it is characteristic of the enormous numbers that we encounter when we attempt to describe systems on a microscopic basis.

In statistical mechanics the term *microstate* is used to characterize an individual, distinguishable distribution. The six particle–two box example possessed a total of 64 such microstates. The term *macrostate* is used to designate a group of microstates with common characteristics. In this case we may choose to consider a macrostate as consisting of all microstates with the same number of particles in compartment A and the same number in compartment B. Thus all microstates that have any 3 particles in compartment A and the other 3 in B would constitute one macrostate. All those microstates with 4 particles in A

TABLE 3-1 Distribution of Six Particles in Two Boxes

Distributions	Compartment A	Compartment B
0–6		1–2–3–4–5–6
1–5	1	2–3–4–5–6
1–5	2	1–3–4–5–6
1–5	3	1–2–4–5–6
1–5	4	1–2–3–5–6
1–5	5	1–2–3–4–6
1–5	6	1–2–3–4–5
2–4	1–2	3–4–5–6
2–4	1–3	2–4–5–6
2–4	1–4	2–3–5–6
2–4	1–5	2–3–4–6
2–4	1–6	2–3–4–5
2–4	2–3	1–4–5–6
2–4	2–4	1–3–5–6
2–4	2–5	1–3–4–6
2–4	2–6	1–3–4–5
2–4	3–4	1–2–5–6
2–4	3–5	1–2–4–6
2–4	3–6	1–2–4–5
2–4	4–5	1–2–3–6
2–4	4–6	1–2–3–5
2–4	5–6	1–2–3–4
3–3	1–2–3	4–5–6
3–3	1–2–4	3–5–6
3–3	1–2–5	3–4–6
3–3	1–2–6	3–4–5
3–3	1–3–4	2–5–6
3–3	1–3–5	2–4–6
3–3	1–3–6	2–4–5
3–3	1–4–5	2–3–6
3–3	1–4–6	2–3–5
3–3	1–5–6	2–3–4
3–3	2–3–4	1–5–6
3–3	2–3–5	1–4–6
3–3	2–3–6	1–4–5
3–3	2–4–5	1–2–6
3–3	2–4–6	1–3–5
3–3	2–5–6	1–3–4
3–3	3–4–5	1–2–6
3–3	3–4–6	1–2–5
3–3	3–5–6	1–2–4
3–3	4–5–6	1–2–3
4–2 5–1 0–6	These are mirror images of the first 3 distributions.	

and 2 in B would constitute a second macrostate; 4 particles in B and 2 in A would represent still a third macrostate.

Table 3-1 shows that the 3-3 macrostate consists of 20 microstates, while the 4-2 and 2-4 macrostates each include 15 microstates. On the other hand, the 6-0 and 0-6 macrostates contain only a single microstate each. It should be noted that different arrangements (permutations) of the same particles within a given box do not constitute additional microstates. Thus for the $0A$-$6B$ microstate, the 123456 arrangement is considered identical to the 234561 arrangement.

The number of combinations for placing N particles in two cells with M in one and $N - M$ in a second is given by the expression

$$C_N^{M,N-M} = \frac{N!}{(N-M)!\,M!} \tag{3-2}$$

Thus for $N = 6$ we see that if $M = 5$,

$$C_6^{5,1} = \frac{6!}{1!\,5!} = 6$$

We observe indeed that there are 6 microstates in the macrostate $5A$-$1B$ and 6 in the macrostate $1A$-$5B$.

SAMPLE PROBLEM 3-1. Deduce a general expression for the number of combinations for placing N particles in three cells with M_1 in one cell, M_2 in a second cell (where $M_1 + M_2 \leq N$). Extend this result to k cells with M_1, $M_2, \ldots, M_k$ particles in each of the cells.

Solution: Let us consider, first, the ways by which the $1A$-$5B$ macrostate in the preceding discussion might be formed. Assume that the first ball picked will go in A and the remaining 5 in B. We may pick any of the 6 for the first, any of the remaining 5 next—and so forth—for a total of $6 \cdot 5 \cdot 4 \cdot 3 \cdot 2 \cdot 1 = 6!$ ways of placing the 6 balls in the two boxes. (Had we chosen to put the second ball drawn in A, then we would have had 6 choices for the first ball in B, 5 for the one in A, 4, 3, 2, and 1 for the remaining choices in B. Again a total of 6!. If we choose any other draw to go in A, clearly we still end up with 6! total permutations possible. Similarly, the result would have been the same for any other macrostate chosen.) However, of these 6! possible orderings, many just involve perturbations of the order in which particles are placed in the boxes, not the particles that finally end up in the two boxes. Thus if we examine the microstate where ball 1 is in box A, and balls 2, 3, 4, 5, and 6 are in B, the orderings

$$\begin{array}{cccccccc} 1 & A & 2 & 3 & 4 & 5 & 6 & B \\ 1 & A & 3 & 4 & 5 & 6 & 2 & B \end{array}$$

are clearly the same microstate. The number of such permutations with particle 1 in A and 2, 3, 4, 5 and 6 in B is

$$1 \cdot (5 \cdot 4 \cdot 3 \cdot 2 \cdot 1) = 1!\,5!$$

In a like manner, every other microstate making up the $1A$-$5B$ macrostate can be formed in $1!\,5!$ permutations. The number of microstates in turn is given by the total number of permutations of the six balls divided by the number of permutations in each microstate that has 1 ball in A and 5 in B. Thus

$$\text{no. microstates} = \frac{6!}{5!\,1!} = C_6^{5,1}$$

This result may easily be generalized to show that the number of microstates in the M_1A-M_2B macrostate ($M_1 + M_2 = 6$) is

$$\text{no. microstates} = C_N^{M_1,M_2} = \frac{N!}{M_1!\,M_2!}$$

If we have cells with M_1, M_2, and M_3 balls in each ($M_1 + M_2 + M_3 = N$), the total number of microstates in the $M_1 - M_2 - M_3$ macrostate is simply

$$\text{no. microstates} = \frac{N!}{M_1!\,M_2!\,M_3!}$$

For k cells we simply add additional factorial terms to the denominator for each additional cell as follows:

$$C_N^{M_1\cdots M_k} = \frac{N!}{M_1!\,M_2!\cdots M_{k-1}!\,M_k!}$$

or, in terms of the continued product, Π,

$$C_N^{M_1\cdots M_k} = \frac{N!}{\prod_{i=1}^{k} M_i!}$$

If each of the six particles has an equal probability of going into either compartment A or B, then any one of the 64 distributions (or microstates) would have an equal probability of occurring each time the six particles are distributed in the two cells. (It is assumed that the order in which particles are arranged in the cells in irrelevant.) However, it can be seen that there is considerably less likelihood of finding macrostate 6A-0B than 3A-3B, because it occurs in just 1 of the 64 possibilities. Its probability is thus $\frac{1}{64}$, as compared with a probability of $\frac{20}{64}$ for 3A-3B or $\frac{15}{64}$ for either 2A-4B or 4A-2B. The probability of finding any given macrostate under these conditions is directly related to the number of microstates it contains: The larger the number of microstates the greater the likelihood of finding that particular macrostate. If the particles are free to fluctuate between the two compartments, we would expect the system to spend a greater fraction of its time in the 3A-3B macrostate than in any of the other six, as shown in Table 3-2.

TABLE 3-2 Macrostates for a Six Particle–Two Box System

Macrostate *A*	Macrostate *B*	No. of Microstates	Fraction of Time in Macrostate
6	0	1	1/64
5	1	6	3/32
4	2	15	15/64
3	3	20	5/16
2	4	15	15/64
1	5	6	3/32
0	6	1	1/64

An examination of Table 3-2 indicates that of the seven possible macrostates, the 3-3 or uniform distribution contains the largest number of microstates and would exist $\frac{5}{16}$ of the time. Inclusion of the 2-4 and 4-2 distributions with the 3-3 arrangement raises to $\frac{25}{32}$ the probability of finding the particles in a "more or less" uniform distribution.

Table 3-3 represents the behavior of a twenty particle–two cell problem which contains 1,048,576 microstates. One begins to get an appreciation for what happens as the number of particles increases. Note that although the most probable macrostate occurs with a probability of only 0.176, the combined probability of the three most likely equals 0.496. Thus such a system would spend approximately half its time in these three macrostates.

TABLE 3-3 Macrostates for a Twenty Particle–Two Box System

Particles in Cell A	Particles in Cell B	No. of Microstates	Fraction of Total Microstates	Fraction of Microstates At Least This Evenly Distributed	Fraction of Microstates Less Evenly Distributed
20	0	1	0.000001		
19	1	20	0.00002		
18	2	190	0.00018	0.99998	0.00002
17	3	1,140	0.00109	0.99962	0.00038
16	4	4,845	0.00462	0.99744	0.00256
15	5	15,504	0.01479	0.98820	0.01180
14	6	38,760	0.03696	0.95862	0.04138
13	7	77,520	0.07393	0.88470	0.11530
12	8	125,970	0.12014	0.73684	0.26316
11	9	167,960	0.16018	0.49656	0.50344
10	10	184,756	0.17620	0.17620	0.82380
9	11	167,960	0.16018		
8	12	125,970	0.12014		
7	13	77,520	0.07393		
6	14	38,760	0.03696		
5	15	15,504	0.01479		
4	16	4,845	0.00462		
3	17	1,140	0.00109		
2	18	190	0.00018		
1	19	20	0.00002		
0	20	1	0.000001		
		$1{,}048{,}576 = 2^{20}$	1.00002		

If this type of analysis is extended to systems containing many particles—say the order of Avogadro's number (10^{23})—we find that the probability of obtaining this "more or less" evenly distributed arrangement becomes overwhelming.

Although slight variations from the most probable distribution are continually occurring, these variations are so small that we have no instruments sensitive enough to measure them. The likelihood of any measurable deviation from an essentially uniform distribution is so small that it may essentially be discounted! For example, Denbigh[1] reports that the probability of observing a 0.001 per cent variation in the density of 1 cm^3 of air is less than 10^{-10^8} and is not likely to be observed in trillions of years. Thus we find that the spontaneous appearance of all the gaseous molecules in one portion of the room is *not totally impossible*; it is just so unlikely that the chances of it occurring are negligibly small.

SAMPLE PROBLEM 3-2. What is the probability of finding all the molecules in a particular cubic inch of space if we have a 1 lb-mole of gas occupying 359 ft^3 of space at standard temperature and pressure? If the molecules are uniformly distributed, how many would each 1-in.^{3} cell contain? How many microstates would make up the latter macrostate? (*Hint*: Stirling's approximation for evaluation of $N!$ when N is large is helpful: $\ln N! \cong N \ln N - N$.)

Solution:

$$359 \text{ ft}^3 = 620{,}352 \text{ in.}^3 = \text{number of cells}$$
$$1 \text{ lb-mole} \cong 2.7 \times 10^{26} \text{ molecules}$$

From equation (3-1): $\Omega = 620{,}352^{2.7\times10^{26}}$ = total number of microstates. If all of the particles are to be in *one particular* cell, there is only one microstate that can exist. The probability, P, of finding the desired macrostate is given by the number of microstates in that macrostate divided by the total number of microstates. Thus

$$P = \frac{1}{(620{,}352)^{2.7\times10^{26}}}$$

If we allow the molecules to collect in any one of the 620,352 cubic inches, then there are 620,352 possible microstates and the probability is given by

$$P = \frac{620{,}352}{(620{,}352)^{2.7\times10^{26}}}$$

A uniform distribution of N particles among M cells would find (N/M) particles in each cell. The number of microstates in such a uniform distribution would be given by

$$C_N^{(N/M)_1\cdots(N/M)_M} = \frac{N!}{\prod_{i=1}^{M}\left(\frac{N}{M}\right)!} = C$$

or upon taking natural logarithms

$$\ln C = \ln (N!) - M \ln \left(\frac{N}{M}\right)!$$

Assuming N is large so Stirling's approximation is applicable, we find

$$\ln C = N \ln N - N - M\left(\frac{N}{M} \ln \frac{N}{M} - \frac{N}{M}\right)$$

[1] K. Denbigh, *The Principles of Chemical Equilibrium*, Cambridge University Press, New York, 1966, p. 59.

or

$$\ln C = N \ln M$$

which gives upon exponentiation

$$C = M^N$$

Thus if we consider 2.7×10^{26} particles in 620,352 cells, we find from Stirling's approximation that the number of ways we can form a uniform particle distribution is

$$C = (620{,}352)^{2.7 \times 10^{26}}$$

The probability of finding this macrostate then is given by

$$P = \frac{C}{\Omega} = \frac{(620{,}352)^{2.7 \times 10^{26}}}{(620{,}352)^{2.7 \times 10^{26}}} = 1.0$$

However, we know this result is seriously in error: The actual probability of finding an *exactly even* distribution is negligible, even though the probability of a *more or less even* distribution is, as we have seen, overwhelming. The Stirling's approximation we have used is not accurate enough to distinguish between the many "almost evenly" distributed arrangements. Rather, the approximation lumps all of these together, and hence we obtain the result that the exactly even distribution is *apparently* always obtained. (Had we examined any of the other nearly even distributions with Stirling's approximation we would have obtained the same result.)

In addition to a spatial distribution, it is necessary to consider the energy distribution of individual molecules in the various energy levels to which the molecules have access. If the energy, U, of the collection is specified along with the number of particles, N, many possibilities exist regarding the energy that any one molecule might possess. For example, the energy possessed by an individual polyatomic molecule may be distributed among its various energy modes, such as translation, vibration, and rotation. A spectrum of energies will be found for each of the modes among the molecules making up the system.

Quantum principles state that in any of these modes only certain discrete energy levels are permissible. Stated differently, this means that a translating molecule cannot possess any velocity but is limited to particular values that depend on system parameters, specifically its volume and energy. Transitions from one energy state to another involve the liberation or absorption of a finite amount of energy that is exactly the amount of energy needed to shift from one quantum state to the next. The process is similar to an electron which jumps from one discrete energy level to an adjacent one by emitting or absorbing a fixed amount of energy.

For any given system of molecules with N, V, and U specified, there is a finite number of permissible energy levels available to each of the molecules. In our earlier examples involving spatial distributions, we have considered a given number of particles and a finite number of volume elements, M, within a given volume, V. The number M was rather arbitrarily selected, and it could be made to approach infinity if one chose to select infinitesimally small elements. Similarly, our consideration of energy distributions will involve the placement of a given number of particles, N, into a large, but finite, number of "energy cells." Each cell (or level, as it is commonly termed) has a specific energy associated with it. It, too, can accommodate varying numbers of particles, as could a volume element. However, if we place a total energy, U, on our system of particles, only distributions in which the sum of particle energies totals U are permitted. Thus, if we define arbitrary energy units (e.u.), such that the permitted energy levels have energies of 0, 1, 2, . . . , 100 e.u., and if we consider the distribution of 10 particles such that the total system energy equals 20 e.u., we immediately recognize that certain distributions are unacceptable. For example, no particle could occupy a level above that with a value of 20 e.u., or the system energy exceeds the prescribed value regardless of which levels the other particles occupy. Such levels are said to be inaccessible for the total energy specified, even though at higher energies some of them might be populated.

In like manner, all 10 particles could not reside in the energy level of 1 e.u., for a system energy of 10 e.u. does not meet the total energy condition of 20 e.u. The reader can immediately identify numerous other distributions that are inconsistent with the stipulated value for U.

The most obvious of the many distributions that would satisfy the energy requirement is the placement of each of the 10 particles in energy level 2 such that the system energy totals 20 e.u. Equally sufficient from a total energy point of view, but far less probable (as we might expect from our earlier discussion), is the distribution of 1 particle in the 20-e.u. level and the remaining 9 in the 0-e.u. level.

As was the case with spatial considerations, the macrostate that contains the greatest number of microstates will have the largest probability of existing. As the number of particles and hence microstates becomes very large, it will have properties that will approach the average properties of all the microstates. Thus the macrostate representing the most probable energy distribution can be used to characterize the equilibrium macroscopic behavior of a system, even though the system may seldom exist in precisely that state. (It should be pointed out again that a many-particle system at equilibrium is continuously fluctuating between many microstates, most of which differ only insignificantly from *the* most probable one.)

SAMPLE PROBLEM 3-3. Six molecules, A through F, are to be distributed among 21 energy levels ranging from 0 to 20 e.u. with no limit on the number of particles per level. Calculate the number of microstates in each possible macrostate if the total energies, U, of the system are 5 and 10 e.u.

Solution: Let us begin by identifying the macrostates that satisfy the conditions for each of the two total energies specified. If we systematically consider all possibilities that exist by first assigning 6 molecules to the 0 level, then 5, 4, 3, 2, 1, and 0 in succession, we minimize the likelihood of omitting any possibilities. Clearly all molecules cannot be in the 0 level or the total energy requirement is not met. However, if we place 5 in the 0 level and 1 in either the 5 or 10 level, we can satisfy the conditions. Since we have six different molecules that could be chosen to place in the 5 or 10 level, there are six microstates comprising each of these macrostates. Recalling Sample Problem 3-2,

$$C_N^{M_1 \cdots M_k} = \frac{N!}{\prod_{i=0}^{k} M_i!}$$

For $N = 6$ and $M_0 = 5$, M_5 or $M_{10} = 1$,

$$C_6^{5,1} = \frac{6!}{5!1!} = 6$$

If we then consider all possibilities in which four molecules remain in the 0 level for each total energy requirement, we shall find two configurations for $U = 5$ e.u. and five for $U = 10$ e.u. Applying the same formula to each distribution permits us to calculate the number of ways each particular configuration (macrostate) can be achieved. Tables S3-3a and S3-3b give all possible arrangements for $U = 5$ and 10 e.u.

TABLE S3-3a Distribution of Six Particles in Various Energy Levels (0, 1, 2, 3, 4, . . . Energy Units)–System Energy = 5 e.u.

	Energy Level													No. of	
Macrostate	0	1	2	3	4	5	6	7	8	9	10	11	12	Microstates	Probablity, %
1	5	0	0	0	0	1								6	2.4
2	4	0	1	1	0	0								30	11.9
3	4	1	0	0	1	0								30	11.9
4	3	1	2	0	0	0								60	23.8
5	3	2	0	1	0	0								60	23.8
6	2	3	1	0	0	0								60	23.8
7	1	5	0	0	0	0								6	2.4
														252	

As Tables 3-3a and 3-3b show, the given internal energy of the collection can be achieved by many different distributions within the permissible energy levels. Just as was the case with spatial distributions, certain energy distributions occur with overwhelming probability (for systems with large numbers of particles and energy levels), because they can be achieved in a great many ways. Under conditions where the system is permitted to remain undisturbed for a sufficient period of time, those distributions that are most probable are the only ones that will be observed macroscopically. As was the case with spatial distributions, the most probable distributions tend to be those with the most uniform distribution of particles among the energy levels.

Although an in-depth treatment of statistical thermodynamics would show entropy to be related to the most probable energy distribution, as well as the spatial distributions discussed earlier, this development intentionally avoids such a detailed treatment because of complicating factors which arise, both conceptually and mathematically, without a commensurate contribution to our understanding of entropy. Since our goal in this treatment is the latter and not one of teaching statistical thermodynamics, we refer the reader to any of several fine textbooks which cover this subject in greater detail.[2]

[2] Knuth, Eldon, *Introduction to Statistical Thermodynamics*, McGraw-Hill Inc., New York (1966). Desloge, Edward, *Statistical Physics*, Holt, Rinehart, & Winston (1966). M. T. Howerton, *Engineering Thermodynamics*, D. van Nostrand Co., Inc. New York (1962).

TABLE S3-3b Distribution of Six Particles in Various Energy Levels (0, 1, 2, 3, 4, . . . Energy Units)–System Energy = 10 e.u.

Macrostate	Energy Level													No. of Microstates	Probability, %
	0	1	2	3	4	5	6	7	8	9	10	11	12		
1	5	0	0	0	0	0	0	0	0	0	1			6	0.2
2	4	1	0	0	0	0	0	0	0	1	0			30	1.0
3	4	0	1	0	0	0	0	0	1	0	0			30	1.0
4	4	0	0	1	0	0	0	1	0	0	0			30	1.0
5	4	0	0	0	1	0	1	0	0	0	0			30	1.0
6	4	0	0	0	0	2	0	0	0	0	0			15	0.5
7	3	2	0	0	0	0	0	0	1	0	0			60	2.0
8	3	1	1	0	0	0	0	1	0	0	0			120	4.1
9	3	1	0	1	0	0	1	0	0	0	0			120	4.1
10	3	1	0	0	1	1	0	0	0	0	0			120	4.1
11	3	0	2	0	0	0	1	0	0	0	0			60	2.0
12	3	0	1	1	0	1	0	0	0	0	0			120	4.1
13	3	0	1	0	2	0	0	0	0	0	0			60	2.0
14	3	0	0	2	1	0	0	0	0	0	0			60	2.0
15	2	3	0	0	0	0	0	1	0	0	0			60	2.0
16	2	2	1	0	0	0	1	0	0	0	0			180	6.1
17	2	2	0	1	0	1	0	0	0	0	0			180	6.1
18	2	2	0	0	2	0	0	0	0	0	0			90	3.1
19	2	1	2	0	0	1	0	0	0	0	0			180	6.1
20	2	1	1	1	1	0	0	0	0	0	0			360	12.2
21	2	1	0	3	0	0	0	0	0	0	0			60	2.0
22	2	0	2	2	0	0	0	0	0	0	0			90	3.1
23	1	4	0	0	0	0	1	0	0	0	0			30	1.0
24	1	3	1	0	0	1	0	0	0	0	0			120	4.1
25	1	3	0	1	1	0	0	0	0	0	0			120	4.1
26	1	2	2	0	1	0	0	0	0	0	0			180	6.1
27	1	1	3	1	0	0	0	0	0	0	0			120	4.1
28	1	0	5	0	0	0	0	0	0	0	0			6	0.2
29	1	2	1	2	0	0	0	0	0	0	0			180	6.1
30	0	5	0	0	0	1	0	0	0	0	0			6	0.2
31	0	4	1	0	1	0	0	0	0	0	0			30	1.0
32	0	4	0	2	0	0	0	0	0	0	0			15	0.5
33	0	3	2	1	0	0	0	0	0	0	0			60	2.0
34	0	2	4	0	0	0	0	0	0	0	0			15	0.5
														2943	

Relationship of Entropy to Microscopic Character of Matter

At this point you may logically ask: What is the connection between these arguments about the probability of a given macrostate and the concept of entropy? In answer to such a question we might make the following observations: As an isolated system moves toward equilibrium, it moves successively from less probable (more ordered) states, where macroscopic potential differences exist, to more probable ones; eventually the system will reach its most probable or equilibrium condition at which point all macroscopic potential differences within the system have vanished. At this point work may no longer be obtained from the (isolated) system. Thus it seems logical to attempt to express entropy in terms of the thermodynamic probability of a system's macrostate, because both approach a maximum as the system approaches equilibrium. Before attempting to relate the two, it is helpful to define the *thermodynamic probability*, p_i, of any macrostate, i, as the number of microstates it possesses:

p_i = number of microstates in the i macrostate

p_{max} = number of microstates in the most probable or equilibrium macrostate

The functional relationship between entropy and thermodynamic probability was first proposed by Boltzmann and may be formulated by recognizing the following conditions, which the functional form must satisfy. We know that entropy, as originally formulated by Clausius and developed by others, is an extensive property. If we combine two systems that are in the same state, the total entropy must be the sum of the entropies of the two systems. However, the thermodynamic probability, (p_{1+2}), for the combined system is the product of the thermodynamic probabilities of the two original systems, (p_1) and (p_2). Thus, although the entropies must be additive, the thermodynamic probabilities of the combined system are multiplicative. The only functional relationship that satisfies these conditions is the logarithmic function. With these considerations in mind it is postulated that the entropy of a given configuration, or macrostate, is related to its thermodynamic probability, p_i, by the simple expression

$$S_i = k \ln p_i \tag{3-3}$$

The proportionality constant, k, is the *Boltzmann constant*, which is equal to the ideal gas constant, R, divided by Avogadro's number. This relationship between thermodynamic probability and entropy is now generally accepted and serves as the basis for the development of statistical thermodynamics.

Although it is not our goal in this book to develop this subject, it should be observed that as our understanding of atomic and molecular behavior is increased, statistical thermodynamics becomes an increasingly valuable tool for the engineer. Its utilization requires an understanding of all the energy modes and the values of energy levels to which particles have access in each of the energy macrostates. Today such information is available for only the simplest of molecules in the gaseous state. If in the future the behavior of more complex molecules can also be described by scientists, then in theory it would be possible to calculate thermodynamic properties directly from a knowledge of the fundamental parameters. Once established, such procedures could eliminate the need for scientists and engineers to devote sizable efforts to obtain experimentally the property data needed to make engineering calculations. However, that day has not yet arrived, and engineers will continue in the foreseeable future to determine much of the property data required for their analyses experimentally.

The value of having introduced equation (3-3) in this development relates to the increased insight one gains with regard to entropy. For each macrostate we can compute an entropy from equation (3-3) provided only that we can determine the number of microstates that comprise the macrostate. Although the number of macrostates for given values of N, V, and U in any real system is enormous, it is, nevertheless, a finite number. Of these, we know that certain macrostates will occur with far greater frequency because of the overwhelming number of microstates which they contain. Figure 3-3 shows the data of Table 3-3 with p_i for each macrostate. The ordinate can be interpreted as either the thermodynamic probability, p_i, or entropy, S_i, for the ith macrostate with an appropriate modification of the scale utilized. Although these data originate from just a twenty particle–two cell distribution, it clearly shows the tendency for the system to cluster about the most probable configurations. As implied earlier, when the number of cells and particles is increased to a level that compares with the number of molecules in a system and the number of energy levels they can occupy, the distribution peaks considerably more sharply about the most probable value.

The macrostates with the largest thermodynamic probability, or largest values of entropy, are those which exist once the system reaches equilibrium. As previously indicated, fluctuations about the equilibrium state occur, but these fluctuations are so small that for practical purposes the system may be considered to be in its equilibrium state. Thus we might associate the value of entropy, labeled S_{max}, with the equilibrium state corresponding to the values of N, V, and U for which the plot applies. A change in any of these parameters would produce a corresponding change in the curve in Fig. 3-3 as well as the system's equilibrium entropy.

One very important point should be learned from the foregoing discussion: that a system with a given N, V, and U possesses a unique value of S only if considered to be in its equilibrium state. If we treat thermodynamics as a study of equilibrium states, entropy retains its state character. However, a system in a perturbed or nonequilibrium condition can be thought of as possessing some value of entropy less than its equilibrium value, as seen in Fig. 3-3. The nonequilibrium system would correspond to a less probable macrostate than that at the peak of the curve, and the entropy of any such state is clearly less than S_{max}. Energy exchange between various energy macrostates or molecules in a nonequilibrium isolated system would move it from less probable to more probable distributions until the system finally reaches the most probable, or equilibrium, state at the peak of the curve.

A three-dimensional plot of entropy, S, plotted against the internal energy, U, and volume, V, of the system is shown in Fig. 3-4. The surface that results represents all the equilibrium states for the possible combinations of U and V for a given value of N. All nonequilibrium states possible would fall beneath the surface, inasmuch as such states would possess entropies less than those corresponding to the equilibrium values.

Two equilibrium states are designated on the surface. Any path connecting

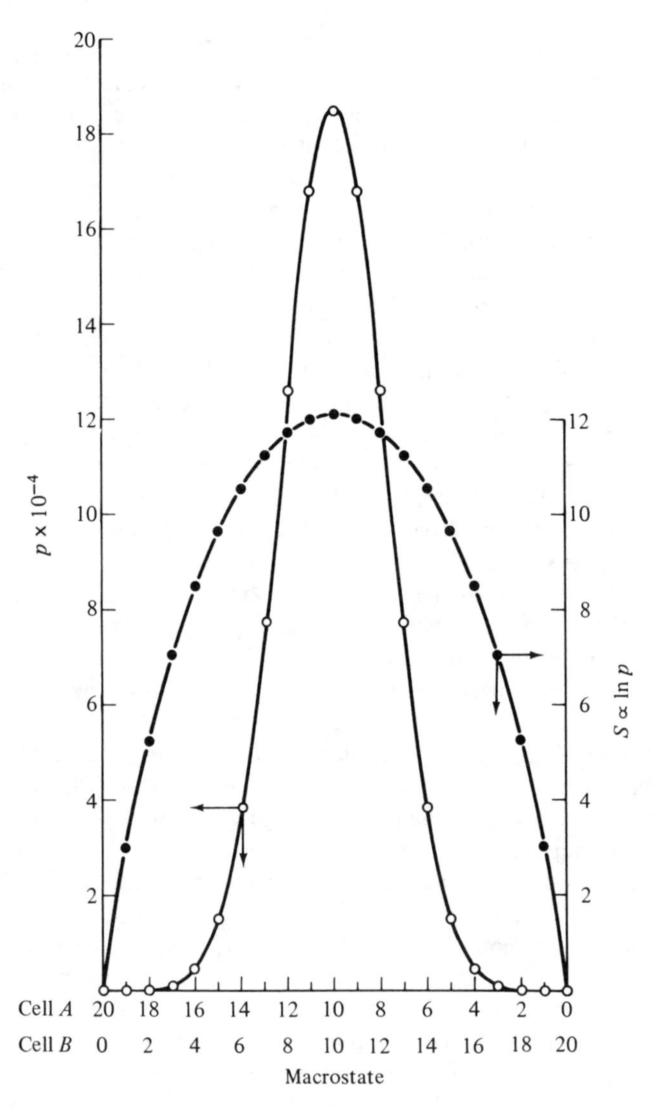

FIGURE 3-3

the two points and lying wholly on the surface would represent a reversible path, because each intermediate state is itself an equilibrium state. There are an infinite number of such paths possible for any system in going from one state to another. (*Note*: Processes that occur along the equilibrium U–V–S surface are also quasistatic processes!)

Similarly, there are an infinite number of paths that depart from the surface which could also be used in taking a system from state 1 to state 2. Such processes represent irreversible and nonquasistatic processes. The difficulties in analyzing such a process should take on added significance in view of the earlier discussion on entropy. Since these processes pass through nonequi-

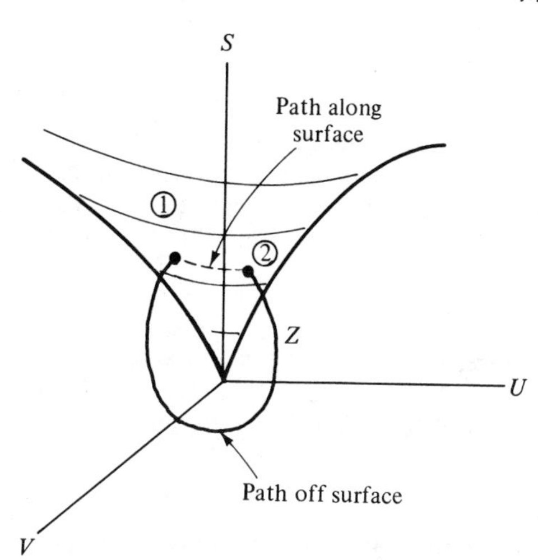

FIG. 3-4. The U–V–S equilibrium surface.

librium states, any given value of U and V can have many different values of S.

In Fig. 3-4 it should be noted that a decrease in U at constant V or a decrease in V at constant U both produce a decrease in S. In view of our previous discussion, this would be expected, because a reduction of either U or V reduces the number of microstates accessible to the system. The greater the energy, the greater the number of ways in which it may be distributed among the various particles (see Sample Problem 3-3). Similarly, as V is increased (equivalent to increasing M, the number of cells, in our earlier discussion), the total number of cells increases, as does the thermodynamic probability corresponding to the most probable macrostate. The latter is directly related to entropy, as equation (3-3) showed.

SAMPLE PROBLEM 3-4. One lb-mole of pure solid copper and three lb-mole of pure solid nickel are brought into intimate contact and held in this fashion at elevated temperatures until the copper and nickel have completely diffused, and the remaining solid is a uniform, random mixture of copper and nickel atoms. If we assume that the original solid copper and nickel were perfect crystals, as is the final mixture, determine the entropy change associated with the mixing of the copper and nickel.

Solution: After the diffusion occurs, the copper and nickel atoms are randomly arranged in the atomic sites available to them. Let us assume that in our molecular ordering the copper atoms are randomly placed among the total number of atomic sites, and that the nickel atoms are then used to fill the remaining sites. For purposes of generality, let us assume that we have C copper atoms and N nickel atoms for a total of $C + N$ atomic sites.

Initially all copper atoms are in the copper matrix and the nickel atoms are in the nickel matrix. If we assume perfect crystals with no atomic movement, then there is only one possible microstate which satifies the initial conditions and the initial entropy is

$$S = k \ln p_i = 0$$

Now after the diffusion, the C copper atoms are randomly dispersed between $C + N$ sites. The total number of ways in which these C atoms can be so arranged is found by observing that the first copper atom can go in any of the $(N + C)$ sites, the second in the $(N + C - 1)$ remaining, and so on until we have

$$(N + C)(N + C - 1) \ldots (N + 1) = \frac{(N + C)!}{N!}$$

arrangements. But of these, $C!$ are simply rearrangements of the C atoms in the same atomic sites, and hence do not represent new microstates. Thus the total number of distinguishable microstates is given by

$$p_i = \frac{(N + C)!}{N!\,C!}$$

and the entropy of the mixture is given by

$$S = k \ln p_i = k \ln \left[\frac{(N + C)!}{N!\,C!}\right]$$

Applying Stirling's approximation and simplifying gives

$$S = k[N + C]\left[\ln (N + C) - \frac{N}{N + C} \ln N - \frac{C}{N + C} \ln C\right]$$

but

$$\frac{N}{(N + C)} = x_{\text{Ni}} \quad \text{and} \quad \frac{C}{(N + C)} = x_{\text{Cu}}$$

where x represents mole fractions. Thus

$$S = k[N + C]\,[\ln (N + C) - x_{\text{Ni}} \ln N - x_{\text{Cu}} \ln C]$$

but

$$x_{\text{Ni}} + x_{\text{Cu}} = 1$$

so that

$$\begin{aligned} S &= -k[N + C]\left[x_{\text{Ni}} \ln \left(\frac{N}{N + C}\right) + x_{\text{Cu}} \ln \left(\frac{C}{N + C}\right)\right] \\ &= -k[N + C][x_{\text{Ni}} \ln x_{\text{Ni}} + x_{\text{Cu}} \ln x_{\text{Cu}}] \end{aligned}$$

but

$$N + C = n_T A, \quad \text{where} \quad \begin{aligned} n_T &= \text{Total number of moles} \\ A &= \text{Avogadro's number} \\ R &= \text{Ideal gas constant} \end{aligned}$$

and

$$R = Ak$$

so that

$$\begin{aligned} S &= -Rn_T[x_{\text{Ni}} \ln x_{\text{Ni}} + x_{\text{Cu}} \ln x_{\text{Cu}}] \\ &= (-1.987)(4.0)[0.75\,(\ln 0.75) + 0.25\,(\ln 0.25)]\,\text{Btu/°R} \\ &= +4.5\ \text{Btu/°R} \end{aligned}$$

and we observe, as expected, that the entropy has increased.

Relationship of Entropy to Macroscopic Concepts

Heat

As was stated earlier, from a historical point of view, entropy was first thought of in a macroscopic sense as relating to heat and its conversion to work. The consistency of the more recent microscopic developments with the classical evolution of the concept was suggested when we examined the relationships among entropy, heat, and work. We just observed that a process in which a system's energy is increased without any change in the number of particles, or in the volume, should lead to an increase in the system's entropy. One method by which such a process could occur would be to transfer energy into a closed, rigid system as heat. Thus a small addition of heat, δQ, should produce a small entropy increase, dS.

$$dS_{N,V} = K\,\delta Q \tag{3-4}$$

where K represents some positive proportionality factor that relates the entropy change to the flow of thermal energy.

We note that equation (3-4) expresses the entropy change in a system that is undergoing a very restricted process—one at constant N and V, one in which energy transfers as heat occur. In Chapter 5 we shall develop methods for calculating dS during more general processes. At that time we will recognize that the entropy change indicated in equation (3-4) is the amount of energy that flows by virtue of the thermal energy (heat) transfer.

Note that in establishing equation (3-4) we have refrained from discussing entropy in an absolute sense. We have simply observed a consistency between our microscopic definition and the relative change in a system's entropy arising from heat exchange. Some insight into both the nature of the proportionality constant and the absolute value of entropy can be gained if one considers what happens to a given system in a microscopic sense when heat addition or removal occurs.

Let us consider as our system a collection of water molecules in the vapor state in a rigid container. We have observed that if such a system is cooled sufficiently (heat removal) it would condense and form liquid water; upon further cooling it would eventually form solid water or ice. Between each of these phase changes a reduction in temperature occurs.

While in the gas phase, the molecules are translating, vibrating, and rotating throughout the container with relatively little interaction. Their distribution is completely random and the fluid possesses no "structure." As heat is removed we would observe a reduction in temperature macroscopically and a slowing down of molecular movement microscopically. Since the total system energy is less, there are fewer possible energy distributions for the system's molecules, and thus the entropy for each successive equilibrium state would be less as cooling proceeds.

At some point in the cooling process the kinetic energy of individual molecules will be reduced to a point where the attractive forces between molecules will become significant relative to translational or kinetic effects and condensation will occur. The liquid state, although still permitting molecules to move about throughout the system, limits molecular movement to a much greater degree than existed in the vapor state. Molecules move in groups or clusters and are always affected by their interactions with neighboring molecules. Some degree of order begins to appear in the system as a result of these interparticle forces. Further cooling slows the molecular movement even more, and the entropy continues to decrease. Finally the point is reached where solidification or freezing occurs.

At this point further removal of energy results in virtually a complete loss of translational energy. Molecules no longer possess the necessary kinetic energy to overcome the short-range intermolecular forces and a crystalline structure results, with each molecule assuming a certain position within a lattice or network arrangement. Molecules continue to vibrate about these positions but not with sufficient energy to move from the lattice point. The absence of the translational mode greatly reduces the number of possible energy distributions and hence we would expect the entropy to decrease rather rapidly while this phase change occurs.

Continued cooling of the solid causes the molecules to vibrate with progressively smaller amplitudes. The cooling process also results in a lowering of the temperature of the ice, as we would expect. If enough energy is removed, the molecules finally cease to vibrate completely and remain in a fixed position at each lattice point. We now have a perfectly ordered lattice in which there is no translational, vibrational, or rotational energy. If we neglect energy associated with the mass of the molecule itself and the intermolecular potential energy, the system now has zero energy. Only one energy distribution meets such a condition, that for which each molecule has zero energy. In addition, since each molecule is at a fixed lattice point (assuming we have a perfect crystalline structure), there is only one configurational arrangement possible. Thus p_{max} becomes 1, and $S = k \ln p_{max} = 0$, or the system's entropy (by our earlier definition) has been reduced to zero.

Had we continued to measure temperature throughout the postulated cooling process, we would have found that on the Fahrenheit scale this condition occurred when we reached $-459.67°F$ ($-273.15°C$ on the centigrade scale). Interestingly enough, if we were to have used any other pure material that forms a (perfect) crystalline lattice in the solid state, we would have observed that all molecular motion would have ceased at exactly the same temperature. This temperature then takes on particular significance in the thermodynamic sense in that it corresponds to that point at which pure crystalline materials have zero entropy. For this reason two new temperature scales (termed *absolute temperature scales*) were conceived, in which temperature differences have the same values as in the Fahrenheit and centigrade scales but in which this unique temperature is labeled zero. These scales are called the *Rankine* and *Kelvin* scales.

These observations led to the *third law of thermodynamics*, which states simply that pure, perfect crystalline substances have zero entropy at the absolute zero of temperature, 0°R or 0°K. It also provides us with a reference base for entropy such that we can later refer to absolute values of entropy as well as to the relative value of entropy changes.

The exact nature of the functional relationship between entropy and heat remains to be discussed. Indeed, it is a rather difficult task to present the justification for the equation that links entropy and heat in a manner completely understandable to most students. However, two specific observations may be made which provide some credence for the function that is used.

It should be remembered that entropy as it relates to an equilibrium state is a state property; that is, its value is independent of the processes by which a given state was achieved. Therefore, the difference in entropy between two states, $\Delta S = S_2 - S_1$, must also be independent of the path taken in moving from state 1 to state 2. If the equation $dS = f(\delta Q)$ is to satisfy such a condition, the right-hand side of the equation must also behave as a state variable such that $\int_1^2 f(\delta Q)$, which equals ΔS, is independent of the path chosen. Heat flow, or δQ, is not a state variable, inasmuch as different amounts of energy exchanged as heat are possible in moving between any two given states. Thus, from a mathematical point of view it is necessary to find an integrating factor that relates heat to entropy. In Section 5.2 it will be shown that reciprocal temperature is the factor we seek such that $K\delta Q$ takes on the character of a state property. Thus the relationship $dS = \delta Q/T$ adequately accounts for that portion of the entropy changes attributable solely to heat transfer.

Although proof of the preceding conclusion is delayed, it may be of value to the student at this point to observe the consistency of this result. At 0°R the entropy of pure crystalline substances becomes zero, because all molecular movement has stopped and a perfectly ordered structure exists. A slight transfer of energy to the system, however, which raises the temperature just above absolute zero could be distributed in a great number of ways in a many-particle system. Thus, for a very small heat addition, δQ, p_{max} goes from unity to a very large number rather quickly. Near the absolute zero of temperature the proportionality factor, $1/T$, becomes very large and thus appears to adequately describe this large entropy change even for a small value of δQ.

Another observation that supports this conclusion is recognition that the entropy change associated with a phase change is directly related to the amounts of the two phases present. If one removes twice as much heat from a saturated vapor, twice as much saturated liquid is produced as long as one stays in a saturated region. Since the entropy of the mixture equals the sum of the entropies of the individual phases, the entropy change produced in the second case would be twice that in the first. During a phase transition, entropy changes bear a constant and direct relation to the amount of heat transfer. Since temperature remains constant during a phase change, its use in the expression $dS = \delta Q/T$ does not lead to an inconsistency.

Work

In the preceding discussion entropy changes were related to a heat effect. Certainly a system's energy can be altered by other than a heat effect and one might justifiably ask: What about entropy changes resulting from energy transfer to a system in the form of work? Consider the compression of gas in an insulated cylinder, a process involving only energy transfer as work. Let us examine the microscopic phenomena associated with this work transfer process. Before the piston begins to move, molecules are striking and rebounding from the piston surface at random. Since the collisions between the molecules and piston will be elastic (unless heat transfer is occurring), the kinetic energy of the molecules will remain constant during collision with the wall, and thus no energy transfer has occurred between the gas molecules and the piston. If the piston begins to move into the fluid (that is, a compression), then the gas molecules will rebound with a higher velocity than they possessed before the collision. Thus energy is transferred into the fluid. *Although the energy in transit across the piston is termed work, once the energy is transferred into the system it clearly becomes part of the internal energy of the system.* Similarly, if the piston moves away from the fluid (an expansion), then the molecules will rebound with a lower velocity than they had when they struck the piston. The energy of the particles is reduced, and a transfer of energy from the molecules to the piston will have occurred.

Thus we observe that during the compression process, energy (in the form of work) is transferred into the system. This energy transfer produces a corresponding increase in the number of energy distributions available to the gas, just as it did in the case of heat addition. However, unlike the constant-volume heating process, the compression is accompanied by a reduction in volume. This change can be thought of as either reducing the number of volume elements throughout which the molecules can distribute themselves or reducing the average size of each element. This reduction tends to compensate for the system's entropy increase, which results from increased energy. Interestingly enough, the two effects exactly compensate one another if the compression occurs reversibly (and with no heat transfer). If the system performs work (reversibly) on its surroundings, the system still experiences no net change in entropy since the decreased energy is accompanied by an increase of volume.

Suppose, however, that a system insulated from its surroundings expands without transferring energy to its surroundings. An example of such a process might be the gas confined by the massless piston shown in Fig. 3-5. Let us assume that no resisting force or atmosphere restricts the motion of the piston to the right. The first molecule that strikes the massless piston after the latch is removed causes the piston to move immediately to the far right part of the container. If the piston does not rebound, then the volume undergoes a step increase. When the piston is struck by the second, third, and following molecules it is no longer moving, and thus no energy interchange occurs between the gas and the piston. The total energy of the gas remains constant while the volume

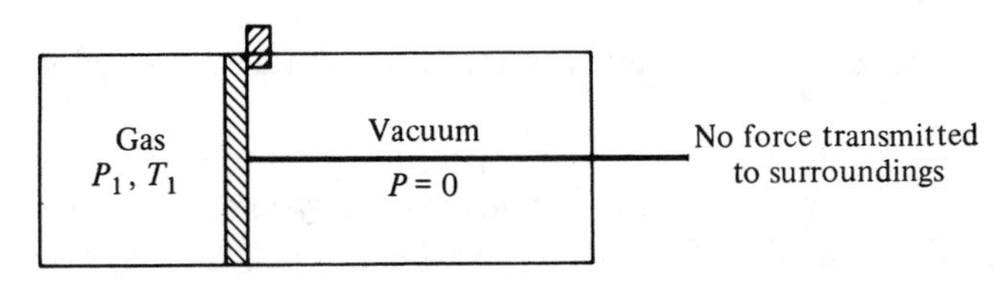

FIG. 3-5. Free expansion of gas.

has increased. The volume increase leads to an increased number of volume microstates. However, since there is no comparable decrease in the number of energy microstates, the total number of microstates increases, as does the entropy of the system—in spite of the lack of an energy transfer.

The process described above is clearly irreversible, because the system can never be returned to its original state without work supplied by the surroundings. Furthermore, it represents a reduction of the pressure potential ($P_2 < P_1$) of the gas without having utilized it to extract energy (as work) from the system. In effect the system, by undergoing this irreversible process, has forever lost the ability to perform a certain amount of work in spite of the fact it still has the same energy. Such a sacrifice is referred to as *lost work*. The symbol *LW* will be used to characterize this quantity, which will be considered in greater detail in Chapter 5. It will be observed that lost work arises any time a process causes the degradation of an energy potential. Furthermore, it will always lead to an increase in entropy, just as it did in this illustration.

The concept of lost work proves useful in later chapters, where the efficiency of energy-conversion processes is of importance. Physically it may be thought of as work which could have been produced had the process been conducted reversibly, but which can no longer be completely realized by any process known to man. It might more appropriately be thought of as "lost potential," inasmuch as energy continues to be conserved throughout any irreversible change. It simply becomes less available to man as he seeks to utilize the energy in the universe to produce change.

The foregoing discussion is intended to relate the macroscopic and microscopic aspects of entropy and to suggest to the student the utility of the entropy concept. By examining certain processes from a microscopic point of view it is hoped that some basis has been established for stating that entropy increases when heat is added to a system or when an irreversible change occurs. In Chapter 5, where the entropy balance is developed, use will be made of these observations, and the student may want to review the contents of this section at that time.

3.6 Temperature and Pressure

Frequent reference has been made to temperature in the preceding sections. It was left to the student's intuition to interpret the temperature in each instance where the term was used. To most people, temperature is thought of as a relative measure of the hotness or coldness of a given body. Similarly, pressure is a familiar concept related to the average force per unit area exerted by a gas on its environs.

An energy potential, such as temperature or pressure, is an intensive property and as such unrelated to the size of the system. They are a measure only of the relative ability of a system to exchange energy. Neither is related in any way to the amount of energy that could be transferred or absorbed. A differential volume at an extremely high pressure, although capable of transferring energy to any system at a lower pressure, can in fact transfer very little.

In previous sections entropy has been shown to be expressible as a function of the other fundamental variables, N, V, and U. The functional relationship between these parameters is termed the *fundamental equation of a system.* Given such a relationship for any system, it would be possible to derive all other thermodynamic information about that system. However, such relationships are not currently available, because of the difficulties in determining the form of the equation and obtaining meaningful values for the conceptual properties, energy and entropy.

A fundamental equation could be expressed with entropy as either a dependent or independent variable. In our earlier discussion it was customary to think of entropy as a dependent parameter with the relationship taking the form $S = S(U, V, N)$. This relationship can be written in differential form as

$$dS = \left(\frac{\partial S}{\partial U}\right)_{N,V} dU + \left(\frac{\partial S}{\partial V}\right)_{N,U} dV + \left(\frac{\partial S}{\partial N}\right)_{U,V} dN \tag{3-5}$$

Because of the state nature of each of the parameters, it is possible to relate differential changes in each of the properties N, V, and U regardless of the process path. Just as the individual parameters are state properties, so are the derivative coefficients of dU, dV, and dN.

It is also possible to express the fundamental relationship in an energy-explicit (or dependent) form: $U = U(S, V, N)$. Assuming an energy-explicit form, the fundamental relation can be expressed in differential form:

$$dU = \left(\frac{\partial U}{\partial S}\right)_{N,V} dS + \left(\frac{\partial U}{\partial V}\right)_{S,N} dV + \left(\frac{\partial U}{\partial N}\right)_{S,V} dN \tag{3-5a}$$

The partial derivatives that appear as coefficients of the differential terms appear so frequently in thermodynamic analyses that they are assigned special symbols and given special designations (purely for convenience sake).

The first of these quantities is termed the thermodynamic temperature and is defined as

$$T \equiv \left(\frac{\partial U}{\partial S}\right)_{N,V} \tag{3-6}$$

The relationship of this thermodynamic temperature to one's intuitive concept of temperature as measured by a thermometer remains to be established. Nevertheless, a careful analysis of temperature as defined here is quite revealing.

First, from a mathematical point of view the derivative of two state variables is itself a state function. The temperature should possess the same state characteristics as the fundamental properties themselves. Second, recalling the relationship between energy and entropy developed earlier, any change in U under constant N and V always produces a change in S of the same sign. Thus the thermodynamic temperature must always have a positive sign.[3] Finally, as observed earlier, the $(\partial U/\partial S)_{N,V}$ would tend toward zero for low energies (the reciprocal of this quantity was shown to become very large at low energy). Thus an absolute zero of temperature is shown to occur at low energies. Figure 3-4 demonstrates the last two points.

From the foregoing observations it may be concluded that the thermodynamic temperature will be a positive state variable with values ranging from zero at extremely low energy levels to very high values as the system energy increases.

Having established the characteristics and properties of the thermodynamic temperature scale we may now ask: How do we measure the thermodynamic temperature? Clearly evaluation of the derivative

$$\left(\frac{\partial U}{\partial S}\right)_{N,V}$$

is an impractical method. We shall defer further consideration of this question until Chapter 7, where we will show how the thermodynamic temperature can be measured and indicate the relationship between the thermodynamic temperature scale and the more commonly encountered Fahrenheit and centigrade scales.

The second of the three derivatives of equation (3-5a) defines the thermodynamic pressure:

$$P \equiv -\left(\frac{\partial U}{\partial V}\right)_{S,N} \tag{3-7}$$

We will see in our later discussion of the mathematics of properties that this pressure is identical to the familiar pressure, which is thought of as the force exerted on a unit area of surface. Although our intuitive feel for pressure may fail us when we consider behavior in solid phases or flowing fluids, the definition of equation (3-7) always retains physical significance.

3.7 *Interrelationship of Properties*

Each of the properties discussed in the previous sections is interrelated with all other physical properties for every state in which matter exists. The study and development of these relationships has been, and continues to be, the subject of much research. It was stated in Chapter 2 that each state of matter is characterized by a unique set of properties. We observe, for example, that when water is taken at atmospheric pressure and 70°F, it always possesses the same

[3] For an interesting discussion concerning the possibility of negative temperatures, the interested reader is refered to F. J. Dyson, What is Heat?, *Sci. Amer.* (Sept. 1954), p. 63.

density or specific volume. Since the state of a substance is not related to the mass of the substance considered, we correctly note that $\underline{V}$, the specific volume, or $1/\underline{V}$, the density, both intensive properties, are parameters that are functions of the temperature and pressure of the substance. Exactly the same observation could be made for steam at a given pressure and temperature, or for any single-component, single-phase substance: that if $\underline{V}$ is measured at a given value of pressure and temperature, it will always have the same value.

Suppose we ask ourselves if the same value of $\underline{V}$ would result if only one property, such as temperature, were specified while pressure was permitted to change. If the pressure on the water was increased to 10 atm and the density very carefully determined, a slight change could be observed. The amount of such a change for liquids and solids, both relatively incompressible phases of matter, would be small but finite. For the gaseous phase, steam, an increase in the pressure on a fixed mass produces an obvious reduction in the volume and hence $\underline{V}$. Thus it is clear that specification of temperature alone does not result in a uniquely defined state for any single-component, single-phase system. Rather, experience indicates that two properties are required to specify a third property for a single-component, single-phase system. However, the choice of properties is somewhat arbitrary. For example, if one chose to fix T and $\underline{V}$ for these systems, the same value of P would always result.

In each of these instances we have observed that specifications of two intensive properties was sufficient to define uniquely a third intensive property. Although pressure, temperature, and volume have been used in this illustration, any other intensive property would behave in an analogous fashion. The specific internal energy for either the water or steam would always be the same, regardless of how many times one were to measure it for given values of P and T. The same would be true for entropy, enthalpy, and the free energies (the latter properties will be introduced later) if measured on a unit mass basis. We thus conclude from many observations that the specification of any two independent intensive properties is sufficient to determine all other intensive properties and hence the state of a single-phase, single-component substance.

If we perform a series of experiments in which two phases of a single component are held in equilibrium with each other, we *observe* that specification of only a single property of one phase *completely specifies the states of the individual phases.* For example, specification of the temperature of a steam–water mixture fixes the pressure exerted by the mixture, as well as the density, entropy, and other properties of the individual phases. If we consider an equilibrium mixture containing three phases of a single component, we would find that *no properties can be arbitrarily specified without losing one phase.* The requirement of three coexisting equilibrium phases for a single component automatically fixes all the properties of the individual phases.

Thus we see for a single-component system, experience indicates that the number of degrees of freedom, f (that is, the number of properties of the individual phases that can be arbitrarily specified before fixing the states of the individual phases), is given by the expression

$$f = 3 - P \tag{3-8}$$

where P is the number of coexisting phases.

As we will show in Chapter 12, equation (3-8) is a special case of the Gibbs phase rule, which expresses the degrees of freedom for a multicomponent–multiphase mixture:

$$f = C + 2 - P \tag{3-9}$$

where C is the number of components.

We shall consider many more implications of the phase rule after our formal derivation in Chapter 12.

Equation (3-9) tells us that a two-phase mixture of a single component can exist only at a single pressure (temperature) once the temperature (pressure) of the mixture has been specified. However, equation (3-9) offers no clue about the form of this relationship. The equilibrium-phase diagram illustrated in Fig. 3-6 allows us to convey concisely the equilibrium pressures and temperatures for which various pairs of phases may be in equilibrium with each other.

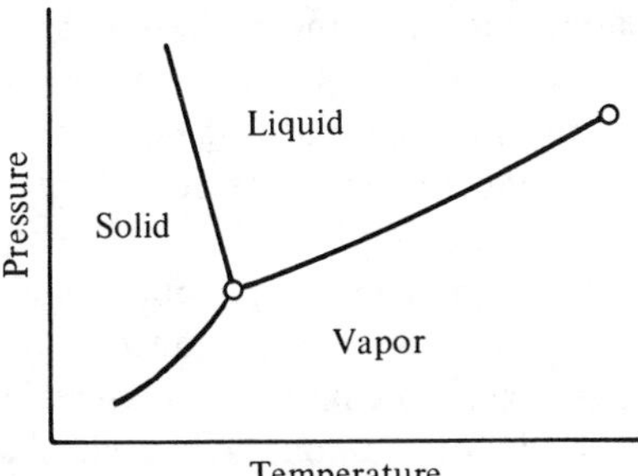

FIG. 3-6. Equilibrium-phase (or P–T) diagram for water.

We may determine the equilibrium phase (phases) of a system by plotting its pressure and temperature on the phase diagram. If the point falls within a single-phase region, the system will exist in that phase. If the point falls on the dividing line between two phases, these phases may coexist at equilibrium (a phase that can exist at equilibrium with one or more other phases is called *saturated*). Thus the solid lines represent the pressures and temperatures at which two saturated phases *may coexist* (there is, however, no guarantee that both phases *do exist*). The intersection of the three lines gives the triple point, the point at which three saturated phases *may coexist*.

The line separating the solid and liquid regions on the P–T diagram is the freezing-point line and shows the effect of pressure on freezing temperature. For almost all substances this line is nearly vertical and indicates that the freezing temperature is relatively independent of pressure. Since the freezing temperature normally increases with pressure, the slope of the freezing-point line is usually positive. Water, however, behaves in the opposite direction and has a negatively sloped freezing-point line, because the freezing temperature of water decreases with increasing pressure. Indeed, it is just this phenomenon that is involved when an ice skate melts a thin film of water under the skate blade. This film of water acts as a lubricant and allows the skate to glide smoothly over the ice. In very cold areas the temperature may drop so low that the skate blade cannot melt the ice. When this occurs, conventional ice skating is not possible.

The line that separates the liquid and vapor phases (and represents the pressures and temperatures at which saturated liquid and vapor may coexist) is known as the vapor-pressure curve. As opposed to the freezing-point line, which may have a positive or negative slope, all vapor-pressure curves have a positive slope (the same is true for the line that separates the solid and vapor regions).

Since the vapor-pressure line has a positive slope, we know that an increased pressure is needed to liquefy a gas, when its temperature is increased. At higher pressures the gaseous phase becomes more dense, and eventually it is impossible to distinguish between the gaseous and liquid phases. The point on the vapor-pressure curve where the two phases become indistinguishable is known as the *critical point*. The pressure and temperatures at the critical point are known as the critical pressure and critical temperature, respectively. The critical temperature is the highest temperature at which a liquid phase may coexist in equilibrium with a separate vapor phase. The critical point is frequently determined by plotting some property (such as density) of the vapor and liquid phases against the equilibrium pressure or temperature and extrapolating to the point where the properties of the phases are equal, as shown in Fig. 3-7. In this manner it is possible to determine the critical conditions without having to observe the disappearance of the liquid phase.

When a system contains a mixture of two saturated phases, it is necessary to know the relative amounts of each phase before the total properties of the mixture can be calculated. For example, the specific volume, $\underline{V}$, of a mixture is given by the total volume of the mixture divided by the total mass of the mixture. The volume of the mixture is the sum of the volumes of the individual phases, as, of course, is the mass:

$$V_{\text{mix}} = \sum_{i=1}^{P} M^i \underline{V}^i \tag{3-10}$$

where i is the phase counter. The specific volume of the mixture is then

$$\underline{V}_{\text{mix}} = \frac{V_{\text{mix}}}{M_{\text{mix}}} = \sum_{i=1}^{P} \frac{M^i}{M_{\text{mix}}} \underline{V}^i = \sum_{i=1}^{P} x^i \underline{V}^i \tag{3-11}$$

where x^i is the mass fraction of mixture in phase i.

Thus the specific volume of the mixture is simply the weighted average of the specific volumes of the individual phases. Although equation (3-11) has been derived for the specific volume of a mixture, the same derivation may be applied to any specific property of the mixture, $\underline{E}_{\text{mix}}$:

$$\underline{E}_{\text{mix}} = \sum_{i=1}^{P} x^i \underline{E}^i \tag{3-12}$$

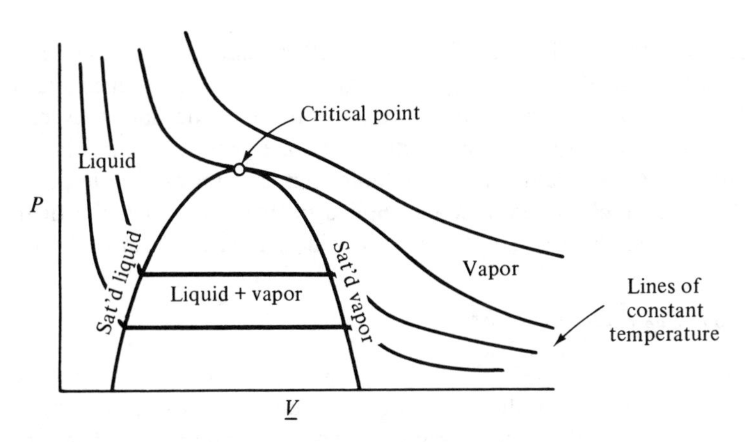

FIG. 3-7. Determination of the critical point.

When the mixture in question consists of only liquid and vapor, the mass fraction of the mixture that is vapor is called the *quality* of the mixture. The fraction liquid is termed the *moisture content* of the mixture.

3.8 Equations of State

The equation of state expresses a relationship between two or more thermodynamic properties. For single-component, single-phase systems the equation of state will always involve three properties, any two of which may be assigned values independently. Although in principle functional relationships involving any three thermodynamic properties (such as T, P, $\underline{V}$, $\underline{U}$, and $\underline{S}$) might be constructed, analytical expressions of the interrelationships among properties have been almost completely limited to P, T, and $\underline{V}$. Because of our incomplete understanding of the interactions that occur among molecules, particularly in the liquid and solid states, empirical methods have been used in developing many of the widely used P–$\underline{V}$–T equations of state. Since pressure, volume, and temperature can all be measured directly, the data needed to evaluate the constants in such equations can be obtained experimentally. Properties such as internal energy or entropy are not directly measurable and must be calculated from other properties and appropriate thermodynamic relations. Equations of state in which $\underline{U}$ or $\underline{S}$ appear as variables have not been independently developed, although we will see in Chapter 6 how these may be derived from the P–$\underline{V}$–T equation of state. (We shall henceforth restrict our use of the term "equation of state" to mean only a relation among P, $\underline{V}$, and T.)

An equation of state may be long and complicated, sometimes involving up to 15 terms—as in the Martin–Hou equation[4]—or short and simple, with as few terms as the one term of the ideal gas equation. The choice of which equation to use in a given application depends greatly on the desired accuracy

[4] J. J. Martin, and Y. C. Hou, *Amer. Inst. Chem. Engrs. J.*, *1*, 142 (1955).

and the endurance of the user. Since the coefficients of almost all equations of state must be evaluated by fitting the equations to various experimental P–$\underline{V}$–T data, the equations can never be more accurate than the data they represent. However, in many instances the equation of state cannot adequately represent the available P–$\underline{V}$–T data, and thus limits our accuracy. This is particularly true when the simpler equations are applied in the region of the critical point. We shall now examine some of the more widely used equations of state.

Ideal Gas Equation of State

The earliest statement of the ideal gas equation of state appears to be *Boyle's law*, which states that the specific volume of a gas at low pressure is inversely proportional to the pressure at constant temperature. That is,

$$PV = f(T) \tag{3-13}$$

Years later, Sir Charles noted that when PV was plotted versus T, a relation of the form shown in Fig. 3-8 was found. Thus a new temperature can be defined such that

$$PV = nRT \tag{3-14}$$

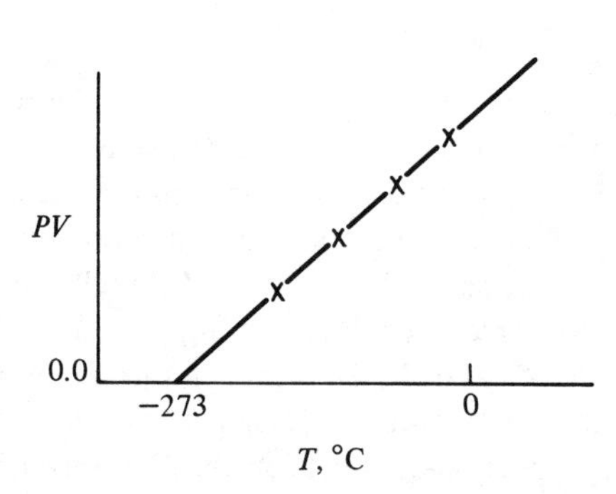

FIG. 3-8. Volumetric behavior of gases at low pressures.

where R is a constant and T is the *ideal gas temperature*. The constant R is the *ideal gas constant* and has different numerical values, depending on the units of P, V, n, and T. Frequently used units of R, and its corresponding numerical values, are presented in Table 3-4.

Equation (3-14) is a formal statement of the ideal gas equation of state. Since it was derived from experimental observations of real gases at low pressures, the ideal gas equation of state is widely used for calculations involving gases and vapors under these conditions. Unfortunately, as the pressure is raised above 5 to 15 atm, most gases no longer behave according to the ideal gas equation of state, and more complicated equations are needed to obtain engineering accuracy.

It is interesting to note that the ideal gas equation of state may be derived from statistical mechanics if the gas molecules are assumed to be infinitesimal spheres which occupy no volume and which undergo completely elastic colli-

TABLE 3-4 Values of the Ideal Gas Constant

Value of R	Units
8.317×10^7	ergs/(g-mol °K)
1.9872	cal/(g-mol °K)
8.3144	J/(g-mol °K)
0.082057	(liter atm)/(g-mol °K)
82.057	(cm^3 atm)/(g-mol °K)
62.361	(liter mm Hg)/(g-mol °K)
0.0848	(kg_f/cm^2 liter)/(g-mol °K)
998.9	(mm Hg ft^3)/(lb-mol °K)
1.314	(atm ft^3)/(lb-mol °K)
1.9869	Btu/(lb-mol °R)
7.805×10^{-4}	(hp hr)/(lb-mol °R)
5.819×10^{-4}	(kw hr)/(lb-mol °R)
0.7302	(atm ft^3)/(lb-mol °R)
555	(mm Hg ft^3)/(lb-mol °R)
10.731	(psi ft^3)/(lb-mol °R)
1545	(lb_f-ft)/(lb-mol °R)
1.851×10^4	(lb_f-in.)/(lb-mol °R)

sions with each other and with the walls of the container. It is also assumed that no intermolecular attraction or repulsion occurs between the gas molecules.

The van der Waals Equation

In an effort to correct the ideal gas equation of state for its two worst assumptions—infinitesimal molecular size and no intermolecular forces—van der Waals proposed the following relation:

$$\left(P + \frac{a}{\underline{V}^2}\right)(\underline{V} - b) = RT \tag{3-15}$$

The term b is meant to account for the finite size of the gas molecules and is sometimes referred to as the "molecular volume." Its value depends on the size and nature of gas molecules. The term $a/\underline{V}^2$ is a correction which was meant to account for the attractive forces that exist between molecules. This attractive force tends to increase the effective pressure on the gas, and therefore is added to the external pressure to get the total effective pressure.

The two constants in the van der Waals equation may be chosen to fit experimental P–$\underline{V}$–T data in any small region. However, since there are only two constants in the equation, we would not expect it to accurately describe P–$\underline{V}$–T data over a great range of P, $\underline{V}$, or T. However, the van der Waals equation does allow for greater accuracy than the ideal gas equation, and therefore may be used when a simple equation of state with somewhat greater accuracy than the ideal gas equation is needed.

It is possible to determine the a and b of the van der Waals equation without recourse to specific P–$\underline{V}$–T data from the following general observation.

Experimental P–$\underline{V}$ data on all real substances *at their critical temperature* indicates that the P–$\underline{V}$ isotherm goes through a horizontal inflection point at the critical pressure of the substance, as shown in Fig. 3-7. That is, both the first and second derivatives of P with respect to $\underline{V}$ vanish at the critical conditions:

$$\left(\frac{\partial P}{\partial \underline{V}}\right)_T = \left(\frac{\partial^2 P}{\partial \underline{V}^2}\right)_T = 0 \tag{3-16}$$

when $P = P_c$ and $T = T_c$.

Substituting these relations into the van der Waals equation we find that a and b are related to the critical P and T of the material by the following relations:

$$a = \frac{27}{64} R^2 \frac{T_c^2}{P_c}$$
$$b = \frac{RT_c}{8P_c} = \frac{\underline{V}_c}{3} \tag{3-17}$$

and

$$\frac{P_c \underline{V}_c}{RT_c} = \frac{3}{8} \tag{3-18}$$

The critical constants (and van der Waals constants calculated from them) for several common gases are presented in Table 3-5.

TABLE 3-5 Critical† and van der Waals Constants for Several Gases

Gas	P_c, atm	T_c, °R	a, atm ft^6 /(lb-mol)^2	b, ft^3 /lb-mol
O_2	50.1	278.6	349	0.510
N_2	33.5	227.1	346	0.618
H_2O	218.3	1165.3	1400	0.486
CO	34.5	240	374	0.630
CO_2	72.9	547.5	924	0.685
CH_4	45.8	343.9	579	0.684
C_2H_6	48.2	549.8	1410	1.04
C_3H_8	42.0	665.9	2370	1.45
C_4H_{10}	37.5	765.2	3670	1.94
NH_3	111.3	729.8	1080	0.598
H_2	12.8	59.9	63	0.427

† Critical constants taken from K. A. Kobe and R. E. Lynn, Jr., *Chem. Rev.*, *52*, 47–236 (1953).

However, it must be stressed that these values will not produce results nearly as accurate as those found when a and b are determined from fitting experimental P–$\underline{V}$–T data over a small region. In fact, the critical volume is, in general, not even predicted accurately by these constants.

If the constants expressed by equation (3-17) are substituted into the van der Waals equation, an equation of state in terms of "reduced" variables is obtained:

$$\left(P_r + \frac{3}{V_r^2}\right)(3V_r - 1) = 8T_r \tag{3-19}$$

where $P_r = P/P_c$ = reduced pressure
$T_r = T/T_c$ = reduced temperature
$V_r = \underline{V}/\underline{V}_c$ = reduced volume

We shall return to this form of the equation at a later time.

The Redlich–Kwong Equation

Unfortunately, the van der Waals equation does not represent experimental P–$\underline{V}$–T data particularly well, except at very low pressures (where the ideal gas equation is not too bad either). The question naturally arises, however, whether any two-constant equation of state can better fit the experimental data. Many forms of two-constant equations have been attempted, but the one that has received the widest acceptance is that due to Redlich and Kwong[5]:

$$P = \frac{RT}{\underline{V} - b} - \frac{a}{T^{1/2}\underline{V}(\underline{V} + b)} \tag{3-20}$$

where a and b are the "Redlich–Kwong" constants. By requiring that the critical isotherm possess a horizontal inflection at the critical point, it is possible to express a and b in terms of only the critical temperature and pressure of the fluid:

$$\begin{aligned} a &= 0.42748\frac{R^2T_c^{5/2}}{P_c} \\ b &= 0.08664\frac{RT_c}{P_c} \end{aligned} \tag{3-21}$$

As with the van der Waals constants, these constants may be substituted into the Redlich–Kwong to obtain the "reduced" form:

$$P_r = \frac{T_r}{\frac{P_c\underline{V}_c}{RT_c}(V_r) - 0.08664} - \frac{0.42748}{T_r^{0.5}V_r\left(\frac{P_c\underline{V}_c}{RT_c}\right)^2\left(V_r + \frac{0.08664}{P_c\underline{V}_c/RT_c}\right)} \tag{3-22}$$

Although the Redlich–Kwong equation is considered by many to be the "best" two-constant equation of state, it is only a two-constant equation and therefore cannot be expected to accurately describe P–$\underline{V}$–T data over extended ranges of pressure and temperature. For greater accuracy—particularly at high pressures and low temperatures—we must use more complex equations with larger numbers of constants.

Beattie–Bridgeman and Benedict–Webb–Rubin Equations of State

Among the more notable of the multiconstant equations of state are those of Beattie–Bridgeman and Benedict–Webb–Rubin:

[5] O. Redlich and J. N. S. Kwong, *Chem. Rev.*, *44*, 233 (1949).

Beattie–Bridgeman equation:

$$P\underline{V}^2 = RT\left[\underline{V} + B_0\left(1 - \frac{b}{\underline{V}}\right)\right]\left(1 - \frac{C}{\underline{V}T^3}\right) - A_0\left(1 - \frac{a}{\underline{V}}\right) \quad (3\text{-}23)$$

Benedict–Webb–Rubin equation:

$$P = \frac{RT}{\underline{V}} + \frac{1}{\underline{V}^2}\left(RTB_0 - A_0 - \frac{C_0}{T^2}\right) + \frac{1}{\underline{V}^3}(RTb - a)$$
$$+ \frac{a\alpha}{\underline{V}^6} + \frac{c}{T^2\underline{V}^3}\left(1 + \frac{\gamma}{\underline{V}^2}\right)\exp\left(\frac{-\gamma}{\underline{V}^2}\right) \quad (3\text{-}24)$$

The Beattie–Bridgeman (B-B) equation of state has five constants; the Benedict–Webb–Rubin (B-W-R) equation has eight constants. These equations, by virtue of their complex nature, are capable of accurately representing P–$\underline{V}$–T data in regions where the van der Waals and Redlich–Kwong equations fail. The B-B and B-W-R equations may be used where the densities are less than 0.8 and 1.2 times the critical density, respectively.

The B-W-R equation, the more complex of the two equations, is capable of somewhat better accuracy than the B-B equation. The mathematical complexity of the B-B and B-W-R equations make all but the simplest hand calculations extremely tedious. Indeed, before the advent of high-speed digital computers even fitting the five or eight constants to experimental data was a task

TABLE 3-6 Constants for the Beattie–Bridgeman Equation†,‡

Gas	A_0	a	B_0	b	$C \times 10^{-4}$
Helium	0.0216	0.05984	0.01400	0.0	0.0040
Neon	0.2125	0.02196	0.02060	0.0	0.101
Argon	1.2907	0.02328	0.03931	0.0	5.99
Hydrogen	0.1975	−0.00506	0.02096	−0.04359	0.0504
Nitrogen	1.3445	0.02617	0.05046	−0.00691	4.20
Oxygen	1.4911	0.02562	0.04624	0.004208	4.80
Air	1.3012	0.01931	0.04611	−0.001101	4.34
CO_2	5.0065	0.07132	0.10476	0.07235	66.00
$(C_2H_5)_2O$	31.278	0.12426	0.45446	0.11954	33.33
C_2H_4	6.152	0.04964	0.12156	0.03597	22.68
Ammonia	2.3930	0.17031	0.03415	0.19112	476.87
CO	1.3445	0.02617	0.05046	−0.00691	4.20
N_2O	5.0065	0.07132	0.10476	0.07235	66.0
CH_4	2.2769	0.01855	0.05587	−0.01587	12.83
C_2H_6	5.8800	0.05861	0.09400	0.01915	90.00
C_3H_8	11.9200	0.07321	0.18100	0.04293	120
n-C_4H_{10}	17.794	0.12161	0.24620	0.09423	350
n-C_7H_{16}	54.520	0.20066	0.70816	0.19179	400

† *Proc. Amer. Acad. Arts Sci.*, *63*, 229–308 (1928); *Z. Physik*, *62*, 95–101 (1930); *J. Chem. Phys.*, *3*, 93–96, (1935); *J. Amer. Chem. Soc.*, *59*, 1587–1589, 1589–1590 (1937); *61*, 26–27 (1939). As reported by G. J. Van Wylen and R. E. Sonntag, *Fundamentals of Classical Thermodynamics*, John Wiley & Sons, Inc., New York, 1965, p. 360.

‡ Pressure in atmospheres, volume in liters/g-mol, temperature in °K; $R = 0.08206$ atm liter/g-mol °K

of considerable size. Today, computers have taken much of the drudgery out of using complicated equations of state, and equations with up to 18 and 20 terms are being developed. However, even with the aid of the digital computer the job of fitting an 18-constant equation to a set of experimental data can be monumental!

Beattie–Bridgeman and Benedict–Webb–Rubin equation-of-state constants for several gases are presented in Tables 3-6 and 3-7.

The Virial Equation of State

The Redlich–Kwong, Beattie–Bridgeman, and Benedict–Webb–Rubin equations are of a semiempirical nature. That is, the form of these equations has generally been chosen such that they can accurately describe experimental data rather than conform to a theoretical description of actual molecular behavior. (The van der Waals equation is not in this class, because the nonideal terms were chosen to account for specific "molecular effects." In addition, we have indicated that the ideal gas equation of state may be derived from statistical mechanics if certain molecular behavior is assumed.) In 1901 Kamerlingh Onnes[6] suggested the *virial equation of state*:

$$P\underline{V} = RT + \frac{B(T)}{\underline{V}} + \frac{C(T)}{\underline{V}^2} + \frac{D(T)}{\underline{V}^3} \tag{3-25}$$

The functions $B(T)$, $C(T)$, $D(T)$, . . . are called the *virial coefficients* and are functions of temperature only. It can be shown (see, for example, Hirschfelder, Curtiss, and Bird[7]) that the virial coefficients are directly related to the molecular forces that exist between groups of molecules. For example, $B(T)$ describes interactions between pairs of molecules, $C(T)$ describes interactions between groups of three molecules, and so on. These interactions are termed *virials* and may be expressed in terms of complicated integrals of the intermolecular forces. Thus, if one knew the intermolecular forces between any combination of molecules as functions of the molecular separations, it would be possible to perform the required integrations and obtain expressions for the virial coefficients without recourse to experimental data. Unfortunately, these calculations are extremely complex and have not been successfully completed except for certain simplified intermolecular force potentials. (Even with simplified force potentials, the calculations have been feasible only for the second, and occasionally third, virial coefficients.) Thus, although the virial equation offers the tempting possibility of describing P–$\underline{V}$–T behavior without recourse to experimental data, computational difficulties prevent us from calculating the virial coefficients without recourse to experimental data. In addition, the particular form of the virial equation makes fitting it to experimental data a fairly difficult task. Therefore, we find that the virial equation, per se, is not frequently encountered in engineering applications. However, as we will see shortly, many of the common-

[6] H. K. Onnes, *Commun. Phys. Lab., Univ. Leiden* (1901).

[7] J. O. Hirschfelder, C. F. Curtiss and R. B. Bird, *Molecular Theory of Gases and Liquids*, John Wiley & Sons, Inc., New York, 1964.

TABLE 3-7 Constants for the Benedict–Webb–Rubin Equation†,‡

Gas	A_0	B_0	$C_0 \times 10^{-6}$	a	b	$c \times 10^{-6}$	$\alpha \times 10^3$	$\gamma \times 10^2$
Methane	1.85500	0.0426000	0.0225700	0.494000	0.00338004	0.00254500	0.124359	0.60000
Ethylene	3.33958	0.0556833	0.131140	0.259000	0.0086000	0.021120	0.178000	0.923000
Ethane	4.15556	0.0627724	0.179592	0.345160	0.0111220	0.0327670	0.243389	1.18000
Propylene	6.11220	0.0850647	0.439182	0.774056	0.0187059	0.102611	0.455696	1.82900
Propane	6.87225	0.0973130	0.508256	0.947700	0.0225000	0.129000	0.607175	2.20000
i-Butane	10.23264	0.137544	0.849943	1.93763	0.0424352	0.286010	1.07408	3.4000
i-Butylene	8.95325	0.116025	0.927280	1.69270	0.0348156	0.274920	0.910889	2.95945
n-Butane	10.0847	0.124361	0.992830	1.88231	0.0399983	0.316400	1.10132	3.4000
i-Pentane	12.7959	0.160053	1.74632	3.75620	0.0668120	0.695000	1.7000	4.63000
n-Pentane	12.1794	0.156751	2.12121	4.07480	0.0668120	0.824170	1.81000	4.75000
n-Hexane	14.4373	0.177813	3.31935	7.11671	0.109131	1.51276	2.80186	6.66849
n-Heptane	17.5206	0.199005	4.74574	10.36475	0.151954	2.47000	4.35611	9.00000

† M. Benedict, G. B. Webb, and L. C. Rubin, *Chem. Eng. Progr.*, *47*, 419 (1951). As reported by G. J. Van Wylen and R. E. Sonntag, *Fundamentals of Classical Thermodynamics*, John Wiley & Sons, Inc., New York, 1965, p. 361.

‡ Units: atmospheres, liters, moles, °K.

ly used equations of state can be reduced to a form that is identical with, or quite similar to, the virial form. In this form the equations of state are particularly convenient to use, and therefore it is frequently useful to convert a given equation of state to the virial form.

SAMPLE PROBLEM 3-5. Convert the Beattie–Bridgeman equation of state into the virial form.

Solution: We begin by writing the Beattie–Bridgeman equation of state in the form in which it is most usually encountered, as shown in equation (3-23):

$$P\underline{V}^2 = RT\left[\underline{V} + B_0\left(1 - \frac{b}{\underline{V}}\right)\right]\left(1 - \frac{C}{\underline{V}T^3}\right) - A_0\left(1 - \frac{a}{\underline{V}}\right)$$

Now perform the multiplications indicated on the right-hand side:

$$P\underline{V}^2 = \left(RT\underline{V} + B_0RT - \frac{RTB_0b}{\underline{V}}\right)\left(1 - \frac{C}{\underline{V}T^3}\right) - A_0 + \frac{A_0a}{\underline{V}}$$

or

$$P\underline{V}^2 = RT\underline{V} + B_0RT - \frac{RTB_0b}{\underline{V}} - \frac{RC}{T^2} - \frac{B_0RC}{\underline{V}T^2} + \frac{RB_0bC}{\underline{V}^2T^2} - A_0 + \frac{A_0a}{\underline{V}}$$

Now collect terms in like powers of V:

$$P\underline{V}^2 = RT\underline{V} + \left(B_0RT - A_0 - \frac{RC}{T^2}\right) + \left(A_0a + RTB_0b - \frac{B_0RC}{T^2}\right)\frac{1}{\underline{V}} + \left(\frac{RB_0bC}{T^2}\right)\frac{1}{\underline{V}^2}$$

Finally, divide by $\underline{V}^2$ to get

$$P = \frac{RT}{\underline{V}} + \left(B_0RT - A_0 - \frac{RC}{T^2}\right)\frac{1}{\underline{V}^2} + \left(A_0a - RTB_0b - \frac{B_0RC}{T^2}\right)\frac{1}{\underline{V}^3} + \frac{RB_0bC}{T^2}\frac{1}{\underline{V}^4}$$

which is the desired fourth-order virial form:

$$P = \frac{RT}{\underline{V}} + \frac{\beta(T)}{\underline{V}^2} + \frac{\gamma(T)}{\underline{V}^3} + \frac{\delta(T)}{\underline{V}^4}$$

where we have set

$$\beta(T) = B_0RT - A_0 - \frac{RC}{T^2}$$

$$\gamma(T) = A_0a - RTB_0b - \frac{B_0RC}{T^2}$$

$$\delta(T) = \frac{RB_0bC}{T^2}$$

Computational Aspects of Equations of State

The recent development of high-speed digital computers has had a marked effect on the types of calculations it is feasible to perform with complex equations of state. Calculations that in the past would have required many days, months, or even years are now routinely performed in a matter of minutes by the computer. As with many other areas of engineering, thermodynamics has been greatly affected by the "computer revolution." Calculations that previously had to be performed by approximate analysis are now handled exactly, or with considerably fewer approximations.

At various points throughout the remainder of this book we shall discuss some of the types of calculations upon which digital computers have had their largest impact on thermodynamic analysis. Since many of these computer-oriented problems have as a common thread some form of equation-of-state calculation, we shall begin our discussion by considering such a calculation. In addition, since many of the problems in later chapters will be considerably more complex, it will be useful to choose one form of equation of state and use this form throughout. In this manner, the programs we develop in the early sections will be useful in the later sections. For purposes of illustration, we shall use the Beattie–Bridgeman equation throughout. This equation has been chosen because it is complicated enough to make hand calculations quite tedious but still simple enough that the programming problems are not overwhelming.

SAMPLE PROBLEM 3-6. Natural gas is to be transported by pipeline from gas fields in Texas to major markets in the Midwest. The gas, which is essentially pure methane, enters the pipeline at a rate of 70 lb_m/sec at a pressure of 3000 $lb_f/in.^2$ and a temperature of 65°F. The pipeline is 12 in. in inside diameter. Calculate the inlet density expressed as lb_m/ft^3 and initial velocity expressed in ft/sec, assuming that methane obeys (1) the ideal gas equation of state and (2) the Beattie–Bridgeman equation of state.

Solution: (a) We begin by discussing the direct solution based on the ideal gas equation of state:

$$P\underline{V} = RT$$

Since $\rho = 1/\underline{V}$, the equation of state becomes

$$\rho = \frac{P}{RT}$$

Before attempting to substitute numbers into this equation, we must examine the units of the various terms to ensure that ρ will have the proper units. Although the unit problem may seem trivial (and actually is quite simple) for the ideal gas equation, we shall find that the unit problem is a significant one when we attempt to deal with the Beattie–Bridgeman equation.

It has been the experience of the authors that the simplest way of overcoming unit problems is simply to convert the units of all physical quantities into a consistent set of units before any calculations are attempted. In this way the need for conversion factors in the governing equation is eliminated and the chance of error significantly reduced. Since the engineering system of units is still used in this country, we shall use this system almost exclusively (except in the later chapters on thermochemistry). Therefore, all dimensions will be expressed in terms of the units feet (ft), pounds force

(lb_f), pounds mass (lb_m), seconds (sec), and degrees Fahrenheit (°F) or degrees Rankine (°R).

In the equation

$$\rho = \frac{P}{RT}$$

we express

$$P = 3000 \text{ psi} = 4.32 \times 10^5 \text{ lb}_f/\text{ft}^2$$
$$T = 65°\text{F} = 525°\text{R}$$
$$R = 1545 \frac{\text{ft-lb}_f}{\text{lb-mol °R}}$$

but the molecular weight of methane is 16 lb_m/lb-mol, so the ideal gas constant (in mass units) is expressed as

$$R = 1545 \frac{\text{ft-lb}_f}{\text{lb-mol °R}} \frac{\text{lb-mol}}{16 \text{ lb}_m} = 96.7 \frac{\text{ft-lb}_f}{\text{lb}_m \text{ °R}}$$

and the density is given by

$$\rho = \frac{4.32 \times 10^5 \text{ lb}_f/\text{ft}^2}{96.7 \frac{\text{ft-lb}_f}{\text{lb}_m \text{ °R}} \cdot 525°\text{R}} = 8.5 \frac{\text{lb}_m}{\text{ft}^3}$$

Note that the units on density automatically give lb_m/ft^3 when all other quantities are expressed as shown.

The entering velocity may be calculated from the expression

$$\dot{M} = \rho u A$$
$$u = \frac{\dot{M}}{\rho A}$$
$$\dot{M} = 70 \text{ lb}_m/\text{sec}$$
$$\rho = 8.5 \text{ lb}_m/\text{ft}^3$$
$$A = \frac{\pi D^2}{4} = \frac{\pi(1 \text{ ft})^2}{4} = 0.785 \text{ ft}^2$$

Therefore,

$$u = \frac{70 \text{ lb}_m/\text{sec}}{8.5 \text{ lb}_m/\text{ft}^3 \cdot 0.785 \text{ ft}^2} = 10.5 \text{ ft/sec}$$

(b) Let us now assume that the methane obeys the Beattie–Bridgeman equation of state:

$$P = \frac{RT}{\underline{V}} + \frac{\beta(T)}{\underline{V}^2} + \frac{\gamma(T)}{\underline{V}^3} + \frac{\delta(T)}{\underline{V}^4}$$

where

$$\beta(T) = RB_0T - \frac{RC}{T^2} - A_0$$

$$\gamma(T) = A_0a - RbB_0T - \frac{RB_0C}{T^2}$$

$$\delta(T) = \frac{RbB_0C}{T^2}$$

We know the values of P and T and can obtain values of the constants A_0, a, B_0, b, and C from Table 3-5. Therefore, the problem is reduced to one of finding that value of $\underline{V}$ which satisfies the Beattie–Bridgeman equation at the temperature and pressure involved. Once $\underline{V}$ is known, the density, ρ, is given by $\rho = 1/\underline{V}$.

The Beattie–Bridgeman and most other commonly used equations of state are pressure-explicit equations of state. That is, given a value of T and $\underline{V}$, pressure can be calculated directly. On the other hand, if T (or $\underline{V}$) is to be calculated from a known P and $\underline{V}$ (or T), the calculation is considerably more involved, because neither T nor $\underline{V}$ can be evaluated directly. That is, most equations of state are implicit in temperature and volume. For certain types of implicit relations an exact solution may be found. For example, the van der Waals and Beattie–Bridgeman equations are simple polynomials in volume. The van der Waals equation is cubic, and the Beattie–Bridgeman equation is quartic. Closed-form solutions for the roots of cubic and quartic polynomials are available. However, the quartic is the highest polynomial for which a closed-form solution is known. Thus, for many equations of state it is not possible to find T or $\underline{V}$ directly. In these instances some form of trial-and-error procedure must be used. If the calculations are to be performed by a computer, this trial-and-error procedure should be of an iterative type—where the results of one trial provide a better value for the next trial. As we will see, it is possible to develop iterative techniques that are extremely efficient; that is, they reach an accurate solution in only a very few iterations. These iterative solutions are so efficient that they are frequently used with even the van der Waals and Beattie–Bridgeman equations, where direct solutions are available but quite cumbersome to use. (How many of you have ever seen the direct solution to a fourth-order polynomial?)

Since we wish to demonstrate the use of digital computations in handling complex equations of state, we shall examine one of the more commonly used iterative techniques for determining volume from the Beattie–Bridgeman equation when pressure and temperature are known. The technique is known as the *Newton* or *Newton–Raphson iteration* and is derived from a Taylor series expansion as follows. Suppose one has the equation

$$f(x) = 0$$

where the form of $f(x)$ is known, and it is wished to find x such that $f(x) = 0$. Also suppose one has a reasonable estimate of the correct x, x^i, at which the value of $f(x)$ is $f(x^i)$. Now write a Taylor series expansion in $f(x)$ about the point x^i. Let x^{i+1} be the root of $f(x)$. That is, $f(x^{i+1}) = 0$:

$$0 = f(x^{i+1}) = f(x^i) + (x^{i+1} - x^i)\frac{\partial f(x^i)}{\partial x} + \frac{(x^{i+1} - x^i)^2}{2!}\frac{\partial^2 f(x^i)}{\partial x^2}$$
$$+ \text{ higher-order terms.}$$

If x^i was fairly close to the correct value, then $(x^{i+1} - x^i)$ is a small number and $(x^{i+1} - x^i)^2$ is a very small number. Let us neglect the terms in $(x^{i+1} - x^i)$ which are raised to the second and higher power. The Taylor series then reduces to

$$0 = f(x^i) + (x^{i+1} - x^i)\frac{\partial f(x^i)}{\partial x}$$

which may be solved for x^{i+1} to give

$$x^{i+1} = x^i - \frac{f(x^i)}{\partial f(x^i)/\partial x}$$

That is, given a reasonable approximation to x, x^i, the procedure gives us (hopefully) a better approximation to x, x^{i+1}. The iteration is simply continued until $f(x)$ is as close to zero as desired.

[A word of caution before we proceed: The first term of the Taylor series expansion that we neglected was of the form

$$\frac{(x^{i+1} - x^i)^2}{2} f''(x^i)$$

As indicated, if x^i is a good approximation to the root, $(x^{i+1} - x^i)^2$ will be very small. However, if the function $f(x)$ has a very high curvature at x^i, $f''(x^i)$ may be a large number, so the term we discarded is not necessarily small. When this occurs the iteration procedure is likely to fail—an experience not uncommon to anyone who has had much experience with the Newton iteration. Fortunately, for our purposes most equations of state are well-behaved functions that are easily solved with the Newton technique. However, if one gets too close to the critical point, the equations of state frequently become less well behaved, and numerical difficulty may be encountered.]

Let us now write the equation of state in the form from which we will solve for $\underline{V}$. Let

$$f(\underline{V}) = 0 = -P + \frac{RT}{\underline{V}} + \frac{\beta(T)}{\underline{V}^2} + \frac{\gamma(T)}{\underline{V}^3} + \frac{\delta(T)}{\underline{V}^4}$$

$$\frac{\partial f(\underline{V})}{\partial \underline{V}} = -\left[\frac{RT}{\underline{V}^2} + \frac{2\beta(T)}{\underline{V}^3} + \frac{3\gamma(T)}{\underline{V}^4} + \frac{4\delta(T)}{\underline{V}^5}\right]$$

but

$$\underline{V}^{i+1} = \underline{V}^i - \frac{f(\underline{V}^i)}{\partial f(\underline{V}^i)/\partial V}$$

so

$$\underline{V}^{i+1} = \underline{V}^i + \frac{\dfrac{RT}{(\underline{V}^i)} + \dfrac{\beta(T)}{(\underline{V}^i)^2} + \dfrac{\gamma(T)}{(\underline{V}^i)^3} + \dfrac{\delta(T)}{(\underline{V}^i)^4} - P}{\dfrac{RT}{(\underline{V}^i)^2} + \dfrac{2\beta(T)}{(\underline{V}^i)^3} + \dfrac{3\gamma(T)}{(\underline{V}^i)^4} + \dfrac{4\delta(T)}{(\underline{V}^i)^5}}$$

For a starting value (to get the iteration under way) we use the ideal gas approximation

$$\underline{V}^\circ = \frac{RT}{P}$$

Now let us unravel the units of the various terms in the equations of state. If we examine Table 3-5 we find that the units of the various constants are not explicitly given. Rather we are told that the units are liters for volume, atmospheres for pressure, °K for temperature, and gram-moles for mass. Thus our first task is to determine the actual units of each constant. Once we know what these units are, we shall convert all units to our previously established system of ft, lb_m, lb_f, sec, and °R or °F.

The equation of state is written as

$$P = \frac{RT}{\underline{V}} + \frac{\beta(T)}{\underline{V}^2} + \frac{\gamma(T)}{\underline{V}^3} + \frac{\delta(T)}{\underline{V}^4}$$

Each term in the sum on the right-hand side must have the units of pressure or the equation is meaningless. The units of $\underline{V}$ are volume/mass [=] liters/g-mol. Thus the units of R, $\beta(T)$, $\gamma(T)$, and $\delta(T)$ must be

$$R\,[=]\,\frac{P\underline{V}}{T}\,[=]\,\frac{\text{atm liter}}{\text{g-mol °K}}$$

$$\beta(T)\,[=]\,P\underline{V}^2\,[=]\,\frac{\text{atm liter}^2}{\text{g-mol}^2}$$

$$\gamma(T)\,[=]\,P\underline{V}^3\,[=]\,\frac{\text{atm liter}^3}{\text{g-mol}^3}$$

$$\delta(T)\,[=]\,P\underline{V}^4\,[=]\,\frac{\text{atm liter}^4}{\text{g-mol}^4}$$

But

$$\beta(T) = RB_0T - \frac{RC}{T^2} - A_0\,[=]\,\frac{\text{atm liter}^2}{\text{g-mol}^2}$$

$$\gamma(T) = A_0a - RB_0bT - \frac{RB_0C}{T^2}\,[=]\,\frac{\text{atm liter}^3}{\text{g-mol}^3}$$

$$\delta(T) = \frac{RB_0bC}{T^2}\,[=]\,\frac{\text{atm liter}^4}{\text{g-mol}^4}$$

As with the equation of state as a whole, the individual groupings of constants in each term must have the units of the term. That is,

$$RB_0T\,[=]\,\frac{RC}{T^2}\,[=]\,A_0\,[=]\,\frac{\text{atm liter}^2}{\text{g-mol}^2}$$

Therefore,

$$A_0\,[=]\,\frac{\text{atm liter}^2}{\text{g-mol}^2}$$

$$B_0\,[=]\,\frac{\text{atm liter}^2}{\text{g-mol}^2 RT}\,[=]\,\frac{\text{liter}}{\text{g-mol}}$$

$$C\,[=]\,\frac{\text{atm liter}^2\,T^2}{\text{g-mol}^2 R}\,[=]\,\frac{\text{liter °K}^3}{\text{g-mol}}$$

we get the units of a and b by examining the units of the $\gamma(T)$ term:

$$\gamma(T)\,[=]\,A_0a\,[=]\,RB_0bT\,[=]\,\frac{\text{atm liter}^3}{\text{g-mol}^3}$$

Thus

$$a\,[=]\,\frac{\text{atm liter}^3}{\text{g-mol}^3 A_0}\,[=]\,\frac{\text{liter}}{\text{g-mol}}$$

$$b\,[=]\,\frac{\text{atm liter}^3}{\text{g-mol}^3 RB_0T}\,[=]\,\frac{\text{liter}}{\text{g-mol}}$$

We now write the values of the five constants as

$$A_0 = 2.2769\,\frac{\text{atm liter}^2}{\text{g-mol}^2}$$

$$a = 0.01855\,\frac{\text{liter}}{\text{g-mol}}$$

$$B_0 = 0.05587\,\frac{\text{liter}}{\text{g-mol}}$$

$$b = -0.01587\,\frac{\text{liter}}{\text{g-mol}}$$

$$C = 1.283 \times 10^5 \frac{\text{liter °K}^3}{\text{g-mol}}$$

These may be converted into the desired units by the following set of conversions:

$$A_0 = 2.2769 \frac{\text{liter}^2 \text{ atm}}{\text{g-mol}^2} 74.7354 \frac{\text{lb}_f\text{-ft}}{\text{liter atm}} 3.35315 \times 10^{-2}$$

$$\times \frac{\text{ft}^3}{\text{liter}} (453.59)^2 \left(\frac{\text{g-mol}}{\text{lb-mol}}\right)^2 (6.25 \times 10^{-2})^2 \left(\frac{\text{lb-mol}}{\text{lb}_m}\right)^2$$

or

$$A_0 = 4.829 \times 10^3 \frac{(\text{lb}_f\text{-ft})\text{ft}^3}{\text{lb}_m}$$

$$a = 0.01855 \frac{\text{liter}}{\text{g-mol}} 0.035315 \frac{\text{ft}^3}{\text{liter}} 453.59 \frac{\text{g-mol}}{\text{lb-mol}}$$

or

$$a = 0.01857 \frac{\text{ft}^3}{\text{lb}_m}$$

$$B_0 = 0.05587 \frac{\text{liter}}{\text{g-mol}} (0.03513)(453.59)(0.0625) \frac{\text{ft}^3/\text{lb}_m}{\text{liter/g-mol}}$$

or

$$B_0 = 0.0593 \frac{\text{ft}^3}{\text{lb}_m}$$

$$b = 0.01587 \frac{\text{liter}}{\text{g-mol}} (0.03513)(453.59)(0.0625) \frac{\text{ft}^3/\text{lb}_m}{\text{liter/g-mol}}$$

or

$$b = -0.01589 \frac{\text{ft}^3}{\text{lb}_m}$$

$$C = 12.83 \times 10^4 \frac{\text{liter °K}^3}{\text{g-mol}} 0.35315 \frac{\text{ft}^3}{\text{liter}} 453.59 \frac{\text{g-mol}}{\text{lb-mol}} 0.0625 \frac{\text{lb-mol}}{\text{lb}_m} 5.82 \frac{\text{°R}^3}{\text{°K}^3}$$

or

$$C = 7.47 \times 10^5 \frac{\text{ft °R}^3}{\text{lb}_m}$$

Finally, the ideal gas constant, temperature, and pressure are given by

$$R = 1545 \frac{\text{ft-lb}_f}{\text{lb-mol °R}} 0.0625 \frac{\text{lb-mol}}{\text{lb}_m} = 96.5 \frac{\text{ft-lb}_f}{\text{lb}_m \text{ °R}}$$

$$T = 65 \text{ °F} = 525 \text{ °R}$$

$$P = 3000 \frac{\text{lb}_f}{\text{in.}^2} = 4.32 \times 10^5 \frac{\text{lb}_f}{\text{ft}^2}$$

A final check of the equation of state will show that the correct units for volume are ft^3/lb_m, just as we want them.

Given below is a Fortran IV computer program designed to perform the desired calculations. Following the program listing is a set of input and output data as processed by the program.

```
C  ---  PROGRAM FOR DETERMINING SPECIFIC VOLUME OF A GAS FROM THE
C  ---  BEATTIE-BRIDGEMAN EQUATION OF STATE WHEN PRESSURE AND
C  ---  TEMPERATURE ARE KNOWN ---
C  ---  DEFINE FUNCTIONS FOR LATER USE ---
        BETA(T) = R*B0*T - R*C/T**2 - A0
        GAMMA(T) = A0*A - R*B0*B*T - R*B0*C/T**2
        DELTA(T) = R*B0*B*C/T**2
        PRESS(T, VBAR) = R*T/VBAR + BETA(T)/VBAR**2 + GAMMA(T)/VBAR**3
       1  + DELTA(T)/VBAR**4
C  ---  READ AND PRINT THE INPUT DATA R, T, P, A0, A, B0, B, C, EPSP ---
   10   READ (105, 20) R, T, P, A0, A, B0, B, C, EPSP
   20   FORMAT (5E16.8)
   30   WRITE (108,40) A0, A, B0, B, C, R, T, P, EPSP
   40   FORMAT (1H1, 36H THE BEATTIE-BRIDGEMAN CONSTANTS ARE, 7H A0 = ,
       1E10.4, 5H A = ,E10.4/ 6H B0 = , E10.4, 5H B = , E10.4, 5H C = , E10.4/
       2 5H R = , E10.4, 5H T = , E10.4, 5H P = , E10.4, 8H EPSP = , E10.4)
C  ---  EVALUATE IDEAL GAS VOLUME AS INITIAL GUESS ---
   50   VBAR = R*T/P
C  ---  BEGIN ITERATION ON VOLUME ---
   60   DO 100 IT = 1, 15
   70   ERR = PRESS(T, VBAR) - P
C  ---  CHECK FOR CONVERGENCE OF ITERATION SCHEME ---
   80   IF (ABS(ERR) .LE. EPSP) GO TO 140
C  ---  UPDATE ESTIMATE OF VBAR ---
   90   VBAR = VBAR - ERR/(-R*T/VBAR**2 - 2.*BETA(T)/VBAR**3 -
       13.*GAMMA(T)/VBAR**4 - 4.*DELTA(T)/VBAR**5)
  100   WRITE (108, 105) IT, VBAR
  105   FORMAT (1H , 6H IT = , I5, 8H VBAR = , E11.4)
C  ---  PRINT COMMENT THAT ITERATION FAILED TO CONVERGE ---
  110   WRITE (108, 120)
  120   FORMAT (1H0, 15H NO CONVERGENCE)
C  ---  RETURN TO START FOR NEW CALCULATION ---
  130   GO TO 10
C  ---  CONVERGENCE OBTAINED, PRINT RESULTS ---
  140   WRITE (108, 150) IT, VBAR
  150   FORMAT (1H , 28H CONVERGENCE OBTAINED AFTER, I5, 19H ITERATION
       1S VBAR = , E 11.4)
C  ---  RETURN TO START FOR NEW CALCULATION ---
  160   GO TO 10
  170   END
THE BEATTIE-BRIDGEMAN CONSTANTS ARE  A0 = .4830E 04 A = .1857E-01
B0 = .5593E-01 B = -.1589E-01 C = .7476E 06
 R = .9650E 02 T =  .5250E 03 P =  .4320E 06 EPSP =  .1000E 03
 IT =   1 VBAR =  .8416E-01
 IT =   2 VBAR =  .9150E-01
 IT =   3 VBAR =  .9255E-01
CONVERGENCE OBTAINED AFTER  4 ITERATIONS VBAR =  .9255E-01
```

3.9 The Law of Corresponding States

We have shown that both the van der Waals and Redlich–Kwong equations of state may be expressed in terms of a single equation in reduced properties that should be valid for all gases:

van der Waals equation:

$$\left(P_r + \frac{3}{V_r^2}\right)(3V_r - 1) = 8T_r \tag{3-26}$$

Redlich–Kwong equation:

$$P_r = \frac{T_r}{(P_cV_c/RT_c)V_r - 0.08664} - \frac{0.42748}{T_r^{0.5}V_r(P_cV_c/RT_c)^2[V_r + (0.8664RT_c/P_cV_c)]} \tag{3-27}$$

Although we recognize that neither of these equations is a quantitatively correct representation of real gases, they do represent the qualitative behavior of most gases. Thus they suggest that it may be possible to develop a "generalized" equation of state in terms of reduced properties that would be applicable to all real gases. That is, if we specify any two reduced properties (P_r, T_r, or V_r), the third one is fixed. Since pressure and temperature are the easiest properties to arbitrarily fix, these are the properties usually specified. Gases at the same reduced pressure and temperature are said to be in *corresponding states*. The *law of corresponding states* states that *all fluids in corresponding states should have the same reduced volume.*

We now introduce the compressibility factor,

$$Z = \frac{P\underline{V}}{RT} \tag{3-28}$$

as a measure of deviation of real gas behavior from ideal gas behavior. The compressibility may be expressed in terms of reduced variables as

$$Z = \frac{P_rV_r}{RT_r}\frac{P_c\underline{V}_c}{T_c} = \frac{P_rV_r}{T_r}\frac{P_c\underline{V}_c}{RT_c} \tag{3-29}$$

The term $P_c\underline{V}_c/RT_c$ is simply the compressibility at the critical point, Z_c, so equation (3-28) becomes

$$Z = Z_c\frac{P_rV_r}{T_r} \tag{3-30}$$

Since the law of corresponding states indicates that V_r is a universal function of P_r and T_r, then equation (3-30) suggests that the compressibility factor should also be a universal function of P_r and T_r for all gases that have the same critical compressibilities. Examination of the critical properties (P_c, $\underline{V}_c$, and T_c) of most real gases indicates that their critical compressibilities fall within a fairly restricted range $0.25 \leq Z_c \leq 0.31$. Thus the law of corresponding states indicates that the compressibility factors for all gases should be essentially a universal function of P_r and T_r. Shown in Fig. 3-9 are some experimental curves of compressibility factors as functions of P_r for various values of T_r.

Examination of the data in Fig. 3-9 shows that it is possible to adequately represent the various Z versus P_r isothermal data by single curves. For many engineering applications, the law of corresponding states expressed in terms of compressibilities, with constant Z_c, is a useful method for correlating the P–$\underline{V}$–T data of many gases onto a single curve. A generalized compressibility chart for $Z_c = 0.27$ is presented in Appendix C.

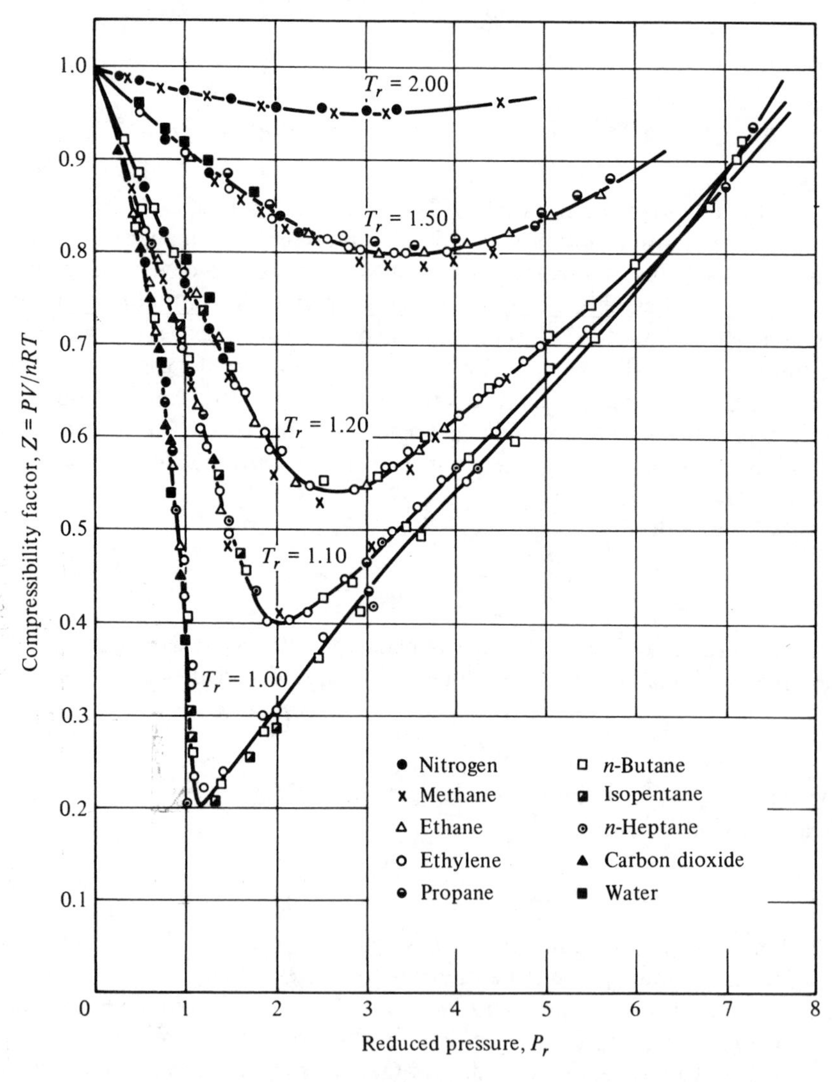

FIG. 3-9. Gas compressibilities as functions of reduced pressure and temperature. [From Gouq-Jen Su, *Ind. Eng. Chem.*, *38*, 803 (1946).]

In an effort to increase the accuracy of results obtained from the law of corresponding states, many attempts have been made to add a third correlating parameter in addition to P_r and T_r. The most widely accepted of these parameters appears to be the critical compressibility. Thus Z is presented as a function of P_r, T_r, and Z_c rather than simply P_r and T_r. Most correlations that do not use Z_c as a parameter correspond to $Z_c = 0.27$, which appears to be a reasonable value for Z_c when no other information is available.

3.10 Other Property Representations

In theory an equation of state could be developed which relates any property of a system to any two other properties. However, the extremely complicated nature of many of these equations would make their use extremely cumbersome even where computers are available to handle the calculations. To present property information in an accurate, but convenient, form for hand or graphical calculations, various property charts and tabulations have been developed. Frequently these charts and tables are prepared from calculations based on one of the complex equations of state. However, once tabulated, or plotted, these data are readily available.

Probably the simplest form of presenting property data is through the property table, or tabulation. In these tables we usually find $\underline{H}$, $\underline{V}$, and $\underline{S}$ presented as functions of P and T. Since the property tables must be of finite size, $\underline{H}$, $\underline{V}$, and $\underline{S}$ can only be presented at a certain number of values of pressure and temperature. If the properties at an intermediate P and/or T are needed, then some form of interpolation must be used. Property tabulations usually allow greater precision than a chart, but the charts have the distinct advantage of presenting the data in a somewhat more convenient, and readily usable, form.

Property tabulations that cover large regions of P and T are available for only a very limited number of gases and liquids. Some of the more readily obtainable property tabulations are those for steam, Freon 12 and Freon 22, NH_3, SO_2, H_2, O_2, and air [see, for example, R.H. Perry et al., *Chemical Engineers' Handbook* (McGraw-Hill, New York, 4th ed., 1963), and E.I. Du Pont de Nemours, *Freon Property Tabulations*].

Because they are so convenient and simple to use, property charts are perhaps the most widely used method of presenting property data. Two properties are plotted along the horizontal and vertical axes with constant values of additional properties superimposed as lines to complete the chart.

We shall now examine some of the more commonly encountered property charts.

P–$\underline{V}$ Diagram

The pressure–volume diagram is a graphical representation of the P–$\underline{V}$–T behavior of a given substance as illustrated in Fig. 3-7. In the figure are shown lines of constant temperature, or isotherms. Movement along an isotherm corresponds to an isothermal compression or expansion and gives the density or specific volume as a function of pressure for any given temperature.

Several regions of interest occur in the P–$\underline{V}$ diagram: The region above the line $T = T_c$ corresponds to temperatures greater than the critical temperature, and therefore represents a fluid state. The region below the line $T = T_c$ corresponds to temperatures below the critical temperature, where it is possible to have more than one phase in equilibrium. The saturated liquid and saturated vapor lines represent the specific volume–pressure relations for the saturated liquid and vapor, respectively. The point where the saturated liquid and saturated vapor lines meet is, of course, the critical point. Note that the critical isotherm is tangent to the saturated vapor–saturated liquid line at the critical point.

The region below the saturated vapor–saturated liquid line is the two-phase region. In this region specification of P and T will not fix $\underline{V}$ of the system. $\underline{V}$ may, however, be determined, as previously demonstrated, from $\underline{V}^{\text{liq}}$ and $\underline{V}^{\text{vap}}$ and the mass fraction of liquid.

The region to the left of the saturated liquid line and below the $T = T_c$ isotherm corresponds to a liquid phase; the region to the right of the saturated vapor line and below the $T = T_c$ isotherm corresponds to a gaseous or vapor phase.

T–S Diagram

The temperature–entropy diagram is one of the more useful of the property diagrams. It finds its greatest application in the analysis of heat and power cycles, and therefore is most readily available for those materials which are used in these cycles. An example of a typical T–S diagram is shown in Fig. 3-10.

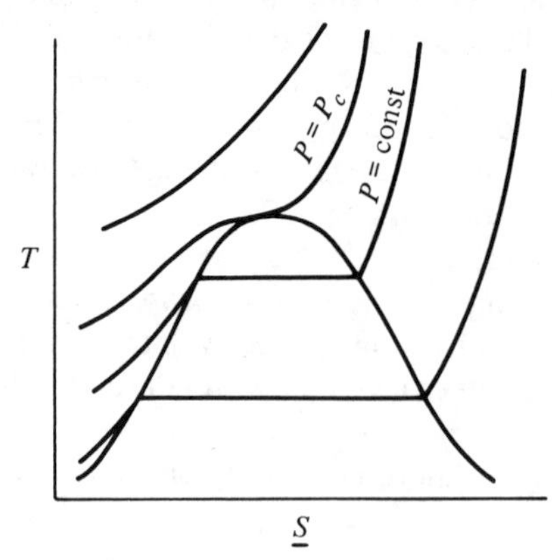

FIG. 3-10. Temperature–entropy diagram.

The dome enclosed by the saturated liquid–saturated vapor line represents the two-phase region. The lines "$P = \text{const}$" represent lines of constant pressure, or isobars. The line $P = P_c$ is the critical isobar and is tangent to the saturation dome at the critical point. Lines of constant enthalpy or constant volume are also frequently included on T–S diagrams.

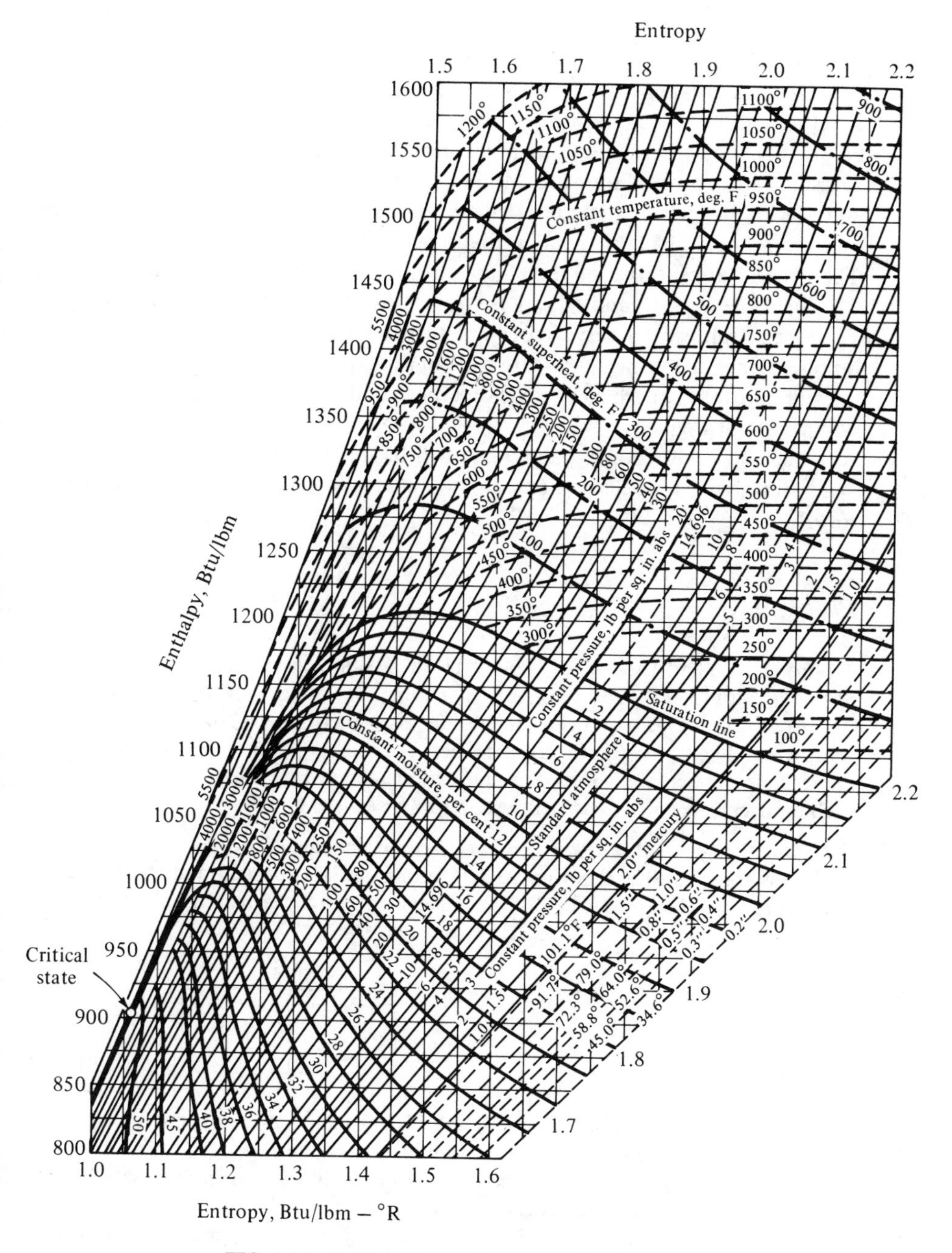

FIG. 3-11. Mollier diagram for steam (H_2O). (From Joseph H. Keenan and Joseph Keyes, *Thermodynamic Properties of Steam*, John Wiley & Sons, Inc., New York, 1936, by permission.

Mollier, or H–S, Diagram

The Mollier diagram is an enthalpy–entropy diagram (the enthalpy, $\underline{H}$, equals the internal energy, $\underline{U}$, plus the product $P\underline{V}$) with lines of constant temperature and pressure (and others on occasion) added for extra usefulness. It is commonly used for calculations involving heating, cooling, expansion or compression, and so on. Complete Mollier diagrams are available for only a few substances, but when they are available they are extremely useful. One of the most commonly available Mollier diagrams is that for the steam–water system. The general appearance of the steam Mollier diagram is shown in Fig. 3-11. The lines $P = P_1, P_2$, and P_3 represent lines of constant pressure. Lines of $T = T_1, T_2, \ldots$ represent lines of constant temperature; lines of $X = X_1, X_2 \ldots$, represent lines of constant moisture content (lines of constant quality). Sometimes lines of constant superheat (where superheat $= T - T_{sat}$) and lines of constant volume are also given.

Several other property diagrams will be introduced and used as the need arises during our later discussions. In most cases these diagrams are self-explanatory.

Problems

3-1 (a) If you had 1 lb-mol of gas at STP (32°F and 1 atm), what is the probability of finding all the molecules in a particular cubic foot of gas? In any one cubic foot?

(b) If it could be assumed that the life of a microstate is 10^{-6} sec, how many years would likely pass (on the average) between such occurrences? Would it make an appreciable difference if we made the life a microstate 10^{-100} sec? Could you detect such an occurrence if it lasted 10^{-6} sec? 10^{-100} sec?

3-2 Pure nitrogen and pure oxygen in two compartments of a cylinder are separated by a membrane as shown in Fig. P3-2. The membrane is broken and the two gases allowed to mix. Has the entropy of the universe increased, decreased, or remained the same? Justify your answer in at least two ways.

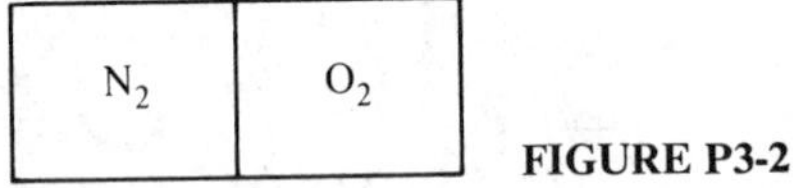

FIGURE P3-2

3-3 Two blocks of copper are initially at different temperatures. The two blocks are held together until the temperatures in both blocks are the same.

(a) What has happened to the entropy of the block that initially was hotter?

(b) Is this a reversible process?

(c) How do we reconcile our answer to (a) with our statement that during an irreversible process the entropy always increases?

(d) What has happened to the entropy of the block that was initially cool?

(e) How does the magnitude of the change of entropy for the hot block compare with the magnitude of the change in entropy for the cold block?

3-4 Let us consider two similar heat exchangers, both working with the same fluids and the same rates. In one exchanger there is an average temperature difference of 10°F between hot and cold streams. In the second exchanger the temperature difference is 20°F. Since the same amounts of heat are to be transferred by each exchanger, the unit with the 20°F ΔT is only half as large as the unit with 10°F ΔT and therefore will cost considerably less to install. However, there are other factors involved in choosing the actual ΔT that will be used.

(a) Will either exchanger operate reversibly?

(b) Which exchanger will have the greater entropy production?

(c) Which exchanger will be more efficient in a thermodynamic sense? (That is, which exchanger will recover more of the available energy from the high-temperature stream?)

This problem should serve to illustrate that with every irreversible process there is a loss in potential and a corresponding increase in the entropy of the universe. The more irreversible a process is (that is, the higher the rate of the process), the greater this loss in potential and the higher the increase in entropy.

3-5 Suppose we have two connected cylinders as illustrated in Fig. P3-5. Initially the cylinder on the left is filled with a gas at a pressure, P. The cylinder on the right is completely evacuated. Suddenly the valve is opened. Explain, using three different concepts, whether there is a change in the entropy of the system and, if so, whether it is an increase or decrease.

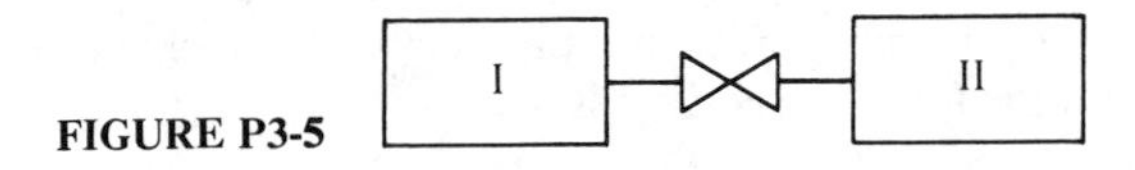

FIGURE P3-5

3-6 A blob of putty is sitting on the top of a wall. If somebody pushes the putty off the wall so that it falls to the ground, what has happened to the entropy of the universe?

3-7 If we had 6×10^{20} particle in two boxes,

(a) What do you think the probability of having exactly 3.0×10^{20} particles in any one box is?

(b) What do you imagine to be the probability of finding a "more or less" even distribution where no more than 50.001 percent of the particles were in any one box?

3-8 Let us extend our concept of ordered versus disordered motion from speaking of atoms to speaking of electrons. Suppose we have a stream of moving electrons (an electric current) in a conductor.

(a) Can you describe some methods in which usable energy could be obtained from these electrons?

(b) If the electron stream passes through a resistor so that its potential (voltage) is decreased, what has happened to the energy of the electron stream?

(c) What has happened to the entropy of the universe?

3-9 We have stated that the entropy of an isolated system at equilibrium must be maximized (for the total volume and energy of the system). Show that this statement is in agreement in at least two different ways with our description of entropy.

3-10 Sugar placed in a cup of hot coffee slowly dissolves. What has happened to the entropy of the universe? Justify your answer in two ways. What happens to the entropy of the coffee–sugar mixture as the mixture cools to room temperature? Justify your answer.

3-11 The Clausius statement of the second law of thermodynamics may be phrased as follows: It is impossible to construct a device that operates in a cycle and produces no effect other than the transfer of heat from a cooler body to a hotter body. Bear in mind that a device operating a cycle periodically returns to its original state and thus exhibits no net change in its properties with time. Show that the Clausius statement is consistent with our statements regarding entropy.

3-12 During the passage of a strong pressure (or shock) wave through a polyatomic gas, it is found that the energy which the shock wave transfers to the gas appears mainly in the translational and rotational forms. Very little increase in the vibrational energy is noted immediately after passage of the shock wave. However, as time passes, random molecular collisions transfer energy from the translational and rotational modes to the vibrational modes, and equilibrium is finally achieved.

(a) Describe the entropy change in the universe which occurs during the period after the passage of the shock wave, but before the vibrational energy approaches its equilibrium value.

(b) Justify your answer to (a) with three different arguments.

3-13 It has been suggested that methane in pressurized cylinders be used as an emergency fuel for a plant heating system which normally uses natural gas (largely methane). A sufficient stockpile of gas cylinders must be kept on hand to provide 100,000 Btu/hr for 24 hr. If methane yields 175,000 Btu/lb-mol upon burning and is available in 2.0-ft^3 cylinders at 3000 psia and 70°F, how many cylinders must be kept in the stockpile? Obtain predictions based on each of the following equations of state:

(a) The ideal gas equation.
(b) The van der Waals equation.
(c) The Beattie–Bridgeman equation.
(d) The law of corresponding states.

3-14 For gaseous chlorine the Redlich–Kwong constants, a and b, are

$$a = 134$$
$$b = 0.039$$

when P is measured in atm, $\underline{V}$ in liters/g-mol, and T in °K. What pressure must a 5-ft^3 cylinder of chlorine withstand if it is to contain 0.1 lb-mol of chlorine at 72°F?

3-15 Throttling devices are usually assumed to be adiabatic, so the flow through them is treated as being isenthalpic. However, these devices do not operate in an isentropic manner. If the state function enthalpy ($\underline{H}$) is constant across the throttle and the throttle operates adiabatically, why is not the entropy, also a state function, constant?

The Energy Balance 4

Introduction

In this chapter we shall develop and apply the *energy balance*—a mathematical equation expressing the conservation of energy, a special case of which is sometimes referred to as the *first law of thermodynamics*. Examination of the general balance equation shown in Chapter 2 reveals that the energy balance will incorporate groups of terms representing a) the addition of energy to the system, b) the removal of energy from the system, and c) the accumulation of energy in the system. In order for the energy balance to be applicable to the variety of systems and processes encountered in thermodynamics, we examine first what terms need to be included in each group and then we combine them to obtain the energy balance.

The application of the energy balance to typical thermodynamic processes is illustrated in the remainder of the chapter. The student will discover how this balance can be used to relate the changes of state of a system to the energy exchanges between the system and its surroundings. He will also learn that, in making this connection for a given process, some shrewdness and experience in the selection of elements to be identified as *the system* can greatly facilitate the computations.

4.1 Work

The dictionary defines work in several ways. It may be referred to as either an exertion directed to produce or accomplish work, toil, or employment, or it may be referred to as the result of an exertion, labor, or activity. Unfortunately neither of these descriptions provides the quantitative description that we as engineers demand. To provide this quantitative description we turn to the physicists' definition of work: Work is the product of an applied force, F, and the distance, ΔX, through which the force moves, multiplied by the cosine of the angle, θ, between the force and the displacement it causes. In thermodynamics we shall restrict our definition of work to include only energy in transition across the system boundaries by virtue of an energy potential other than temperature (for example, pressure or voltage). *Energy transfers that occur wholly within a system are not considered as work.*

Let us consider a man who supports a 100 lb_f weight without raising it. He has transferred no work to the weight, because the force he applies was not accompanied by motion. On the other hand, a man who lifts a 1 lb_m book (taken as the system) and places it on a table has transferred work to the system, since this force is accompanied by motion.

When either the applied force or the angle between the force and the displacement changes during the process, we may evaluate the work performed during the whole process as the sum of a number of smaller processes during which both F and θ are constant. In the limit this sum becomes an integral, and we obtain the relation

$$W = \int_{X_1}^{X_2} |F| \cos\theta \, |dX| \tag{4-1}$$

The following sign convention will be adopted in regard to work: Work will be considered positive when done by a system on its surroundings, negative when done by the surroundings on the system. Thus it is seen that a choice of system and surroundings must be made before a sign can be given to a unit of work.

Let us now examine the quantitative calculation of the work performed in several commonly encountered processes.

Friction Processes

Suppose a block rests on a flat surface as illustrated in Fig. 4-1. When the force F exerted on the block (the system) is large enough to overcome the

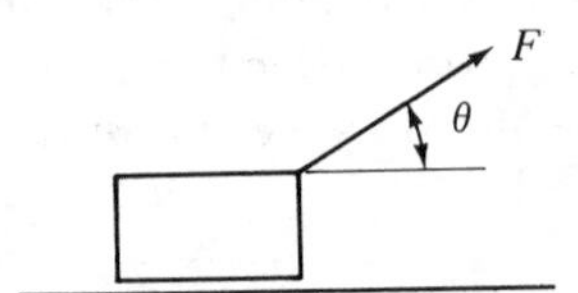

FIG. 4-1. Work against friction.

frictional resistance, the force causes a horizontal motion through a differential distance dX. The differential work this force has performed on the system is then

$$\delta W = F \cos\theta \, dX \tag{4-1a}$$

If the force continues to move the block until it has undergone movement through a total distance $X_2 - X_1$, we may evaluate the total work performed on the system by integrating equation (4-1) to yield

$$W = \int_{X_1}^{X_2} \delta W = \int_{X_1}^{X_2} F \cos \theta \, dX \tag{4-2}$$

If F and θ remain constant, equation (4-2) may be integrated to yield

$$W = F \cos \theta (X_2 - X_1)$$

or (4-3)

$$W = F \cos \theta (\Delta X)$$

where the Greek letter Δ (capital delta) signifies the difference between a property at the end and the beginning of a process. This convention for the symbol Δ is followed throughout this text.

Lifting Process

If a constant force F is applied to lift a weight (the system) through a distance $\Delta Z = Z_2 - Z_1$, the work required to lift the system is

$$W = \int_{Z_1}^{Z_2} F \cos \theta \, dZ = F \int_{Z_1}^{Z_2} \cos \theta \, dZ \tag{4-4}$$

However, $\theta = 0°$, because F is directed parallel to the direction of the motion it produces. Therefore,

$$W = F \, \Delta Z \tag{4-5}$$

and the work done on the system has resulted in an increase in its potential energy.

Accelerating Processes

If a system with mass M is accelerated from a velocity u_1 to a velocity u_2, we may evaluate the work necessary to perform this acceleration as follows:

$$W = \int_{X_1}^{X_2} F \cos \theta \, dX \tag{4-6}$$

But $\cos \theta = 1$, and from Newton's law, $F = Ma/g_c$, where a is the acceleration of the body.

However, by definition,

$$a = du/dt \tag{4-7}$$

and

$$u = dX/dt \tag{4-8}$$

Therefore,

$$F = \frac{M}{g_c}\frac{du}{dt} = \frac{M}{g_c}\frac{du}{dX}\frac{dX}{dt} = \frac{M}{g_c}\frac{du}{dX}u \tag{4-9}$$

$$W = \frac{M}{g_c}\int_{X_1}^{X_2} u\frac{du}{dX}\,dX = \frac{M}{g_c}\int_{u_1}^{u_2} u\,du \tag{4-10}$$

since M is constant. Upon integrating,

$$W = \frac{M}{2g_c}(u_2^2 - u_1^2) = \frac{M}{2g_c}\Delta\left(u^2\right) \tag{4-11}$$

where the work has resulted in an increase in the system's kinetic energy.

Spring Processes

Suppose we examine a weightless, frictionless, ideal spring (the system) as illustrated in Fig. 4-2. The spring is attached to a solid wall at one end and is allowed to float freely with no constraint on the other end. The no-load position of the free end is chosen as a coordinate reference point to which the value $X = 0$ is attached. If a force F is applied to the spring, it will be compressed and shortened according to Hooke's law,

$$F = KX \tag{4-12}$$

where K is the spring constant of the spring, F the force applied to the spring, and X the displacement of the spring caused by the force F (Fig. 4-3).

The work necessary to compress the spring from the point $X = 0$ to $X = X$ may be evaluated from

$$W = \int_{X=0}^{X=X} F\,dX = \int_{X=0}^{X=X} KX\,dX \tag{4-13}$$

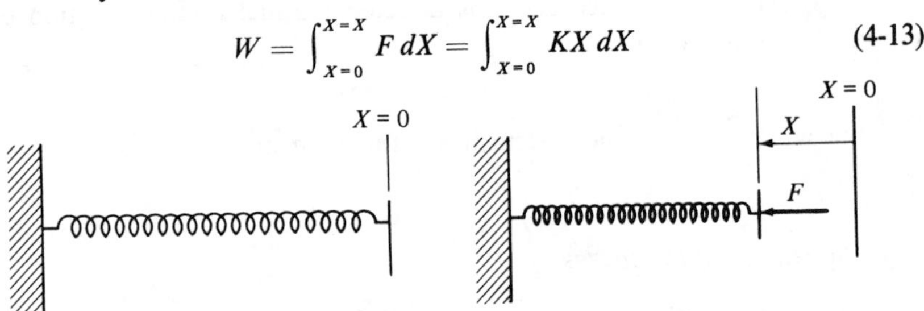

FIG. 4-2. Uncompressed spring. **FIG. 4-3.** Compressed spring.

or

$$W = K\int_{X=0}^{X} X\,dX \tag{4-13a}$$

which gives

$$W = \left(\frac{KX^2}{2}\right)_{X=0}^{X=X} = \frac{KX^2}{2} \tag{4-14}$$

The work done on the system in this case results in an increase in the system's internal energy.

Compression or Expansion Processes—Closed Systems

Assume that a gas (the system) of volume V and pressure P is contained within a cylinder and piston as illustrated in Fig. 4-4. The piston is assumed

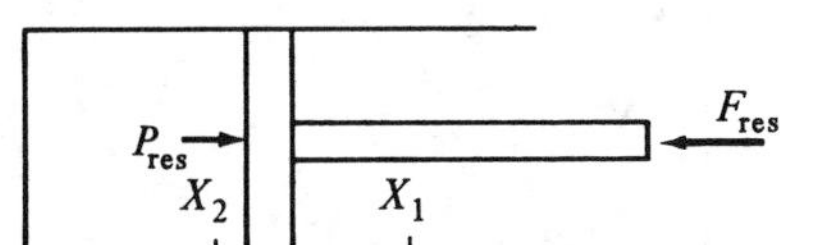

FIG. 4-4. Work of compression.

to move frictionlessly in the cylinder and has a force F applied to it in order to constrain the gas within the cylinder. If the force F is increased so that the gas is compressed, we may evaluate the work of compression performed by the surroundings on the system as follows:

$$W = \int_{X_1}^{X_2} F_{res}\, dX \tag{4-15}$$

where F_{res} represents the *resisting force exerted on the system boundary.* But the force exerted on the system boundary can be converted to an effective pressure by defining P_{res} such that

$$F_{res} = P_{res}A \tag{4-16}$$

where A is the area of the piston and P_{res} is the resisting pressure on the system boundary. Substitution of equation (4-16) into (4-15) then gives

$$W = \int_{X_1}^{X_2} P_{res}A\, dX \tag{4-17}$$

Since $A\, dX = dV$, the volume swept out by the piston as it advances, equation (4-17), can be rewritten as

$$W = \int_{V_1}^{V_2} P_{res}\, dV \tag{4-18}$$

During the compression the volume of the gas is decreasing, so dV is a negative quantity. Therefore, the amount of work transferred by the system to the surroundings is negative. That is, the surroundings are transferring work to the system. If the gas were expanding, the volume change of the system would be positive, so W is positive and the system is transferring work to the surroundings.

As we shall see later, the evaluation of this integral is often an extremely difficult task, because the relation of P to V may not be explicitly known. Even in some cases where an analytical expression between P and V is known, the evaluation of equation (4-17) may be time consuming or impractical—especially if the expression relating P to V is obtained from one of the complex equations of state. However, if a plot of P vs V for the process at hand can be constructed as shown in Fig. 4-5, the evaluation of $\int_{V_1}^{V_2} P\, dV$ is reduced to the evaluation of the area under the curve between the limits of V_1 and V_2. For example, in Fig. 4-6 the work done on the gas in compressing it from V_1 to V_2 is equal to $P\, dV$, which is identically equal to the shaded area under the curve. It should be noted that the area is negative if the integration is from state 1 to state 2. Since in many cases the P–V curve for a gas may be available or readily plotted, evaluation of the area under the P–V curve (called *graphical integration*) is often a far simpler and quicker procedure than analytical evaluation of the integral.

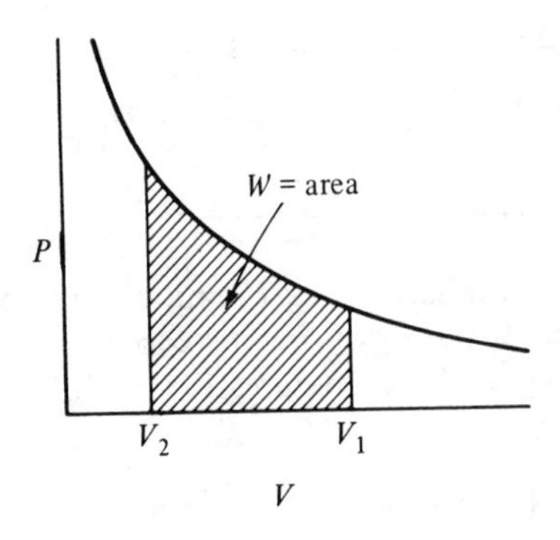

FIG. 4-5. Work of compression—P-V diagram.

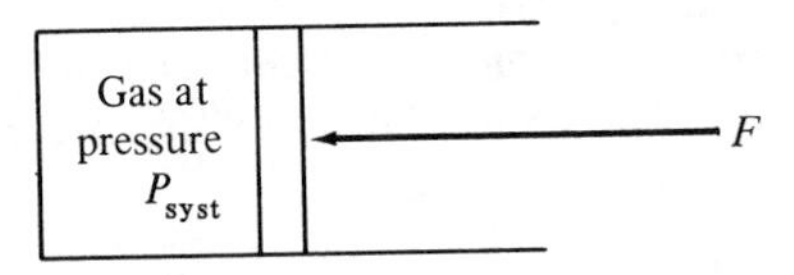

FIG. 4-6. Piston–cylinder arrangement.

4.2 Reversible and Irreversible Work

Processes with Infinitesimal Driving Forces

Suppose a piston and cylinder are arranged as shown in Fig. 4-6. The piston is assumed to move *frictionlessly* within the cylinder. A force F is applied to the piston and is assumed to exactly counterbalance the force the gas exerts on the piston head. Thus the pressure within the cylinder is always equal to the resisting pressure on the system boundary. If the force exerted on the piston decreases by a differential amount, the differential pressure difference that now exists will cause the piston to move to the right, so the volume of the gas within the cylinder increases by a differential amount dV. This is a quasi-static process, since all motion takes place infinitesimally slowly, and is a *reversible process, since no dissipation of a potential occurs*. The work performed by the gas within the cylinder on the surroundings may then be evaluated from equation (4-18) as

$$\delta W = P_{\text{res}}\, dV_{\text{syst}} \tag{4-18a}$$

where the subscript "syst" refers to the system. However, for the frictionless nonaccelerating process,

$$P_{\text{res}} = P_{\text{syst}} \tag{4-19}$$

Therefore,

$$\delta W = P_{\text{syst}}\, dV_{\text{syst}} \tag{4-20}$$

As the force is continually lowered, the gas continues to expand. The total work performed by the gas on its surroundings may then be determined by integrating equation (4-20):

$$W = \int_{V_1}^{V_2} P_{\text{syst}}\, dV_{\text{syst}} \tag{4-20a}$$

Thus for a frictionless (reversible) process we may evaluate the total work performed as a function only of system properties—a relation we shall find extremely useful.

The integration of equation (4-20a) requires two other pieces of information—an equation of state for the substance under consideration, and a relation between the temperature of the system and either its pressure or volume. With these additional items we may then integrate equation (4-20a). As may be expected, these two pieces of information will, in general, *not* be available—particularly the pressure (or volume)–temperature relation of the process.

SAMPLE PROBLEM 4-1. An ideal gas is to undergo an isothermal, reversible compression in a frictionless piston–cylinder from 1 to 10 atm. Calculate the initial and final molar volumes of the gas and the work necessary to perform the compression if the gas is initially at 100°F.

Solution: The ideal gas equation of state is given by

$$P\underline{V} = RT$$

or

$$\underline{V} = \frac{RT}{P}$$

where $T = 560°\text{R}$
$R = 0.7302$ atm ft^3/lb-mol °R

That is,

$$\underline{V} = \frac{(0.7302)(560)}{P} \frac{\text{ft}^3\ \text{atm}}{\text{lb-mol}}$$
$$= \frac{410}{P} \frac{\text{ft}^3\ \text{atm}}{\text{lb-mol}}$$

Therefore, at $P = 1$ atm,

$$\underline{V} = 410\ \text{ft}^3/\text{lb-mol}$$

At $P = 10$ atm,

$$\underline{V} = 41.0\ \text{ft}^3/\text{lb-mol}$$

Since the compression is reversible, the work of compression may be obtained from the integral

$$\underline{W} = \int_{\underline{V}_1}^{\underline{V}_2} P_{\text{syst}}\, d\underline{V}_{\text{syst}}$$

But for an ideal gas, $P_{\text{syst}} = RT/\underline{V}_{\text{syst}}$. Therefore,

$$\underline{W} = \int_{\underline{V}_1}^{\underline{V}_2} \frac{RT}{\underline{V}}\, d\underline{V} = \int_{\underline{V}_1}^{\underline{V}_2} RT \frac{d\underline{V}}{\underline{V}}$$

But we have been told that the process is isothermal; therefore, T is a constant and RT may be removed from under the integral sign, to give

$$\underline{W} = RT \int_{\underline{V}_1}^{\underline{V}_2} \frac{d\underline{V}}{\underline{V}} = RT \ln \frac{\underline{V}_2}{\underline{V}_1}$$

or

$$\underline{W} = 1.987 \frac{\text{Btu}}{\text{lb-mol °R}} (560\text{°R}) \ln \frac{41.0}{410}$$
$$= -2.56 \times 10^3 \text{ Btu/lb-mol}$$

The minus sign indicates that the surroundings must perform work on the system to effect the desired compression.

SAMPLE PROBLEM 4-2. If the piston–cylinder of Sample Problem 4-1 is made adiabatic, that is, perfectly insulated against heat transfer, it can be shown that the following relation exists between the pressure and temperature of the gas within the cylinder (see Section 6.7 for a derivation of this result):

$$\frac{T_2}{T_1} = \left(\frac{P_2}{P_1}\right)^{1-1/k}$$

Using this relation, obtain an expression for the amount of work needed to adiabatically and reversibly compress an ideal gas from 1 atm and 100°F to 10 atm. The coefficient k is a constant whose value depends on the ideal gas in question.

Solution: From the ideal gas equation of state we may eliminate temperature from the above equation to give

$$\frac{T_2}{T_1} = \frac{P_2\underline{V}_2}{P_1\underline{V}_1} = \left(\frac{P_2}{P_1}\right)^{1-1/k}$$

or

$$\frac{\underline{V}_2}{\underline{V}_1} = \left(\frac{P_2}{P_1}\right)^{-1/k} = \left(\frac{P_1}{P_2}\right)^{1/k}$$

or at any general point,

$$\frac{\underline{V}}{\underline{V}_1} = \left(\frac{P_1}{P}\right)^{1/k}$$

or

$$P = P_1\left(\frac{\underline{V}_1}{\underline{V}}\right)^k$$

But the work for a reversible compression is given by

$$\underline{W} = \int_{\underline{V}_1}^{\underline{V}_2} P\, d\underline{V}$$

Substituting the relation between P and $\underline{V}$ then gives

$$\underline{W} = \int_{\underline{V}_1}^{\underline{V}_2} P_1\left(\frac{\underline{V}_1}{\underline{V}}\right)^k d\underline{V}$$

or

$$\underline{W} = P_1(\underline{V}_1)^k \int_{\underline{V}_1}^{\underline{V}_2} \underline{V}^{-k}\, d\underline{V}$$
$$= \frac{P_1(\underline{V}_1)^k}{1-k}[\underline{V}^{1-k}]_{\underline{V}_1}^{\underline{V}_2}$$
$$= \frac{P_1(\underline{V}_1)^k}{1-k}[(\underline{V}_2)^{1-k} - (\underline{V}_1)^{1-k}]$$

or

$$\underline{W} = \frac{P_2\underline{V}_2 - P_1\underline{V}_1}{1-k}$$

which gives us the desired relation. $\underline{V}_2$ is determined from the relation

$$\underline{V}_2 = \underline{V}_1\left(\frac{P_1}{P_2}\right)^{1/k}$$

Processes with Finite Driving Forces

Even if the piston of Fig. 4-7 is not frictionless, or if more than a differential pressure difference exists between the gas confined by the piston and the effective pressure of the surroundings, we can still substitute P_{syst} for P_{res} in the determination of W. For example, let us examine Fig. 4-7. Initially the

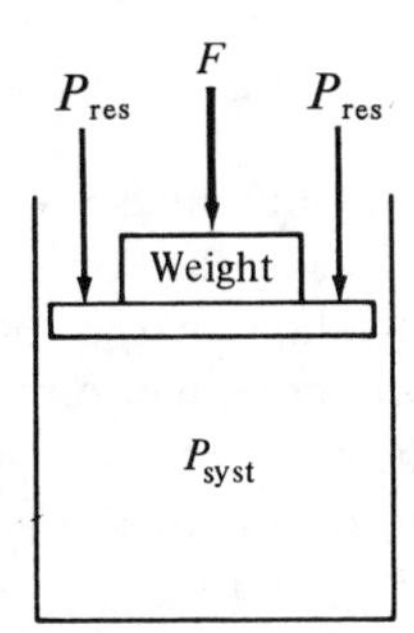

FIG. 4-7. Expansion with finite pressure difference.

effective pressure of the piston and weights (P_{surr}) is identical to the pressure of the system, where the system is herein chosen to be the cylinder and gas but not including the piston. Let us for the moment assume that the piston is frictionless. If we suddenly remove a portion of the weights on the piston, the effective pressure of the surroundings instantaneously drops before the volume of the system has had a chance to adjust itself (see Fig. 4-8). However, if we examine a force balance around the system boundary, we see that P_{res} must still equal P_{syst} if no unbalanced force is to exist across the system boundary. (Since the boundary has no mass, it cannot support a finite pressure imbalance.) On the other hand, a finite pressure difference now exists across the piston resulting in

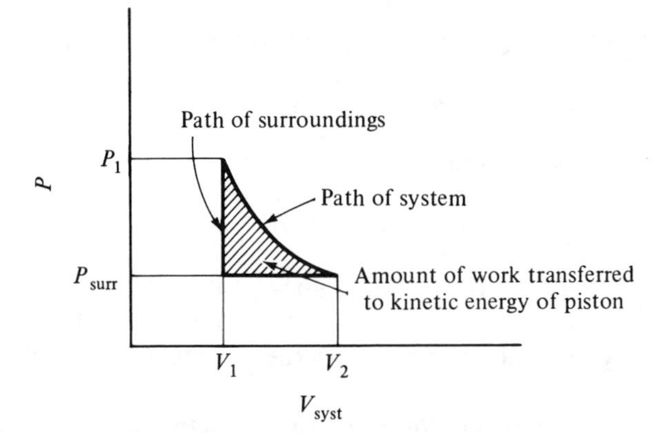

FIG. 4-8. The variation of P_{syst} with P_{res}.

an unbalanced force on the piston, and Newton's law tells us that this unbalanced force will cause the piston to accelerate. *If this acceleration is not too rapid*, we may still assume that the gas within the cylinder is undergoing a quasi-static process. Therefore, the work the system (the gas) does is identical to the amount of work the gas did in the reversible case (after all, the gas cannot tell what it is pushing against as long as *it* is undergoing a quasi-static process). That is, the work done by the gas can be expressed as

$$W = \int P_{\text{syst}}\, dV \tag{4-21}$$

However, of this only

$$W_{\text{surr}} = \int P_{\text{surr}}\, dV \tag{4-22}$$

goes to pushing back the surroundings. What has happened to the rest? Very simply, it has gone into increasing the kinetic energy (acceleration) of the piston! It is now a simple matter to show that as long as the piston is frictionless, once set in motion like this it will continue to move for time immemorial (provided it does not shoot out of the cylinder), in a periodic oscillation whose frequency and amplitude will depend on the mass of the piston and its weights and the initial difference between P_{syst} and P_{res}.

If the piston is now assumed to be frictional, we have a situation very similar to that in the previous case. Suppose a frictional force F_f must be overcome to move the piston up or down in the cylinder. Then we can say that a frictional pressure drop $P_f = F_f/A$ is required before motion can occur. That is, unless the absolute magnitude of $|P_{\text{syst}} - P_{\text{res}}| > P_f$, no motion can occur. When this inequality is satisfied, the piston will move (in the direction toward the lower pressure). Suppose $|P_{\text{syst}} - P_{\text{res}}| \gg P_f$. Then, as before, the piston will experience an imbalanced force and will accelerate. If the acceleration is slow enough to assume that the gas within the cylinder is undergoing a quasi-static process, then once again the *work performed by the gas* is given by

$$W = \int P_{\text{syst}}\, dV$$

The work performed in pushing back the surrounding atmosphere is given by

$$W_{\text{surr}} = \int P_{\text{surr}}\, dV$$

The work lost through friction is

$$W_f = \int P_f\, dV$$

and the remainder $(W - W_{\text{surr}} - W_f)$ goes into accelerating the piston head. Because of the effects of friction, the piston in this case will not continue to move indefinitely but will, after a time, settle at some position where $|(P_{\text{syst}} - P_{\text{res}})| \leq P_f$.

We might at this point stop to consider whether either of the preceding processes is reversible. Let us consider the frictionless piston first. Since there were no frictional losses, we observed that the piston would oscillate indefinitely in the cylinder. If we solve the equations of motion that govern these oscillations we find that at the lowest point of each swing the piston will be at exactly the same position (and the gas in the same state) as at the start of the process. *Since both system and surroundings can be simultaneously returned to their initial states* (and, in fact, do so periodically of their own accord), *this is a reversible process.*

On the other hand, if we consider the frictional piston we find that the friction prevents the piston from ever returning to its original state without the surroundings supplying work to it. *Thus the frictional process is irreversible. However, if we consider only the gas expanding within the cylinder, we find that it has undergone a reversible process because there has been no degradation of potential within the gas.* The degradation (or irreversibility) has occurred outside the gas at the interface between the piston and cylinder. The gas cannot tell what it is pushing against—be it a restraining force, friction, or inertia—and hence undergoes a reversible process whenever there is no internal degradation of potential. We can make an important observation: *Although a complete process is irreversible, there may be many reversible components within the process.* Since the thermodynamic analysis of a reversible process is generally much simpler than that for an irreversible one, this is a worthwhile point to remember when picking a system. We shall return to the question of reversible and irreversible processes and the location of the irreversibilities in Chapter 5 when we consider a rigorous derivation of the entropy balance.

SAMPLE PROBLEM 4-3. A 1440-lb_m piston is initially held in place by a removable latch above a cylinder as shown in Fig. S4-3. The cylinder has an area of 1 ft^2; the volume of the gas within the cylinder initially is 2 ft^3 and at a pressure of 10 atm. The working fluid may be assumed to obey the ideal gas equation of state. The cylinder has a total volume of 5 ft^3 and the top end is open to the surrounding atmosphere, whose pressure is 1 atm.

(a) If the piston rises frictionlessly in the cylinder when the latches are removed and the gas within the cylinder is always kept at the same temperature, what will be the velocity of the piston as it leaves the cylinder?

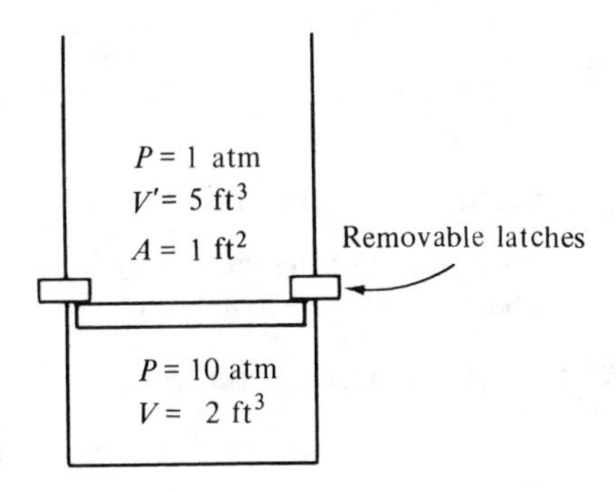

FIGURE S4-3

(b) What will be the maximum height to which the piston will rise?

Solution: (a) First we must find the effective pressure of the surroundings:

$$P_{surr} = P_{atm} + \frac{\text{weight of piston}}{\text{area of piston}}$$

$$= 14.7 \text{ psia} + \frac{1440 \text{ lb}_f}{144 \text{ in.}^2}$$

$$= 24.7 \text{ psia}$$

Next, we determine the total work performed by the gas on both the piston and the atmosphere. We have shown for isothermal expansions of ideal gases that the total work performed by the gas is

$$W = nRT \ln (\underline{V}_2/\underline{V}_1)$$

But

$$nRT = P_1 V_1$$

Thus

$$W = (10 \text{ atm} \cdot 2 \text{ ft}^3) \ln (5.0/2.0)$$

$$= 18.4 \text{ atm ft}^3 = 3.88 \times 10^4 \text{ ft-lb}_f$$

Of this, however,

$$W_{surr} = \int P_{surr} \, dV_{syst}$$

$$= 24.7 \frac{\text{lb}}{\text{in.}^2} \cdot 3 \text{ ft}^3$$

$$= 1.07 \times 10^4 \text{ ft-lb}_f$$

must be supplied to move the "effective" surrounding pressure. Therefore,

$$(3.88 - 1.07) \times 10^4 = 2.81 \times 10^4 \text{ ft-lb}_f$$

of energy is supplied to accelerate the piston. That is, the change in the kinetic energy of the piston is equal to 2.81×10^4 ft-lb_f, which tells us that

$$\frac{1}{2} \frac{Mu^2}{g_c} = 2.81 \times 10^4 \text{ ft-lb}_f$$

or

$$\frac{1}{2} \frac{1440 \text{ lb}_m \text{ sec}^2}{32.2 \text{ ft}} u^2 = 2.81 \times 10^4 \text{ ft-lb}_f$$

or

$$u^2 = 1255 \text{ ft}^2/\text{sec}^2$$

or

$$u = 35.4 \text{ ft/sec at its exit from the cylinder}$$

(b) At the maximum rise of the piston its potential energy (PE) would be equal to the total energy imparted to it. Therefore, at its maximum height,

$$\text{PE} = \frac{MgZ}{g_c} = WZ = 2.81 \times 10^4 \text{ ft-lb}_f$$

But its weight is 1440 lb_f; therefore,

$$Z_{max} = \frac{2.81 \times 10^4}{1440} = 19.5 \text{ ft above cylinder top}$$

(Checking of the units is left to the reader as an exercise).

4.3 Heat

By the late eighteenth and early nineteenth centuries our present view of heat as energy in transit by virtue of a temperature difference had pretty well evolved. Although it is clearly possible to have energy transfers wholly within a system by virtue of internal temperature differences, we shall not refer to these energy flows as heat. Rather we shall restrict our definition of heat (in a thermodynamic context) to include only *those energy transfers which occur between the system and its surroundings by virtue of a temperature difference*. Thus to determine the existence or nonexistence of heat flow, we must examine the boundary between the system and its surroundings. It is also meaningless to attempt to describe heat by reference to a single system: Heat can only be transferred from one system to another. (It should be remembered that work is also energy in transit across the system boundary and therefore has meaning only in the context of an interaction between the system and its surroundings.) Once heat has crossed the boundary between the system and surroundings, it is no longer considered heat but becomes part of the thermal component of the internal energy of the receiving body. A process during which no heat is transferred between the system and surroundings is termed *adiabatic*.

Heat is measured in terms of the energy changes it causes in either the system or the surroundings. Since these energy changes are most easily described in terms of temperature changes (at either constant volume or pressure), we define the unit of heat in terms of the temperature change (at constant V or P) it can effect. Two units of heat (or energy) are commonly encountered: The British thermal unit (Btu) is defined as the amount of heat (energy) that must be added to 1 lb_m of water at 1 atm pressure to raise its temperature from 59.5 to 60.5°F. The calorie (cal) is the amount of heat (energy) that must be added to 1 g of water at 1 atm pressure to raise its temperature from 14.5 to 15.5°C. For the general case of a given fluid at any temperature we may express the amount of heat needed to raise the temperature of the system of mass M from a temperature of T to $T + dT$ at constant pressure (or volume) as

$$\delta Q = MC_P\,dT \qquad \text{for constant pressure}$$

or

$$\delta Q = MC_V\,dT \qquad \text{for constant volume} \tag{4-23}$$

(It is assumed that the mass of the system, along with its potential and kinetic energy, do not change during the heat addition.) In equation (4-23) C_P (C_V) is the constant-pressure (volume) heat capacity of the system. It can be seen that, by the definition of the units of heat, we have defined the constant-pressure heat capacity (C_P) of water at 1 atm to be 1 Btu/lb_m °F at 60 °F or 1 cal/g °C at 15 °C. The constant-pressure (constant-volume) heat capacity of a body can be considered as simply the ratio of the amount of heat needed to raise the temperature of a unit mass of the body by one temperature unit to the amount of heat needed to raise the temperature of a unit mass of water by one temperature unit when the water is at 15 °C or 60 °F and a constant pressure of 1 atm. In some instances it is convenient to express heat capacities on a molar basis rather than the mass basis shown here—the two are related by the molecular weight of the material.

Heat capacities (either C_P or C_V) are, in general, functions of both T and P, or V. However, over small temperature ranges C_P and C_V are usually relatively constant, so finite heat transfers can be expressed as

$$Q = MC_P\Delta T \qquad \text{for constant pressure}$$

or (4-24)

$$Q = MC_V\Delta T \qquad \text{for constant volume}$$

provided that the temperature differences are "small enough" so that the heat capacities are relatively constant. For larger temperature ranges we must integrate $C_P\,dT$ (or $C_V\,dT$) or use an average heat capacity

$$(C_P)_{\text{av}} = \frac{\int_{T_1}^{T_2} C_P\,dT}{T_2 - T_1}$$
$$(C_V)_{\text{av}} = \frac{\int_{T_1}^{T_2} C_V\,dT}{T_2 - T_1} \tag{4-25}$$

Heat capacities are usually measured directly in an instrument known as a calorimeter. Two types of calorimeters are commonly used: the constant-volume (or bomb) type for measuring C_V; and the constant pressure (or flow) type for measuring C_P. The calorimeter experiment is extremely simple (in theory): A known amount of heat is supplied to a known mass of fluid at constant pressure (or volume) and the temperature change is measured. The average heat capacity over the temperature range is then given by solving equations (4-24) and (4-25) for the average heat capacities:

$$(C_P)_{\text{av}} = \frac{Q}{M\,\Delta T} \qquad \text{for constant pressure}$$
$$(C_V)_{\text{av}} = \frac{Q}{M\,\Delta T} \qquad \text{for constant volume} \tag{4-26}$$

By taking the limit as ΔT approaches zero the temperature-dependent values of C_P and C_V may be determined. (Since it is not an easy matter to deliver a precisely known amount of heat to a mass of fluid, these experiments are tedious and difficult in practice.)

4.4 Conservation of Energy

The fact that many forms of mechanical energy are completely interconvertible (theoretically) has been well known for many years. However, the recognition that thermal and mechanical energy are interconvertible is not so apparent. Indeed, as we have seen, it was not until the late 1700s that we began to develop an understanding of the correct relation between heat and work.

As further evidence of the interconvertibility of heat and energy accumulated, several crude experiments were undertaken to determine the "mechanical equivalent of heat." In 1843 Joule performed the first accurate and reproducible determination of this quantity using the simple equipment illustrated in Fig. 4-9. The weight W is allowed to fall through a distance ΔZ,

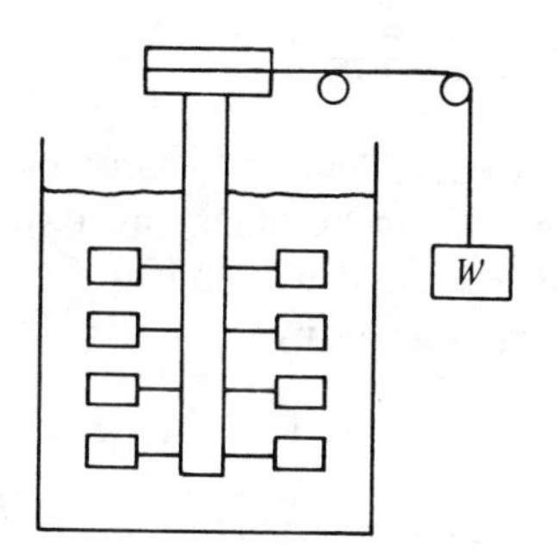

FIG. 4-9. Joule's experiment.

performing an amount of work $W\,\Delta Z$ on the paddle wheel. The paddle wheel in turn performs this same amount of work on the water, where viscous forces convert the mechanical energy into internal energy. By measuring the temperature rise of the water, Joule was able to determine the amount of thermal energy absorbed, and, from the amount of work the weight performed, was able to determine the mechanical equivalent of heat. The best value obtained by Joule was[1]

$$1 \text{ Btu} = 772.5 \text{ ft-lb}_f$$

or

$$1 \text{ g cal} = 4.155 \text{ J}$$

Later measurements have fixed these values more accurately as

$$\begin{aligned} 1 \text{ Btu} &= 777.647 \text{ ft-lb}_f \\ &= 1054.35 \text{ J (w sec)} \\ 1 \text{ g cal} &= 4.184 \text{ J} \end{aligned}$$

The concept of the equivalency and interconvertibility of heat, work, and all other forms of energy has been derived from the work of Rumford and Joule. This concept is expressed in the *law of conservation of energy*, which

[1] M. W. Zemansky and H. C. Van Ness, *Basic Engineering Thermodynamics*, McGraw-Hill Book Company, Inc., New York, 1966, p. 74.

states: Energy can neither be created nor destroyed—only changed in form. It would seem at first glance that atomic and nuclear processes in which mass is converted to energy according to Einstein's law, $E = mc^2$, disobey the law of conservation of energy. However, nuclear processes only indicate that the traditional concept of the separation of mass and energy is coming under attack. It now appears that mass and energy themselves are nothing but two different forms of the same fundamental phenomenon, and therefore are in many ways no more different than thermal and mechanical energy. However, in this book we shall not deal with atomic or nuclear reactions and therefore will not be concerned with the interconversion of energy and mass. For our purposes we shall consider mass and energy to be separate properties, each of which obeys its respective law of conservation.

4.5 State and Path Functions

We have defined work as equivalent to a force moving through a distance. Any time we have motion of the boundary of a system against a restraining force, work is performed. Earlier it was shown that the amount of work performed on a closed system is given by

$$W = \int_{V_1}^{V_2} P_{res}\, dV \tag{4-18}$$

where P_{res} is the effective pressure exerted on the system boundary. For simplicity, let us again consider our system to be the gas enclosed within the cylinder of Fig. 4-10.

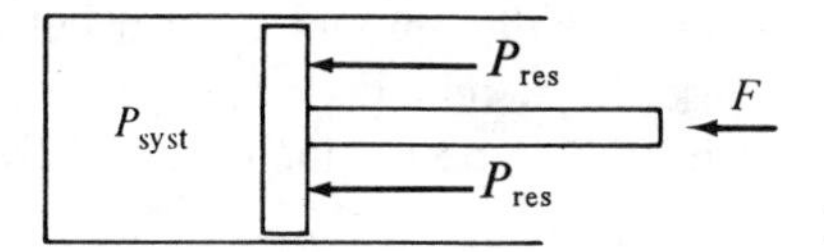

FIG. 4-10. Piston–cylinder arrangement.

The piston is assumed to be weightless and considered as part of the system. However, the piston is *not* assumed to be frictionless. The process is assumed to be quasi-static, so the pressure within the gas is at all points equal. When the piston moves, the forces of friction (when it is present) will prevent the pressure of the gas within the system from equaling that of the surroundings. Thus, when we evaluate the work performed by the system on the surroundings, we must be very careful to specify only that (mechanical) energy which is actually transferred to the surroundings. That portion which might have been transferred to the surroundings but is dissipated by friction within the system is termed *lost work* and will be discussed further in Chapter 5.

The work that is actually transferred to the surroundings is W and is evaluated from

$$W = \int_{V_1}^{V_2} P_{res}\, dV \tag{4-18}$$

Thus we see that the work done on, or by, a system may very well be a function of things not associated with the terminal states of the system, as, for example, the effects of friction. We find, then, that in going between any two states the work a system can produce may have many different values depending on the path the system or the surroundings follows between the two states. Therefore, we say that the work a system can produce in going between two states is a *function of the path* or *a path function*. Similarly, it may be demonstrated that the heat absorbed by a system in going between two states is also a path function.

In addition to path functions, which depend on the path followed between states, there are many system properties that are independent of path. These properties are called *state properties* or *functions of state*. All physical properties —density, pressure, temperature, viscosity, etc.—are examples of state functions. In addition, any property that can be expressed uniquely as a function of only these properties must itself be a function of state. Thus we find that all the thermodynamic properties $\underline{U}$, $\underline{H}$, $\underline{S}$, . . . which may be expressed as functions of temperature and pressure alone are functions of state!

The distinction between functions of path and functions of state may be quantitatively described by looking at the integral of the function in question around a closed path such as that shown in Fig. 4-11. If the function P in

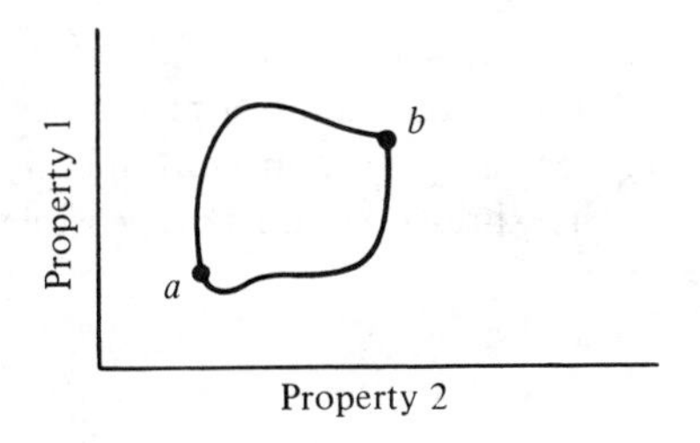

FIG. 4-11. The path integral.

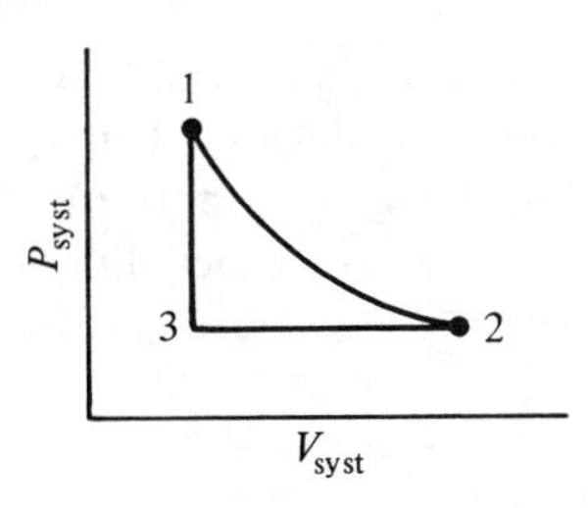

FIG. 4-12. The P–V diagram.

question is a state function, the closed-path integral $\oint dP$ will equal zero independent of the path traversed between a and b or between b and a. If a single closed path can be found for which the closed-path integral is not identicaly zero, the function is a path function.

Let us examine several ways in which we may reversibly expand an ideal gas contained within a piston–cylinder (system: the gas) between states 1 and 2 on the P_{syst}–V_{syst} diagram shown in Fig. 4-12. For illustration purposes, let $T_1 = T_2$. The direct path 1-2 is chosen to be the reversible, isothermal expansion between P_1 and P_2. As we have shown in Section 4-2, the work transferred along this path is given by

$$W_{1\text{-}2} = nRT_1 \ln\left(\frac{V_2}{V_1}\right)$$

Along the indirect path 1-3-2, the gas is assumed to undergo a reversible two-step process. During the first step (1-3) the gas is cooled at constant volume until the pressure drops to the outlet pressure, P_2. In the second step (3-2) the gas is reheated, but at constant pressure, back to the inlet temperature, T_1. Since we are assuming that both steps are reversible, $P_{res} = P_{syst}$, and the work transferred to the surroundings is given by

$$W_{1\text{-}3\text{-}2} = W_{1\text{-}3} + W_{1\text{-}2} \tag{4-27}$$

$$= \int_{V_1}^{V_3} P_{syst}\, dV_{syst} + \int_{V_3}^{V_2} P_{syst}\, dV_{syst} \tag{4-27a}$$

Since no volume change occurs in the first step, $W_{1\text{-}3}$ is zero, and

$$W_{1\text{-}3\text{-}2} = \int_{V_3}^{V_2} P_{syst}\, dV_{syst} \tag{4-28}$$

Since P_{syst} is constant at P_2 during the second step, the integral reduces to

$$W_{1\text{-}3\text{-}2} = P_2(V_2 - V_3) = P_2(V_2 - V_1) \tag{4-28a}$$

Examination of the two work expressions, $W_{1\text{-}2}$ and $W_{1\text{-}3\text{-}2}$, clearly indicates that they are different—even though both processes are reversible. *Thus we must conclude that work is a path function whose value depends on the path chosen between the initial and final states.* Similarly, if we examine the closed-path integral we find that the integral does not vanish since $W_{1\text{-}2}$ does not equal $W_{1\text{-}3\text{-}2}$. Once again we are led to the conclusion that W is a path function.

The mathematical differences between path and state functions will become more significant when we deal with the mathematics of properties. We know that changes in state functions may be uniquely and conveniently expressed in terms of exact differentials. Such expressions are not possible for path functions that are inexact differentials.

4.6 The Energy Balance

In the following sections we shall discuss the derivation and use of the energy balance. The balance is derived under very general conditions. We shall then consider some of the simplications that are commonly justified.

In formulating the energy balance for a system, the same procedure as was used in Chapter 2 for mass may be used. For an open system, the overall energy balance may be written as

$$\text{energy}_{in} - \text{energy}_{out} + \text{energy}_{gen} = \text{energy}_{acc} \tag{4-29}$$

In a previous section we indicated that energy can neither be created nor destroyed (at least for nonnuclear processes). Therefore, $\text{energy}_{gen} = 0$ and the energy balance becomes

$$\text{energy}_{in} - \text{energy}_{out} = \text{energy}_{acc} \tag{4-29a}$$

Let us assume that the system has a differential amount of mass, δM_{in}, flowing into the system and a differential amount of mass, δM_{out}, leaving the system, as illustrated in Fig. 4-13. The mass entering the system has an energy per unit mass of $\underline{E}_{in}$; the material leaving contains $\underline{E}_{out}$ energy per unit mass. Let us also assume that a differential amount of heat, δQ, flows into the system and that the system performs a differential amount of work, $\delta W'$, on the surroundings. The total energy accumulated by the system will then be $(\underline{E}M)_{syst}$ at the end, $-(\underline{E}M)_{syst}$ at the beginning, or $d(\underline{E}M)_{syst}$, for the differential process just described. The energy balance may then be written as

$$(\underline{E}\delta M)_{in} + \delta Q - (\underline{E}\delta M)_{out} - \delta W' = d(\underline{E}M)_{syst} \tag{4-30}$$

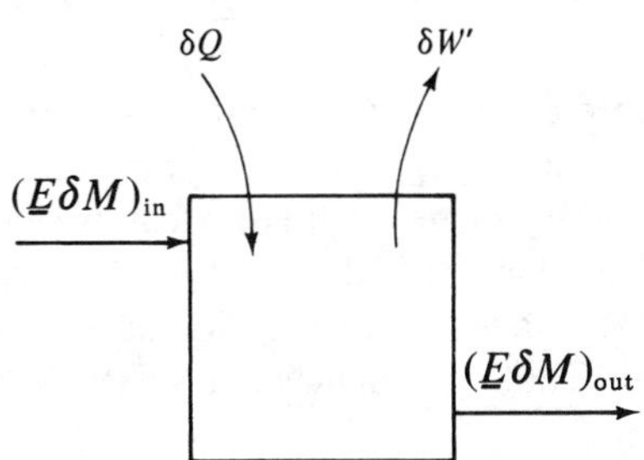

FIG. 4-13. The open system.

For a process with no chemical reaction the mass balance is written as

$$\delta M_{in} - \delta M_{out} = dM_{syst} \tag{4-31}$$

Particular attention should be given at this point to the different meanings of the two infinitestimal operators δ and d: δ is used to indicate the transfer of a differential amount of something such as mass, heat, or work; d is used to signify the *change* in a (state) property [that is, the value of a (state) property now, minus its value at some previous time]. The difference between δ and d becomes especially important when they are integrated. The integral of δ is simply the total amount of whatever is entering (or leaving) the system, or the total internal production. However, the integral of d is the total change of the system property and is represented by the difference operator Δ. That is,

$$\int_{A_1}^{A_2} dA = \Delta A = A_2 - A_1$$

while (4-32)

$$\int \delta W = W$$

Sign Convention for δQ and δW

In the previous derivation of the energy balance, we have assumed that heat, δQ, is being added to the system while the system is performing work, $\delta W'$, on the surroundings. That is, δQ is positive when heat is transferred into the system. If the system is losing heat to the surroundings, then δQ will be a negative number—but the sign before the δQ term in the energy balance should remain +. Similarly, $\delta W'$ has a positive value when the system is performing work on the surroundings. However, when the surroundings perform work on the system, $\delta W'$ has a negative value, which when combined with the minus sign in the energy balance yields a positive contribution to the energy of the system.

Total Energy

The total energy of the system, $\underline{E}$, may be split into three separate terms: the internal energy, $\underline{U}$, the kinetic energy, $u^2/2g_c$, and the (gravitational) potential energy, gZ/g_c. Therefore, $\underline{E}$ becomes

$$\underline{E} = \underline{U} + \frac{u^2}{2g_c} + \frac{gZ}{g_c} \tag{4-33}$$

where all terms represent energy per unit mass (mole). In some fields of study other forms of potential energy are also important, and this equation can be generalized by including them. In this text, however, they will not be needed.

Now let us examine $\delta W'$, the total work transferred from the system to the surroundings. We have shown that whenever a force acts over a distance, work is performed. That is, whenever a system's boundaries move, work is done.

When we examine an open system in close detail, the following process is seen to occur at the inlet and outlet (as illustrated in Fig. 4-14). As a differ-

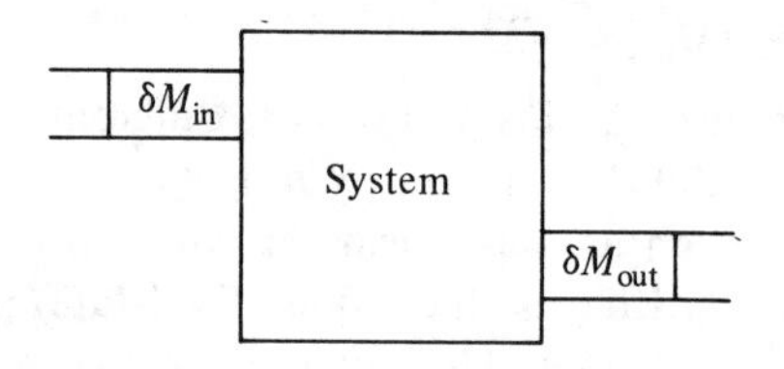

FIG. 4-14. The effect of mass flows.

ential amount of mass, δM_{in}, is forced into the system, the surroundings perform an amount of work equal to $(P\,dV)_{in} = (P\underline{V}\delta M)_{in}$ on the increment of mass entering the system. Similarly, as δM_{out} is forced out of the system, the system performs an amount of work equal to $(P\underline{V}\,\delta M)_{out}$ on the surroundings. If an amount of work δW is performed by the system on its surroundings in addition to the work required to force mass in and out of the system (such as the work needed to lift a weight or turn a shaft), then the total amount of work performed by the system on its surroundings is

$$\delta W' = \delta W + (P\underline{V}\,\delta M)_{out} - (P\underline{V}\,\delta M)_{in} \tag{4-34}$$

The term δW represents the amount of work that must be supplied to (or removed from) the system by a source other than the material entering or leaving the system and is commonly referred to as the "shaft work." If the expression for $\delta W'$ is substituted into the energy balance, equation (4-29), we obtain

$$(\underline{E} + P\underline{V})_{in}\,\delta M_{in} - (\underline{E} + P\underline{V})_{out}\,\delta M_{out} + \delta Q - \delta W = d(\underline{E}M)_{syst} \tag{4-35}$$

or upon substitution of equation (4-33) for $\underline{E}$;

$$\left(\underline{U} + P\underline{V} + \frac{u^2}{2g_c} + \frac{gZ}{g_c}\right)_{in}\delta M_{in} - \left(\underline{U} + P\underline{V} + \frac{u^2}{2g_c} + \frac{gZ}{g_c}\right)_{out}\delta M_{out}$$
$$+ \delta Q - \delta W = d\left[M\left(\underline{U} + \frac{u^2}{2g_c} + \frac{gZ}{g_c}\right)\right]_{syst} \tag{4-36}$$

Equation (4-36) is the *open-system energy balance*. It is extremely general (although only for processes in which only internal, potential and kinetic energies changes are important). As we shall see, the energy balance can be greatly simplified for many special cases.

In equation (4-36), the term $\underline{U} + P\underline{V}$ appears twice. Since this term is encountered frequently, it has attained the status of a property and has been given its own name and symbol. $\underline{U} + P\underline{V}$ is called the *specific enthalpy* and given the symbol $\underline{H}$. Therefore,

$$\underline{H} = \underline{U} + P\underline{V} \tag{4-37}$$

or, multiplying by the total mass, M, the *total enthalpy* is

$$H = U + PV \tag{4-37a}$$

Substitution of the enthalpy into the energy balance gives

$$\left(\underline{H} + \frac{u^2}{2g_c} + \frac{gZ}{g_c}\right)_{\text{in}} \delta M_{\text{in}} - \left(\underline{H} + \frac{u^2}{2g_c} + \frac{gZ}{g_c}\right)_{\text{out}} \delta M_{\text{out}} + \delta Q - \delta W$$
$$= d\left[M\left(\underline{U} + \frac{u^2}{2g_c} + \frac{gZ}{g_c}\right)\right]_{\text{syst}} \tag{4-38}$$

Since $\underline{U}$, P, and $\underline{V}$ are functions only of the state of the material involved, $\underline{H}$ is also a state function, whose value therefore does not depend on its past history. Values of enthalpy are commonly found in tabulations of thermodynamic properties.

4.7 Flow or Shaft Work and Its Evaluation

We have said that work is equivalent to a force moving through a distance. A system is capable of performing (or consuming) work in three primary fashions: (1) the boundaries of the system may move against a restraining force; (2) a shaft may add (or remove) work through the system's boundary or (3) energy may transfer across the system boundary by virtue of a thermodynamic potential other than temperature (for example, an electrical potential).

The work associated with a process depends on the choice of the system. As was noted earlier, an energy transfer—regardless of the potential causing it —is not considered to be either heat or work if it occurs wholly within the system. Thus it is always possible to define a system such that all flows are internal and thus neither heat nor work in a thermodynamic sense. These observations will be examined in the discussion that follows.

Let us consider first a simple closed system as illustrated in Fig. 4-15 (choose the gas as the system). As the gas expands, it does work by pushing back the surroundings. If a shaft within the system supplies an amount of work

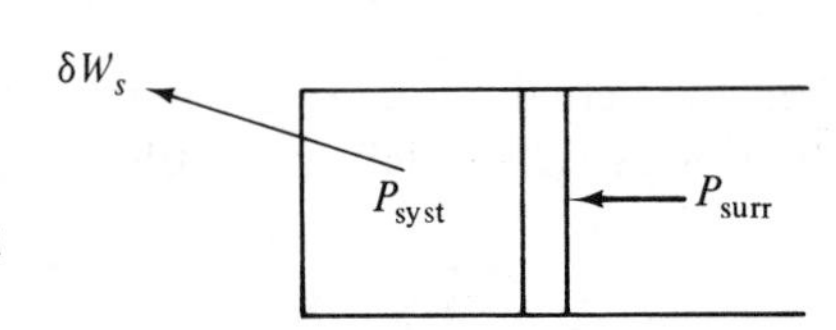

FIG. 4-15. Work produced by a closed system.

δW_s to the surroundings, then the total work may be evaluated as

$$\delta W_{\text{total}} = \delta W_s + P_{\text{res}}\, dV_{\text{syst}} \tag{4-39}$$

where $P_{\text{res}}\, dV$ represents the work needed to push back the surroundings.

Let us now consider a gas flowing through a turbine or a compressor (Fig. 4-16). Suppose we choose as our system 1 lb_m of gas as it flows through the

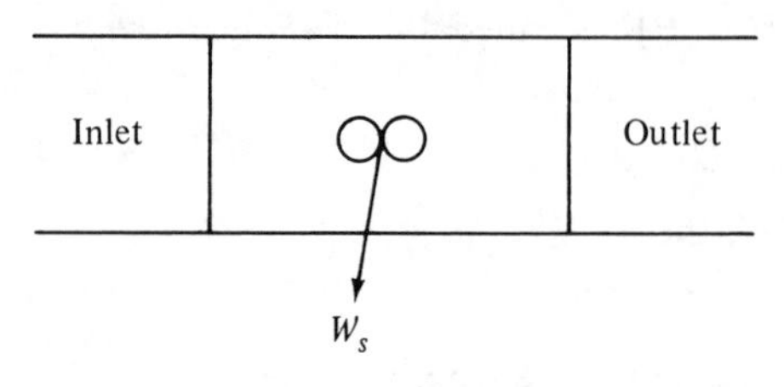

FIG. 4-16. Work produced in an open system.

turbine (the turbine blades and shaft are chosen to be outside the system). The gas represents a closed system. As the gas flows through the turbine the total amount of work it does is

$$\underline{W}' = \int P_{\text{syst}}\, d\underline{V}_{\text{syst}} \tag{4-40}$$

where the prime refers to the total amount of work delivered by the mass of gas to its surroundings. This work is divided into three parts: (1) that work which the gas does on the gas in front of it as it moves down the pipeline, given by $P_{\text{out}}\underline{V}_{\text{out}}$; (2) that work which the gas receives from the gas behind it as it is forced down the pipeline, given by $P_{\text{in}}\underline{V}_{\text{in}}$; and (3) that work which is transmitted to the shaft, $\underline{W}_s$. Thus

$$\underline{W}' = \int P_{\text{syst}}\, d\underline{V}_{\text{syst}} = \underline{W}_s + P_{\text{out}}\underline{V}_{\text{out}} - P_{\text{in}}\underline{V}_{\text{in}} \tag{4-41}$$

Solution for the work transmitted to the shaft, W_s, then gives

$$\begin{aligned} \underline{W}_s &= \int P_{\text{syst}}\, d\underline{V}_{\text{syst}} - (P_{\text{out}}\underline{V}_{\text{out}} - P_{\text{in}}\underline{V}_{\text{in}}) \\ &= \int P_{\text{syst}}\, d\underline{V}_{\text{syst}} - \Delta(P\underline{V}) \end{aligned} \tag{4-42}$$

Integration by parts of the first integral in equation (4-42) gives

$$\int_1^2 P_{\text{syst}}\, d\underline{V}_{\text{syst}} = -\int_1^2 \underline{V}\, dP + \int_1^2 d(P\underline{V}) \tag{4-43}$$

or

$$\int_1^2 P_{\text{syst}}\, d\underline{V}_{\text{syst}} = -\int_1^2 \underline{V}\, dP + \Delta P\underline{V} \tag{4-44}$$

which may be substituted into equation (4-42) to give

$$\underline{W}_s = -\int_1^2 \underline{V}\, dP \tag{4-45}$$

W_s is then the actual work *transferred from the gas to the shaft*. (If the shaft passes frictionlessly through the walls of the turbine, then $\underline{W}_s$ is the work transmitted to the *surroundings of the turbine*. If the shaft encounters frictional losses as it passes through the turbine housing, then the amount of work transmitted to the turbine's surroundings will be decreased by a comparable amount.)

If we now consider the turbine and its contents as an open system, then the work term in the energy balance is only that work which the system transmits to its surroundings via the connecting shaft. (Had the volume of the turbine been changing, a $P_{\text{surr}}\, dV_{\text{syst}}$ term would also be needed.) For the reversible process this would be $\underline{W}_s$ of equation (4-45). For a nonreversible process it would be $\underline{W}_s$ minus the frictional losses.

Thus we learn that when we are attempting to analyze the *reversible* flow of a material *through* a piece of (rigid) equipment, there are two different work terms that may be encountered, depending on how the system is chosen. The first is the work produced by a unit mass of material (the system) as it flows through the equipment and is given by

$$\underline{W}' = \int P_{\text{syst}}\, d\underline{V}_{\text{syst}} \tag{4-40}$$

The second is work transmitted by the gas via a mechanical linkage (shaft) to the surroundings of the equipment and is given by

$$\underline{W}_s = -\int \underline{V}_{\text{syst}}\, dP_{\text{syst}} \tag{4-46}$$

where the system is taken as the turbine and its contents. The two work terms differ by $\Delta(P\underline{V})$, the net work the unit of material expends pushing fluid out of its way as it passes through the piece of equipment. Because this amount of work, $\Delta P\underline{V}$, is the work associated only with the flow of the working fluid, it is frequently termed *flow energy* or *flow work*. For processes in which no mass flow occurs, there is no work needed to move the fluid around and thus such a term does not arise.

The shaft work, $\underline{W}_s$, is the maximum amount of *recoverable* work that may be obtained from the flow of a fluid through a piece of process equipment (for irreversible flows, the actual $\underline{W}$ is less). For example, if we are analyzing the main turbine in an electrical generating station, the maximum amount of work that may be recovered from the working fluid and transmitted to the electric generators would be $\underline{W}_s$, not $\underline{W}'$. Similarly, if we were analyzing the work requirements of a compressor, $\underline{W}_s$, not $\underline{W}'$ would be the minimum amount of work that would have to be supplied to the compressor motor in the form of electrical energy. Thus we see that, for flow processes, it is the shaft work, $\underline{W}_s$, not $\underline{W}'$, that is the term of primary importance to us. This is the reason we have chosen to write the general energy balance in terms of $\underline{W}_s$ rather than $\underline{W}'$. Of course, if the (flow) system boundaries are expanding or contracting against a surrounding pressure, then a work term for this must also be added to $\underline{W}_s$. Similarly, for irreversible processes, the actual work transfer, $\underline{W}$, is the term of primary importance.

4.8 Special Cases of the Energy Equation

Now let us examine some of the simplifications of the energy equation for certain special, but still common and important cases.

Closed System

The gas enclosed by a cylinder and piston is a typical example of a closed system (see Fig. 4-7). Since no mass enters or leaves the closed system, $\delta M_{in} = \delta M_{out} = dM_{syst} = 0$. Therefore, the energy equation becomes

$$\delta Q - \delta W = Md\left(\underline{U} + \frac{u^2}{2g_c} + \frac{gZ}{g_c}\right)_{syst} \tag{4-47}$$

If the system does not accelerate greatly its velocity, u, does not vary greatly and we may neglect changes in $u^2/2g_c$. If its height does not vary appreciably, we may also neglect changes in gZ/g_c in comparison to $\underline{U}$, and the closed-system energy balance further reduces to

$$\delta Q - \delta W = M\,d\underline{U}_{syst} = dU_{syst} \tag{4-48}$$

Equation (4-48) is frequently referred to as the *first law of thermodynamics* and is presented as a general form of the energy balance. However, as we have seen, equation (4-48) is merely a special case of the more general, open system energy balance. *It may only be used for closed systems that neither accelerate nor change their height appreciably.*

Open Systems

As you will learn in later courses, most equipment and process designers strive to design equipment and processes that will operate at steady state for extended periods of time. Unsteady processes are frequently avoided because they are usually less economical than steady-state processes. In addition, flow processes are usually much simpler to operate and maintain than batch, or unsteady, processes.

Since all flow processes may be considered as open systems (with the process equipment and its contents as the system), most engineering design problems may be treated as open systems—usually operating at steady state.

Of course, any process can be treated using a closed-system analysis if we choose as the system a unit of mass as it flows through the equipment. However, when this is done we must be able to evaluate the work performed on, or by, the system as it flows through the process. Unfortunately, this evaluation normally requires knowledge of all the system properties at every point during the process—knowledge that is frequently not available. The open-system approach, on the other hand, requires knowledge of only the "shaft work" supplied to, or by, the processing machinery. Usually this quantity can be easily calculated or directly measured. For this reason the flow-system approach to thermodynamic calculations is usually preferred to the closed-system approach. When sufficient data for both a closed- and an open-system analysis are available, the results from both calculations *must be identical* (except for determination of the system work and heat transfer).

A *steady-state process* is defined as one in which the system properties (either point or average properties) do not change with time. That is, if we examine either a single point in the system or the system as a whole, its properties will not vary with time. Note, this in no way implies that the properties at all points must be identical, only that the properties of each point are invariant with time. Since the mass of a system is a property of the system, steady state implies that $\delta M_{\text{in}} = \delta M_{\text{out}}$ and $dM_{\text{syst}} = 0$. Since total energy is also a system property, steady state also implies that $d[\underline{U} + (u^2/2g_c) + (gZ/g_c)]_{\text{syst}} = 0$. Thus the energy balance for a steady-state process reduces to

$$\left(\underline{H} + \frac{u^2}{2g_c} + \frac{gZ}{g_c}\right)_{\text{in}} \delta M - \left(\underline{H} + \frac{u^2}{2g_c} + \frac{gZ}{g_c}\right)_{\text{out}} \delta M + \delta Q - \delta W = 0 \tag{4-49}$$

We may divide equation (4-49) by δM to get

$$\left(\underline{H} + \frac{u^2}{2g_c} + \frac{gZ}{g_c}\right)_{\text{in}} - \left(\underline{H} + \frac{u^2}{2g_c} + \frac{gZ}{g_c}\right)_{\text{out}} + \frac{\delta Q}{\delta M} - \frac{\delta W}{\delta M} = 0 \tag{4-49a}$$

where

$$\frac{\delta Q}{\delta M} = \underline{Q} = \frac{\text{heat absorbed}}{\text{unit mass flow through equipment}}$$

$$\frac{\delta W}{\delta M} = \underline{W} = \frac{\text{work liberated}}{\text{unit mass flow through equipment}} \tag{4-50}$$

Thus equation (4-49a) reduces to

$$-\Delta\left(\underline{H} + \frac{u^2}{2g_c} + \frac{gZ}{g_c}\right) + \underline{Q} - \underline{W} = 0 \tag{4-51}$$

Compressors and Expanders—Steady State

Compressors and expanders are machines that are capable of changing the pressure on a substance. Compressors (pumps) are used to increase the pressure on a substance, and they consume work in so doing. There are two main classes of compressors: positive-displacement compressors, which include the piston, diaphragm, and gear compressors; and centrifugal compressors. Positive-displacement compressors are usually capable of high compression ratios per stage, but being noncontinuous they cannot handle high flow rates. Centrifugal compressors, on the other hand, are not capable of producing high pressure ratios per stage but can handle extremely large flow volumes, because they operate continuously. Most of the pumps and compressors used in normal process operations are of the centrifugal type—fitting into our previous observation that most processes tend to be of the continuous type.

Expanders operate in the reverse fashion from compressors—decreasing the pressure on the working fluid while producing work. All compressors can in theory be run backward to operate as expanders. However, in practice only continuous machinery is used in work-producing expanders. Turbine expanders are by far the most commonly encountered type of expansion machines.

For most compressors and expanders it is found that the $\Delta(u^2/2g_c)$ and $\Delta(gZ/g_c)$ terms are usually quite small in comparison to the $\Delta \underline{H}$ term. Therefore, these terms are often assumed to be negligible and are neglected. Under these assumptions the steady-state energy equation reduces to

$$\Delta \underline{H} = \underline{Q} - \underline{W} \tag{4-52}$$

Equation (4-52) can be considered the open-system analogy to equation (4-48).

If the compressor is run adiabatically, that is, with no heat exchange with the surroundings, then $\underline{Q} = 0$, and equation (4-52) reduces to

$$\Delta \underline{H} = -\underline{W} \tag{4-53}$$

Throttling Devices—Steady State

A throttling device is a device used to irreversibly reduce the pressure of a flowing fluid without obtaining any shaft work. The valve that reduces the water pressure in your kitchen faucet from the pressure of the water main to atmospheric pressure is a good example of a throttling device. As with most compressors and expanders, most throttling devices produce a negligible change in $u^2/2g_c$ and gZ/g_c; therefore, the energy balance reduces to

$$-\Delta \underline{H} + \underline{Q} - \underline{W} = 0 \tag{4-52}$$

However, since throttling devices provide no work to the surroundings, $\underline{W} = 0$, and equation (4-52) reduces to

$$\Delta \underline{H} = \underline{Q} \tag{4-54}$$

Usually the fluid passing through a throttling device is moving so rapidly that it does not remain within the device long enough to absorb, or give off, much heat. Therefore, for many cases it is reasonable to assume that $\underline{Q} = 0$ (unless special effort is taken to provide heating or cooling within the device), and the energy balance becomes

$$\Delta \underline{H} = 0 \tag{4-55}$$

or

$$\underline{H}_{\text{in}} = \underline{H}_{\text{out}} \tag{4-55a}$$

Nozzles—Steady State

A nozzle is a device specifically designed to increase the kinetic energy of a high-pressure fluid at the expense of its pressure and temperature. Since a nozzle is designed to increase the velocity of the working fluid (and often to a very high value), we cannot neglect the $u^2/2g_c$ term. However, except for extremely long vertical nozzles, the gZ/g_c term is still small, and the $\underline{Q}$ is usually neglected for the same reasons as with the throttling devices. Normally no shaft work is produced, so $\underline{W} = 0$ and the energy balance reduces to

$$\Delta \underline{H} + \Delta \frac{u^2}{2g_c} = 0 \tag{4-56}$$

or

$$\Delta \underline{H} = -\Delta \frac{u^2}{2g_c} \tag{4-57}$$

4.9 Heat Capacities

We have defined the constant-volume and constant-pressure heat capacities by the relations

$$\begin{aligned} \delta Q &= MC_V\,dT \qquad \text{for constant-volume processes} \\ \delta Q &= MC_P\,dT \qquad \text{for constant-pressure processes} \end{aligned} \tag{4-58}$$

(where it is assumed that the system's mass, as well as its potential and kinetic energy, were constant throughout the heat-transfer process).

Let us now show how C_V and C_P are related respectively to internal energy, and enthalpy. Consider as a system a unit of mass as it proceeds through a given process. Since the system is closed, its mass is constant, and the energy balance reduces to

$$\delta Q - \delta W = Md\left(\underline{U} + \frac{u^2}{2g_c} + \frac{gZ}{g_c}\right) \tag{4-59}$$

If we also assume that the potential and kinetic energies of the system are constant, then equation (4-59) simplifies to

$$\delta Q - \delta W = M\,d\underline{U} \tag{4-60}$$

Now, if the process is performed at constant volume (or specific volume since the mass is constant), the work term ($\delta W = P_{\text{res}}\,dV_{\text{syst}}$) vanishes and we get

$$\delta Q = M\,d\underline{U} \tag{4-61}$$

But at constant volume we have shown that

$$\delta Q = MC_V\,dT \tag{4-62}$$

Eliminating δQ between equations (4-61) and (4-62) gives

$$MC_V\,dT = M\,d\underline{U} \qquad \text{at constant } \underline{V} \tag{4-63}$$

or

$$C_V = \left(\frac{\partial \underline{U}}{\partial T}\right)_{\underline{V}} \tag{4-63a}$$

If, on the other hand, the process is performed at constant pressure as in a flow calorimeter, the system's pressure is the same as the gas that surrounds it as it moves through the process. Thus, $P_{\text{syst}} = P_{\text{surr}}$ and

$$\delta W = P_{\text{res}}\,dV_{\text{syst}} = P_{\text{syst}}\,dV_{\text{syst}} = d(P_{\text{syst}}V_{\text{syst}}) = Md(P_{\text{syst}}\underline{V}_{\text{syst}}) \tag{4-64}$$

Substitution of this work expression into equation (4-60) gives

$$\delta Q - Md(P\underline{V}) = M\,d\underline{U} \tag{4-65}$$

At constant pressure

$$\delta Q = MC_P\,dT \tag{4-66}$$

Elimination of δQ between equations (4-65) and (4-66) gives

$$\begin{aligned} MC_P\,dT &= M\,d\underline{U} + Md(P\underline{V}) = Md(\underline{U} + P\underline{V}) \\ &= M\,d\underline{H} \qquad \text{at constant pressure} \end{aligned} \tag{4-67}$$

That is,

$$C_P = \left(\frac{\partial \underline{H}}{\partial T}\right)_P \tag{4-68}$$

Thus we find that the heat capacities may be directly related to the derivatives of $\underline{U}$ and $\underline{H}$ with respect to temperature.

We will show later that the heat capacities of gases that obey the ideal gas equation of state have two very useful properties:

1. C_P and C_V are both independent of P and $\underline{V}$ at constant temperature.
2. C_P and C_V are related through the equation

$$C_P = C_V + R \tag{4-69}$$

In addition, it is frequently assumed that C_P and C_V are independent of T as well as P and $\underline{V}$ for an ideal gas.

4.10 Sample Problems

SAMPLE PROBLEM 4-4. A turbine is driven by 10,000 lb_m/hr of steam, which enters the turbine at 650 psia and 850°F with a velocity of 200 ft/sec. The steam leaves the turbine exhaust at a point 10 ft below the turbine inlet with a velocity of 1200 ft/sec. The shaft work produced by the turbine is measured as 943 hp, and the heat loss from the turbine has been calculated to be 10^5 Btu/hr.

A small portion of the exhaust steam from the turbine is passed through a throttling valve and discharges at atmospheric pressure. Velocity changes across the valve may be neglected.

(a) What is the temperature of the steam leaving the valve?

(b) What is the quality or degrees of superheat (whichever is applicable) of the steam leaving the throttling valve?

Solution: The process is as illustrated in Fig. S4-4. Since we know the pressure of the gas leaving the throttling valve (point 3), determination of one other property of the throttled stream will serve to fix its state and thus its temperature and quality (or superheat). Since we are supplied with considerable information about the energy

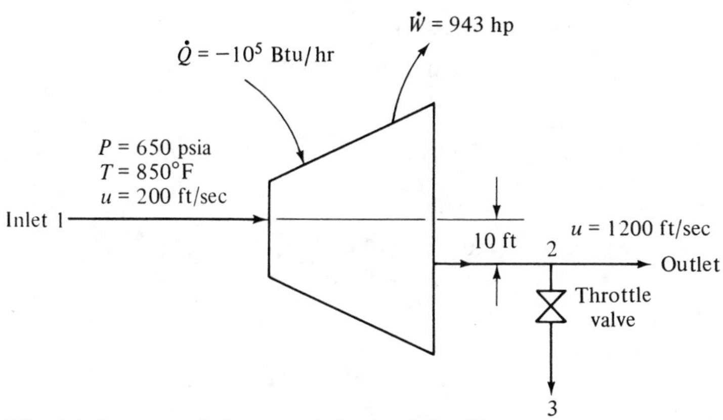

(The dot above a symbol represents "per unit time.")

FIGURE S4-4

requirements of the turbine, it is logical that the property of the gas at point 3 we should attempt to determine is the enthalpy.

We begin our determination of the enthalpy at point 3 by considering an energy balance around the throttling valve:

$$\left(\underline{H}_2 + \frac{u_2^2}{2g_c} + \frac{gZ_2}{g_c}\right)\delta M_{\text{in}} - \left(\underline{H}_3 + \frac{u_3^2}{2g_c} + \frac{gZ_3}{g_c}\right)\delta M_{\text{out}} + \delta Q - \delta W = d[M(\underline{E})_{\text{valve}}]$$

The valve is operating at steady state, so $d(M\underline{E}) = 0$ and $\delta M_{\text{in}} = \delta M_{\text{out}}$. We are told to neglect $\Delta(u^2/2g_c)$ and recognize that ΔZ will be small enough to neglect also. Therefore, the energy balance becomes

$$(\underline{H}_2 - \underline{H}_3)\,\delta M + \delta Q - \delta W = 0$$

However, we showed earlier that $\delta W = 0$ for a throttle valve, and for most cases δQ is also equal to zero. The energy balance then gives

$$\underline{H}_3 = \underline{H}_2$$

Thus, if we can determine $\underline{H}_2$, we will know a second property, $\underline{H}_3$ at point 3. $\underline{H}_2$ is determined from an energy balance around the turbine:

$$\left(\underline{H}_1 + \frac{u_1^2}{2g_c} + \frac{gZ_1}{g_c}\right)\delta M_1 - \left(\underline{H}_2 + \frac{u_2^2}{2g_c} + \frac{gZ_2}{g_c}\right)\delta M_2 + \delta Q - \delta W = d(M\underline{E})_{\text{turb}}$$

Base all future calculations on $\delta M_1 = 1\ \text{lb}_\text{m}$.

Since the turbine is operating at steady state, $d(M\underline{E})_{\text{syst}} = 0$ and $\delta M_2 = \delta M_1$. Also, since the turbine exit is only 10 ft below the inlet, we know that the $g\Delta Z/g_c$ term will be small and may be neglected. The energy balance then becomes

$$\left(\underline{H}_1 + \frac{u_1^2}{2g_c}\right) - \left(\underline{H}_2 + \frac{u_2^2}{2g_c}\right) + \underline{Q} - \underline{W} = 0$$

or

$$-\Delta\underline{H} - \Delta u^2/2g_c + \underline{Q} - \underline{W} = 0$$

but from the data given in the statement,

$$\underline{Q} = \frac{\dot{Q}}{\dot{M}} = \frac{-100{,}000 \text{ Btu/hr}}{10{,}000 \text{ lb}_m/\text{hr}} = -10 \frac{\text{Btu}}{\text{lb}_m}$$

$$\underline{W} = \frac{\dot{W}}{\dot{M}} = 943 \text{ hp}/10{,}000 \text{ lb}_m/\text{hr} = 240 \text{ Btu/lb}_m$$

$$\underline{H}_1 = 1430 \text{ Btu/lb}_m$$

$$\frac{\Delta u^2}{2g_c} = \frac{(1200)^2 - (200)^2 \text{ ft}^2/\text{sec}^2}{(2)(32)\ (\text{lb}_m/\text{lb}_f)(\text{ft/sec}^2)} = \frac{43{,}500 \text{ ft-lb}_f}{\text{lb}_m}$$

or, converting to Btu,

$$\frac{\Delta u^2}{2g_c} = 28 \text{ Btu/lb}_m$$

(A quick examination of the magnitude of the two $u^2/2g_c$ terms indicates that the 200-ft/sec term accounts for only about 3 percent of the total change in kinetic energy or about 1 Btu/lb$_m$. That is, for all practical purposes the kinetic energy associated with velocities below 200 ft/sec is negligible in comparison with the enthalpy or internal energy.) Thus the energy balance gives

$$\Delta \underline{H} = -28\text{Btu/lb}_m - 10 \text{ Btu/lb}_m - 240 \text{ Btu/lb}_m = -278 \text{ Btu/lb}_m$$

But

$$\Delta \underline{H} = \underline{H}_2 - \underline{H}_1 = -278 \text{ Btu/lb}_m$$

Therefore,

$$\begin{aligned}\underline{H}_2 &= \underline{H}_1 - 278 \text{ Btu/lb}_m = 1430 - 278 \text{ Btu/lb}_m \\ &= 1152 \text{ Btu/lb}_m\end{aligned}$$

but

$$\underline{H}_3 = \underline{H}_2 = 1152 \text{ Btu/lb}_m$$

We may now examine the steam tables to locate the point whose pressure is 14.696 psia and whose enthalpy is 1152 Btu/lb$_m$. This point is just above the saturation dome, and therefore we know that the gas must be superheated—about 2°F—so the temperature of the gas at point 3 is 214°F.

SAMPLE PROBLEM 4-5. Steam flowing through an insulated 1-in.-diameter pipe has a pressure of 800 psia and a quality of 95 per cent. If the steam is adiabatically throttled to 14.696 psia through a horizontal valve in the line, what should the diameter of the downstream line be if there is to be no change in the velocity of the stream?

Solution: The process under consideration is illustrated in Fig. S4-5. We may write a mass balance around the throttling valve as follows:

$$\dot{M}_{\text{in}} = \dot{M}_{\text{out}} = \dot{M}$$

$$\dot{M} = \rho_1 u_1 A_1 = \rho_2 u_2 A_2$$

where ρ_1 = density at point 1
u_1 = velocity at point 1
A_1 = area of pipe at point 1

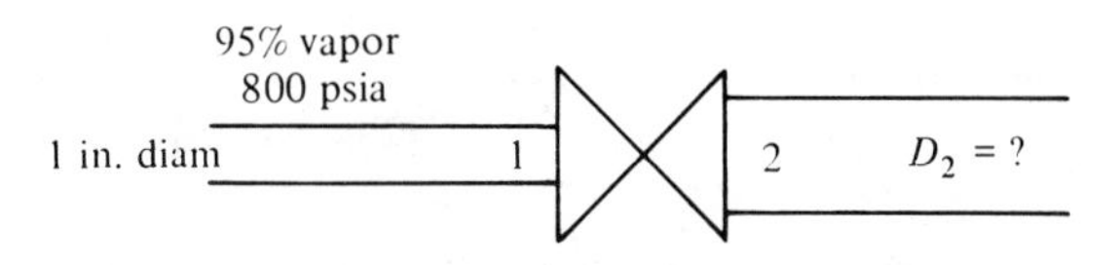

FIGURE S4-5

However, we are asked to find the downstream diameter such that $u_1 = u_2$. Therefore,

$$\rho_1 A_1 = \rho_2 A_2$$

or

$$\rho_1 \pi D_1^2 = \rho_2 \pi D_2^2$$

or

$$D_2 = D_1 \sqrt{\frac{\rho_1}{\rho_2}} = D_1 \sqrt{\frac{\underline{V}_2}{\underline{V}_1}}$$

since $\rho = 1/\underline{V}$. Therefore, our problem reduces to determining the downstream specific volume, since we can determine the upstream volume from the saturation pressure and moisture content.

If we can determine one more property of the downstream fluid (other than its pressure), we can fix the state of the fluid, from which we can determine its specific volume. An energy balance around the throttling valve will supply us with the needed property value.

Assuming horizontal, nonaccelerating, adiabatic, steady-state flow through the throttling valve, the energy balance reduces to

$$\underline{H}_2 = \underline{H}_1$$

but

$$\underline{H}_1 = 1162 \text{ Btu/lb}_m \text{ (from Mollier diagram)}$$

Therefore,

$$\underline{H}_2 = 1162 \text{ Btu/lb}_m$$

Knowing P_2 and $\underline{H}_2$, we can find T_2 and $\underline{V}_2$ from the steam tables.

$$T_2 = 238°\text{F}$$

$$\underline{V}_2 = 28.0 \text{ ft}^3/\text{lb}_m \text{ (from } P_2 \text{ and } \underline{H}_2)$$

Now let us determine the inlet volume, from which we can evaluate D_2:

$$\underline{V}_1 = X_1 \underline{V}^V + (1 - X_1)\underline{V}^L$$

where X_1 = vapor fraction at 1. But at 800 psia the saturation properties of vapor and liquid are

$$\underline{V}^V = 0.5685 \text{ ft}^3/\text{lb}_m$$

$$\underline{V}^L = 0.0209 \text{ ft}^3/\text{lb}_m$$

$$X_1 = 0.95 \text{ mass fraction vapor}$$

Therefore,

$$\underline{V}_1 = (0.95)(0.5685) + (0.05)(0.0209)$$
$$= 0.541 \text{ ft}^3/\text{lb}_m$$

$$D_1 = 1 \text{ in.}$$

and

$$D_2 = D_1\sqrt{\frac{\underline{V}_2}{\underline{V}_1}} = 7.2 \text{ in.}$$

SAMPLE PROBLEM 4-6. An evacuated chamber with perfectly insulated walls is connected to a steam main through which steam at 100 psia and 350°F is flowing. The valve is opened and steam flows rapidly into the chamber until the pressure within the chamber is 100 psia. If no heat is lost to the surroundings or transferred back into the main, find the temperature of the steam in the chamber when the flow stops.

Solution: When the valve is opened, steam will flow into the chamber until the pressure within the chamber is equal to that in the line (Fig. S4-6). However, since no

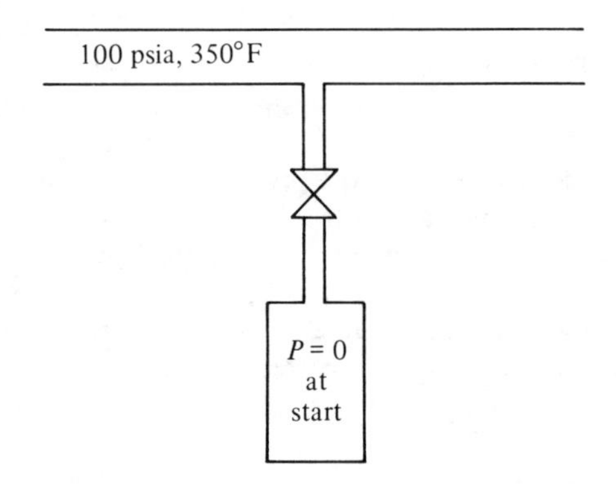

FIGURE S4-6

heat is transferred from the chamber, the temperature within the chamber may not (indeed will not) be equal to that in the main. Therefore, it is necessary to determine another property of the steam in the chamber before the state can be fixed and its temperature determined. Thus we should like to know the enthalpy (or, as we shall see, more conveniently the internal energy) of the material in the tank at the end of the process. At this point we expect that some form of energy balance will probably be needed. But before we apply the energy balance we must choose the system. At least two systems are possible:

1. We could choose the closed system: all the gas that ends up in the chamber at the end of the process, or
2. We could pick the open system: the chamber, the valve, and their contents at any moment.

For this illustration we shall choose the open-system approach. (The student should solve this problem using the closed-system approach to prove that both methods of attack yield the same solution.) With the open-system approach we may write an energy balance around the chamber and its contents as

$$\left(\underline{H} + \frac{u^2}{2g_c} + \frac{gZ}{g_c}\right)_{\text{in}} \delta M_{\text{in}} - \left(\underline{H} + \frac{u^2}{2g_c} + \frac{gZ}{g_c}\right)_{\text{in}} \delta M_{\text{out}} + \delta Q - \delta W$$
$$= d\left[M\left(\underline{U} + \frac{u^2}{2g_c} + \frac{gZ}{g_c}\right)\right]_{\text{syst}}$$

However, no mass flows out of the system; that is, $\delta M_{out} = 0$. Also, we may neglect $u^2/2g_c + gZ/g_c$ in relation to $\underline{H}$ and $\underline{U}$. Therefore, the energy balance simplifies to

$$(\underline{H}\delta M)_{in} + \delta Q - \delta W = d(M\underline{U})_{syst}$$

But $\delta Q = 0$ because of the insulation, and $\delta W = 0$ because the system boundaries do not move and no shaft brings work through the system boundaries. The energy balance now becomes

$$(\underline{H}\delta M)_{in} = d(M\underline{U})_{syst}$$

We must now integrate the energy balance over all the mass that enters. Since H_{in} is a function only of the T and P in the main, it does not vary with mass and may be removed from under the integral sign, to give

$$\underline{H}_{in} \int \delta M_{in} = \int d(M\underline{U})_{syst}$$

or

$$(\underline{H}M)_{in} = \int d(M\underline{U}) = (M\underline{U})_2 - (M\underline{U})_1$$

But M_{in} = total mass added to the system $= M_2 - M_1$. Therefore,

$$\underline{H}(M_2 - M_1) = (M\underline{U})_2 - (M\underline{U})_1$$

But $M_1 = 0$ since chamber was initially evacuated; therefore, the integrated energy balance becomes

$$\underline{H}_{in}M_2 = M_2\underline{U}_2$$

or

$$\underline{H}_{in} = \underline{U}_2$$

That is, the internal energy of the material in the chamber at the end of the process is equal to the enthalpy of the incoming material.

Now we must find that temperature at which the internal energy of steam at 100 psia is equal to the enthalpy of the incoming steam, which is $\underline{H}_{in} = 1200$ Btu/lb$_m$. Unfortunately, the Mollier diagram and steam tables do not have entries with the internal energy. Therefore, we must determine $\underline{U}$ from the relation

$$\underline{U} = \underline{H} - P\underline{V}$$

and the given values of $\underline{H}$ and $\underline{V}$ at various temperatures (Table S4-6; $P = 100$ psia). Now we must interpolate between the values of $\underline{U}$ to find T_2. The interpolation is performed in Fig. S4-6a, from which we see that the final temperature is 570°F. Of course, if our Mollier diagram or steam tables had tabulated values of the internal energy (and some do), then this interpolation would be unnecessary.

TABLE S4-6

T, °F	$\underline{H}$, Btu/lb$_m$	$\underline{V}$, ft^3/lb$_m$	$P\underline{V}$, Btu/lb$_m$	$\underline{U}$
450	1253.7	5.266	97.2	1156.5
500	1278.6	5.589	103.5	1175.1
600	1327.9	6.217	115.0	1212.9
700	1377.5	6.836	126.5	1251.0

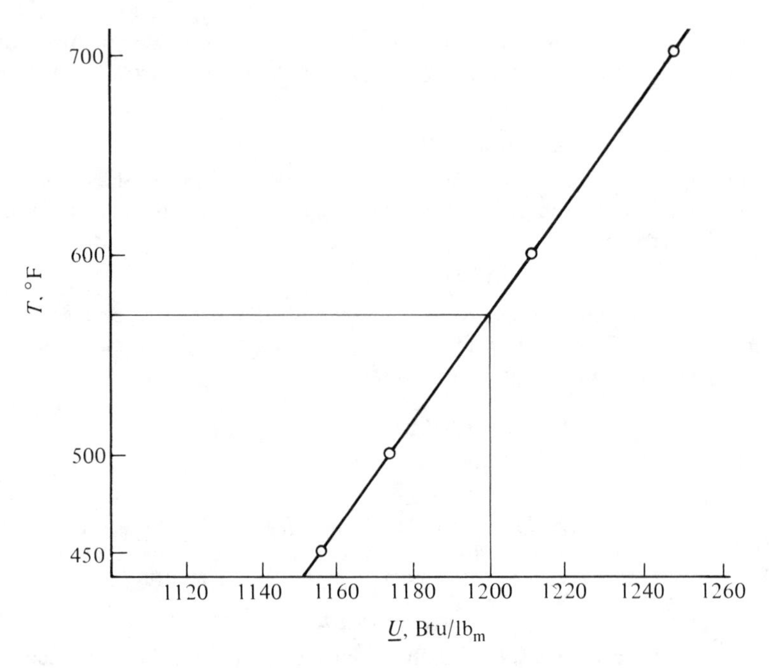

FIGURE S4-6a

SAMPLE PROBLEM 4.7. Methane is to be transported in high-pressure-gas cylinders. The cylinders have previously been used for methane shipment, and they are being refilled. Before refilling, the cylinders contain methane at a pressure of 500 psia and a temperature of 75°F. Methane for refilling is available from a large pipeline at a pressure of 4500 psia and 120°F. The cylinders are refilled by connecting them to the methane pipeline and allowing methane to flow into the cylinder until the pressure within the cylinder is 4500 psia. Each cylinder has a volume of 1 ft^3.

(a) If the cylinders are filled slowly, so that the final temperature of the methane is room temperature (75°F), how much methane is added to each cylinder? How much methane does each cylinder contain at the end of the filling operation?

(b) If the cylinder is filled rapidly so that the process is essentially adiabatic, how much methane is added to each cylinder? How much methane is contained in the cylinder at the end of the filling operation?

You may assume that methane obeys the Beattie–Bridgeman equation of state for the temperatures and pressures involved, and that the internal energy is expressed as a function of T and $\underline{V}$ by (we shall derive this expression in Chapter 6)

$$\underline{U}(T, \underline{V}) = CVA(T - T_0) + \frac{CVB}{2}(T^2 - T_0^2) + \frac{CVC}{3}(T^3 - T_0^3)$$
$$- \frac{1}{\underline{V}}\left(A_0 + \frac{3RC}{T^2}\right) + \frac{1}{2\underline{V}^2}\left(A_0 a - \frac{3RB_0C}{T^2}\right) + \frac{RB_0bC}{\underline{V}^3T^2}$$

where A_0, a, B_0, b, C, and R are the constants in the Beattie–Bridgeman equation; T_0 is a reference temperature such that $U(T_0, \infty)$ equals zero; and CVA, CVB, and CVC are the constants in the zero-pressure constant-volume heat capacity expression. For methane these have the values

$$CVA = 1.394$$

$$CVB = 10.03 \times 10^{-3}$$

$$CVC = -1.327 \times 10^{-6}$$

Thus for methane

$$C_V^* = 1.394 + 10.03 \times 10^{-3}T - 1.327 \times 10^{-6}T^2 \qquad \text{(where } T = {}^\circ\text{R)}$$

$$C_V^* = \text{Btu/}^\circ\text{R lb-mol.}$$

Solution: (a) Each cylinder has a volume of 1 ft^3. When the cylinders are returned, they contain methane at 500 psia and 75°F, but we do not know how much. We could find how much they contain from the specific volume of the gaseous methane and the relation

$$V = M\underline{V}$$

or

$$M = \frac{V}{\underline{V}}$$

where $V = 1\ ft^3$
$\underline{V}$ = specific volume at 500 psia and 75°F

Since we know the pressure and temperature of the methane, the specific volume is fixed and may be determined from the Beattie–Bridgeman equation of state. Using the program we wrote for Sample Problem 3-6, we feed in the appropriate values for the Beattie–Bridgeman constants, along with the pressure and temperature, and obtain the value for $\underline{V}$:

$$\underline{V}_{\text{initial}} = 0.6755\ ft^3/lb_m$$

Thus

$$M_{\text{initial}} = 1\ ft^3/(0.6755\ ft^3/lb_m) = 1.480\ lb_m$$

At the end of the filling operation we still know the pressure and temperature ($P = 4500$ psia, $T = 75$°F), and thus can use exactly the same logic as in the initial case to determine the final mass in the tank at the end of filling:

$$\underline{V}_{\text{final}} = 0.07979\ ft^3/lb_m$$

$$M_{\text{final}} = 12.52\ lb_m$$

and thus the total mass added during filling is

$$\Delta M = M_{\text{final}} - M_{\text{initial}} = 11.04\ lb_m$$

(b) In the second part of the problem things get somewhat more complicated, because we do not know the final temperature of the methane in the cylinder. Let us now attempt to develop a set of equations from which the final temperature and specific volume can be calculated. We begin by writing the energy balance. Let us choose as a system the tank and its contents at any time (Fig. S4-7). Since no mass flows out of the tank, $\delta M_{\text{out}} = 0$; also $\delta Q = \delta W = 0$. Finally, neglect potential and kinetic energy

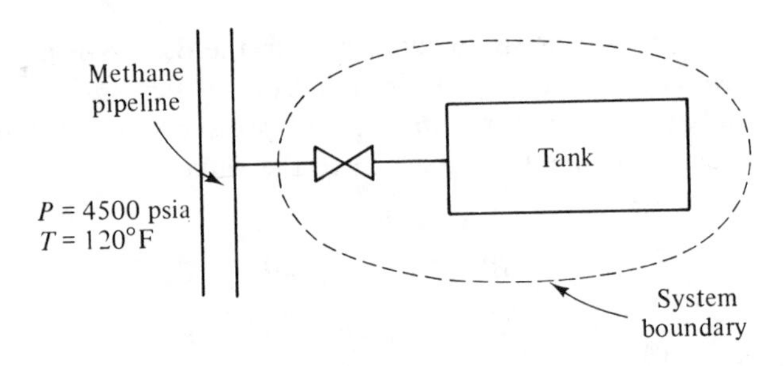

FIGURE S4-7

changes so that the energy equation reduces to

$$(\underline{H}\,\delta M)_{\text{in}} = d(M\underline{U})_{\text{syst}}$$

However, as mass flows into the system the properties of the gas in the pipeline do not change. Thus when we integrate over the total mass charged to the tank, the left-hand side becomes $\int \underline{H}_{\text{in}}\,\delta M = H_{\text{in}}(M_{\text{final}} - M_{\text{initial}})$. The right-hand side simply becomes

$$\int d(M\underline{U})_{\text{syst}} = \Delta(M\underline{U})_{\text{syst}} = (M\underline{U})_{\text{final}} - (M\underline{U})_{\text{initial}}$$

Thus the integrated energy equation becomes

$$\underline{H}_{\text{in}}(M_{\text{final}} - M_{\text{initial}}) = (M\underline{U})_{\text{final}} - (M\underline{U})_{\text{initial}}$$

M_{initial} is known from part (a). $\underline{U}_{\text{initial}}$ can be found from the $\underline{U}(T, \underline{V})$ expression because we know the initial T and $\underline{V}$. $\underline{H}_{\text{in}}$ can be found by evaluating $\underline{U}_{\text{in}}$ from the $U(T, \underline{V})$ expression and the known inlet conditions, and simply adding $(P\underline{V})_{\text{inlet}}$ to U_{inlet}. ($\underline{V}_{\text{inlet}}$ will have to be found from the Beattie–Bridgeman equation.) Thus there are still two unknowns in one equation, M_{final} and $\underline{U}_{\text{final}}$. We need more equations.

Write the volume expression

$$M_{\text{final}} = \frac{V}{\underline{V}_{\text{final}}}$$

This adds another equation but also another unknown. However, we know that

$$\underline{U}_{\text{final}} = \phi(\underline{V}_{\text{final}}, P_{\text{final}})$$

We know that $P_{\text{final}} = 4500$, so in theory this equation adds no new unknowns, and we have a determinate system of three equations in three unknowns. Unfortunately, we do not know the form of the functional relationship

$$U = \phi(P, \underline{V})$$

but know instead that

$$\underline{U} = \underline{U}(T, \underline{V})$$

Thus we must add a fourth unknown to the system. The fourth equation is provided by the equation of state

$$P = P(T, \underline{V})$$

Thus we have four simultaneous, nonlinear equations in four unknowns which we must solve to obtain the outlet conditions. The four equations are summarized below:

$$M_{\text{final}}(\underline{H}_{\text{in}} - \underline{U}_{\text{final}}) = M_{\text{initial}}(\underline{H}_{\text{in}} - \underline{U}_{\text{initial}}) \quad \text{(a)}$$

$$\underline{V}_{\text{final}} = \frac{V}{M_{\text{final}}} \quad \text{(b)}$$

$$\underline{U}_{\text{final}} = \underline{U}(T_{\text{final}}, \underline{V}_{\text{final}}) \quad \text{(c)}$$

$$P_{\text{final}} = P(T_{\text{final}}, \underline{V}_{\text{final}}) \quad \text{(d)}$$

We may simplify this system of equations somewhat by eliminating V_{final} through equation (b) to give a system of three equation in three unknowns:

$$M_{\text{final}}(\underline{H}_{\text{in}} - \underline{U}_{\text{final}}) = M_{\text{initial}}(\underline{H}_{\text{in}} - \underline{U}_{\text{initial}}) \quad \text{(a')}$$

$$\underline{U}_{\text{final}} = \underline{U}\left(T_{\text{final}}, \frac{V}{M_{\text{final}}}\right) \quad \text{(b')}$$

$$P_{\text{final}} = P\left(T_{\text{final}}, \frac{V}{M_{\text{final}}}\right) \quad \text{(c')}$$

where

$$\underline{U}(T, \underline{V}) = CVA(T - T_0) + \frac{CVB}{2}(T^2 - T_0^2) + \frac{CVC}{3}(T^3 - T_0^3)$$
$$- \frac{1}{\underline{V}}\left(A_0 + \frac{3RC}{T^2}\right) + \frac{1}{2\underline{V}^2}\left(A_0 a - \frac{3RB_0C}{T^2}\right) + \frac{RB_0bC}{\underline{V}^3T^2}$$

$$P(T, \underline{V}) = \frac{RT}{\underline{V}} + \frac{\beta(T)}{\underline{V}^2} + \frac{\gamma(T)}{\underline{V}^3} + \frac{\delta(T)}{\underline{V}^4}$$

where $\beta(T) = RB_0T - \dfrac{RC}{T^2} - A_0$

$$\gamma(T) = A_0a - RB_0bT - \frac{RB_0C}{T^2}$$

$$\delta(T) = \frac{RB_0bC}{T^2}$$

Because equations (a′), (b′), and (c′) represent a set of simultaneous, nonlinear algebraic equations, the likelihood of finding an exact solution is extremely small. (Remember that in Sample Problem 3-6 we had only one of these equations.) Thus we shall develop a numerical procedure for obtaining an approximate solution to this set of equations. The procedure to be developed is essentially an extension to multivariable systems of the Newton iteration used in Sample Problem 3-6; it is called the Newton–Raphson technique and may be derived as follows.

It is desired to find the values of x, y, and z, such that

$$f(x, y, z) = 0$$

$$g(x, y, z) = 0$$

$$h(x, y, z) = 0$$

where the functional forms $f(x, y, z)$, $g(x, y, z)$, and $h(x, y, z)$ are known. Suppose reasonable approximations x^k, y^k, z^k to the exact x, y, and z are known and it is wished to obtain new approximations x^{k+1}, y^{k+1}, and z^{k+1} which are better approximations to x, y, and z. This procedure, once established, can then be used to obtain approximations x^n, y^n, and z^n, which are as close to x, y, and z as desired.

Let us expand the functions $f(x, y, z)$, $g(x, y, z)$, and $h(x, y, z)$ in Taylor series around the point x^k, y^k, and z^k:

$$f(x^{k+1}, y^{k+1}, z^{k+1}) = f(x^k, y^k, z^k) + (x^{k+1} - x^k)\frac{\partial f(x^k, y^k, z^k)}{\partial x}$$
$$+ (y^{k+1} - y^k)\frac{\partial f(x^k, y^k, z^k)}{\partial y} + (z^{k+1} - z^k)\frac{\partial f(x^k, y^k, z^k)}{\partial z}$$
$$+ \text{higher-order terms}$$

$$g(x^{k+1}, y^{k+1}, z^{k+1}) = g(x^k, y^k, z^k) + (x^{k+1} - x^k)\frac{\partial g(x^k, y^k, z^k)}{\partial x}$$
$$+ (y^{k+1} - y^k)\frac{\partial g(x^k, y^k, z^k)}{\partial y} + (z^{k+1} - z^k)\frac{\partial g(x^k, y^k, z^k)}{\partial z}$$
$$+ \text{higher-order terms}$$

$$h(x^{k+1}, y^{k+1}, z^{k+1}) = h(x^k, y^k, z^k) + (x^{k+1} - x^k)\frac{\partial h(x^k, y^k, z^k)}{\partial x}$$
$$+ (y^{k+1} - y^k)\frac{\partial h(x^k, y^k, z^k)}{\partial y} + (z^{k+1} - z^k)\frac{\partial h(x^k, y^k, z^k)}{\partial z}$$
$$+ \text{higher-order terms}$$

However, we want $f(x^{k+1}, y^{k+1}, z^{k+1}) = g(^{k+1}, y^{k+1}, z^{k+1}) = h(x^{k+1}, y^{k+1}, z^{k+1}) = 0$. Thus set them equal to zero. Also if x^k, y^k, z^k was a good approximation to (x, y, z), then the higher-order terms can be neglected. Finally, call $x^{k+1} - x^k = \Delta x$, $y^{k+1} - y^k = \Delta y$, and $z^{k+1} - z^k = \Delta z$. Thus the three equations become

$$-f(x^k, y^k, z^k) = \Delta x\frac{\partial f(x^k, y^k, z^k)}{\partial x} + \Delta y\frac{\partial f(x^k, y^k, z^k)}{\partial y} + \Delta z\frac{\partial f(x^k, y^k, z^k)}{\partial z}$$

$$-g(x^k, y^k, z^k) = \Delta x\frac{\partial g(x^k, y^k, z^k)}{\partial x} + \Delta y\frac{\partial g(x^k, y^k, z^k)}{\partial y} + \Delta z\frac{\partial g(x^k, y^k, z^k)}{\partial z}$$

$$-h(x^k, y^k, z^k) = \Delta x\frac{\partial h(x^k, y^k, z^k)}{\partial x} + \Delta y\frac{\partial h(x^k, y^k, z^k)}{\partial y} + \Delta z\frac{\partial h(x^k, y^k, z^k)}{\partial z}$$

Since we know the functional forms of $f(x, y, z)$, $g(x, y, z)$, and $h(x, y, z)$, the nine partial derivatives and the three functions can be evaluated at x^k, y^k, z^k, so these equations reduce to three simultaneous linear equations in the three unknowns, Δx, Δy, Δz. These three equations will be solved by use of a package subroutine supplied by the computing center. (Alternatively, a small subroutine using determinants could be written by the user.) Once Δx, Δy, and Δz are known, x^{k+1}, y^{k+1}, and z^{k+1} are found from

$$x^{k+1} = x^k + \Delta x$$
$$y^{k+1} = y^k + \Delta y$$
$$z^{k+1} = z^k + \Delta z$$

The calculations are then repeated until some desired degree of convergence of x, y, and z is obtained.

Let us now put our three equations into the desired form, and evaluate the nine partial derivatives:

$$M_{\text{final}}(\underline{H}_{\text{in}} - \underline{U}_{\text{final}}) - M_{\text{initial}}(\underline{U}_{\text{in}} - \underline{H}_{\text{initial}}) = 0 \quad \text{(a)}$$

$$-\underline{U}_{\text{final}} + \Big[CVA(T_{\text{final}} - T_0) + \frac{CVB}{2}(T_{\text{final}}^2 - T_0) + \frac{CVC}{3}(T_{\text{final}}^3 - T_0)$$
$$- \frac{M_{\text{final}}}{V}\left(A_0 + \frac{3RC}{T_{\text{final}}^2}\right) + \frac{M_{\text{final}}^2}{2V^2}\left(A_0 a - \frac{3RB_0C}{T_{\text{final}}^2}\right) + \frac{M_{\text{final}}^3}{V^3 T_{\text{final}}^2} RB_0bC\Big] = 0 \quad \text{(b)}$$

$$-P_{\text{final}} + \left[\frac{RT_{\text{final}}M_{\text{final}}}{V} + \frac{M_{\text{final}}^2}{V^2}\beta(T_{\text{final}}) + \frac{M_{\text{final}}^3}{V^3}\gamma(T_{\text{final}}) + \frac{M_{\text{final}}^4}{V^4}\delta(T_{\text{final}})\right] = 0 \qquad \text{(c)}$$

Call equation (a) = f, equation (b) = g, and equation (c) = h. The three unknowns are $M_{\text{final}} = x$, $\underline{U}_{\text{final}} = y$, and $T_{\text{final}} = z$. The partial derivaties are then

$$\frac{\partial f}{\partial x} = \underline{H}_{\text{in}} - \underline{U}_{\text{final}} \qquad \frac{\partial f}{\partial y} = -M_{\text{final}} \qquad \frac{\partial f}{\partial z} = 0$$

$$\frac{\partial g}{\partial x} = -\frac{1}{V}\left(A_0 + \frac{3RC}{T_{\text{final}}^2}\right) + \frac{M_{\text{final}}}{V^2}\left(A_0 a - \frac{3RB_0C}{T_{\text{final}}^2}\right) + \frac{3M_{\text{final}}^2}{V^3}\frac{RB_0bC}{T_{\text{final}}^2}$$

$$\frac{\partial g}{\partial y} = -1$$

$$\frac{\partial g}{\partial z} = CVA + CVB(T_{\text{final}}) + CVC(T_{\text{final}}^2) + \frac{M_{\text{final}}}{V}\frac{6RC}{T_{\text{final}}^3} + \frac{M_{\text{final}}^2}{V^2}\frac{3RB_0C}{T_{\text{final}}^3} - \frac{2M_{\text{final}}^3}{V^3}\frac{RbB_0C}{T_{\text{final}}^3}$$

$$\frac{\partial h}{\partial x} = \frac{RT_{\text{final}}}{V} + \frac{2M_{\text{final}}}{V^2}\beta(T_{\text{final}}) + \frac{3M_{\text{final}}^2}{V^3}\gamma(T_{\text{final}}) + \frac{4M_{\text{final}}^3}{V^4}\delta(T_{\text{final}})$$

$$\frac{\partial h}{\partial y} = 0$$

$$\frac{\partial h}{\partial z} = \frac{RM_{\text{final}}}{V} + \frac{M_{\text{final}}^2}{V^2}\beta'(T_{\text{final}}) + \frac{M_{\text{final}}^3}{V^3}\gamma'(T_{\text{final}}) + \frac{M_{\text{final}}^4}{V^4}\delta'(T_{\text{final}})$$

where $\beta(T)$, $\gamma(T)$, and $\delta(T)$ are the virial coefficients and the prime represents d/dT.

A program for performing the desired calculations is presented below.

```
C    ---  PROGRAM FOR DETERMINING THE AMOUNT OF GAS ADDED TO A
C    ---  TANK DURING FILLING
          REAL M, MSTART, MEND
          DIMENSION AA(3, 3), BB(3), CC(3)
C    ---  DEFINE FUNCTIONS TO BE USED LATER ---
          BETA(T) = R*B0*T - R*C/(T*T) - A0
          GAMMA(T) = A0*A - R*B0*B*T - R*B0*C/(T*T)
          DELTA(T) = R*B0*B*C/(T*T)
          BPRIME(T) = R*B0 + 2.0*R*C/(T*T*T)
          GPRIME(T) = 2.0*R*B0*C/(T*T*T) - R*B0*B
          DPRIME(T) = -2.0*R*B0*B*C/ (T*T*T)
          PRESS(T, VBAR) = R*T/VBAR + BETA(T) / (VBAR*VBAR) + GAMMA(T)/
         1VBAR**3 + DELTA(T) / VBAR**4
          UBAR(T, VBAR) = CVA*(T - T0) + CVB*(T*T - T0*T0)/2.0 + CVC*(T**3
         1 - T0**3)/3.0 - (1.0/VBAR)*(A0 + 3.0*R*C/(T*T)) + (0.5/(VBAR*
         2VBAR))*( A0*A - 3.0*R*B0*C/(T*T)) + R*B0*B*C/((VBAR**3)*T*T)
          HBAR(T, VBAR) = UBAR(T, VBAR) + PRESS(T, VBAR)*VBAR
          COMMON / SAFETY / R, A0, A, B0, B, C
          IFLAG = 0
C    ---  START CALCULATIONS BY READING DATA ---
```

```
   10   READ (105,20) R, TLINE, PLINE, A0, A, B0,B,C, EPSP, VOL, PSTART,
       1 TSTART, CVA, CVB, CVC, T0, EPST, EPSM, EPSU
   20   FORMAT (5E16.8)
   30   WRITE ( 108,40 ) A0, A, B0, B, C, R, EPSP, VOL,
       1 PSTART, TSTART, PLINE, TLINE, T0, CVA, CVB, CVC, EPSM, EPST, EPSU
   40   FORMAT (1H1, 36H THE BEATTIE-BRIDGEMAN CONSTANTS ARE, 7H A0 =
       1E10.4, 5H A = , E10.4// 6H B0 = , E10.4, 5H B = , E10.4, 5H C = , E10.4/
       2 5H R = , E10.4, 8H EPSP = , E10.4, 10H VOLUME = , E10.4,
       3 10H PSTART = , E10.4//11H TSTART = , E10.4 / 9H PLINE = ,
       4 E10.4, 9H TLINE = , E10.4//23H REFERENCE TEMP, T0 = , E10.4,
       5 7H CVA = , E10.4, 7H CVB = , E10.4// 8H CVC = , E10.4,
       6 8H EPSM = , E10.4, 8H EPST = , E10.4, 8H EPSU = , E10.4)
C  ---  CALCULATE INITIAL SPECIFIC VOLUME AND SPECIFIC VOLUME IN
C  ---  PIPELINE ---
C  ---  USE SUBROUTINE WHICH USES NEWTON ITERATION ON BEATTIE-
C  ---  BRIDGEMAN EQUATION ---
   50   VSTART = VOLUME (PSTART, TSTART, EPSP)
   60   VLINE = VOLUME (PLINE, TLINE, EPSP)
C  ---  FIND INITIAL CONDITIONS IN TANK
   70   MSTART = VOL/ VSTART
   80   USTART = UBAR( TSTART, VSTART)
   90   HIN = HBAR( TLINE, VLINE)
  100   ULINE = UBAR( TLINE, VLINE)
C  ---  SET INITIAL CONDITIONS FOR NEWTON-RAPHSON ITERATION, ASSUME
C  ---  TANK CONDITIONS ARE SAME AS IN PIPELINE ---
  110   U = ULINE
  120   T = TLINE
  130   M = VOL/VLINE
C  ---  START NEWTON-RAPHSON ITERATION ---
  140   DO 330 IT = 1, 30
C  ---  DEFINE THE FUNCTIONS F(X,Y,Z), G(X,Y,Z), H(X,Y,Z) ---
  150   FXYZ = M*(HIN - U) - MSTART*(HIN - USTART)
  155   VBAR = VOL/M
  160   GXYZ = UBAR(T, VBAR) - U
  170   HXYZ = PRESS(T, VBAR) - PLINE
C  ---  DEFINE THE NINE PARTIAL DERIVATIVES ---
  180   DFDX = HIN - U
  190   DFDY = -M
  200   DFDZ = 0.0
  210   DGDX = -(A0 + 3.0*R*C/(T*T))/VOL + (M/(VOL*VOL))*(A0*A -
       13.0*R*B0*C/(T*T)) + (3.0*M*M/(VOL**3))*R*B0*C/(T*T)
  220   DGDY = -1.0
  230   DGDZ = CVA + CVB*T + CVC*T*T + (M/VOL)*6.0*R*C/(T*T*T) +
       1(M/VOL)*(M/VOL)*(3.0*R*B0*C)/(T*T*T) - 2.0*((M/VOL)
       2**3)*R*B*B0*C/(T*T*T)
  240   DHDX = R*T/VOL + 2.0*M*BETA(T)/(VOL*VOL) + 3.0*M*M*GAMMA(T)/
       1(VOL*VOL*VOL) + 4.0*M*M*M*DELTA(T)/(VOL**4)
  250   DHDY = 0.0
  260   DHDZ = R*M/VOL + ((M/VOL)**2)*BPRIME(T) + ((M/VOL)**3)*GPRIME
       1(T) + ((M/VOL)**4)*DPRIME(T)
```

```
C   ---  LOAD COEFFICIENTS INTO LINEAR EQUATION SOLVER MATRIX ---
   270   AA(1,1) = DFDX
         AA(1,2) = DFDY
         AA(1,3) = DFDZ
         CC(1) = -FXYZ
   280   AA(2,1) = DGDX
         AA(2,2) = DGDY
         AA(2,3) = DGDZ
         CC(2) = -GXYZ
   290   AA(3,1) = DHDX
         AA(3,2) = DHDY
         AA(3,3) = DHDZ
         CC(3) = -HXYZ
C   ---  USE MATRIX INVERSION ROUTINE TO OBTAIN SOLUTION ---
   300   CALL LINEQ(AA,CC,BB, 3,3, IFLAG)
C   ---  UPDATE VALUES OF M,U, AND T, AND CHECK FOR CONVERGENCE ---
   310   M = M + BB(1)
         U = U + BB(2)
         T = T + BB(3)
C   ---  DELTA M = BB(1), DELTA U = BB(2), DELTA T = BB(3) ---
   320   IF(ABS(BB(1))·LT·EPSM·AND·ABS(BB(2))·LT·EPSU·AND·ABS(BB(3))
        1·LT· EPST) GO TO 360
C   ---  CONTINUE ITERATIONS ---
   330   WRITE(108, 325) IT, M, U, T
   325   FORMAT (1H 0, 7H AFTER, I2, 24H ITERATIONS WE HAVE M = , E10.4,
        1 5H U = , E10.4, 9H AND T = , E10.4)
C   ---  ITERATION HAS FAILED, START NEW CALCULATION ---
   340   WRITE (108, 345)
   345   FORMAT (1H0, 38H ITERATION HAS FAILED, NO CONVERGENCE)
   350   GO TO 10
C   ---  FINISH PROGRAM BY PRINTING RESULTS ---
   360   WRITE( 108, 370) U,M,T, (VOL/M)
   370   FORMAT (1H0, 40H CONVERGENCE OBTAINED, RESULTS ARE U = , E15.5,
        15H M = , E15.5//6H T = , E15.5, 8H VBAR = , E15.5)
   380   WRITE (108, 380) (M - MSTART)
   390   FORMAT (1H0, 22H MASS ADDED TO TANK = , E15.5)
   400   GO TO 10
   410   END

         FUNCTION VOLUME ( P,T, EPSP)
         COMMON/SAFETY / R, A0, A, B0, B, C
         BETA(T) = R*B0*T - R*C/T**2 - A0
         GAMMA(T) = A0*A - R*B0*B*T - R*B0*C/T**2
         DELTA(T) = R*B0*B*C/T**2
         PRESS(T, VBAR) = R*T/VBAR + BETA(T)/VBAR**2 + GAMMA(T)/VBAR**3
        1 + DELTA(T)/VBAR**4
    50   VBAR = R*T/P
    60   DO 100 IT = 1, 15
    70   ERR = PRESS(T, VBAR) - P
    80   IF ( ABS( ERR) ·LE· EPSP) GO TO 140
```

```
90     VBAR = VBAR - ERR/( -R*T/VBAR**2 - 2.*BETA(T)/VBAR**3 -
     13.*GAMMA(T)/VBAR**4 - 4.*DELTA(T)/VBAR**5)
100    CONTINUE
110    WRITE ( 108, 120)
120    FORMAT ( 1H0, 15H NO CONVERGENCE )
130    CALL EXIT
140    VOLUME = VBAR
150    RETURN
170    END
```

```
THE BEATTIE-BRIDGEMAN CONSTANTS ARE  A0 = .4830E 04 A = .1857E-01
B0 = .5593E-01 B = -.1589E-01 C = .7476E 06
R = .9650E 02 EPSP = .1000E 03 VOLUME = .1000E 01 PSTART = .7200E 05
TSTART = .5350E 03
PLINE = .6470E 06 TLINE = .5800E 03
REFERENCE TEMP, T0 = .4600E 03 CVA = .1085E 04 CVB = .7800E 01
CVC = -.1032E-02 EPSM = .1000E-02 EPST = .1000E-01 EPSU = .1000E 03
AFTER  1 ITERATIONS WE HAVE M = .7892E 01 U = .5366E 06 AND T = .5842E 03
AFTER  2 ITERATIONS WE HAVE M = .1067E 02 U = .5675E 06 AND T = .5863E 02
AFTER  3 ITERATIONS WE HAVE M = .8897E 01 U = .5550E 06 AND T = .5849E 03
AFTER  4 ITERATIONS WE HAVE M = .1000E 02 U = .5574E 06 AND T = .5844E 03
AFTER  5 ITERATIONS WE HAVE M = .9272E 01 U = .5545E 06 AND T = .5842E 03
AFTER  6 ITERATIONS WE HAVE M = .9747E 01 U = .5563E 06 AND T = .5842E 03
AFTER  7 ITERATIONS WE HAVE M = .9434E 01 U = .5550E 06 AND T = .5842E 03
AFTER  8 ITERATIONS WE HAVE M = .9638E 01 U = .5558E 06 AND T = .5842E 03
AFTER  9 ITERATIONS WE HAVE M = .9504E 01 U = .5552E 06 AND T = .5842E 03
AFTER 10 ITERATIONS WE HAVE M = .9592E 01 U = .5556E 06 AND T = .5842E 03
AFTER 11 ITERATIONS WE HAVE M = .9534E 01 U = .5553E 06 AND T = .5842E 03
AFTER 12 ITERATIONS WE HAVE M = .9572E 01 U = .5555E 06 AND T = .5842E 03
AFTER 13 ITERATIONS WE HAVE M = .9547E 01 U = .5554E 06 AND T = .5842E 03
AFTER 14 ITERATIONS WE HAVE M = .9563E 01 U = .5555E 06 AND T = .5842E 03
AFTER 15 ITERATIONS WE HAVE M = .9553E 01 U = .5554E 06 AND T = .5842E 03
AFTER 16 ITERATIONS WE HAVE M = .9560E 01 U = .5554E 06 AND T = .5842E 03
AFTER 17 ITERATIONS WE HAVE M = .9555E 01 U = .5554E 06 AND T = .5842E 03
AFTER 18 ITERATIONS WE HAVE M = .9558E 01 U = .5554E 06 AND T = .5842E 03
AFTER 19 ITERATIONS WE HAVE M = .9556E 01 U = .5554E 06 AND T = .5842E 03
AFTER 20 ITERATIONS WE HAVE M = .9557E 01 U = .5554E 06 AND T = .5842E 03
CONVERGENCE OBTAINED, RESULTS ARE U = ,    .55543E 06 M =    .95566E 01
T =        .58418E 03 VBAR =        .10464E 00
MASS ADDED TO TANK =        .80765E 01
```

Problems

4-1 (a) Write the *complete* first-law energy balance for the system shown in Fig. P4-1.

(b) If the black box is a compressor operating at steady state and streams 2 and 5 are cooling water while streams 1, 3, and 4 are the working fluid (streams 3 and 4 are at different pressure), reduce the energy balance to its simplest form.

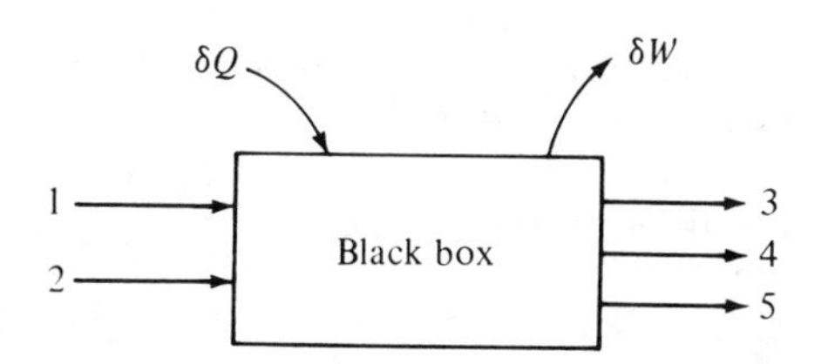

FIGURE P4-1

4-2 $$\left[\left(\underline{U} + P\underline{V} + \frac{u^2}{2g_c} + \frac{gZ}{g_c}\right)\delta M\right]_{\text{in}} - \left[\left(\underline{U} + P\underline{V} + \frac{u^2}{2g_c} + \frac{gZ}{g_c}\right)\delta M\right]_{\text{out}} + \delta Q - \delta W = d\left[\left(\underline{U} + \frac{u^2}{2g_c} + \frac{gZ}{g_c}\right)M\right]_{\text{syst}}$$

Reduce the above mathematical statement of the first law of thermodynamics to its simplest possible form for application to each of the following processes and systems:

(a) A piece of hot steel suddenly immersed in cold water; system: the piece of steel.

(b) Cold water being heated in the tubes of a heat exchanger, horizontal flow at a constant rate; system: the tubes and the water in them.

(c) A freely falling body passing through a differential increment of height; system: the body.

(d) Steam flowing steadily through a horizontal, insulated nozzle; system: the nozzle and its contents.

(e) Same as (d); system: 1 lb of steam flowing.

(f) A rubber balloon being inflated; system: the rubber.

(g) A storage battery discharging across a resistance; system: the resistance.

(h) An automobile accelerating on smooth, level pavement (assume frictionless); system: the automobile.

(i) A frictionless windmill driving an electric generator; system: the windmill.

(j) A tennis ball dropped from shoulder height bounces on a sidewalk until it finally comes to rest; system: the tennis ball.

(k) A bullet embeds itself in a bowling ball that is rolling frictionlessly along a horizontal surface; system: the bowling ball.

(l) Hot oil vapors are cooled at a steady rate by cooling water in a horizontal double-pipe heat exchanger that is well insulated; system: the exchanger and its contents.

(m) A heavy steel block slides slowly down an inclined plane until it comes to rest; system: the block.

(n) Water drips slowly out of a hole in the bottom of an enclosed tank; system: the tank and its contents.

(o) A gas is confined in a vertical cylinder fitted with a frictionless piston and an evacuated space above the piston. The piston rises as the cylinder is heated; system: the gas.

(p) Same as (o); system: the gas and the piston.

(q) Same as (p) except that the space above the piston is not evacuated.

(r) Water flows steadily through a long, horizontal pipe; system: the pipe and its contents.

(s) Gas flows slowly into an insulated tank that was initially evacuated; system: the tank and its contents.

(t) Two metal blocks, isolated from the rest of the universe and initially at different temperatures, are brought together until their temperatures are equal; system: the hot block.

(u) Same as (t); system: the cold block.

(v) Same as (u); system: both blocks.

(w) A surge tank is "riding" on a compressed air line. The line pressure starts to fall and air flows slowly out of the tank for several minutes; system: the tank.

(x) Same as (w); system: all the air that will remain in the tank when the flow stops.

4-3 A gas that is at a temperature T_0 is flowing in a pipe (Fig. P4-3). A small amount is bled off into an evacuated cylinder. The bleeding continues until the pressure in the tank is equal to P, the pressure in the pipe. If the cylinder is perfectly insulated and the gas is ideal (with C_P and C_V independent of temperature and pressure), find the temperature, T, which exists in the cylinder at the end of the process by two separate approaches:

(a) Assume that the cylinder and its contents constitute an open system.

(b) Assume that the mass of gas M, which will end up in the cylinder, constitutes a closed system.

You should express your answers to (a) and (b) in terms of T_0, C_P, C_V, and P alone. Do the answers to (a) and (b) differ? Does this surprise you? Which approach is the simpler one to apply *in this case*, (a) or (b)?

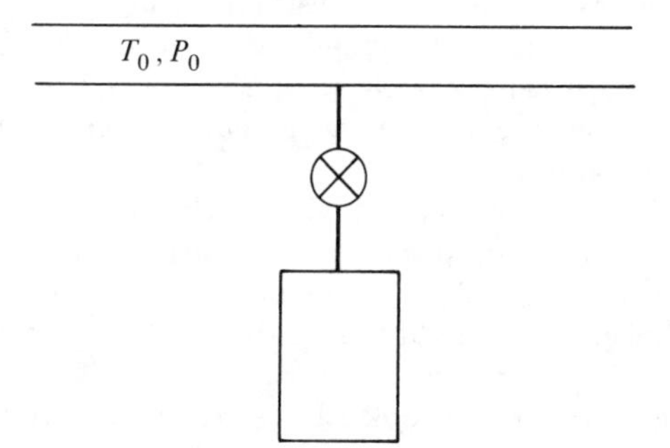

FIGURE P4-3

4-4 A compressor takes in CO_2 at a rate of 1000 ft^3/hr at 70°F and atmospheric pressure. The discharge conditions are 115 psia and 85°F. The compressor is cooled by a jacket that removes heat to cooling water. By noting the rise in temperature of a measured quantity of the water, it has been found that 7500 Btu of heat is removed every hour. The motor that drives the compressor is drawing 2.8 kw of electrical power from the line. Determine the efficiency of the electric motor:

$$(C_P)_{CO_2} = 9.3 \text{ Btu/lb-mol °F}$$

4-5 High-pressure steam at a rate of 1000 lb$_m$/hr initially at 500 psia and 700°F is expanded in a turbine to produce work. Two exit streams are removed from the turbine as illustrated in Fig. P4-5. Stream 2 is at 200 psia and 400°F and has a flow rate equal to one third of the inlet flow. Stream 3 is at 100 psia and is known to be a mixture of saturated vapor and liquid. A small representative fraction of stream 3 is passed through a throttling valve and expanded to 1 atm. The temperature after expansion is found to be 240°F. If the measured work obtained from the compressor is 54 hp, estimate the heat loss from the turbine expressed as Btu/hr.

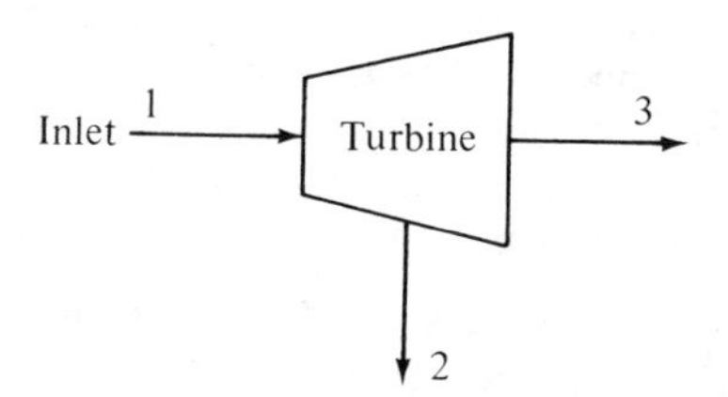

FIGURE P4-5

4-6 An adiabatic steam turbine is used to drive an electrical generator, which, in turn, supplies power to four water pumps. The steam enters the turbine at 1000 lb_m/hr with an enthalpy of 1200 Btu/lb_m and leaves with an enthalpy of 600 Btu/lb_m. Each pump pumps 50 gal/min of water from an open tank 100 m above sea level to another 800 m above sea level. Kinetic energy changes and friction in the water lines may be neglected. What is the overall efficiency of the operation, that is, the fraction of the energy (enthalpy) recovered from the steam that is converted to the potential energy of the water?

4-7 A turbine is used in a nitrogen refrigeration process to derive some work from the expansion of nitrogen; 1000 lb_m of nitrogen per hour enters at 1500 psia and −40°F at 200 ft/sec. The nitrogen leaves the turbine exhaust at a point 10 ft below the inlet with a velocity of 1200 ft/sec. The measured shaft work of the turbine is 12.6 hp, and heat is transferred from the surroundings to the turbine at the rate of 10,000 Btu/hr. A small portion of the exhaust nitrogen from the turbine is passed through a throttling valve and discharged at atmospheric pressure. Velocity change in passing through the valve may be neglected. What is the temperature of the stream leaving the valve?

4-8 An insulated tank initially contains 1000 lb_m of steam and water at 500 psia. Fifty per cent of the tank volume is occupied by liquid and 50 per cent by vapor. Fifty pounds of moisture-free vapor is slowly withdrawn from the tank so that the pressure and temperature are always uniform throughout the tank. Analyze the situation carefully and calculate the pressure in the tank after the 50 lb_m of steam is withdrawn.

4-9 A well-insulated cylinder, fitted with a frictionless piston, initially contained 20 lb_m of liquid water and 1 lb_m of water vapor at a pressure of 200 psia (Fig. P4-9). The valve to the steam line was opened and 5 lb_m of superheated steam at 250 psia was admitted to the cylinder. The valve was closed and the contents of the cylinder allowed to come to equilibrium. If the final volume of the contents of the cylinder was six times the initial volume, determine the temperature of the superheated steam that was admitted to the cylinder.

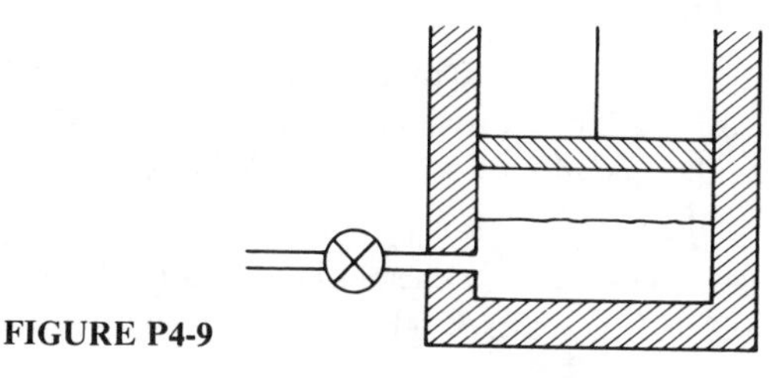

FIGURE P4-9

4-10 A fireless steam engine is used to haul boxcars around an explosives plant. The engine has a well-insulated 100-ft^3 tank. This tank is periodically charged with high-pressure steam which is used to power the engine until the tank is depleted. At the end of a given run the tank contains saturated steam at atmospheric pressure. It is then connected to a supply line carrying steam at 800 psia and 620°F. A valve in the supply line is opened until no more steam flows into the tank. The valve is then closed. The filling operation takes place very rapidly. Using the steam tables (and stating any assumptions), determine the amount of steam in the tank just after filling.

4-11 A well-insulated vertical cylinder is fitted with a frictionless piston and contains 2 lb_m of steam at 520°F and 114 psia. A line leads from the bottom of the cylinder to an insulated tank having a volume of 8 ft^3 and initially containing 1 lb_m of steam at 50 psia. A valve is opened and steam flows slowly into the tank until the tank and cylinder pressures are equal, at which time the valve is closed. It may be assumed that the volume of the line is negligible and that there is no transfer of *heat* through the valve.

(a) How many pounds of steam flow into the tank?

(b) What is the final temperature in the tank?

(c) Calculate Q, W, and ΔU (all in Btu), taking as the system (1) the atmosphere, (2) the piston, (3) the piston and atmosphere, and (4) all the steam. Determine ΔH in two different ways.

(d) What is the internal energy of (1) the steam in the tank under initial conditions? (2) the steam initially in the cylinder which is to flow into the tank? (3) the steam in the tank under final conditions?

4-12 An insulated cylinder with a volume of 2 ft^3 contains a piston (volume negligible) attached to a shaft as shown in Fig. P4-12. The piston is initially at the extreme left of the cylinder, which contains air at 1 atm. The valve is opened and steam enters the cylinder, forcing the piston to the right. When the piston first touches the right end of the cylinder, the valve is closed. At this point 43.6 Btu of shaft work has been performed and the cylinder contains 0.5 lb_m of steam. What is the pressure of steam in the cylinder? (*Note:* The piston cannot be considered frictionless.)

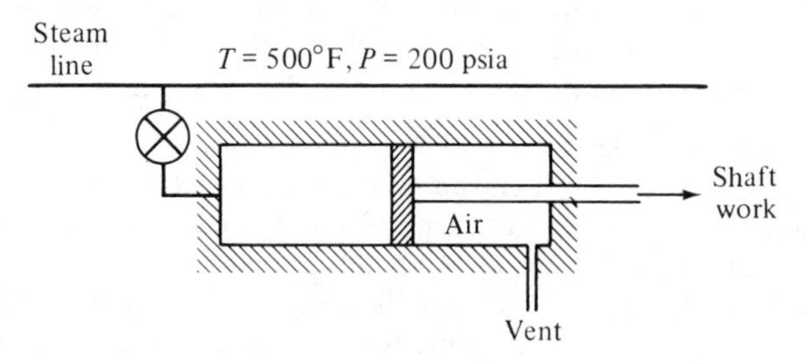

FIGURE P4-12

4-13 An adiabatic turbine is supplied with steam at 240 psia and 540°F and exhausts to a pressure of 15 psia. If the quality of the exhaust steam is 96 per cent and there is negligible change in velocity, compute the quantity of work done by 1 lb_m of steam flowing through the turbine.

4-14 The exhaust steam from a turbine is to be used in a chemical process. The turbine is supplied with steam at 200 psia and 500°F, heat loss is 30.4 Btu/lb_m of steam supplied, and the work is 138.5 Btu/lb_m of steam. Find the enthalpy and quality of the exit steam if the exit pressure is 35 psia.

4-15 A 10-hp turbine is supplied with steam at 300 psia and 600°F and exhausts at 20 psia and 98 per cent quality. Find the heat loss in Btu/hr, if 200 lb_m of steam is supplied per hour.

4-16 The steam–water power cycle shown in Fig. P4-16 has been recommended for the production of useful work needed elsewhere in the plant. Heat to the cycle is supplied in the boiler and superheater and is removed in the condenser. The turbine produces an amount of work W_1. However, $-(W_2 + W_3)$ of this is needed to power the pumps. On the basis of the information supplied in Fig. P4-16, determine the thermal efficiency of this cycle,

$$\eta = \frac{\text{net work produced}}{Q_1 + Q_2}$$

Supply all pressures, temperatures, and enthalpies which do not appear on the diagram. You may assume that there are no pressure changes across any equipment or piping except the turbine and pumps. Also you may assume $\underline{V}$ of the liquid passing through the pump to be independent of pressure.

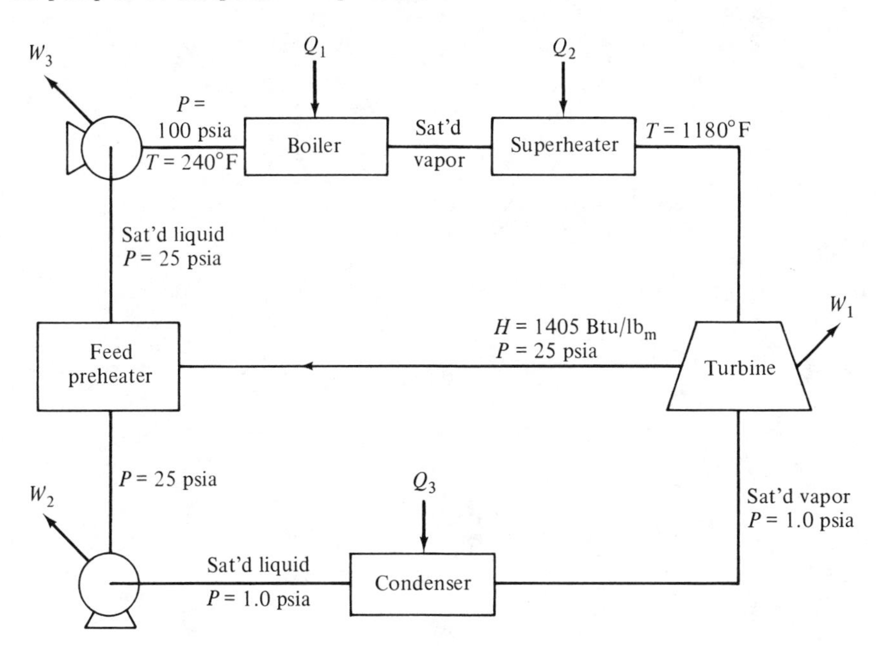

FIGURE P4-16

4-17 In a certain paper mill two steam boilers are to be operated in parallel. Each has a volumetric capacity of 1000 ft³ and each contains 18,000 lb_m of steam and water. The first boiler registers a pressure of 200 lb/in.² absolute, but, owing to an error, it is connected to the second boiler when the pressure in the latter is but 75 lb/in.² absolute. What will be the pressure in the system after equilibrium has been attained, on the assumption that no steam is withdrawn, that no heat is added to the system during the change, and that there is no interchange of heat between the boiler shells and their contents?

4-18 (a) One mole of gas is confined on one side of a piston at 5 atm and 200°F as shown in Fig. P4-18. If the piston has a mass of 1 lb_m and the gas expands adiabatically and reversibly, calculate the piston's velocity when the pressure has fallen to 1 atm. Assume the gas to be ideal with $C_P = 6$ Btu/lb-mol °F.

(b) How much work was done by the gas in (a)?

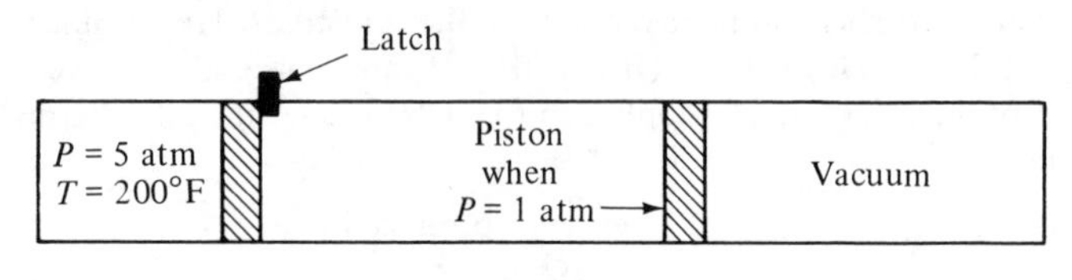

FIGURE P4-18

The Entropy Balance 5

Introduction

In Chapter 4 the first of three basic thermodynamic relationships, the energy balance, was developed. Its utility in the analysis of engineering problems will become increasingly evident as we progress. However, the energy balance alone is insufficient in analyzing many thermodynamic systems and processes, because nature has imposed certain restrictions on energy transfer and conversion over and above those embodied in the conservation-of-energy principle. The quantitative treatment of these restrictions has been facilitated by introducing the entropy concept.

Our development thus far has departed from the historical approach to the treatment of entropy in that we have attempted to produce some physical basis for the property entropy. The presentation has combined our present understanding of matter on the microscopic level with the power of statistical mechanics to achieve this objective. However, the utility of the entropy concept depends on relating it to the changes with which we as engineers are involved. Since our concerns are generally macroscopic in nature, it is necessary to associate entropy with macroscopic phenomena, such as heat and work, so that it can be used in the analysis of processes involving these energy flows and interconversions. Development of the entropy balance provides a quantitative relationship of considerable engineering utility, as will be demonstrated in this chapter.

The idea of accounting for entropy changes in a system by constructing an entropy balance is one that arises rather naturally from our earlier treatments of mass and energy. Entropy, like mass and energy, is an extensive property that may be used to characterize the state of thermodynamic systems at equilibrium. A system's entropy can undergo change as a result of a process just as its energy, mass, or volume may change. The general procedure cited in Section 2.5 for monitoring such changes can be applied to entropy in the same manner as it was applied to mass or energy. Thus, if we consider an open system as shown in Fig. 5-1, we may write an entropy balance as

$$\text{entropy}_{\text{in}} - \text{entropy}_{\text{out}} + \text{entropy}_{\text{gen}} = \text{entropy}_{\text{acc}} \tag{5-1}$$

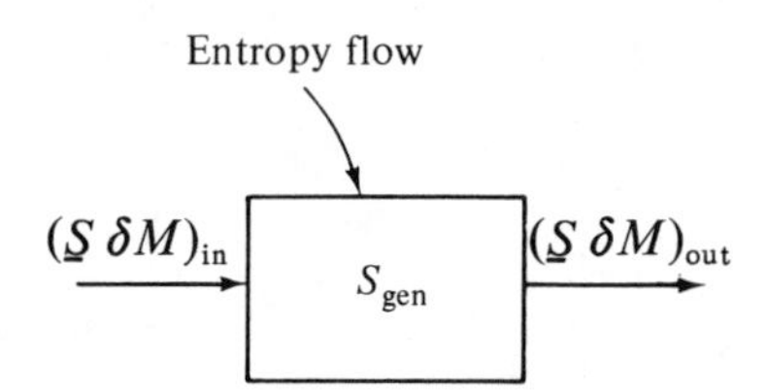

FIG. 5-1. An open system.

As in the case of the energy equation, we break the inlet and outlet terms into an entropy flow that is associated with the mass flow, $\underset{\sim}{S}\,\delta M$, and an entropy flow that is not associated with mass flow. Since entropy is a nonconservative property, unlike mass and energy, it will be necessary to include a generation term in the entropy balance to account for entropy production arising from system irreversibilities. The accumulation term is simply the change in entropy of the system and is expressed as $d(M\underset{\sim}{S})_{\text{syst}}$. Thus the entropy balance becomes

$$(\underset{\sim}{S}\,\delta M)_{\text{in}} - (\underset{\sim}{S}\,\delta M)_{\text{out}} + \delta(\text{entropy flow}) + \delta(\text{entropy})_{\text{gen}} = d(M\underset{\sim}{S})_{\text{syst}} \tag{5-2}$$

Before we proceed, it is necessary to investigate the character of the entropy-flow and generation terms.

5.1 Entropy Flow

In developing the energy balance, a seemingly arbitrary distinction was made between heat and work as energy flows. Justification for this differentiation becomes apparent if we reconsider certain points raised in Section 3.5 regarding the effect that these flows have on a system's entropy.

It must be remembered that both heat and work refer to flows across system boundaries. Just as we choose to designate water as rain when moving from water droplets in clouds to the earth (once it is in our rivers and lakes, it is no longer rain), we designate energy as heat or work only as it crosses the boundary of our system. Energy crosses the boundary of a system in one of these two categories, but once in or out of the system it is no longer regarded as heat or work, but simply becomes part of the system's (or the surroundings') total energy.

If we could station ourselves at the system's boundary and watch only the energy flowing across it, we would discover that we were unable to distinguish energy flows in reversible processes from those in irreversible ones. Since the irreversibilities that affect a system's entropy are those occurring within the system, we would not even see them. Furthermore, we would have no basis for relating them to our observations of energy flow at the boundary. In formulating the flow terms for the entropy balance, it is imperative that we recognize this important point: that *the flow terms relate just to what crosses the boundary and not to any changes occurring within the system, be they reversible or irreversible.*

Energy flowing as heat was shown to produce entropy changes within a system which were *directly attributable* only to the flow of heat, whether the overall process is reversible or not. Energy transfer as work was discussed only for a simple compression or expansion process and was *shown not to produce an entropy change* as a direct result of the energy transfer. In a more general sense, one might extend that observation to all energy-transfer processes between system and surroundings in which the energy transfer is categorized as work. In each case the energy content of the system is changed, but the change is compensated for by a change in some other extensive property, such as volume, so that no net change in entropy occurs. This statement is not intended to imply that entropy changes cannot result within a system which is either doing or receiving work, *only that such entropy change is not a direct consequence of the (mechanical) energy flow. Thus, in terms of energy flows between system and surroundings, only heat has an entropy flow associated directly with it.*

In Chapter 3 it was reasoned that the flow of entropy directly related to the flow of heat should be proportional to the amount of heat flow. This was expressed as

$$\delta S = K\,\delta Q \tag{5-3}$$

where (1) K was a positive term which became very large as the system's energy content was reduced to very low levels, and (2) K remained constant during phase changes. K can be shown to equal $1/T$, where T is the absolute temperature, by the following reasoning.

Consider the pressure–volume behavior of a substance subjected to a cyclical reversible process consisting of two isothermal steps, *A-B* and *C-D*, and two adiabatic steps, *B-C* and *D-A*, as shown in Fig. 5-2. The system begins in state A and after the four steps is returned to the same state. All state properties, including entropy, should thus return to their original values at the completion of the process.

The curves labeled t_1, t_2, and t_3 represent isotherms of progressively lower temperature (that is, $t_1 > t_2 > t_3$); the curves labeled S_1 and S_2 represent adiabatic lines which for reversible changes are also lines of constant entropy. (The function t is assumed to be related to the thermodynamic temperature but not necessarily identical to it.) As discussed earlier, in the absence of irreversibility the only means of changing a closed system's entropy is by adding or removing heat. Thus step *A-B* involves addition of heat to the system at a constant temperature, t_1; step *B-C* involves an adiabatic expansion; step *C-D*

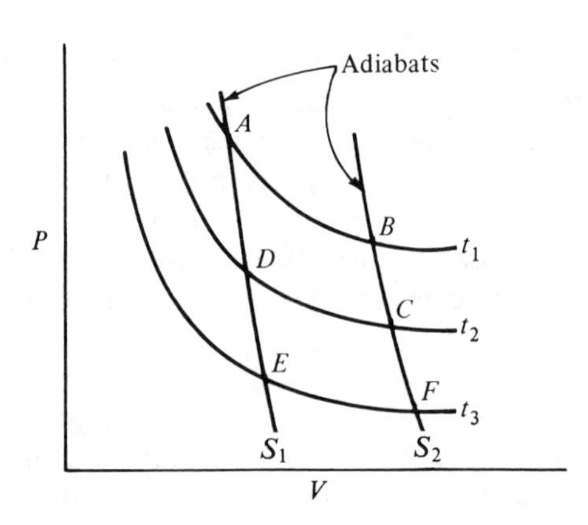

FIGURE 5-2

involves heat removal at a lower temperature, t_2; and step *D-A* involves an adiabatic compression that returns the system to its original state. Such a cycle represents a Carnot engine (a device for converting thermal energy to mechanical energy), which will be discussed further in Chapter 7.

In Chapter 4 it was shown that the area beneath a P–V curve equals the work associated with the process. In a cyclical process the area enclosed by the path *ABCD* equals the net work, W, for the cycle. Application of the energy balance to the system yields

$$W = Q_1 + Q_2 \tag{5-4}$$

where Q_1 and Q_2 represent the total heat effects at t_1 and t_2, respectively, Q_1 being a positive quantity (heat absorbed) and Q_2 a negative quantity (heat rejected). Clearly, $|Q_1| > |Q_2|$ if work is produced by the system (since W is positive by convention if work is done by the system).

Consider what happens if instead of discharging heat at t_2 it is discharged at t_3. The enclosed area becomes larger; therefore, W increases and (for a given Q_1) $|Q_3|$ must be less than $|Q_2|$. In general, for a reversible, cyclic process consisting of two isothermal and two adiabatic legs, Q_i is a function of t_i. Moreover, it can be shown that $|Q_i|$ is a monotonic function of t_i ($|Q_1| > |Q_2| > |Q_3|$ in this case).

In either of the cycles *ABCD* or *ABFE*, entropy changes occur only along the isothermal legs of the path. Since the only entropy change encountered by the gas along the isothermal paths are those associated with the entropy flows, the entropy changes along these paths are simply equal to the entropy flows, so that:

$$\Delta S_{t_1} = \int_A^B K_1\, \delta Q_1 \qquad \text{and} \qquad \Delta S_{t_2} = \int_C^D K_2\, \delta Q_2 \tag{5-5}$$

If entropy is to be a state function, $\Delta S_{t_1} \equiv -\Delta S_{t_2}$ and:

$$\int_A^B K_1\, \delta Q_1 + \int_C^D K_2\, \delta Q_2 = 0 \tag{5-6}$$

if the total entropy change around the cycle is to vanish. For a process in which only a differential amount of heat is involved at t_1 and t_2 (and thus S_1 and S_2 differ by only a differential amount, dS), equation (5-6) can be written as

$$K_1\,\delta Q_1 + K_2\,\delta Q_2 = 0 \tag{5-7}$$

or

$$\frac{\delta Q_1}{\delta Q_2} = -\frac{K_2}{K_1} \tag{5-8}$$

or

$$\frac{|\delta Q_1|}{|\delta Q_2|} = \frac{K_2}{K_1}$$

It can now be argued that because $|\delta Q_i|$ is a function of t_i, the ratio $|\delta Q_1|/|\delta Q_2|$ is a ratio of functions of the temperatures.

$$\frac{|\delta Q_1|}{|\delta Q_2|} = \frac{K_2}{K_1} = \frac{f(t_1)}{f(t_2)} \tag{5-9}$$

Rewriting as

$$\frac{|\delta Q_1|}{f(t_1)} = \frac{|\delta Q_2|}{f(t_2)} \tag{5-10}$$

where $K_1 = 1/f(t_1)$ and $K_2 = 1/f(t_2)$, it becomes apparent that K_1 and K_2 must be a function of the temperatures t_1 and t_2. *The thermodynamic temperature scale is now defined such that*

$$T = f(t) \tag{5-11}$$

Although many alternative definitions are possible—and in fact several have been suggested—no other definition has received wide acceptance, and thus we shall continue to use the definition suggested in equation (5-11). We will show in Chapter 6 that this thermodynamic temperature is identical to that discussed in Chapter 3:

$$T = \left(\frac{\partial U}{\partial S}\right)_V \tag{5-12}$$

In Chapter 7 the equivalence of the thermodynamic temperature and the ideal gas temperature, $T = P\underline{V}/R$, will be shown.

Substitution of equation (5-11) into the definition of K then yields

$$K = \frac{1}{T} \tag{5-13}$$

so that the entropy flow term is given by:

$$\delta S = \frac{\delta Q}{T} \tag{5-14}$$

which is the desired result. The temperature scale, which is used in calculations based on equation (5-14), must be either the Rankine or Kelvin scale, each of which is an absolute, or thermodynamic, temperature scale.

Substitution of the thermodynamic temperature scale for $f(t)$ in equation (5-10) then yields

$$\frac{\delta Q_1}{T_1} + \frac{\delta Q_2}{T_2} = 0 \tag{5-15}$$

Equation (5-15) may be generalized for a reversible process to give

$$\oint \frac{\delta Q}{T} = \oint dS = 0 \tag{5-16}$$

which demonstrates that $1/T$ is the appropriate integrating factor for relating changes in the state property, entropy, to the path function, heat.

Thus we learn that heat flow produces an entropy flow which is given by

$$\delta\,(\text{entropy flow}) = \frac{\delta Q}{T} \tag{5-17}$$

where T is the absolute temperature of the system at the point where heat flow occurs. When the temperature of the system at the boundary is not constant over the areas where heat transfer is occurring, equation (5-17) must be rewritten:

$$\delta\,(\text{entropy flow}) = \int_{\text{surface area}} \frac{\delta q\, dA}{T} \tag{5-17a}$$

where q = finite heat transferred per unit area of boundary

δq = differential heat transferred per unit area of boundary

The total entropy flow for a process attributable to heat is given by the double integral

$$\text{entropy flow} = \int_{\text{process}} \int_{\text{surface area}} \frac{\delta q\, dA}{T} \tag{5-18}$$

As has been shown, all nonthermal forms of energy transfer between system and surroundings (excluding the transfer of mass) do not have an entropy flow associated with them. However, dissipatory processes (such as friction, turbulence, etc.), which occur simultaneously *within the system* while it is receiving or delivering work or heat, can result in entropy changes. Indeed, they can occur in a system for which neither a heat nor a work effect exist. Entropy generation by such means will be taken up later when generation terms are discussed and should not be confused with the entropy flow currently under discussion. Since work and heat are defined as energy in transition across a system boundary, energy flows wholly within a system do not qualify as either heat or work, so the preceding discussion does not apply to such flows.

SAMPLE PROBLEM 5-1. Two blocks that are initially at different temperatures and completely isolated from the remainder of the universe are brought together so that thermal energy flows from the hot block to the cold block. Analyze the entropy change of the universe associated with this process by choosing as your system (a) the two blocks individually, and (b) the two blocks together. Assume that the heat transfer within each individual block is reversible so that any irreversibilities which occur, occur only at the interface between the two blocks.

Solution: (a) Let us examine the individual blocks first. The entropy balance for either block as a system may be written as

$$(\underline{S}\,\delta M)_{\text{in}} - (\underline{S}\,\delta M)_{\text{out}} + \frac{\delta Q}{T} + \delta(\text{entropy})_{\text{gen}} = d(M\underline{S})_{\text{syst}}$$

Because the system is closed $dM = \delta M_{\text{in}} = \delta M_{\text{out}} = 0$. If the heat transfer in each block is reversible, the entropy generation within each block is zero and the entropy balance reduces to

$$M\,d\underline{S} = \frac{\delta Q}{T}$$

Applied individually to the low and high-temprature blocks, the entropy balance then gives

$$M\,d\underline{S}_L = \frac{\delta Q_L}{T_L}; \qquad M\,d\underline{S}_H = \frac{\delta Q_H}{T_H}$$

The change in entropy of the universe as a result of this process then is given by

$$dS_{\text{univ}} = dS_H + dS_L = \frac{\delta Q_H}{T_H} + \frac{\delta Q_L}{T_L}$$

The total change in entropy of the universe is obtained by integrating the above expression to give

$$\Delta S_{\text{univ}} = \int \frac{\delta Q_H}{T_H} + \frac{\delta Q_L}{T_L}$$

or, since $\delta Q_L = -\delta Q_H > 0$,

$$\Delta S_{\text{univ}} = \int\left(\frac{1}{T_H} - \frac{1}{T_L}\right)\delta Q_H$$

When performing the integration, it is important to keep in mind that T_H and T_L refer to the temperatures at which δQ_H and δQ_L are transferred. Since $T_H > T_L$, our integral for ΔS_{univ} indicates that $\Delta S_{\text{univ}} > 0$, as we know it must be.

(b) Taking both blocks together as the system, the entropy balance may again be written:

$$(\underline{S}\,\delta M)_{\text{in}} - (\underline{S}\,\delta M)_{\text{out}} + \frac{\delta Q}{T} + \delta(\text{entropy})_{\text{gen}} = d(M\underline{S})_{\text{syst}}$$

As in part (a), the closed system indicates that $\delta M_{\text{in}} = \delta M_{\text{out}} = dM = 0$, and the entropy balance reduces to

$$\frac{\delta Q}{T} + \delta(\text{entropy})_{\text{gen}} = M\,d\underline{S}_{\text{syst}}$$

Since no heat is exchanged between the surroundings and the two blocks, $\delta Q = 0$. (Although thermal energy is transferred between the blocks, this is an internal transfer and thus not a flow term.) Thus the entropy balance further reduces to

$$M\,d\underline{S} = (\text{entropy})_{\text{gen}}$$

Since the system is the only portion of the universe undergoing changes, $\Delta S_{\text{univ}} = \Delta S_{\text{syst}}$, or

$$\Delta S_{\text{univ}} = \int dS_{\text{syst}} = \int \delta(\text{entropy generation})$$

and the total entropy change of the universe is now associated with internal irreversible flows.

5.2 Entropy Generation and Lost Work

The need for an entropy-generation term to account for the entropy changes brought about by irreversible energy flows internal to the system is clearly demonstrated by Sample Problem 5-1. The entropy change of the universe was shown to be a positive, nonzero number when the two blocks were considered separately as systems. Since the entropy change of the universe must be independent of the way by which it is calculated, it must be the same for parts (a) and (b). When the two blocks of Sample Problem 5-1 were taken as a combined system, it became an isolated system, because no exchange of energy or mass occurred between system and surroundings. Therefore, the only way in which the entropy change of the system (and hence the universe) could be accounted for was by entropy generation within the system.

In this example we could clearly identify a flow of thermal energy from the high-temperature part of the system to the lower-temperature portion. Dissipation of the temperature potential results in a permanent decrease in the amount of work that can ultimately be produced in our universe. As is shown in Chapter 7, a portion of the resulting thermal energy flow could have been converted to work by a heat engine. Failure to have done so results in an increase in the system's entropy. Our accounting scheme, the entropy balance, is one that carefully accounts for all transfers across the system boundary. However, our definition of the system in the second case incorporated the flow resulting from an internal, finite driving force within the system, and therefore it escaped detection. Thus we must devise a term to account for such irreversible or dissipative phenomena when they occur within the system.

Although the observation that the entropy-generation term is positive is based on a single calculation in Sample Problem 5-1, we shall soon show that this is universally true. Only in the case of a reversible process where any internal gradients are infinitesimal and friction absent can the entropy-generation term become zero; *it can never be negative.*

Although our discussion of entropy generation thus far has related to a process in which an energy transfer occurs within a system as the result of a temperature difference, it can be shown that entropy will be generated whenever any energy potential is decreased without work having been produced or another energy potential having experienced an equivalent increase. For example, in Section 3.5 we examined the adiabatic, free expansion of a compressible substance and observed that the entropy of the gas always increased when it expanded in the absence of a resisting force. The net result of this process, which was certainly irreversible, was to reduce one of the system potentials (its pressure) for producing mechanical energy without having produced the work which was theoretically extractable. An irreversible flow occurred within the system (chosen to be the container and the gas it contains) because of the pressure imbalance that existed during the rapid expansion. As in the case of internal thermal-energy transfer, the entropy change occurring during this internal transfer of mass is accounted for by the entropy-generation term.

In the case of the gas undergoing an adiabatic free expansion, the net effect of the process was to end up with the same amount of gas at the same total energy but at a lower pressure. Since pressure is an energy potential which causes an energy flow that results in work, it is obvious that at a lower pressure the gas is less capable of converting its energy into work than it was at its original pressure.

The term *lost work* has been chosen to indicate this loss in the ability of the gas to produce work. It should be emphasized that the conservation-of-energy principle is not violated in such a process because the energy has not really been lost—only the ability to convert a portion of it to work. Lost work is thus defined as work that could have been performed but was not because of dissipative effects or irreversibilities. Whenever an irreversible change within a system leads to the lowering of an energy potential (such as pressure, temperature, electrical potential, etc.), without transferring as much energy to the surroundings, in the form of work, as possible, lost work results and entropy production occurs.

Friction is another example of an irreversible, or dissipatory, process in that it reduces a system's ability to deliver work as shown in Fig. 5-3 where the system is taken to be the gas + cylinder + piston. As

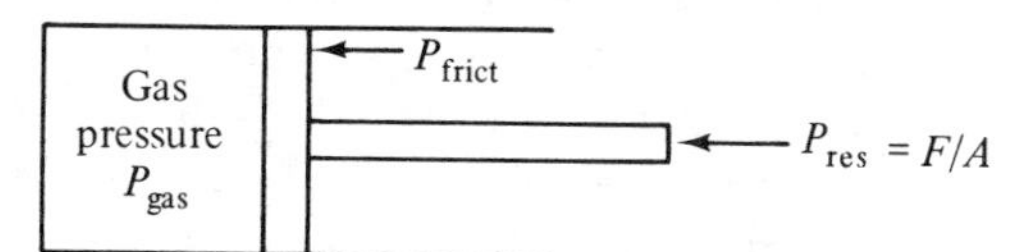

FIG. 5-3. The effect of friction.

the gas expands, it exerts a force $F = P_{gas} \times A_{piston}$ on the piston head. In the absence of friction this force can be transferred directly to the surroundings through the shaft and will produce useful work. The presence of friction between the piston and the wall diminishes the net force that is transferrable to the surroundings and hence the maximum amount of work that may be produced. In a sense it is equivalent to a process that occurs with a finite difference in pressure if we compare the pressure of the gas, P_{gas}, with the pressure, P_{res}, which is resisting the motion of the system boundary. Thus, although the friction slows the expansion down so that the process may be considered quasi-static, it requires the existence of a finite pressure difference between system and surroundings to produce an energy transfer.

The frictional force may be thought of as friction pressure, P_{frict}, acting to oppose movement of the piston. In this case, it would act to resist motion to the right and is thus shown as acting toward the left. The sum of the pressures opposing the motion of the piston must be less than the gas pressure, P_{gas}, if an expansion is to occur. Assuming that the piston has negligible mass, it can support only an infinitesimal pressure (or force) imbalance without experiencing an acceleration. Thus the pressures acting on the piston must be in equilibrium, so that

$$P_{gas} = P_{res} + P_{frict} \tag{5-19}$$

If the gas pressure is held constant, an increase in the friction losses, P_{frict}, causes a corresponding decrease in the usable force, P_{res}, transmitted to the surroundings. Therefore, less work can be transferred to the surroundings.

In the expansion process pictured in Fig. 5-3 the maximum work will be transferred to the surroundings when the frictional losses vanish so that $P_{res} = P_{gas}$. Since the work delivered to the surroundings is given by

$$\int_{V_1}^{V_2} P_{res}\, dV$$

the maximum work that can be transferred to the surroundings is given by

$$\int_{V_1}^{V_2} P_{gas}\, dV$$

where V_1 and V_2 equal the initial and final volumes of the confined gas. In those instances where friction is present, $P_{res} = P_{gas} - P_{frict}$, which is less than P_{gas}. The decrease in work transferred to the surroundings is given by

$$\int_{V_1}^{V_2} P_{frict}\, dV$$

This lost work will be degraded into thermal energy by friction at the interface between the piston and the cylinder. If the piston and cylinder are insulated against transfer of heat to the surroundings, this thermal energy will be transferred back into the gas, causing its entropy to increase (assuming that it is not absorbed by the piston and cylinder—that is, assuming that the piston and cylinder have either negligible masses or negligible heat capacities). This transfer qualifies as heat if we consider the gas alone as the system. However, if the system is taken as gas + piston + cylinder, the transfer, although still occurring, is internal and not considered as heat. However, in terms of this system it has clearly reduced the amount of work extractable from the expansion and therefore qualifies as lost work.

Let us examine the expansion of the previous example from the point of view of the gas within the piston–cylinder. We find that the gas has no way of detecting whether the expansion is proceeding reversibly, or irreversibly, as long as the gas pressure is exactly balanced by the sum $P_{res} + P_{frict}$. The irreversibilities that exist in the total process occur outside the gas and therefore do not affect the gas except insofar as these irreversibilities may lead to a heat flow back to the gas.

In analyzing this expansion process, we might choose to define a system such that the irreversibilities occur outside the system (that is, choose the gas alone as system) or in such a way that the irreversibilities occur within the system (for example, the gas + piston + cylinder as a system). If we assume that the piston + cylinder are perfectly insulated against heat loss to the surroundings, then the gas in either case will undergo an increase in entropy. In the former case the entropy increase results from the transfer of thermal energy (caused by friction) from the surroundings. In the latter case no heat flow is present, because the thermal-energy transfer (from piston to gas) is internal to the system and not between the system and the surroundings. In the second instance the entropy change is attributable wholly to entropy generation, which results from the internal irreversibilities.

The entropy increase in the first case can be calculated as

$$\Delta S_{\text{gas}} = \int \frac{\delta Q}{T} \tag{5-20}$$

where δQ is the heat flow to the gas as a result of the interfacial friction. As shown before, δQ is exactly equal to the mechanical energy dissipated at the wall:

$$\delta Q = P_{\text{frict}}\, dV$$

The temperature in the denominator of equation (5-20) is the system temperature at the point where the heat transfer takes place. Thus the entropy change is given by

$$(\Delta S)_{\text{gas}} = \int \frac{\delta Q}{T} = \int_{V_1}^{V_2} \frac{P_{\text{frict}}\, dV}{T_{\text{syst}}} \tag{5-21}$$

In the second case, however, there is no δQ term, but the same entropy increase must be experienced by the gas. In this case the system produced less work than it could have in the absence of friction. The lost work, LW, represents this difference:

$$LW = \int_{V_1}^{V_2} (P_{\text{gas}} - P_{\text{res}})\, dV = \int_{V_1}^{V_2} P_{\text{frict}}\, dV \tag{5-22}$$

or, upon differentiating,

$$\delta LW = P_{\text{frict}}\, dV \tag{5-22a}$$

where δ again represents the differential amount of the quantity LW.

From our earlier calculation we know that the entropy change experienced by the gas is given by

$$\Delta S = \int_{V_1}^{V_2} \frac{P_{\text{frict}}\, dV}{T_{\text{syst}}} \tag{5-23}$$

but since $\delta LW = P_{\text{frict}}\, dV$, equation (5-23) reduces to

$$\Delta S = \int_{V_1}^{V_2} \frac{\delta LW}{T_{\text{syst}}} \tag{5-24}$$

Thus the entropy change associated with an internal irreversibility or entropy generation is given for a differential process by

$$\delta(\text{entropy})_{\text{gen}} = \frac{\delta LW}{T_{\text{syst}}} \tag{5-25}$$

If irreversibilities are distributed throughout the volume of the system and the temperature varies within the system, the entropy generation must be evaluated by integrating $(\delta lw/T)dV$ over the entire volume of the system along the entire path between initial and final states:

$$\text{entropy generation between states 1 and 2} = \int_{\text{process}} \int_{\text{volume}} \frac{\delta lw}{T}\, dV \tag{5-26}$$

where lw = (finite) lost work per unit volume of system
δlw = differential lost work per unit volume of system

(Note: lw is used for lost work on a unit volume basis as contrasted to LW, used earlier, which is the total lost work.) The limits on the integral imply that the quantity $\delta lw/T$ must be evaluated at all points in the system where irreversibilities occur.

Although we were successful in evaluating the lost work and entropy production for the relatively simple piston–cylinder problem, in general it cannot be precisely calculated because of the extremely complex nature of the dissipative phenomena. Although we can measure changes in other properties by which we can calculate entropy changes arising from irreversibilities, an a priori prediction of LW is extremely difficult and generally impractical.

This difficulty in treating irreversible processes in an analytic manner necessitates the use of empirical methods for predicting such effects. For example, consider the flow that occurs within a pipeline due to a pressure gradient. Since lost work (and entropy production) must be evaluated with respect to a specified system, let the pipe and its contents be our system. The drag and turbulence that occur in the pipeline lead to lost work much as friction does in the piston–cylinder example. Unfortunately, a quantitative evaluation of this dissipation is a difficult task. Fluid mechanics offers a mathematical expression for the dissipation, but its evaluation frequently requires a greater knowledge of the fluid velocity distribution than is available. Consequently, empirical procedures are generally used to estimate these effects. When the topic of the thermodynamics of fluid flow is discussed later we shall consider the question of empirical correlations for lost work in flow systems in greater detail.

Direct evaluation of lost work effects in process equipment, such as turbines and compressors, is far beyond our present capabilities. Therefore, in analyzing processes involving operations of this nature, it is often necessary to approximate the real situation with a reversible one in which $LW = 0$. Past experience with many devices, such as compressors and turbines, often permits the engineer to relate performance under reversible conditions to actual operation under real conditions. The relation is usually expressed by means of an efficiency factor. For devices such as pumps and compressors which utilize work from the surroundings, efficiency is defined as

$$\text{efficiency} = \frac{W_{\text{rev}}}{W_{\text{act}}} \times 100\% \tag{5-27}$$

For turbines and other expansion devices that supply work to the surroundings the definition is inverted to give

$$\text{efficiency} = \frac{W_{\text{act}}}{W_{\text{rev}}} \times 100\% \tag{5-28}$$

In either case, W_{rev} is the work produced or consumed if the process is operated reversibly, and W_{act} is the actual work that is involved. A knowledge of the efficiency of a particular type of device, coupled with an analysis based on a reversible process, thus enables one to estimate with reasonable engineering accuracy the actual energy requirements for a specific piece of equipment.

SAMPLE PROBLEM 5-2. Consider an insulated oven with an electrical heating element. The oven is turned on, and the temperature begins to rise. Indicate whether LW or Q are present for the following systems:

(a) The heating element alone.

(b) The oven and its contents (including the element).

(c) The gas within the oven.

Solution: (a) The heating element alone: In the heating element, electrical energy (a form of mechanical energy) is transformed into thermal energy and then transferred to the surroundings as heat. Thus the heating element has a Q term. In addition to the Q term, a LW term is present because of the internal degradation of electrical energy to thermal energy resulting from the electrical resistance of the element. Thus both Q and LW are present when the heating element is chosen as a system.

(b) If we choose the oven and its contents as a system, LW is clearly present, since electrical energy is converted to thermal energy within the system. If we assume that the oven is insulated so that no heat is lost to the surroundings, then no heat transfer occurs, because the thermal energy transfers that exist are wholly within the system.

(c) If we now consider the gas within the oven as a system, Q is present in the form of the energy transferred from the heating element to the gas. If we assume that ther is a negligible temperature gradient within the gas, then the gas is undergoing reversible heat transfer and no LW occurs.

5.3 The Entropy Balance

Incorporation of $\int_{\text{surface}} (\delta q/T)\, dA$ for the entropy flow and $\int_{\text{volume}} (\delta lw/T)\, dV$ for the entropy-generation terms in equation (5-3) yields the complete entropy balance for a general process

$$(\underset{\sim}{S}\,\delta M)_{\text{in}} - (\underset{\sim}{S}\,\delta M)_{\text{out}} + \int_{\text{surface}} \frac{\delta q}{T}\, dA + \int_{\text{volume}} \frac{\delta lw}{T}\, dV = d(M\underset{\sim}{S})_{\text{syst}} \quad (5\text{-}29)$$

If the system temperature is everywhere uniform, equation (5-29) reduces to

$$(\underset{\sim}{S}\,\delta M)_{\text{in}} - (\underset{\sim}{S}\,\delta M)_{\text{out}} + \frac{\delta Q}{T} + \frac{\delta LW}{T} = d(M\underset{\sim}{S})_{\text{syst}} \quad (5\text{-}30)$$

This equation can be applied to any thermodynamic system just as the energy balance is applied. The balance has been written in differential notation with δ used to denote differential flow and production terms. Sign conventions are the same as in the energy balance; that is, $d(M\underset{\sim}{S})_{\text{syst}}$ represents the entropy change of the system over a differential time period. δQ represents heat flowing into the system (if positive) and to the surroundings (if negative) for a differential time period. δM represents a mass flow over the differential time period and is always taken as positive (since its sign is already accounted for in the "in" and "out" terms). δLW represents the energy dissipation over the differential period and will always be positive.

The temperature in the flow term $\delta q/T$ is the temperature of the system at the point along the boundary where the transfer occurs. By the same token, the temperature in the $\delta lw/T$ term is the temperature at that point within the system where the irreversibility is occurring. For isothermal (uniform temperature) systems, application of the entropy balance is greatly simplified. For nonisothermal systems it is necessary to sum (or integrate) the $\delta q/T$ and $\delta lw/T$ terms over all portions of the system where either δq or δlw is present. It is also important to stress that the terms $\underline{S}_{in}$ and $\underline{S}_{out}$ relate to the stream entropies at the point where the streams cross the system's boundary.

As with the energy balance, the entropy balance we have derived is an extremely general relation designed to fit a wide variety of processes. As such, it contains terms that may frequently be ignored in any given problem. Let us now examine some of the more commonly encountered simplifications of the entropy balance. (For purposes of simplicity, we use the isothermal form of the entropy balance in the ensuing discussion. However, it must be remembered that the integral form must be used when heat transfer or lost work exists in a nonisothermal system.)

Closed Systems

For a closed system the δM_{in} and δM_{out} terms vanish, as does dM_{syst}, and the entropy balance reduces to

$$M\,d\underline{S}_{syst} = \frac{\delta Q}{T} + \frac{\delta LW}{T} \tag{5-31}$$

For a reversible process $\delta LW = 0$ and we obtain a further simplification:

$$dS_{syst} = M\,d\underline{S}_{syst} = \frac{\delta Q}{T} \tag{5-32}$$

Open Systems

When an open system operates at steady state $d(MS)_{syst} = 0$ and $\delta M_{in} = \delta M_{out}$, so the entropy balance reduces to

$$(\underline{S}_{in} - \underline{S}_{out})\,\delta M + \frac{\delta Q}{T} + \frac{\delta LW}{T} = 0 \tag{5-33}$$

For an adiabatic process $\delta Q = 0$, whereas for a reversible process $\delta LW = 0$. These facts may be used to modify equations (5-31) and (5-33) when applicable.

SAMPLE PROBLEM 5-3. Two insulated horizontal cylinders are separated by removable insulation as shown in Fig. S5-3. Each cylinder is divided into two

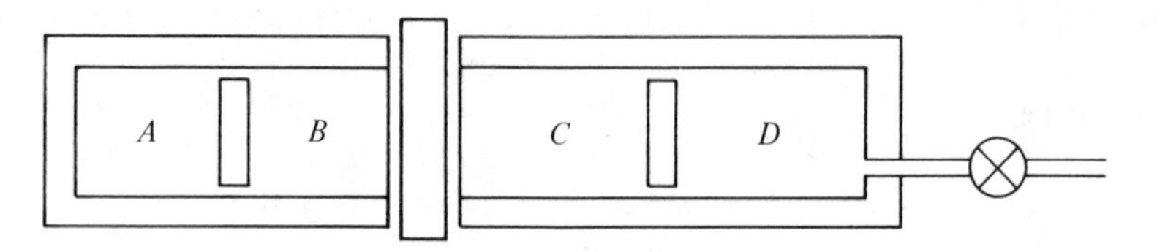

FIGURE S5-3

parts by a freely floating, insulated piston having negligible mass. Initially, chamber A has a volume of 5 ft^3 and contains steam at 490 psia and 520°F; chamber B, also 5 ft^3, contains steam at 740°F; chamber C contains 10 lb$_m$ of saturated water vapor at 320°F; chamber D contains air. The removable insulation is raised, and heat flows slowly from chamber B to C, until the steam in chamber A is a saturated vapor. At the same time, air is bled slowly out of D to maintain a constant temperature in C. What is the final pressure in C?

Solution: We may summarize the changes that occur in each region as follows: Chamber A undergoes an adiabatic reversible expansion (no irreversibilities occur in chamber A) in which it performs work on chamber B. Chamber B undergoes a non-adiabatic compression during which it receives work from chamber A and loses heat to chamber C. Since the piston separating chambers A and B is frictionless, $P_A = P_B$ at all times. Chamber C undergoes an isothermal expansion at 320°F, during which it receives heat from chamber B and transfers work to chamber D. Chamber D serves to receive work from chamber C, allowing the temperature in C to remain constant. The initial conditions in the various chambers are summarized in Table S5-3.

TABLE S5-3

Chamber	P, psia	T, °F	$\underline{H}$, Btu/lb$_m$	$\underline{S}$, Btu/lb$_m$ °R	$\underline{V}$, ft^3/lb$_m$	V, ft^3	M, lb$_m$	$\underline{U} = \underline{H} - P\underline{V}$, Btu/lb$_m$
A	490	520	1248.4	1.5116	1.0535	5	?	1152
B	490	740	1378	1.630	1.390	5	?	1252
C	90	320	1185	1.6116	4.898	?	10	Unneeded
D (air)	90							

We may calculate the mass in chambers A and B from the definition of $\underline{V}$:

$$\underline{V} = \frac{V}{M}$$

or

$$M = \frac{V}{\underline{V}}$$

Thus

$$M_A = 5 \text{ ft}^3/1.0535 \text{ ft}^3/\text{lb}_m = 4.75 \text{ lb}_m$$
$$M_B = 5 \text{ ft}^3/1.390 \text{ ft}^3/\text{lb}_m = 3.60 \text{ lb}_m$$

The volume in C is calculated to be

$$V_C = \underline{V}_C M_C = 48.9 \text{ ft}^3$$

We may now begin determining the conditions when the steam in chamber A reaches saturation conditions. Since the steam in chamber A is to be a saturated vapor at the end of its expansion, determination of any property will serve to fix its final state. Choosing the steam in chamber A as a system, the energy and entropy balances are written as follows (assuming negligible changes in potential and kinetic energy):

Energy:

$$(\underline{H}\,\delta M)_{\text{in}} - (H\,\delta M)_{\text{out}} + \delta Q - \delta W = d(M\underline{U})_{\text{syst}}$$

Entropy:

$$(\underline{S}\,\delta M)_{\text{in}} - (\underline{S}\,\delta M)_{\text{out}} + \int_{\text{area}} \frac{\delta q}{T}\,dA + \int_{\text{vol}} \frac{\delta lw}{T}\,dV = d(M\underline{S})_{\text{syst}}$$

Since the system is closed, $\delta M_{\text{in}} = \delta M_{\text{out}} = dM_{\text{syst}} = 0$, and the energy and entropy balances reduce to

$$\delta Q - \delta W = M\,d\underline{U}$$

$$\int_{\text{area}} \frac{\delta q}{T}\,dA + \int_{\text{vol}} \frac{\delta lw}{T}\,dV = M\,d\underline{S}$$

Since the expansion in chamber A is adiabatic and reversible, $\delta q = \delta lw = 0$, and the balances further reduce to

$$-\delta W = M\,d\underline{U}$$

$$0 = d\underline{S}$$

Integration of the entropy balance over the entire expansion then yields

$$(\underline{S}_A)_{\text{final}} - (\underline{S}_A)_{\text{init}} = 0$$

$$(\underline{S}_A)_{\text{final}} = (\underline{S}_A)_{\text{init}} = 1.5116$$

Thus the entropy balance allows us to determine the needed property which fixes the state of the steam in A when it reaches saturation.

If we integrate the energy balance over the entire expansion we can determine the work done by the gas in chamber A as

$$-W = -\int \delta W = M[(\underline{U}_A)_{\text{final}} - (\underline{U}_A)_{\text{init}}]$$

$(\underline{U}_A)_{\text{init}}$ is obtained from $\underline{U} = \underline{H} - P\underline{V}$, and $(\underline{U}_A)_{\text{final}}$ can be determined from the final state of the steam in A as follows. Since the steam in A undergoes an isentropic expansion, $(\underline{S}_A)_{\text{final}} = (\underline{S}_A)_{\text{init}} = 1.5116$. At saturation conditions this entropy corresponds to a final state summarized as follows:

$$T = 420°\text{F}$$

$$P = 300 \text{ psia}$$

$$\underline{H} = 1203 \text{ Btu/lb}_{\text{m}}$$

$$\underline{V} = 1.54 \text{ ft}^3/\text{lb}_{\text{m}}$$

$$\underline{U} = \underline{H} - P\underline{V} = 1117 \text{ Btu/lb}_{\text{m}}$$

The final volume of chamber A is given by

$$(V_A) = M\underline{V} = (4.75)(1.54)\,(\text{ft}^3/\text{lb}_{\text{m}})\,\text{lb}_{\text{m}}$$

$$(V_A)_{\text{final}} = 7.32 \text{ ft}^3$$

The work performed by the steam in chamber A is given by

$$-W = M\,\Delta\underline{U} = (4.75)(1117 - 1152) \text{ Btu}$$

or

$$W = 166 \text{ Btu}$$

We may now begin to analyze the final state of the material in chamber B.

Since $P_A = P_B$ at all times, $(P_B)_{\text{final}} = (P_A)_{\text{final}} = 300$ psia. Thus determination of one other property of the system will completely specify its state. We may easily determine the specific volume of the steam in chamber B by noting that $(V_A)_{\text{final}} + (V_B)_{\text{final}} = (V_A)_{\text{init}} + (V_B)_{\text{init}} = 10 \text{ ft}^3$ and remembering that $(\underline{V}_B)_{\text{final}} = (V_B)_{\text{final}}/M_B = (V_B)_{\text{final}}/3.6 \text{ lb}_m$. Since $(V_A)_{\text{final}} = 7.32 \text{ ft}^3$, $(V_B)_{\text{final}} = 2.68 \text{ ft}^3$ and

$$(\underline{V}_B)_{\text{final}} = \frac{(V_B)_{\text{final}}}{3.6 \text{ lb}_m} = \frac{2.68}{3.6} \text{ ft}^3/\text{lb}_m$$

or

$$(\underline{V}_B)_{\text{final}} = 0.745 \text{ ft}^3/\text{lb}_m$$

If we attempt to locate $\underline{V} = 0.745 \text{ ft}^3/\text{lb}_m$ and $P = 300$ psia in the steam tables, we find that this corresponds to a mixture of saturated liquid and saturated vapor. The proportion of vapor and liquid is determined from the mixture rule:

$$\underline{V}_{\text{mix}} = \underline{V}^V X^V + \underline{V}^L X^L$$

But

$$X^V = 1 - X^L$$
$$\underline{V}_{\text{mix}} = 0.745 \text{ ft}^3/\text{lb}_m$$
$$\underline{V}^V = 1.54 \text{ ft}^3/\text{lb}_m$$
$$\underline{V}^L = 0.091 \text{ ft}^3/\text{lb}_m$$

or

$$0.745 = 1.54(1 - X^L) + 0.091 X^L$$

Solving for X^L yields

$$X^L = 0.52$$
$$X^V = 0.48$$

$(\underline{U}_B)_{\text{final}}$ is then determined from the mixture rules:

$$(\underline{U}_B)_{\text{mix}} = X^L(\underline{U})_{\text{sat liq}} + (1 - X^L)(\underline{U})_{\text{sat vap}}$$
$$\underline{U}_{\text{sat vap}} = 1117 \text{ Btu/lb}_m \text{ (from chamber } A)$$
$$\underline{U}_{\text{sat liq}} = (\underline{H} - P\underline{V})_{\text{sat liq}} = 393 \text{ Btu/lb}_m$$

so

$$(\underline{U}_B)_{\text{mix}} = (0.52)(393) + (0.48)(1117) = 740 \text{Btu/lb}_m$$

We are now in a position to determine the heat transfer between chambers B and C. The energy balance is written about chamber B. Since the system is closed, $\delta M_{\text{in}} = \delta M_{\text{out}} = dM_{\text{syst}} = 0$. Neglecting potencial and kinetic energy changes, the energy balance reduces to

$$\delta Q - \delta W = M\, d\underline{U}_{\text{syst}}$$

This may be integrated to yield

$$Q - W = M\, \Delta\underline{U} = M[(\underline{U}_B)_{\text{final}} - (\underline{U}_B)_{\text{init}}]$$

We know $(\underline{U}_B)_{\text{final}}$, $(\underline{U}_B)_{\text{init}}$, and M. In addition, the only work transferred between chamber B and its surroundings is the work transferred from chamber A to chamber B. Since work is considered positive when transferred from the system to the surroundings,

$$W_B = -W_A = -166 \text{ Btu}$$

We may now calculate the heat transferred to chamber B:

$$\begin{aligned} Q_B &= W_B + M_B(\underline{U}_B)_{\text{final}} - (\underline{U}_B)_{\text{init}} \\ &= -166 + [(3.6)(740) - 1252] \text{ Btu} \\ &= -2010 \text{ Btu} \end{aligned}$$

Having calculated the heat loss from chamber B we can now begin to work on chamber C. Since chamber C undergoes an isothermal expansion, $T = \text{const} = 320°\text{F}$. Thus once again, determination of another system property will fix the remaining properties and thereby allow us to determine the final pressure in chamber C as desired.

Assuming that chamber C is a closed system the entropy and energy balances around system C reduce to

$$\text{Energy:} \quad \delta Q - \delta W = M\, d\underline{U}$$

$$\text{Entropy:} \quad \frac{\delta Q}{T} + \frac{\delta LW}{T} = M\, d\underline{S}$$

These may be integrated to yield

$$Q - W = M\,\Delta\underline{U}$$

$$\frac{Q}{T} + \frac{LW}{T} = M\,\Delta\underline{S}$$

Since the steam in chamber C undergoes a reversible process, $LW = 0$, so that

$$\frac{Q_C}{T_C} = M\,\Delta\underline{S}_C = M_C[(\underline{S}_C)_{\text{final}} - (\underline{S}_C)_{\text{init}}]$$

But

$$\begin{aligned} Q_C &= -Q_B = 2010 \text{ Btu} \\ T &= 320°\text{F} = 780°\text{R} \\ (\underline{S}_C)_{\text{init}} &= 1.6116 \text{ Btu/lb}_\text{m}°\text{R} \\ M &= 10 \text{ lb}_\text{m} \end{aligned}$$

Thus we may solve for the final entropy in chamber C to get

$$\begin{aligned} (\underline{S}_C)_{\text{final}} &= (\underline{S}_C)_{\text{init}} + \frac{Q_C}{T_C M_C} \\ &= \left[1.6116 + \frac{2010}{(780)(10)}\right] \text{Btu/lb}_\text{m}\ °\text{R} \end{aligned}$$

or

$$(\underline{S}_C)_{\text{final}} = 1.8715 \text{ Btu/lb}_\text{m}\ °\text{R}$$

From the final entropy and temperature, the final pressure in chamber C is found to be $(P_C)_{\text{final}} = 10$ psia.

SAMPLE PROBLEM 5-4. A steam line supplies steam at 620 psia and 700°F to an adiabatic reversible turbine (expander) which discharges to an insulated collector fitted with a frictionless piston that maintains a constant pressure of 23 psia. Additional steam is fed to the collector via a throttling valve so as to hold the temperature in the collector constant at 270°F, as shown in Fig. S5-4. If the collecting vessel has a cross-sectional area of 37.18 ft², how many pounds of steam flow through the turbine per foot of piston rise? Neglect potential and

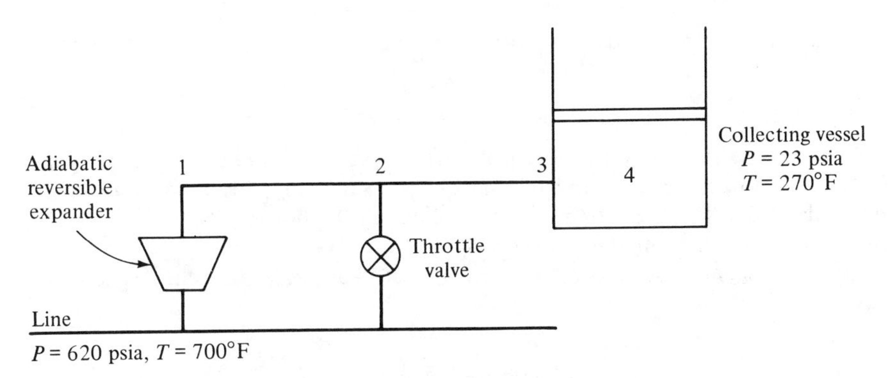

FIGURE S5-4

kinetic energy changes in the steam and heat and frictional losses in the connecting lines.

Solution: We begin our analysis by attempting to determine the state of the steam entering the collecting vessel. Since the pressure in the collector is 23 psia and there are no frictional losses in the connecting lines, the pressure of the material in the line entering the collecting vessel will also be 23 psia. Thus determination of one more property will fix the state of the entering stream. The entropy balance may be written around the collecting vessel and gives

$$(\underline{S}\,\delta M)_{\text{in}} - (\underline{S}\,\delta M)_{\text{out}} + \int_{\text{area}} \frac{\delta q}{T}\,dA + \int_{\text{vol}} \frac{\delta lw}{T}\,dV = d(M\underline{S})_{\text{syst}}$$

Since no mass flows out of the system, $\delta M_{\text{out}} = 0$. The collecting vessel is well-insulated, so $\delta q = 0$. Since no irreversibilities occur within the system (the piston and connecting lines are frictionless), $lw = 0$ and the entropy equation reduces to

$$(\underline{S}\,\delta M)_{\text{in}} = d(M\underline{S})_{\text{syst}}$$

We may integrate this expression for 1 lb_{m} of steam entering the collector. Since the state of both the steam entering the collector and that within it are not functions of the mass in the cylinder, the specific entropies of the entering steam and those within the collector are also independent of the mass within the collector. Thus they may be removed from under the integral sign, and the integrated entropy balance reduces to

$$(\underline{S}M)_{\text{in}} = (\Delta M)\underline{S}_{\text{syst}}$$

However, the conservation of mass indicates that $M_{\text{in}} = \Delta M$, so the entropy balance becomes

$$\underline{S}_{\text{in}} = \underline{S}_{\text{syst}}$$

That is, the entropy of steam does not change as it enters the collector. However, we have already indicated that $P_{\text{syst}} = P_{\text{in}}$, so two of the properties of the steam do not change as the steam enters the collector. Since two properties fix the state of a system, two properties remaining unchanged means that the state of the steam is unchanged and all other properties are unchanged.

We now determine the properties of the steam within the collecting vessel. We know that $P = 23$ psia and $T = 270°F$; therefore,

$$\underline{H} = 1175 \text{ Btu/lb}_m$$
$$\underline{V} = 18.6 \text{ ft}^3/\text{lb}_m$$

The cross-sectional area of the collector is 37.18 ft^2, so the volume increase corresponding to 1 ft of piston rise is 37.18 ft^3. Since $\underline{V} = 18.6$ ft^3/lb$_m$, 1 ft of piston rise corresponds to $37.18/18.6 = 2.0$ lb$_m$ of steam entering the piston. The enthalpy of the steam in the line entering the collector is then $\underline{H}_3 = 1175$ Btu/lb$_m$.

The energy balance around the mixing "T" (point 2 in the diagram), is now written as

$$\sum \left[\left(\underline{H} + \frac{u^2}{2g_c} + \frac{gZ}{g_c}\right)\delta M\right]_{\text{in}} - \left[\left(\underline{H} + \frac{u^2}{2g_c} + \frac{gZ}{g_c}\right)\delta M\right]_{\text{out}} + \delta Q - \delta W$$
$$= d\left[M\left(\underline{U} + \frac{u^2}{2g_c} + \frac{gZ}{g_c}\right)\right]_{\text{syst}}$$

However, we may neglect $u^2/2g_c$, gZ/g_c, and δQ. $\delta W = 0$ since no motion of, or through, the system boundaries occurs. In addition, if the mixing process in the "T" is at steady state, the entire right-hand portion of the equation vanishes, and the energy equation reduces to

$$\sum (\underline{H}\,\delta M)_{\text{in}} - (\underline{H}\,\delta M)_{\text{out}} = 0$$

Since the enthalpies of the various streams do not change with M, the energy balance can be integrated to

$$\sum (\underline{H}M)_{\text{in}} - (\underline{H}M)_{\text{out}} = 0$$

or

$$\sum (\underline{H}M)_{\text{in}} = (\underline{H}M)_{\text{out}}$$

The incoming streams to the mixing "T" come from the turbine and throttling valve. Thus the summation reduces to

$$(\underline{H}M)_{\text{turb}} + (\underline{H}M)_{\text{valve}} = (\underline{H}M)_{\text{out}}$$

A mass balance around the "T" gives

$$M_{\text{turb}} + M_{\text{valve}} = M_{\text{out}}$$

Since we are attempting to solve for M_{turb}, eliminate M_{valve} between the energy and mass balances to give

$$(\underline{H}M)_{\text{turb}} + \underline{H}_{\text{valve}}(M_{\text{out}} - M_{\text{turb}}) = (\underline{H}M)_{\text{out}}$$

or, solving for M_{turb}

$$M_{\text{turb}} = \frac{(\underline{H}_{\text{out}} - \underline{H}_{\text{valve}})M_{\text{out}}}{\underline{H}_{\text{turb}} - \underline{H}_{\text{valve}}}$$

Since the material leaving the mixing "T" is the same as that entering the collecting chamber, $M_{\text{out}} = 2$ lb$_m$ per foot of piston rise, and $\underline{H}_{\text{out}} = 1175$ Btu/lb$_m$. Thus if we can determine $\underline{H}_{\text{valve}}$ and $\underline{H}_{\text{turb}}$, we can solve directly for the mass passing through the turbine per foot of piston rise.

We may determine $\underline{H}_{\text{valve}}$ from an energy balance around the valve. Neglecting potential and kinetic energies across the valve and assuming steady-state operation, the energy balance around the valve reduces to

$$(\underline{H}_{\text{in}})\,\delta M - (\underline{H}_{\text{out}})\,\delta M + \delta Q - \delta W = 0$$

However, for the valve $\delta W = 0$, and if we assume that the valve is adiabatic, $\delta Q = 0$. Thus the energy balance becomes

$$(\underline{H}_{\text{in}})_{\text{valve}} = (\underline{H}_{\text{out}})_{\text{valve}}$$

But $(H_{\text{in}})_{\text{valve}} = 1350 \text{ Btu/lb}_m$, so

$$(\underline{H}_{\text{out}})_{\text{valve}} = 1350 \text{ Btu/lb}_m$$

However,

$$\underline{H}_{\text{valve}} = (\underline{H}_{\text{out}})_{\text{valve}} = 1350 \text{ Btu/lb}_m$$

We now need to determine the exit enthalpy from the turbine. Since we know that the outlet pressure from the turbine is 23 psia, determination of any other system property will fix the state of the outlet stream and give us its enthalpy. The energy and entropy balances may be written around the turbine as a system. If we neglect potential and kinetic changes and assume steady-state operation, these equations reduce to

Energy:

$$(\underline{H}_{\text{in}} - \underline{H}_{\text{out}})_{\text{turb}}\, \delta M + \delta Q - \delta W = 0$$

Entropy:

$$(\underline{S}_{\text{in}} - \underline{S}_{\text{out}})_{\text{turb}}\, \delta M + \int_{\text{area}} \frac{\delta q}{T}\, dA + \int_{\text{vol}} \frac{\delta lw}{T}\, dV = 0$$

Since the turbine is adiabatic and reversible $\delta q = \delta lw = 0$ and the entropy balance gives

$$(\underline{S}_{\text{out}})_{\text{turb}} = (\underline{S}_{\text{in}})_{\text{turb}}$$

Since we know P_{in} and T_{in} we can obtain $\underline{S}_{\text{in}}$ and then $\underline{S}_{\text{out}}$. From $\underline{S}_{\text{out}}$ and $P_{\text{out}} = 23$ psia we may now look up the final enthalpy and find

$$(\underline{H}_{\text{out}})_{\text{turb}} = \underline{H}_{\text{turb}} = 1065 \text{ Btu/lb}_m$$

(Had we continued to work with the energy equation we finally would have obtained $-\underline{W} = \Delta \underline{H}$, so we could now determine the work produced by the turbine if this were desired.) Substitution of $\underline{H}_{\text{turb}}$ and $\underline{H}_{\text{valve}}$ into the expression for M_{turb} then gives

$$M_{\text{turb}} = \frac{(1175 - 1350) \text{ Btu/lb}_m}{(1065 - 1350) \text{ Btu/lb}_m} \frac{2.0 \text{ lb}_m}{\text{ft of piston travel}}$$

Therefore,

$$M_{\text{turb}} = 1.24 \frac{\text{lb}_m}{\text{ft of piston rise}}$$

and 1.24 lb_m flow through the turbine per foot of piston rise in the collector.

SAMPLE PROBLEM 5-5. A vapor-compression desalination process will be used to produce drinking water aboard an ocean-going ship. The process is as illustrated in Fig. S5-5. Sea water (3.5 per cent salt) enters the unit and is preheated by countercurrent contact with the drinking water and waste brine. The preheated sea water then enters the evaporator, where a portion is boiled off by condensing steam. The condensing steam is obtained by compressing the vapor formed in the evaporator. The cooled condensate is the drinking water.

The evaporator normally operates at 1 atm pressure. The temperature is 218°F because of the boiling-point elevation. The vapor entering the compressor

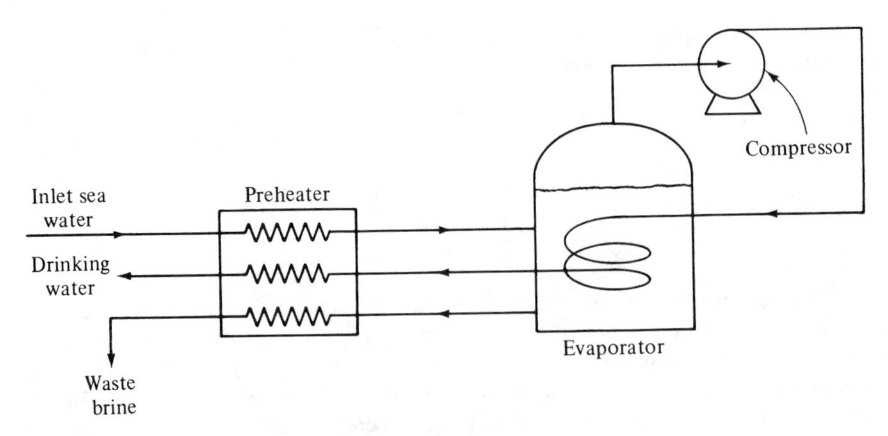

FIGURE S5-5a

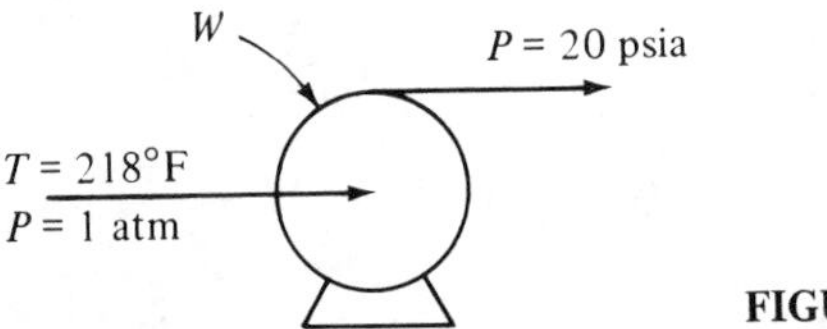

FIGURE S5-5b

is compressed adiabatically to 20 psia. If the compressor operates at an efficiency of 60 per cent determine (a) the outlet temperature from the compressor and (b) the work of compression per pound of drinking water formed.

Solution: We may picture the compressor as a separate unit that receives superheated steam at 218°F and 1 atm and discharges it at 20 psia, as shown in Fig. S5-5b. Choosing the turbine and its contents as a system the energy balance may be written. However, if potential and kinetic energy changes are neglected and the process is assumed to be at steady state, the energy balance reduces to

$$\underline{Q} - \underline{W} = \Delta \underline{H} = \underline{H}_{\text{out}} - \underline{H}_{\text{in}}$$

Since we are told that the compression is adiabatic, $Q = 0$ and the energy balance becomes

$$-\underline{W} = \Delta \underline{H}$$

Thus evaluation of either $\underline{W}$ or $\Delta \underline{H}$ fixes the other. Our problem is to evaluate $\underline{W}$ and the outlet temperature T_2. However, since we already know that the outlet pressure $P_2 = 20$ psia, determination of any other system property will fix the state, and hence temperature, of the outlet stream. Determination of $\Delta \underline{H}$ will give us the outlet enthalpy, since the inlet enthalpy can be obtained from the inlet tempeature and pressure. Therefore, our problem reduces to one of finding either $\underline{W}$ or $\Delta \underline{H}$.

We are told that the compressor may be assumed to be 60 per cent efficient. Therefore, the ratio of the actual work of compression to that needed if the compression were reversible is

$$\frac{\underline{W}_{\text{rev}}}{\underline{W}_{\text{act}}} = 0.60$$

or

$$\underline{W}_{\text{act}} = \frac{\underline{W}_{\text{rev}}}{0.60}$$

If the compression were reversible, then $lw = 0$ and the entropy balance (assuming steady state) could be written as

$$(\underline{S}_{\text{in}} - \underline{S}_{\text{out}})\,\delta M + \int_{\text{area}} \frac{\delta q}{T}\,dA = 0$$

Since the process is adiabatic (whether it's reversible or not, it can always be adiabatic), $\delta q = 0$ and the entropy balance for the reversible process reduces to

$$\underline{S}_{\text{in}} = \underline{S}_{\text{out}}$$

Thus the proces is isentropic. From the inlet conditions we determine the inlet entropy. From the entropy and the outlet pressure we obtain the outlet conditions (assuming, of course, that the process were reversible). That is,

$$(\underline{H}_2)_{\text{rev}} = 1176 \text{ Btu/lb}_{\text{m}}$$

The inlet enthalpy is

$$\underline{H}_1 = 1153 \text{ Btu/lb}_{\text{m}}$$

Therefore if the process were reversible, the enthalpy change would have been

$$\Delta\underline{H}_{\text{rev}} = (1176 - 1153) \text{ Btu/lb}_{\text{m}} = 23 \text{ Btu/lb}_{\text{m}}$$

Since the energy balance is independent of the reversibility of the process,

$$-\underline{W}_{\text{rev}} = (\Delta\underline{H})_{\text{rev}} = 23 \text{ Btu/lb}_{\text{m}}$$

But from the efficiency of the process we know

$$\underline{W}_{\text{act}} = \frac{\underline{W}_{\text{rev}}}{0.60} = \frac{-23 \text{ Btu/lb}_{\text{m}}}{0.60} = -38.3 \text{ Btu/lb}_{\text{m}}$$

so that the actual work requirements of the compressor are 38.3 Btu/lb_{m} of drinking water.

Again, since the energy balance holds independent of any reversibility arguments

$$(\Delta\underline{H})_{\text{act}} = -\underline{W}_{\text{act}} = 38.3 \text{ Btu/lb}_{\text{m}}$$

The inlet enthalpy is still $\underline{H}_1 = 1153 \text{ Btu/lb}_{\text{m}}$, so that the actual outlet enthalpy is

$$(\underline{H}_2)_{\text{act}} = (1153 + 38.3) \text{ Btu/lb}_{\text{m}} = 1191.3 \text{ Btu/lb}_{\text{m}}$$

The actual outlet enthalpy and the outlet pressure are enough to fix the outlet state and allow determination of the outlet temperature, which is

$$T_2 = 300°\text{F}$$

So the temperature of the steam leaving the compressor is 300°F.

5.4 *Irreversible Thermodynamics*

The topic of irreversible thermodynamics deals with the rate at which entropy is produced during many different irreversible flow processes. The flows are recognized as arising from potential gradients such as temperature, composition, and pressure. The interrelationship between potential gradients (forces) and fluxes is developed and related to the rate of entropy production. The study of irreversible thermodynamics thus bridges the gap between classical equilibrium (reversible) thermodynamics and the nonequilibrium rate processes. We shall briefly examine the general concepts of irreversible thermodynamics in Chapter 14.

Problems

5-1 A friend claims to have invented a flow device to increase the superheat of steam, and he solicits your financial backing. He is secretive about details but boasts that he can feed steam at 20 psia and 250°F and obtain steam at 450°F and 1 atm. The device also yields liquid water at 212°F and 1 atm. It receives no additional heat or work from the surroundings, but heat losses may be anticipated. The ratio of product steam to product water is 10 : 1.

(a) Will you invest your money in the project? Justify the soundness of your conclusion, on thermodynamic grounds.

(b) Describe how the device *may* operate; that is, what kind of equipment would be needed inside the box?

(c) If the ratio of product steam to product water were increased, how would this affect the operation? What, if any, are the thermodynamic limitations to such an increase?

5-2 Steam is supplied at 490 psia and 780°F to a well-insulated reversible turbine. The turbine exhausts at 40 psia, the exhaust steam going directly to the heating coils of an evaporator where the steam is condensed. The liquid condensate from the heating coils is trapped (that is, put through a trap which allows only liquid to pass) at 40 psia and flows from the trap into an open barrel, where it is weighed for the purpose of making an energy balance around the evaporator. The atmospheric pressure is 14.7 psia, and you may assume that the condensate leaves the evaporator at 212°F.

(a) Find the work done by the turbine per pound of liquid water weighed in the barrel.

(b) How many Btu's of heat are transferred in the heating coils per pound of liquid water weighed in the barrel?

5-3 A certain process requires 1000 lb_m/hr of process steam at 20 psia with not less than 96 per cent quality and not more than 12°F superheat. Steam is available at 260 psia and 500°F.

(a) It has been suggested that the exhaust steam from a turbine operating from the available steam supply be utilized for the purpose. What maximum horsepower would be available from such a turbine if the heat losses from the turbine were 5000 Btu/hr?

(b) The nature of the process makes it necessary for the process steam to be available when the turbine is down for service. It has been suggested that an alternative source of process steam could be obtained by throttling the available steam supply (260 psia and 500°F) through an adiabatic throttling valve to the required pressure and then cooling to the required condition of quality or superheat. What minimum amount of heat would have to be removed from the throttled steam (Btu/hr) to attain the required conditions for the process steam?

(c) If the expansion in (a) was conducted adiabatically and reversibly, how much superheat must the steam have originally to ensure no liquid in the turbine exhaust at 20 psia if the original pressure is 260 psia?

5-4 A turbine is supplied with 5,000 lb_m/hr of steam at 900 psia and 820°F. The exhaust pressure is 80 psia. A sample of the exhaust steam is passed through an adiabatic throttling calorimeter, where it expands to atmospheric pressure and a temperature of 240°F. Heat losses from the turbine are estimated to be 140,000 Btu/hr. How much work is being done by the turbine? What is the quality of the steam exhausting from the turbine? Show whether there is any lost work in the turbine.

5-5 The two cylinders in Fig. P5-5 are in a well-insulated box having a removable insulating partition *C*. Both cylinders are fitted with heavy, frictionless pistons of 1 ft^2 cross-sectional area each. Space *A* contains 1 lb_m of saturated water vapor at 500 psia and 467°F. Space *B* contains 50 lb_m of saturated liquid water at 50 psia and 281°F. The partition *C* is withdrawn and later replaced when the volume in space *A* is one half its original value. Assuming negligible heat capacity of the materials of construction, find Q, W, ΔU, and ΔS for:

(a) The vapor in space *A* as the system.
(b) The liquid in space *B* as the system.
(c) Both pistons and the fluids in both cylinders as the system.

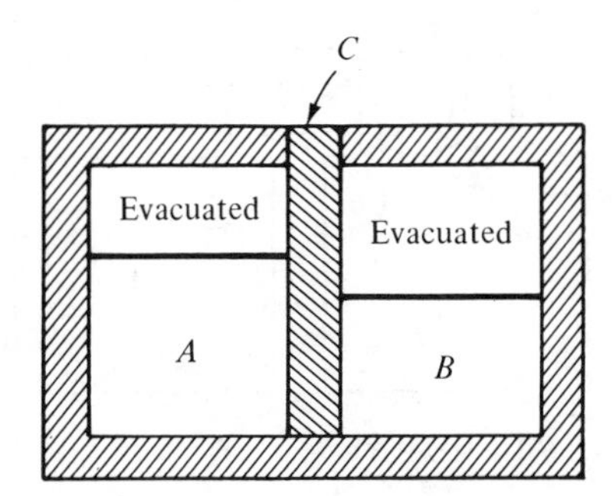

FIGURE P5-5

5-6 Maneuvering of an orbiting space capsule is accomplished by use of small thruster rockets mounted on the side of the capsule. In a typical design liquid hydrogen and oxygen are fed to a combustion chamber, where they combine to form hot steam. The steam is then expanded in a nozzle and produces the required thrust. A steam rocket of this type, in which the steam enters the nozzle at 700 psia and 800°F, is being designed. During ground tests the nozzle will discharge its exhaust steam at 15 psia.

(a) What is the exit velocity of the steam?

(b) If the nozzle has an exit cross-sectional area of 1 $in.^2$, what is the volumetric flow of steam into the atmosphere (in ft^3/sec)?

(c) What is the mass flow rate of steam through the nozzle (in lb_m/sec)?

You may make the following assumptions:

(1) The flow of steam through the nozzle is adiabatic and reversible (this is confirmed by experiment).

(2) The velocity of the steam entering the nozzle is negligible in comparison to the velocity of the exiting steam.

(3) The flow within the nozzle is one-dimensional, so you need use only *one* velocity component in the energy balance.

5-7 Cylinder 1 (Fig. P5-7) contains 10 lb_m of steam at 345 psia and 800°F. Cylinders 1, 3, and 4 are equal in volume. Cylinder 2 has 2 ft^2 of cross-sectional area. Each pipe has a volume of 0.5 ft^3. All the cylinders and pipes are completely insulated and frictionless. The piston in cylinder 2 weighs 500 lb. Cylinder 3 and pipes *A* and *B* are completely evacuated. Cylinder 4 contains steam at 14.7 psia and 300°F.

(a) Open valve *A*, leaving *B* and *C* closed. (1) How far up does the piston rise when equilibrium is attained? (2) What are the conditions of the steam in cylinders 1 and 2 if the valve *A* is shut when the piston is 1 ft above the equilibrium position?

(b) Open valve *B*, leaving *A* and *C* closed. What are the final conditions of the steam?

(c) Open valve *C*, leaving *A* and *B* closed. What are the final conditions of the steam?

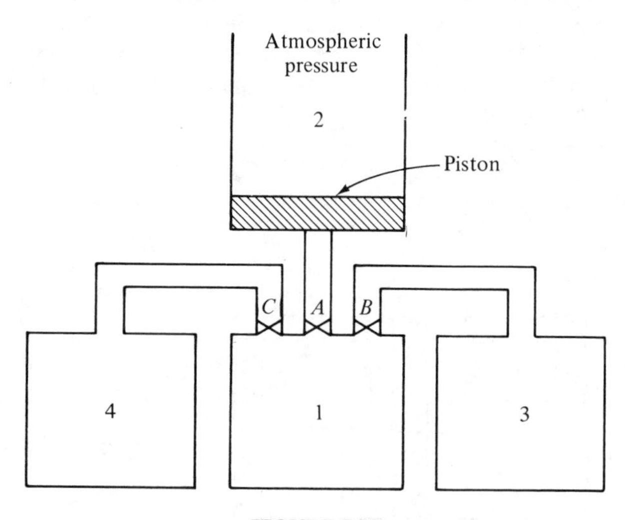

FIGURE P5-7

5-8 Airplanes are launched from aircraft carriers by means of a steam catapult as shown in Fig. P5-8. The catapult is a well-insulated cylinder that contains steam and is fitted with a frictionless piston. The piston is connected to the airplane by a cable. As the steam expands, the movement of the piston causes movement of the plane. A

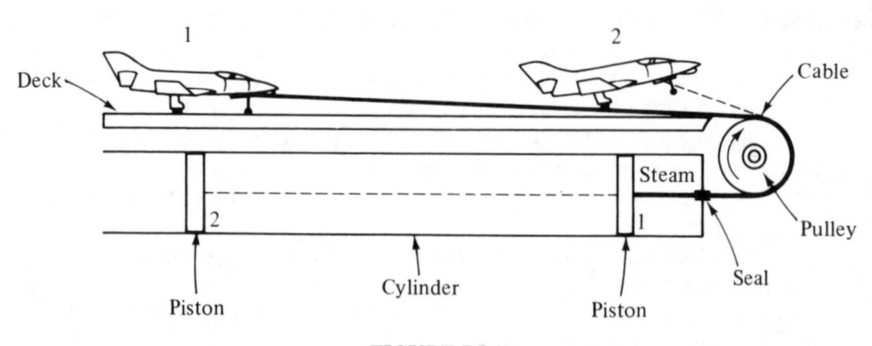

FIGURE P5-8

catapult design calls for 600 lb_m of steam at 2000 psia and 800°F to be expanded to 50 psia. Will this catapult be adequate to accelerate a 30-ton fighter aircraft from rest to 200 mph? Neglect the mass of the piston head and connecting cables as well as the thrust produced by the plane's engines.

5-9 Steam flows through a 1 in. pipe. At the inlet of the pipe the steam is at 240 psia and 600°F and the velocity is 80 ft/sec. At the outlet of the pipe, the pressure is 60 psia. Heat losses along the pipe are estimated to be 50 Btu/lb_m of steam flowing. At the outlet of the pipe the steam enters a well-insulated, reversible nozzle. The steam leaves the nozzle saturated at atmospheric pressure.

(a) Determine the conditions of the steam entering the nozzle.

(b) What is the velocity of the steam leaving the nozzle?

Use the same assumptions as in Problem 5-6 regarding flow through the nozzle.

5-10 In a certain chemical plant a 40 per cent solution of organic salts is being evaporated to an 80 per cent solution (all percentages on a mass basis). The organic salts cause only a very small boiling-point rise, which can be considered negligible. The condenser operates in such a manner that the pressure in the vapor line to the condenser is equivalent to a temperature of 130°F. At the present time the only steam available for this operation is that which comes directly from the boiler. Since the tubes in the evaporator cannot withstand high pressure, the boiler steam is throttled to 10 psig before entering the evaporator. A manufacturer of turbo expander–compressors suggests improving the steam economy by installing one of his machines, shown by the dashed lines in Fig. P5-10. Per pound of feed solution handled, calculate the saving in boiler steam which might be made if the turbo expander–compressor operates with an efficiency of 80 per cent (based on adiabatic reversible operation) at both ends.

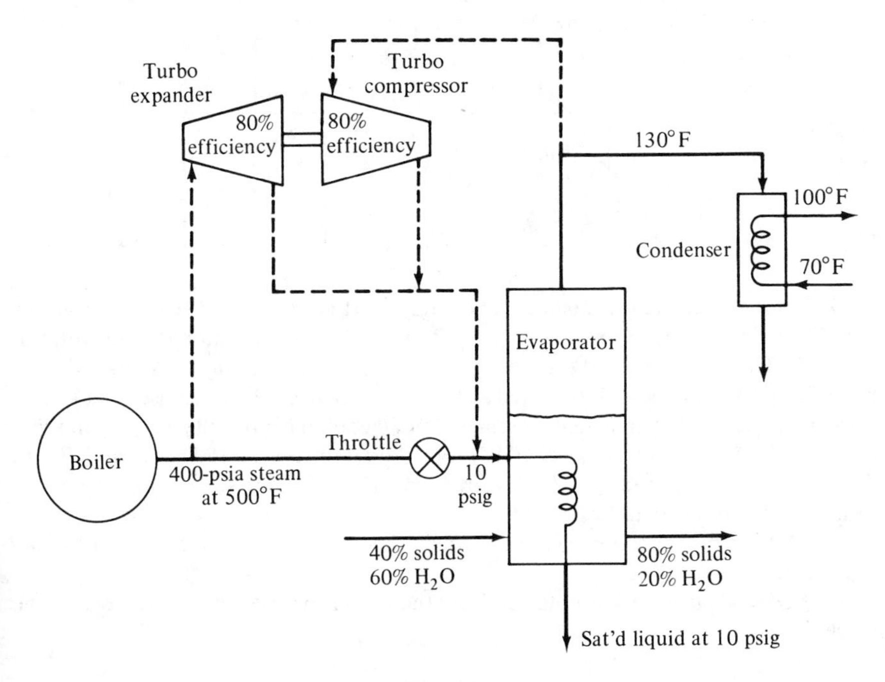

FIGURE P5-10

5-11 Low-pressure steam at 100 psia and 400°F is to be used to fill an insulated tank with a movable piston whose resisting force is equivalent to 500 psia. An adiabatic compressor is used to boost the steam pressure from 100 to 500 psia, as shown in Fig. P5-11. If the tank is initially empty and the compressor operates at 70 percent efficiency, what will be the final pressure in the tank when its volume is 10 ft^3, and how much work will have been supplied to the compressor?

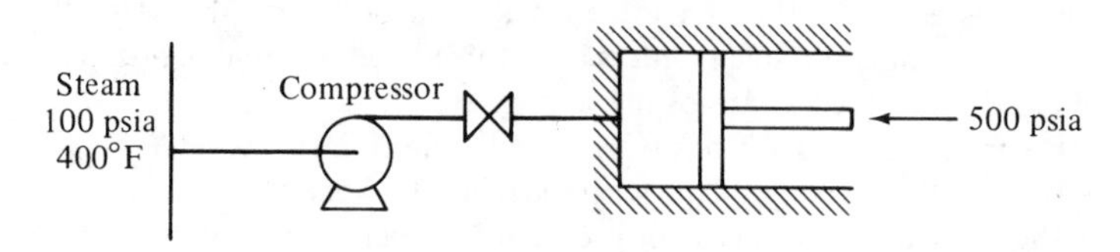

FIGURE P5-11

5-12 It is desired to determine the unknown volume of a tank (tank B) in a complex chemical process (Fig. P5-12). It is not possible either to measure directly the volume of the tank or to weigh its contents. However, tank B may be evacuated and then connected to a second tank (tank A) of known volume which contains steam at known P and T. (The lines and valve connecting the two tanks are assumed to have negligible volume and to be perfectly insulated.) The valve in the line connecting the two tanks is opened, and the pressures in the tanks are allowed to equilibrate rapidly. The valve is then closed (before any heat transfer can occur) and the pressure in one of the tanks is measured. Explain how you would determine the volume of tank B.

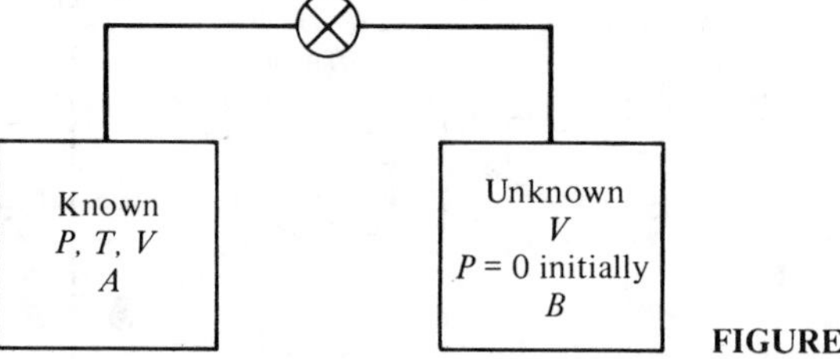

FIGURE P5-12

5-13 Water is circulated through a nuclear reactor (Fig. P5-13) and leaves at 1500 psia and 500°F (as *liquid*). The water is passed into a flash evaporator that operates at 300 psia. The vapor from the evaporator is used to drive a turbine (adiabatic and reversible). The condenser is operated at 1 psia. Water from the condenser and evaporator are mixed and recirculated. (Assume that all equipment is well insulated and that pressure drops in lines are negligible.) For liquid water at 1500 psia and 500°F, $\underline{H} = 487.53$ Btu/lb_m and $\underline{S} = 0.68515$ Btu/lb_m °F. Reference state: $\underline{H} = \underline{S} = 0$ for saturated liquid water at 32°F.

(a) How much vapor (*in pounds*) enters the turbine per pound of water leaving the reactor?

(b) How much work is obtained from the turbine per pound of water leaving the reactor?

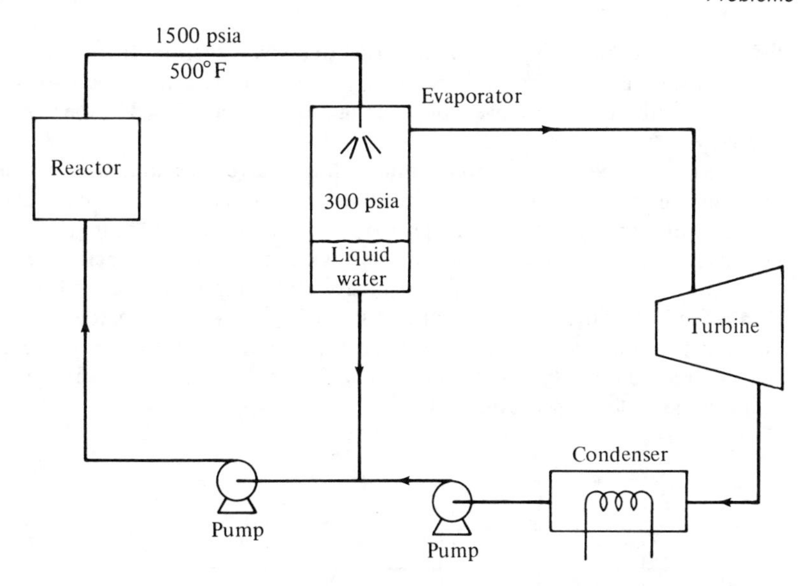

FIGURE P5-13

5-14 A steam turbine is being tested after installation in a refinery steam plant. Steam at 1000 psia is generated in a boiler and fed to the turbine. Just before the steam enters the turbine, a small portion is bled through a well-insulated throttling valve and expanded to 20.0 psia. The temperature of the throttled steam is found to be 740°F. The unthrottled portion of the steam is passed into the turbine, where it is adiabatically expanded to 20 psia. Tests on the turbine exhaust vapor indicate that it is a saturated vapor.

(a) What is the temperature of the steam entering the turbine?
(b) How much work is recovered per lb_m of steam expanded in the turbine?
(c) What is the efficiency of the turbine?

5-15 An insulated cylinder is fitted with a freely floating piston and contains 1 lb_m of steam at 120 psia and 90 percent quality; the space above the piston contains air to maintain the pressure on the steam (Fig. P5-15). Additional air is forced into the upper chamber, forcing the piston down and increasing the steam pressure until the steam has 100 percent quality.

(a) Determine the steam pressure at 100 percent quality.
(b) How much work must be done on the steam during the compression?

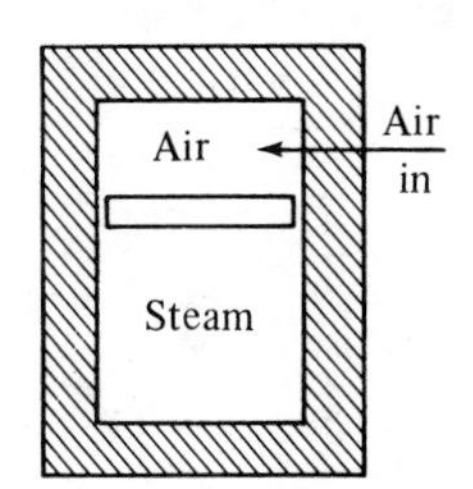

FIGURE P5-15

5-16 The cycle shown in Fig. P5-16 is used to convert thermal energy (in the form of heat transferred to the boiler–superheater) into work. It is assumed that the turbine operates adiabatically and reversibly and that the work consumed by the pump is small, so that $\underline{H}_e \approx \underline{H}_d$.

(a) The cycle is meant to operate with the valve *open* so that all the properties at point a equal those at b. (1) Under these conditions, what is the circulation rate (lb_m/sec) of water in the cycle to provide the required power output in the turbine (10,000 hp)? (2) At what rate is heat to be provided in the boiler–superheater?

(b) The valve between a and b is inadvertently left partially closed and the steam adiabatically throttled to a pressure of 120 psia at b. (The properties at a, d, and e are assumed to be unchanged.) (1) What is the water circulation rate needed to provide the required power output in the turbine (10,000 hp)? (2) At what rate is heat to be provided in the boiler–superheater?

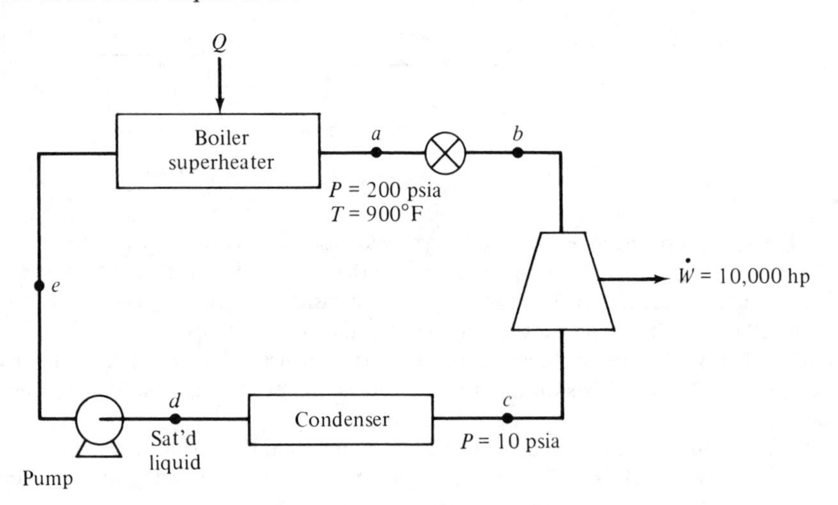

FIGURE P5-16

The Property Relations and the Mathematics of Properties

6

6.1 The Property Relations

In previous sections we have developed the single-component energy and entropy balances as follows:

Energy:

$$\left(\underline{H} + \frac{u^2}{2g_c} + \frac{gZ}{g_c}\right)_{\text{in}} \delta M_{\text{in}} - \left(\underline{H} + \frac{u^2}{2g_c} + \frac{gZ}{g_c}\right)_{\text{out}} \delta M_{\text{out}} + \delta Q - \delta W = d\left[M\left(\underline{U} + \frac{u^2}{2g_c} + \frac{gZ}{g_c}\right)\right]_{\text{syst}} \tag{6-1}$$

Entropy:

$$(\underline{S}\,\delta M)_{\text{in}} - (\underline{S}\,\delta M)_{\text{out}} + \int_{\text{surface area}} \frac{\delta q\, dA}{T} + \int_{\text{volume}} \frac{\delta lw\, dV}{T} = d(M\underline{S})_{\text{syst}} \tag{6-2}$$

These expressions may be applied to both open and closed systems through proper use of the δM_{in} and δM_{out} terms. However, these equations are frequently inconvenient to use to evaluate changes in the state variables $\underline{H}$, $\underline{U}$, or $\underline{S}$. By combining these equations in such a way as to eliminate the terms that are path functions (δQ, δW, and δlw) we shall describe the changes in state properties of a differential mass as it undergoes a given process in terms of other state functions, such as P, $\underline{V}$, T, C_P, and C_V. The differential mass is a closed system so that the energy and entropy equations may be written

Energy:

$$\delta Q - \delta W = d\left[M\left(\underline{U} + \frac{u^2}{2g_c} + \frac{gZ}{g_c}\right)\right]_{\text{syst}} \tag{6-3}$$

Entropy:

$$\int_{\text{surface area}} \frac{\delta q\, dA}{T} + \int_{\text{volume}} \frac{\delta lw\, dV}{T} = dS_{\text{syst}} \tag{6-4}$$

For simplicity let us assume that $u^2/2g_C$ and gZ/g_C are negligible in relation to $\underline{U}$ and therefore may be neglected. (These assumptions are not critical, and these terms can easily be included in everything that follows.) Thus the energy balance reduces to the familiar closed-system form

$$\delta Q - \delta W = dU \tag{6-5}$$

As the state of the system changes, it may transfer work to its surroundings by expansion. Let us choose our system boundaries such that all work which is transferred to the surroundings is transferred as expansion, or $P\,dV$, work; that is, assume that no shaft crosses the boundaries of the system as it flows through the process. (This in no way says that the gas cannot expand against a shaft and do work, but only that the system boundaries be chosen to exclude the shaft from entering the system.) Thus we say that the work transferred to the surroundings is given by

$$\delta W = P_{\text{res}}\, dV_{\text{syst}} \tag{6-6}$$

Now suppose that due to the effect of friction, the pressure within the system is different from that outside the system. As an example, let us take the piston and the gas confined within a cylinder as our system. Assume that the gas is expanding against the forces of friction and the surroundings as illustrated in Fig. 6-1. P_{syst} is the pressure within the system, P_{res} the effective resisting pressure

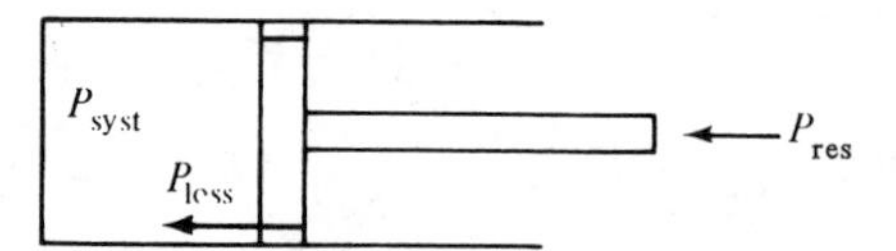

FIG. 6-1. Effect of friction.

of the surroundings on the piston, and P_{loss} the effective pressure loss due to friction which equals (frictional force)/(piston area). P_{loss} is assumed to have a plus sign when in the direction of P_{res}.

If the piston is massless, or not accelerating, a simple force balance gives

$$P_{\text{syst}} = P_{\text{res}} + P_{\text{loss}} \tag{6-7}$$

which may be substituted into equation (6-6) for δW to give

$$\delta W = (P_{\text{syst}} - P_{\text{loss}})\, dV_{\text{syst}} \tag{6-8}$$

Since we are considering a differential mass of fluid as our system, we may assume that the temperature is uniform across the system so that equation (6-4) may be rearranged to give

$$\int_{\text{area}} \delta q\, dA + \int_{\text{volume}} \delta lw\, dV = T_{\text{syst}}\, dS_{\text{syst}} \tag{6-9}$$

or

$$\delta Q + \delta LW = T_{\text{syst}}\, dS_{\text{syst}} \tag{6-10}$$

But as we have shown in Chapter 5,

$$\delta LW = P_{\text{loss}}\, dV_{\text{syst}} \tag{6-11}$$

which may be substituted into equation (6-10) to yield

$$\delta Q = TdS - P_{\text{loss}}\, dV_{\text{syst}} \tag{6-12}$$

Substitution of equations (6-8) and (6-12) into (6-5) for the work and heat terms, respectively, gives

$$[T_{\text{syst}}\, dS_{\text{syst}} - P_{\text{loss}}\, dV_{\text{syst}}] - [P_{\text{syst}} - P_{\text{loss}}]\, dV_{\text{syst}} = dU_{\text{syst}} \tag{6-13}$$

or dividing by M, the system mass, and canceling the two $P_{\text{loss}}\, dV_{\text{syst}}$ terms, we obtain

$$d\underline{U}_{\text{syst}} = T_{\text{syst}}\, d\underline{S}_{\text{syst}} - P_{\text{syst}}\, d\underline{V}_{\text{syst}} \tag{6-14}$$

Equation (6-14) is the *fundamental property relationship* and represents the third of the three basic equations of thermodynamics. (The energy and entropy balances are the other two.) Its great value lies in the fact that it relates only state functions to each other. For this reason, it is an exact differential equation and has properties that we shall find extremely useful in our work. It is left as an exercise to attempt this analysis for the case where the piston is not part of the system. (What happens to Q in this case?)

If instead of assuming that the system does work on its surroundings we assumed that the surroundings did work on the system, our development would have been identical. Since P_{loss} is positive when in the direction of P_{res}, we would have simply found that P_{loss} was a negative number. Equations (6-8) through (6-14) would not, however, have been affected in the least!

Although we have used the simple piston process to help us visualize the manner in which the lost-work term, $P_{\text{loss}}\, dV_{\text{syst}}$, is eliminated by combining the energy and entropy process, this elimination can be demonstrated for any process in general. Therefore, the property relation, equation (6-14), is itself a general equation which applies for any process in which the only forms of energy change are heat transfer and the transfer of *expansion* work to or from the surroundings. In our later work on the thermodynamics of chemical equilibria, we will see how equation (6-14) may be expanded to take into account the energy changes associated with changes in chemical composition. However, at this point we shall restrict ourselves to single-component, or constant-composition, processes. Similarly, electrical and other forms of work could have been considered and would have led to other work terms similar in form to the $P\, dV$ term in this abbreviated form of the fundamental property relationship.

6.2 The Convenience Functions and Their Property Relations

Although it is entirely possible to express all thermodynamic relations in terms of the fundamental properties $\underline{U}$, $\underline{V}$, P, $\underline{S}$, and T, we find that certain combinations of these fundamental properties occur so frequently that it is to our advantage to define these groupings as new thermodynamic functions. Since these functions are combinations of state properties, they must also be functions of state themselves.

Since these properties are not fundamental to the study of thermodynamics but only introduced for our convenience, they have been given the name *convenience functions*. We have already introduced the first of the convenience functions, the enthalpy, H. It has been defined as

$$H = U + PV \tag{6-15}$$

We now add two more convenience functions: the *Helmholtz free energy*, A, and the *Gibbs free energy*, or *Gibbs free enthalpy*, G. These are defined as

$$A = U - TS \tag{6-16}$$

$$G = H - TS \tag{6-17}$$

Equations (6-15) through (6-17) may be divided by total mass to yield their specific counterparts:

$$\underline{H} = \underline{U} + P\underline{V} \tag{6-15a}$$

$$\underline{A} = \underline{U} - T\underline{S} \tag{6-16a}$$

$$\underline{G} = \underline{H} - T\underline{S} \tag{6-17a}$$

We may derive property relations for H, A, and G similar to equation (6-14) as follows. Differentiate equation (6-15) to give

$$dH = dU + P\,dV + V\,dP \tag{6-18}$$

Substitute equation (6-14) for dU to give

$$dH = T\,dS + V\,dP \tag{6-19}$$

Similarly, differentiation of equations (6-16) and (6-17) followed by elimination of the dU and dH terms yields

$$dA = -S\,dT - P dV \tag{6-20}$$

$$dG = -S\,dT + V\,dP \tag{6-21}$$

which along with the relations for dH and dU,

$$dH = T\,dS + V\,dP \tag{6-19}$$

$$dU = T\,dS - P\,dV \tag{6-14}$$

are the desired property relations for U, H, A, and G.

Mass independent forms of the property relations may be obtained by dividing equations (6-14), (6-19), (6-20), and (6-21) by the total mass to obtain

$$d\underline{U} = T\,d\underline{S} - P\,d\underline{V} \tag{6-14a}$$

$$d\underline{H} = T\,d\underline{S} + \underline{V}\,dP \tag{6-19a}$$

$$d\underline{A} = -\underline{S}\,dT - P\,d\underline{V} \tag{6-20a}$$

$$d\underline{G} = -\underline{S}\,dT + \underline{V}\,dP \tag{6-21a}$$

These property relations will be of fundamental importance to us in the development of the mathematics of property changes. Since U, H, A, and G are functions of state, equations (6-14) and (6-19) to (6-21) are exact differentials and therefore are susceptible to mathematical manipulations that yield still other useful relations among properties.

6.3 The Maxwell Relations

We begin our discussion and derivation of the Maxwell equations by first discussing some of the useful mathematical properties of the state variables. We have indicated in earlier chapters that any of the intensive state variables, T, $\underline{U}$, P, $\underline{S}$, $\underline{H}$, $\underline{V}$, $\underline{G}$, and $\underline{A}$, can be expressed in terms of any two other state variables. In mathematical terminology we express this fact through an equation of the form

$$F = F(A, B) \tag{6-22}$$

where F, A, and B can be any three intensive state properties and $F(A, B)$ is the functional relation among the three properties. If F, A, and B were, for example, P, $\underline{V}$, and T, then $F(A, B)$ would represent an equation of state.

Since F can be expressed solely as a function of A and B, the laws of calculus tell us that

$$dF = \left[\frac{\partial F(A, B)}{\partial A}\right]_B dA + \left[\frac{\partial F(A, B)}{\partial B}\right]_A dB \tag{6-23}$$

or dropping the (A, B) for convenience,

$$dF = \left(\frac{\partial F}{\partial A}\right)_B dA + \left(\frac{\partial F}{\partial B}\right)_A dB \tag{6-23a}$$

Since F is an exact function of A and B, equations (6-23) and (6-23a) are exact differentials. For purposes of simplicity, we now express the partial derivatives in equation (6-23a) as X and Y according to the following relationships:

$$X = \left(\frac{\partial F}{\partial A}\right)_B \qquad Y = \left(\frac{\partial F}{\partial B}\right)_A \tag{6-24}$$

Then

$$dF = X\,dA + Y\,dB \tag{6-25}$$

Conversely, if an equation of the form of equation (6-25) is encountered, X and Y must be the partial derivatives indicated in equation (6-24). For example, consider the $d\underline{U}$ property relation:

$$d\underline{U} = T\,d\underline{S} - P\,d\underline{V}$$

We see that T and P are analogous to X and Y in equation (6-25) and thus must be expressible as

$$\begin{aligned} T &= \left(\frac{\partial \underline{U}}{\partial \underline{S}}\right)_{\underline{V}} \\ P &= -\left(\frac{\partial \underline{U}}{\partial \underline{V}}\right)_{\underline{S}} \end{aligned} \tag{6-26}$$

Thus we see that the thermodynamic temperature, introduced in Section 3-6, is, in fact, the absolute temperature defined in Chapter 5 to relate δQ and dS.

Similar expressions for T may be derived from the equation for dH, and for P from dA. Thus we see that T and P themselves are not fundamental thermodynamic properties, but can be considered as convenient representations for the derivatives of the state variables $\underline{U}$, $\underline{S}$, and $\underline{V}$.

However, exceedingly more useful relations than equation (6-26) can be derived from (6-25). If we cross differentiate the variables in equation (6-24), we obtain

$$\begin{aligned} \left(\frac{\partial X}{\partial B}\right)_A &= \left[\frac{\partial}{\partial B}\left(\frac{\partial F}{\partial A}\right)_B\right]_A \\ \left(\frac{\partial Y}{\partial A}\right)_B &= \left[\frac{\partial}{\partial A}\left(\frac{\partial F}{\partial B}\right)_A\right]_B \end{aligned} \tag{6-27}$$

However, the rules of calculus tell us that the order of differentiation of a state function cannot change the value of the derivatives. That is,

$$\left[\frac{\partial}{\partial B}\left(\frac{\partial F}{\partial A}\right)_B\right]_A \equiv \left[\frac{\partial}{\partial A}\left(\frac{\partial F}{\partial B}\right)_A\right]_B$$

Therefore, we reach the conclusion that

$$\left(\frac{\partial X}{\partial B}\right)_A = \left(\frac{\partial Y}{\partial A}\right)_B \tag{6-28}$$

Thus in the $d\underline{U}$ property relation if we let $F = \underline{U}$, $X = T$, $Y = -P$, and A and B equal $\underline{S}$ and $\underline{V}$, respectively, equation (6-28) yields

$$-\left(\frac{\partial T}{\partial \underline{V}}\right)_{\underline{S}} = \left(\frac{\partial P}{\partial \underline{S}}\right)_{\underline{V}} \qquad \text{(from } d\underline{U}\text{)} \tag{6-29}$$

Similar expressions derived from the $d\underline{H}$, $d\underline{A}$, and $d\underline{G}$ property relations yield (it is left to the student as an exercise to derive the relations)

$$-\left(\frac{\partial \underline{S}}{\partial P}\right)_T = \left(\frac{\partial \underline{V}}{\partial T}\right)_P \qquad \text{(from } d\underline{G}\text{)} \tag{6-30}$$

$$\left(\frac{\partial T}{\partial P}\right)_{\underline{S}} = \left(\frac{\partial \underline{V}}{\partial \underline{S}}\right)_P \qquad \text{(from } d\underline{H}\text{)} \tag{6-31}$$

$$\left(\frac{\partial \underline{S}}{\partial \underline{V}}\right)_T = \left(\frac{\partial P}{\partial T}\right)_{\underline{V}} \qquad \text{(from } d\underline{A}\text{)} \tag{6-32}$$

Equations (6-29) to (6-32) and their reciprocal relations,

$$\left(\frac{\partial \underline{V}}{\partial T}\right)_{\underline{S}} = -\left(\frac{\partial \underline{S}}{\partial P}\right)_{\underline{V}} \tag{6-33}$$

$$-\left(\frac{\partial P}{\partial \underline{S}}\right)_T = \left(\frac{\partial T}{\partial \underline{V}}\right)_P \tag{6-34}$$

$$\left(\frac{\partial P}{\partial T}\right)_{\underline{S}} = \left(\frac{\partial \underline{S}}{\partial \underline{V}}\right)_P \tag{6-35}$$

$$\left(\frac{\partial \underline{V}}{\partial \underline{S}}\right)_T = \left(\frac{\partial T}{\partial P}\right)_{\underline{V}} \tag{6-36}$$

are termed the *Maxwell equations*. The utility of these equations will become evident when we attempt to derive expressions for the evaluation of property changes in terms of the easily measured variables P, T, $\underline{V}$, C_P, and C_V, rather than the unwieldy and nonmeasurable quantity entropy. (Mass dependent forms of the Maxwell equations can be found by multiplying by the total mass, and introducing the total properties.)

6.4 Mathematics of Property Changes

With the derivation of the Maxwell relations we are now ready to consider other property changes. In the following sections we shall establish formulas that allow us to calculate changes in $\underline{U}$, $\underline{H}$, $\underline{G}$, and $\underline{A}$ solely in terms of P, $\underline{V}$, T, C_V, and C_P—variables that can be controlled or measured.

We begin our discussion by developing an expression for $d\underline{U}$ in terms of P, $\underline{V}$, T, C_V, and C_P. The property relation tells us that

$$d\underline{U} = T\,d\underline{S} - P\,d\underline{V} \tag{6-14a}$$

However, the occurrence of entropy in the property relation makes actual evaluation of $d\underline{U}$ from this equation difficult, because we have no direct means of measuring entropies.

As we showed previously, the property relation is a specific example of the more general form

$$d\underline{U} = \left(\frac{\partial \underline{U}}{\partial \underline{S}}\right)_{\underline{V}} d\underline{S} + \left(\frac{\partial \underline{U}}{\partial \underline{V}}\right)_{\underline{S}} d\underline{V} \tag{6-37}$$

which implies that $\underline{U}$ is an exact function of $\underline{S}$ and $\underline{V}$, or

$$\underline{U} = \underline{U}(\underline{S}, \underline{V}) \tag{6-38}$$

However, we also know that $\underline{U}$ can be expressed as an exact function of any two intensive state properties. Therefore, consider

$$\begin{aligned} \underline{U} &= \underline{U}(P, \underline{V}) \\ \underline{U} &= \underline{U}(T, \underline{V}) \\ \underline{U} &= \underline{U}(T, P) \\ \underline{U} &= \underline{U}(\underline{S}, \underline{A}) \end{aligned} \tag{6-39}$$

or any other combination of properties. Since we really wish to express $\underline{U}$ as a function of P, $\underline{V}$, and T, it appears that one of the three first alternatives is most likely to yield the desired result. However, experience has shown that one of these alternatives will yield a formulation for $d\underline{U}$ which is easier to evaluate than the other two; this does not mean that correct formulas cannot be developed by starting with the other two alternatives, only that these formulas will not be as useful to us as those derived from the best alternative.

As will be seen shortly, the best form of equation (6-39) is

$$\underline{U} = \underline{U}(T, \underline{V}) \tag{6-39}$$

Thus we may write

$$d\underline{U} = \left(\frac{\partial \underline{U}}{\partial T}\right)_{\underline{V}} dT + \left(\frac{\partial \underline{U}}{\partial \underline{V}}\right)_T d\underline{V} \tag{6-40}$$

However, the heat capacity at constant volume, C_V, has been shown to be expressible as

$$C_V = \left(\frac{\partial \underline{U}}{\partial T}\right)_{\underline{V}}$$

which may be substituted into equation (6-40) to give

$$d\underline{U} = C_V\, dT + \left(\frac{\partial \underline{U}}{\partial \underline{V}}\right)_T d\underline{V} \tag{6-40a}$$

The $(\partial \underline{U}/\partial \underline{V})_T$ term in equation (6-40a) must be derived from the $d\underline{U}$ property relation:

$$d\underline{U} = T\, d\underline{S} - P\, d\underline{V} \tag{6-14a}$$

Division of equation (6-14a) by $d\underline{V}$ at constant T then gives us

$$\left(\frac{\partial \underline{U}}{\partial \underline{V}}\right)_T = T\left(\frac{\partial \underline{S}}{\partial \underline{V}}\right)_T - P \tag{6-41}$$

Since we wish to replace all terms involving entropy with terms involving only P, $\underline{V}$, or T, we use a Maxwell relationship to replace $(\partial \underline{S}/\partial \underline{V})_T$. From equation (6-32) we observe that

$$\left(\frac{\partial \underline{S}}{\partial \underline{V}}\right)_T = \left(\frac{\partial P}{\partial T}\right)_{\underline{V}} \tag{6-42}$$

Therefore equation (6-41) becomes

$$\left(\frac{\partial \underline{U}}{\partial \underline{V}}\right)_T = T\left(\frac{\partial P}{\partial T}\right)_{\underline{V}} - P \tag{6-43}$$

Equation (6-43) is substituted into (6-40) to give

$$d\underline{U} = C_V\, dT + \left[T\left(\frac{\partial P}{\partial T}\right)_{\underline{V}} - P\right] d\underline{V} \tag{6-44}$$

Equation (6-44) is the desired relation for $d\underline{U}$ in terms of the directly measurable properties, C_V, P, T, and $\underline{V}$.

Had we attempted to determine $d\underline{U}$ in terms of dP and dT, or dP and $d\underline{V}$, rather than $d\underline{V}$ and dT, we would have found that the expression for $d\underline{U}$ was more complicated than equation (6-44) and therefore more difficult to use. Since in the long run we shall really want to integrate equation (6-44) to obtain $\Delta\underline{U}$, any of the three expressions for $d\underline{U}$ could be used (knowledge of any two of the three properties P, $\underline{V}$, or T fixes the third through the equation of state), but clearly the simplest expression is the most desirable one.

SAMPLE PROBLEM 6-1. Express $d\underline{H}$ as a function of C_P, C_V, P, $\underline{V}$, and T.

Solution: As with $d\underline{U}$, we can express $d\underline{H}$ in terms of dT and dP or $d\underline{V}$, or $d\underline{V}$ and dP, but experience tells us that dT and dP will yield the simplest expression. The property relation for $d\underline{H}$ is

$$d\underline{H} = Td\underline{S} + \underline{V}dP$$

Assume that

$$\underline{H} = \underline{H}(T, P)$$

The total differential for enthalpy may be expressed as

$$d\underline{H} = \left(\frac{\partial \underline{H}}{\partial T}\right)_P dT + \left(\frac{\partial \underline{H}}{\partial P}\right)_T dP$$

But

$$\left(\frac{\partial \underline{H}}{\partial T}\right)_P = C_P$$

Therefore

$$d\underline{H} = C_P\,dT + \left(\frac{\partial \underline{H}}{\partial P}\right)_T dP$$

We can determine $(\partial \underline{H}/\partial P)_T$ from division of the $d\underline{H}$ property relation by dP at constant T to yield

$$\left(\frac{\partial \underline{H}}{\partial P}\right)_T = T\left(\frac{\partial \underline{S}}{\partial P}\right)_T + \underline{V}$$

From the Maxwell relations we obtain

$$\left(\frac{\partial \underline{S}}{\partial P}\right)_T = -\left(\frac{\partial \underline{V}}{\partial T}\right)_P$$

which is substituted into the last equation to yield

$$\left(\frac{\partial \underline{H}}{\partial P}\right)_T = -T\left(\frac{\partial \underline{V}}{\partial T}\right)_P + \underline{V}$$

The desired expression for $d\underline{H}$ then becomes

$$d\underline{H} = C_P\,dT + \left[\underline{V} - T\left(\frac{\partial \underline{V}}{\partial T}\right)_P\right] dP$$

Evaluation of Entropy Changes

We now turn our attention to developing expressions for changes in entropy in terms of C_P, C_V, P, $\underline{V}$, and T. Once these expressions are found, determinations of changes in $\underline{G}$ and $\underline{A}$ can be calculated from the corresponding changes in $\underline{H}$, $\underline{U}$, and $\underline{S}$.

Let us find an expression for $d\underline{S}$ in terms of dT and $d\underline{V}$. Let

$$\underline{S} = \underline{S}(T, \underline{V}) \tag{6-45}$$

Therefore,

$$d\underline{S} = \left(\frac{\partial \underline{S}}{\partial T}\right)_{\underline{V}} dT + \left(\frac{\partial \underline{S}}{\partial \underline{V}}\right)_T d\underline{V} \tag{6-46}$$

Since we have derivatives of entropy we first check the Maxwell relations to see if either of the terms can be eliminated. We find that $(\partial \underline{S}/\partial \underline{V})_T = (\partial P/\partial T)_{\underline{V}}$ but can find no expression for $(\partial \underline{S}/\partial T)_{\underline{V}}$. Therefore we must still find a way of evaluating this term. Let us examine the internal-energy property relation:

$$d\underline{U} = T\,d\underline{S} - P\,d\underline{V} \tag{6-14a}$$

Division of this expression by dT at constant $\underline{V}$ yields

$$\left(\frac{\partial \underline{U}}{\partial T}\right)_{\underline{V}} = T\left(\frac{\partial \underline{S}}{\partial T}\right)_{\underline{V}} \tag{6-47}$$

But $(\partial \underline{U}/\partial T)_{\underline{V}} = C_V$. Therefore, we find that

$$\left(\frac{\partial \underline{S}}{\partial T}\right)_{\underline{V}} = \frac{C_V}{T} \tag{6-48}$$

which is the necessary expression for $(\partial \underline{S}/\partial T)_{\underline{V}}$. Substitution of equation (6-48) and the Maxwell relation for $(\partial \underline{S}/\partial \underline{V})_T$ into equation (6-46) yields the desired relation for dS in terms of $d\underline{V}$ and dT:

$$d\underline{S} = \frac{C_V}{T}\,dT - \left(\frac{\partial P}{\partial T}\right)_{\underline{V}} dV \tag{6-49}$$

SAMPLE PROBLEM 6-2. Find $d\underline{S}$ in terms of dT and dP.

Solution: Let $\underline{S} = \underline{S}(T, P)$. Therefore,

$$d\underline{S} = \left(\frac{\partial \underline{S}}{\partial T}\right)_P dT + \left(\frac{\partial \underline{S}}{\partial P}\right)_T dP$$

but from the Maxwell relation,

$$\left(\frac{\partial \underline{S}}{\partial P}\right)_T = -\left(\frac{\partial \underline{V}}{\partial T}\right)_P$$

so

$$d\underline{S} = \left(\frac{\partial \underline{S}}{\partial T}\right)_P dT - \left(\frac{\partial \underline{V}}{\partial T}\right)_P dP$$

We evaluate $(\partial \underline{S}/\partial T)_P$ from the enthalpy property relation:

$$d\underline{H} = T\,d\underline{S} + \underline{V}\,dP$$

by dividing by dT at constant P. That is,

$$\left(\frac{\partial \underline{H}}{\partial T}\right)_P = T\left(\frac{\partial \underline{S}}{\partial T}\right)_P$$

But

$$\left(\frac{\partial \underline{H}}{\partial T}\right)_P = C_P$$

Therefore,

$$\left(\frac{\partial \underline{S}}{\partial T}\right)_P = \frac{1}{T}C_P$$

so

$$d\underline{S} = \frac{C_P}{T}dT - \left(\frac{\partial \underline{V}}{\partial T}\right)_P dP$$

which is the desired relation for $d\underline{S}$ in terms of dP and dT.

It is interesting to note that the equations previously derived for $d\underline{U}$ and

$d\underline{H}$ in terms of dT and $d\underline{V}$, and dT and dP, respectively, may be easily derived from their respective property and entropy relations as follows:

$$d\underline{U} = T\,d\underline{S} - P\,d\underline{V} \tag{6-14a}$$

But

$$d\underline{S} = \frac{C_V}{T}\,dT + \left(\frac{\partial P}{\partial T}\right)_{\underline{V}} d\underline{V} \tag{6-49}$$

Therefore,

$$d\underline{U} = C_V\,dT + \left[T\left(\frac{\partial P}{\partial T}\right)_{\underline{V}} - P\right] d\underline{V} \tag{6-50}$$

Similarly for $d\underline{H}$,

$$d\underline{H} = T\,d\underline{S} + \underline{V}\,dP \tag{6-51}$$

But

$$d\underline{S} = \frac{C_P}{T}\,dT - \left(\frac{\partial \underline{V}}{\partial T}\right)_P dP \tag{6-52}$$

Therefore,

$$d\underline{H} = C_P\,dT + \left[\underline{V} - T\left(\frac{\partial \underline{V}}{\partial T}\right)_P\right] dP \tag{6-53}$$

which are the relations we previously derived for $d\underline{U}$ and $d\underline{H}$.

6.5 Other Useful Expressions

In addition to the expressions for $d\underline{U}$, $d\underline{H}$, and $d\underline{S}$ previously developed, we shall find use for certain other relations which may be derived using techniques similar to those used for $d\underline{H}$, $d\underline{U}$, and $d\underline{S}$.

Joule–Thomson Coefficient

We have previously noted that a fluid which undergoes an adiabatic throttling process at steady state, while experiencing negligible changes in potential energy and kinetic energy, has undergone an isenthalpic (constant-enthalpy) process. Very often this isenthalpic throttling process is used to reduce the temperature of the fluid as part of a refrigeration process. Therefore, it is frequently desirable to know the degree of temperature reduction that can be obtained during the expansion. A quantitative measure of this temperature change is expressed by the *Joule–Thomson coefficient*, μ, which is defined by the relation

$$\mu = \left(\frac{\partial T}{\partial P}\right)_{\underline{H}} \tag{6-54}$$

μ may be expressed in terms of C_P, P, $\underline{V}$, and T as follows:

$$d\underline{H} = C_P\,dT + \left[\underline{V} - T\left(\frac{\partial \underline{V}}{\partial T}\right)_P\right] dP \tag{6-53}$$

But for an isenthalpic process $d\underline{H} = 0$; therefore,

$$0 = (C_P\,dT)_{\underline{H}} + \left(\left[\underline{V} - T\left(\frac{\partial \underline{V}}{\partial T}\right)_P\right] dP\right)_{\underline{H}} \tag{6-55}$$

or

$$(C_P\,dT)_{\underline{H}} = \left(\left[T\left(\frac{\partial \underline{V}}{\partial T}\right)_P - \underline{V}\right] dP\right)_{\underline{H}} \tag{6-56}$$

Therefore,

$$\mu = \left(\frac{\partial T}{\partial P}\right)_{\underline{H}} = \frac{T(\partial \underline{V}/\partial T)_P - \underline{V}}{C_P} \tag{6-57}$$

A similar expression, the *Euken coefficients*, ζ, can be derived for a closed-system process at constant internal energy:

$$\zeta = \left(\frac{\partial T}{\partial \underline{V}}\right)_{\underline{U}} = -\frac{T(\partial P/\partial T)_{\underline{V}} - P}{C_V} \tag{6-58}$$

Variations in C_P and C_V with Respect to P and $\underline{V}$, Respectively

We may evaluate the derivative $(\partial C_P/\partial P)_T$ as follows. The enthalpy property relation is written

$$d\underline{H} = C_P\,dT + \left[\underline{V} - T\left(\frac{\partial \underline{V}}{\partial T}\right)_P\right] dP \tag{6-53}$$

Since equation (6-53) is an exact differential, we may apply a Maxwell-relation type of cross differentiation to obtain

$$\left(\frac{\partial C_P}{\partial P}\right)_T = \left(\frac{\partial[\underline{V} - T(\partial \underline{V}/\partial T)_P]}{\partial T}\right)_P \tag{6-59}$$

or

$$\left(\frac{\partial C_P}{\partial P}\right)_T = \left(\frac{\partial \underline{V}}{\partial T}\right)_P - \left(\frac{\partial \underline{V}}{\partial T}\right)_P - T\left(\frac{\partial^2 \underline{V}}{\partial T^2}\right)_P \tag{6-60}$$

Therefore, the change of C_P with pressure is given by

$$\left(\frac{\partial C_P}{\partial P}\right)_T = -T\left(\frac{\partial^2 \underline{V}}{\partial T^2}\right)_P \tag{6-61}$$

Similarly, an expression for $(\partial C_V/\partial \underline{V})_T$ can be derived from the expression for $d\underline{U}$ as

$$\left(\frac{\partial C_V}{\partial \underline{V}}\right)_T = T\left(\frac{\partial^2 P}{\partial T^2}\right)_{\underline{V}} \tag{6-62}$$

These expressions may be derived alternatively as follows:

$$C_P = T\left(\frac{\partial \underline{S}}{\partial T}\right)_P$$

Therefore,

$$\left(\frac{\partial C_P}{\partial P}\right)_T = T\frac{\partial}{\partial P}\left[\left(\frac{\partial \underline{S}}{\partial T}\right)_P\right]_T = T\frac{\partial}{\partial T}\left[\left(\frac{\partial \underline{S}}{\partial P}\right)_T\right]_P \tag{6-63}$$

but from the Maxwell relations

$$\left(\frac{\partial \underline{S}}{\partial P}\right)_T = -\left(\frac{\partial \underline{V}}{\partial T}\right)_P$$

Therefore,

$$\left(\frac{\partial C_P}{\partial P}\right)_T = -T\left(\frac{\partial^2 \underline{V}}{\partial T^2}\right)_P \tag{6-64}$$

which is identical to equation (6-61). The reader should rederive equation (6-62) in the same manner.

General Derivative Formulas

Examination of the eight thermodynamic variables P, $\underline{V}$, T, $\underline{S}$, $\underline{U}$, $\underline{H}$, $\underline{A}$, and $\underline{G}$ indicates that there are 168 different partial derivatives (ignoring reciprocals) which involve any three of these variables. For example, $(\partial P/\partial \underline{V})_T$, $(\partial \underline{H}/\partial \underline{S})_G$, $(\partial \underline{V}/\partial \underline{A})_S$, . . . can be listed as just a few. As may be imagined, the number of equations relating these partial derivatives is extremely great. However, if we add the additional restriction that all derivatives be expressed in terms of P, $\underline{V}$, T, C_P, and C_V, the number is greatly reduced.

We shall now develop techniques for expressing each of these 168 derivatives in terms of P, $\underline{V}$, T, C_P, and C_V. We begin by stating without proof several extremely important properties of functions whose differentials are exact. (For proofs of the various properties the reader is referred to any standard textbook on advanced calculus.) Let us assume that F is an exact function of A and B:

$$F = F(A, B) \tag{6-65}$$

Then the following properties of the derivatives of F, A, and B are true:

$$dF = \left(\frac{\partial F}{\partial A}\right)_B dA + \left(\frac{\partial F}{\partial B}\right)_A dB \tag{6-66}$$

$$\left(\frac{\partial A}{\partial B}\right)_F = \frac{1}{(\partial B/\partial A)_F}$$

$$\left(\frac{\partial B}{\partial A}\right)_F = \frac{(\partial B/\partial C)_F}{(\partial A/\partial C)_F} \qquad \text{where } C \text{ is a third property other than } A \text{ or } B$$

$$\left(\frac{\partial A}{\partial B}\right)_F = -\frac{(\partial A/\partial F)_B}{(\partial B/\partial F)_A} = -\frac{(\partial F/\partial B)_A}{(\partial F/\partial A)_B}$$

The last property is frequently called the *triple-product relation* and is extremely useful because it allows us to move an unfavorable constraint (such as S, G, A, H, or U) into the derivatives.

The general technique for reducing a complicated derivative is then as follows:

1. If the derivative contains $\underline{U}$, $\underline{H}$, $\underline{A}$, or $\underline{G}$, bring these functions to the numerator of their respective derivatives. Then eliminate the $\underline{U}$, $\underline{H}$, $\underline{A}$, or $\underline{G}$

via the appropriate property relation. If a $\underline{U}$, $\underline{H}$, $\underline{A}$, or $\underline{G}$ still appears as a constraint on the derivative, use the triple product to bring it inside the derivatives. The offending quantity is then eliminated via its property relation as illustrated below.

$$\left(\frac{\partial T}{\partial V}\right)_{\underline{U}} = -\frac{(\partial T/\partial \underline{U})_{\underline{V}}}{(\partial \underline{V}/\partial \underline{U})_T} = -\frac{(\partial \underline{U}/\partial \underline{V})_T}{(\partial \underline{U}/\partial T)_{\underline{V}}}$$

Now substitute $dU = T\,dS - P\,dV$ to obtain

$$\left(\frac{\partial T}{\partial \underline{V}}\right)_U = \frac{T(\partial \underline{S}/\partial \underline{V})_T - P(\partial \underline{V}/\partial \underline{V})_T}{T(\partial \underline{S}/\partial T)_{\underline{V}}} = \frac{T(\partial \underline{S}/\partial \underline{V})_T - P}{T(\partial \underline{S}/\partial T)_{\underline{V}}}$$

2. At this point we have eliminated all references to $\underline{U}$, $\underline{H}$, $\underline{A}$, or $\underline{G}$ from the derivatives. We now eliminate the entropy. First bring the entropy to the numerator of all derivatives (the triple product may be used to eliminate entropy as a constraint). Next eliminate any entropy derivatives that appear in the Maxwell equations. Then eliminate derivatives of entropy with respect to temperature via the definitions of C_P and C_V:

$$C_P = T\left(\frac{\partial \underline{S}}{\partial T}\right)_P$$

$$C_V = T\left(\frac{\partial \underline{S}}{\partial T}\right)_{\underline{V}}$$

Any entropy derivatives that still exist may be eliminated by introducing $1/dT$ into the numerator and denominator of the derivative (see the third property of general derivatives), and then using the definitions of C_P and/or C_V to eliminate the entropy, as indicated in the following example.

$$\left(\frac{\partial T}{\partial \underline{V}}\right)_U = \frac{T(\partial \underline{S}/\partial \underline{V})_T - P}{T(\partial \underline{S}/\partial T)_{\underline{V}}}$$

but $(\partial \underline{S}/\partial \underline{V})_T = (\partial P/\partial T)_{\underline{V}}$ via the Maxwell relation, and $T(\partial \underline{S}/\partial T)_{\underline{V}} = C_V$. Thus

$$\left(\frac{\partial T}{\partial \underline{V}}\right)_U = \frac{T(\partial P/\partial T)_{\underline{V}} - P}{C_V}$$

SAMPLE PROBLEM 6-3. Evaluate $(\partial \underline{H}/\partial P)_{\underline{G}}$ in terms of P, $\underline{V}$, T, C_P, C_V, and absolute entropy, $\underline{S}$ (no entropy derivatives, however).

Solution: We begin by substituting the $d\underline{H}$ property relation for the $d\underline{H}$ term in the numerator:

$$\left(\frac{\partial \underline{H}}{\partial P}\right)_{\underline{G}} = T\left(\frac{\partial \underline{S}}{\partial P}\right)_{\underline{G}} + \underline{V}\left(\frac{\partial P}{\partial P}\right)_{\underline{G}} = T\left(\frac{\partial \underline{S}}{\partial P}\right)_{\underline{G}} + \underline{V}$$

Now bring $\underline{G}$ into the derivatives by application of the triple-product expansion:

$$\left(\frac{\partial \underline{H}}{\partial P}\right)_{\underline{G}} = -T\frac{\left(\frac{\partial \underline{G}}{\partial P}\right)_S}{\left(\frac{\partial \underline{G}}{\partial \underline{S}}\right)_P} + \underline{V}$$

Now eliminate $\underline{G}$ with the $d\underline{G}$ property relation:

$$\left(\frac{\partial \underline{H}}{\partial P}\right)_{\underline{G}} = -T\left[\frac{\left(-\underline{S}\left(\frac{\partial T}{\partial P}\right)_{\underline{S}} + \underline{V}\right)}{\left(-\underline{S}\frac{\partial T}{\partial \underline{S}}\right)_P + \underline{V}\left(\frac{\partial P}{\partial \underline{S}}\right)_P}\right] + \underline{V}$$

But $(\partial P/\partial \underline{S})_P = 0$, so we get

$$\left(\frac{\partial \underline{H}}{\partial P}\right)_{\underline{G}} = \left[-T\frac{\left(-\underline{S}\left(\frac{\partial T}{\partial P}\right)_{\underline{S}} + \underline{V}\right)}{-\underline{S}\left(\frac{\partial T}{\partial \underline{S}}\right)_P}\right] + \underline{V}$$

$$= -T\left[\left(\frac{\partial T}{\partial P}\right)_{\underline{S}}\left(\frac{\partial \underline{S}}{\partial T}\right)_P - \frac{\underline{V}}{\underline{S}}\left(\frac{\partial \underline{S}}{\partial T}\right)_P\right] + \underline{V}$$

Now use the triple product on the first term in parentheses:

$$= T\left[\left(\frac{\partial \underline{S}}{\partial P}\right)_T + \frac{\underline{V}}{\underline{S}}\left(\frac{\partial \underline{S}}{\partial T}\right)_P\right] + \underline{V}$$

But $(\partial \underline{S}/\partial P)_T = -(\partial \underline{V}/\partial T)_P$ from the Maxwell equations, and

$$\left(\frac{\partial \underline{S}}{\partial T}\right)_P = \frac{C_P}{T}$$

Thus the derivative is finally expressed as

$$\left(\frac{\partial \underline{H}}{\partial P}\right)_{\underline{G}} = -T\left(\frac{\partial \underline{V}}{\partial T}\right)_P + \underline{V}\left(\frac{C_P}{\underline{S}} + 1\right)$$

The absolute entropy can not be reduced, so this is the simplest form possible.

Let us now derive one more useful expression—that for the difference between C_P and C_V. We have shown that C_V may be expressed as

$$C_V = T\left(\frac{\partial \underline{S}}{\partial T}\right)_{\underline{V}}$$

However, we also know that

$$d\underline{S} = \frac{C_P}{T}\,dT - \left(\frac{\partial \underline{V}}{\partial T}\right)_P dP$$

Thus the expression for C_V becomes

$$C_V = T\left[\frac{C_P}{T}\left(\frac{\partial T}{\partial T}\right)_{\underline{V}} - \left(\frac{\partial \underline{V}}{\partial T}\right)_P\left(\frac{\partial P}{\partial T}\right)_{\underline{V}}\right] \tag{6-67}$$

which reduces to

$$C_V = C_P - T\left(\frac{\partial \underline{V}}{\partial T}\right)_P\left(\frac{\partial P}{\partial T}\right)_{\underline{V}} \tag{6-68}$$

or

$$C_P - C_V = T\left(\frac{\partial \underline{V}}{\partial T}\right)_P\left(\frac{\partial P}{\partial T}\right)_{\underline{V}} \tag{6-68a}$$

which is the desired relation. Several other identical forms of equation (6-68a) are possible if we employ the triple product to expand one or both of the remaining derivatives. However, these forms obviously convey no new information.

For a gas that obeys the ideal gas equation of state,

$$\left(\frac{\partial \underline{V}}{\partial T}\right)_P = \frac{R}{P}; \quad \left(\frac{\partial P}{\partial T}\right)_{\underline{V}} = \frac{R}{\underline{V}}$$

Thus:

$$C_P - C_V = \frac{TR^2}{P\underline{V}} = R \tag{6-69}$$

6.6 Thermodynamic Properties of an Ideal Gas

We have defined an ideal gas as one that obeys the ideal equation of state:

$$P\underline{V} = RT$$

Using this equation of state, we may calculate thermodynamic property changes which occur with an ideal gas as follows:

Internal Energy

$$d\underline{U} = C_V\,dT + \left[T\left(\frac{\partial P}{\partial T}\right)_{\underline{V}} - P\right] d\underline{V} \tag{6-50}$$

but from the ideal gas equation of state,

$$T\left(\frac{\partial P}{\partial T}\right)_{\underline{V}} = T\frac{R}{\underline{V}} = P$$

Therefore, equation (6-50) becomes

$$d\underline{U} = C_V\,dT \tag{6-70}$$

That is, the internal energy change of an ideal gas is independent of the volume change and may be expressed as a function of temperature alone.

Enthalpy

$$d\underline{H} = C_P\,dT + \left[\underline{V} - T\left(\frac{\partial \underline{V}}{\partial T}\right)_P\right] dP \tag{6-53}$$

But for an ideal gas $T(\partial \underline{V}/\partial T)_P = \underline{V}$, so that

$$d\underline{H} = C_P\,dT \tag{6-71}$$

Therefore, the enthalpy, as well as the internal energy, of an ideal gas is a function of its temperature only, and not a function of its pressure.

Entropy: T and V Known

$$d\underline{S} = \frac{C_V}{T}\,dT + \left(\frac{\partial P}{\partial T}\right)_{\underline{V}} dV \tag{6-49}$$

But for an ideal gas,

$$\left(\frac{\partial P}{\partial T}\right)_{\underline{V}} = \frac{R}{\underline{V}}$$

Therefore,

$$d\underline{S} = C_V \frac{dT}{T} + R \frac{d\underline{V}}{\underline{V}} \tag{6-72}$$

or

$$d\underline{S} = C_V \, d \ln T + R \, d \ln \underline{V} \tag{6-72a}$$

Entropy: T and P Known

$$d\underline{S} = \frac{C_P}{T} dT - \left(\frac{\partial \underline{V}}{\partial T}\right)_P dP \tag{6-52}$$

But for an ideal gas,

$$\left(\frac{\partial \underline{V}}{\partial T}\right)_P = \frac{R}{P}$$

Therefore,

$$d\underline{S} = C_P \frac{dT}{T} - R \frac{dP}{P} \tag{6-73}$$

or

$$d\underline{S} = C_P \, d \ln T - R \, d \ln P \tag{6-73a}$$

Thus we see that changes in $\underline{U}$, $\underline{H}$, and $\underline{S}$ can be calculated easily and directly for an ideal gas provided that we know C_P or C_V (remember that $C_P - C_V = R$ for an ideal gas) and the initial and final temperature and pressure. For $\underline{U}$ and $\underline{H}$ this simplifies even further to requiring that we know only the initial and final temperatures and C_P or C_V.

Joule–Thomson Coefficient, μ

We have shown that the Joule–Thomson coefficient may be evaluated from

$$\mu = \left(\frac{\partial T}{\partial P}\right)_{\underline{H}} = \frac{T(\partial \underline{V}/\partial T)_P - \underline{V}}{C_P} \tag{6-57}$$

But we have shown that for an ideal gas

$$T\left(\frac{\partial \underline{V}}{\partial T}\right)_P - \underline{V} = 0$$

Therefore, the Joule–Thomson coefficient for an ideal gas is identically equal to zero and indicates that an ideal gas will neither increase nor decrease its temperature during an isenthalpic pressure change.

The Euken Coefficient, ζ

The Euken coefficient is given by

$$\zeta = \left(\frac{\partial T}{\partial \underline{V}}\right)_{\underline{U}} = \frac{T(\partial P/\partial T)_{\underline{V}} - P}{C_V} \tag{6-58}$$

However, for an ideal gas

$$T\left(\frac{\partial P}{\partial T}\right)_{\underline{V}} - P = 0$$

Therefore, the Euken coefficient is zero, and an ideal gas can neither increase nor decrease in temperature during a constant internal-energy process.

Variation of C_P and C_V with P and $\underline{V}$, Respectively

$$\left(\frac{\partial C_P}{\partial P}\right)_T = -T\left(\frac{\partial^2 \underline{V}}{\partial T^2}\right)_P \tag{6-61}$$

$$\left(\frac{\partial C_V}{\partial \underline{V}}\right)_T = T\left(\frac{\partial^2 P}{\partial T^2}\right)_{\underline{V}} \tag{6-62}$$

But for an ideal gas

$$\left(\frac{\partial^2 \underline{V}}{\partial T^2}\right)_P = \left(\frac{\partial^2 P}{\partial T^2}\right)_{\underline{V}} = 0$$

Therefore, for an ideal gas

$$\left(\frac{\partial C_P}{\partial P}\right)_T = \left(\frac{\partial C_V}{\partial \underline{V}}\right)_T = 0 \tag{6-74}$$

Thus, we find for an ideal gas that C_P is independent of P, and C_V is independent of V. Therefore, we can say that C_P and C_V are, at most, functions only of the temperature, and not independent functions of P or V for ideal gases.

If the ideal gas is composed of rigid, noninteracting, monatomic particles, it is possible, by use of statistical mechanical arguments, to show that the heat capacities of the ideal gas are independent of temperature, as well as pressure and volume, and therefore are true constants. (For very high temperatures we must also exclude ionization effects.) However, if the ideal gas is composed of real molecules (for example, oxygen at room temperature and 1 atm obeys the ideal gas equation of state) which are polyatomic, then the heat capacities will not be independent of temperature and are not true constants.

6.7 Evaluations of Change in $\underline{U}$, $\underline{H}$, and $\underline{S}$ for Various Processes

We shall now examine the techniques for integratingthe equations for $d\underline{U}$, $d\underline{H}$, and $d\underline{S}$. These integrations will allow us to evaluate the changes in $\underline{U}$, $\underline{H}$, $\underline{S}$, $\underline{G}$, and $\underline{A}$ which occur when a substance undergoes a given process. Since the equations for $d\underline{U}$, $d\underline{H}$, and $d\underline{S}$ take their simplest form when applied to an ideal gas, we shall study this case first. Then we shall examine the methods used to integrate more complex equations generated when the working fluid is not an ideal gas.

Ideal Gases

In Section 6.6 we showed that for an ideal gas $d\underline{U}$, $d\underline{H}$, and $d\underline{S}$ could be expressed as

$$d\underline{U} = C_V\, dT \tag{6-70}$$

$$d\underline{H} = C_P\, dT \tag{6-71}$$

$$d\underline{S} = C_V\, d\ln T + R\, d\ln \underline{V} \tag{6-72a}$$

$$d\underline{S} = C_P\, d\ln T - R\, d\ln P \tag{6-73a}$$

For an ideal gas whose C_P and C_V are independent of T, C_P and C_V are true constants, and these four relations may be integrated directly to yield

$$\int_{\underline{U}_1}^{\underline{U}_2} d\underline{U} = \Delta\underline{U} = \underline{U}_2 - \underline{U}_1 = \int_{T_1}^{T_2} C_V\, dT = C_V(T_2 - T_1) \tag{6-75}$$

$$\int_{\underline{H}_1}^{\underline{H}_2} d\underline{H} = \Delta\underline{H} = \underline{H}_2 - \underline{H}_1 = \int_{T_1}^{T_2} C_P\, dT = C_P(T_2 - T_1) \tag{6-76}$$

$$\int_{\underline{S}_1}^{\underline{S}_2} d\underline{S} = \Delta\underline{S} = \underline{S}_2 - \underline{S}_1 = \int_{T_1}^{T_2} C_V\, d\ln T + \int_{\underline{V}_1}^{\underline{V}_2} R\, d\ln V$$

$$= C_V \ln\frac{T_2}{T_1} + R\ln\frac{\underline{V}_2}{\underline{V}_1} \tag{6-77}$$

$$\int_{\underline{S}_1}^{\underline{S}_2} d\underline{S} = \Delta\underline{S} = \underline{S}_2 - \underline{S}_1 = \int_{T_1}^{T_2} C_P\frac{dT}{T} - \int_{P_1}^{P_2} R\, d\ln P$$

$$= C_P \ln\frac{T_2}{T_1} - R\ln\frac{P_2}{P_1} \tag{6-78}$$

Thus we can evaluate the change in any property directly in terms of P, $\underline{V}$, T, C_P and C_V.

As we mentioned earlier, almost all real gases at room (or higher) temperatures, and relatively low pressures—less than 5 atm—obey the ideal gas equation of state. However, for many of these gases the heat capacities may vary appreciably with temperature (although not at all with P or V), and therefore may not be considered constants. Heat capacities at low pressures are usually distinguished from all others by the superscript *, that is, C_P^* and C_V^*. Usually $C_P^*(T)$ and $C_V^*(T)$ are presented in the form of a polynomial expansion in (T):

$$C_P^*(T) = A + BT + CT^2 + \cdots$$

$$C_V^*(T) = A' + B'T + C'T^2 + \cdots$$

At all temperatures, however,

$$C_P^*(T) - C_V^*(T) = R$$

which implies that $B' = B$, $C' = C$, and $A - A' = R$.
From this point on, unless otherwise indicated, an *ideal gas* will be taken as one whose heat capacities are true constants.

SAMPLE PROBLEM 6-4. Derive an expression for $\Delta\underline{U}$, $\Delta\underline{H}$, and $\Delta\underline{S}$ for a gas at low pressure whose heat capacities are given by

$$C_P^* = A + BT + CT^2$$
$$C_V^* = (A - R) + BT + CT^2$$

Solution: Assume that the gas obeys the ideal gas equation of state, so

$$\Delta\underline{U} = \int_{T_1}^{T_2} C_V^*\, dT = \int_{T_1}^{T_2} (A - R + BT + CT^2)\, dT$$

Therefore,

$$\Delta\underline{U} = (A - R)(T_2 - T_1) + \frac{B}{2}(T_2^2 - T_1^2) + \frac{C}{3}(T_2^3 - T_1^3)$$

$$\Delta\underline{H} = \int_{T_1}^{T_2} C_P^*\, dT = \int_{T_1}^{T_2} (A + BT + CT^2)\, dT$$

or

$$\Delta\underline{H} = A(T_2 - T_1) + \frac{B}{2}(T_2^2 - T_1^2) + \frac{C}{3}(T_2^3 - T_1^3)$$

$$\Delta\underline{S} = \int_{T_1}^{T_2} C_P^* \frac{dT}{T} - \int_{P_1}^{P_2} R\, d\ln P$$

Therefore,

$$\Delta\underline{S} = \int_{T_1}^{T_2} \left(\frac{A}{T} + B + CT\right) dT - R \ln \frac{P_2}{P_1}$$

or

$$\Delta\underline{S} = A \ln \frac{T_2}{T_1} + B(T_2 - T_1) + \frac{C}{2}(T_2^2 - T_1^2) - R \ln \frac{P_2}{P_1}$$

Equations (6-77) and (6-78) gives us some very useful information about the P–$\underline{V}$–T relations of an ideal gas undergoing an isentropic process. Since many compressors and expanders operate in nearly an adiabatic and reversible fashion, processes that may be treated as isentropic will be encountered fairly often in our work. For an isentropic process $\Delta\underline{S} = 0$. Therefore, equations (6-77) and (6-78) yield

$$\Delta\underline{S} = 0 = C_V \ln \frac{T_2}{T_1} + R \ln \frac{\underline{V}_2}{\underline{V}_1} \tag{6-79}$$

or

$$\Delta\underline{S} = 0 = C_P \ln \frac{T_2}{T_1} - R \ln \frac{P_2}{P_1} \tag{6-80}$$

Therefore,

$$\ln \frac{T_2}{T_1} = -\frac{R}{C_V} \ln \frac{\underline{V}_2}{\underline{V}_1} \tag{6-81}$$

$$\ln \frac{T_2}{T_1} = \frac{R}{C_P} \ln \frac{P_2}{P_1} \tag{6-82}$$

Equations (6-81) and (6-82) allow us to calculate the outlet temperature from an isentropic process if we know the inlet conditions (T and P or $\underline{V}$) and the outlet P or $\underline{V}$. Once we have determined the outlet temperature T_2, we may use equation (6-70) or (6-71) to determine $\Delta\underline{U}$ or $\Delta\underline{H}$. Thus it is seen that determination of the outlet temperature is the first step in calculating the work requirement of an isentropic process. As we shall see later, this becomes an extremely complicated and tedious task when we have to deal with nonideal gases.

By equating the right-hand sides of equations (6-81) and (6-82), we can relate the pressure and volume during an isentropic process as follows:

$$-\frac{C_P}{C_V}\ln\frac{\underline{V}_2}{\underline{V}_1} = \ln\frac{P_2}{P_1}$$

Let $C_P/C_V = k =$ the ratio of specific heats. Therefore,

$$0 = \ln\frac{P_2}{P_1} + \ln\left(\frac{\underline{V}_2}{\underline{V}_1}\right)^k = \ln\frac{P_2\underline{V}_2^k}{P_1\underline{V}_1^k} \tag{6-83}$$

or

$$P_2\underline{V}_2^k = P_1\underline{V}_1^k \tag{6-84}$$

which relates the outlet pressure and volume to the inlet conditions.

Real Gases with Use of Equations of State

We now consider the calculation of property changes for real gases. As we shall see, many of the terms that conveniently dropped out when we were dealing with ideal gases no longer vanish. These terms are responsible for many of the difficulties encountered in calculations involving real gases.

The differential equations for $d\underline{U}$, $d\underline{H}$, and $d\underline{S}$ may be written for any gas (real or ideal) as follows:

$$d\underline{U} = C_V\,dT + \left[T\left(\frac{\partial P}{\partial T}\right)_{\underline{V}} - P\right]d\underline{V} \tag{6-44}$$

$$d\underline{H} = C_P\,dT - \left[T\left(\frac{\partial \underline{V}}{\partial T}\right)_P - \underline{V}\right]dP \tag{6-53}$$

$$d\underline{S} = \frac{C_P}{T}\,dT - \left(\frac{\partial \underline{V}}{\partial T}\right)_P dP \tag{6-52}$$

For real gases, the simplifications which result from the ideal gas equation of state are frequently not valid. Rarely will the dP or $d\underline{V}$ terms in the $d\underline{H}$ or $d\underline{U}$ expression vanish or be easily evaluated in the $d\underline{S}$ expression. Second, C_P and C_V are functions of T and P or $\underline{V}$ for real gases, so even evaluation of the integral $\int C_P\,dT$ is complicated. Third, we shall usually not know the exact path (that is, P–T or $\underline{V}$–T relation) along which the process in question is proceeding, so we shall not be able to evaluate the integrals needed to find $\Delta\underline{U}$, $\Delta\underline{H}$, or $\Delta\underline{S}$, even if the derivatives are expressible in terms of P, $\underline{V}$, and T.

Therefore, we develop an alternative approach which takes advantage of the fact that $\underline{U}$, $\underline{H}$, and $\underline{S}$ are state functions. Thus any changes that occur in these properties depend only on the initial and final states of the substance and are not functions of the path between the states. Thus, if we evaluate $\Delta \underline{U}$, $\Delta \underline{H}$, and $\Delta \underline{S}$ for any specific path between the initial and final states of the system, the result must be the same for all other paths between the same initial and final states. However, we recognize that the change in a state property over any path may be evaluated as the sum of the changes of the property over all segments of the original path. That is, by evaluating $\Delta \underline{H}_{1-2}$ and $\Delta \underline{H}_{2-3}$ (where $\Delta \underline{H}_{1-2} = \underline{H}_2 - \underline{H}_1$), the total change in enthalpy between state 1 and state 3 is obtained from the sum

$$\Delta \underline{H}_{1-3} = \Delta \underline{H}_{1-2} + \Delta \underline{H}_{2-3}$$

Thus if we wish to evaluate a property change along any arbitrary path, we may break this path into a series of smaller paths (or subpaths) along which the change in the desired property can be evaluated.

Since in general we will know C_P or C_V only for very low pressures, where $C_P = C_P^*$ and $C_V = C_V^*$, it is frequently necessary to choose the integration paths in such a way that changes in temperature occur only at low pressures. Thus, if we wished to evaluate $\Delta \underline{H}$ or $\Delta \underline{S}$ between points 1 and 2 on a P–T diagram, we would use a three-part path as indicated in Fig. 6-2.

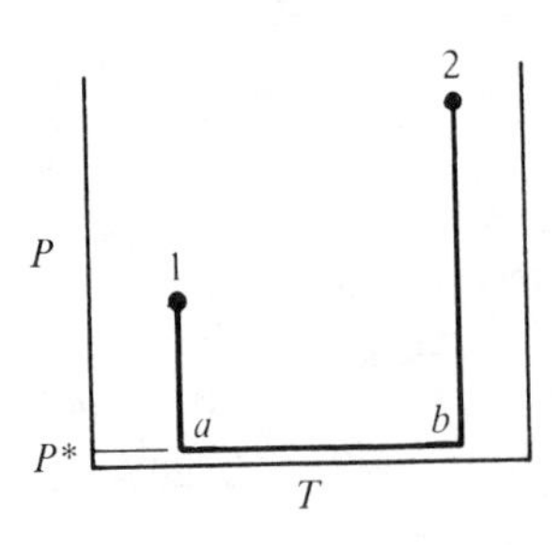

FIG. 6-2. Three-step process.

On the first leg of the path from point 1 to point a, T is held constant at T_1, and the pressure is reduced from P_1 to P^* (where P^* is some very low pressure at which C_P is known). Since T is held constant, $dT = 0$ and $\Delta \underline{H}_{1-a} = \underline{H}_a - \underline{H}_1$ may be evaluated by integrating equation (6-53):

$$\int_1^a d\underline{H} = \Delta \underline{H}_{1-a} = \underline{H}_a - \underline{H}_1 = \int_{P_1}^{P^*} \left[\underline{V} - T_1 \left(\frac{\partial \underline{V}}{\partial T} \right)_P \right] dP \tag{6-85}$$

Similarly

$$\Delta \underline{S}_{1-a} = \int_{P_1}^{P^*} -\left(\frac{\partial \underline{V}}{\partial T} \right)_P dP \tag{6-86}$$

Along the second leg, from a to b, the pressure is held constant at P^* and the temperature changed from T_1 to T_2. Since P is constant, $dP = 0$, and $\Delta \underline{H}_{a-b}$ and $\Delta \underline{S}_{a-b}$ are calculated from

$$\Delta \underline{H}_{a-b} = \underline{H}_b - \underline{H}_a = \int_{T_1}^{T_2} C_P^* \, dT \tag{6-87}$$

$$\Delta \underline{S}_{a-b} = \underline{S}_b - \underline{S}_a = \int_{T_1}^{T_2} \frac{C_P^*}{T} \, dT \tag{6-88}$$

where C_P^* indicates C_P at P^*.

Along the last leg of the path from b to point 2, the temperature is held constant at T_2 so $dT = 0$, and the pressure P is raised from P^* to P_2. Thus $\Delta\underline{H}_{b-2}$ and $\Delta\underline{S}_{b-2}$ become

$$\Delta\underline{H}_{b-2} = \underline{H}_2 - \underline{H}_b = \int_{P^*}^{P}\left[\underline{V} - T_2\left(\frac{\partial \underline{V}}{\partial T}\right)_P\right] dP \tag{6-89}$$

$$\Delta\underline{S}_{b-2} = \underline{S}_2 - \underline{S}_b = \int_{P^*}^{P} -\left(\frac{\partial \underline{V}}{\partial T}\right)_P dP \tag{6-90}$$

The overall $\Delta\underline{H}_{1-2}$ or $\Delta\underline{S}_{1-2}$ is then simply the sum of the $\Delta\underline{H}$'s and $\Delta\underline{S}$'s of the three path segments:

$$\begin{aligned} \Delta\underline{H}_{1-2} &= \underline{H}_2 - \underline{H}_1 = \Delta\underline{H}_{1-a} + \Delta\underline{H}_{a-b} + \Delta\underline{H}_{b-2} \\ \Delta\underline{S}_{1-2} &= \underline{S}_2 - \underline{S}_1 = \Delta\underline{S}_{1-a} + \Delta\underline{S}_{a-b} + \Delta\underline{S}_{b-2} \end{aligned} \tag{6-91}$$

Substitution of the expressions for the three path-segment changes then gives us the total change in enthalpy or entropy. Similar expression for $\Delta\underline{U}$ or $\Delta\underline{S}$ can be developed using T and $\underline{V}$, rather than T and P, as the integration variables.

SAMPLE PROBLEM 6-5. In Sample Problem 4-7 you were told that the internal energy of a gas that obeyed the Beattie–Bridgeman equation of state could be expressed as

$$\underline{U}(T, \underline{V}) = CVA(T - T_0) + \frac{CVB}{2}(T^2 - T_0^2) + \frac{CVC}{3}(T^3 - T_0^3)$$
$$- \frac{1}{\underline{V}}\left(A_0 + \frac{3RC}{T^2}\right) + \frac{1}{2\underline{V}^2}\left(A_0 a - \frac{3RB_0C}{T^2}\right) + \frac{RbB_0C}{\underline{V}^3T^2}$$

if the zero pressure C_V was expressible as

$$C_V^* = CVA + CVB(T) + CVC(T^2)$$

The reference state was chosen as $\underline{U} = 0$ when $\underline{V} \to \infty$ and $T = T_0$. Derive the above expression for $\underline{U}(T, \underline{V})$.

Solution: Set up a two-step path from the reference conditions, T_0, $\underline{V} = \infty$ to any given condition $(T, \underline{V})$ as shown in Fig. S6-5. Along path 1, $\underline{V}$ is constant, so the change in $\underline{U}$ is given by

$$d\underline{U} = C_V\, dT$$

Since $\underline{V} = \infty$, $P = 0$ and $C_V = C_V^*$. Therefore, $d\underline{U}$ becomes

$$d\underline{U} = C_V^*\, dT$$

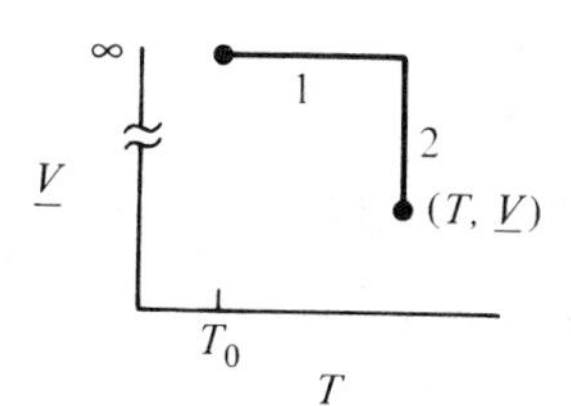

FIGURE S6-5

which may be integrated between the reference state and state a:

$$\underline{U}_a - \underline{U}_{\text{ref}} = \int_{T_{\text{ref}}}^{T} C_V^* \, dT$$

But $\underline{U}_{\text{ref}} = 0$ and $T_{\text{ref}} = T_0$, and C_V^* is expressed as

$$C_V^* = CVA + CVB(T) + CVC(T^2)$$

Thus

$$\underline{U}_a = \int_{T_0}^{T} [CVA + CVB(T) + CVC(T^2)] \, dT$$

$$= CVA(T - T_0) + \frac{CVB}{2}(T^2 - T_0^2) + \frac{CVC}{3}(T^3 - T_0^3)$$

Along path 2, T is constant, and $d\underline{U}$ is given by

$$d\underline{U} = \left(\frac{\partial \underline{U}}{\partial \underline{V}}\right)_T d\underline{V}$$

But

$$\left(\frac{\partial \underline{U}}{\partial \underline{V}}\right)_T = \left[T\left(\frac{\partial P}{\partial T}\right)_{\underline{V}} - P\right]$$

so

$$d\underline{U} = \left[T\left(\frac{\partial P}{\partial T}\right)_{\underline{V}} - P\right] d\underline{V}$$

For the Beattie–Bridgeman equation of state,

$$P = \frac{RT}{\underline{V}} + \frac{\beta(T)}{\underline{V}^2} + \frac{\gamma(T)}{\underline{V}^3} + \frac{\delta(T)}{\underline{V}^4}$$

Therefore,

$$\left(\frac{\partial P}{\partial T}\right)_{\underline{V}} = \frac{R}{\underline{V}} + \frac{\beta'(T)}{\underline{V}^2} + \frac{\gamma'(T)}{\underline{V}^3} + \frac{\delta'(T)}{\underline{V}^3}$$

where the primes represent d/dT. Thus

$$T\left(\frac{\partial P}{\partial T}\right)_{\underline{V}} = \frac{RT}{\underline{V}} + \frac{T\beta'(T)}{\underline{V}^2} + \frac{T\gamma'(T)}{\underline{V}^3} + \frac{T\delta'(T)}{\underline{V}^4}$$

$$T\left(\frac{\partial P}{\partial T}\right)_{\underline{V}} - P = \frac{1}{\underline{V}^2}[T\beta'(T) - \beta(T)] + \frac{1}{\underline{V}^3}[T\gamma'(T) - \gamma(T)]$$

$$+ \frac{1}{\underline{V}^4}[T\delta'(T) - \delta(T)]$$

Now integrate from $\underline{V}_\infty$ to $\underline{V}$ to get

$$\underline{U}(T, \underline{V}) - \underline{U}_a = \int_{\underline{V}_\infty}^{\underline{V}} \left[T\left(\frac{\partial P}{\partial T}\right)_{\underline{V}} - P\right] d\underline{V}$$

or

$$\underline{U}(T, \underline{V}) - \underline{U}_a$$

$$= \int_{\underline{V}_\infty}^{\underline{V}} \left\{\frac{1}{\underline{V}^2}[T\beta'(T) - \beta(T)] + \frac{1}{\underline{V}^3}[T\gamma'(T) - \gamma(T)] + \frac{1}{\underline{V}^4}[T\delta'(T) - \delta(T)]\right\} d\underline{V}$$

Performing the integration and substituting limits gives

$$\underline{U}(T, \underline{V}) - \underline{U}_a = [T\beta'(T) - \beta(T)]\left[-\left(\frac{1}{\underline{V}} - \frac{1}{\infty}\right)\right]$$
$$+ [T\gamma'(T) - \gamma(T)]\left[-\frac{1}{2}\left(\frac{1}{\underline{V}^2} - \frac{1}{\infty^2}\right)\right]$$
$$+ [T\delta'(T) - \delta(T)]\left[-\frac{1}{3}\left(\frac{1}{\underline{V}^3} - \frac{1}{\infty^3}\right)\right]$$

Since $1/\infty = 0$, this simplifies to

$$\underline{U}(T, \underline{V}) = \underline{U}_a - \frac{1}{\underline{V}}[T\beta'(T) - \beta(T)] - \frac{1}{2\underline{V}^2}[T\gamma'(T) - \gamma(T)]$$
$$- \frac{1}{3\underline{V}^3}[T\delta'(T) - \delta(T)]$$

or

$$\underline{U}(T, \underline{V}) = \underline{U}_a + \frac{1}{\underline{V}}[\beta(T) - T\beta'(T)] + \frac{1}{2\underline{V}^2}[\gamma(T) - T\gamma'(T)]$$
$$+ \frac{1}{3\underline{V}^3}[\delta(T) - T\delta'(T)]$$

The virial coefficients are then expressed in terms of the Beattie–Bridgeman constants as

$$\beta(T) = RB_0T - \frac{RC}{T^2}A_0$$

so

$$\beta'(T) = RB_0 + \frac{2RC}{T^3}$$
$$\beta(T) - T\beta'(T) = -\frac{3RC}{T^3} - A_0$$

Similarly,

$$\gamma(T) = A_0a - RbB_0T - \frac{RB_0C}{T^2}$$
$$\gamma'(T) = -RbB_0 + \frac{2RB_0C}{T^3}$$
$$\gamma(T) - T\gamma'(T) = A_0a - \frac{3RB_0C}{T^2}$$

Finally,

$$\delta(T) = \frac{RB_0bC}{T^2}$$
$$\delta'(T) = -\frac{2RB_0bC}{T^3}$$
$$\delta(T) - T\delta'(T) = \frac{3RB_0bC}{T^2}$$

Substitution into the $\underline{U}(T, \underline{V})$ expression yields

$$\underline{U}(T, \underline{V}) = \underline{U}_a - \frac{1}{\underline{V}}\left(A_0 + \frac{3RC}{T^2}\right) + \frac{1}{2\underline{V}^2}\left(A_0a - \frac{3RB_0C}{T^2}\right) + \frac{RB_0bC}{\underline{V}^3T^2}$$

But

$$\underline{U}_a = CVA(T - T_0) + \frac{CVB}{2.0}(T^2 - T_0^2) + \frac{CVC}{3}(T^3 - T_0^3)$$

so the final expression for $\underline{U}(T, \underline{V})$ is given by

$$\underline{U}(T, \underline{V}) = CVA(T - T_0) + \frac{CVB}{2.0}(T^2 - T_0^2) + \frac{CVC}{3}(T^3 - T_0^3)$$
$$- \frac{1}{\underline{V}}\left(A_0 + \frac{3RC}{T^2}\right) + \frac{1}{2\underline{V}^2}\left(A_0 a - \frac{3RB_0C}{T^2}\right) + \frac{RB_0bC}{\underline{V}^3T^2}$$

which is the desired form.

SAMPLE PROBLEM 6-6. Helium enters a reversible isothermal compressor at 540°R and 12 atm and is continuously compressed to 180 atm. Calculate the work per mole of helium needed to run the compressor and the amount of heat per mole of helium that must be removed from the compressor if

(a) Helium behaves as an ideal gas.
(b) Helium behaves according to the equations of state

$$P\underline{V} = RT - \frac{a}{T}P + bP$$

where $a = 11.13$ °R ft³/lb-mol
$b = 0.2445$ ft³/lb-mol

Solution: We may picture the process as shown in Fig. S6-6a. We may write energy and entropy balances around the compressor and its contents as follows:

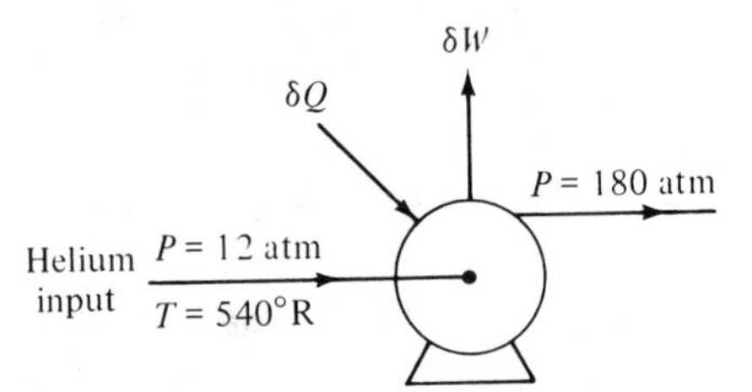

FIGURE S6-6a

Energy:

$$\left(\underline{H} + \frac{u^2}{2g_c} + \frac{gZ}{g_c}\right)_{\text{in}} \delta M_{\text{in}} - \left(\underline{H} + \frac{u^2}{2g_c} + \frac{gZ}{g_c}\right)_{\text{out}} \delta M_{\text{out}} + \delta Q - \delta W = d(M\underline{E})_{\text{syst}}$$

Entropy:

$$(\underline{S}\,\delta M)_{\text{in}} - (\underline{S}\,\delta M)_{\text{out}} + \frac{\delta Q + \delta LW}{T_{\text{syst}}} = d(M\underline{S})_{\text{syst}}$$

Since the compressor operates under steady-state conditions, $d(M\underline{E})_{\text{syst}} = d(M\underline{S})_{\text{syst}} = 0$, and $\delta M_{\text{in}} = \delta M_{\text{out}}$. Kinetic and potential energy changes in the compressor will be assumed to be negligible. Since the compressor operates reversibly, $\delta LW = 0$. Therefore, the energy and entropy equations reduce to

Energy:

$$\Delta\underline{H} = \underline{Q} - \underline{W}$$

Entropy:

$$\Delta \underset{\sim}{S} = \frac{\underset{\sim}{Q}}{T_{\text{syst}}}$$

or

$$\underset{\sim}{Q} = T_{\text{syst}}(\Delta \underset{\sim}{S})$$

But

$$T_{\text{syst}} = 540°\text{R}$$

Therefore,

$$\underset{\sim}{Q} = 540°\text{R}\ (\Delta \underset{\sim}{S})$$

Thus, if we can find $\Delta \underset{\sim}{S}$ for the process, we can evaluate $\underset{\sim}{Q}$. Then if we can find $\Delta \underset{\sim}{H}$, we can use $\underset{\sim}{Q}$ and the energy equation to determine $\underset{\sim}{W}$. On a P–T diagram the compression process is given as shown in Fig. S6-6b.

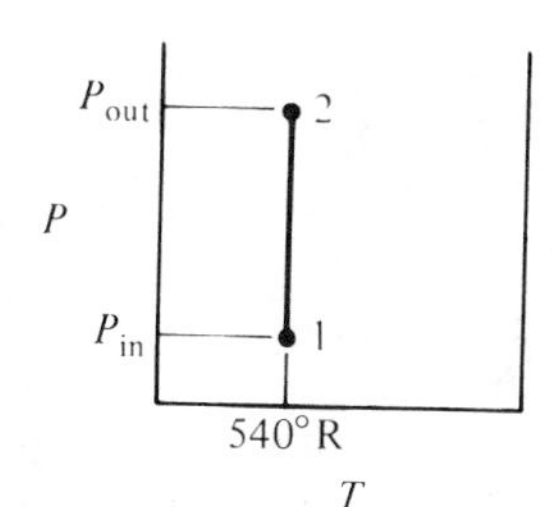

FIGURE S6-6b

Since no temperature change occurs, it is not necessary to use a three-step process; a simple one-step integration from P_{in} to P_{out} at constant temperature $T = 540°\text{R}$ will be sufficient. For an isothermal process we have shown that $\Delta \underset{\sim}{H}$ and $\Delta \underset{\sim}{S}$ may be expressed by the relations

$$(\Delta \underset{\sim}{H}_{1-2})_T = \int_{P_1}^{P_2} \left[\underset{\sim}{V} - T\left(\frac{\partial \underset{\sim}{V}}{\partial T}\right)_P \right] dP$$

$$(\Delta \underset{\sim}{S}_{1-2})_T = -\int_{P_1}^{P_2} \left(\frac{\partial \underset{\sim}{V}}{\partial T}\right)_P dP$$

We must now evaluate these integrals for the desired equations of state.

(a) Ideal gas: $P\underset{\sim}{V} = RT$.

$$\left(\frac{\partial \underset{\sim}{V}}{\partial T}\right)_P = \frac{R}{P} = \frac{\underset{\sim}{V}}{T}$$

Thus

$$T\left(\frac{\partial \underset{\sim}{V}}{\partial T}\right)_P = \underset{\sim}{V}$$

so

$$d\underset{\sim}{H}_T = 0 \qquad \text{and} \qquad \Delta \underset{\sim}{H}_{1-2} = 0$$

That is, the change in enthalpy for the isothermal compression of an ideal gas is zero.

$$\Delta \underset{\sim}{S}_{1-2} = -\int_{P_1}^{P_2} \frac{R}{P}\, dP = -R \ln \frac{P_2}{P_1}$$

$$= -1.987\ \text{Btu/lb-mol °R} \ln \frac{180}{12} = -5.36\ \text{Btu/lb-mol °R}$$

$$\underline{Q} = T\,\Delta\underline{S}_{1-2} = -(540)(5.36)\ \text{Btu/lb-mol}$$
$$= -2900\ \text{Btu/lb-mol}$$

But the energy balance gives

$$\Delta\underline{H} = \underline{Q} - \underline{W} = 0$$

Therefore,

$$\underline{Q} = \underline{W} = -2900\ \text{Btu/lb-mol}$$

(b) Real gas: $P\underline{V} = RT + [(-a/T + b)]P$.

$$\underline{V} = \frac{RT}{P} + \left(b - \frac{a}{T}\right)$$

So

$$\left(\frac{\partial \underline{V}}{\partial T}\right)_P = \frac{R}{P} + \frac{a}{T^2}$$

Therefore,

$$\Delta\underline{S}_{1-2} = -\int_{12}^{180}\left(\frac{R}{P} + \frac{a}{T^2}\right)dP$$
$$= R\ln\frac{180}{12} - \frac{a}{T^2}(180 - 12)$$
$$= -5.36\frac{\text{Btu}}{\text{lb-mol °R}} - \frac{11.13}{(540)^2}$$
$$\times \frac{(180 - 12)(144)(14.7)}{778\ \text{lb-mol °R}^2}\ \frac{\text{°R ft}^3\ \text{lb}_f\text{-in.}^2\ \text{atm Btu}}{\text{lb-mol °R}^2\ \text{in.}^2\ \text{ft}^2\ \text{atm lb}_f\text{-ft}}$$
$$= -5.36\frac{\text{Btu}}{\text{lb-mol °R}} - 0.0175\frac{\text{Btu}}{\text{lb-mol °R}}$$
$$= -5.38\frac{\text{Btu}}{\text{lb-mol °R}}$$

$$\underline{Q} = T\,\Delta\underline{S} = -2910\ \text{Btu/lb-mol}$$

Enthalpy:

$$\underline{V} - T\left(\frac{\partial \underline{V}}{\partial T}\right)_P = -\frac{a}{T} + b - \frac{a}{T} + \frac{RT}{P} - \frac{RT}{P}$$
$$= b - \frac{2a}{T}$$

Therefore,

$$\Delta\underline{H}_{1-2} = \int_{P_1}^{P_2}\left(b - \frac{2a}{T}\right)dP = \left(b - \frac{2a}{T}\right)(P_2 - P_1)$$

Therefore,

$$\Delta\underline{H}_{1-2} = \left[0.2445\frac{\text{ft}^3}{\text{lb-mol}} - \frac{(2)(11.13)\ \text{°R ft}^3}{540\ \text{°R lb-mol}}\right](180 - 12)\ \text{atm}$$

or

$$\Delta\underline{H}_{1-2} = (0.2032)(180 - 12)\frac{\text{ft}^3\ \text{atm}}{\text{lb-mol}}$$
$$= 34.1\ \frac{\text{ft}^3\ \text{atm}}{\text{lb-mol}}$$
$$= 34.1 \times \frac{(14.7)(144)}{778}\ \frac{\text{ft}^3\ \text{atm psi psf Btu}}{\text{lb-mol atm psi lb}_f\text{-ft}}$$
$$= 93.5\ \text{Btu/lb-mol}$$

But

$$\Delta \underline{H} = \underline{Q} - \underline{W}$$

or

$$\underline{W} = \underline{Q} - \Delta \underline{H} = -2{,}910 - 93$$
$$= -3003 \text{ Btu/lb-mol}$$

Isentropic Processes with Real Gases

As we previously mentioned, many compressors, expanders, and other devices operate in a manner which may be treated as isentropic for design purposes. Since it is often necessary to calculate the work produced or required by these machines, we must devise methods for evaluating the changes in $\underline{H}$ and $\underline{U}$ which occur during isentropic processes. By and large, compressor or expander designs specify the inlet conditions (pressure and temperature) and the desired outlet pressure. Since the change in entropy of the fluid as it passes through an isentropic process is zero, the entropy of the fluid leaving an isentropic process must equal the entropy of the fluid entering the process. Thus if we know any two properties of the fluid entering the process, we can fix the fluid's state, and therefore determine its entropy. From the inlet entropy, we know the outlet entropy, which, together with the outlet pressure, fixes the state of the exiting fluid. Therefore, all other outlet properties must have fixed, but not necessarily known, values. As we are about to see, determination of these values is a straightforward, but often long and tedious, job.

We shall discuss the determination of the final temperature first. Once the final temperature is known, the three-step process discussed in Section 6-6 may be used to evaluate changes in $\underline{U}$ and $\underline{H}$, and thereby allow us to calculate work transferred to the surroundings during the process.

Since the equations that must be solved to determine the exit temperature are frequently extremely complicated, a trial-and-error solution for the final temperature is usually indicated. We may envision the logic of this trial-and-error solution as follows:

1. Estimate a final temperature T_2.
2. Using the three-step process of Section 6-6, and equations (6-86), (6-88), and (6-90), calculate the total change in entropy that would occur in a process between P_1, T_1 and P_2, T_2.
3. From the $\Delta \underline{S}$ of step 2, estimate a new value for T_2 to bring $\Delta \underline{S}$ closer to zero.
4. Repeat steps 2 and 3 until the change in T of step 3 is within the desired accuracy.

Another method of solution is to simply guess a whole series of T_2's and then calculate the $\Delta \underline{S}$'s for a process between P_1, T_1 and each P_2, T_2. A plot of $\Delta \underline{S}$ versus T_2 is then interpolated to $\Delta \underline{S} = 0$, which gives T_2.

The second method of solutions has the advantage that it does not require us to devise a systematic method of correcting our last guess on T_2 (step 3 of the first method) and, therefore, for one-shot calculations is probably easier to use than the first method. However, for calculations that are likely to be repeated frequently (as in design of a whole series of similar compressors), the first method will require less calculations per case studied once the correction algorithm is devised and, therefore, is preferred. In addition, the first method can be programmed for computer solution; the second cannot.

SAMPLE PROBLEM 6-7. If the cooling water to the compressor of Sample Problem 6-6 fails so that the compressor now operates adiabatically (and reversibly) rather than isothermally, determine the outlet temperature of the gas and the amount of work per mole of helium that must be supplied to the compressor to maintain an outlet pressure of 180 atm if

(a) Helium behaves as an ideal gas.

(b) Helium behaves according to the equation of state

$$P\underline{V} = RT + \left(b - \frac{a}{T}\right)P$$

where $a = 11.13$ °R ft³/lb-mol

$b = 0.2445$ ft³/lb-mol

C_P^* may be assumed to be constant:

$$C_P^* = 4.97 \text{ Btu/lb-mol °R}$$

Solution: Application of the energy and entropy balances around the compressor yields the following relations:

Energy:

$$\left(\underline{H} + \frac{u^2}{2g_C} + \frac{gZ}{g_C}\right)_{\text{in}} \delta M_{\text{in}} - \left(\underline{H} + \frac{u^2}{2g_C} + \frac{gZ}{g_C}\right)_{\text{out}} \delta M_{\text{out}} + \delta Q - \delta W = d(M\underline{E})_{\text{syst}}$$

Entropy:

$$(\underline{S}\,\delta M)_{\text{in}} - (\underline{S}\,\delta M)_{\text{out}} + \int_{\text{area}} \frac{\delta q dA}{T} + \int_{\text{volume}} \frac{\delta lw dV}{T} = d(M\underline{S})_{\text{syst}}$$

Since the compressor operates at steady state, $d(M\underline{E})_{\text{syst}} = d(M\underline{S})_{\text{syst}} = \delta M_{\text{in}} - \delta M_{\text{out}} = 0$. The compressor also operates adiabatically and reversibly, so $\delta q = \delta lw = 0$. Neglecting potential and kinetic energy changes in the compressor, the energy and entropy equations reduce to

Energy:

$$\Delta\underline{H}_{1-2} = -\underline{W}$$

Entropy:

$$\Delta\underline{S}_{1-2} = 0$$

Therefore, we know that the entropy of the material leaving the compressor is identical to the entropy of the material entering the compressor. We may now use this condition to calculate the temperature of the material leaving the compressor, which may then be used to calculate $\Delta\underline{H}_{1-2}$ across the compressor, which in turn tells us the work that must be supplied.

(a) Ideal gas: $P\underline{V} = RT$. We have shown previously that the final T and P of an ideal (with constant C_P) undergoing an isentropic process are related according to the expression

$$C_P \ln \frac{T_2}{T_1} = R \ln \frac{P_2}{P_1}$$

That is,

$$\ln\frac{T_2}{T_1} = \frac{R}{C_P}\ln\frac{P_2}{P_1}$$

or

$$T_2 = T_1\left(\frac{P_2}{P_1}\right)^{R/C_P}$$

But

$$T_1 = 540°\text{R}$$
$$P_2 = 180 \text{ atm}$$
$$P_1 = 12 \text{ atm}$$
$$C_P = 4.97 \text{ Btu/lb-mol °R}$$
$$R = 1.987 \text{ Btu/lb-mol °R}$$

Therefore,

$$T_2 = 540°\text{R}\,\frac{180}{12}\,\frac{1.987}{4.97} = 1595\,°\text{R}$$

and the total change in enthalpy may be calculated as

$$\Delta\underline{H}_{1-2} = \int_{T_1}^{T_2} C_P\,dT = (4.97)(1595 - 540) \text{ Btu/lb-mol}$$
$$= 5250 \text{ Btu/lb-mol}$$

(b) Real gas: $P\underline{V} = RT + [b - a/T)]P$. We must now attempt to find T_2 such that $\Delta\underline{S}_{1-2}$ between P_1, T_1 and P_2, T_2 is zero. According to the first procedure outlined in the text, the following trial-and-error solution will be attempted.

1. Guess an initial value of T_2.
2. Calculate $\Delta\underline{S}$ from P_1, T_1 to P_2, T_2.
3. Correct last guess for T_2.
4. Repeat steps 2 and 3.

For our initial guess of T_2, let us choose $T_2 = 1550°$R. (Usually a wise initial choice for T_2 is something near the value determined for the ideal gas case.)

We must now set up our three-step path for the evaluation of $\Delta\underline{S}$. A P–T diagram for the path is given in Fig. S6-7a and the total entropy change $\Delta\underline{S}_{1-2}$ is given by

$$\Delta\underline{S}_{1-2} = \Delta\underline{S}_{1-a} + \Delta\underline{S}_{a-b} + \Delta\underline{S}_{b-2}$$

where

$$\Delta\underline{S}_{1-a} = \int_{P_1}^{P^*} -\left(\frac{\partial\underline{V}}{\partial T}\right)_P dP = \int_{P_1}^{P^*} -\left(\frac{R}{P} + \frac{a}{T^2}\right) dP = -R\ln\frac{P^*}{P_1} - \frac{a}{T_1^2}(\overset{0}{P^*} - P_1)$$

$$\Delta\underline{S}_{a-b} = \int_{T_1}^{T_2} \frac{C_P^*}{T}\,dT = C_P^*\ln\frac{T_2}{T_1}$$

$$\Delta\underline{S}_{b-2} = \int_{P^*}^{P_2} -\left(\frac{\partial V}{\partial T}\right)_P dP = -R\ln\frac{P_2}{P^*} - \frac{a}{T_2^2}(P_2 - \overset{0}{P^*})$$

where P^* is a very low pressure approaching zero. Therefore,

$$\Delta\underline{S}_{1-2} = -R\ln\frac{P_2}{P_1} + C_P^*\ln\frac{T_2}{T_1} + \frac{aP_1}{T_1^2} - \frac{aP_2}{T_2^2}$$

but

$$-R\ln\frac{P_2}{P_1} = -1.987 \text{ Btu/lb-mol°R}\ln\frac{180}{12} = -5.36 \text{ Btu/lb-mol °R}$$

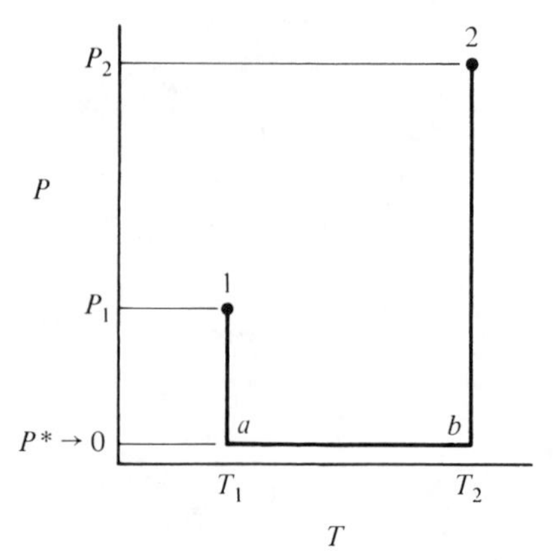

FIGURE S6-7a

and

$$\frac{aP_1}{T_1^2} = \frac{(11.13)(12.)(2.72)}{(540)^2} \text{ Btu/lb-mol °R} = 1.24 \times 10^{-3} \text{ Btu/lb-mol °R}$$

$$aP_2 = (11.13)(180)(2.72) \text{ Btu °R/lb-mol} = 5.45 \times 10^3 \text{ Btu °R/lb-mol}$$

Therefore,

$$\Delta \underline{S}_{1-2} = -5.36 \text{ Btu/lb-mol °R} + C_P^* \ln \frac{T_2}{540} - \frac{5.45 \times 10^3}{(T_2)^2} \text{ Btu °R/lb-mol}$$

We set up Table S6-7 in order to evaluate $\Delta \underline{S}_{1-2}$ as a function of T_2. We now see that our first guess of T_2 was too low and we must raise it—but by how much? We may frequently obtain a pretty good estimate for the next guess of T_2 as follows.

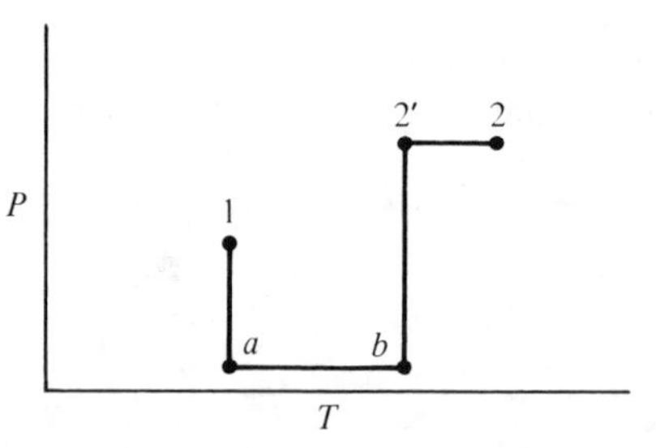

FIGURE S6-7b

TABLE S6-7.

T_2	$\frac{T_2}{540}$	$C_P^* \ln \frac{T_2}{540}$, $\frac{\text{Btu}}{\text{lb-mol °R}}$	$\frac{5.45 \times 10^3}{(T_2)^2}$, $\frac{\text{Btu}}{\text{lb-mol °R}}$	$\Delta \underline{S}_{1-2}$, $\frac{\text{Btu}}{\text{lb-mol °R}}$
1550	2.37	5.25	2.26×10^{-3}	−0.11
1590	2.94	5.37	2.16×10^{-3}	~0

Let us examine the three-step path actually followed during the previous calculation with $T_{2'}$ as the final temperature (Fig. S6-7b). Let point 2 be the actual point where

$\Delta\underline{S}_{1-2} = 0$. We have evaluated $\Delta\underline{S}_{1-2'} = -0.11$ Btu/lb-mol °R. Therefore $\Delta\underline{S}_{2'-2}$ should equal 0.11 Btu/lb-mol °R. But $(\Delta\underline{S})_P$ is given by

$$(\Delta\underline{S})_P = \int_{T_{2'}}^{T_2} \frac{C_P}{T}\, dT$$

Now as an approximation, assume that $C_P = C_P^*$. Therefore,

$$\Delta\underline{S}_{2'-2} = C_P^* \ln \frac{T_2}{T_{2'}}$$

or

$$0.11 \text{ Btu/lb-mol °R} = 4.97 \text{ Btu/lb-mol °R} \ln T_2/1550\text{°R}$$

Therefore,

$$\ln \frac{T_2}{1550} = 0.0235$$

or

$$\frac{T_2}{1550} = 1.0238$$

so

$$T_2 = 1590\text{°R}$$

which we now use as our second guess for T_2. Performing the calculations a second time, we find that our second guess, $T = 1590$°F, satisfies the $\Delta\underline{S}_{1-2} = 0$ condition to within the accuracy of our calculations and therefore is the final temperature.

With the final temperature, we may now calculate the change in enthalpy of the gas as it passes through the process from the familiar three-part path:

$$\Delta\underline{H}_{1-2} = \Delta\underline{H}_{1-a} + \Delta\underline{H}_{a-b} + \Delta\underline{H}_{b-2}$$

where

$$\Delta\underline{H}_{1-a} = \int_{P_1}^{P^*} \left[\underline{V} - T\left(\frac{\partial \underline{V}}{\partial T}\right)_P\right] dP = \left(b - \frac{2a}{T_1}\right)(P^* - P_1) \quad (P^* \to 0)$$

$$\Delta\underline{H}_{a-b} = \int_{T_1}^{T_2} C_P\, dT = C_P^*\,(1590 - 540)\text{°R}$$

$$\Delta\underline{H}_{b-2} = \int_{P^*}^{P_2} \left[\underline{V} - T\left(\frac{\partial \underline{V}}{\partial T}\right)_P\right] dP = \left(b - \frac{2a}{T_2}\right)(P_2 - P^*) \quad (P^* \to 0)$$

Substitution of the values for a, b, P_1, P_2, T_1, and T_2 yields the value

$$\Delta\underline{H}_{1-2} = 5360 \text{ Btu/lb-mol}$$

so

$$\underline{W} = -5360 \text{ Btu/lb-mol}$$

SAMPLE PROBLEM 6-8. A lean natural gas stream (essentially pure methane) is available at 100°F and 200 psia. The gas is to be compressed to 5000 psia before transmission by underground pipeline. The compression will be adiabatic, and the methane may be assumed to obey the Beattie–Bridgeman equation of state with

$$C_P^* = 4.52 + 0.00737T$$

where C_P = Btu/lb-mol °R
T = °R

Determine the compressor outlet temperature and work of compression (ft-lb_f/lb_m) if the compressor

(a) operates reversibly.

(b) operates irreversibly with an efficiency of 75 per cent.

Solution: Let us attack part (a) first. If the compressor is to operate adiabatically and reversibly, then a simple entropy balance tells us that the methane undergoes an isentropic process during the compression and hence its entropy after compression is identical to its entering entropy. Since we know the inlet temperature and pressure, we know the state of the incoming methane and hence its entropy.

Since we know the entropy and pressure of the exiting methane, its state is fixed, and all other properties (T, $\underline{H}$, $\underline{V}$, etc.) are in principle known. Our problem is, of course, to remove the words "in principle."

One's first thoughts for solution would probably be to use the same technique as outlined in Sample Problem 6-7. That is, guess a final T, express $\underline{S}$ as a function of T via a three-step process on a P–T diagram, and find T such that $\Delta\underline{S} = 0$. Unfortunately, this procedure is subject to severe computational difficulties. In particular, during the evaluation of $\underline{S}$ along the two isothermal legs of the three-step process the integral $\int (\partial\underline{V}/\partial T)_P\, dP$ is needed. Because the Beattie–Bridgeman equation of state is pressure explicit—that is, written in the form $P = P(T, \underline{V})$—the expression for the derivative $(\partial\underline{V}/\partial T)_P$ is quite complicated and its integral fairly difficult to evaluate. Since most of the commonly used equations of state (van der Waals, Redlich–Kwong, virial, Benedict–Webb–Rubin, etc.) are pressure, rather than volume, explicit, this is a common difficulty. (A very similar problem arises in the evaluation of $\Delta\underline{H}$ with the standard three-step process. Here the integral $\int [\underline{V} - T(\partial\underline{V}/\partial T)_P]\, dP$ is needed. The difficulty is even more severe here because most of the pressure-explicit equations of state cannot be solved explicitly for volume. In this case it is not possible to perform the integration directly in terms of pressure.) The difficulties associated with both of these calculations can be avoided if we express the changes in entropy and enthalpy in terms of volume and its derivatives rather than pressure and its derivatives. For example, express

$$\underline{S} = S(T, \underline{V})$$

$$d\underline{S} = \frac{C_V}{T}\, dT + \left(\frac{\partial P}{\partial T}\right)_{\underline{V}} d\underline{V}$$

For enthalpy let

$$\underline{H} = \underline{U} + P\underline{V}$$

where

$$\underline{U} = \underline{U}(T, \underline{V})$$

$$d\underline{U} = C_V\, dT + \left[T\left(\frac{\partial P}{\partial T}\right)_{\underline{V}} - P\right] d\underline{V}$$

Now if we put our three-step processes on a T–$\underline{V}$ (rather than T–P) diagram, the evaluation of $\underline{S}$ or $\underline{U}$ can be simply expressed in terms of the initial temperature and volume and the final temperature and volume. Changes in enthalpy are found by adding $\Delta P\underline{V}$ to $\Delta\underline{U}$. Thus if we knew the outlet density, or specific volume we could express $\Delta\underline{S}$ between inlet and outlet as a function of the outlet temperature and solve for T such that $\Delta\underline{S} = 0$. Unfortunately, we do not know the outlet specific volume but instead know the outlet pressure. Thus we must add a second equation to the entropy condition. This equation must relate the outlet temperature, pressure, and volume and is of course nothing more than the equation of state. Thus the two equations

$$\underline{S}(T_2, \underline{V}_2) = \underline{S}(T_1, \underline{V}_1)$$
$$P(T_2, \underline{V}_2) = P_2 = 5000 \text{ psia}$$

in the unknowns T_2 and $\underline{V}_2$ must be solved simultaneously. Since both of the equations in T_2 and $\underline{V}_2$ are highly nonlinear, we shall not attempt an exact solution but will use the Newton–Raphson procedure to perform an iterative solution.

Let us now develop the specific forms of the two equations that must be satisfied. Consider the three-step process pictured on the $\underline{V}$–T diagram of Fig. S6-8, where path

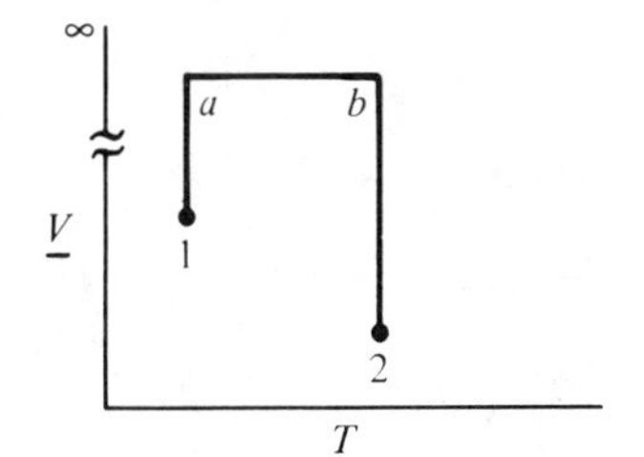

FIGURE S6-8

$1 - a$ is at $T = T_1$, path $b - 2$ is at $T = T_2$, and path $a - b$ is at $\underline{V} = \infty$. The total entropy change from 1 to 2 is given by

$$\Delta\underline{S}_{1-2} = \Delta\underline{S}_{1-a} + \Delta\underline{S}_{a-b} + \Delta\underline{S}_{b-2}$$

But

$$\Delta\underline{S}_{1-a} = \int_{\underline{V}_1}^{\infty} \left(\frac{\partial P}{\partial T}\right)_{\underline{V}} d\underline{V}$$

$$\Delta\underline{S}_{a-b} = \int_{T_1}^{T_2} \frac{C_{\underline{V}}^* \, dT}{T}$$

$$\Delta\underline{S}_{b-2} = \int_{\underline{V}_2}^{\infty} \left(\frac{\partial P}{\partial T}\right)_{\underline{V}} d\underline{V}$$

From the virial form of the Beattie–Bridgeman equation,

$$P = \frac{RT}{\underline{V}} + \frac{\beta(T)}{\underline{V}^2} + \frac{\gamma(T)}{\underline{V}^3} + \frac{\delta(T)}{\underline{V}^4}$$

where $\beta(T)$, $\gamma(T)$, and $\delta(T)$ have their standard meanings. Differentiation of P with respect to T at constant $\underline{V}$ gives

$$\left(\frac{\partial P}{\partial T}\right)_{\underline{V}} = \frac{R}{\underline{V}} + \frac{\beta'(T)}{\underline{V}^2} + \frac{\gamma'(T)}{\underline{V}^3} + \frac{\delta'(T)}{\underline{V}^4}$$

where the prime indicates d/dT. Thus

$$\Delta\underline{S}_{1-a} = \int_{\underline{V}_1}^{\infty} \left(\frac{\partial P}{\partial T}\right)_{\underline{V}} d\underline{V} = \int_{\underline{V}_1}^{\infty} \left[\frac{R}{\underline{V}} + \frac{\beta'(T)}{\underline{V}^2} + \frac{\gamma'(T)}{\underline{V}^3} + \frac{\delta'(T)}{\underline{V}^4}\right] d\underline{V}$$

$$= \left[R \ln \underline{V} - \frac{\beta'(T)}{\underline{V}} - \frac{\gamma'(T)}{2\underline{V}^2} - \frac{\delta'(T)}{3\underline{V}^3}\right]_{\underline{V}_1}^{\infty}$$

$$= R \ln \frac{\infty}{\underline{V}_1} - \beta'(T_1)\left(\frac{1}{\infty} - \frac{1}{\underline{V}_1}\right) - \frac{\gamma'(T_1)}{2}\left(\frac{1}{\infty^2} - \frac{1}{\underline{V}_1^2}\right) - \frac{\delta'(T_1)}{3}\left(\frac{1}{\infty^3} - \frac{1}{\underline{V}_1^3}\right)$$

or

$$\Delta \underline{S}_{1-a} = R \ln \frac{\infty}{\underline{V}_1} + \frac{\beta'(T_1)}{\underline{V}_1} + \frac{\gamma'(T_1)}{2\underline{V}_1^2} + \frac{\delta'(T_1)}{3\underline{V}_1^2}$$

Similarly,

$$\Delta \underline{S}_{b-2} = \int_{\infty}^{\underline{V}_2} \left(\frac{\partial P}{\partial T}\right)_{\underline{V}} d\underline{V}$$

$$= R \ln \frac{\underline{V}_2}{\infty} - \frac{\beta'(T_2)}{\underline{V}_2} - \frac{\gamma'(T_2)}{2\underline{V}_2^2} - \frac{\delta'(T_2)}{3\underline{V}_2^3}$$

Since the ∞ volumes along both legs are equivalent, when $\Delta \underline{S}_{1-a} + \Delta \underline{S}_{b-2}$ are added the $R \ln (\infty/\infty)$ terms cancel and we obtain

$$\Delta \underline{S}_{1-a} + \Delta \underline{S}_{b-2} = R \ln \frac{\underline{V}_2}{\underline{V}_1} + \frac{\beta'(T_1)}{\underline{V}_1} + \frac{\gamma'(T_1)}{2\underline{V}_1^2} + \frac{\delta'(T_1)}{3\underline{V}_1^3} - \frac{\beta'(T_2)}{\underline{V}_2} - \frac{\gamma'(T_2)}{2\underline{V}_2^2} - \frac{\delta'(T_2)}{3\underline{V}_2^3}$$

The last portion of the three-part path is then $\Delta \underline{S}_{a-b}$ and is given by

$$\Delta S_{a-b} = \int_{T_1}^{T_2} \frac{C_V^*}{T} dT = \int_{T_1}^{T_2} \left[\frac{CVA}{T} + CVB + CVC(T)\right] dT$$

$$= CVA \ln \frac{T_2}{T_1} + CVB(T_2 - T_1) + \frac{CVC}{2}(T_2^2 - T_1^2)$$

The total change in entropy between any two points is given by the sum

$$\Delta \underline{S}_{1-2} = \Delta \underline{S}_{1-a} + \Delta \underline{S}_{a-b} + \Delta \underline{S}_{b-2}$$

so

$$\Delta \underline{S}_{1-2} = R \ln \frac{\underline{V}_2}{\underline{V}_1} + \frac{\beta'(T_1)}{\underline{V}_1} + \frac{\gamma'(T_1)}{2\underline{V}_1^2} + \frac{\delta'(T_1)}{3\underline{V}_1^3} - \frac{\beta'(T_2)}{\underline{V}_2} - \frac{\gamma'(T_2)}{2\underline{V}_2^2} - \frac{\delta'(T_2)}{3\underline{V}_2^3} + CVA \ln \frac{T_2}{T_1} + CVB(T_2 - T_1) + \frac{CVC}{2}(T_2^2 - T_1^2)$$

But we know from the isentropic conditions that $\Delta \underline{S}_{1-2} = 0$, which gives us one equation in the two unknowns $(T_2, \underline{V}_2)$. As previously indicated, the second equation we need is simply the equation of state:

$$P_2 - \frac{RT_2}{\underline{V}_2} - \frac{\beta(T_2)}{\underline{V}_2^2} - \frac{\gamma(T_2)}{\underline{V}_2^3} - \frac{\delta(T_2)}{\underline{V}_2^4} = 0$$

Since P_2 is known (in our specific problem $P_2 = 5000$ psia), the equation of state adds no new unknowns, and we have two equations in two unknowns. Examination of these equations indicates that they are coupled and highly nonlinear. Therefore, a direct solution seems quite unlikely, and we shall instead rely on the Newton–Raphson iteration procedure to provide an efficient iterative solution. Our two equations (in T_2 and $\underline{V}_2$) are already in the form

$$F(T_2, \underline{V}_2) = 0 \qquad \text{(entropy condition)}$$

$$G(T_2, \underline{V}_2) = 0 \qquad \text{(equation of state)}$$

The two-unknown iteration scheme may be derived in exactly the same manner as the three-unknown scheme of Sample Problem 4-7. Only the results will be presented here. Given a kth guess $(T_2^k, \underline{V}_2^k)$ we obtain a (hopefully) better guess $(T_2^{k+1}, \underline{V}_2^{k+1})$ from the equations

$$-F(T_2^k, \underline{V}_2^k) = (T_2^{k+1} - T_2^k)\frac{\partial F(T_2^k, \underline{V}_2^k)}{\partial T} + (\underline{V}_2^{k+1} + \underline{V}_2^k)\frac{\partial F(T_2^k, \underline{V}_2^k)}{\partial \underline{V}}$$

$$-G(T_2^k, \underline{V}_2^k) = (T_2^{k+1} - T_2^k)\frac{\partial G(T_2^k, \underline{V}_2^k)}{\partial T} + (\underline{V}_2^{k+1} - \underline{V}_2^k)\frac{\partial G(T_2^k, \underline{V}_2^k)}{\partial \underline{V}}$$

Thus all we need do is choose an initial guess and iterate until satisfactory convergence is obtained. We may obtain a reasonable initial guess as follows:

1. Evaluate $C_V(T_1) = CVA + CVB(T_1) + CVC(T_1^2)$ and $C_P(T_1) = C_V(T_1) + R$.
2. Assume that the gas is ideal and that C_P is constant with T. Then the final temperature and volume may be estimated from the isentropic, ideal gas equations:

$$T_2 = T_1\left(\frac{P_2}{P_1}\right)^{R/C_P}$$

$$\underline{V}_2 = \underline{V}_1\left(\frac{P_1}{P_2}\right)^{C_V/C_P}$$

3. These estimates are then used as the starting values. The partial derivatives of $F(T_2, \underline{V}_2)$ and $G(T_2, \underline{V}_2)$ are given by

$$\frac{\partial F(T_2, \underline{V}_2)}{\partial T} = -\frac{\beta''(T_2)}{\underline{V}_2} - \frac{\gamma''(T_2)}{2\underline{V}_2^2} - \frac{\delta''(T_2)}{3\underline{V}_2^3} + \frac{CVA}{T_2} + CVB + CVC(T_2)$$

$$\frac{\partial F(T_2, \underline{V}_2)}{\partial \underline{V}} = \frac{R}{\underline{V}_2} + \frac{\beta'(T_2)}{\underline{V}_2^2} + \frac{\gamma'(T_2)}{\underline{V}_2^3} + \frac{\delta'(T_2)}{\underline{V}_2^4}$$

$$\frac{\partial G(T_2, \underline{V}_2)}{\partial T} = -\left[\frac{R}{\underline{V}_2} + \frac{\beta'(T_2)}{\underline{V}_2^2} + \frac{\gamma'(T_2)}{\underline{V}_2^3} + \frac{\delta'(T_2)}{\underline{V}_2^4}\right]$$

$$\frac{\partial G(T_2, \underline{V}_2)}{\partial \underline{V}} = \frac{RT_2}{\underline{V}_2^2} + \frac{2\beta(T_2)}{\underline{V}_2^3} + \frac{3\gamma(T_2)}{\underline{V}_2^4} + \frac{4\delta(T_2)}{\underline{V}_2^5}$$

A program to carry out the necessary calculations is listed below.

```
C     --- PROGRAM TO DETERMINE THE OUTLET CONDITIONS FROM AN
C         ISENTROPIC COMPRESSION
C         WHEN THE INLET CONDITIONS AND OUTLET PRESSURE ARE KNOWN ---
          DIMENSION AA(2, 2), BB(2), CC(2)
          COMMON / SAFETY / R, A0, A, B0, B, C
C     --- DEFINE FUNCTIONS TO BE USED LATER ---
          BETA(T) = R*B0*T - R*C/(T*T) - A0
          GAMMA(T) = A0*A - R*B0*B*T - R*B0*C/(T*T)
          DELTA(T) = R*B0*B*C/(T*T)
          BPRIME(T) = R*B0 + 2.0*R*C/(T*T*T)
          GPRIME(T) = 2.0*R*B0*C/(T*T*T) - R*BO*B
          DPRIME(T) = -2.0*R*BO*B*C/ (T*T*T)
          PRESS(T,VBAR) = R*T/VBAR + BETA(T) / (VBAR*VBAR) + GAMMA(T)/
         1 VBAR**3 + DELTA(T) / VBAR**4
          UBAR(T,VBAR) = CVA*(T-T0) + CVB*(T*T - T0*T0)/ 2.0 + CVC*(T**3
         1 - T0**3)/3.0 - (1.0/VBAR)*(A0 + 3.0*R*C/(T*T)) + (0.5/(VBAR*
         2 VBAR))*( A0*A - 3.0*R*B0*C/(T*T)) + R*B0*B*C/((VBAR**3)*T*T)
          BPPRME(T) = -6.0*R*C/(T**4)
          GPPRME(T) = -6.0*R*B0*C/(T**4)
          DPPRME(T) = 6.0*R*B0*B*C/(T**4)
          HBAR(T,VBAR) = UBAR(T, VBAR) + PRESS(T, VBAR)*VBAR
```

```
      DSBAR(T, VBAR) = R* ALOG(VBAR/V1) + BPRIME(T1)/V1 + GPRIME(T1)/
     1(2.0*V1*V1) + DPRIME(T1)/ (3.*V1*V1*V1) - BPRIME(T)/VBAR - GPRIME
     2(T)/ (2.*VBAR*VBAR) - DPRIME(T)/(3.0*(VBAR**3)) + CVA*ALOG(T/T1)
     3 + CVB*(T -T1) + 0.5*CVC*(T*T - T1*T1)
C   --- START EXECUTION OF ACTUAL PROGRAM ---
 10   READ(105,20) R, T1, P1, A0, A, B0, B, C, EPSP, EPST, EPSV, PEND,
     1 CVA, CVB, CVC, T0
 20   FORMAT (5E16.8)
 30   WRITE(108, 40) A0, A; B0, B, C, R, EPSP, EPST, EPSV, T1, P1, PEND
     1, CVA, CVB, CVC, T0
 40   FORMAT (1H1, 36H THE BEATTIE-BRIDGEMAN CONSTANTS ARE, 7H A0 = ,
     1E10.4,5H A = ,E10.4// 6H B0 = ,E10.4,5H B = ,E10.4,5H C = , E10.4/
     21H0, 5H R = , E10.4, 8H EPSP = , E10.4, 8H EPST = , E10.4, 8H EPS
     3V = , E10.4 /1H0, 6H T1 = , E10.4, 6H P1 = , E10.4, 8H PEND = ,
     4 E10.4 /1HO, 7H CVA = , E10.4, 7H CVB = , E10.4, 7H CVC = , E10.4,
     5 6H T0 = , E10.4)
C   --- CALCULATE INLET SPECIFIC VOLUME USING VOLUME SUBROUTINE ---
 50   V1 = VOLUME(P1, T1, EPSP)
 60   HIN = HBAR(T1, V1)
C   --- MAKE INITIAL GUESSES AT VALUES OF T2 AND V2 ---
 70   CV = CVA + CVB*T1 + CVC*T1*T1
 80   CP = CV + R
 90   T2 = T1*(PEND/P1)**(R/CP)
 100  V2 = V1*(P1/PEND)**(CV/CP)
C   --- ENTER NEWTON-RAPHSON ITERATION ---
 110  DO 200 IT = 1, 15
 120  FTV = DSBAR(T2, V2)
 130  GTV = PEND - PRESS(T2, V2)
C   --- EVALUATE THE FOUR PARTIAL DERIVATIVES ---
 140  DFDT = CVA/T2 + CVB + CVC*T2 - BPPRME(T2)/ V2 - 0.5*GPPRME(T2)/
     1(V2*V2) - DPPRME(T2)/ (3.0*V2*V2*V2)
      DFDV = R/V2 + BPRIME(T2)/(V2*V2) + GPRIME(T2)/(V2*V2*V2) +
     1DPRIME(T2)/(V2**4)
      DGDT = -DFDV
      DGDV = R*T2/( V2*V2) + 2.0*BETA(T2)/ ( V2*V2*V2) + 3.0*GAMMA(T2)/
     1 ( V2**4) + 4.0*DELTA(T2)/ ( V2**5 )
C   --- LOAD DERIVATIVES IN TO COEFFICIENT MATRIX AND SOLVE FOR DEL(T)
C     AND DEL(V)
 150  AA(1,1) = DFDT
      AA(1,2) = DFDV
      CC(1) = -FTV
      AA(2,1) = DGDT
      AA(2,2) = DGDV
      CC(2) = -GTV
C   --- USE LIBRARY SUBROUTINE, "LINEQ" TO SOLVE EQUATIONS.
 160  CALL LINEQ( AA, CC, BB, 2,2, IFLAG)
C   --- UPDATE T2 AND V2, DEL T2 = BB(1), DEL V2 = BB(2) ---
 170  T2 = T2 + BB(1)
```

```
   180    V2 = V2 + BB(2)
C     - - - CHECK FOR CONVERGENCE - - -
   190    IF( ABS(BB(1)) .LT. ESPT . AND . ABS( BB(2)).LT. EPSV) GO TO 250
   200    WRITE (108, 210) IT, T2, V2
   210    FORMAT (1H0, 7H AFTER, I5, 17H ITERATIONS, T = , E15.5, 9H AND
         1V = , E15.5)
   220    WRITE (108, 230)
   230    FORMAT (1H0, 18H ITERATION FAILED)
   240    GO TO 10
   250    HOUT = HBAR(T2, V2)
   260    DELH = HOUT - HIN
   270    WRITE(108, 280) T2, V2, DELH
   280    FORMAT (1H0, 28H CONVERGENCE OBTAINED, T2 = , E15.5, 10H AND V2
         1 = , E15.5,/20H ENTHALPY CHANGE = , E15.5)
   290    GO TO 10
   300    END
          FUNCTION VOLUME (P, T, EPSP)
          COMMON/ SAFETY/ R, A0, A, B0, B, C
          BETA(T) = R*B0*T - R*C/T**2 - A0
          GAMMA(T) = A0*A - R*B0*B*T - R*B0*C/T**2
          DELTA(T) = R*B0*B*C/T**2
          PRESS(T,VBAR) = R*T/VBAR + BETA(T)/VBAR**2 + GAMMA(T)/VBAR**3
         1 + DELTA(T)/VBAR**4
    50    VBAR = R*T/P
    60    DO 100 IT = 1, 15
    70    ERR = PRESS(T, VBAR) - P
    80    IF (ABS(ERR) .LE. EPSP) GO TO 140
    90    VBAR = VBAR - ERR/(-R*T/VBAR**2 - 2.*BETA(T)/VBAR**3 - 3.*
         1 GAMMA(T)/VBAR**4 - 4.*DELTA(T)/VBAR**5)
   100    CONTINUE
   110    WRITE (108, 120)
   120    FORMAT (1H0, 15H NO CONVERGENCE)
   130    CALL EXIT
   140    VOLUME = VBAR
   150    RETURN
   170    END
THE BEATTIE-BRIDGEMAN CONSTANTS ARE A0 = .4830E 04 A = .1857E-01
B0 = .5593E-01 B = -.1589E-01 C = .7476E 06
R = .9650E 02 EPSP = .1000E 03 EPST = .1000E-01 EPSV = .1000E-02
T1 = .5600E 03 P1 = .2880E 05 PEND = .7200E 06
CVA = .1232E 03 CVB = .3570E 00 CVC = .0000E 00 T0 = .4600E 03
AFTER 1 ITERATIONS, T = .10809E 04 AND V = .16104E 00
AFTER 2 ITERATIONS, T = .10801E 04 AND V = .16221E 00
CONVERGENCE OBTAINED, T2 = .10801E 04 AND V2 = .16222E 00
ENTHALPY CHANGE = .25375E 06
```

Having completed the reversible case, we are ready to tackle the irreversible operation. Since the process is now irreversible $\Delta \underline{S}_{1-2} > 0$, but by an unknown amount. Thus the entropy condition is no longer useful in fixing the state of the fluid at the compressor discharge. Instead, we fall back on the energy balance and the efficiency factor. If the compression is adiabatic, $\underline{Q} = 0$ and the energy balance reduces to (assuming steady state and negligible changes in potential and kinetic energy)

$$\Delta\underline{H} = -\underline{W}$$

From the isentropic calculation we have $(\Delta\underline{H})_{rev} = (-\underline{W})_{rev} = 2.54 \times 10^4$ ft lb_f/lb_m. For the irreversible process we know that the efficiency factor is related to the actual and reversible works by

$$\text{efficiency} = \frac{(\underline{W})_{rev}}{(\underline{W})_{act}}$$

or

$$\underline{W}_{act} = \frac{(\underline{W})_{rev}}{\text{efficiency}}$$

or, again applying the energy balance.

$$(\Delta\underline{H})_{act} = \frac{(\Delta\underline{H})_{rev}}{\text{efficiency}}$$

Thus for the irreversible process we know the outlet enthalpy and pressure but still lack the outlet temperature. If we could express $\Delta\underline{H}$ as a simple function of T_1, P_1 and T_2, P_2 we would have to solve only a single equation for the T_2 which gave $(\Delta\underline{H})_{act} = (\Delta\underline{H})_{rev}/\text{efficiency}$. Unfortunately, as we have previously indicated, the pressure-explicit form of the Beattie–Bridgeman equation does not lend itself to expressions in T and P but rather to expressions in T and $\underline{V}$. Thus $\Delta\underline{H}$ is most conveniently expressed in terms of T_1, $\underline{V}_1$ and T_2, $\underline{V}_2$. This adds a second unknown, which of course requires a second equation. The second equation is simply the equation of state. Therefore, we must once again solve two equations in the two unknowns T_2 and $\underline{V}_2$. The expression for $\Delta\underline{H}_{1-2}$ is obtained from the previously developed equation for $\Delta\underline{U}$ (see Sample Problem 6-5) by the relationship

$$\Delta\underline{H} = \Delta\underline{U} + \Delta P\underline{V}$$

But

$$\underline{U}_1 = CVA(T_1 - T_0) + \frac{CVB}{2}(T_1^2 - T_0^2) + \frac{CVC}{3}(T_1^3 - T_0^3)$$
$$- \frac{1}{\underline{V}_1}\left(A_0 + \frac{3RC}{T_1^2}\right) + \frac{0.5}{\underline{V}_1^2}\left(A_0 a - \frac{3RB_0C}{T_1^2}\right) + \frac{RB_0bC}{\underline{V}_1^3 T_1^2}$$
$$P\underline{V}_1 = RT_1 + \frac{\beta(T_1)}{\underline{V}_1} + \frac{\gamma(T_1)}{\underline{V}_1^2} + \frac{\delta(T_1)}{\underline{V}_1^3}$$

These may be added to give the total expression for $\underline{H}(T_1, \underline{V}_1)$. The expression for $\underline{H}(T_2, \underline{V}_2)$ is found in an identical manner. The two equations that must be solved simultaneously are then

$$0 = -\Delta\underline{H}_{act} + [\underline{H}(T_2, \underline{V}_2) - \underline{H}(T_1, \underline{V}_1)] = F(T_2, \underline{V}_2)$$
$$0 = P_2 - P(T_2, \underline{V}_2) \qquad = G(T_2, \underline{V}_2)$$

Since these equations are again quite nonlinear we will not attempt a direct solution but will return once again to the Newton–Raphson iteration. As a starting guess we will simply use the results of the isentropic calculation. The partial derivatives of the second function are identical to the partial derivatives of the $G(T_2, \underline{V}_2)$ in the isentropic analysis and hence will not be presented here. The partial derivatives of the first function $F(T_2, \underline{V}_2)$ are

$$\frac{\partial F(T_2, \underline{V}_2)}{\partial T_2} = \frac{\partial \underline{H}(T_2, \underline{V}_2)}{\partial T_2} = \frac{\partial[\underline{U}(T_2, \underline{V}_2) + P(T_2, \underline{V}_2)(\underline{V}_2)]}{\partial T_2}$$

$$= CVA + CVB(T_2) + CVC(T_2^2) + \frac{6RC}{\underline{V}_2 T_2^3} + \frac{3RB_0C}{\underline{V}_2^2 T_2^3}$$

$$- \frac{2RB_0bC}{\underline{V}_2^3 T_2^3} + R + \frac{\beta'(T_2)}{\underline{V}_2} + \frac{\gamma'(T_2)}{\underline{V}_2^2} + \frac{\delta'(T_2)}{\underline{V}_2^3}$$

and, similarly,

$$\frac{\partial F(T_2, \underline{V}_2)}{\partial \underline{V}_2} = \frac{1.0}{\underline{V}_2^2}\left(A_0 + \frac{3RC}{T_2^2}\right) - \frac{1}{\underline{V}_2^3}\left(A_0 a - \frac{3RB_0C}{T_2^2}\right)$$

$$- \frac{3RB_0bC}{\underline{V}_2^4 T_2^2} - \frac{\beta(T_2)}{\underline{V}_2^2} - \frac{2\gamma(T_2)}{\underline{V}_2^3} - \frac{3\delta(T_2)}{\underline{V}_2^4}$$

A program to perform the required calculations is given below.

```
C    --- PROGRAM TO DETERMINE OUTLET CONDITIONS FROM NON-REVERSIBLE
C        COMPRESSION WHEN EFFICIENCY AND REVERSIBLE OUTLET
C        CONDITIONS ARE KNOWN ---
         COMMON /SAFETY/R, A0, A, B0, B, C
         DIMENSION AA(2,2), BB(2), CC(2)
C    --- DEFINE FUNCTIONS TO BE USED LATER ---
         BETA(T) = R*B0*T - R*C/(T*T) - A0
         GAMMA(T) = A0*A - R*B0*B*T - R*B0*C/(T*T)
         DELTA(T) = R*B0*B*C/(T*T)
         BPRIME(T) = R*B0 + 2.0*R*C/(T*T*T)
         GPRIME(T) = 2.0*R*B0*C/(T*T*T) - R*B0*B
         DPRIME(T) = -2.0*R*B0*B*C/ (T*T*T)
         PRESS(T,VBAR) = R*T/VBAR + BETA(T)/(VBAR*VBAR) + GAMMA(T)/
        1VBAR**3 + DELTA(T) / VBAR**4
         UBAR(T, VBAR) = CVA*(T-T0) + CVB*(T*T-T0*T0)/2.0 + CVC*(T**3
        1 - T0**3)/3.0 - (1.0/VBAR)*(A0 + 3.0*R*C/(T*T)) + (0.5/(VBAR*
        2 VBAR))*(A0*A - 3.0*R*B0*C/(T*T)) + R*B0*B*C/((VBAR**3)*T*T)
         HBAR(T,VBAR) = UBAR(T, VBAR) + PRESS(T, VBAR)*VBAR
         DSBAR(T, VBAR) = R*ALOG(VBAR/V1) + BPRIME(T1)/V1 + GPRIME(T1)/
        1(2.0*V1*V1) + DPRIME(T1)/ (3.*V1*V1*V1) - BPRIME(T)/VBAR - GPRIME
        2(T)/(2.*VBAR*VBAR) - DPRIME(T)/(3.0*(VBAR**3)) + CVA*ALOG(T/T1)
        3 + CVB*(T - T1) + 0.5*CVC*(T*T - T1*T1)
   10    IFLAG = 0
   20    READ(105, 30) R, T1, P1, A0, A, B0, B, C, EPSP, EPST, EPSV, PEND,
        1 CVA, CVB, CVC, T0, T2, V2, DHREV, EFF
   30    FORMAT (5E16.8)
   40    WRITE(108, 50) A0, A, B0, B, C, R, EPSP, EPST, EPSV, T1, P1, PEND
        1, CVA, CVB, CVC, T0,T2, V2, DHREV, EFF
   50    FORMAT (1H1, 36H THE BEATTIE-BRIDGEMAN CONSTANTS ARE, 7H A0 =,
        1E10.4,5H A = ,E10.4//6H B0 = ,E10.4,5H B = ,E10.4,5H C = ,E10.4/
        21H0, 5H R = ,E10.4, 8H EPSP = ,E10.4 8H EPST = ,E10.4, 8H EPS
```

```
         3V = ,E10.4 /1H0, 6H T1 = ,E10.4,6H P1 = ,E10.4, 8H PEND = ,
         4 E10.4/1HO, 7H CVA = ,E10.4,7H CVB = ,E10.4, 7H CVC = ,E10.4,
         5 6H T0 = ,E10.4/1H0, 6H T2 = ,E10.4, 6H V2 = ,E10.4,
         69H DHREV = ,E10.4, 7H EFF = ,F10.5)
C    - - - CALCULATE INITIAL SPECIFIC VOLUME - - -
   60    V1 = VOLUME(P1,T1,EPSP)
C    - - - CALCULATE ACTUAL ENTHALPY CHANGE - - -
   70    DHACT = (HBAR(T2,V2) - HBAR(T1,V1))/ EFF
C    - - - START NEWTON-RAPHSON ITERATION - - -
   80    DO 170 IT = 1, 15
   90    FTV = (HBAR(T2,V2) - HBAR(T1,V1)) - DHACT
  100    GTV = PEND - PRESS(T2,V2)
C    - - - CALCULATE THE FOUR PARTIAL DERIVATIVES - - -
  110    DFDT = CVA + CVB*T2 + CVC*T2*T2 + 6.0*R*C/((V2*V2)*(T2*T2*T2)) +
        1 3.0*R*B0*C/((V2*V2)*(T2*T2*T2)) - 2.0*R*B0*B*C/((V2*V2*V2)*
        2(T2*T2*T2)) + R + BPRIME (T2)/V2 + GPRIME(T2)/(V2*V2) + DPRIME(T2)/
        3(V2*V2*V2)
         DFDV = (A0 + 3.*R*C/(T2*T2))/(V2*V2) - (A0*A - 3.*R*B0*C/
        1 (T2*T2))/(V2*V2*V2) - 3.0*R*B0*B*C/((V2**4)*(T2*T2)) - BETA(T2)/
        2(V2*V2) - 2.0*GAMMA(T2)/(V2*V2*V2) - 3.0*DELTA(T2)/(V2**4)
         DGDT = -(R/V2 + BPRIME(T2)/(V2*V2) + GPRIME(T2)/(V2*V2*V2) +
        1DPRIME(T2)/(V2**4))
         DGDV = R*T2/(V2*V2) + 2.0*BETA(T2)/(V2*V2*V2) + 3.0*GAMMA(T2)/
        1 (V2**4) + 4.0*DELTA(T2)/(V2**5)
C    - - - LOAD DERIVATIVES INTO COEFFICIENT MATRIX AND SOLVE FOR DEL(T)
C          AND DEL(V)
  120    AA(1,1) = DFDT
         AA(1,2) = DFDV
         CC(1) = -FTV
         AA(2,1) = DGDT
         AA(2,2) = DGDV
         CC(2) = -GTV
  130    CALL LINEQ (AA, CC, BB, 2,2, IFLAG)
C    - - - UPDATE T2 AND V2, DEL T2 = BB(1), DEL V2 = BB(2) - - -
  140    T2 = T2 + BB(1)
  150    V2 = V2 + BB(2)
C    - - - CHECK FOR CONVERGENCE - - -
  160    IF (ABS(BB(1)) .LT.EPST.AND.ABS(BB(2)).LT.EPSV) GO TO 220
  170    WRITE (108, 180) IT, T2, V2
  180    FORMAT (1H0, 7H AFTER, 15, 17H ITERATIONS, T = ,E15.5, 9H AND
        1V = , E15.5)
  190    WRITE(108, 200)
  200    FORMAT (1H0, 18H ITERATION FAILED)
  210    GO TO 10
  220    WRITE(108, 230) T2, V2, DSBAR(T2,V2)
  230    FORMAT (1H0, 28H CONVERGENCE OBTAINED, T2 = , E15.5, 10H AND V2
        1 = , E15.5,/20H ENTROPY CHANGE = ,E15.5)
  240    GO TO 10
```

```
  250   END
        FUNCTION VOLUME (P, T, EPSP)
        COMMON / SAFETY / R, A0, A, B0, B, C
        BETA(T) = R*B0*T - R*C/T**2 - A0
        GAMMA(T) = A0*A - R*B0*B*T - R*B0*C/T**2
        DELTA(T) = R*B0*B*C/T**2
        PRESS(T, VBAR) = R*T/VBAR + BETA(T)/VBAR**2 + GAMMA(T)/VBAR**3
       1 + DELTA(T)/VBAR**4
   50   VBAR = R*T/P
   60   DO 100 IT = 1, 15
   70   ERR = PRESS(T, VBAR) - P
   80   IF (ABS(ERR).LE.EPSP) GO TO 140
   90   VBAR = VBAR - ERR/(-R*T/VBAR**2 - 2.*BETA(T)/VBAR**3 - 3,
       1*GAMMA(T)/VBAR**4 - 4.*DELTA(T)/VBAR**5)
  100   CONTINUE
  110   WRITE (108, 120)
  120   FORMAT (1H0, 15H NO CONVERGENCE)
  130   CALL EXIT
  140   VOLUME = VBAR
  150   RETURN
  170   END
THE BEATTIE-BRIDGEMAN CONSTANTS ARE A0 = .4830E 04 A = .1857E-01
B0 = .5593E-01 B = -.1589E-01 C = .7476E 06
R = .9650E 02 EPSP = .1000E 03 EPST = .1000E-01 EPSV = .1000E-02
T1 = .5600E 03 P1 = .2880E 05 PEND = .7200E 06
CVA = .1232E 03 CVB = .3570E 00 CVC = .0000E 00 TO = .4600E 03
T2 = .1080E 04 V2 = .1622E 00 DHREV = .2537E 06 EFF = .75000
AFTER 1 ITERATIONS, T = .12073E 04 AND V = .18254E 00
AFTER 2 ITERATIONS, T = .12062E 04 AND V = .18207E 00
AFTER 3 ITERATIONS, T = .12062E 04 AND V = .18207E 00
CONVERGENCE OBTAINED, T2 = .12062E 04 AND V2 = .18207E 00
ENTROPY CHANGE = .74018E 02
```

6.8 Fugacities and the Fugacity Coefficient

From the Gibbs free-energy-property relation we learn that

$$\left(\frac{\partial \underline{G}}{\partial P}\right)_T = \underline{V} \tag{6-92}$$

or

$$d\underline{G} = \underline{V}\, dP \qquad \text{at constant } T \tag{6-92a}$$

For a fluid that obeys the ideal gas equation of state, $\underline{V} = RT/P$ and equation (6-92) becomes

$$(d\underline{G})_T = RT\frac{dP}{P} = RT\, d\ln P \tag{6-93}$$

For fluids that do not obey the ideal gas equation of state, equation (6-93) is not correct. However, its form is quite useful and thus we may define a new quantity—*the fugacity*—such that the form of (6-93) is retained even for gases that do not obey the ideal gas equation of state. The fugacity, f, is thus defined so that

$$(d\underline{G})_T = \underline{V}\, dP = RT\, d \ln f \qquad \text{at constant } T \tag{6-94}$$

Since equation (6-83) is a differential expression, it can only fix the fugacity to within a constant term. The value of the constant is determined by also requiring that

$$f \longrightarrow P \qquad \text{as } P \longrightarrow 0 \tag{6-95}$$

or

$$\frac{f}{P} \longrightarrow 1 \qquad \text{as } P \longrightarrow 0 \tag{6-95a}$$

Equations (6-94) and (6-95) or (6-95a) then completely define the fugacity function.

The fugacity coefficient, ν, is defined as the ratio of the fugacity to the pressure:

$$\nu = \frac{f}{P} \tag{6-96}$$

The fugacity coefficient for a fluid may be expressed in terms of the P–$\underline{V}$–T behavior of the fluid as follows. We begin with equation (6-94):

$$RT\left(\frac{\partial \ln f}{\partial P}\right)_T = \underline{V} \tag{6-94}$$

Subtract $RT(\partial \ln P/\partial P)_T = RT/P$ from both sides of equation (6-94), which may be rearranged to give

$$RT\left(\frac{\partial \ln f/P}{\partial P}\right)_T = \underline{V} - \frac{RT}{P} \tag{6-97}$$

But $f/P = \nu$, so (6-97) reduces to

$$RT\left(\frac{\partial \ln \nu}{\partial P}\right)_T = \underline{V} - \frac{RT}{P} \tag{6-98}$$

Now integrate from $P = 0$ to the pressure of interest at constant temperature:

$$RT \int_{P=0}^{P} d \ln \nu = \int_{P=0}^{P} \left(\underline{V} - \frac{RT}{P}\right) dP \tag{6-99}$$

or

$$RT[(\ln \nu)_P - (\ln \nu)_{P=0}] = \int_{P=0}^{P} \left(\underline{V} - \frac{RT}{P}\right) dP \tag{6-100}$$

But from the definition of the fugacity, at $P = 0$, $\nu = 1$, so equation (6-100) reduces to

$$RT \ln \nu = \int_{P=0}^{P} \left(\underline{V} - \frac{RT}{P}\right) dP \tag{6-101}$$

or

$$\ln \nu = \int_{P=0}^{P} \left(\frac{\underline{V}}{RT} - \frac{1}{P} \right) dP \tag{6-101a}$$

For a gas that obeys the ideal gas equation of state, $1/P = \underline{V}/RT$, so the right-hand side of equation (6-101a) vanishes and we find that $\nu = 1$ independent of P or T. Thus for an ideal gas the fugacity is simply equal to the pressure. For a gas that does not obey the ideal gas equation of state, the fugacity coefficient may be found by performing the integration in equation (6-101a).

6.9 Calculation of Fugacities from an Equation of State

Most of the useful equations of state represent pressure, P, as the dependent variable as a function of the independent variables temperature, T, and specific volume, $\underline{V}$. Since fugacity is a state function, it should be possible to use such equations to compute fugacity as a function of $\underline{V}$ and T. Our defining equation for fugacity,

$$\left(\frac{\partial \ln f}{\partial P} \right)_T = \frac{\underline{V}}{RT} \tag{6-94}$$

is not well suited to this task, however, because it treats P and T as the independent variables. We can, however, rearrange equation (6-94) to obtain the alternative expression,

$$\left(\frac{\partial \ln f}{\partial \underline{V}} \right)_T = \left(\frac{\partial \ln f}{\partial P} \right)_T \left(\frac{\partial P}{\partial \underline{V}} \right)_T = \frac{\underline{V}}{RT} \left(\frac{\partial P}{\partial \underline{V}} \right)_T \tag{6-102}$$

Thus, in principle, we could insert an equation-of-state expression for P here and, by integration, evaluate f as a function of $\underline{V}$ and T. Since $f \longrightarrow P$ as $P \longrightarrow 0$ ($\underline{V} \longrightarrow \infty$), the range of the integration should be $\underline{V} = \infty$ to $\underline{V}$. We note that, because of their logarithmic nature, some terms in this expression may not be bounded in the limit $\underline{V} \longrightarrow \infty$, and thus will seek an alternative form in which these problems are eliminated. Since the fugacity coefficient, $\nu = f/P$, approaches unity (rather than zero) as $P \longrightarrow 0$, we shall attempt to derive an expression for ν rather than f directly.

First we rearrange equation (6-102) to the form

$$\left(\frac{\partial \ln f}{\partial \underline{V}} \right)_T = \left(\frac{\partial}{\partial \underline{V}} \right)_T \frac{P\underline{V}}{RT} - \frac{P}{RT} \tag{6-103}$$

and note that

$$\left(\frac{\partial}{\partial \underline{V}} \right)_T \ln \frac{P\underline{V}}{RT} = \left(\frac{\partial \ln P}{\partial \underline{V}} \right)_T + \frac{1}{\underline{V}} \tag{6-104}$$

which can be rearranged to give

$$\left(\frac{\partial \ln P}{\partial \underline{V}} \right)_T = -\frac{1}{\underline{V}} + \left(\frac{\partial}{\partial \underline{V}} \right)_T \ln \frac{P\underline{V}}{RT} \tag{6-105}$$

Subtraction of equation (6-105) from (6-103) will yield the desired expression:

$$\left(\frac{\partial}{\partial \underline{V}}\right)_T \ln \frac{f}{P} = \frac{1}{\underline{V}} - \frac{P}{RT} + \left(\frac{\partial}{\partial \underline{V}}\right)_T \left(\frac{P\underline{V}}{RT} - \ln \frac{P\underline{V}}{RT}\right) \tag{6-106}$$

Integration over the range $\underline{V} = \infty$ to $\underline{V}$, noting that in the limit $\underline{V} \longrightarrow \infty$, $f/P \longrightarrow 1$, and $P\underline{V}/RT \longrightarrow 1$, produces

$$\ln \frac{f}{P} = \frac{1}{RT} \int_{\infty}^{\underline{V}} \left(\frac{RT}{\underline{V}} - P\right) d\underline{V} + \frac{P\underline{V}}{RT} - 1 - \ln \frac{P\underline{V}}{RT} \tag{6-107}$$

Although the first term in the integral might be integrated directly, it is convenient to leave it within the integral so as to assure convergence at $\underline{V} \longrightarrow \infty$.

Equation (6-107) is useful with any equation of state in which pressure is given as a function of $\underline{V}$ and T. The only limit on its application is the range of validity of the equation of state, for the derivation above has made no approximations that are not generally valid. For equations such as the Beattie–Bridgeman or Redlich–Kwong it can be used to compute vapor-phase fugacities, but for equations such as the Benedict–Webb–Rubin, which represent the liquid region as well as the vapor region, the equation can be used for vapor or liquid fugacities.

6.10 Evaluation of $\Delta \underline{H}$, v, and $\Delta \underline{S}$ Using the Law of Corresponding States

Enthalpy Changes

In the evaluation of $\Delta \underline{H}_{1-2}$ using the three-step process previously described, we showed that evaluation of the integrals

$$\Delta \underline{H}_{1-a} = \underline{H}_a - \underline{H}_1 = \int_{P_1}^{P^*} \left[\underline{V} - T_1 \left(\frac{\partial \underline{V}}{\partial T}\right)_P\right] dP \tag{6-85}$$

$$\Delta \underline{H}_{b-2} = \underline{H}_2 - \underline{H}_b = \int_{P^*}^{P_2} \left[\underline{V} - T_2 \left(\frac{\partial \underline{V}}{\partial T}\right)_P\right] dP \tag{6-89}$$

was necessary. Since these integrals depend solely on the P–$\underline{V}$–T behavior of the substance in question, we might hope to express the integrands only in terms of compressibility, Z, and the reduced variables, P_r and T_r. We could then use the generalized compressibility charts to evaluate these integrals as functions of P_r and T_r. We shall now show how this evaluation may be performed.

In the generalized charts, the compressibility factor, $Z = P\underline{V}/RT$, is expressed as a function of P_r and T_r. Therefore, before we can use these charts for the evaluation of the equations for $\Delta \underline{H}_{1-a}$ and $\Delta \underline{H}_{b-2}$, we must first express the integrals in terms of only reduced variables and compressibilities. This is accomplished as follows. Rearrangement of the definition of Z gives an expression for $\underline{V}$:

$$\underline{V} = \frac{RTZ}{P} \tag{6-108}$$

which may be differentiated with respect to T at constant P to give

$$\left(\frac{\partial \underline{V}}{\partial T}\right)_P = \frac{TR}{P}\left(\frac{\partial Z}{\partial T}\right)_P + \frac{RZ}{P} \tag{6-109}$$

Combination of equations (6-108) and (6-109) yields

$$\underline{V} - T\left(\frac{\partial \underline{V}}{\partial T}\right)_P = \frac{RT}{P}\left[Z - Z - T\left(\frac{\partial Z}{\partial T}\right)_P\right] \tag{6-110}$$

Simplification and multiplication by dP yields the expression

$$\left[\underline{V} - T\left(\frac{\partial \underline{V}}{\partial T}\right)_P\right] dP = \frac{RT^2}{P}\left(\frac{\partial Z}{\partial T}\right)_P dP \tag{6-111}$$

The T's and P's of equation (6-111) can be converted to reduced variables by division by T_c and P_c. The reduced version of equation (6-111) is then substituted into equation (6-85) to give

$$\underline{H} - \underline{H}^* = \int_{P^*}^{P}\left[\underline{V} - T\left(\frac{\partial \underline{V}}{\partial T}\right)_P\right] dP = \int_{P^*_r}^{P_r} \frac{RT_c T_r^2}{P_r}\left(\frac{\partial Z}{\partial T_r}\right)_{P_r} dP_r \tag{6-112}$$

Division by RT_c gives

$$\frac{\underline{H} - \underline{H}^*}{RT_c} = \int_{P^*_r}^{P_r} \frac{T_r^2}{P_r}\left(\frac{\partial Z}{\partial T_r}\right)_{P_r} dP_r \tag{6-113}$$

At a given T_r and P_r we may evaluate $(\partial Z/\partial T_r)_{P_r}$ by graphically differentiating the data presented in the generalized compressibility chart. With the values of $(\partial Z/\partial T_r)_{P_r}$ so determined, we can graphically integrate equation (6-113) by plotting $(T_r^2/P_r)(\partial Z/\partial T_r)_{P_r}$ versus P_r from some very low pressure P_r^* to the desired pressure P_r. The value of the integral is then the area under the curve. In this manner we can determine the isothermal enthalpy change for a process between zero pressure and any desired reduced pressure P_r. Although the term $1/P_r$ in equation (6-113) would appear to cause trouble as $P_r \longrightarrow 0$, it is known that $(\partial Z/\partial T)_{P_r}$ approaches zero as $P_r \longrightarrow 0$, so that the integrand is indeterminant. However, l'Hôspital's rule tells us that

$$\lim_{P_r \to 0}\left(\frac{\partial Z/\partial T_r}{P_r}\right) = \frac{\partial}{\partial P_r}\frac{(\partial Z/\partial T_r)_{P_r}}{1.0}\bigg|_{P_r \to 0} = \frac{\partial}{\partial P_r}\left(\frac{\partial Z}{\partial T_r}\right)_{P_r}\bigg|_{P_r \to 0} \tag{6-114}$$

Since

$$\frac{\partial}{\partial P_r}\left(\frac{\partial Z}{\partial T_r}\right)_{P_r}$$

is bounded as $P_r \longrightarrow 0$, the integrand is finite and the integral can be evaluated.

In order to eliminate the need for evaluating the integral of equation (6-113) whenever an enthalpy calculation must be performed, values of this integral have been tabulated as functions of T_r and P_r. These tables are found in many standard reference books. In addition to tables, graphs of $(\underline{H} - \underline{H}^*)/RT_c$, as functions of P_r and T_r, have also been assembled. A set of these graphs have been included in Appendix C.

In terms of the generalized $(\underline{H} - \underline{H}^*)$ charts, we can now express $\Delta \underline{H}$ for any process by means of our three-step path as

$$\Delta \underline{H}_{1-2} = \underline{H}_2 - \underline{H}_1 = -(\underline{H} - \underline{H}^*)_1 + (\underline{H} - \underline{H}^*)_2 + (\underline{H}_2^* - \underline{H}_1^*) \tag{6-115}$$

where

$$\underline{H}_2^* - \underline{H}_1^* = \Delta\underline{H}_{a-b} = \int_{T_1}^{T_2} C_P^* \, dT$$

is the ideal change in enthalpy, while $(\underline{H}^* - \underline{H})_1 = \Delta\underline{H}_{1-a}$ and $(\underline{H}^* - \underline{H})_2 = -\Delta\underline{H}_{b-2}$.

The Fugacity Coefficient, $\boldsymbol{\nu}$

We have shown that the fugacity coefficient may be evaluated from

$$\ln \nu = \int_{P=0}^{P} \left(\frac{\underline{V}}{RT} - \frac{1}{P}\right) dP \tag{6-101a}$$

Equation (6-101a) may be rearranged to give

$$\ln \nu = \int_{P=0}^{P} \left(\frac{P\underline{V}}{RT} - 1\right) d \ln P \tag{6-116}$$

or, if we introduce compressibilities and reduced pressures,

$$\ln \nu = \int_{P_r^*}^{P_r} (Z - 1)\, d \ln P_r \tag{6-117}$$

In the limit of $P \longrightarrow 0$, both $(Z - 1)$ and $P_r \longrightarrow 0$, so that the integrand of equation (6-117) becomes indeterminant. However, applications of l'Hôspital's rule tells us that

$$\lim_{P \to 0} \frac{Z - 1}{P_r} \longrightarrow \left(\frac{\partial Z}{\partial P_r}\right)_{P_r = 0} \tag{6-118}$$

Examination of the compressibility charts indicates that as $P_r \longrightarrow 0$, $\partial Z/\partial P_r$ is finite, so the integrand is bounded and the integral is finite. Values of f/P, or ν, may then be graphically determined from the area under a plot of $(Z - 1)/P_r$ versus P_r.

As with equation (6-113), the integral on the right side of equation (6-117) may be evaluated once and for all as a function of P_r and T_r and then stored for future use either as a table or chart of f/P versus P_r and T_r. A chart of (f/P) versus P_r and T_r is also included in Appendix A.

Entropy Changes

In the evaluation of $\Delta\underline{S}_{1-2}$ using the three-step process, we found it necessary to evaluate the integrals

$$\Delta\underline{S}_{1-a} = -\int_{P_1}^{P^*} \left(\frac{\partial \underline{V}}{\partial T}\right)_P dP \tag{6-86}$$

$$\Delta\underline{S}_{b-2} = -\int_{P^*}^{P_2} \left(\frac{\partial \underline{V}}{\partial T}\right)_P dP \tag{6-90}$$

As with the enthalpy, we shall attempt to express these integrals in terms of reduced variables and compressibilities. However, it will be more to our advantage to take a slightly different approach in evaluating the two isothermal portions of the three-part $\Delta \underline{S}$ path. We begin by rearranging the enthalpy property relation

$$d\underline{H} = T\, d\underline{S} + \underline{V}\, dP$$

to solve for $d\underline{S}$:

$$d\underline{S} = \frac{d\underline{H}}{T} - \frac{\underline{V}\, dP}{T} \tag{6-119}$$

At constant temperature, equation (6-119) may be integrated to yield

$$\Delta \underline{S}_{1-a} = \frac{\Delta \underline{H}_{1-a}}{T} - \frac{1}{T}\int_{P_1}^{P^*} \underline{V}\, dP \tag{6-120}$$

where the path $1 - a$ represents one of the isothermal legs of the three-step process. However, we have shown in Section 6-9 how to evaluate $\Delta \underline{H}_{1-a}$. Therefore, all we must do now is evaluate the integral $\int_{P_1}^{P^*} \underline{V}\, dP$.

In Section 6-8 we have defined the fugacity such that

$$RT\left(\frac{\partial \ln f}{\partial P}\right)_T = \underline{V} \tag{6-121}$$

Thus the second term in equation (6-120) may be expressed in terms of fugacities as

$$\frac{1}{T}\int_{P_1}^{P^*} \underline{V}\, dP = R\int_{P_1}^{P^*} d\ln f \tag{6-122}$$

or

$$\frac{1}{T}\int_{P_1}^{P^*} \underline{V}\, dP = R\ln\frac{f^*}{f_1} \tag{6-122a}$$

which may be substituted into equation (6-120) to give

$$\Delta \underline{S}_{1-a} = \frac{\Delta \underline{H}_{1-a}}{T} - R\ln\frac{f^*}{f_1} \tag{6-123}$$

where f^* is the fugacity evaluated at some low pressure, P^*. However, according to equation (6-95a) $f^* \longrightarrow P^*$ as $P^* \longrightarrow 0$. Therefore, equation (6-123) becomes

$$\Delta \underline{S}_{1-a} = \frac{\Delta \underline{H}_{1-a}}{T_1} - R\ln\frac{P^*}{f_1} \tag{6-124}$$

Application of equation (6-124) for the second isothermal leg of the three-step path yields

$$\Delta \underline{S}_{b-2} = \frac{\Delta \underline{H}_{b-2}}{T_2} - R\ln\frac{f_2}{P^*} \tag{6-124a}$$

The change in entropy for the isobaric portion of the process is simply

$$\Delta \underline{S}_{a-b} = \int_{T_1}^{T_2} \frac{\underline{C}_P}{T}\, dT$$

which may be combined with equations (6-124) and (6-124a) to give the total change in entropy for the whole process as

$$\Delta \underline{S}_{1-2} = \Delta \underline{S}_{1-a} + \Delta \underline{S}_{a-b} + \Delta \underline{S}_{b-2} \tag{6-125}$$

$$\Delta \underline{S}_{1-2} = \frac{\Delta \underline{H}_{1-a}}{T_1} - R\ln\frac{P^*}{f_1} + \int_{T_1}^{T_2} \frac{\underline{C}_P}{T}\, dT + \frac{\Delta \underline{H}_{b-2}}{T_2} - R\ln\frac{f_2}{P^*} \tag{6-125a}$$

However, the two terms in $\ln (f/P^*)$ may be combined to yield

$$\Delta \underline{S}_{1-2} = \frac{\Delta \underline{H}_{1-a}}{T_1} + \frac{\Delta \underline{H}_{b-2}}{T_2} + \int_{T_1}^{T_2} \frac{C_P}{T}\,dT - R \ln \frac{f_2}{f_1} \tag{6-126}$$

from which the overall entropy change may be evaluated.

SAMPLE PROBLEM 6-9. Ethylene is to be compressed adiabatically and reversibly from 200 psia and 140°F to 1500 psia. Compute the minimum work per pound of ethylene. $C_P^* = 5.34 + 0.0064T$ (°R), critical temperature = 508°R, and critical pressure = 645.0 psia.

Solution: (See Fig. S6-9a.) As illustrated in Sample Problem 6-8, the energy and entropy balances taken around the compressor and its contents as a system reduce to

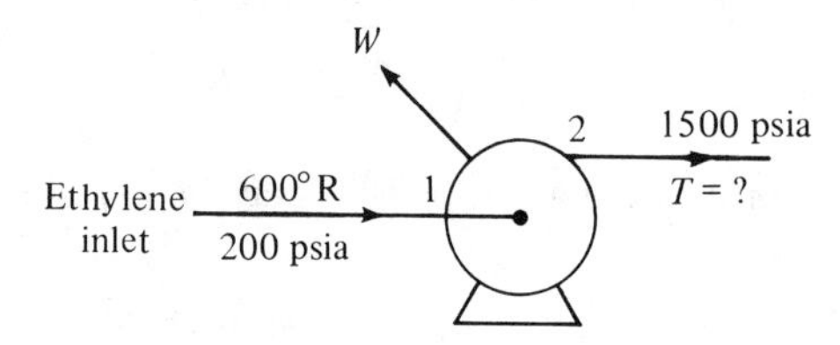

FIGURE S6-9a

Energy:

$$\Delta \underline{H} = -\underline{W}$$

Entropy:

$$\Delta \underline{S} = 0$$

As in Sample Problem 6-8, we must use the constant entropy condition to determine the temperature of the material exiting from the compressor. From this temperature, we may then calculate the change in enthalpy and hence the work needed to drive the compressor. Since no information is given to us regarding the P–$\underline{V}$–T behavior of ethylene, we are forced to use the generalized charts for the required volumetric behavior.

We may establish our three-step process for the evaluation of $\Delta \underline{S}$ and $\Delta \underline{H}$ across the compressor as indicated in Fig. S6-9b. The total entropy change $\Delta \underline{S}_{1-2}$ is split into its three component parts:

$$\Delta \underline{S}_{1-2} = \Delta \underline{S}_{1-a} + \Delta \underline{S}_{a-b} + \Delta \underline{S}_{b-2}$$

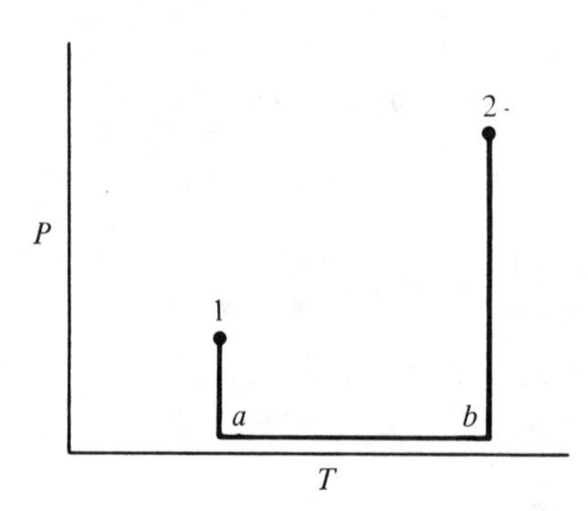

FIGURE S6-9b

where $\Delta \underline{S}_{1-a}$ is evaluated at constant $T = T_1$

$\Delta \underline{S}_{a-b}$ is evaluated at constant $P = P^* \longrightarrow 0$

$\Delta \underline{S}_{b-2}$ is evaluated at constant $T = T_2$

As given in equation (6-124), $\Delta \underline{S}_{1-a}$ and $\Delta \underline{S}_{b-2}$ may be expressed as

$$\Delta \underline{S}_{1-a} = \frac{\Delta \underline{H}_{1-a}}{T_1} - R \ln \frac{f^*}{f_1}$$

$$\Delta \underline{S}_{b-2} = \frac{\Delta \underline{H}_{b-2}}{T_2} - R \ln \frac{f_2}{f^*}$$

while $\Delta \underline{S}_{a-b}$ is given by

$$\Delta \underline{S}_{a-b} = \int_{T_1}^{T_2} \frac{C_P}{T}\, dT = \int_{T_1}^{T_2} \frac{5.34 + 0.0064T}{T}\, dT$$

$$= 5.34 \ln \frac{T_2}{T_1} + 0.0064(T_2 - T_1) \frac{\text{Btu}}{\text{lb-mol °R}}$$

Therefore, the overall change in entropy is given by

$$\Delta \underline{S}_{1-2} = \frac{\Delta \underline{H}_{1-a}}{T_1} - R \ln \frac{f^*}{f_1} + \int_{T_1}^{T_2} \frac{C_P}{T}\, dT + \frac{\Delta \underline{H}_{b-2}}{T_2} - R \ln \frac{f_2}{f^*}$$

or

$$\Delta \underline{S}_{1-2} = \frac{\Delta \underline{H}_{1-a}}{T_1} + \frac{\Delta \underline{H}_{b-2}}{T_2} - R \ln \frac{f_2}{f_1} + \int_{T_1}^{T_2} \frac{C_P}{T}\, dT$$

$\Delta \underline{H}_{1-a}$ and f_1 may be calculated directly from the generalized charts and P_{1r} and T_{1r}. However, $\Delta \underline{H}_{b-2}$ and f_2 cannot be calculated before T_2 is known. Thus we must use a trial-and-error procedure to find the T_2 at which $\Delta \underline{S}_{1-2} = 0$. Because of the complex nature of the relation between $\Delta \underline{S}_{1-2}$ and T_2, it seems that the second type of trial-and-error procedure suggested earlier will be more appropriate. Therefore, we now guess a series of T_2's and calculate the $\Delta \underline{S}$'s corresponding to these T_2's.

Inlet Data:

$$P_1 = 200 \text{ psi} \qquad P_{1r} = 0.268$$

$$T_1 = 140°\text{F}, 600°\text{R} \qquad T_{1r} = 1.18$$

Therefore, $(f/P)_{\text{inlet}} = 0.95$. Therefore, $f_1 = 190$ psia.

$$\frac{\underline{H}^* - \underline{H}_1}{T_c} = 0.40 \text{ Btu/lb-mol °R} = \frac{\underline{H}_a - \underline{H}_1}{T_c} = \frac{\Delta \underline{H}_{1-a}}{T_c}$$

But

$$\frac{\Delta \underline{H}_{1-a}}{T_1} = \frac{\Delta \underline{H}_{1-a}/T_c}{T_1/T_c} = \frac{\Delta \underline{H}_{1-a}/T_c}{T_{1r}} = \frac{0.40}{1.18} \frac{\text{Btu}}{\text{lb-mol °R}} = 0.34 \frac{\text{Btu}}{\text{lb-mol °R}}$$

Outlet Data:

$$P_2 = 1500 \text{ psi} \qquad P_{2r} = 2.01$$

Therefore, we calculate $\Delta \underline{S}_{1-2}$ as shown in Table S6-9.

We now plot $\Delta \underline{S}$ versus T (see Fig. S6-9c) and extrapolate to $\Delta \underline{S} = 0$, giving $T = 918°$R. Now we must evaluate $\Delta \underline{H}_{1-2}$. But

$$\Delta \underline{H}_{1-2} = \Delta \underline{H}_{1-a} + \Delta \underline{H}_{a-b} + \Delta \underline{H}_{b-2}$$

$$\Delta \underline{H}_{1-a} = T_c \frac{\underline{H}_a^* - \underline{H}_1}{T_c} = (508)(0.40) \frac{\text{Btu °R}}{\text{lb-mol °R}}$$

$$= 203 \text{ Btu/lb-mol}$$

TABLE S6-9 Entropy Calculation

T_2, °R	T_{2r}	f_2, psia	$1.987 \ln \frac{f_2}{f_1}$, $\frac{\text{Btu}}{\text{lb-mol °R}}$	$5.34 \ln \frac{T_2}{T_1}$	$0.0064(T_2 - T_1)$	$\int_{T_1}^{T_2} \frac{C_P}{T} dT$	$\frac{\Delta \underline{H}_{b-2}}{T_c}$	$\frac{\Delta \underline{H}_{b-2}}{T_2}$	$\frac{\Delta \underline{H}_{b-1}}{T_1} + \frac{\Delta \underline{H}_{b-2}}{T_2}$	$\Delta \underline{S}_{1-2}$ $\frac{\text{Btu}}{\text{lb-mol °R}}$
800	1.57	1320	3.87	1.55	1.28	2.83	−2.0	−1.27	−0.93	−1.97
900	1.78	1400	3.92	2.16	1.92	4.08	−1.45	−0.815	−0.475	−0.32
1000	1.97	1440	4.06	2.73	2.56	5.29	−1.20	−0.61	−0.27	0.96
1200	2.36	1500	4.1	3.17	3.84	7.01	−0.85	−0.36	−0.02	2.89

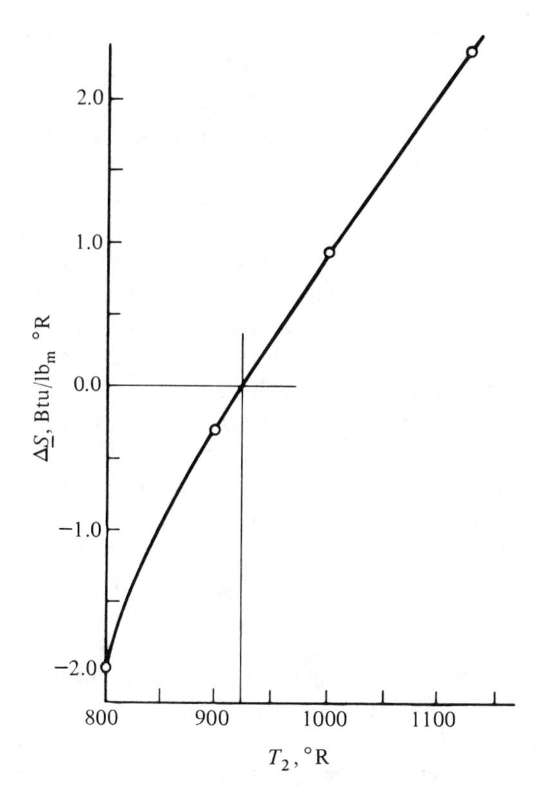

FIGURE S6-9c

$$\Delta \underline{H}_{a-b} = \int_{600}^{918} C_P\,dT = (5.34)(918 - 600) + (0.0032)(918^2 - 600^2)$$

$$= 1670 + 1920 = 3590 \text{ Btu/lb-mol}$$

$$\Delta \underline{H}_{b-2} = -T_c \frac{\underline{H}_b^* - \underline{H}_2}{T_c} = (508)(1.40)\frac{\text{Btu °R}}{\text{lb-mol °R}} = -711 \text{ Btu/lb-mol}$$

Therefore,

$$\Delta \underline{H}_{1-2} = [203 + 3590 - 711] \text{ Btu/lb-mol} = 3080 \text{ Btu/lb-mol}$$

$$\underline{W} = -\Delta \underline{H}_{1-2} = -3080 \text{ Btu/lb-mol}$$

6.11 Equilibrium Between Phases

Up to this point we have talked of computing thermodynamic properties and property changes for systems that undergo no phase change. How are we to compute changes when a gas condenses? What, in fact, are the conditions at which a vapor will liquify reversibly? To answer these questions, we must establish the conditions necessary for equilibrium between phases.

We first recall that one speaks of equilibrium of a body or system with respect to the constraints that are imposed on the body. Thus a hot body may be in equilibrium if it is perfectly insulated from its surroundings. (The constraint is its isolation from any body with which it can exchange thermal energy.) If, however, the same hot body is placed in contact with a cooler body, it will come to equilibrium at a lower temperature than before. (The constraint is now the contact with another body, with which the hot body can exchange thermal energy.) Thus, as we reopen the question of equilibrium, we must be careful to specify the constraints that we wish to impose on the body.

Having specified the constraints to be imposed on a system at equilibrium, we have observed in Chapter 3 that (1) the entropy of the system and its surroundings is maximized, (2) there are no potential differences between the system and its surroundings, and (3) it is impossible to obtain work from an engine operating reversibly between the system and its surroundings. (The surroundings in each of the above cases is construed as that portion of the universe with which the system is free to interact.) These are not all independent criteria, but each provides some insight not directly available from the others.

These ideas, although useful, are not sufficiently quantitative for our purposes in describing phase and chemical equilibrium. We shall therefore make use of them to obtain quantitative criteria for equilibrium—that is, the relationship between the thermodynamic state variables of bodies in equilibrium with each other.

The ensuing development has as its immediate purpose the development of criteria for phase equilibrium of pure materials. At the outset, however, let it be noted that the treatment does not specifically rely on this circumstance. The conclusions obtained will be applicable to the seemingly more complex situations of multicomponent phase equilibrium (Chapter 10) and chemical reaction equilibrium (Chapter 11) as well.

The constraints that will be common to all considerations of phase and chemical equilibrium in this book are:

1. The absence of barriers to thermal-energy transfer between the system and its surroundings.

2. The absence of barriers to mechanical-energy transfer between the system and its surroundings through a volume change.

Except when we consider chemical reaction equilibrium specifically, we shall impose an additional constraint:

3. No chemical reaction can take place between the chemical species present.

If any of these constraints is altered, or additional constraints imposed, we shall so specify at that time.

By any of the criteria of equilibrium mentioned earlier it is evident that constraints 1 and 2 require that the temperature and pressure of the system be the same as that of the surroundings and that there be no temperature gradients within the system. Were this not so we could, for instance, obtain work from a heat engine operating reversibly between the high-and low-temperature regions or from an expansion of the high-pressure material to lower pressure. Because of these conclusions, it is common to merely specify that temperature and pressure are fixed by the surroundings—which we see is an equivalent assumption.

A further important observation about equilibrium of such systems can be made by asking what happens to the total Gibbs free energy of a material in a constant pressure, isothermal, steady-flow process which is proceeding reversibly—that is, the maximum amount of work is being exchanged between the flow system and the surroundings. The flowing material might, for example, consist of liquid and gaseous A, or of liquid and gaseous mixtures of A and B, or of a gaseous mixture of chemically reacting A, B, and C. The conclusions reached will be generally valid; the above examples are offered only as an aid in visualizing the process.

For the purpose of illustration, let the flow system be divided into a sequence of smaller stages as shown in Fig. 6-3. To determine what changes occur,

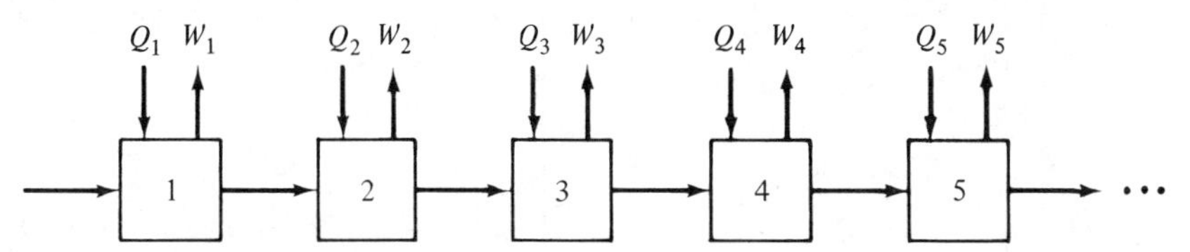

FIG. 6-3. Staged, isothermal, reversible flow process.

let us write the energy balance and the entropy balance around the first stage, remembering that reversibility is assured if $LW = 0$. (Kinetic and potential energy changes are assumed to be negligible, although a similar development can be followed for cases where they are not.)

Using the notation that H and S are the total enthalpy and total entropy of the flowing stream and that Q and W represent the heat and work transfers for the same basis, the energy and entropy balances are

Energy:

$$H_{in} - H_{out} + Q_1 - W_1 = 0$$

Entropy:

$$S_{in} - S_{out} + \frac{Q_1}{T} = 0 \tag{6-127}$$

If the two equations are combined, eliminating Q_1, we have

$$W_1 = -[H_{out} - TS_{out} - (H_{in} - TS_{in})] \tag{6-128}$$

or, since $G = H - TS$,

$$W_1 = -(G_{out} - G_{in}) = -\Delta G_1 \tag{6-129}$$

where $\Delta G_1 = G_{out} - G_{in}$ denotes the total Gibbs-free-energy change of the flowing material across the first stage. By repetition of the arguments, we see that the work produced in any portion of the steady, isothermal, reversible flow process is equal to the negative of the change in total Gibbs free energy across that portion of the process. As long as the total Gibbs free energy decreases, work can be recovered from the process.

If we replace the finite number of stages by an infinite series of infinitesimal stages, then the requirement for positive work production in each stage is simply that $dG < 0$. If $dG > 0$, then work must be supplied by the surroundings to continue the process; $dG = 0$ implies that work is neither produced nor consumed. Thus, if we plot G as a function of X, a process parameter (as shown in Fig. 6-4), we can distinguish the region in which work is produced as the

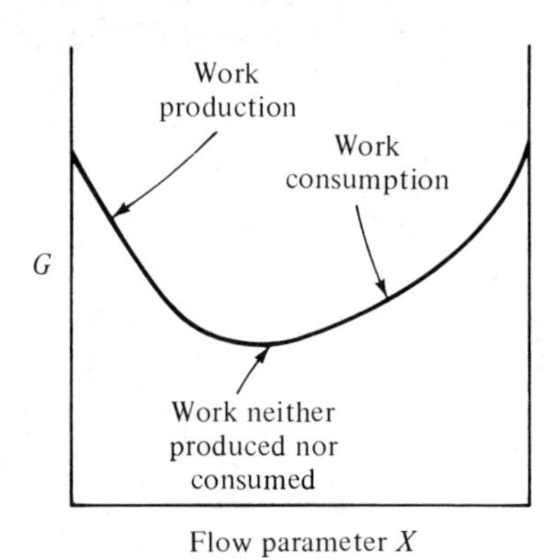

FIG. 6-4. Gibbs free energy as a function of the process parameter X.

region of negative slope. A region of positive slope indicates that work must be supplied to continue the process. The minimum, corresponding to $dG = 0$, is the point at which work is neither produced nor consumed. The process parameter, X, might be any useful measure of the change occurring within the system, such as overall specific volume in the case of phase equilibrium or fractional conversion in the case of chemical reaction.

Let us now consider what happens if we simply allow the process to occur spontaneously. As long as the system is to the left of the minimum in the G–X curve, the process will proceed with a decrease in the total Gibbs free energy of the system because the reversible work for that process would be positive. When the minimum in the curve is reached the process can proceed no further, because further change can occur only if the surroundings supply work to the system. Our assumption of spontaneity implies that the surroundings cannot supply the needed work; hence the change must cease. Similarly, the process cannot proceed in the reverse direction, because this, too, would correspond to increasing G. Similarly, if the system started in a state to the right of the minimum in Fig. 6-4, it could move only toward the minimum point (or to the left) if the change is to occur spontaneously. Therefore, we find that the process will proceed only up to that point at which the minimum in the Gibbs-free-energy curve is reached. Once this point is attained, the change can proceed in neither direction, and the system will remain in this state indefinitely (unless work is supplied from the surroundings). The system at this point is in equilibrium under the constraints we have imposed, because there is no way one can obtain further work from this system. Thus we conclude that, *at fixed temperature and pressure, the Gibbs free energy of the system is a minimum at equilibrium*. For specific cases of phase or chemical equilibrium, we shall determine the conditions that correspond to the minimum in the G–X curve by setting $(dG/dX)_{T,P} = 0$.

Let us now apply this to the specific instance of phase equilibrium for pure materials. If we denote the two phases present as I and II, then

$$G = M^{\mathrm{I}}\underline{G}^{\mathrm{I}} + M^{\mathrm{II}}\underline{G}^{\mathrm{II}} \tag{6-130}$$

Now if we transfer a small amount of material from one phase to the other without changing the temperature or pressure, the total change in Gibbs free energy of the mixture is given by

$$dG = \underline{G}^{\mathrm{I}}dM^{\mathrm{I}} + \underline{G}^{\mathrm{II}}dM^{\mathrm{II}} \tag{6-131}$$

(We have made use of the fact that the specific Gibbs free energies of each phase depend only on the temperature and pressure, and hence do not change.) However, a mass balance around the mixture gives

$$dM^{\mathrm{I}} = -\,dM^{\mathrm{II}} \tag{6-132}$$

so equation (6-131) reduces to

$$dG = (\underline{G}^{\mathrm{I}} - \underline{G}^{\mathrm{II}})\,dM^{\mathrm{I}} \tag{6-131a}$$

If the transfer process is performed at constant temperature and pressure, then equilibrium between the phases corresponds to that condition in which it is not possible to obtain work from a transfer of mass between the phases. Thus dG must vanish at the equilibrium conditions, and equation (6-131a) becomes

$$\underline{G}^{\mathrm{I}} - \underline{G}^{\mathrm{II}} = 0 \tag{6-133}$$

or

$$\underline{G}^{\mathrm{I}} = \underline{G}^{\mathrm{II}} \tag{6-133a}$$

at equilibrium.

We can say in summary that equilibrium between different phases (of pure materials) is characterized by the equalities

$$P^{\mathrm{I}} = P^{\mathrm{II}} \quad ; \quad T^{\mathrm{I}} = T^{\mathrm{II}} \quad ; \quad \underline{G}^{\mathrm{I}} = \underline{G}^{\mathrm{II}} \tag{6-134}$$

Because of the desire to cover more general cases to come in Chapters 10 and 11, we have spoken of fixing both P and T independently. Since specification of P and T fully determines the state of a pure material, we cannot assure the existence of two phases for arbitrary choices of the two. The relationship of the points on the P–T equilibrium curve is discussed in a following section. We can say, nonetheless, that whenever two phases of a pure material are in equilibrium, all of equations (6-134) must be satisfied.

Let us now return to the original question of this section: How can we compute property changes for systems undergoing a phase change? We have learned that the phase change can be performed at constant temperature and pressure with no change in the Gibbs free energy. If we measure the heat associated with a phase change (in an isothermal flow calorimeter, for example), we can compute $\Delta\underline{H}$ and, since $\Delta\underline{G} = 0$, $\Delta\underline{S} = \Delta\underline{H}/T$. If, in addition, we measure the volume change associated with the phase change, we find that $\Delta\underline{U} = \Delta\underline{H} - \Delta(P\underline{V}) = \Delta\underline{H} - P\,\Delta\underline{V}$ and, since $\Delta\underline{G} = 0$, $\Delta\underline{A} = \Delta\underline{G} - \Delta(P\underline{V}) = -P\,\Delta\underline{V}$.

In short, for a pure material, knowledge of the volume and enthalpy change associated with a phase change at known temperature and pressure is sufficient to permit calculation of the change in all other thermodynamic properties of the material.

6.12 Evaluation of Liquid- and Solid-Phase Fugacities

As we have seen, much useful information can be obtained from the fugacity. This will continue to be the case when we deal with mixtures in Chapters 10 and 11. Accordingly, we now open the question of calculating liquid- and solid-phase fugacities.

Equation (6-94) upon integration at constant temperature becomes

$$RT \ln \frac{f^{\text{II}}}{f^{\text{I}}} = \underline{G}^{\text{II}} - \underline{G}^{\text{I}} \tag{6-135}$$

We have noted that when the superscripts I and II denote the properties of two phases in equilibrium, the right-hand side of this equation is identically zero. Consequently, it must also be true that the fugacities are equal at equilibrium,

$$f^{\text{I}} = f^{\text{II}} \tag{6-136}$$

This relationship enables us to compute the fugacity of pure liquids and solids assuming that we may evaluate the fugacity of a vapor which is in equilibrium with the liquid or solid under consideration. We have seen in Section 6-8 how the fugacity of a gas at elevated temperature and pressure may be obtained by integrating equations (6-102) to (6-107) at constant temperature from a low pressure reference state to the pressure in question. If the elevated pressure is the vapor pressure, P', we have shown that the gas and liquid fugacities are equal at that pressure. Since the definition of fugacity in equation (6-94) is not limited to the gaseous state, we may use that equation for the liquid state as well,

$$\left(\frac{\partial \ln f^L}{\partial P}\right)_T = \frac{\underline{V}^L}{RT} \tag{6-94}$$

and continue integration into the liquid range. The computation of liquid fugacity then follows the procedure outlined by the following integrated form of equation (6-94):

$$\ln \frac{f^L}{P^*} = \int_{P^*}^{P'} \frac{\underline{V}^V}{RT}\, dP + \int_{P'}^{P} \frac{\underline{V}^L}{RT}\, dP \tag{6-137}$$

To circumvent problems that might arise in the limit of $P^* \longrightarrow 0$, we can subtract the equation

$$\ln \frac{P'}{P^*} = \int_{P^*}^{P'} \frac{1}{P}\, dP$$

to obtain the working equation

$$\ln \frac{f^L}{P'} = \int_{P^*}^{P'} \left(\frac{\underline{V}^V}{RT} - \frac{1}{P} \right) dP + \int_{P'}^{P} \frac{\underline{V}^L}{RT}\, dP \tag{6-138}$$

It should be clear, as well, that the same procedures can be applied to calculation of the fugacities of pure solids by noting that for a solid and gas in equilibrium (through sublimation of the solid), $f^S = f^V$. Equation (6-138) can be applied by changing all L superscripts to S and understanding P' to mean the vapor pressure of the pure solid at the temperature in question.

SAMPLE PROBLEM 6-10. To illustrate the procedures above, calculate the fugacity of liquid water at 400 psia and 300°F, using data from the steam tables found in Appendix B.

Solution: The steam tables found in Appendix B give data for the specific volume of water at 300°F as indicated in the first two columns of Table S6-10. The remaining columns were computed from the handbook data using values for R and T as follows:

$$R = 10.731 \frac{\text{psia ft}^3}{\text{lb-mole °R}} \frac{1}{18.016} \frac{\text{lb-mole}}{\text{lb}_m}$$
$$= 0.5956 \text{ psia ft}^3/\text{lb}_m \text{ °R}$$
$$T = 459.7 + 300 = 759.7 \text{ °R}$$

TABLE S6-10

Pressure, psia	$\underline{V}$, ft^3/lb$_m$	$\frac{\underline{V}}{RT}$, psia^{-1}	$\frac{\underline{V}}{RT} - \frac{1}{P}$ psia^{-1}
1	452.3	0.9996	-4×10^{-4}
2	226.0	0.4995	-5×10^{-4}
5	90.25	0.1995	-5×10^{-4}
10	45.00	0.0995	-5×10^{-4}
20	22.36	0.0494	-6×10^{-4}
40	11.040	0.0244	-6×10^{-4}
60	7.259	0.0160	-6×10^{-4}
67.013	6.466 (vapor)	0.0143	-6×10^{-4}
67.013	0.01745 (liquid)	3.85×10^{-5}	—
200	0.01744	3.85×10^{-5}	—
400	0.01742	3.85×10^{-5}	—

Because the difference $\underline{V}/RT - 1/P$ is so nearly constant, the first integral in equation (6-138) can be approxmiated as follows:

$$\int_0^{P'} \left(\frac{\underline{V}^V}{RT} - \frac{1}{P} \right) dP = \left(\frac{\underline{V}^V}{RT} - \frac{1}{P} \right)_{\max} (P' - 0)$$
$$= (-6 \times 10^{-4})(67.013)$$
$$= -0.0402$$

A more careful numerical procedure gives the value -0.0386.

The second integral of equation (6-138) also has a constant integrand, with the result

$$\int_{P'}^{P} \frac{\underline{V}^L}{RT}\,dP = (3.85 \times 10^{-5})(400 - 67.013)$$
$$= 0.0128$$

Thus the fugacity of water at both the vapor pressure, P', and 400 psia, P, can now be calculated.

$$\ln \frac{f^{\text{sat vap}}}{P'} = -0.0402$$
$$f^{\text{sat vap}} = (67.013)\exp(-0.0402) = 64.4 \text{ psia}$$

and

$$\ln \frac{f^L}{P'} = -0.0402 + 0.0128$$
$$f^L = (67.013)\exp(-0.0274) = 65.1 \text{ psia}$$

The striking thing about this result is that the fugacity of the compressed liquid is not very different from that of the saturated vapor or saturated liquid. The reason for this is evident from an inspection of equation (6-137) in light of the data above. The volume of the liquid is so much less than the volume of the vapor that the second integral is much smaller than the first.

Sample Problem 6-10 has shown how we may compute the fugacity of a pure liquid from specific volume data and has revealed that the fugacity does not change greatly as the liquid is compressed beyond the saturation pressure. In fact, we can say, with reasonable accuracy, that

$$f^L(P, T) = f^{\text{sat liq}}(T) = f^{\text{sat vap}}(T) \tag{6-139}$$

In other words, the fugacity of a compressed liquid is equal to its fugacity at its vapor pressure at the same temperature. This approximation will be enormously helpful in our later work, for we will not need to have liquid specific volume data to obtain values of the fugacity for compressed liquids.

The approximation in equation (6-139) will involve negligible error provided the second integral in equation (6-138) is small. In other words, it will be valid if $\underline{V}^L \ll \underline{V}^V$ and P is not $\gg P'$. The former is true except near the critical conditions and the latter is true provided the liquid is not compressed well beyond the vapor pressure.

One of the most interesting applications of equation (6-107), together with the Benedict–Webb–Rubin (BWR) equation of state, is its use in calculating vapor–liquid equilibrium data. Several authors[1] have recently called attention to the usefulness of similar techniques in predicting phase equilibrium for multicomponent mixtures as well. This is a very powerful tool in that actual experimental data on mixtures is generally very limited. To provide a background for such techniques, Sample Problem 6-11 indicates how that equation

[1] See, for example, R. V. Orye, *Ind. Eng. Chem. Process Design Develop.* 8, 579–588 (1969).

of state can be used to compute vapor–liquid equilibrium data for pure materials.

SAMPLE PROBLEM 6-11. Using the BWR equation of state, compute the vapor–liquid equilibrium curve for normal pentane over the temperature range −75 to +175°C.

Solution: In order that we understand the computation to follow, let us examine the isotherm predicted by the BWR equation of state. It is shown as the solid line in Fig. S6-11. The two-phase region is indicated by the dashed line with the two points of inter-

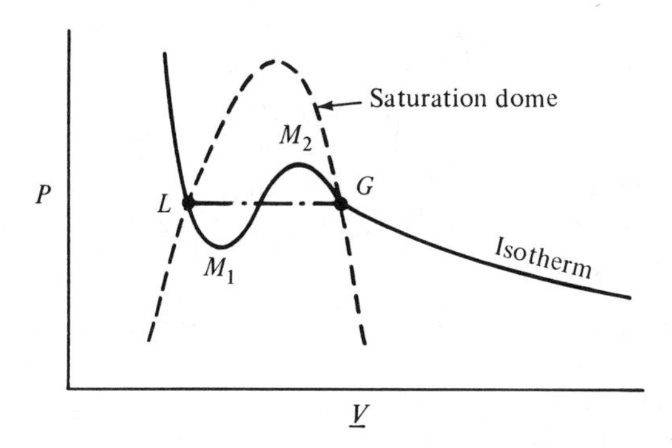

FIGURE S6-11

section (L and G) connected by an isobar (—·—). To the right of G, the BWR line predicts the behavior of superheated gas, and to the left of L it predicts the compressed liquid behavior. Between the two it does not predict the true stable states of liquid–vapor mixtures, for we know that they lie along the isobar LG. The question posed by this problem is just this: Given the isotherm but no knowledge of the liquid–vapor region, how can we select that isobar LG (from all that might be drawn between the local maximum and minimum at M_1 and M_2) which corresponds to the correct liquid–vapor equilibrium? The answer is simple: We require that not only the pressures but also f^L and f^G be equal:

$$\begin{aligned} P(T, \underline{V}^L) &= P(T, \underline{V}^G) \\ f(T, \underline{V}^L) &= f(T, \underline{V}^G) \end{aligned} \tag{a}$$

in which, for a given choice of temperature, we must find the liquid and gas specific volumes which satisfy both equations.

For a given choice of specific volume and temperature, it is a straightforward matter to compute the pressure from the BWR equation. We have, moreover, shown in this section how to integrate the BWR equation to compute the fugacity as a function of $\underline{V}$ and T. Since fugacity and pressure are state properties and the BWR equation correctly predicts the liquid state for $\underline{V} \leq \underline{V}^L$ and the gaseous state for $\underline{V} \geq \underline{V}^G$, we need not be concerned with what path is used by the BWR equation in predicting the state-property changes between points L and G.

To solve equation (a) we need to have explicit expressions for each term involved. The first consists of the BWR equation itself. It is often convenient to write the equation in terms of density rather than specific volume. Noting that $\rho = 1/\underline{V}$, we can transform the BWR equation given in Chapter 3 to the form

$$P = RT\rho + \frac{B_0RT - A_0 - C_0}{T^2}\rho^2 + (bRT - a)\rho^3 + a\alpha\rho^6 + \frac{C\rho^3}{T^2}(1 + \gamma\rho^2)e^{-\gamma\rho^2} \tag{b}$$

With appropriate choices of density (ρ^L and ρ^G) this is used to construct the first of equations (a). It can be substituted into equation (6-107) and the indicated integration carried out to arrive at a BWR expression for the fugacity,

$$RT\ln f = RT\ln\rho RT + 2\left(\frac{B_0RT - A_0 - C_0}{T^2}\right)\rho + \frac{3}{2}(bRT - a)\rho^2 + \frac{6}{5}a\alpha\rho^5 + \frac{C}{T^2}\left[\frac{1}{\gamma} + \left(-\frac{1}{\gamma} + \frac{\rho^2}{2} + \gamma\rho^4\right)e^{-\gamma\rho^2}\right]$$

which, with appropriate choices of density (ρ^L and ρ^G), forms the second of equations (a). The computer program below solves equations (a) by application of a second-order Newton–Raphson algorithm.

```
      DIMENSION CK(6),RHO(2),P(2),PP(2),F(2),FF(2),RM(2),PPP(2)
      DIMENSION TTL(4)
C ---  INPUT AND OTHER PRELIMINARIES---
C ---  IDENTIFICATION OF SPECIAL FUNCTIONS---
      COP(X)=(C*X**3/T**2)*(1.+G*X**2)*EXP(-G*X**2)
      COPP(X)=(C*X**2/T**2)*(3.+3.*G*X**2-2. *G**2*X**4)*EXP(-G*X**2)
      COPPP(X)=(3.+3.*G*X**2-9.*G**2*X**4+2.*G**3*X**6)*EXP(-G*X**2)
      COF(X)=(C/T**2)*(1./G+(-1./G+.5*X**2+G*X**4)*EXP(-G*X**2))
      DATA HG,AT/6H MM HG,6H ATM  /
C ---  INPUT OF BWR AND CRITICAL CONSTANTS---
      READ(6,200) BO,AO,CO,B,A,C,AL,G,PC,TC
C ---  TO, TF, AND TC ARE IN DEGREE CENTIGRADE
  200 FORMAT(E15.0)
      READ(6,204)TTL
  204 FORMAT(4A6)
      PC = PC*14.696
      TC = 1.8*(TC+273.16)
      R = 10.731
      WRITE(8,203) TTL
  203 FORMAT(1H ,51H THIS PROGRAM COMPUTES THE VAPOR-LIQUID
     1EQUILIBRIUM,
     1/ 49H AS A FUNCTION OF TEMPERATURE FOR PURE COMPOUNDS /
     248H USING THE BENEDICT-WEBB-RUBIN EQUATION OF STATE //
     316H THE COMPOUND IS ,4A6, //)
C ---  INPUT OF TEMPERATURE RANGE AND INCREMENTS---
      READ(6,201) TO,TINC, TF
  201 FORMAT(F10.0)
      TO = (TO+273.16)*1.8
      TINC = TINC*1.8
```

```
      TF = (TF+273.16)*1.8
      T = TO - TINC
      IF ((TF/TC).GT. 0.95) TF = 0.95*TC
      IF (TO .GT. TC) GO TO 60
      GO TO 90
60    WRITE(8,201) T,TINC,TF
      CALL EXIT
90    WRITE(8,202)
202   FORMAT(//,47X,8H VAPOR ,9X,8H LIQUID/13H TEMPERATURE ,4X,
     110H PRESSURE ,6X,10H FUGACITY ,7X,9H DENSITY, 7X,9H DENSITY/4X,
     29H (DEG C) ,19X,28H COEFFICIENT (LBMOLE/CUFT) ,15H (LBMOLE/CUFT)
      RM(2) = 1.E-5
      RM(1) = 2.5
C --- INDEXING OF TEMP. AND CALCULATION OF BWR COEFF. GROUPS---
17    T = T + TINC
      CK(1) = R*T
      CK(2) = BO*R*T-AO-CO/T**2
      CK(3) = R*R*T-A
      CK(4) = 0.0
      CK(5) = 0.0
      CK(6) = A*AL
C --- COMPUTATION OF DENSITIES, RM(1) AND RM(2), AT THE PRESSURE
C     MAX AND MIN(M1 AND M2 ON FIGURE) SO AS TO IDENTIFY THE
C     RANGE FOR SEARCH FOR LIQUID(RHO(1)) AND VAPOR (RHO(2))
C     DENSITIES.---
      DO 20 I = 1,2
22    PP(I) = CK(1)+COPP(RM(I))
      PPP(I) = (2.*C*RM(1)/T**2)*COPPP(RM(I))
      DO 21 J = 2,6
      EJ = J
      PP(I) = PP(I)+EJ*CK(J)*RM(I)**(J-1)
21    PPP(I) = PPP(I)+EJ*(EJ-1.)*CK(J)*RM(I)**(J-2)
      RM(I) = RM(I)-PP(I)/PPP(I)
      IF (PP(I) - 1.0E-04)25,25,22
25    IF (ABS(PP(I)/(PPP(I)*RM(I)))-1.E-04)20,20,22
20    CONTINUE
      IF (RM(1)-RM(2)*1.01)23,23,24
23    WRITE(8,220)RM,PP,T
220   FORMAT(20H EXTREMUM PROBLEMS ,/,2E12.5)
      GO TO 17
C --- APPLICATION OF THE SECOND-ORDER NEWTON-RAPHSON ALGORITHM TO
C     EQUATIONS A. NOTE THAT F(I) = RTLNF AT THIS STAGE.---
24    RHO(1) = 1.2*RM(1)
      RHO(2) = 0.1*RM(2)
      JC = 0
      JK = 0
12    JC = JC + 1
      DO 10 I = 1,2
      P(I) = CK(1)*RHO(I)+COP(RHO(I))
```

```
      F(I) = CK(1)*ALOG(CK(1)*RHO(I))*COF(RHO(I))
      PP(I) = CK(1)+COPP(RHO(I))
      FF(I) = CK(1)/RHO(I)+COPP(RHO(I))/RHO(I)
      DO 19 J = 2,6
      EJ = J
      P(I) = P(I)+CK(J)*(RHO(I)**J)
      F(I) = (I)+(EJ/(EJ-1.))*CK(J)*(RHO)(I)**(J-1))
      PP(I) = PP(I)+EJ*CK(J)*(RHO(I)**(J-1))
   19 FF(I) = FF(I)+EJ*CK(J)*(RHO)(I)**(J-2))
      F(I) = EXP(F(I)/(R*T))
      FF(I) = FF(I)*F(I)/(R*T)
   10 CONTINUE
      DENOM = PP(2)*FF(1)-PP(1)*FF(2)
      DRHO1 = ((P(2)-P(1))*FF(2)-(F(2)-F(1))*PP(2))/DENOM
      DRHO2 = ((P(2)-P(1))*FF(1)-(F(2)-F(1))*PP(1))/DENOM
      RHO(1) = RHO(1) - DRHO1
      RHO(2) = RHO(2) - DRHO2
C --- APPLICATION OF CONVERGENCE TESTS---
C --- ADJUSTMENT OF INITIAL GUESS IF *CORRECTED* DENSITIES ARE NOT
C     POSITIVE AND WITHIN THEIR PROPER RANGES---
      JK = 0
      IF (RM(2)-RHO(2))30,30,31
   30 RHO(2) = .99*RM(2)
      JK = 1
   31 IF (RHO(1)-RM(1))32,32,33
   32 RHO(1) = 1.01*RM(1)
      JK = 2
   33 IF (RHO(2))34,34,35
   34 RHO(2) = RM(2)*10.**(-JC-1)
      JK = 3
   35 IF (JK) 18,18,12
C --- TEST FOR CONVERGENCE OF COMPUTATION WITH RESPECT TO ALL
C     STATE PARAMETERS---
   18 IF (ABS(DRHO1/RHO(1))-1.E-06)11,11,41
   11 IF (ABS(DRHO2/RHO(2))-1.E-04)13,13,41
   13 IF (ABS((F(2)-F(1))/F(2))-1.E-03)14,14,44
   14 IF (ABS((P(2)-P(1))/P(2))-1.E-03)15,15,44
   41 JK = 4
      GO TO 12
   44 JK = 5
      GO TO 12
C --- COMPUTATION OF OUTPUT PARAMETERS AND OUTPUT---
   15 FC = F(2)/P(2)
      P(2) = P(2)/14.696
      TP = (T/1.8)-273.16
      IF (P(2)-1.)42,43,43
   42 P(2) = P(2)*760
      WRITE(8,210)TP,P(2),HG,FC,RHO(2),RHO(1)
      GO TO 45
```

```
   43     WRITE(8,210)TP,P(2),AT,FC,RHO(2),RHO(1)
   210    FORMAT(F10.2,F11.3,A6,3E14.5)
C  ---    RETURN FOR NEXT TEMPERATURE VALUE OR EXIT---
   45     IF (TF-T-0.1*TINC)16,16,17
   16     CALL EXIT
          END
```

```
THIS PROGRAM COMPUTES THE VAPOR-LIQUID EQUILIBRIUM
AS A FUNCTION OF TEMPERATURE FOR PURE COMPOUNDS
USING THE BENEDICT-WEBB-RUBIN EQUATION OF STATE

THE COMPOUND IS NORMAL PENTANE
```

TEMPERATURE (DEG C)	PRESSURE	FUGACITY COEFFICIENT	VAPOR DENSITY (LBMOLE/CUFT)	LIQUID DENSITY (LBMOLE/CUFT)
−75.00	0.106 MM HG	.99997E 00	.53461E-06	.64669E 00
−50.00	3.401 MM HG	.99931E 00	.15265E-04	.61356E 00
−25.00	31.834 MM HG	.99561E 00	.12899E-03	.58422E 00
0.00	150.828 MM HG	.98536E 00	.56109E-03	.55751E 00
25.00	476.993 MM HG	.96631E 00	.16593E-02	.53248E 00
50.00	1.537 ATM	.93833E 00	.38734E-02	.50832E 00
75.00	3.176 ATM	.90265E 00	.77764E-02	.48421E 00
100.00	5.827 ATM	.86095E 00	.14161E-01	.45916E 00
125.00	9.810 ATM	.81471E 00	.24284E-01	.43179E 00
150.00	15.505 ATM	.76480E 00	.40601E-01	.39941E 00
175.00	23.411 ATM	.71081E 00	.70114E-01	.35408E 00

6.13 *Clausius–Clapeyron Equations*

The pressure–temperature plot for a pure component introduced in Fig. 3-6 included three curves, which represented liquid–vapor, solid–vapor, and liquid–solid equilibrium. The Clausius–Clapeyron equation provides information on the slope of these curves, $(dP/dT)_{\text{equilibrium}}$, which is obtained by noting that at all points on these equilibrium curves $\underline{G}^{\text{I}} = \underline{G}^{\text{II}}$, where the superscripts denote different phases. Thus, for a change dT and dP along this curve

$$d\underline{G}^{\text{I}} = d\underline{G}^{\text{II}}$$

or

$$\underline{V}^{\text{I}}\,dP - \underline{S}^{\text{I}}\,dT = \underline{V}^{\text{II}}\,dP - \underline{S}^{\text{II}}\,dT \tag{6-140}$$

Solving this equation for dP/dT and making use of the fact that

$$\underline{G}^{\text{I}} = \underline{H}^{\text{I}} - T\underline{S}^{\text{I}} = \underline{H}^{\text{II}} - T\underline{S}^{\text{II}} = \underline{G}^{\text{II}} \tag{6-141}$$

or

$$\underline{S}^{\text{II}} - \underline{S}^{\text{I}} = \frac{\underline{H}^{\text{II}} - \underline{H}^{\text{I}}}{T} \tag{6-141a}$$

to eliminate entropy yields the result known as the *Clausius–Clapeyron equation*,

$$\left(\frac{dP}{dT}\right)_{eq} = \frac{\underline{S}^{II} - \underline{S}^{I}}{\underline{V}^{II} - \underline{V}^{I}} \tag{6-142}$$

$$= \frac{\underline{H}^{II} - \underline{H}^{I}}{(\underline{V}^{II} - \underline{V}^{I})T} \tag{6-142a}$$

If phase II is considered to be vapor and phase I to be liquid, then $\underline{H}^{II} - \underline{H}^{I} = \Delta\underline{H}_V$, the enthalpy change on vaporization, and $\underline{V}^{II} - \underline{V}^{I} = \Delta\underline{V}_V$, the volume change on vaporization. An approximate form of this equation may be obtained by noting that usually $\underline{V}^L \ll \underline{V}^V$ and that, if the vapor is an ideal gas,

$$\Delta\underline{V}_V = \frac{RT}{P} \tag{6-143}$$

Substituting these simplifications into equation (6-142) reduces the Clausius–Clapeyron equation to

$$d \ln P = -\frac{\Delta\underline{H}_V}{R} d\left(\frac{1}{T}\right) \tag{6-144}$$

Equation (6-144) in turn may be integrated over a range of temperatures in which $\Delta\underline{H}_V$ is constant to give

$$\ln \frac{P_2}{P_1} = -\frac{\Delta\underline{H}_V}{R}\left(\frac{1}{T_2} - \frac{1}{T_1}\right) \tag{6-145}$$

All the above conditions are reasonably well met provided we are not too close to the critical point.

Many examples can be cited to illustrate the utility of the foregoing relationships. Because of the form of equation (6-145), vapor-pressure data are often plotted as $\ln P'$ versus $1/T$, with nearly straight lines resulting. Another example is given below.

SAMPLE PROBLEM 6-12. Freon 13 is being used as a refrigerant in a commercial refrigerating system. At one point in the system saturated liquid Freon 13 enters an evaporator (a heat exchanger) under a pressure of 150 psia and a temperature of -10°F. The Freon 13 absorbs 48.9 Btu/lb_m of liquid entering and leaves the evaporator at the same temperature and pressure but with a quality of 96 per cent. It has become necessary to lower the temperature of the Freon 13 in the evaporator. Would the pressure have to be increased or decreased to carry out the evaporation at -20°F? By how much?

The specific volume of the saturated liquid at 150 psia and -10°F is estimated to be 5 per cent that of the saturated vapor at the same temperature and pressure. Other constants of Freon 13 are as follows:

$$T_c = 83.9°\text{F}$$

$$P_c = 561.3 \text{ psia}$$

$$\text{mol wt} = 104.5$$

Solution: If the slope of the equilibrium curve at 150 psia and -10°F can be estimated, then we can make a reasonable estimate of the pressure at -20°F. The Clausius–Clapeyron equation (6-142a) is well suited to this task. We first calculate the enthalpy change upon vaporization. The heat transfer was carried out in a steady flow process (with no shaft work, kinetic or potential energy changes) for which the energy balance reduces to

$$\Delta \underline{H} = \underline{Q} = 48.9 \text{ Btu/lb}_m \text{ liquid entering}$$

Since this was required to evaporate 96 per cent of the liquid,

$$\Delta \underline{H}_V = (48.9/0.96) \times 104.5 \text{ (Btu/lb}_m\text{) (lb}_m\text{/lb-mol)} = 5310 \text{ Btu/lb-mol}$$

Because the saturated liquid volume is 5 per cent that of the saturated vapor, we may write

$$\Delta \underline{V}^V = \underline{V}^V - 0.05\underline{V}^V = 0.095\ \underline{V}^V$$

and use the generalized compressibility charts to estimate $\underline{V}^V$:

$$T_r = \frac{450}{543.9} = 0.828$$

$$P_r = \frac{150}{561.3} = 0.267$$

$$Z = 0.81$$

$$\underline{V}^V = \frac{ZRT}{P} = \frac{(0.81)(10.731)(450)}{150} \frac{\text{ft}^3 \text{ psi °R}}{\text{lb-mol psi °R}} = 26.1 \text{ ft}^3\text{/lb-mol}$$

The slope of the equilibrium curve is obtained from the Clausius–Clapeyron equation:

$$\left(\frac{dP}{dT}\right)_{\text{eq}} = \frac{\Delta \underline{H}_V}{\Delta \underline{V}_V T} = \frac{(5310)(778)}{(0.95)(26.1)(450)} \frac{\text{Btu lb-mol/ft-lb}_f}{\text{lb-mol Btu ft}^3 \text{ °R}}$$
$$= 370 \text{ lb}_f\text{/ft}^2 \text{ °R}$$
$$= 2.57 \text{ psia/°R}$$

Note that the conversion factor 778 ft-lb$_f$/Btu has been employed here so as to yield familiar pressure units. The units Btu/ft^3 are pressure units but not convenient ones.

To answer the question originally posed, it is now evident that a pressure decrease of 25.7 psia will be necessary to decrease the temperature by 10°F. The approximate forms of the Clausius–Clapeyron equation, (6-144) and (6-145), are not sufficiently accurate here because the vapor is not an ideal gas ($Z = 0.81$).

Problems

6-1 (a) Evaluate the following derivatives in terms of only P, $\underline{V}$, T, C_P, C_V, and their derivatives.

(1) $\left(\frac{\partial^2 \underline{V}}{\partial T^2}\right)_P$ (2) $\left(\frac{\partial \underline{S}}{\partial P}\right)_T$

(3) $\left(\frac{\partial \underline{U}}{\partial \underline{V}}\right)_T$ (4) $\left(\frac{\partial P}{\partial \underline{V}}\right)_{\underline{S}}$

(5) $\left(\frac{\partial T}{\partial P}\right)_{\underline{H}}$ (6) $\left(\frac{\partial \underline{H}}{\partial P}\right)_T$

(7) $\left(\frac{\partial T}{\partial P}\right)_{\underline{S}}$ (8) $\left(\frac{\partial \underline{V}}{\partial T}\right)_{\underline{U}}$

(b) Evaluate the derivatives for a gas that obeys the ideal gas equation of state.

(c) Evaluate the derivatives of part (a), (2) through (8), for a gas that obeys the van der Waals equation of state:

$$P = \frac{RT}{\underline{V} - b} - \frac{a}{\underline{V}^2}$$

6-2 Reduce the following to forms involving only P, $\underline{V}$, T, C_P, C_V, and their derivatives:

(a) $\left(\frac{\partial \underline{U}}{\partial \underline{S}}\right)_{\underline{V}}$ (b) $\left(\frac{\partial \underline{V}}{\partial T}\right)_{\underline{U}}$

(c) $\left(\frac{\partial P}{\partial T}\right)_{\underline{S}}$ (d) $\left(\frac{\partial \underline{U}}{\partial P}\right)_{T}$

(e) $\left(\frac{\partial \underline{U}}{\partial P}\right)_{\underline{S}}$ (f) $\left(\frac{\partial \underline{S}}{\partial P}\right)_{T}$

(g) $\left(\frac{\partial T}{\partial P}\right)_{\underline{H}}$ (h) $\left(\frac{\partial \underline{H}}{\partial P}\right)_{T}$

6-3 For the purpose of solving a problem involving a refrigeration cycle, some changes in thermodynamic properties of Freon 12 are needed. Show clearly how the change in entropy and enthalpy for a process proceeding from 330 to 350 psia at 400°F (isothermal) could be calculated using the data given for specific volumes (in ft^3/lb_m)

Pressure, psia	380°F	390°F	400°F	410°F	420°F
300	0.22522	0.22893	0.23260	0.23625	0.23987
340	0.19589	0.19927	0.20261	0.20592	0.20921
380	0.17271	0.17583	0.17892	0.18197	0.18500

6-4 (a) During certain types of nozzle calculations it is necessary to know the rate at which the enthalpy of the nozzle fluid varies with density. Since the nozzle flow is usually approximated as isentropic, the derivative in question is

$$\left(\frac{\partial \underline{H}}{\partial \rho}\right)_S$$

Express this derivative in terms of only P, $\underline{V}$, T, C_P, C_V, and their derivatives. Evaluate for a gas that obeys the ideal gas equation of state.

(b) It has been suggested that the temperature of an organic liquid might be reduced by adiabatically expanding the liquid through a reversible turbine. The coefficient of thermal expansion of the liquid $[\beta = (1/\underline{V})(\partial \underline{V}/\partial T)_P]$ is

$$\beta = 4.0 \times 10^{-3}\ {}^\circ R^{-1}$$

and the liquid has a specific gravity of 0.8, $C_P = 0.09$ Btu/lb_m °R, and a molecular weight of 100. Determine the isentropic temperature decrease with pressure, $(\partial T/\partial P)_S$, at $T = 200$°F.

6-5 (a) Evaluate

$$\left(\frac{\partial \underline{U}}{\partial \underline{S}}\right)_{\underline{A}}$$

in terms of P, $\underline{V}$, T, C_P, C_V, and their derivatives. Your answer may include absolute values of $\underline{S}$ if it is not associated with a derivative.

(b) Find an expression for the partial derivative

$$\left(\frac{\partial \underline{U}}{\partial P}\right)_T$$

in terms of P, $\underline{V}$, T, C_P, C_V, and their derivatives.

(c) Evaluate the above derivative for a gas that obeys the Redlich–Kwong equation of state.

6-6 Estimate the work required to produce droplets 0.001 ft in diameter per pound of water. Assume that the process is conducted adiabatically and reversibly and that the fluid is incompressible. [*Note:* The property relationship can be extended to include surface effects by adding a term, $\sigma\, d\underline{A}$, so that

$$d\underline{U} = T\,d\underline{S} - P\,d\underline{V} + \sigma\,d\underline{A}$$

where σ is surface tension (4.93×10^{-3} lb_f/ft) and $\underline{A}$ is surface area per pound mass.]

6-7 (a) Prove that

$$\left(\frac{\partial \underline{H}}{\partial P}\right)_T = -\mu C_P$$

(b) The data below give μ (°C/atm) for air as a function of temperature and pressure:

Pressure, atm	0°C	25°C	50°C	75°C
1	0.2663	0.2269	0.1887	0.1581
20	0.2494	0.2116	0.1777	0.1490
60	0.2143	0.1815	0.1527	0.1275
100	0.1782	0.1517	0.1283	0.1073

Source: J. R. Roebuck, *Proc. Amer. Acad. Arts Sci., 60*, 535–596 (1925).

Estimate the change of heat capacity with pressure at 50°C. The observed value was 7.3×10^{-3} cal/mol deg atm. The molar heat capacity of air ($P = 100$ atm) is

$$C_P = \begin{cases} 7.91 \text{ cal/deg mol} & \text{at } 50°\text{C} \\ 7.79 \text{ cal/deg mol} & \text{at } 75°\text{C} \end{cases}$$

[*Hint:* Differentiate the equation of part (a) with respect to T at constant P.]

6-8 Calculate the final temperature after a Joule–Thomson expansion from 1000 atm and 300°K to 1 atm.

$$C_P = 7 \text{ cal/g-mol °K}$$

$$\underline{V} = \frac{RT}{P} + (10^{-4})T^2$$

where $\underline{V}$ is in cm³/g-mol and P is in atm.

6-9 It is desired to develop a general-purpose computer program for constructing a pressure–enthalpy chart from a knowledge only of an equation of state for the gas, its C_P^* as a function of T, and the vapor-pressure curve. In particular, a method for determining the lines of constant entropy would be extremely useful.

(a) Prove that the slope of the isentrope is given by:

$$\left(\frac{\partial P}{\partial \underline{H}}\right)_S = \frac{1}{\underline{V}}$$

(b) Indicate clearly how the expression derived above may be integrated at constant entropy.

[*Hint:* How would one relate $\underline{V}$ to P at constant entropy knowing only the information given in the problem statement? Be specific in indicating what must be done, but do not attempt to carry out the calculations for a specific equation of state.]

6-10 A single-acting, two-stage compressor is handling hydrogen at 1 atm and 60°F. Exhaust is at 250 psia. A heat exchanger cools the gas between the stages to 60°F; otherwise operation is assumed to be adiabatic. For hydrogen k (heat capacity ratio) = 1.42.

(a) What pressure between the two stages minimizes the reversible work for the compressor?

(b) Calculate the minimum work per 100 lb-mol of gas.

Suppose now that the compressor is equipped with jacketed cylinders so that the operation is essentially isothermal at all times.

(c) What pressure between the two stages minimizes the reversible work?

(d) Compare the amount of heat removed from the jacketed cylinders in the second case with the amount of heat removed from the intercooler (heat exchanger) in the first case.

(e) Compare the work per 100 lb-mol for the two cases.

6-11 A portable compressed air tank is frequently used to inflate flat tires as shown in Fig. P6-11. The tank is initially filled with compressed air at 150 psia and 70°F; the tire

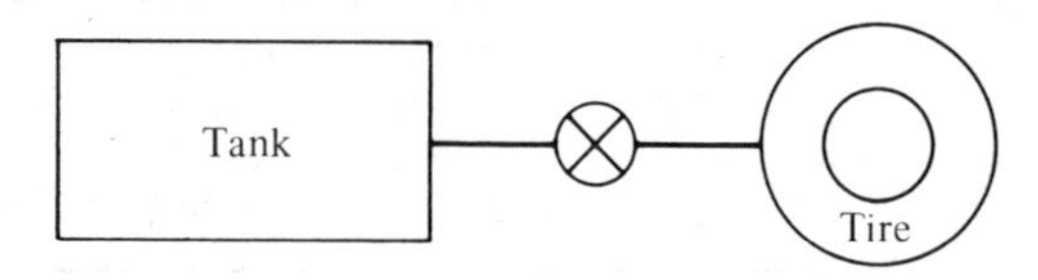

FIGURE P6-11

may be assumed to be fully evacuated (a *really* flat tire). After connecting the air hose to the tire, the valve is opened and air rushes into the tire. In the specific circumstances under study the valve is closed when the pressure in the tank has fallen to 75 psia. The volume of the tank is $\frac{1}{2}$ ft^3; that of the tire is 1 ft^3. Assume that the tire is rigid, so its walls do not move during the filling operation, and that the filling takes place rapidly enough that the process may be considered adiabatic; $k_{air} = 1.40$.

(a) Taking the gas that remains in the tank as a system, has this system undergone a reversible or irreversible expansion? Justify your answer.

(b) What are the final temperatures in the tire and tank?

(c) What is the final pressure in the tire (psig)?

6-12 An air-actuated control device is designed as shown in Fig. P6-12. Chambers I

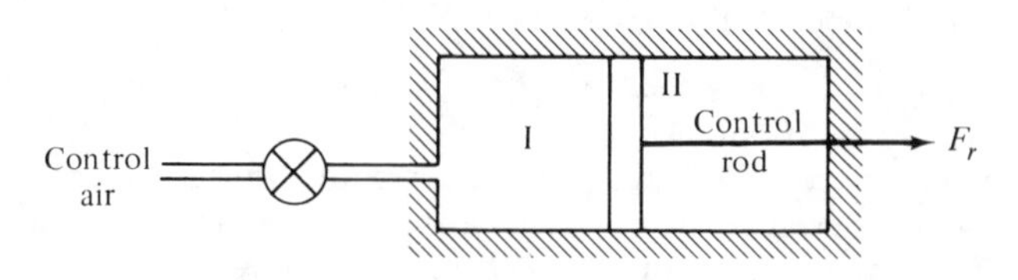

FIGURE P6-12

and II are separated by a perfectly insulated frictionless piston which is connected to a movable control rod. Motion of the control rod is controlled by addition or removal of air from chamber I. Chamber II is completely airtight, and the whole device is insulated from the surroundings. Chamber I initially occupies 10 per cent of the total volume and contains air at 100 psia and 140°F; chamber II initially occupies 90 per cent of the total volume and contains air at 20 psia and 70°F. If the resisting force on the control rod, $F_{r'}$, almost counterbalances the pressure difference between chambers I and II so that the motion of the control rod can be considered reversible, determine the conditions that exist in the chambers when the control rod ceases moving. What fraction of the total volume is now occupied by chamber I? $(C_P^*)_{air} = 6.95$ Btu/lb-mol° R, and you may assume that air is an ideal gas under the problem conditions.

6-13 A diesel engine requires no spark plug when it is operating. Ignition of the fuel is carried out by adiabatically compressing the fuel–air mixture until its temperature exceeds the ignition temperature and the fuel burns. Assume that the fuel–air mixture is pumped into the cylinder at 5 atm and 50°F and is adiabatically and reversibly compressed in the closed cylinder until its volume is $\frac{1}{10}$ the initial volume. Assuming that no ignition has occurred at this point, determine the final T and P, as well as the work needed to compress each mole of air–fuel mixture. C_V^* for the mixture may be taken as 12.5 Btu/lb-mol, and the gas may be assumed to obey the equation of state

$$P\underline{V} = RT + AP$$

where A is a constant whose value is

$$A = 3 \text{ ft}^3/\text{lb-mol}$$
$$R = 0.73 \text{ atm ft}^3/\text{lb-mol °R}$$

Do not assume that C_V is independent of $\underline{V}$.

6-14 A rigid container of negligible mass holds 18 lb_m of propane (C_3H_8) gas at 30 psia and 60°F. If the container is heated until the pressure becomes 45 psia, determine

(a) The final temperature.
(b) The amount of heat transferred.
(c) The change of internal energy of the propane.
(d) The change of entropy of the propane.
(e) The volume of the container.

Propane gas follows the PVT relation

$$\frac{PV}{nRT} = 1 - \frac{0.4T_cP}{P_cT}$$

where $P_c = 617.4$ psia and $T_c = 666$°R. The constant-volume heat capacity of propane may be represented by the equation

$$C_V^* = 4 + 0.021T(\text{°R}) \text{ Btu/lb-mol °R}$$

6-15 Natural gas (essentially pure methane) is being compressed prior to transmission by pipeline. The gas enters the compressor at 70°F and 200 psia and is isothermally compressed to 3000 psia. Evaluate $\Delta\underline{H}$, $\underline{Q}$, and $\underline{W}$ for the compression if

(a) The natural gas behaves as an ideal gas.
(b) The natural gas behaves according to the van der Waals equation of state.

(*Hint:* Because of the nature of the equation of state, you may find it simpler to let $\Delta\underline{H} = \Delta\underline{U} + \Delta P\underline{V}$, and then calculate $\Delta\underline{U}$ and $\Delta P\underline{V}$.)

6-16 Gas enters an isothermal gas turbine at 70 atm and 80°F and discharges at 20 atm. Part of the work from the turbine is used to drive an adiabatic isentropic compressor in which all the gas discharged from the turbine is partially recompressed and leaves the compressor at 200°F.

(a) At what pressure does the gas leave the compressor?

(b) What fraction of the total work delivered by the turbine must be supplied to the compressor?

Equation of state: $P\underline{V} = RT + a\underline{V}$, where $a = 5.6$ atm. $C_V^* = 5.0$ Btu/lb-mol °R. Assume that $\underline{Q}_{turb} = T(\underline{S}_0 - \underline{S}_i)$.

6-17 Gas enters an adiabatic, isentropic nozzle at 40 atm and 500°F and with negligible velocity and is continuously expanded to 1 atm. (See Problem 5-8 for a description of nozzle-flow assumptions you may use.)

(a) At what velocity (ft/sec) will the gas leave the nozzle?

(b) For a gas flow rate of 10 lb_m/hr, what cross-sectional area (ft^2) will be required at the nozzle exit?

Equation of state: $P(\underline{V} - b) = RT$, where $b = 0.5$ ft³/lb-mol. C_P at 1 atm = 7.0 Btu lb-mol °F. Molecular weight of gas = 30.

6-18 Two insulated tanks of identical volumes are connected by a pipe with a valve. One tank contains 11 lb_m of propane (C_3H_8) at 250 psia and 160°F. The other tank is completely evacuated. The valve is opened for a short time until the flow of propane ceases; the valve is then closed. Estimate, using reasonable assumptions, the mass of propane in each tank at the end of the process.

Propane gas follows the PVT relation

$$\frac{PV}{nRT} = 1 - \frac{0.4T_cP}{P_cT}$$

where $P_c = 617.4$ psia and $T_c = 666$°R. The constant-volume heat capacity of propane may be represented by the equation

$$C_V = 4 + 0.021T(°\text{R}) \text{ Btu/lb-mol °R}$$

6-19 Natural gas as received from gas wells usually contains moderate amounts of valuable heavy hydrocarbons (C_2 through C_{10}) which are removed from the gas before it is sold as fuel. The lean natural gas is then compressed to high pressures so that it can be economically transported through gas pipelines to the major fuel markets. A lean natural gas is available at 50°F and 1000 psia. The gas must be compressed to 5000 psia for economical pipeline transportation. Determine the work that must be supplied to the compressor per lb-mass of gas under the following assumptions:

(a) The compression is isothermal and reversible and the natural gas behaves according to the following equations of state: (1) The ideal gas law, and (2)

$$1 - Z = \frac{P_r}{T_r}\left(-0.115 + \frac{0.132}{T_r} + \frac{0.356}{T_r^2}\right)$$

where $Z = P\underline{V}/RT$.

(b) The compression is adiabatic and reversible, and the natural gas behaves according to the same two equations of state as given in (a).

You may assume that the natural gas behaves as if it is pure methane with the following properties:

$$P_c = 45.8 \text{ atm}$$
$$T_c = 191.1°\text{K}$$
$$C_P^* = 4.52 + 0.00737T(°\text{R}) \text{ Btu/lb-mol °R}$$

6-20 A tank of oxygen (volume 2 ft^3) is stored outdoors for a long time in a place where the temperature is 40°F. The pressure in the tank at this temperature is measured to be 1500 psia. The tank is brought indoors and stored near a furnace, where its temperature rises to 190°F.

(a) What is the new pressure in the tank?

(b) How much heat was transferred to the contents of the tank to heat them from 40 to 190°F?

(c) With the tank at 190°F, suppose the valve is opened slightly and some gas allowed to flow into the room. Assuming the velocity of the stream to be small, what will be the temperature of the oxygen as it leaves the tank (the first increment)?

For oxygen

$$P_c = 49.7 \text{ atm}$$
$$T_c = 154.3°\text{K}$$
$$\underline{V}_c = 1.19 \text{ ft}^3/\text{lb-mol}$$
$$C_P^* = 7.02 \text{ cal/°K g-mol}$$
$$\text{Melting point} = 54.36°\text{K}$$
$$\text{Boiling point} = 90.19°\text{K}$$

6-21 Chloropentafluoroethane (C_2ClF_5) is used in certain refrigeration cycles. At one point the gas at 80 psia and 160°F is compressed to 320 psia. Per 1000 ft^3 of gas at the suction side, calculate the heat and work effects for the following cases:

(a) The compressor operates reversibly and isothermally.

(b) The compressor operates reversibly and adiabatically.

If the compressed gas from case (a) is expanded adiabatically through a throttle valve to a pressure of 80 psia, what will be the exit temperature?

The critical properties of C_2ClF_5 are as follows:

$$P_c = 453 \text{ psia}$$
$$T_c = 175.9°\text{F}$$
$$\underline{V}_c = 0.02687 \text{ ft}^3/\text{lb}_m$$

The heat capacity of the gas at low pressure is

$$C_P^* = 0.044976 + 0.00033T(°\text{R}) - 1.52 \times 10^{-7}T^2 \text{ Btu/lb}_m°\text{R}$$

6-22 A diesel engine operates without a spark plug by using the high-temperature gas generated during the compression stage to ignite the fuel. During a typical compression, pure air that is originally at 70°F and 0.95 atm is reversibly and adiabatically compressed to $\frac{1}{20}$ of its original volume by the piston, as shown in Fig. P6-22.

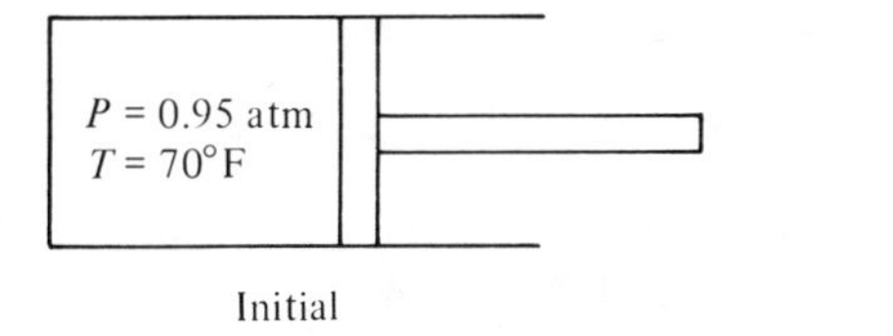

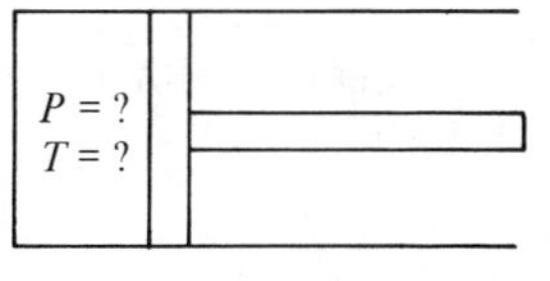

FIGURE P6-22

(a) If air is assumed to obey the ideal gas equation of state, find the pressure and temperature in the cylinder at the end of the compression. $(C_P)_{air} = 7.0$ Btu/lb-mol. What is the work of compression?

(b) Using the generalized charts, determine a (hopefully) better approximation to the final temperature and pressure in the cylinder.

6-23 Ethylene is contained in a steel vessel at 1490 psia and 70°F. A valve is opened and the tank pressure drops immediately to atmospheric pressure. If the temperature of the surroundings is 60°F, calculate the temperature of the gas remaining in the tank immediately after the expansion.

Critical properties for ethylene:

$$T_c = 282.5^\circ\text{K}$$

$$P_c = 50 \text{ atm}$$

$$C_P^* = 2.71 + 16.20 \times 10^{-3}T(^\circ\text{R}) - 2.80 \times 10^{-6}T^2 \text{ But/lb-mol } ^\circ\text{R}$$

6-24 Air is to be cooled before entering the distillation tower, where pure nitrogen and oxygen are produced. The cooling will take place in a gas-refrigeration process as follows. Air enters the refrigeration plant at 1 atm and 75°F. It is then compressed to 1100 psia and water cooled to 50°F and then cooled to −150°F by heat exchange with the separated nitrogen and oxygen. In the final step, the compressed air is then throttled to 1 atm and fed to the distillation tower. Assuming that the throttling operation is adiabatic, determine the temperature of the air that enters the tower. (You may assume negligible changes in potential and kinetic energy.)

The following data may prove useful:

$$C_P^* = 7.0 \text{ Btu/lb-mol } ^\circ\text{R in this temperature region}$$

$$P_c = 37.2 \text{ atm}$$

$$T_c = -140^\circ\text{C}$$

$$\underline{V}_c = 2.86 \text{ cm}^3/\text{g}$$

6-25 If the compressor in Problem 6-24 operates adiabatically and reversibly, what is the temperature of the outlet air stream? What is the work of compression, and how much heat is removed in the water cooler, where the compressed oxygen is cooled to 50°F? Use the data of Problem 6-24.

6-26 Carbon dioxide is to be expanded isentropically through an insulated nozzle of proper design. The carbon dioxide will be supplied to the nozzle entrance at 800 psia and 100°F, and with negligible velocity. The discharge pressure will be 500 psia. At a pressure of 1 atm the heat capacity of carbon dioxide is given by the equation

$$C_P = 6.85 + 0.00474T(^\circ\text{R}) - (7.64 \times 10^{-7})T^2 \text{ Btu/lb-mol } ^\circ\text{R}$$

Determine the velocity (ft/sec) at the nozzle outlet

(a) Assuming CO_2 to be an ideal gas.
(b) Assuming CO_2 to be a van der Waals gas.
(c) Using generalized plots.
(d) Using the BWR equation.

6-27 Natural gas (assumed to be pure methane) flows from a pipeline into an initially empty insulated underground gas-storage reservoir. The gas flowing in the pipeline is at 75°F and 3000 psia. Determine the temperature of the gas in the reservoir when the pressure reaches 3000 psia. (You may assume that the filling operation is adiabatic.)

$$T_c = 191.1°K$$
$$P_c = 45.8 \text{ atm}$$
$$C_P^* = 5.52 + 0.00737T(°R) \text{ Btu/lb-mol °R}$$

6-28 The fugacity, f, has been defined in such a manner as to correct for the nonideal behavior of real gases. The compressibility factor, Z, has been defined for the same reason. Therefore, it has been suggested that the equation defining Z ($P\underline{V} = ZRT$) may be rewritten as

$$f\underline{V} = RT$$

Comment in depth on the correctness of this suggestion.

6-29 Show that G is a minimum at equilibrium for the closed, isothermal, constant-pressure system shown in Fig. P6-29. Note that both pistons are free moving, massless,

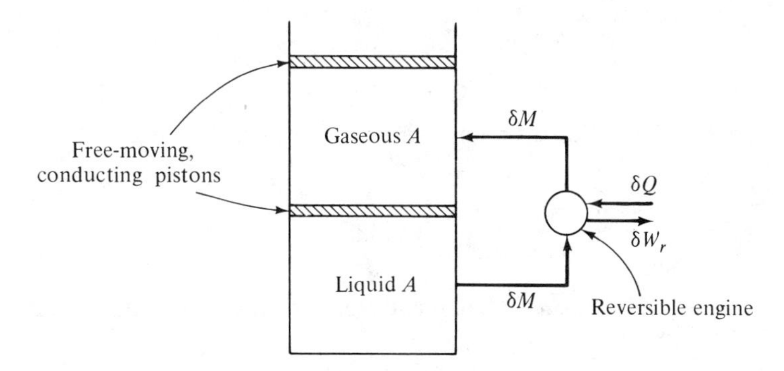

FIGURE P6-29

and perfectly conducting. In writing the energy balance, be sure to account for the work done in changing the total volume of the system as well as the work of the reversible engine. The equilibrium conditions correspond to the conditions where no work may be obtained *from the engine.*

6-30 Show that A is a minimum at equilibrium for the closed, isothermal, constant-total-volume system that results if the upper piston in Fig. P6-29 is assumed to be fixed in position. Show also that the specific Gibbs free energies of the two phases are equal at equilibrium.

6-31 Listed below are properties of saturated carbon dioxide at temperatures below the triple point.

		Volume, ft^3/lb_m	
T, °F	P, psia	Solid	Vapor
−140	3.18	0.01008	24.320
−120	8.90	0.01018	9.179
−100	22.22	0.01032	3.804
−90	33.98	0.01040	2.525
−80	50.85	0.01048	1.700

Compute the $\Delta\underline{H}$ and $\Delta\underline{U}$ of sublimation of solid carbon dioxide at −100°F in Btu/lb_m.

6-32 Water and ice are in equilibrium at 0°C and 1 atm pressure. The latent heat of fusion of ice to water under these conditions is 1438 cal/g-mol. The volume of water is 18 cm^3/g-mol, of ice 19.7 cm^3/g-mol. The heat capacity of water is 18 cal/g-mol °C, and of ice 9 cal/g-mol °C.

(a) Compute $\Delta \underline{G}$ for the change of subcooled water to ice at −20°C and 1 atm.
(b) Compute the pressure at which ice and water are in equilibrium at −20°C.

6-33 Normal butane (C_4H_{10}) is stored as a compressed liquid at 190°F and 14 atm pressure. In order to use the butane in a low-pressure gas-phase process, it is throttled to 1.5 atm and passed to a vaporizer. From the vaporizer the butane emerges as a gas at 160°F and 1.5 atm. Using only the data below and any assumptions you believe are reasonable, calculate the heat that must be supplied the vaporizer per pound of butane passing through it. Describe quantitatively the condition of the butane entering the vaporizer.

For normal butane

$$T_c = 765.3°\text{R}$$
$$P_c = 550.7 \text{ psia}$$

Heat capacity of the liquid at any constant pressure, $C_P = 0.555 + 0.0005T(°\text{F})$ Btu/lb_m °F

Vapor pressure given by

$$\ln P' (\text{atm}) = \frac{-4840}{T(°R)} + 9.92$$

Pressure has negligible effect on liquid enthalpy. Volume of saturated liquid = 0.025 + $0.0004T(°\text{F})$ $\text{ft}^3/\text{lb}_\text{m}$.

6-34 Liquid butane is pumped to a vaporizer as a saturated liquid under a pressure of 237 psia. The butane leaves the exchanger as a wet vapor of 90 per cent quality and at practically the same pressure as it entered. From the following information, estimate the heat load on the vaporizer per pound of butane entering.

$$P_c = 37.48 \text{ atm}$$
$$T_c = 305.3°\text{F}$$
$$\ln P' = -4840/T + 9.92$$

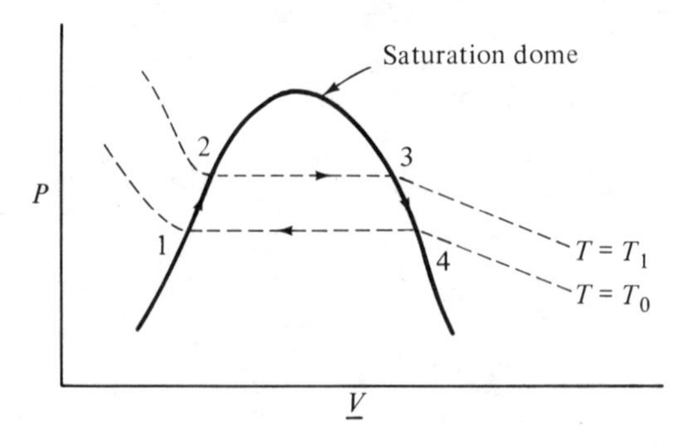

FIGURE P6-35

where P is in atm and T is in °R. At 237 psia the specific volume of the saturated liquid is estimated as $\frac{1}{10}$ that of saturated vapor.

6-35 A certain cycle operates as shown by the path 1–2–3–4 in Fig. P6-35. Derive an expression involving only the heat of vaporization and the temperature T_1 and T_0 for the maximum work per cycle that can be removed per unit weight of working fluid. You may assume that heat of vaporization to be constant over the range T_0 to T_1. Note that for a closed cycle

$$\oint P \, d\underline{V} = -\oint \underline{V} \, dP$$

Thermodynamics of Energy Conversion 7

Introduction

Virtually every process with which the engineer is associated involves energy, either as a product or as a means of producing a product. In the preceding chapters the basic relationships that govern energy transfer have been developed. In addition, the thermodynamic potentials that serve as the driving forces for energy flows, as well as the interrelationships among these potentials, have been developed. In this chapter these relationships are utilized to study the methods and machines devised by engineers for the utilization of our energy resources in meeting our everyday power and refrigeration needs.

The various forms in which energy may be manifested, such as thermal, mechanical, electrical, and chemical, have been cited in the earlier chapters. Energy flows caused solely by temperature gradients were categorized as heat, while energy flows resulting from thermodynamic potentials other than temperature (for example, pressure, electrical potential, etc.) were classified as work. As we have previously indicated, the energy flows we call work all have one important common trait: They are, in theory, completely interconvertible. To be sure, these interconversions are not easily achieved without irreversibilities which lead to degradation of some of the mechanical energy to the thermal form. Frequently mechanical forms are intentionally converted to thermal forms, as in the case of an electrical heating unit or a mechanical stirrer.

Nature has endowed the universe with a variety of energy sources; some, such as our water reservoirs, permit the direct production of mechanical energy (hydroelectric power) without the intermediate combustion step required with conventional fuels. Although we continue to develop such resources by adding dams and hydroelectric facilities along our rivers, the bulk of our usable electric power continues to come from fossil fuels. In the past 5 to 10 years nuclear power plants have begun to make serious inroads into the electric power market. However, these facilities, much as the fossil fuel-powered generating stations, use the high temperatures produced in the reactor core as the driving force by which energy is extracted.

The efficient conversion of thermal energy into mechanical energy has challenged scientists and engineers for centuries. This challenge continues today because a large portion of our energy reserves exists in forms such that the energy is extracted in a thermal form by first burning (or fissioning) the fuel. A portion of the thermal energy so obtained is then frequently converted to a mechanical form, in which it has far greater utility. In spite of our best efforts we have been unable to devise a procedure by which the thermal energy released in the combustion process can be completely converted to a mechanical form. In the next few sections we shall examine the reason for such a limitation. We shall also study the various schemes that have been devised to perform this conversion in such a way that the maximum possible work is obtained from a unit of thermal energy.

Recent advances in the fields of magnetohydrodynamics, fuel cells, and thermoelectric generators raise the hope that someday we may be able to economically generate electric energy without the need for heat engines. However, it appears that the bulk of energy-conversion devices will continue to rely on heat engines for many years to come.

7.1 Noncyclic Heat Engines: The Steam Engine

Since the thermal potential, temperature, cannot be used directly to produce work, it is necessary to utilize the thermal potential to produce an increase in some other energy potential, such as pressure, before work can be extracted. For example, the heating of a closed vessel of water (producing steam) would produce an increased pressure as its energy content was increased. The higher pressure possessed by the system could then be used to push a piston, as is done in the simple steam engine. The relationships developed in Chapter 6 will be used to provide an insight into the variation of the pressure and temperature of the steam as the energy content is changed.

The steam engine is a device capable of converting thermal energy to mechanical energy (as evidenced by the operation of the steam locomotive). However, a closer examination of the conversion process reveals that only a small fraction of the thermal energy supplied to the boiler actually gets converted to work. Consider the steam engine shown in Fig. 7-1 as the system. Water is pumped from the feedwater tank into the boiler as a liquid at low temperature and high pressure. The water is then heated and converted to steam before passing into the piston–cylinder converter. By maintaining the pressure

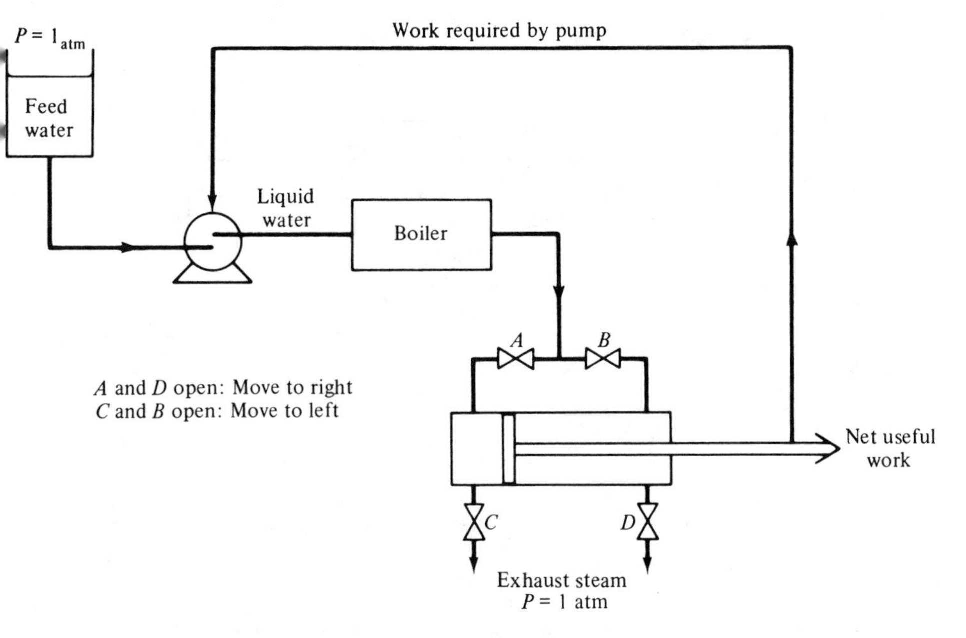

FIG. 7-1. Steam engine.

on the other side of the cylinder at atmospheric pressure, work can be obtained from the pressure (force) difference across the piston. Since the steam is exhausted to the atmosphere at, or slightly above, atmospheric pressure, the steam leaves the engine with a much higher enthalpy than when it entered as water. Thus a significant fraction of the thermal energy supplied leaves the engine in the exhaust steam rather than in the form of useful work.

SAMPLE PROBLEM 7-1. An emergency pump in an oil refinery is to be powered by a steam turbine as shown in Fig. S7-1a. Feedwater will enter the pump at 1 atm and 100°F. The water is pumped to a pressure of 100 psia and passed into the boiler, where it is converted to saturated steam. From the boiler the steam enters the turbine, where it is expanded to 1 atm. A portion of the work produced by the turbine is used to power the feedwater pump; the remainder is the useful work of the engine. If the turbine is assumed to operate in an adiabatic and reversible fashion, determine the net work produced per unit of thermal energy supplied to the boiler.

Solution: The feedwater enters the engine (the combination of pump, boiler, and turbine is termed "the engine") at 100°F and 1 atm. All calculations are based on 1 lb_m of water passing through the engine. The work required by the pump to increase the pressure of the water from 1 atm to 100 psia is given by

$$\underline{W}_{\text{pump}} = -\int_{P_1}^{P_2} \underline{V}\, dP$$

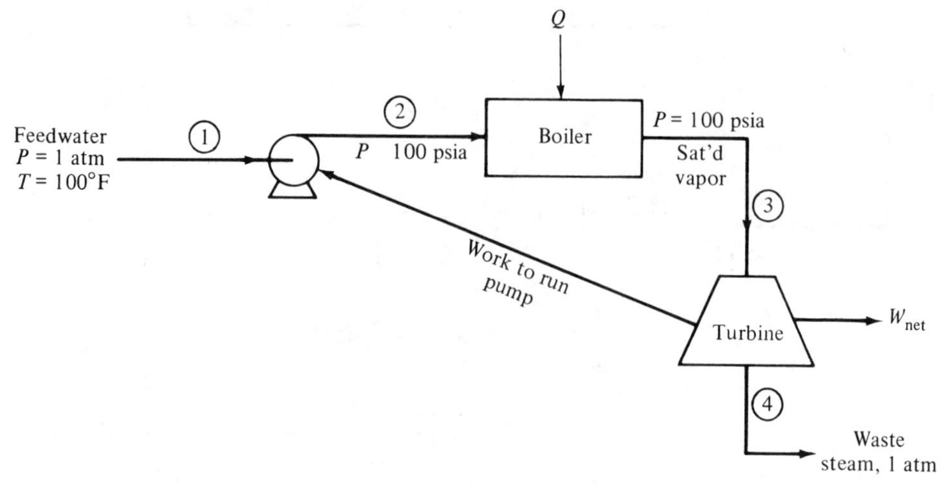

FIGURE S7-1a

but for liquid water $\underline{V}$ is constant, and the integral reduces to

$$\underline{W}_{\text{pump}} = -\underline{V}\int_{P_1}^{P_2} dP = -\underline{V}(P_2 - P_1)$$

$\underline{V} = 0.017$ ft³/lb$_m$ for liquid water, and $\underline{W}$ becomes

$$\underline{W}_{\text{pump}} = -0.017\,\frac{\text{ft}^3}{\text{lb}_m}(100 - 14.7)\frac{\text{lb}_f}{\text{in.}^2} \times \frac{144}{778}\frac{\text{in.}^2}{\text{ft}^2}\frac{\text{Btu}}{\text{lb}_f\text{-ft}}$$
$$= 0.265 \text{ Btu/lb}_m$$

An energy balance taken around the pump, assuming steady state and negligible changes in potential and kinetic energy gives

$$\underline{Q} - \underline{W} = \Delta\underline{H}$$

Assuming the flow of water through the pump to be essentially adiabatic, $\underline{Q} = 0$, and the energy balance reduces to

$$\Delta\underline{H} = -\underline{W} = +0.265 \text{ Btu/lb}_m$$

or, if we label the various positions in the engine as indicated in the diagram,

$$\underline{H}_2 - \underline{H}_1 = 0.265 \text{ Btu/lb}_m$$

Thus if we can find the enthalpy, $\underline{H}$, of the liquid entering the engine, we can calculate the enthalpy of the liquid leaving the pump and entering the boiler. We begin by noting that we can find the enthalpy of a saturated liquid at 100°F in the steam tables. That is,

$$\underline{H}_{\text{sat liq}} = 67.97 \text{ Btu/lb}_m$$

But

$$\underline{H}_1 = \underline{H}_{\text{sat liq}} + \int_{P_{\text{sat}}}^{P_1}\left(\frac{\partial \underline{H}}{\partial P}\right)_T dP$$

$$P_{\text{sat}} = 1.932 \text{ psia}$$

$$P_1 = 14.7 \text{ psia}$$

$$\left(\frac{\partial \underline{H}}{\partial P}\right)_T = \underline{V} - T\left(\frac{\partial \underline{V}}{\partial T}\right)_P$$

Evaluation of the integral term shows its value to be essentially negligible. Therefore,

$$\underline{H}_1 = 67.97 \text{ Btu/lb}_m$$

$$\underline{H}_2 = 68.24 \text{ Btu/lb}_m$$

Now let us examine the boiler (Fig. S7-1b). Since the outlet steam is a saturated vapor at 100 psia, we obtain from the steam tables (Appendix B)

$$\underline{H}_3 = 1187.3 \text{ Btu/lb}_m$$

$$\underline{S}_3 = 1.6028 \text{ Btu/lb}_m \text{ °R}$$

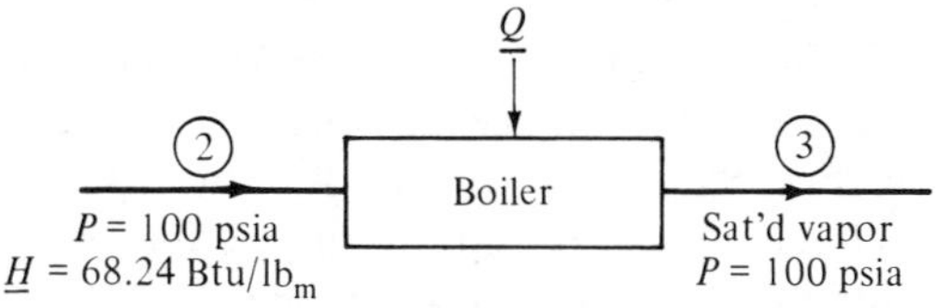

FIGURE S7-1b

An energy balance around the boiler, assuming steady state and negligible changes in potential and kinetic energy, reduces to

$$\underline{Q} - \underline{W} = \Delta \underline{H}$$

Since there is no work added or removed from the boiler, $\underline{W} = 0$, and the energy balance gives

$$\underline{Q} = \Delta \underline{H} = \underline{H}_3 - \underline{H}_2 = (1187.3 - 68.2) \text{ Btu/lb}_m$$

or

$$\underline{Q} = 1119.1 \text{ Btu/lb}_m$$

and represents the total thermal energy input per pound of water passing through the process.

Since the turbine is assumed to be adiabatic and reversible, the flow through it is isentropic. That is,

$$\Delta \underline{S} = \underline{S}_4 - \underline{S}_3 = 0$$

But

$$\underline{S}_3 = 1.603 \text{ Btu/lb}_m \text{ °R}$$

Therefore,

$$\underline{S}_4 = 1.603 \text{ Btu/lb}_m \text{ °R}$$

However, we also know that $P_4 = 1$ atm. Since we know two properties of the steam leaving the turbine, we can determine all the remaining properties. In particular we find (on the Mollier diagram, Appendix A)

$$\underline{H}_4 = 1047 \text{ Btu/lb}_m$$

Again the energy balance around the turbine gives

$$\underline{Q} - \underline{W} = \Delta \underline{H}$$

Since the turbine is adiabatic, $\underline{Q} = 0$, and the energy balance reduces to

$$-\underline{W} = \Delta \underline{H} = \underline{H}_4 - \underline{H}_3 = (1047 - 1187) \text{ Btu/lb}_m$$

or

$$\underline{W} = 140 \text{ Btu/lb}_m$$

However, of this, 0.3 Btu/lb_m is used to run the pump. Thus the net work is

$$\underline{W}_{net} = (142 - 0.3)\ Btu/lb_m = 139.7\ Btu/lb_m$$

and the ratio of net work to thermal energy input (frequently called the thermal efficiency) is given by

$$\eta = \frac{\underline{W}_{net}}{\underline{Q}} = \frac{139.7\ Btu/lb_m}{1119.1\ Btu/lb_m} = 0.126$$

so only 12.6 per cent of the thermal energy supplied to the boiler is converted to useful work. The remaining 87.4 per cent is lost with the exhaust steam.

As shown in Sample Problem 7-1, a large fraction of the thermal energy supplied to the noncyclic steam engines is wasted in the exhaust steam. One might question if it is necessary to discharge the working fluid at this relatively high energy level. Would it not be possible to recycle the exhaust steam through the engine in an effort to avoid losing this energy?

7.2 Cyclic Processes—The Carnot Cycle

The recycling of the discharge steam suggested in Section 7.1 requires two additional considerations. First, since the boiler operates at a higher pressure than the exhaust of the turbine (or whatever expander is used), the pressure of the working fluid must be increased before it returns to the boiler. Thus an expenditure of work is required for this step. This compression work must be subtracted from the work produced by the turbine to calculate the net work obtained from the engine (this loss is partially recovered, because no feedwater pump is required).

Second, from our earlier discussions of reversible processes it is apparent that the net work obtainable from the process will be maximized if both the expander and the compressor (pump) operate reversibly. In addition, let us assume that the pump and expander operate adiabatically; thus the flows through the compressor and expander occur at constant entropy.

Let us now apply energy and entropy balances to each of the three components our system is known to require: (1) the boiler, (2) the expander, and (3) the compressor. Assume steady state, negligible changes in potential and kinetic energy across each unit, and one unit mass of material flowing.

1. The boiler (Fig. 7-2) vaporizes the fluid reversibly at constant temperature, T_H, and pressure.
Energy balance:

$$\underline{Q}_H = \Delta \underline{H} = \underline{H}_2 - \underline{H}_1$$

Entropy balance:

$$\underline{Q}_H = T_H(\Delta \underline{S}_H) = T_H(\underline{S}_2 - \underline{S}_1)$$

2. The expander (turbine) operates reversibly and adiabatically (Fig. 7-3).
Energy balance:

$$-\underline{W}_T = \Delta \underline{H} = \underline{H}_3 - \underline{H}_2$$

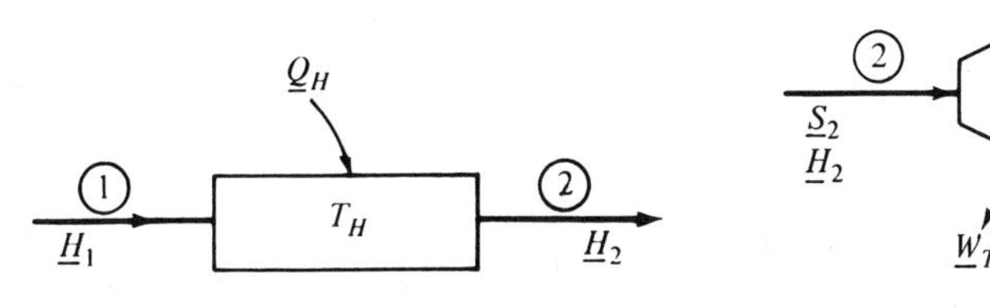

FIGURE 7-2 FIGURE 7-3

Entropy balance:

$$\Delta \underline{S} = 0 = \underline{S}_3 - \underline{S}_2$$

3. The compressor operates reversibly and adiabatically to return fluid to the boiler (Fig. 7-4). The asterisk indicates the state in which the fluid must enter the pump if it is to be returned to state 1 by the pump.
Energy balance:

$$-\underline{W}_P = \Delta \underline{H} = \underline{H}_1 - \underline{H}_*$$

Entropy balance:

$$\Delta \underline{S} = 0 = \underline{S}_1 - \underline{S}_*$$

Examination of the three operations shows that the pump can, in theory, return the working fluid to its original energy level. However, if we examine the three entropy balances we see that two of the processes do not affect the entropy, while the third process (the boiler) increases the entropy. This increase must be offset by a corresponding decrease elsewhere in the system if the working fluid is to complete the cycle in its original state (a necessary condition for cyclical operation—otherwise the process cannot operate at steady state).

The only method by which the fluid's entropy can be decreased in a closed cycle is by removing energy as heat. Thus it is essential to include a heat-removal step. This heat removal must not be immediately before or after that of the boiler. (If it were, its effect would have to be just opposite that of the boiler, and the fluid leaving the two units would be in the same state as that entering, and the cycle would be useless.) Therefore, the heat removal must occur between the expander and the pump. Since the temperature of the working fluid leaving the expander is less than its entering temperature, heat removal takes place at a lower temperature than heat addition in the boiler.

The low-temperature heat exchanger in this type of cycle is called the *condenser* and will be assumed to operate reversibly, isothermally, and at steady state (Fig. 7-5). It will remove sufficient heat so that the entropy of the exiting fluid is equal to the entropy of fluid entering the boiler.

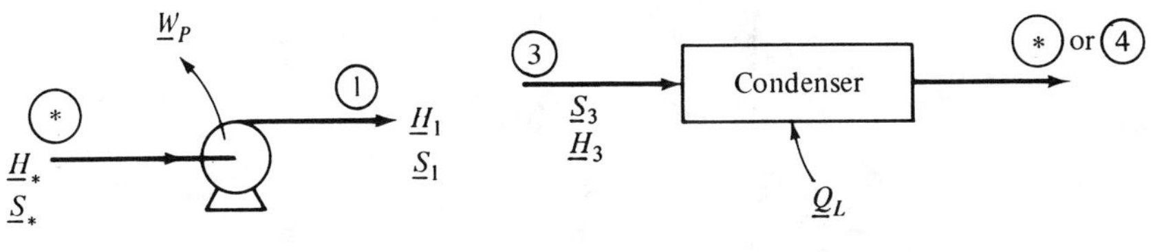

FIGURE 7-4 FIGURE 7-5

Energy balance:

$$\underline{Q}_L = \underline{H}_4 - \underline{H}_3$$

Entropy balance:

$$\underline{S}_4 - \underline{S}_3 = \frac{\underline{Q}_L}{T_L}$$

T_L refers to the temperature of the fluid at which heat is removed in the condenser.

We may picture the completed cycle as shown in Fig. 7-6.

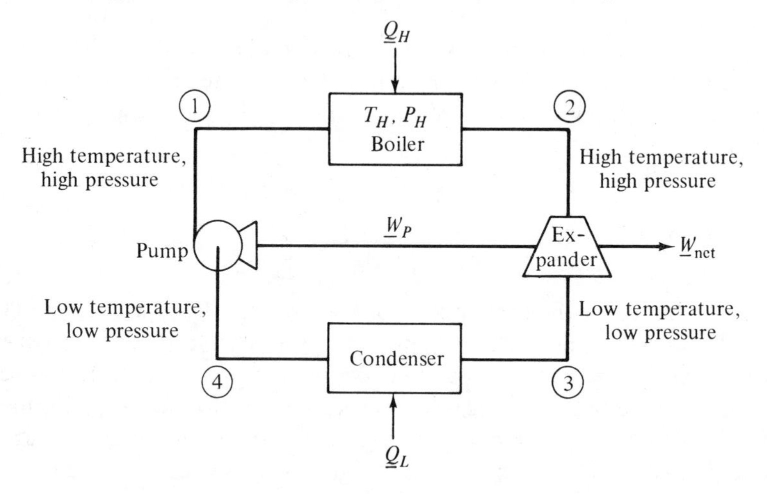

FIG. 7-6. Cyclical heat engine.

The idealized cycle we have developed, in which all processes occur reversibly and which contains isothermal heat absorption and rejection steps and two isentropic steps, is termed a *Carnot cycle*. As we shall demonstrate shortly, the Carnot cycle is the most efficient (efficiency $= W_{\text{net}}/Q_H$) cycle that can operate between any two given temperatures T_H and T_L. We may plot the paths of the various processes encountered during the Carnot cycle on a temperature–entropy, or T–$\underline{S}$, diagram, as shown in Fig. 7-7. Since we have assumed that our

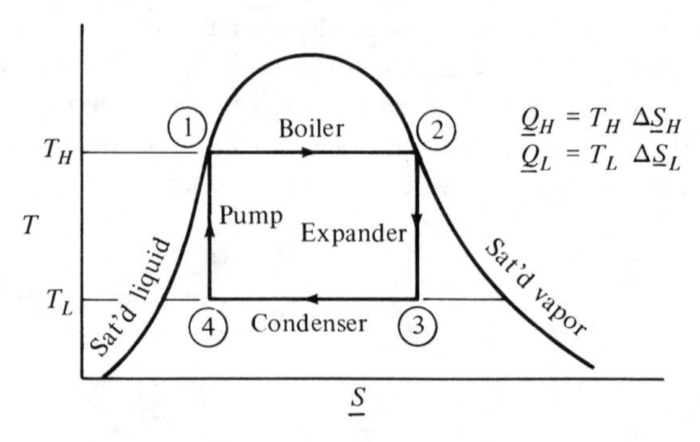

FIG. 7-7. Carnot cycle.

cycle is operating within the two-phase (vapor–liquid) region, the saturation dome has been included in the T–$\underline{S}$ diagram. Since the operation between states 3 and 4 corresponds to a partial condensation of the working fluid, we can see why this heat exchanger is commonly termed the "condenser." This cyclic process we have developed has the ability to absorb heat from a high-temperature source and convert a portion of this energy to work, provided there is a low-temperature heat sink to which the process can discard heat.

Up to this point no mention has been made of the fluid that is to circulate in our Carnot cycle. Obviously in any real device such a consideration will play an important role. However, for the present let us assume that from the many possible working fluids, we can find a suitable choice for whatever temperature range is of interest. Although the Carnot cycle is assumed to consist of two isothermal steps during which all the heat is absorbed and rejected, this restriction is in general unnecessary. In subsequent sections we shall consider cycles where temperature variations occur at both the heat-absorption and heat-rejection steps of the cycle.

An energy balance applied to the Carnot cycle as a whole reduces to the relationship

$$\underline{W}_{\text{net}} = \underline{Q}_H + \underline{Q}_L \tag{7-1}$$

where $\underline{W}_{\text{net}}$ is the net work ($\underline{W}_{\text{turb}} + \underline{W}_{\text{pump}}$) supplied by the cycle per unit mass of fluid circulating. $\underline{Q}_H$ and $\underline{Q}_L$ are the thermal-energy transfers at the high and low temperatures, respectively per unit mass of fluid circulating. The cycle (a closed system by definition) is assumed to be operating reversibly and in the steady state.

Since the entropy balance reduces to $\delta\underline{Q} = T\,d\underline{S}$ for a heat exchanger operating under these conditions, application of the entropy balance to the boiler and the condenser produces

$$\underline{Q}_H = T_H\,\Delta\underline{S}_H \tag{7-2}$$

$$\underline{Q}_L = T_L\,\Delta\underline{S}_L \tag{7-3}$$

Thus it is seen that the areas under curves 1-2 and 3-4 in Fig. 7-7 represent $\underline{Q}_H$ and $\underline{Q}_L$, respectively. The area under curve 1-2 is positive; the area under curve 3-4 is negative, because $\Delta\underline{S}_{3\text{-}4}$ is negative. Addition of the two quantities, as suggested by equation (7-1), yields $\underline{W}_{\text{net}}$. Graphically this results in a cancellation of the area beneath curve 3-4 and shows that $\underline{W}_{\text{net}}$ for a Carnot engine is represented by the enclosed area of the cycle. (We must, however, avoid drawing the same conclusion for any cycle that does not utilize isentropic work processes, for the cancellation of areas beneath the low-temperature heat-removal step does not result if the compression or expansion is irreversible.)

Application of the entropy balance to the cycle then yields

$$\frac{\underline{Q}_H}{T_H} + \frac{\underline{Q}_L}{T_L} = 0 \tag{7-4}$$

or

$$\underline{Q}_L = -\frac{T_L}{T_H}\underline{Q}_H \tag{7-5}$$

which may be substituted into the energy balance, equation (7-1), to give

$$\underline{W}_{\text{net}} = \underline{Q}_H - \frac{T_L}{T_H}\underline{Q}_H = \underline{Q}_H\left(1 - \frac{T_L}{T_H}\right) \tag{7-6}$$

where the subscript "net" indicates that this is the actual work that can be provided by the cycle to the surroundings. Equation (7-6) can be rearranged to give the net work obtainable per unit of thermal-energy input:

$$\frac{W_{\text{net}}}{Q_H} = \frac{\underline{W}_{\text{net}}}{\underline{Q}_H} = 1 - \frac{T_L}{T_H} = \frac{T_H - T_L}{T_H} \tag{7-7}$$

The ratio (W_{net}/Q_H) is termed the *thermal efficiency*, η, of the cycle, and is the fraction of thermal energy which is supplied to the cycle that is converted to useful work. The higher η, the higher the fraction of Q_H that is converted. Note that although we call η an efficiency, it *must* have a value less than unity (except in extraordinary circumstances) even though the cycle is perfectly reversible. *The thermal efficiency of any cyclic heat engine is limited by the necessity of rejecting thermal energy to the surroundings.*

Equation (7-7) indicates that the thermal efficiency of a Carnot cycle is a function solely of the absolute temperature levels at which the cycle absorbs and rejects heat. Complete conversion of heat to work is attainable only if the cycle absorbs heat at an infinite temperature or discards it at an absolute temperature of zero. Clearly neither of these conditions is attainable in any practical cycle, so thermal efficiencies of unity are not possible. (We shall discuss this matter in somewhat greater detail in Section 7.3.) Ordinarily, the low temperature is limited to ambient temperatures or to temperature levels at which large amounts of cooling water are available. Thus we find that power plants are usually built near large rivers or lakes (and occasionally the ocean) whose water acts as a convenient low-temperature sink.

The maximum temperature of a cycle is generally determined by our ability to handle the fluid at the temperature involved. When water is used as the working fluid, the upper temperature is limited by the vapor pressure of the water. With other fluids, such as liquid metals, vapor pressure is not a serious problem, but the limitation frequently is one of finding materials that can withstand the corrosive effects of the liquid metals at high temperatures. Thus we find that it is generally impossible to utilize fully the high temperatures at which thermal energy is often available. Flame temperatures frequently exceed 2000°F (2460°R), but steam temperatures in boilers seldom exceed 1000°F. Nuclear-reactor operating temperatures are limited only by our ability to contain the fission process. (However, this limitation is normally more restrictive than the maximum boiler temperatures cited above, and thus nuclear reactors usually operate at maximum temperatures lower than conventional fossil-fuel units.)

SAMPLE PROBLEM 7-2. Preliminary cost estimates for a 1 million-kW nuclear-powered thermal power station require an estimate of the cooling-water demands of the facility. For the purposes of this estimate, the heat engine will be approximated by a Carnot cycle operating between 650 and 150°F. If the maximum temperature rise in the cooling water is restricted to 50°F, at what rate (gal/min) must the cooling water be supplied?

Solution: The net power produced by the plant is to be 1 million kw = 10^9 w = 5.69×10^7 Btu/min. Therefore, $\dot{W}_{\text{net}} = 5.69 \times 10^7$ Btu/min, where the dot over the symbol indicates per unit time. The thermal efficiency of this cycle is given by

$$\eta = \frac{T_H - T_L}{T_H} = \frac{1110°\text{R} - 610°\text{R}}{1110°\text{R}} = \frac{500°\text{R}}{1110°\text{R}} = 0.45$$

but the thermal efficiency is also

$$\eta = \frac{\dot{W}_{\text{net}}}{\dot{Q}_H}$$

Therefore,

$$\dot{Q}_H = \frac{\dot{W}_{\text{net}}}{\eta} = \frac{5.69 \times 10^7}{0.45} \text{ Btu/min} = 1.26 \times 10^8 \text{ Btu/min}$$

The heat rejected to the cooling water is obtained from the overall energy balance:

$$\dot{W}_{\text{net}} = \dot{Q}_H + \dot{Q}_L$$

Therefore,

$$\dot{Q}_L = -(\dot{Q}_H - \dot{W}_{\text{net}}) = -6.9 \times 10^7 \text{ Btu/min}$$

Since all this heat must be transferred to the cooling water,

$$\dot{Q}_{\text{cw}} = -\dot{Q}_L = 6.9 \times 10^7 \text{ Btu/min}$$

An energy balance around the cooling water gives

$$\dot{Q}_{\text{cw}} = \dot{M}_{\text{cw}}\,\Delta \underline{H}_{\text{cw}} = \dot{M}_{\text{cw}}(C_P\,\Delta T)_{\text{cw}}$$

but C_P for H_2O = 1.0 Btu/lb_m °F, ΔT_{cw} = 50°F. Therefore, the flow rate of cooling water is given by

$$\dot{M}_{\text{cw}} = \frac{6.9 \times 10^7 \text{ Btu/min}}{50.0 \text{ Btu/lb}_\text{m}} = 1.38 \times 10^6 \text{ lb}_\text{m}\text{/min} = 1.66 \times 10^5 \text{ gal/min}$$

That is, 166,000 gallons of cooling water are needed every minute. On a daily basis, this becomes 240 million gallons per day. As a basis of comparison, a small city of about 10,000 normally consumes approximately 1 million gallons per day of drinking water! Consider the effect of raising the temperature of this flow rate of water 50°F on the receiving body of water. This effect is termed "thermal pollution" and is rapidly becoming a serious problem in our inland waters.

7.3 *The Second Law of Thermodynamics*

In our earlier discussions of entropy (in Chapter 3) we observed that spontaneously occurring processes always proceeded in a certain predictable direction. A cup of hot coffee placed in a cool room always tended to cool to the temperature of the room. Once it was at room temperature, we observed that the cup of coffee would never return spontaneously to its original hot state. These observations concerning the one-wayedness of naturally occurring processes were embodied in our statement that all spontaneous processes proceed in such a way that the entropy of the universe never decreases. The consequences of this statement are usually termed the *second law of thermodynamics*. Since our statement that the entropy of the universe never decreases is rather broad, its consequences are also far-reaching.

Although many statements of the second law have been proposed, most of these can be broadly classed as variations of two basic premises—that of Kelvin and Planck, and that of Clausius. Let us examine these statements:

1. The Kelvin–Planck statement: It is impossible to construct a cyclic process whose only effect is to absorb heat at a single temperature and convert it to an equivalent amount of work.
2. The Clausius statement: It is impossible to construct a cyclic process whose only effect is to transfer heat from a lower temperature to a higher one.

We have seen in our development of the Carnot cycle that the Kelvin–Planck statement of the second law is accurate. Although the Carnot cycle is by no means the only conceivable cycle, no other cycle has ever been devised which violates the Kelvin–Planck statement or which outperforms the Carnot cycle. Therefore, we have been unable to disprove it, and it appears that we must live within its consequences. Once the Kelvin–Planck statement is accepted, we may "prove" many other statements and corollaries of the second law.

The Clausius statement of the second law may be proved by examining the cycle shown in Fig. 7-8. Cycle 1 is a proposed cycle which violates the Clausius statement by being able to transfer a unit of heat from the low-temperature sink to the high-temperature sink without receiving work. Let the Carnot engine absorb a unit of heat from the high-temperature reservoir, convert a fraction of it to work, and reject the remainder to the low-temperature sink. Cycle 1 is now allowed to take the amount of heat rejected and transfer it back to the high-temperature reservoir, so the net effect of the two cycles is to absorb some heat from the hot reservoir and completely convert it to work—a feat that violates the Kelvin–Planck statement. Thus the scheme cannot operate as proposed. Since we know that the Carnot engine is theoretically possible, it

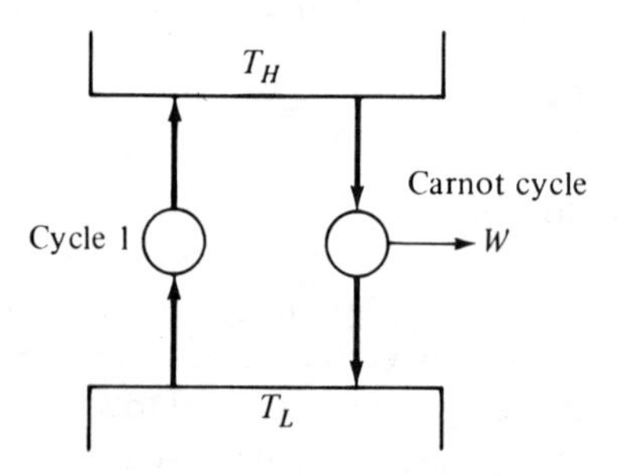

FIG. 7-8. Clausius statement.

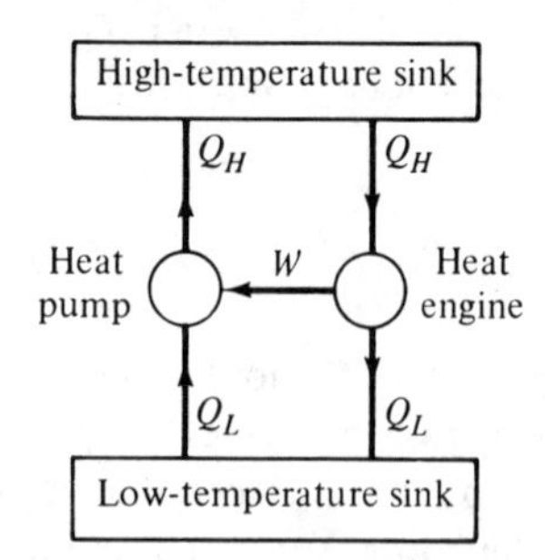

FIG. 7-9. Reversible Carnot engines.

must not be possible to construct a cycle that operates in the fashion proposed for cycle 1. Thus the Clausius statement must be true.

Using reasoning similar to that we used in the proof of the Clausius statement, we now demonstrate that the Carnot cycle is the most efficient cycle that can operate between any two thermal sinks. Before proceeding let us observe that the Carnot engine is capable of operating in reverse, such that it behaves as a heat pump rather than a heat engine. When operated in reverse, the Carnot heat pump consumes exactly the same amount of work to return a unit of heat to the hot reservior as a Carnot heat engine produces when it absorbs this amount of heat. Thus if we place a Carnot heat engine and a heat pump back to back, as shown in Fig. 7-9, the net effect is that nothing happens.

Now let us assume that some heat engine which is more efficient than the Carnot engine is substituted for the Carnot engine. Since this engine is assumed to be more efficient than the Carnot engine, it produces more work than the Carnot engine. Therefore, it provides more work than the Carnot heat pump needs to return Q_L to the high-temperature sink. Thus the net effect of the two units is to absorb heat from the high-temperature sink and completely convert it to work, a feat that again violates the Kelvin–Planck statement of the second law. Thus it must not be possible to construct a heat engine which is more efficient than a Carnot engine, and the proposition is demonstrated.

In an analogous fashion we may show (1) that a reversible cycle is always more efficient than an irreversible cycle operating between the same temperatures, and (2) that a Carnot heat pump requires no more work than any other cycle to transfer a given amount of heat from a low to a high temperature.

7.4 The Thermodynamic Temperature Scale

In Chapter 3 temperature was defined by the expression

$$T = \left(\frac{\partial U}{\partial S}\right)_V = \left(\frac{\partial \underline{U}}{\partial \underline{S}}\right)_{\underline{V}} \qquad (7\text{-}8)$$

In Chapters 5 and 6 it was shown that a temperature so defined was identical to the thermodynamic temperature defined by the relationship

$$\oint \frac{\delta Q}{T} = 0 \qquad (7\text{-}9)$$

Since we cannot measure either internal energy or entropy directly, equation (7-8) does not provide a viable relation for measuring thermodynamic temperatures directly. We may avoid this difficulty by considering the heat absorption and heat rejection of a Carnot heat engine. These thermal-energy transfers are independent of the working fluid and depend only on the temperatures of the hot and cold sink according to the relation

$$-\frac{Q_H}{Q_L} = \frac{T_H}{T_L} \qquad (7\text{-}10)$$

Thus by arbitrarily assigning a numerical value to some temperature (such as the freezing or boiling temperature of water), the temperature of any other body can (in theory) be determined by measuring the ratio of Q_H/Q_L between the reference and unknown temperatures. In this manner a thermodynamic temperature scale can be established.

If we could operate a Carnot cycle between the freezing and boiling points of water at 1 atm pressure, we would find that

$$-\frac{Q_{\text{steam}}}{Q_{\text{ice}}} = \frac{T_{\text{steam}}}{T_{\text{ice}}} = 1.3662 \tag{7-11}$$

If in addition we now specify that there be 180 degrees between the steam and ice temperatures we obtain

$$T_{\text{steam}} - T_{\text{ice}} = 180 \tag{7-12}$$

Equations (7-11) and (7-12) allow us to solve for T_{steam} and T_{ice} to obtain

$$\begin{aligned} T_{\text{ice}} &= 491.67°\text{R} \\ T_{\text{steam}} &= 671.67°\text{R} \end{aligned} \tag{7-13}$$

The temperature scale so defined is the familiar Rankine temperature scale. If instead of 180 degrees between T_{steam} and T_{ice} we had chosen 100 degrees, we would have defined the Kelvin temperature scale, in which

$$\begin{aligned} T_{\text{ice}} &= 273.15°\text{K} \\ T_{\text{steam}} &= 373.15°\text{K} \end{aligned} \tag{7-14}$$

The commonly used Fahrenheit and centigrade temperature scales are not thermodynamic temperature scales, in that they do not obey equation (7-10). They may, however, be obtained directly from the Rankine and Kelvin scales by the relations

$$\begin{aligned} °\text{C} &= °\text{K} - 273.15 \\ °\text{F} &= °\text{R} - 459.67 \end{aligned} \tag{7-15}$$

Although we now have, in theory, a way of measuring the thermodynamic temperature of any body by means of a Carnot engine, in practice this is not a very satisfactory means of measuring temperature. A truly reversible Carnot engine cannot be constructed, and, even if it could, the measurement of Q_H and Q_L would still make this a very difficult technique for measuring temperatures.

Many of the experimental difficulties associated with using the Carnot engine for measuring temperature can be avoided by using the *ideal gas thermometer*. The ideal gas thermometer is based on the ideal gas equation of state:

$$PV = nRT \tag{7-16}$$

or, solving for T,

$$T = \frac{PV}{nR} \tag{7-17}$$

Thus by measuring the pressure and volume of a known amount of an ideal gas held at the unknown temperature we may determine T directly. The ideal gas constant R is chosen such that the temperature of some reference state (again, frequently the boiling or freezing point of water) has an arbitrary value.

The fact that the ideal gas temperature scale defined by equation (7-14) is identical to the thermodynamic temperature scale may be shown as follows. For purposes of distinguishing between the ideal and thermodynamic temperatures in the subsequent discussion, the ideal gas temperature will be represented by t and T will continue to be used for the thermodynamic temperature. Let us begin with the property relationship based on the thermodynamic temperature:

$$dU = T\,dS - P\,dV \tag{7-18}$$

Equation (7-18) divided by dV at constant T yields

$$\left(\frac{\partial U}{\partial V}\right)_T = T\left(\frac{\partial S}{\partial V}\right)_T - P \tag{7-19}$$

The entropy derivative in equation (7-19) may be eliminated by using a Maxwell relationship:

$$\left(\frac{\partial S}{\partial V}\right)_T = \left(\frac{\partial P}{\partial T}\right)_V \tag{7-20}$$

so equation (7-19) becomes

$$\left(\frac{\partial U}{\partial V}\right)_T = T\left(\frac{\partial P}{\partial T}\right)_V - P \tag{7-21}$$

If it is assumed that t varies monotonically with T such that for a given value of T, t remains constant, $(\partial U/\partial V)_T = (\partial U/\partial V)_t$, we can rewrite equation (7-21) as

$$\left(\frac{\partial U}{\partial V}\right)_t = T\left(\frac{\partial P}{\partial T}\right)_V - P \tag{7-22}$$

For a gas which obeys the ideal gas equation of state, the kinetic theory indicates $(\partial U/\partial V)_t = 0$. Thus equation (7-22) becomes

$$0 = T\left(\frac{\partial P}{\partial T}\right)_V - P \tag{7-23}$$

The derivative in equation (7-23) may then be written in terms of t as

$$T\left(\frac{\partial P}{\partial T}\right)_V = \frac{T}{t}\left(\frac{\partial t}{\partial T}\right)_V t\left(\frac{\partial P}{\partial t}\right)_V$$

But from the equation of state,

$$t\left(\frac{\partial P}{\partial t}\right)_V = \frac{nRt}{V} = P \tag{7-24}$$

so equation (7-23) reduces to

$$P\left[\frac{T}{t}\left(\frac{\partial t}{\partial T}\right)_V - 1\right] = 0 \tag{7-25}$$

which may be rearranged to

$$\frac{\partial t}{t} = \frac{\partial T}{T} \tag{7-26}$$

or

$$\partial \ln t = \partial \ln T \tag{7-26a}$$

Equation (7-26a) is then integrated to give

$$\ln T = \ln t + \ln k \tag{7-27}$$

or

$$T = kt \tag{7-27a}$$

where k is a constant of integration. Thus the ideal gas and thermodynamic temperature scales vary by at most a multiplicative constant, and hence are equivalent.

7.5 The International Temperature Scale

Although at first glance the ideal gas thermometer would appear to be simple and straightforward to use, in practice it is extremely difficult and time-consuming. In addition, it is not useful at very low temperatures, where no gas behaves ideally, or at very high temperatures, where the containing vessel would melt. To avoid these difficulties, the ninth General Conference of Weights and Measures in 1948 adopted the "international temperature scale." This scale is based on assigning values to six specific temperatures and using carefully specified interpolating formulas for all intermediate temperatures. The fixed points, as well as the interpolating formulas, have been chosen so that the international temperature scale conforms as closely as possible to the ideal gas temperature scale. The fixed points on this scale are listed in Table 7-1. The fixed points are all at a pressure of 1 standard atmosphere.

TABLE 7-1 Fixed Points on the International Temperature Scale

Points	Temperature, °C
(1) Boiling point of pure oxygen	−182.970
(2) Freezing point of water saturated with air	0.000
(3) Boiling point of pure water	100.000
(4) Boiling point of pure sulfur	444.600
(5) Melting point of pure silver	960.800
(6) Melting point of pure gold	1063.000

For interpolating between the six fixed points the scale is divided into four regions, and the following interpolation formulas specified:

1. From the boiling point of oxygen to the freezing point of water a platinum resistance thermometer is used. The resistance is related to temperature by

$$R_T = R_0[1 + AT + BT^2 + C(T - 100)T^3] \tag{7-28}$$

R_o is the resistance at the freezing point of water. The constants A, B, and C are determined by calibration at the boiling points of oxygen, water, and sulfur.

2. Between the freezing point of water and the freezing point of antimony the resistance of the platinum resistance thermometer is given by

$$R_T = R_0[1 + AT + BT^2] \tag{7-29}$$

where R_o is the resistance at the ice point and A and B are obtained from calibration of the boiling points of water and sulfur.

3. From the freezing point of antimony to the freezing point of gold a platinum versus platinum–rhodium thermocouple is used. The temperature is related to the thermocouple potential by

$$E = A + BT + CT^2 \tag{7-30}$$

where the constants A, B, and C are determined by calibration at the freezing points of antimony, silver, and gold.

4. Above the freezing point of gold a black-body thermometer is used. The temperature is related to the radiant intensity at a fixed frequency by the equation

$$J_T = J_{\text{au}} \frac{\exp[(C_2/\lambda)(T_{\text{au}} + T_0) - 1}{\exp[(C_2/\lambda)(T + T_0) - 1} \tag{7-31}$$

where C_2 = 1.438 cm °K
T_0 = the ice point in °K = 273.16°K
λ = wavelength of radiation in cm

7.6 Practical Considerations in Heat Engines

In our previous discussion of the Carnot cycle it was assumed that all steps occurred reversibly. Such operation will produce the maximum conversion of thermal to mechanical energy for a given source and sink temperature. However, reversible operation is both impossible and impractical to obtain and represents only a theoretical maximum against which a real system's performance can be compared. For example, if one attempted to operate the heat exchangers reversibly, only an infinitesimal temperature difference would exist between the hot and cold sinks and the respective portions of the working fluid. Thus infinitely large heat exchangers (boiler and condenser), which cost an infinite amount of money, would be required. Since this is clearly intolerable, it is necessary to provide a finite temperature difference between the working fluid and the hot and cold sinks. This permits the heat exchanger to be reasonably scaled and priced. Therefore, the high temperature at which the cyclic fluid operates is less than the temperature of the hot sink, and its low temperature is greater than the temperature of the cold sink by an amount necessary to achieve reasonable heat-transfer rates. The overall temperature difference between the fluid temperatures at the two extremes of the cycle is thus less than that between the hot and cold sinks, and the overall efficiency of any heat engine (not just the

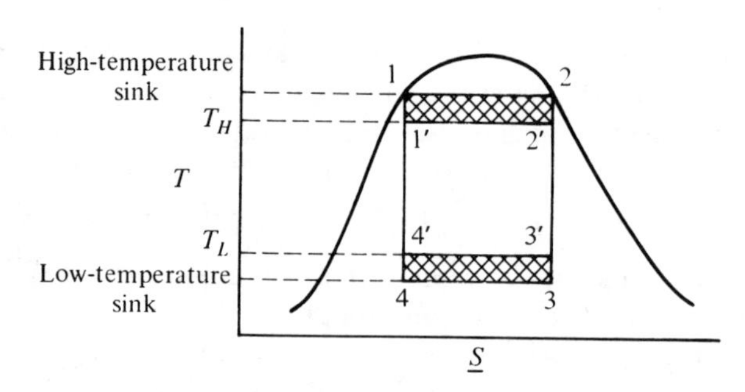

FIG. 7-10. Carnot engine with finite temperature differences.

Carnot engine) is reduced commensurately. The crosshatched area in Fig. 7-10 represents work that could have been obtained from Carnot engines operating around the cycles 1-2-2′-1′ and 4′-3′-3-4 had it not been necessary to provide the temperature driving forces in both exchangers.

In addition, it must be remembered that even the best expanders and compressors available today do not operate reversibly. Consequently some dissipation of mechanical to thermal energy will occur in the expander and the compressor; the compressor requires more work than the reversible minimum, and the expander produces less work than the reversible maximum. Since irreversible processes result in the fluid experiencing an entropy increase (if conducted adiabatically), the two work paths must both display increasing entropies, as seen in Fig. 7-11 for cycle 1-2-3′-4′. At first glance one is tempted

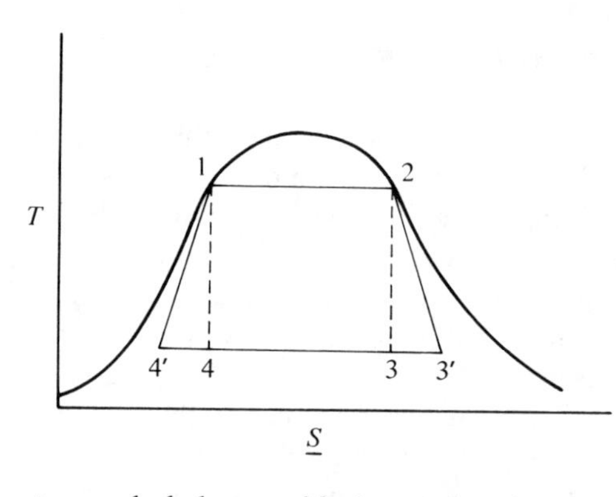

FIG. 7-11. Effect of irreversible pumps and expanders on the Carnot cycle.

to conclude by considering enclosed areas that $\underline{W}_{\text{net}}$ for cycle 1-2-3′-4′ is greater than for the Carnot cycle, 1-2-3-4, but, as cautioned earlier, *for cycles with irreversible work steps, the enclosed area does not represent* $\underline{W}_{\text{net}}$. Instead, we recall that for any cyclic heat engine,

$$\underline{W}_{\text{net}} = \underline{Q}_H + \underline{Q}_L$$

and that $\underline{Q}_H = T_H \, \Delta \underline{S}_H$ = area under curve 1-2. Now contrast the area under curves 3-4 and 3′-4′, both of which are negative. Clearly the area under 3′-4′ is the larger and hence $\underline{W}_{\text{net}}$ for cycle 1-2-3′-4′ is appreciably less than for the

Carnot cycle, 1-2-3-4, even though both cycles receive exactly the same amount of heat at T_H.

7.7 The Rankine Cycle

In previous discussion the Carnot engine has always been sketched on a T–$\underline{S}$ diagram within the liquid-vapor region. This practice has been followed because an isothermal absorption or rejection of a finite amount of heat would result in violation of the isothermal requirement if a single phase (gas or liquid) was passed through the exchanger. In addition to the highly idealized nature of Carnot cycles, other considerations arise if one attempts to operate a heat engine with the fluid path in the two-phase region as shown in Fig. 7-7.

The work requirements for reversibly compressing (or pumping) a fluid between two pressure levels is directly related to its specific volume as shown in Chapter 4:

$$\underline{W} = -\int_{P_L}^{P_H} \underline{V}\, dP \tag{7-32}$$

Thus if one condenses the fluid completely before it enters the pump, the work required to elevate its pressure is considerably less because of the much lower value of $\underline{V}^L$ than $\underline{V}^V$ (for fluids far removed from their critical points). In addition to this consideration, the pumping of two-phase gas–liquid mixtures poses serious mechanical difficulty with most pumps or compressors.

As shown in Fig. 7-12, the entry point to the compressor is shifted to the saturated liquid curve in an effort to improve performance of the cycle. An adiabatic reversible compression of the liquid from the condenser pressure to the boiler pressure takes the fluid along the isentropic path 5-6. Since the tem-

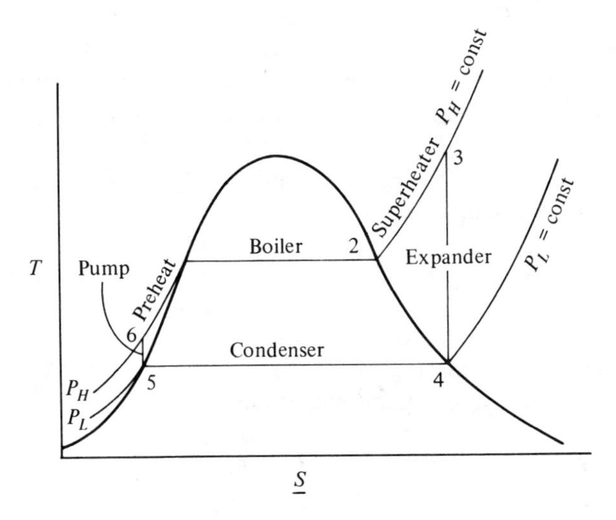

FIG. 7-12. Rankine cycle.

perature drops sharply along lines of constant pressure as one moves away from the saturated liquid curve into the liquid region, the fluid emerges at point 6 in a subcooled state and its temperature is considerably below that corresponding to saturation at P_H. Consequently, the fluid must first experience a sensible heat addition (path 6-1) in the boiler before vaporization commences.

The discharge (from the expander) of the Carnot cycle discussed earlier also occurs in the two-phase region. This, too, is a highly undesirable situation, because the liquid droplets in the two-phase mixture cause serious erosion and vibration problems in the high-velocity portions of the expander. One would thus prefer that the expanding fluid remain in the vapor phase until the expansion is completed. For an isentropic expansion this may be achieved by superheating (at constant pressure) the working fluid before it enters the expander (path 2-3). By controlling the amount of superheat it is possible to predetermine the state of the fluid in the expander exhaust, because both its entropy and pressure are then fixed. For our purposes we shall assume that enough superheat is added so that the fluid exits from the expander as a saturated vapor. The modified cycle shown in Fig. 7-12 is called a Rankine cycle and, although still subject to the consideration of irreversibilities discussed earlier, represents a far more practical (although slightly less efficient) cycle than the Carnot cycle.

In discussing the efficiencies of heat engines it is useful to think in terms of the average temperatures at which heat is received and rejected. Since the cycle with superheat has a higher average temperature of heat addition than the comparable cycle without superheat, its efficiency would be higher. On the other hand, a Carnot cycle operating between the maximum superheat temperature and T_L would have a still higher efficiency, since the Carnot cycle would have a higher average temperature for heat addition. In general we may conclude that anything which increases the average temperature of heat addition increases the cycle efficiency–and vice versa.

Since the liquid in the Rankine cycle enters the boiler below its saturation temperature, the fluid must first be preheated to its boiling point. Because the fluid is absorbing energy at a lower temperature than during the phase change, this addition of sensible heat reduces the average temperature of heat addition, and hence the thermal efficiency. However, the ease of pumping a pure liquid, rather than a two-phase mixture, easily justifies this slight decrease in the thermal efficiency of the Rankine cycle.

It should be remembered that the Rankine cycle (as well as all modifications to be discussed in subsequent sections) is subject to the same irreversible compression and expansion losses as were discussed in Section 7.6 for the Carnot cycle. Thus the T–$\underline{S}$ diagram for actual Rankine cycle would deviate from Fig. 7-12 in that the isentropic work steps would both be replaced by segments along which the entropy increased as shown in Fig. 7-13. Similarly, a small amount of liquid subcooling is generally provided in the condenser to assure that no vapor will pass into the pump. This, too, can be observed at point 6 in Fig. 7-13.

The Rankine cycle with slight modifications (which will be discussed in

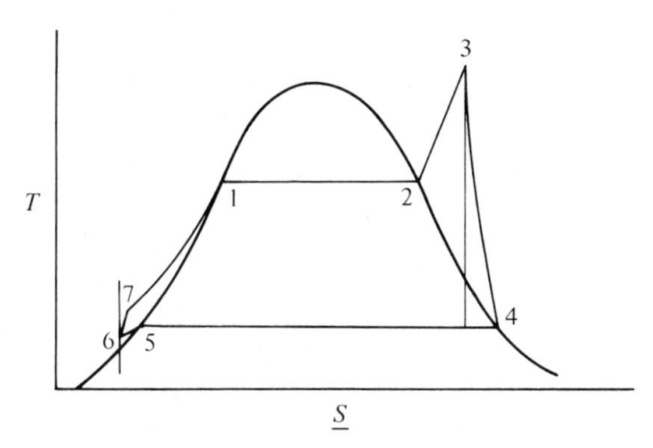

FIG. 7-13. Rankine cycle with irreversibilities.

the next section) forms the basis for most of the thermal power generated in this country. The most recently designed facilities are capable of producing 1 kw hr of electricity per 8000 Btu of thermal energy supplied, for an overall efficiency of 42 per cent. Older installations usually require about 10,000 Btu of thermal energy per kw hr, for an overall efficiency of 34 per cent. These figures illustrate that even our best performance still results in a loss to our rivers and lakes of slightly more than 1 Btu for every Btu of electrical energy we produce.

SAMPLE PROBLEM 7-3. A steam generating station operates on the Rankine cycle with superheat. The boiler tubes are restricted to a maximum pressure of 100 psia, and cooling water is available at 120°F. The condenser is designed for a 30°F temperature difference, so the low temperature in the cycle is 150°F. The superheater and condenser operate so that the expander and pump handle only single-phase materials. The pump and expander may be considered to be adiabatic and reversible.

(a) Determine the superheat (°F) needed, and the thermal efficiency of the cycle.

(b) Compare this efficiency with that of a Carnot cycle operating between 150°F and (1) the boiler temperature, and (2) the maximum superheater temperature.

(c) If new boiler tubes that can withstand a pressure of 200 psia are installed, determine the new thermal efficiency of the cycle.

Solution: (a) We may picture the cycle as shown in Fig. S7-3a or, in terms of the actual equipment used, Fig. S7-3b.

The steam leaving the boiler section (point 2) is a saturated vapor at 100 psia, so $T = 328$°F. It is then superheated to a point where it can be adiabatically and reversibly expanded in the turbine to the point where it is a saturated vapor at 150°F and 3.72 psia. Since the expansion in the turbine is adiabatic and reversible, the entropy balance tells us that it is also isentropic. Thus we can find the temperature that the superheater must achieve by following the isentrope from the outlet of the turbine to a pressure of 100 psia. This gives us a superheater temperature of 860°F, so the steam is $(860 - 328)$°F $= 532$°F superheated.

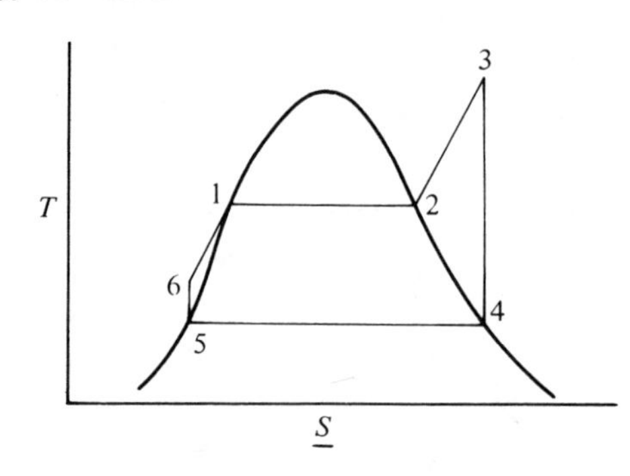

FIGURE S7-3a

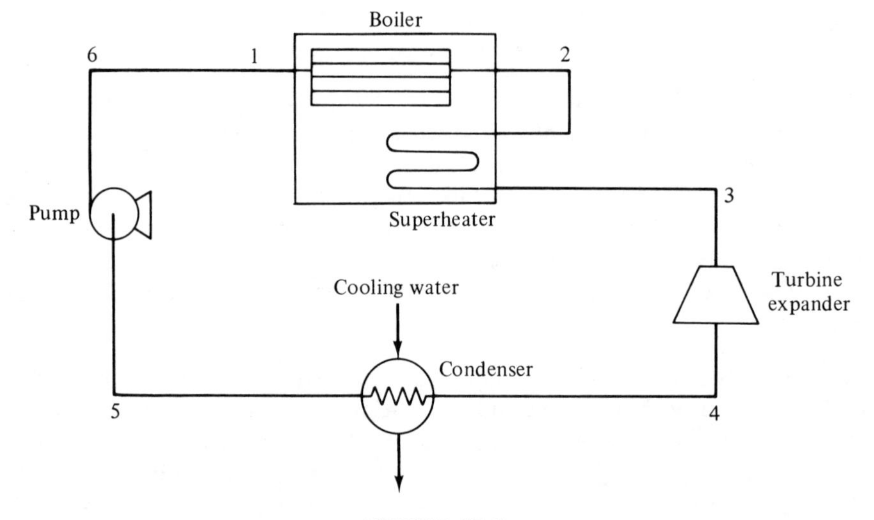

FIGURE S7-3b

Since we now know the conditions of the steam entering and leaving the turbine we can determine the amount of work produced during the expansion. If we neglect potential and kinetic energy changes across the turbine and assume steady-state, adiabatic operation, the energy balance reduces to

$$-\underline{W}_{\text{turb}} = \Delta \underline{H}_{3-4} = \underline{H}_4 - \underline{H}_3$$

but

$$\underline{H}_4 = 1125 \text{ Btu/lb}_\text{m}$$

$$\underline{H}_3 = 1458 \text{ Btu/lb}_\text{m}$$

Therefore,

$$\underline{W}_{\text{turb}} = (1458 - 1125) \text{ Btu/lb}_\text{m}$$
$$= 333 \text{ Btu/lb}_\text{m}$$

The liquid leaving the condenser is a saturated liquid at 150°F. Its enthalpy is $\underline{H}_5 = 117.9$ Btu/lb$_\text{m}$. From here it is compressed adiabatically and reversibly by the pump to a pressure of 100 psia. The work that the pump consumes can be found from the expression

$$\underline{W}_{\text{pump}} = -\int_{P_5}^{P_6} \underline{V}\, dP$$

However, since the liquid is essentially incompressible, $\underline{V}$ is not a function of P and can be removed from under the integral sign to yield

$$\underline{W}_{\text{pump}} = -\underline{V}\int_{P_5}^{P_6} dP = -\underline{V}(P_6 - P_5)$$

$$= -(0.0163)(100 - 3.72)\left(\frac{144}{778}\right)\frac{\text{ft}^3}{\text{lb}_\text{m}}\frac{\text{lb}_\text{f}}{\text{in.}^2}\frac{\text{in.}^2}{\text{ft}^2}\frac{\text{Btu}}{\text{lb}_\text{f}\text{-ft}}$$

$$= -0.3 \text{ Btu/lb}_\text{m}$$

We now determine the enthalpy of the liquid entering the boiler as follows. An energy balance around the pump, assuming negligible changes in potential and kinetic energy and steady-state, adiabatic operation, gives

$$-\underline{W}_{\text{pump}} = \Delta\underline{H}_{5-6} = \underline{H}_6 - \underline{H}_5$$

or

$$\underline{H}_6 = \underline{H}_5 - \underline{W}_{\text{pump}}$$

Since $\underline{H}_5 = 117.9$ Btu/lb$_\text{m}$, $\underline{H}_6 = 118.2$ Btu/lb$_\text{m}$.

We now evaluate the heat input in the boiler–superheater from an energy balance. Assuming negligible changes in potential and kinetic energy and steady-state operation, the energy balance around the boiler–superheater gives

$$\Delta\underline{H}_{6-3} = \underline{Q} - \underline{W}$$

But $\underline{W} = 0$, so

$$\underline{Q} = \Delta\underline{H}_{6-3} = \underline{H}_3 - \underline{H}_6$$

$$= (1458 - 118.2) \text{ Btu/lb}_\text{m}$$

$$= 1339.8 \text{ Btu/lb}_\text{m}$$

The net work produced by the cycle is the sum of the work produced in the turbine and the pump. That is,

$$\underline{W}_{\text{net}} = \underline{W}_{\text{turb}} + \underline{W}_{\text{pump}} = (333 - 0.3) \text{ Btu/lb}_\text{m}$$

$$\underline{W}_{\text{net}} = 332.7 \text{ Btu/lb}_\text{m}$$

and the thermal efficiency of the cycle is given by

$$\eta = \frac{\underline{W}_{\text{net}}}{\underline{Q}_{\text{boiler}}} = \frac{332.7}{1339.8} = 0.248 \text{ or } 24.8\%$$

(b) For the Carnot cycle $\eta = (T_H - T_L)/T_H$. (1) Therefore, for a Carnot cycle operating between the boiler temperature $T = 328°\text{F} = 788°\text{R}$ and the condenser temperature $T = 150°\text{F} = 610°\text{R}$, the thermal efficiency is

$$\eta = \frac{788 - 610}{788} = 0.226$$

(2) The maximum temperature in the superheater is $860°\text{F} = 1320°\text{R}$. The Carnot efficiency of a cycle with this maximum temperature is then

$$\eta = \frac{1320 - 610}{1320} = 0.537$$

Thus we see that the Rankine cycle with superheat is slightly more efficient than the Carnot cycle operating at the boiler temperature. However, a Carnot cycle operating at the maximum superheater temperature is considerably more efficient than the Rankine cycle.

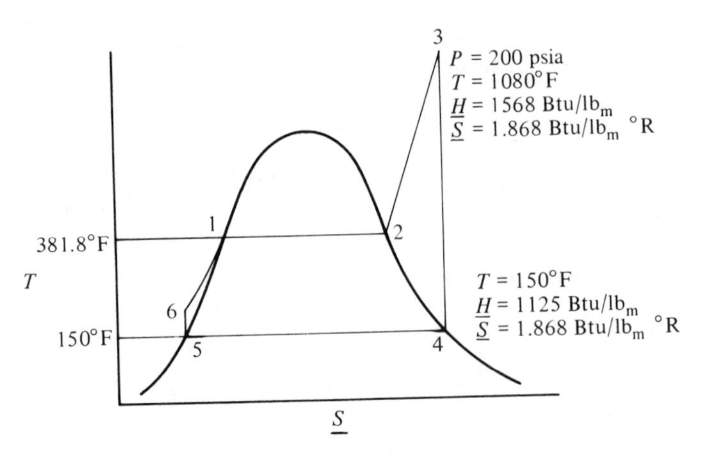

FIGURE S7-3c

(c) If the old 100-psia boiler tubes are replaced by tubes capable of withstanding 200 psia, we must repeat the calculation of part (a), with the new boiler pressure. The T–$\underline{S}$ diagram for the new cycle is given in Fig. S7-3c. We may now fill in the property values at points 3, 4, and 5 as given in Fig. S7-3c. The work produced by the turbine is then

$$-\underline{W}_{\text{turb}} = \Delta\underline{H}_{3-4} = (1125 - 1568)\ \text{Btu/lb}_{\text{m}}$$

$$\underline{W}_{\text{turb}} = 443\ \text{Btu/lb}_{\text{m}}$$

The work consumed by the pump is

$$-\underline{W}_{\text{pump}} = \underline{V}(P_6 - P_5) = 0.0163(200 - 3.72)\left(\frac{144}{778}\right)\frac{\text{Btu}}{\text{lb}_{\text{m}}}$$

$$= 0.6\ \text{Btu/lb}_{\text{m}}$$

The enthalpy of the liquid entering the boiler is

$$\underline{H}_6 = \underline{H}_5 - \underline{W}_{\text{pump}} = 118.5\ \text{Btu/lb}_{\text{m}}$$

The heat addition in the boiler–superheater is given by

$$\underline{Q}_{\text{boiler}} = \Delta\underline{H}_{6-3} = (1568 - 118.5)\ \text{Btu/lb}_{\text{m}}$$

$$= 1449.5\ \text{Btu/lb}_{\text{m}}$$

The net work produced by the cycle is

$$\underline{W}_{\text{net}} = \underline{W}_{\text{turb}} + \underline{W}_{\text{pump}} = (443 - 0.6)\ \text{Btu/lb}_{\text{m}}$$

$$= 442.4\ \text{Btu/lb}_{\text{m}}$$

Thus the thermal efficiency is

$$\eta = \frac{\underline{W}_{\text{net}}}{\underline{Q}_{\text{boiler}}} = \frac{442.4}{1449.5} = 0.305$$

which is significant improvement over the previous values.

7.8 Improvements in the Rankine Cycle

As we saw in Sample Problem 7-3, the thermal efficiency of the Rankine cycle can be significantly increased by using higher boiler pressures. Unfortunately, this requires ever-increasing superheats. [In part (c) of Sample Problem 7-3 a superheat of approximately 700°F was needed for a boiler pressure of only 200 psia.] Since the maximum temperature in the superheater is limited by the temperature the tubes can stand, superheater temperatures are usually restricted to less than about 1100°F. Since the major fraction of the heat supplied to Rankine cycle is supplied in the boiler, not the superheater, we must increase the boiler temperatures (and hence pressures) if significant efficiency improvements are to be obtained. From Sample Problem 7-3 we saw that boiler pressures much above 200 psia are not possible in a simple Rankine cycle if the expanding steam is to remain as a single phase throughout the expansion and a maximum temperature of 1100°F is allowed in the superheater.

The problem of excessive superheater temperatures may be solved while still avoiding two-phase mixtures in the expansion by "reheating" the expanding steam part way through the expansion as shown in Fig. 7-14. Thus the steam

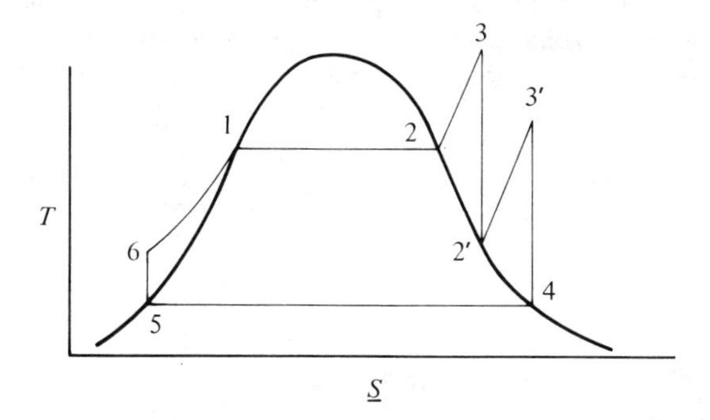

FIG. 7-14. Rankine cycle with reheat.

leaving the boiler section (point 2) is superheated to an acceptable temperature and then expanded (and work removed) until it intersects the saturation curve. The steam is then "reheated" in a second superheater section and expanded in a second turbine (with more work removed) until the saturation curve is again encountered. The steam is then condensed and pumped back into the boiler.

Although the preceding example considered only one reheat operation, clearly we may use as many as are desired. In this fashion it is possible to use higher boiler pressures (thereby obtaining increased efficiencies) without having to increase the maximum superheater temperature above the working limits of the superheater tubes.

SAMPLE PROBLEM 7-4. The tubes in the boiler of Sample Problem 7-3 are replaced by tubes that can withstand pressures up to 900 psia but no temperatures above 860°F. The condenser is still assumed to operate at 150°F.

(a) If a standard Rankine cycle with no reheat is used, what is the maximum boiler pressure and the cycle efficiency?

(b) If the boiler pressure is raised to 900 psia and reheat added to allow single-phase expansions: (1) How many reheat cycles are needed? (2) What is the maximum temperature in the last superheater? (3) What is the overall cycle efficiency? (*Note:* You may assume that the pump and turbine are adiabatic and reversible.)

Solution: (a) Since 860°F is the maximum superheater temperature of Sample Problem 7-3 when the boiler pressure is 100 psia, that is still the maximum boiler pressure that may now be used. If a boiler pressure above this value is used, it will not be possible to superheat the vapor enough to avoid condensation in the turbine. The cycle efficiency is thus the same as part (a) of the Sample Problem 7-3: $\eta = 24.8$ percent.

(b) We may trace out the reheat cycles needed with a boiler pressure of 900 psia as as shown in Fig. S7-4. The steam leaves the boiler at point 2 as a saturated vapor at 900 psia and is superheated to 860°F at constant P (point 3). The steam is now adiabatically and reversibly (that is, isentropically) expanded until it hits the saturation curve which occurs at a pressure of 90 psia. It is then reheated at 90 psia until it hits either the 860°F isotherm or the isentrope that intersects the saturation curve at 150°F. It hits the isentrope first at a temperature of 830°F. From this point the steam is expanded in a second turbine until it intersects the saturation curve at a temperature of 150°F. Thus one reheat is necessary, and the overall cycle will be as shown in Fig. S.7-4

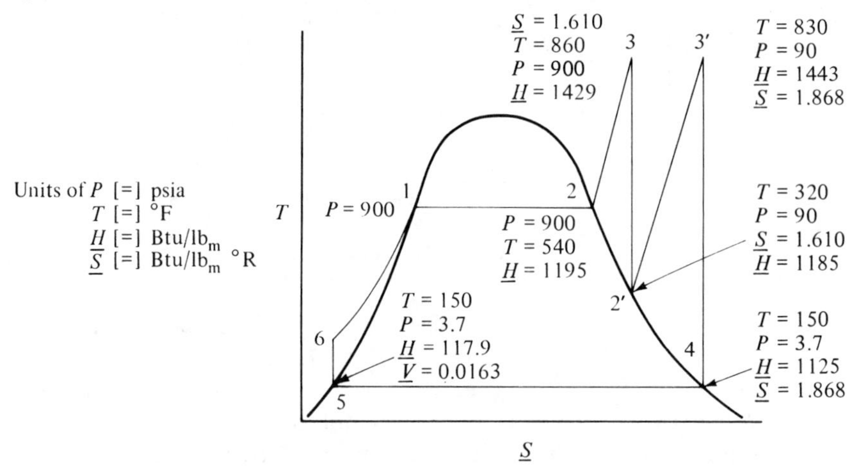

FIGURE S7-4

The net work produced in the cycle is the sum of the works of the two turbines, plus the work of the pump; the heat input is the sum of that added in the preheater–boiler and that added in the two superheaters. Let us begin the analysis of this cycle by considering the two turbines. An energy balance around each of the turbines (assuming no change of potential or kinetic energy, and adiabatic, steady-state flow) gives

$$-\underline{W}_T = \Delta \underline{H}$$

Thus the work produced by the first turbine is

$$\underline{W}_{T_1} = \underline{H}_3 - \underline{H}_{2'} = (1429 - 1185)\ \text{Btu/lb}_\text{m} = 244\ \text{Btu/lb}_\text{m}$$

while that produced in the second is

$$\underline{W}_{T_2} = \underline{H}_{3'} - \underline{H}_4 = (1443 - 1125)\ \text{Btu/lb}_\text{m} = 318\ \text{Btu/lb}_\text{m}$$

The work produced in the pump is given by

$$\underline{W}_P = -\underline{V}\,\Delta P = -0.0163\,(900 - 3.7)\left(\frac{144}{778}\right)\frac{\text{ft}^3}{\text{lb}_\text{m}}\frac{\text{lb}_\text{f}}{\text{in.}^2}\frac{\text{in.}^2}{\text{ft}^2}\frac{\text{Btu}}{\text{ft lb}_\text{f}}$$

$$= -2.70 \text{ Btu/lb}_\text{m}$$

Therefore, the net work produced by the cycle is

$$\underline{W}_{\text{net}} = (244 + 318 - 2.7) \text{ Btu/lb}_\text{m} = 559.3 \text{ Btu/lb}_\text{m}$$

We may obtain the enthalpy of the liquid water entering the boiler from an energy balance around the pump:

$$-\underline{W}_P = \Delta\underline{H} = \underline{H}_6 - \underline{H}_5$$

or

$$\underline{H}_6 = \underline{H}_5 - \underline{W}_P = (117.9 + 2.7) \text{ Btu/lb}_\text{m} = 120.6 \text{ Btu/lb}_\text{m}$$

The heat added to the boiler and first superheater is then obtained from an energy balance around this section. Assuming negligible potential and kinetic energy changes and steady-state flow with no work, the energy balance becomes

$$\underline{Q}_{B-S} = \Delta\underline{H}_{B-S} = \underline{H}_3 - \underline{H}_6 = (1429 - 120.6) \text{ Btu/lb}_\text{m}$$

or

$$\underline{Q}_{B-S} = 1308.4 \text{ Btu/lb}_\text{m}$$

Similarly, the heat added in the reheater is given by

$$\underline{Q}_R = \Delta\underline{H} = \underline{H}_{3'} - \underline{H}_{2'} = (1443 - 1185) \text{ Btu/lb}_\text{m}$$

$$= 258 \text{ Btu/lb}_\text{m}$$

The total heat addition is $\underline{Q}_H = \underline{Q}_{B-S} + \underline{Q}_R$, or

$$\underline{Q}_H = (1308.4 + 258) \text{ Btu/lb}_\text{m} = 1566.4 \text{ Btu/lb}_\text{m}$$

and the thermal efficiency is given by

$$\eta = \frac{\underline{W}_{\text{net}}}{\underline{Q}_H} = \frac{559.3}{1566} = 0.35 \text{ or } 35\%$$

an increase of almost 50 per cent above the efficiency of the simple Rankine cycle with no reheat operating at the same maximum temperature. Even greater increases can be obtained by operating at still higher boiler pressures, but the same superheater temperature, by using two, three, or even four reheat stages.

The thermal efficiency of the Rankine and reheat cycles can be further increased by the use of "regenerative" heat exchange, as shown in Fig. 7-15. In the regenerative cycle, a portion of the partially expanded steam is drawn off, between the high- and low-pressure turbines (or an intermediate tap if only a single turbine is used). This steam is used to preheat the condensed liquid before it is returned to the boiler–superheater. In this way we can reduce the amount of heat added at the low temperatures, so we increase the average temperature at which heat is added to the cycle and thereby increase the thermal efficiency. Although in theory any number of preheaters may be used, the gain in efficiency drops off fairly quickly with the number of preheaters, so it is extremely rare to find more than five preheaters, three being a more typical value.

Two types of feed preheaters are commonly used: the open preheater and

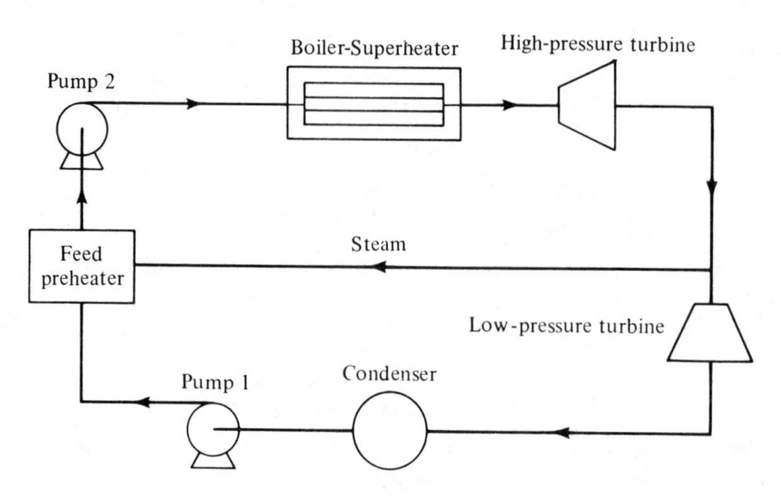

FIG. 7-15. Regenerative cycle.

the closed preheater. In the open preheaters shown in Fig. 7-15, the steam is simply mixed with the feedwater in a tank. The steam condenses and heats the water. The mixture is then pumped to the next stage. In the closed preheater, on the other hand, the steam and feedwater are kept separate. The steam condenses in tubes that pass through the preheater. The condensate is removed through traps and allowed to enter the feed stream at a lower pressure, where it may do so without the need for additional pumps. A two-stage closed preheat scheme is shown in Fig. 7-16.

The advantages of the open preheaters are excellent heat-transfer characteristics, ease of operation and design, and inexpensive construction. They have, however, the disadvantage of requiring an extra feed pump for each preheater stage.

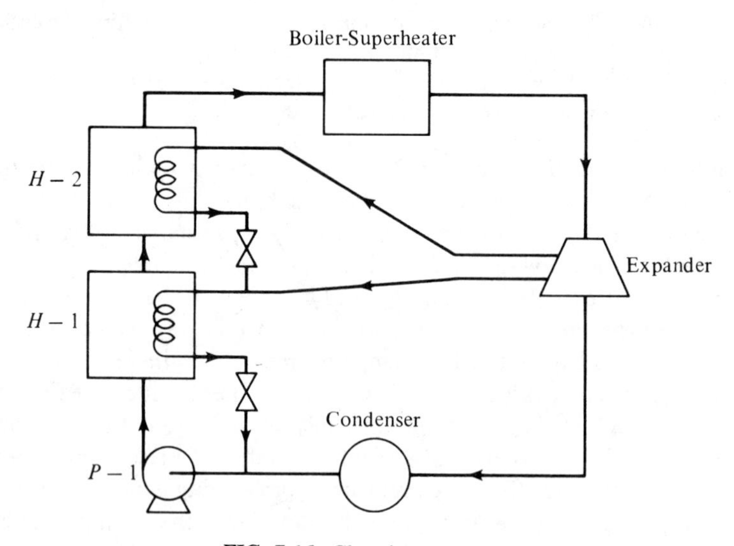

FIG. 7-16. Closed preheaters.

SAMPLE PROBLEM 7-5. The reheat cycle of Sample Problem 7-4 is to be augmented by three stages of feed preheat as shown in Fig. S7-5. The boiler pressure is 900 psia, and a maximum temperature of 860°F is allowable. The

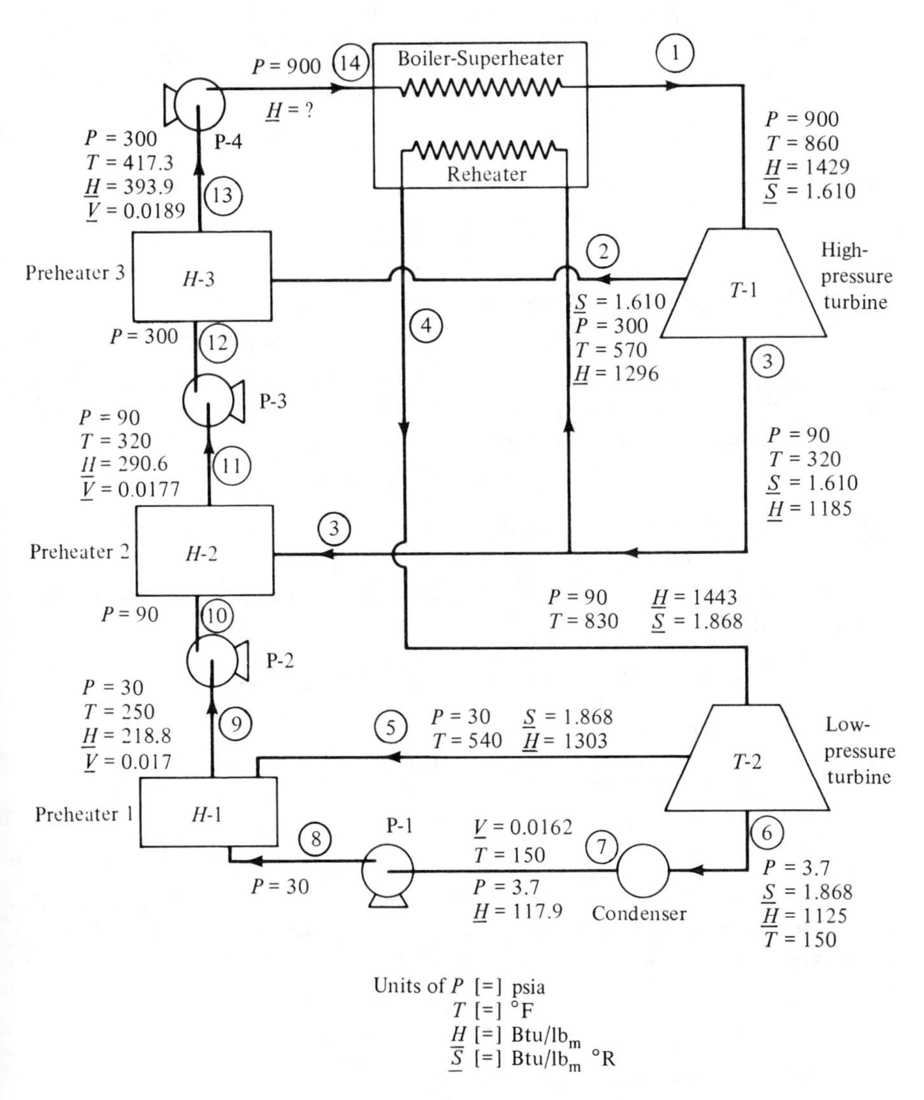

FIGURE S7-5

steam for the feed preheaters is available at pressures of 300, 90, and 30 psia, and the liquids leaving the preheaters are to be saturated liquids at the pressure of the preheater steam. The condenser will operate as a total condenser discharging saturated liquid at a temperature of 150°F. Compare the thermal efficiency of this reheat cycle with and without regenerative preheating of the feed. You may assume that the pumps and turbines are adiabatic and reversible.

Solution: In Sample Problem 7-4 we learned that the efficiency of the cycle without feed preheating was 35 per cent. Let us now calculate the efficiency with feed preheat. We may now determine the values of the properties of many of the streams from the steam tables. These streams and their properties are shown in Fig. S7-5.

With these property values we may now begin to analyze the cycle. Because of the number of streams leaving the turbines, and the number of pumps, we will calculate W^*_{net} from an overall energy balance around the cycle rather than from the sum of the individual works. Thus

$$W^*_{\text{net}} = \sum Q^* = Q^*_{\text{boiler}} + Q^*_{\text{reheater}} + Q^*_{\text{cond}}$$

The thermal efficiency is then

$$\eta = \frac{Q^*_{\text{boiler}} + Q^*_{\text{reheater}} + Q^*_{\text{cond}}}{Q^*_{\text{boiler}} + Q^*_{\text{reheater}}}$$

An energy balance around the condenser gives

$$Q^*_C = \underline{Q}_C = \Delta \underline{H}_C = (117.9 - 1125) \text{ Btu/lb}_m = 1007.1 \text{ Btu/lb}_m$$

(*Note:* For this problem the underbar refers to 1 lb_m through any particular piece of equipment. The superscript asterisk will be used to refer to 1 lb_m through the condenser. The two quantities are related as follows:

$$Q^* = M^* \underline{Q}$$

where $M^* = \dfrac{\text{lb}_m \text{ through particular equipment}}{\text{lb}_m \text{ through the condenser}}$.)

We may now determine the enthalpy of the liquid entering H−1 from an energy balance around P−1:

$$\Delta \underline{H} = \underline{H}_8 - \underline{H}_7 = -\underline{W}_{P-1} = \int \underline{V}\, dP = \underline{V}\, \Delta P$$

or, substituting values,

$$\Delta \underline{H} = (0.0162)(30 - 3.7)\left(\frac{144}{778}\right) \frac{\text{ft}^3}{\text{lb}_m} \frac{\text{lb}_f}{\text{in.}^2} \frac{\text{in.}^2}{\text{ft}^2} \frac{\text{Btu}}{\text{lb}_f\text{-ft}}$$
$$= 0.08 \text{ Btu/lb}_m$$

Therefore, $\underline{H}_8 = 118.0 \text{ Btu/lb}_m$.

We may now determine the flow rate of steam to H−1 from an energy balance around preheater 1. Assuming negligible changes in potential and kinetic energy and steady, adiabatic operation we get

$$\sum M^* \underline{H} = 0$$

$$M^*_8 \underline{H}_8 + M^*_5 \underline{H}_5 - M^*_9 \underline{H}_9 = 0$$

A mass balance around the preheater gives

$$M^*_9 = M^*_8 + M^*_5$$

which is substituted in the energy balance. The energy balance is then solved for M^*_5 to give

$$M^*_5 = M^*_8 \frac{\underline{H}_9 - \underline{H}_8}{\underline{H}_5 - \underline{H}_9}$$

But $M^*_8 = 1$, so M^*_5 becomes

$$M_5^* = \frac{218.8 - 118.0}{1303 - 218.8} = 0.0930 \text{ lb}_m/\text{lb}_m \text{ through condenser}$$

$$M_9^* = 1.093$$

We now proceed in an identical manner to determine the flow rates leaving the other two preheaters. The enthalpy of the fluid leaving P−2 is obtained from

$$\underline{H}_{10} - \underline{H}_9 = \Delta\underline{H} = -\underline{W}_{P-2} = \underline{V}\,\Delta P = (0.017)(90 - 30)\left(\frac{144}{778}\right) \text{ Btu/lb}_m$$
$$= 0.2 \text{ Btu/lb}_m$$

Thus

$$\underline{H}_{10} = (218.8 + 0.2) \text{ Btu/lb}_m = 219 \text{ Btu/lb}_m$$

The mass flow rate of steam into preheater H−2 is obtained from

$$M_3^* = M_{10}^* \frac{\underline{H}_{11} - \underline{H}_{10}}{\underline{H}_3 - \underline{H}_{11}}$$

But

$$M_{10}^* = M_9^* = 1.093$$

Therefore,

$$M_3^* = 1.093 \frac{(290.6 - 218.8) \text{ Btu/lb}_m}{(1185 - 290.6) \text{ Btu/lb}_m} = 0.087$$

$$M_{11}^* = M_3^* + M_{10}^* = 1.180$$

Continue through P−3:

$$\underline{H}_{12} - \underline{H}_{11} = \Delta\underline{H}_{P-3} = -\underline{W}_{P-3} = \underline{V}\,\Delta P$$
$$= (0.0177)(300 - 90)\left(\frac{144}{778}\right) \text{ Btu/lb}_m$$
$$= 0.7 \text{ Btu/lb}_m$$
$$\underline{H}_{12} = (290.6 + 0.7) \text{ Btu/lb}_m = 291.3 \text{ Btu/lb}_m$$

The steam flow rate into H−3 is given by

$$M_2^* = M_{12}^* \frac{\underline{H}_{13} - \underline{H}_{12}}{\underline{H}_2 - \underline{H}_{13}}$$

But

$$M_{12}^* = M_{11}^* = 1.180$$

Therefore,

$$M_2^* = 1.180 \frac{(393.9 - 291.3) \text{ Btu/lb}_m}{(1296 - 393.9) \text{ Btu/lb}_m} = 0.135$$

$$M_{13}^* = M_2^* + M_{12}^* = 1.315 \text{ lb}_m/\text{lb}_m \text{ through condenser}$$

Since $M_{14}^* = M_{13}^*$, 1.315 lb_m of water flows through the boiler per lb_m of fluid through the condenser.

We can determine the enthalpy of the liquid entering the boiler–superheater from the relation

$$\underline{H}_{14} - \underline{H}_{13} = \Delta\underline{H}_{P-4} = -\underline{W}_{P-4} = +\underline{V}\Delta P$$
$$= (0.0189)(900 - 300)\left(\frac{144}{778}\right) \text{ Btu/lb}_m$$
$$= 2.1 \text{ Btu/lb}_m$$

Therefore,

$$\underline{H}_{14} = (393.9 + 2.1) \text{ Btu/lb}_m = 396 \text{ Btu/lb}_m$$

The heat load in the boiler–superheater is obtained from the energy balance around the boiler–superheater:

$$\underline{Q}_B = \Delta\underline{H}_B = \underline{H}_1 - \underline{H}_{14} = 1429 - 396.0 \text{ Btu/lb}_\text{m}$$

or

$$\underline{Q}_B = 1033 \text{ Btu/lb}_\text{m}$$

and

$$\underline{Q}_B^* = M_{14}^* \underline{Q}_B = (1.315)(1033) = 1362 \text{ Btu/lb}_\text{m} \text{ through condenser}$$

The heat absorbed in the reheater is obtained from an energy balance around the reheater:

$$\underline{Q}_R = \Delta\underline{H}_R = \underline{H}_4 - \underline{H}_3 = 1443 - 1185 \text{ Btu/lb}_\text{m}$$

or

$$\underline{Q}_R = 258 \text{ Btu/lb}_\text{m}$$

The mass flow rate through the reheater is obtained from a mass balance around T−2:

$$M_4^* = M_5^* + M_6^*$$
$$= 0.0930 + 1.0 = 1.093 \frac{\text{lb}_\text{m} \text{ through reheater}}{\text{lb}_\text{m} \text{ through condenser}}$$

Therefore, the total heat absorbed in the reheater is given by

$$\underline{Q}_R^* = M_4^* \underline{Q}_R = (1.093)(258)$$
$$= 281 \text{ Btu/lb}_\text{m} \text{ through condenser}$$

The heat load from the condenser is obtained from an energy balance around the condenser:

$$\underline{Q}_C = \Delta\underline{H}_C = \underline{H}_7 - \underline{H}_6 = (117.9 - 1125) \text{ Btu/lb}_\text{m}$$
$$= -1007.1 \text{ Btu/lb}_\text{m}$$

Since $M_C^* = 1$,

$$\underline{Q}_C^* = \underline{Q}_C = -1007.1 \text{ Btu/lb}_\text{m}$$

The net work produced by the cycle is given by

$$W_\text{net}^* = \underline{Q}_B^* + \underline{Q}_R^* + \underline{Q}_C^*$$
$$= 636 \text{ Btu/lb}_\text{m} \text{ through condenser}$$

The thermal efficiency of the whole cycle is then

$$\eta = \frac{W_\text{net}^*}{\underline{Q}_B^* + \underline{Q}_R^*} = \frac{636}{1643} = 0.387$$

This compares with an efficiency of 35 per cent for the cycle without feed preheating. This is not nearly as impressive an increase in the efficiency as was obtained by using reheat. However, it must be remembered that in large generating plants which produce millions of kilowatts of electricity, an efficiency increase of a single per cent represents annual fuel savings of hundreds of thousands, or even millions, of dollars. Thus no increase is too small to be considered.

7.9 Binary Cycles

Our discussion of heat engines up to this point has presumed that water (steam) would be used as the cyclic fluid, which indeed is the case with virtually every commercial power plant. As previously indicated, water, in spite of its many advantages, does possess several important limitations, such as its relatively high vapor pressure at elevated temperatures. Examination of a T–$\underline{S}$ diagram for water (Fig. 7-14) indicates a more serious problem.

As the critical temperature is approached, the two-phase dome narrows progressively and $\Delta \underline{H}_V$ approaches zero at the critical temperature. Thus it is possible to absorb less energy per pound of fluid circulating through the boiler at higher temperatures. If power levels are to be maintained, it is thus necessary to circulate considerably more fluid, and therefore, higher frictional losses and pumping requirements are encountered. Thus higher-temperature water loses much of it attractiveness as a working fluid for heat engines.

In an effort to achieve higher operating temperatures, the binary cycle shown in Fig. 7-17 has been suggested. It consists essentially of two heat engines

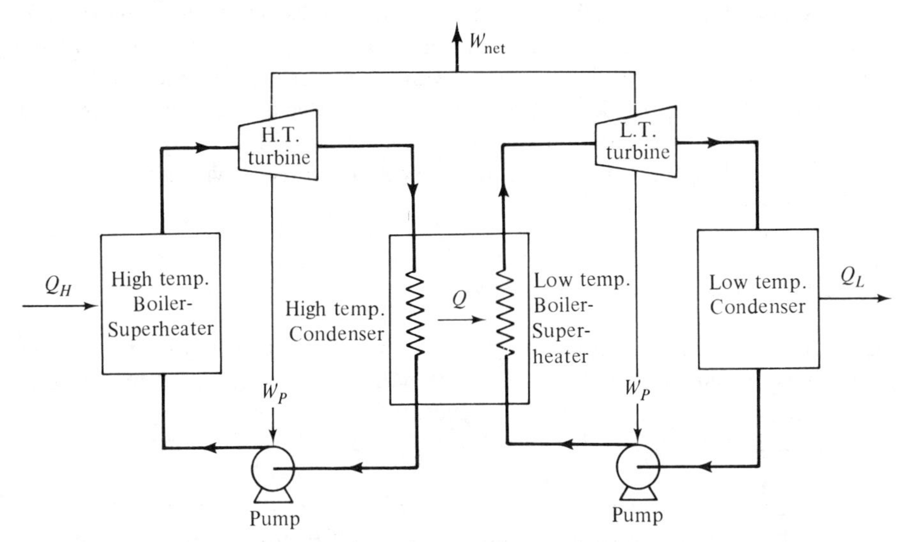

FIG. 7-17. Binary cycle.

with the condenser of the high-temperature cycle serving as the boiler for the lower-temperature cycle (which typically would operate with water). Heat is added from a fuel source to the high-temperature boiler–superheater. The working fluid in the high-temperature cycle is usually a liquid metal whose critical temperature is well above the melting point of the presently used boiler-tube materials. The vaporized metal is partially cooled by expansion in a turbine to an intermediate temperature (about 500 to 700°F). The metal is then condensed by transferring its heat to the boiler in the low-temperature cycle, which would normally use water as its working fluid. The superheated steam then passes through a normal power cycle, which discards heat to the coldest available sink.

The first binary cycles used mercury as the high-temperature working fluid. Since mercury is extremely expensive and toxic, these cycles were not satisfactory and the idea was abandoned. However, in recent years the development of nuclear reactors cooled by liquid sodium and potassium, along with the technology to handle these materials, has once again spurred interest in binary cycles. Since the maximum temperature attainable in nuclear reactors is limited only by the ability of the materials of construction to withstand the temperatures (and the corrosive environment), there is, in theory at least, the possibility of increasing significantly the maximum temperatures that may be used in future binary cycles.

Such binary cycles would permit use of thermal energy at the much higher temperatures and thus achieve improved conversion efficiencies. The lower vapor pressures and higher critical temperatures avoid the problems posed by water at high temperatures. At the same time, by condensing the metallic working fluids at the upper operating temperatures for a steam cycle, we avoid the solidification problems posed by the metals if they were condensed at the low temperatures to which the heat is discarded.

Although liquid-metal technology has been greatly advanced by the nuclear reactor and space programs, the day has not yet arrived when alkali-metal cycles are feasible in large-scale power conversion. Smaller units have been suggested to provide electrical energy from nuclear sources on orbiting space stations, but the design of components that will operate reliably and efficiently for long periods of time in extremely corrosive environments has yet to be achieved.

7.10 Internal Combustion Engines

Up until this point our discussion has been restricted to external combustion engines in which the fuel is burned externally and the heat transferred to a separate, internal working fluid. In the internal combustion engine the fuel is burned internally, and the combustion products are used as the working fluid.

Since combustion products leave internal combustion engines at moderately high temperatures (usually about 300 to 400°F), the thermal efficiency in these engines is usually fairly low. Consequently, internal combustion engines have not found much application in the large-scale generation of electrical energy. On the other hand, by exhausting its waste heat with the combustion products, the internal combustion engine avoids the need for an external source of cooling and is ideally suited for nonstationary propulsion applications such as automobiles and airplanes.

Although a myriad of internal combustion engines have been developed and used, our discussion will be limited by space requirements to a description of the three most important engines: the Otto engine, the diesel engine, and the gas turbine.

The Otto Engine

The first internal combustion engine was produced in Germany by Otto in the mid 1860s. This engine with only minor modifications is still found in almost every automobile in use today. The cycle consists of four strokes: (1) intake, (2) compression, (3) work production, and (4) exhaust.

Its operation can be better understood by examining the P–V relationship of the working fluid during the various strokes shown in Fig. 7-18. During the intake stroke, 1-2, the intake valve is opened and the fuel–air mixture is introduced into the cylinder at almost constant pressure. The intake valve is then closed and the fuel–air mixture compressed along path 2-3. The fuel–air mixture is ignited at point 3 and burns very rapidly at almost constant volume to point 4. After completion of the combustion, work production occurs as the piston expands along path 4-5. At the end of the work stroke the exhaust valve is opened, and the spent combustion products are exhausted along line 5-6-1 as the cycle is completed.

Since the processes that occur in the Otto engine are extremely complex, theoretical prediction of the thermal efficiency of the Otto cycle is very difficult. However, a semiquantitative estimate of the effect of changes in the operating conditions can be studied by examining the air-standard Otto cycle. In the air-standard cycle we assume that the working fluid is air and that it operates in a closed cycle in which the combustion and exhaust strokes are replaced by constant-volume heat-addition and heat-rejection stages. The compression and expansion stages are assumed to be adiabatic and reversible. Finally, the air is assumed to be an ideal gas with constant $C_V = 5.0$ Btu/lb-mol °F, and $k = C_P/C_V = 1.40$. The path followed in the air-standard Otto cycle is illustrated in Fig. 7-19.

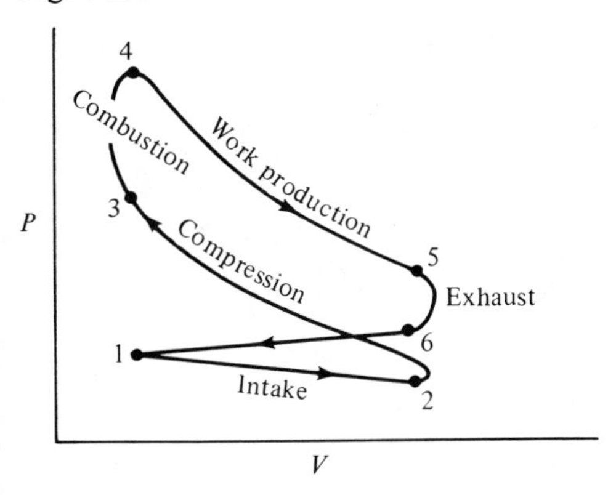

FIG. 7-18. Otto cycle.

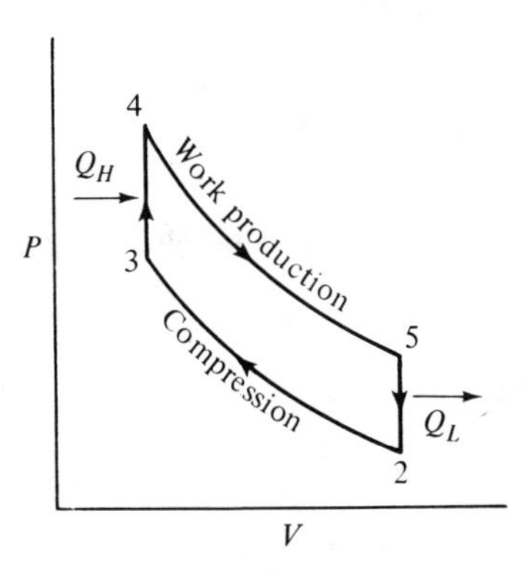

FIG. 7-19. Air-standard Otto cycle.

SAMPLE PROBLEM 7-6. Determine the thermal efficiency of the air-standard Otto cycle as a function of the specific heat ratio, k, and the compression ratio, $\rho = V_2/V_3$.

Solution: The thermal efficiency is given by

$$\eta = \frac{\underline{W}_{\text{net}}}{\underline{Q}_H} = \frac{\underline{Q}_H + \underline{Q}_L}{\underline{Q}_H}$$

An energy balance around the heat-addition and heat-rejection steps gives

$$\underline{Q}_H = \Delta \underline{U}_H = C_V(T_4 - T_3)$$

$$\underline{Q}_L = \Delta \underline{U}_L = C_V(T_2 - T_5)$$

since no work is involved in these nonflow operations. The thermal efficiency then becomes

$$\eta = \frac{\underline{Q}_H + \underline{Q}_L}{\underline{Q}_H} = \frac{C_V[(T_4 - T_3) + (T_2 - T_5)]}{C_V(T_4 - T_3)} = 1 + \frac{T_2 - T_5}{T_4 - T_3}$$

Since the air is assumed to be an ideal gas, $T = PV/nR$, which can be substituted into the thermal efficiency to give

$$\eta = 1 + \frac{P_2 V_2 - P_5 V_5}{P_4 V_4 - P_3 V_3}$$

or, since $V_3 = V_4$ and $V_2 = V_5$,

$$\eta = 1 + \frac{V_2}{V_3}\left[\frac{P_2 - P_5}{P_4 - P_3}\right] = 1 + \rho\left[\frac{P_2 - P_5}{P_4 - P_3}\right] = 1 + \rho\frac{P_5(P_2/P_5 - 1)}{P_3(P_4/P_3 - 1)}$$

Since the compression and expansion are assumed to be adiabatic and reversible, they are isentropic, and the pressure and volume are related by the expression $PV^k = \text{const}$, so

$$P_4 V_4^k = P_5 V_5^k$$

$$P_2 V_2^k = P_3 V_3^k$$

However, since $V_3 = V_4$ and $V_2 = V_5$, this gives

$$P_4 V_3^k = P_5 V_2^k$$

$$P_2 V_2^k = P_3 V_3^k$$

which gives

$$\frac{P_4}{P_3} = \frac{P_5}{P_2}$$

Thus the thermal efficiency is given by

$$\eta = 1 + \rho\frac{P_5(P_2/P_5 - 1)}{P_3(P_4/P_3 - 1)} = 1 - \frac{P_5}{P_4}\rho$$

But from the relation between P and V,

$$\frac{P_5}{P_4} = \frac{P_2}{P_3} = \left(\frac{V_3}{V_2}\right)^k = \rho^{-k}$$

so that we finally obtain for η:

$$\eta = 1 - \rho^{1-k}$$

Sample Problem 7-6 demonstrates that the thermal efficiency increases with increasing compression ratios. The compression ratio in an Otto cycle is usually limited by the ability of the fuel to withstand the compression stroke without preigniting or "knocking." Preignition occurs if the fuel is compressed to the point where its temperature exceeds the autoignition temperature. The ability of a fuel to avoid preignition is characterized by its "octane" number. The higher the octane number, the less likely a fuel is to preignite. Thus higher-octane fuels must be used in higher-compression engines. The fuels currently used in this country have fairly high octane numbers which allow use of high compression ratios. Automobile engines that run on "regular" gasoline typically have compression ratios of about 8.5/1; engines that use "high test" or "ethyl" gasoline have compression ratios of about 10.5/1.

The Diesel Engine

Whereas preignition is harmful and usually avoided in the Otto cycle, the diesel cycle uses the high temperatures that can be attained in the compression step to avoid the need for an external combustion initiator (typically a spark plug in the Otto engine). In the diesel cycle the fuel is not mixed with the air until the end of the compression stage. The fuel is then added slowly, so that the combustion occurs, ideally, at constant pressure. After combustion is complete, the combustion products are expanded and exhausted in the same manner as in the Otto engine. The idealized air-standard diesel engine is illustrated in Fig. 7-20.

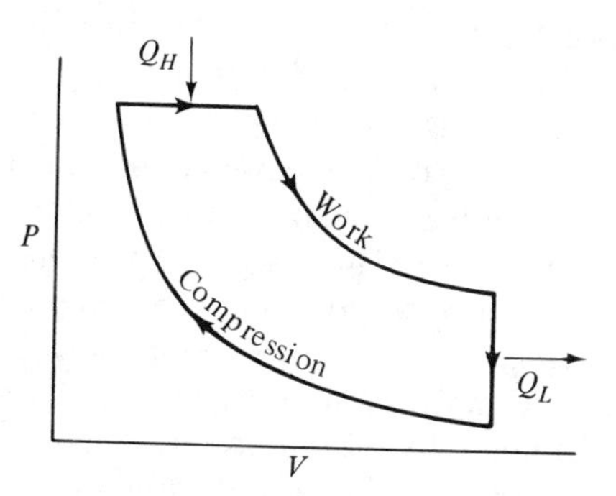

FIG. 7-20. Air-standard diesel cycle.

Although the thermal efficiency of the diesel cycle is slightly lower than that of an Otto cycle with the same compression ratio, diesel engines normally operate with much higher compression ratios than the Otto engines, resulting in actual operating efficiencies higher than those of the Otto engine.

Gas-Turbine Engines

The need for reciprocating pistons and moving valves places restrictions on the speed at which diesel and Otto engines may be driven. The problems in fabricating and operating large pistons also places a practical limit on the size of these engines. By replacing the reciprocating piston expansion device with a nonreciprocating turbine, these restrictions are greatly reduced. Since the piston is no longer available to perform the precombustion compression, a rotary compressor is added to the engine.

The gas-turbine engine is usually built along the lines suggested in Fig. 7-21. The inlet air is compressed in the compressor and mixed with burning fuel in the combustion chamber. The hot, compressed combustion products are then expanded in the turbine, and finally exhausted to the atmosphere. A portion of the work produced in the turbine is used to power the compressor, which is usually mounted on the same shaft as the turbine.

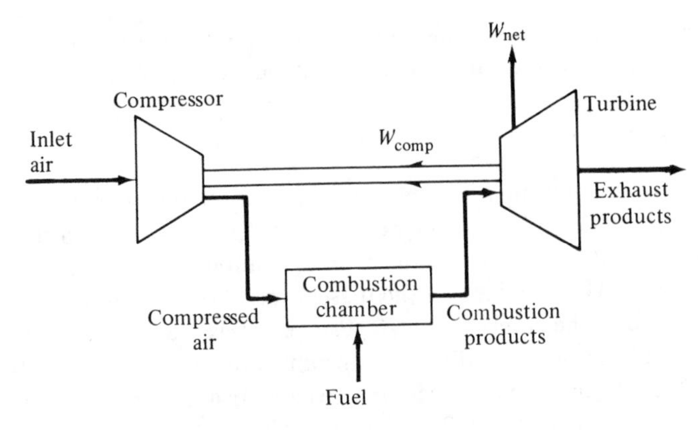

FIG. 7-21. Gas turbine engine.

Since the gas-turbine engine has no reciprocating parts, vibration problems are much less than in reciprocating equipment. Thus gas-turbine engines find some of their greatest use in applications that require extremely high rotational speeds, such as in turbine-powered aircraft. In addition, since the gas-turbine engine has very few moving parts, it is extremely reliable and requires only a minimum of service. Thus they also find great application in unattended field applications, such as powering the compressors on interstate gas pipelines. (In commercial aircraft, where failure must be kept as low as possible, turbine engines are completely overhauled after 2000 hours of flying, as opposed to 200 to 400 hours for piston-powered engines.)

An increasingly important variation of the gas-turbine engine is the "jet engine," illustrated in Fig. 7-22. The incoming air is compressed and reacts with the fuel in the combustion chamber. The hot gases are partially expanded in the turbine, where enough work is extracted to run the compressor. The gas is then completely expanded in the thrust nozzle, where the velocity of the exit gases is greatly increased. The engine produces thrust according to Newton's second law of motion: Every action produces an equal and opposite reaction.

When the speed of the jet engine approaches or exceeds the speed of sound in the surrounding air (approximately 1100 ft/sec), a portion of the inlet compression may be handled by means of a diffuser, as shown in Fig. 7-23. This device reduces the velocity of the incoming air and increases its pressure (a process to be studied in greater detail in Chapter 8). At very high velocities, 3000 mph and higher, this diffuser compression completely eliminates the need for the compressor and turbine. The resulting engine is termed a *ram jet* and is

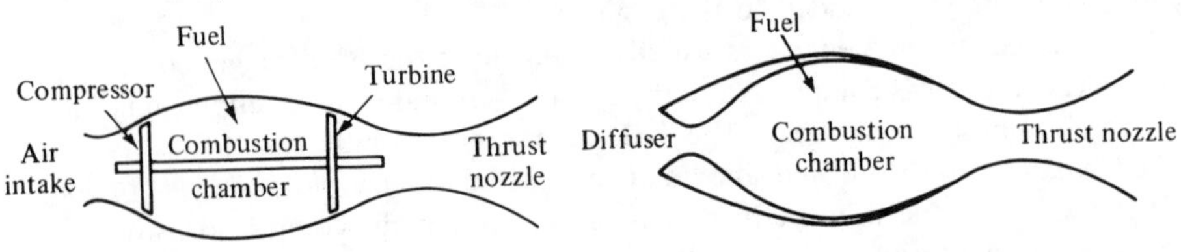

FIG. 7-22. Jet engine.

FIG. 7-23. Ram jet engine.

one of the simplest forms of the internal combustion engine. Unfortunately, it cannot be used as the sole power source for a craft, because an auxiliary engine is needed to accelerate the craft to a velocity high enough to permit operation of the ram jet.

7.11 Vapor-Compression Refrigeration Cycles

As we indicated earlier, the Carnot cycle can be operated in the reverse direction from the cycles we have just been discussing. By supplying work to the cycle, heat can be absorbed from a low-temperature sink and discharged to a high-temperature sink. In this fashion the Carnot cycle may be used to produce a refrigeration effect. As with the heat engines, the Carnot refrigeration cycle represents the most efficient (Btu of cooling/Btu of work supplied) cycle and therefore is used as a standard against which we compare the more practical cycles that are actually used.

The Carnot heat pump, as this version of the cycle is termed, also consists of two isothermal and two isentropic processes and utilizes the same components as the heat engine but operates in reverse sequence, as shown in Fig. 7-24. The

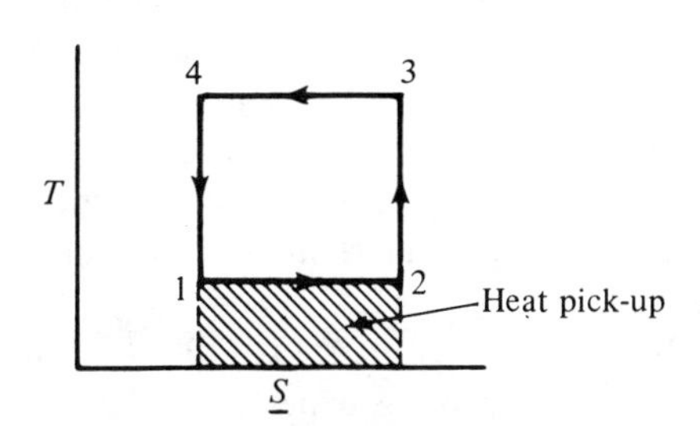

FIG. 7-24. Carnot heat pump.

fluid picks up heat isothermally in a low-temperature heat exchanger (here assumed to be an evaporator), during which time its entropy is increased (step 1-2). The fluid's temperature is then increased by an isentropic compression (step 2-3). The fluid then discharges heat isothermally to a high-temperature sink and experiences an entropy reduction (step 3-4) that precisely offsets its increase at the lower temperature. The fluid is then expanded isentropicly to the pressure and temperature at which it began the cycle (step 4-1). During this step reversible expansion work is extracted.

Application of the energy balance to the cycle, a closed, steady-state system, yields

$$W_{\text{net}} = Q_H + Q_L \tag{7-33}$$

or, on the basis of a unit mass of refrigerant flowing,

$$\underline{W}_{\text{net}} = \underline{Q}_H + \underline{Q}_L \tag{7-33a}$$

Similarly application of the entropy balance to the cycle yields for a unit mass of refrigerant:

$$\frac{\underline{Q}_H}{T_H} + \frac{\underline{Q}_L}{T_L} = 0 \tag{7-34}$$

$\underline{Q}_H$ can be shown to be represented by the area under the curve corresponding to the fluid path through the high-temperature heat exchanger (step 3-4) by simply applying the entropy balance to the process. Since step 3-4 occurs isothermally and reversibly,

$$\frac{\underline{Q}_H}{T_H} + \underline{S}_{\text{in}} - \underline{S}_{\text{out}} = 0 \tag{7-35}$$

or

$$\underline{Q}_H = T_H(\underline{S}_{\text{out}} - \underline{S}_{\text{in}}) = T_H \, \Delta\underline{S}_H \tag{7-35a}$$

$\Delta\underline{S}_{3-4}$ is negative, as is $\underline{Q}_H$, for a process where heat leaves the system at the high temperature. Similarly, $\underline{Q}_L = T_L \, \Delta\underline{S}_L$ and is represented by the area beneath curve 1-2 in Fig. 7-24. $\underline{Q}_L$ is positive, because heat is absorbed in the evaporator. Equation (7-33a) indicates that the net work, $\underline{W}_{\text{net}}$, required is the sum of $\underline{Q}_H$ and $\underline{Q}_L$. Graphically this is represented by the area enclosed by the cycle in Fig. 7-24, because the areas beneath curve 1-2 exactly cancel. This result is precisely the same as observed for the Carnot heat engine with the *important exception* that the net area is *negative* ($|\underline{Q}_H| > |\underline{Q}_L|$ and $\underline{Q}_H < 0$) in the case of the heat pump. Whereas a heat engine's performance is improved by maintaining T_H as high as possible and T_L as low as possible, precisely the reverse is seen to be true for the Carnot heat pump. Thus the most efficient refrigeration cycle would contract the area by raising T_L and decreasing T_H. However, T_L cannot exceed the temperature of the sink from which heat is to be extracted (or heat will not flow into the fluid), and T_H cannot be less than the sink to which heat must be discarded. Thus the ideal refrigeration cycle operates with T_H infinitesimally greater than the hot sink and T_L infinitesimally lower than the cold sink.

The coefficient of performance (COP) of a refrigeration cycle is defined as the amount of cooling produced per unit of work supplied to the cycle. We may develop an expression for the COP of a Carnot refrigerator by eliminating $\underline{Q}_H$ between equations (7-33a) and (7-34) to give

$$\underline{W}_{\text{net}} = \underline{Q}_L\left(1 - \frac{T_H}{T_L}\right) \tag{7-36}$$

Rearrangement yields the coefficient of performance

$$\text{COP} = \frac{-\underline{Q}_L}{\underline{W}_{\text{net}}} = \frac{-1}{1 - T_H/T_L} = \frac{T_L}{T_H - T_L} \tag{7-37}$$

Thus the coefficient of performance of a Carnot refrigerator can be expressed as a function of only the hot and cold sink temperatures. (It should be emphasized that these are *absolute* temperatures.)

Although it is possible in theory to build a gas-phase Carnot cycle, it would be possible to transfer finite amounts of heat to a fluid at constant temperatures only if the fluid flowed at an infinite rate or an isothermal work transfer occurred. We may avoid this difficulty by simply operating the cycle in a two-phase (gas–liquid) region as shown in Fig. 7-25. The working fluid, or refrigerant, boils (evaporates) at constant temperature and pressure as heat is absorbed. Heat

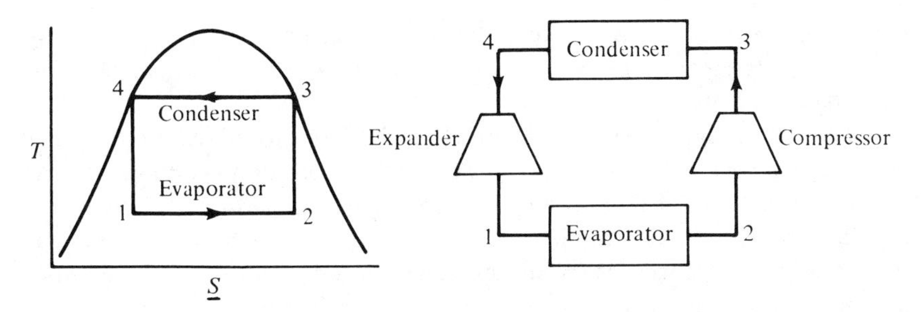

FIG. 7-25. Two-phase Carnot refrigerator.

is rejected in the condenser as the refrigerant condenses—again at constant temperature and pressure. Since boiling and condensing are excellent heat transfer mechanisms, these steps cause little difficulty.

Examination of Fig. 7-25 indicates that both the compression and expansion steps in this Carnot refrigerator occur within the two-phase region. As was discussed for heat engines, the compression of a gas–liquid two-phase mixture is generally avoided. The problem can be eliminated by simply allowing the refrigerant to evaporate completely in the evaporator, producing a saturated vapor as shown in Fig. 7-26. Isentropic compression of the refrigerant to the condenser pressure then produces a superheated vapor at point 3 in Fig. 7-26.

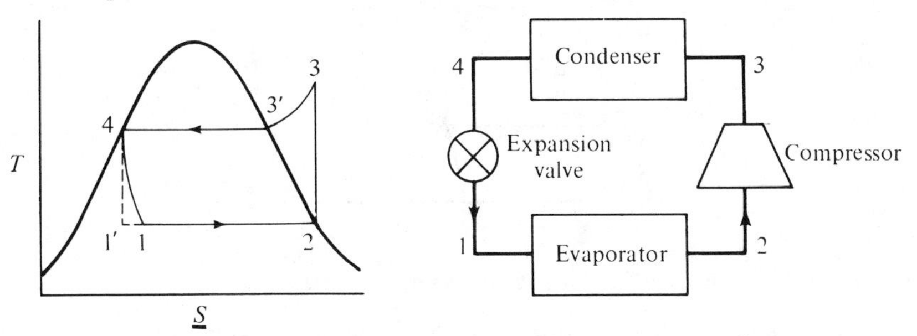

FIG. 7-26. Vapor-compression refrigeration.

This superheating is necessary in order that the path from point 3 to point 3′ be a simple constant-pressure cooling step involving no work. This constant-pressure cooling occurs in the first portion of the condenser. Since this heat removal involves single phase gas heat exchange, it is a rather inefficient process and thus should be kept to a minimum.

Since the fluid passing through the expander is mostly a liquid, its specific volume is relatively low, so the amount of work produced by the expander is not appreciable. Thus little would be lost, in the way of efficiency, if the reversible expander were replaced by a simple isenthalpic throttling expansion. More importantly, the throttling expansion devices are much less expensive than their work-producing counterparts and are essentially maintenance free. Consequently, most refrigeration cycles use a throttling expansion rather than the slightly more efficient work-producing expansion. Since the throttling device, either a fine capillary tube or an expansion valve, involves no moving parts, it is possible to operate in the two-phase region without experiencing serious operating problems.

The throttling process is irreversible and causes an increase in the fluid's entropy (path 4-1), as shown in Fig. 7-26. Less refrigeration is obtained in cycle 1-2-3-4 than in 1′-2-3-4 because of the reduced $\underline{Q}_L$. (The area under curve 1-2 is less than under 1′-2, but the net work required is still greater.) However, this sacrifice is compensated for by the practical considerations mentioned above, and the resulting cycle, the vapor-compression refrigeration cycle, is used in essentially all electrically driven home air conditioners, freezers, and refrigerators.

Since enthalpy values at the various points in the cycle are frequently needed in order to perform calculations on a cycle under consideration, it is common to prepare refrigeration-cycle plots on pressure–enthalpy, or P–$\underline{H}$ plots, as well as on T–$\underline{S}$ diagrams. A typical P–$\underline{H}$ diagram for the vapor-compression refrigeration cycle is illustrated in Fig. 7-27.

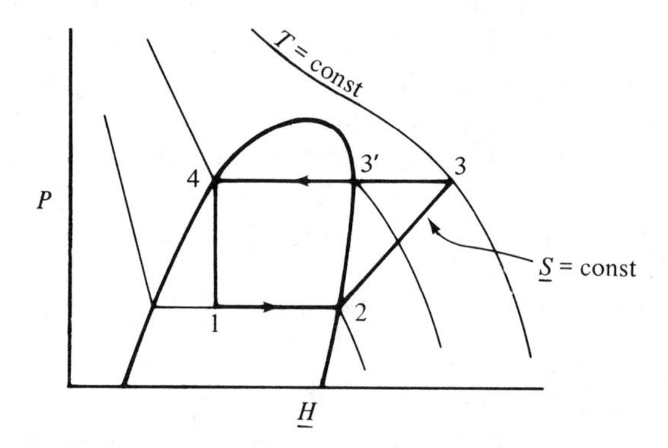

FIG. 7-27. Vapor compression refrigeration cycle.

The standard unit of cooling capacity is called a "ton" and represents the rate of cooling that would be provided by 2000 lb_m (1 ton) of ice melting per day. In terms of Btu, 1 ton of cooling is equivalent to 12,000 Btu of cooling per hour:

$$1 \text{ ton} = 12{,}000 \text{ Btu/hr}$$

SAMPLE PROBLEM 7-7. A heat pump (refrigerator) is to pick up 96,000 Btu/hr from a cold room maintained at a constant temperature of 20°F, and will discard heat to cooling water supplied at an initial temperature of 40°F.

Calculate the minimum work (Btu/hr) required to accomplish the refrigeration for each of the following cases:

(a) If the supply of cooling water is unlimited.

(b) If the cooling-water supply is limited to a maximum of 2100 lb_m/hr.

(c) Same as (b), with additional specification that the refrigerant enter the condenser as a saturated vapor and leave as a saturated liquid. Assume that the pressure drop across the condenser is negligible.

(d) Same as (c), with the additional specification that a minimum "driving force" of 20°F be maintained in both the cold room and the condenser.

(e) Same as (d), with the additional specification that the adiabatic, reversible expander in the heat-pump cycle be replaced with an adiabatic throttling valve. (*Note:* The outlet water temperature will be 100°F for this case.)

Solution: Since we are attempting to calculate the minimum work requirements of the refrigeration scheme, we shall assume that the cycle is reversible in all ways except as it would violate the statement of the problem.

(a) If the cooling-water supply is unlimited, we may assume that the high-temperature sink is at a constant temperature of 40°F = 500°R. The cold room is maintained at a constant temperature of 20°F = 480°R. The absolute minimum work requirements for the desired cooling would be that for a Carnot refrigerator working between the temperatures 480 and 500°R. The COP for a Carnot refrigerator between these temperatures would be

$$\text{COP} = \frac{T_L}{T_H - T_L} = -\frac{Q_L}{\underline{W}_{\text{net}}} = -\frac{\dot{Q}_L}{\dot{W}_{\text{net}}} = \frac{480}{20} = 24.0$$

Since $\dot{Q}_L = 96{,}000$ Btu/hr,

$$\dot{W}_{\text{net}} = \frac{-96{,}000 \text{ Btu/hr}}{24.0} = -4000 \text{ Btu/hr}$$

(b) If the cooling-water supply is limited to 2100 lb_m/hr, the cooling-water temperature will increase as it passes through the condenser. Thus, if we are to keep the work load to a minimum, the refrigerant temperature in the condenser should equal the cooling-water temperature at all points within the condenser. (Realistically, of course, this would be extremely difficult to achieve, but it might be approached by using a large number of small compressors and condensers, as shown in Fig. S7-7.)

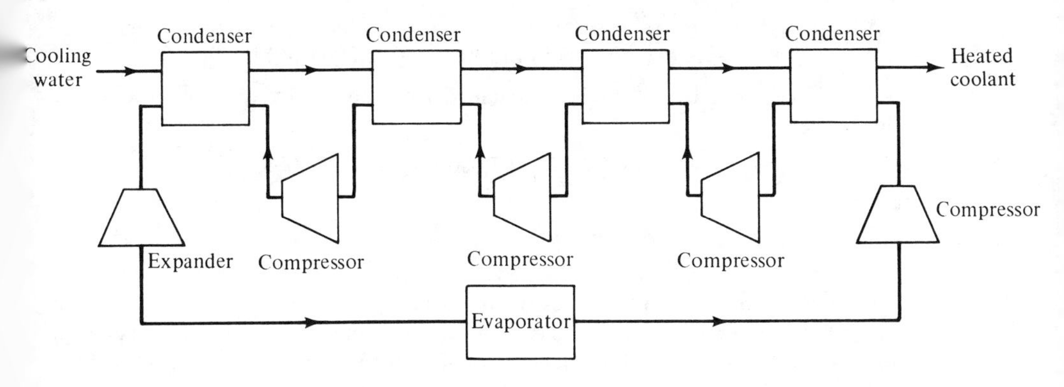

FIGURE S7-7

If the cycle is to be completely reversible, we know that the entropy of the universe will remain constant during the refrigeration process. Since the refrigerator operates in a closed cycle, its entropy does not change. Thus the only portions of the universe that change because of the refrigerator are the cold room, whose entropy decreases, and the cooling water, whose entropy increases. Since there is no net change in the entropy of the universe, the entropy gain of the cooling water must equal the entropy loss in the cold room. Since the cold room is at a constant temperature, T_L,

$$\Delta\dot{S}_{CR} = -\Delta\dot{S}_L = \frac{-\dot{Q}_L}{T_L} = \frac{-96{,}000 \text{ Btu/hr}}{480°\text{R}} = -200 \text{ Btu/hr °R}$$

(Note the minus sign, since heat is removed from the cold room but added to the cycle.)

The entropy change in the cooling water is given by

$$d\dot{S}_{H_2O} = \frac{\dot{M}C_P\,dT}{T}$$

or, assuming that $C_P = \text{const}$,

$$\Delta\dot{S}_{H_2O} = \dot{M}C_P \ln\frac{T_{out}}{T_{in}} = -\Delta\dot{S}_L$$

where $\Delta\dot{S}_L$ = entropy change of refrigerant in condenser. But

$$\dot{M} = 2100 \text{ lb}_m\text{/hr}$$

$$C_P = 1 \text{ Btu/lb}_m \text{ °R}$$

$$T_{in} = 500°\text{R}$$

Therefore,

$$-\Delta\dot{S}_L = \Delta\dot{S}_{H_2O} = 2100\left(\ln\frac{T_{out}}{500}\right) \text{ Btu/hr °R}$$

Since $\Delta\dot{S}_H + \Delta\dot{S}_L = 0$,

$$\Delta\dot{S}_H = -\Delta\dot{S}_L = 200 \text{ Btu/hr °R} = 2100\left(\ln\frac{T_{out}}{500}\right) \text{ Btu/hr °R}$$

Solving for T_{out} gives

$$T_{out} = 550°\text{R}$$

The heat rejected to the cooling water is then given by an energy balance around the condenser:

$$\begin{aligned}\dot{Q}_H &= -\dot{M}(\Delta\underline{H})_{H_2O}\\ &= -\dot{M}C_P(550 - 500) \text{ Btu/hr} = -105{,}000 \text{ Btu/hr}\end{aligned}$$

The work required by the cycle is obtained from an energy balance around the whole cycle:

$$\dot{W}_{net} = \dot{Q}_H + \dot{Q}_L = -105{,}000 \text{ Btu/hr} + 96{,}000 \text{ Btu/hr}$$

or

$$\dot{W}_{net} = -9000 \text{ Btu/hr}$$

so 9000 Btu/hr of work must be supplied to the cycle.

(c) If we add the additional restriction that only one condenser be used, the high temperature in the refrigerator cycle is a constant and must be *no lower* than the maximum temperature of the cooling water. Thus, if T_H is the maximum temperature of the cooling water, it is also the condenser temperature of the refrigeration cycle. Since the refrigeration cycle can still operate as a Carnot cycle,

$$\frac{\dot{Q}_H}{T_H} = \frac{-\dot{Q}_L}{T_L} = -200 \text{ Btu/hr °R}$$

Therefore,

$$\dot{Q}_H = -T_H(200) \text{ Btu/hr}$$

But, from an energy balance around the condenser,

$$\dot{Q}_H = -\dot{Q}_{H_2O} = -\dot{M}C_P(T_H - 500) = -2100\,(T_H - 500)\text{ Btu/hr}$$

Equating the two expressions for $\dot{Q}_H$ gives

$$-T_H(200) = -2100(T_H - 500)$$

or, solving for T_H and $\dot{Q}_H$,

$$T_H = 552.5°\text{R}$$

$$\dot{Q}_H = -110{,}300\text{ Btu/hr}$$

The net work is again obtained from an overall energy balance:

$$\dot{W}_{\text{net}} = \dot{Q}_H + \dot{Q}_L = (-110{,}300 + 96{,}000)\text{ Btu/hr} = -14{,}300\text{ Btu/hr}$$

(d) We now add the requirement of at least a 20°R temperature difference everywhere in both the evaporator and the condenser. Thus the evaporator temperature is 20°R less than the cold room, or 0°F = 460°R.

The condenser temperature will be 20°R above the maximum temperature of the cooling water. Let the maximum temperature in the cooling water be T_w, the outlet temperature. The condenser temperature, T_H, will then be $T_w + 20$. We now proceed in the same way as in part (c):

$$\dot{Q}_H = -T_H\frac{\dot{Q}_L}{T_L} = -T_H\left(\frac{96{,}000}{460}\right) = -208.5T_H\text{ Btu/hr}$$

$$-\dot{Q}_{H_2O} = \dot{Q}_H = -\dot{M}C_P(T_w - 500) = -2100(T_H - 520)\text{ Btu/hr}$$

Eliminating $\dot{Q}_H$ gives

$$-2100(T_H - 520)\text{ Btu/hr} = -208.5T_H\text{ Btu/hr}$$

Solving for T_H and Q_H then yields

$$T_H = 577°\text{R} \quad , T_w = 557°\text{R}$$

$$\dot{Q}_H = -120{,}000\text{ Btu/hr}$$

The work produced is given by

$$\dot{W}_{\text{net}} = \dot{Q}_H + \dot{Q}_L = (-120{,}000 + 96{,}000)\text{ Btu/hr} = -24{,}000\text{ Btu/hr}$$

(e) If we are told the outlet water temperature, we can get $\dot{Q}_H$ from the relation

$$-\dot{Q}_{H_2O} = \dot{Q}_H = -2100(T_w - 500°\text{R})\text{ Btu/hr}$$

independent of the cycle workings. Thus

$$\dot{Q}_H = -2100(560 - 500)\text{ Btu/hr} = -126{,}000\text{ Btu/hr}$$

The work production is then given by

$$\dot{W}_{\text{net}} = (-126{,}000 + 96{,}000)\text{ Btu/hr} = -30{,}000\text{ Btu/hr}$$

so 30,000 Btu/hr of work must be provided to the cycle.

We can summarize the results of the five calculations as shown in Table S7-7.

A quick examination of the results of Sample Problem 7-7 serves to illustrate the extreme differences that can exist between the absolute minimum work requirements for a particular cooling job and the minimum requirements when more realistic operating conditions are assumed. In any real refrigeration scheme more work than suggested in part (e) would be required because of the mechanical inefficiencies associated with the compressor. With this factor taken into account the actual work requirements for this cooling load would be closer to 50,000 or 60,000 Btu/hr. Indeed, it is effects of just this

TABLE S7-7

Case	Restrictions	Work Requirements, Btu/hr
(a)	Pure Carnot cycle, $T_H = 500°R$, $T_L = 480°R$	4,000
(b)	Cooling water limited to 2100 lb_m/hr, $T_L = 480°R$	9,000
(c)	Limited cooling H_2O, single condenser, $T_L = 480°R$	14,300
(d)	Same as (c) but with 20°R ΔT in both condenser and evaporator	24,000
(e)	Same as (d) except use throttle instead of reversible expansion	30,000

type that cause the household air conditioner (which, after all, pumps heat across only a 20 to 30°F temperature difference) to be so expensive to operate, when a simple analysis such as that of part (a) indicates a very small work requirement.

SAMPLE PROBLEM 7-8. A central home air conditioner is being designed to provide 36,000 Btu/hr (or 3 tons) of cooling capacity. The air conditioner will operate on a standard vapor-compression refrigeration cycle with throttling expansion. Freon 22 will be used as the working fluid. The evaporator temperature will be held at 50°F, and the condenser will operate at 140°F. Determine the flow rate of the working fluid, the compressor horsepower, and the total heat load of the condenser. Assume reversible compression.

Solution: If we assume that the compression is adiabatic and reversible, it is isentropic. If the throttling expansion is also adiabatic, then it is isenthalpic. If, in addition, we assume that the evaporator and condenser operate at constant pressure we can plot the proposed refrigeration cycle on a P–$\underline{H}$ diagram for Freon 22 as shown in Fig. S7-8.

The values of the various properties have been taken from the P–$\underline{H}$ diagram for Freon 22 which appears in Appendix D. With these values for the properties we are now in a position to evaluate the behavior of the various cycle components.

Evaporator: The heat absorbed by the working fluid in the evaporator can be obtained from an energy balance around the evaporator. Assuming negligible changes in potential and kinetic energy and steady-state operation, the energy balance reduces to

$$\underline{Q}_L = \Delta\underline{H}_{\text{evap}} = \underline{H}_2 - \underline{H}_1 = 109 - 52 \text{ Btu/lb}_m$$

or

$$\underline{Q}_L = 57 \text{ Btu/lb}_m$$

Since the conditioner is to absorb 36,000 Btu/hr, the required flow rate of refrigerant is given by

$$\dot{M} = \frac{\dot{Q}_L}{\underline{Q}_L} = \frac{36{,}000 \text{ Btu/hr}}{57.0 \text{ Btu/lb}_m} = 632 \text{ lb}_m/\text{hr}$$

Compressor: The work production of the compressor can be obtained from an

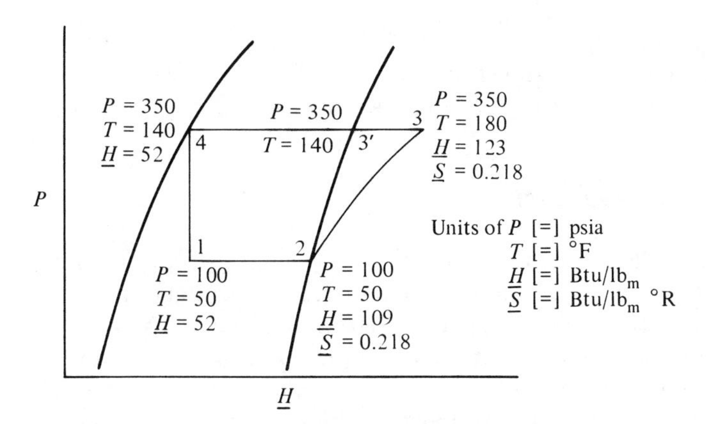

FIGURE S7-8

energy balance around the compressor. Assuming no changes in potential and kinetic energy and adiabatic steady-state flow, the energy balance reduces to

$$\underline{W} = -\Delta\underline{H} = -(\underline{H}_3 - \underline{H}_2) = -(123 - 109) \text{ Btu/lb}_m$$

or

$$\underline{W} = -14 \text{ Btu/lb}_m$$

The total work production is given by

$$\dot{W} = \dot{M}\underline{W} = (632)(-14) \text{ Btu/hr} = -8.85 \times 10^3 \text{ Btu/hr}$$

Since 1 hp = 2545 Btu/hr, the compressor motor will consume 3.47 hp.

Condenser: The total heat load of the condenser can be directly evaluated from an overall energy balance around the refrigerator:

$$\dot{W} = \dot{Q}_H + \dot{Q}_L$$

or

$$\dot{Q}_H = \dot{W} - \dot{Q}_L = -(8.85 \times 10^3 + 36 \times 10^3) \text{ Btu/hr}$$

Therefore, the condenser load is given by

$$\dot{Q}_H = -44.9 \times 10^3 \text{ Btu/hr}$$

In our earlier discussion of the second law we showed that the efficiency (or, similarly, the coefficient of performance) of a Carnot cycle was independent of the working fluid chosen. Unfortunately, this statement is no longer true for the vapor-compression refrigeration cycle. Thus the choice of refrigerant can significantly affect the performance of a particular cycle. In addition to favorable thermodynamic properties, a good refrigerant should possess several other characteristics. It should be (1) nonflammable, (2) nontoxic, (3) noncorrosive, (4) nonexplosive, (5) not too expensive, and (6) have suitable vapor pressures at the condenser and evaporator temperatures so that neither extremely high nor extremely low pressures are involved.

When vapor-compression refrigeration was first introduced, sulfur dioxide, carbon dioxide, and ammonia were the commonly used refrigerants. However, none of these fluids proved entirely satisfactory, and they have been almost completely replaced by the halogenated (chlorine and fluorine) hydrocarbons—commonly known by the duPont trade name Freon. The Freons have proved to be nearly ideal refrigerants for many applications.

7.12 Cascade Cycles

The recent development of the cryogenics industry has spurred great interest in the attainment of very low temperatures. Examination of the thermodynamic properties of the commonly used refrigerants quickly shows that it is not possible to achieve these extremely cold temperatures (less than $-300°F$) in a single vapor-compression cycle (practically any refrigerant that did not freeze in the evaporator would be above its critical point in the condenser). These problems can be avoided by means of a *cascade cycle*, as shown in Fig. 7-28.

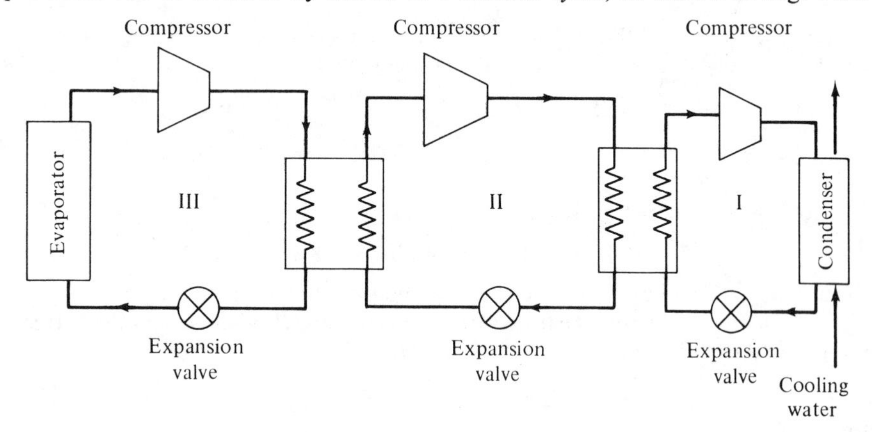

FIG. 7-28. Three-step cascade cycle.

Although cascade cycles can theoretically be built with any number of steps, three or four appear to be the maximum practical limit. Note the similarity of the cascade cycle to the binary power cycle we discussed previously.

Cycle III would use a refrigerant with a normal boiling point around the lowest temperature to be achieved. The heat is absorbed in the evaporator of cycle III. The condenser of cycle III rejects its heat to the evaporator of cycle II. The refrigerant in cycle II would be one with a normal boiling point about two thirds of the way from ambient to the evaporator temperature in cycle III. The condenser for cycle II discharges its heat into the evaporator of cycle I, which finally rejects the heat to cooling water or some other convenient thermal sink. The refrigerant for cycle I would be one with a normal boiling point about halfway between the cooling-water temperature and the evaporator temperature in cycle II. In a three-step cascade cycle designed to provide refrigeration for the liquefaction of natural gas the refrigerants would commonly be: cycle I, propane; cycle II, ethylene; and cycle III, methane.

SAMPLE PROBLEM 7-9. A plant for processing quick-frozen poultry requires the freezing chambers to be held at a temperature of $-160°F$. A two-stage cascade refrigeration unit is being designed to provide the required cooling. The low-temperature evaporator is to operate at $-180°F$; the high-temperature condenser will be at 100°F. The low-temperature cycle will use Freon 503 as its working fluid and will discharge its heat of condensation at $-30°F$. The high-temperature cycle will use Freon 22 as the refrigerant and will absorb heat at $-50°F$. If the refrigerator is to absorb 20,000 Btu/hr from the freezing chamber, determine the compressor horsepower for both cycles. You may assume that the compressions are adiabatic and reversible and that the expansions are isenthalpic.

Solution: We may plot the low temperature cycle on a P–$\underline{H}$ diagram (assuming isentropic compression and isenthalpic expansion) for Freon 503 as shown in Fig. S7-9a. The heat absorbed in the evaporator per lb_m of refrigerant is then given by

$$\underline{Q}_L = \Delta\underline{H}_L = (48 - 3.0) \text{ Btu/lb}_m = 45 \text{ Btu/lb}_m$$

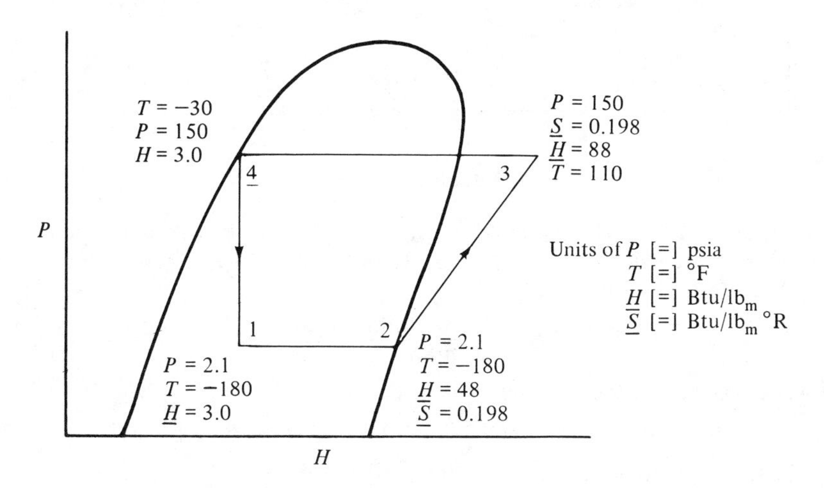

FIGURE S7-9a

Since it is desired to absorb 20,000 Btu/hr the refrigerant rate in this cycle is given by

$$\dot{M} = \frac{\dot{Q}_L}{\underline{Q}_L} = \frac{20{,}000 \text{ Btu/hr}}{45 \text{ Btu/lb}_m} = 445 \text{ lb}_m\text{/hr}$$

The compressor load in this cycle is given by

$$-\underline{W}_C = \Delta\underline{H}_C = (88 - 48) \text{ Btu/lb}_m = 40 \text{ Btu/lb}_m$$

The total compressor load is then

$$-\dot{W}_C = -\underline{W}_C\dot{M} = (40)(445) \text{ Btu/hr} = 17{,}800 \text{ Btu/hr}$$

or, converting to horsepower,

$$-\dot{W}_C = 7.0 \text{ hp}$$

The heat rejected in the low-temperature condenser is then given by

$$-\underline{Q}_H = -\Delta\underline{H} = (88 - 3.0)\ \text{Btu/lb}_m = 85\ \text{Btu/lb}_m$$

and

$$-\dot{Q}_H = \dot{M}\underline{Q}_H = (85)(445)\ \text{Btu/hr} = 37{,}800\ \text{Btu/lb}_m$$

We can now attack the high-temperature Freon 22 cycle. The P–$\underline{H}$ diagram for the cycle (again assuming isentropic compression, and isenthalpic expansion) is as shown in Fig. S7-9b. The heat absorbed in the evaporator is given by

$$\underline{Q}_L = \Delta\underline{H} = (99 - 39)\ \text{Btu/lb}_m = 60\ \text{Btu/lb}_m$$

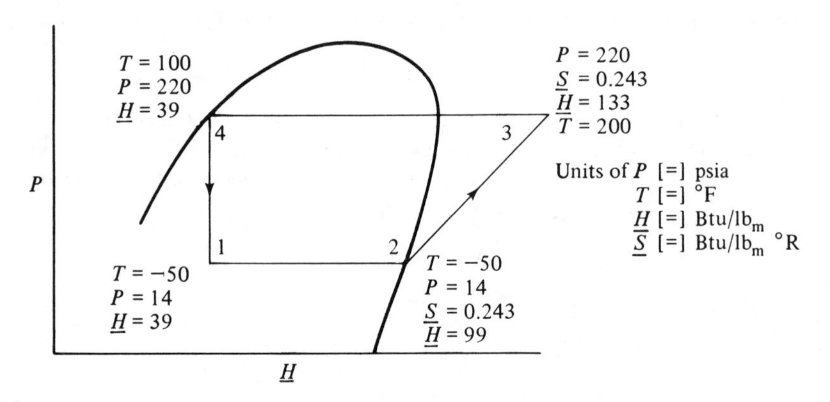

FIGURE S7-9b

However, the total heat absorbed in this evaporator equals the heat discharged by the low-temperature condenser-37,800 Btu/hr. Thus the flow rate of refrigerant in the Freon 22 cycle is given by

$$\dot{M} = \frac{\dot{Q}_L}{\underline{Q}_L} = \frac{37{,}800\ \text{Btu/hr}}{60\ \text{Btu/lb}_m} = 630\ \text{lb}_m\text{/hr}$$

The work supplied to the compressor is given by

$$-\underline{W}_C = \Delta\underline{H}_C = (133 - 99)\ \text{Btu/lb}_m = 34\ \text{Btu/lb}_m$$

Thus the total work requirement of the compressor is given by

$$\dot{W}_C = \underline{W}_C\dot{M} = (-34)(630)\ \text{Btu/hr} = -21{,}400\ \text{Btu/hr}$$

Therefore, the Freon 22 compressor horsepower is

$$-\dot{W}_C = 8.4\ \text{hp}$$

The total compressor power necessary is the sum of the two compressor requirements and equals 15.4 hp.

7.13 Liquefaction of Gases—Cryogenic Temperatures

The first successful liquefaction of a "permanent gas" was achieved by a Frenchman, Georges Claude, who liquefied air around the turn of the century. Claude's cycle used air as the working fluid and consisted of an isothermal compression

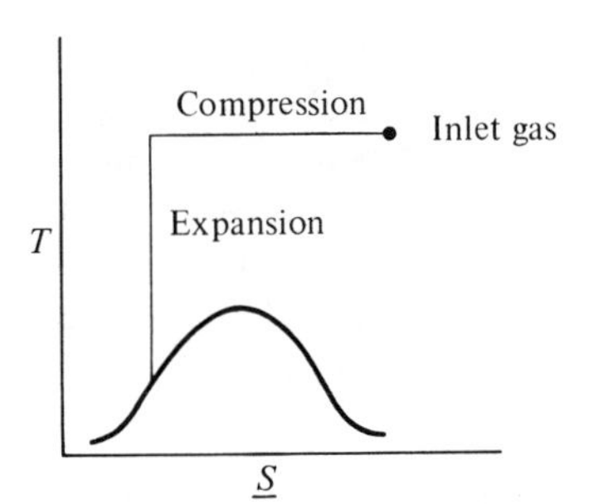

FIG. 7-29. Claude cycle for air liquefaction.

of the gas to a high pressure, followed by isentropic expansion to a low pressure. During the isentropic expansion the gas temperature decreases and finally the gas condenses, as shown in Fig. 7-29.

Although the scheme shown in Fig. 7-29 is the most efficient gas-liquefaction process, it is not a practical process for large-scale applications. The compression pressures would be extremely high. In addition, no expansion devices are available that are capable of efficiently handling the degree of expansion that would be necessary. The first practical scheme for liquefying was developed in the early 1900s in Germany by Carl von Linde. The basic Linde process is shown in Fig. 7-30. The incoming gas is mixed with the returning unliquefied gas, and both are compressed to about 1000 psia. The compressed gas is precooled by an external refrigeration scheme. The precooled gas is further cooled by countercurrent heat exchange with the unliquefied gas. After the heat exchanger the gas passes through a throttling Joule–Thomson expansion, where the gas temperature is reduced and the gas partially liquefied. The

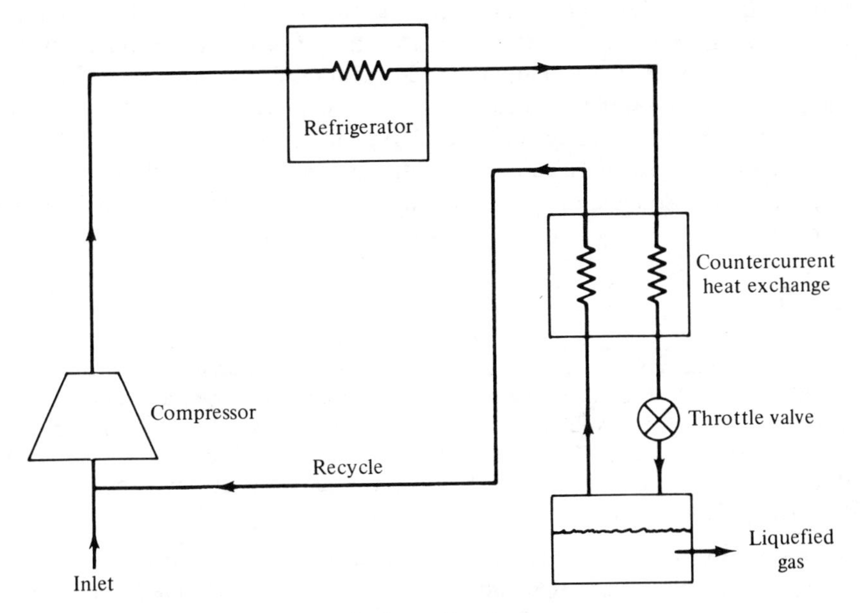

FIG. 7-30. Linde gas-liquefaction process.

liquid is removed and the gas returned through the heat exchanger. The Joule–Thomson expansion requires no moving parts and hence can easily be performed over large pressure ratios. However, the Joule–Thomson expansion will provide the desired temperature decrease only if the gas is first cooled to near its critical temperature, thus the need for the precooling and countercurrent heat exchange. A T–$\underline{S}$ plot of the Linde process is shown in Fig. 7-31.

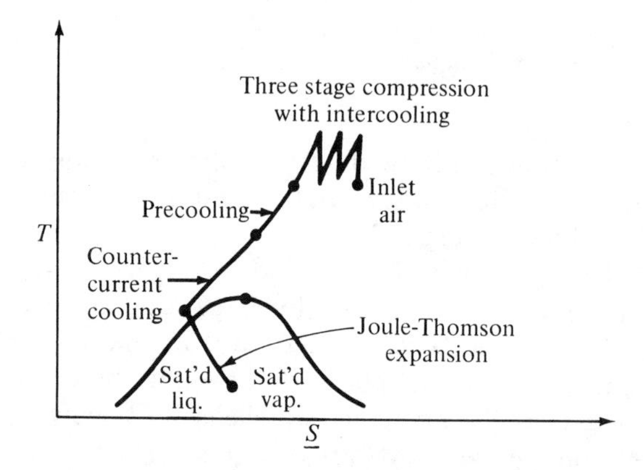

FIG. 7-31. The Linde gas liquefaction process.

Gas liquefaction has become of increasing commercial importance since the 1950s, owing partially to its use in the space program but even more to the natural gas industry. Large quantities of natural gas are now being transported in the liquefied state by tankers from gas fields to areas of high consumption. Storage facilities, both above and below ground, are being used by suppliers and users of liquefied natural gas to store the fuel. To many users, liquefied natural gas supplies large amounts of refrigeration capacity as it is returned to the gaseous state prior to use as a fuel or a raw material.

The recently commercialized basic oxygen steelmaking process makes use of large quantities of purified oxygen obtained from the distillation of liquefied air. It seems reasonable to expect that many other manufacturing facilities will also be affected by cryogenic operations in the not-too-distant future.

Problems

7-1 A steam locomotive is powered by a simple steam engine similar to the one illustrated in Fig. 7-1. The boiler operates at 150 psia and discharges a saturated vapor. The steam is expanded to 20 psia in the piston and discharged to the atmosphere. If feedwater is available at 1 atm and 75°F and the engine is operating at steady state, determine the thermal efficiency of the engine. What is the quality of the steam exhausted from the piston?

7-2 Develop an expression for the thermal efficiency of the air-standard diesel cycle as a function of the compression ratio, $\rho = V_{\text{inlet}}/V_{\text{compressed}}$ and the expansion ratio $V_{\text{hot}}/V_{\text{exhaust}}$.

7-3 Prove that no refrigeration cycle can have a higher coefficient of performance than the reversed Carnot cycle.

7-4 Our space program requires a "portable" power-generating unit that can produce electrical energy for extended periods of time during orbital flight. A modified Rankine cycle (see Fig. P7-4) that utilizes a metallic fluid has been proposed for this purpose. Heat will be extracted at very high temperatures directly from a nuclear reactor. Heat from the condenser will be discharged by radiation to space. If sodium is circulated as the working fluid between the pressures of 126 psia (in the boiler) and 14.7 psia (in the condenser) as shown in Fig. P7-4,

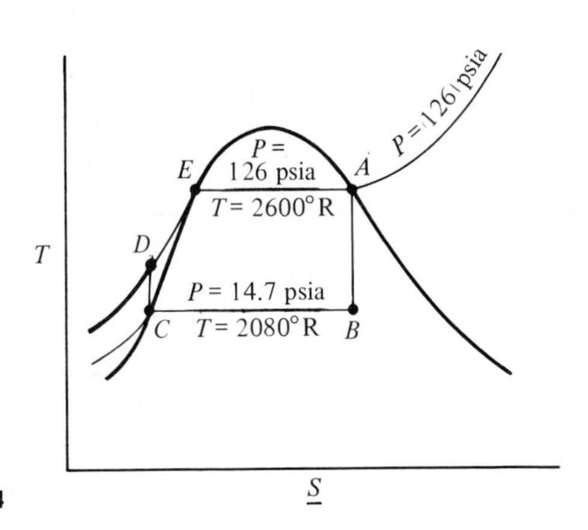

FIGURE P7-4

(a) Calculate the maximum work that could be produced per Btu of heat picked up in the reactor. Assume that the fluid temperature at the exit of the pump (point D) is 2100°R.

(b) What is the quality of the fluid entering the condenser (point B)?

(c) How many Btus of work per Btu of heat picked up could a Carnot cycle deliver operating between the two saturation temperatures (2600°R and 2080°R)?

For sodium: At P = 126 psia, saturation temperature = 2600°R; $(C_P)_{\text{liq}}$ = 0.28 Btu/lb$_m$ °R; latent heat of vaporization = 500 Btu/lb$_m$. At P = 14.7 psia, saturation temperature = 2080°R; latent heat of vaporization = 560 Btu/lb$_m$.

In Fig. P7-4, fluid at point A is saturated vapor at 2600°R, fluid at point C is saturated liquid at 2080°R, fluid at point D is subcooled liquid at 2100°R, and fluid at point E is saturated liquid at 2600°R.

7-5 A steam-heat engine cycle is operated with a boiler, superheater, turbine, condenser, open feedwater heater, and necessary pumps. The steam from the superheater passes to the turbine at 310 psia and 600°F. The turbine exhausts to the condenser at a vacuum of 28.4 in. Hg when the atmospheric pressure is 29.9 in. Hg. The condenser produces saturated liquid which is pumped to the open feedwater heater, which is operating at 50 psia with steam that is bled from an intermediate stage of the turbine. Also, some of the 50 psia steam bled from the turbine is sent out to a chemical process area for use in an evaporator. This steam is returned to the heat-engine cycle as saturated liquid at 50 psia, where it is combined with the feedwater heater product and pumped into the boiler. The turbine is 80 per cent efficient compared to a reversible turbine, and all pumps are 70 per cent efficient on the same basis. All equipment is well insulated and pressure drops are negligible in the piping, condenser, boiler, superheater, and feedwater heater.

(a) For the power cycle alone, without considering the amount of process steam withdrawn for the evaporator, determine the net work delivered per Btu of heat supplied to the boiler and superheater.

(b) Repeat the calculation of (a), assuming that the feedwater heater is not used.

(c) If steam delivered to the turbine is valued at 55 cents per 1000 lb_m what would be a fair charge per 1000 lb_m of steam used by the process department for its evaporator?

7-6 A standard Rankine cycle with superheat (so no moisture condenses in the turbine) is to be operated with two closed feed preheaters as shown in Fig. P7-6. The boiler will

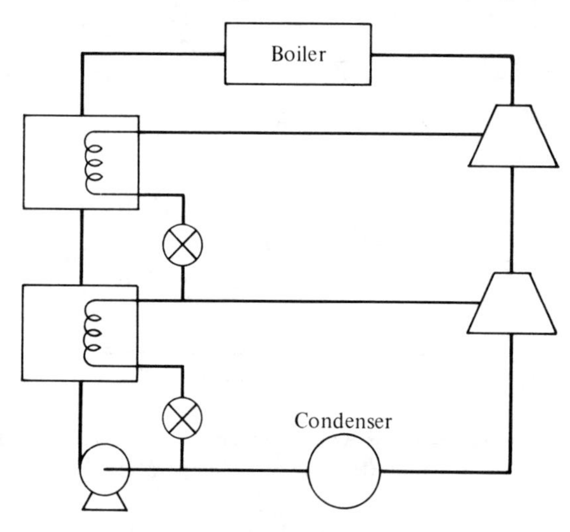

FIGURE P7-6

operate at 200 psia, and the turbine takeoffs will occur at 70 and 15 psia. The condenser will operate at 100°F and will discharge saturated liquid. The feed preheaters discharge liquid at the same temperature as the heating steam is condensing, and the condensed steam leaves each preheater as a saturated liquid. If the pumps and compressors operate adiabatically and all process equipment is reversible, determine the thermal efficiency of this cycle.

7-7 The boiler of a Rankine-cycle steam plant furnishes steam at 350 psia and 800°F to the turbine at the rate of 1000 lb_m/min. From the turbine the steam passes to the condenser, which operates at a pressure of 4 in. Hg using cooling water at 70°F. Assume that the turbine is adiabatic and reversible, that the condensate leaving the condenser is saturated liquid, and that the pump returning the condensate to the boiler is adiabatic and reversible. Calculate

(a) The work produced by the turbine (Btu/lb_m of steam).

(b) The pump work.

(c) The heat transferred in the boiler (Btu/lb_m of steam).

(d) The maximum (Carnot) work per Btu transferred in the boiler for a cycle receiving all its heat at the temperature of the boiler, 2000°F, and discharging at 70°F.

(e) The turbine work (Btu/lb_m of steam) if 500 lb_m/min of process steam is removed from the turbine at the 40-psia stage.

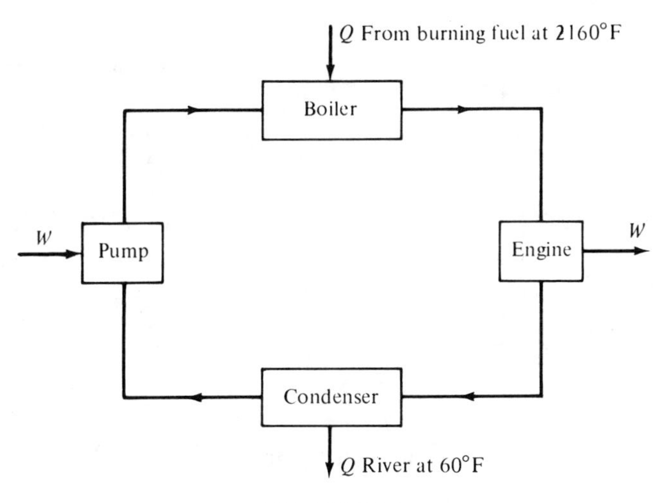

FIGURE P7-8

7-8 A Rankine-cycle steam plant (Fig. P7-8) is operating with a steam pressure of 275.3 psig, a steam temperature of 600°F, and a condenser vacuum of 28.4 in. Hg. Barometer indicates 29.9 in. Hg. Saturated liquid leaves the condenser.

(a) Assuming that all steam lines are well insulated and that the work of expansion and of compression is adiabatic and reversible, compute for 1 lb_m of steam: (1) the engine work, (2) the heat discarded in the condenser, (3) the pump work, and (4) the heat absorbed in the boiler.

(b) Calculate the net work produced per Btu transferred to the steam in the boiler. Compare this with the calculated work per Btu transferred in the boiler for a reversible Carnot engine receiving all of its heat at 2160°F and discharging its heat at 60°F.

(c) How many pounds of steam would be required per horsepower-hour of plant output and what would be the fuel requirement of the plant if it were delivering 10,000 hp while using fuel oil with a heating value of 19,000 Btu/lb_m and a furnace and boiler having an efficiency of 75 per cent? What power would be required to drive the pump?

7-9 Steam enters a bleeder turbine at 310 psia and 700°F and exhausts at 1.5 in. Hg absolute. The saturated condensate from the condenser is heated in an open feedwater heater before being returned to the boiler. Heating is accomplished with steam extracted from the 50-psia stage of the turbine. The pressure in the heater is 50 psia and the liquid leaving the heater is saturated at this pressure. Assume adiabatic reversible expansion of the steam in the turbine and negligible work to pump any liquid into the boiler. Determine the work produced by the turbine per Btu of heat transferred to the steam in the boiler

(a) When using the feedwater heater as described above.

(b) When operating with no feedwater heater (that is, the condensate from the condenser is pumped directly to the boiler.)

How do you account for the difference in work for the two cases?

7-10 (a) In a certain deep-freeze unit the heat transferred into the cold chest from the surroundings is 3000 Btu/day. The chest is maintained at 0°F while the surroundings are at 80°F. What is the power rating (in horsepower) of the smallest motor that can be used to drive an ideal refrigeration cycle to hold the chest at 0°F?

(b) Suppose the cycle uses Freon 12 as the working fluid and has a condenser

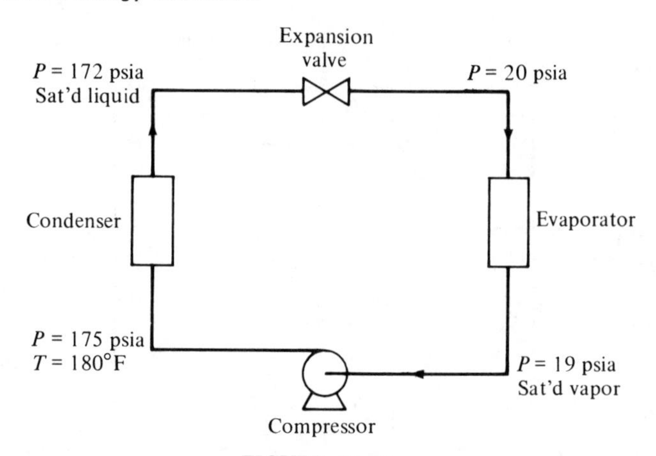

FIGURE P7-10

pressure of 172.35 psia and an evaporator pressure of 19.189 psia. What is the horsepower rating of the smallest motor that can be used to drive this cycle?

(c) The actual refrigeration cycle shown in Fig. P7-10 is to be used to maintain the chest at 0°F. (1) What circulation rate (lb_m/hr) of Freon 12 is required? (2) What is the smallest motor that could be used to drive the pump? (Give the horsepower rating.)

7-11 Water is supplied at 60°F for cooling and for ice manufacture. Ice is to be delivered at 20°F. The cooling water is heated from 60 to 80°F.

(a) Compute the minimum theoretical horsepower required for an ideal reversible operation producing 5 tons of ice per hour under the above conditions.

(b) Draw on the T–$\underline{S}$ diagram the path followed by the water in forming ice and by the cooling water under the above conditions.

In an actual vapor compression plant using ammonia as a refrigerant the conditions are as given in Fig. P7-11.

(c) Compute the horsepower required by the ammonia compressor for producing 5 tons of ice per hour. Assume the mechanical efficiency of the compressor to be 80 per cent and adiabatic compression of the ammonia vapor.

The saturated properties of ammonia are as follows:

		Saturated Liquid		Vapor	
P, psia	T, °F	$\underline{H}$, Btu/lb_m	$\underline{S}$, Btu/lb_m °R	$\underline{H}$, Btu/lb_m	$\underline{S}$, Btu/lb_m °R
23	−10	32.11	0.0738	608.5	1.3558
180	90	143.3	0.2954	432	1.1850
180	248			735.3	1.3558

$$(C_P)_{\text{brine}} = 0.8 \text{ Btu/lb}_m \text{ °F}$$
$$(C_P)_{\text{water}} = 1.0 \text{ Btu/lb}_m \text{ °F}$$
$$(C_P)_{\text{ice}} = 0.5 \text{ Btu/lb}_m \text{ °F}$$

Latent heat ($\Delta\underline{H}$) of melting ice at 32°F = 143.8 Btu/lb_m.

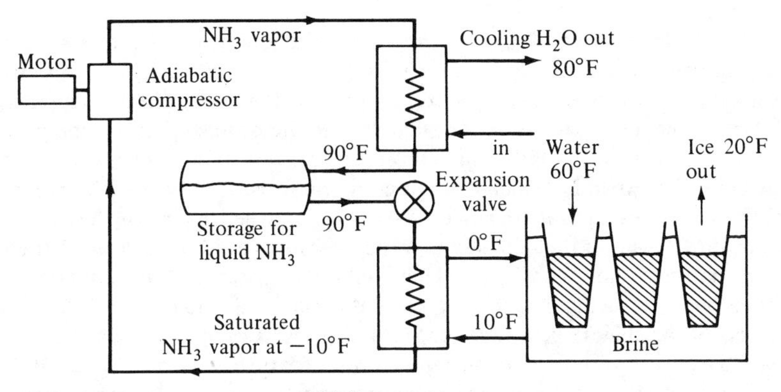

FIGURE P7-11

7-12 A heat pump is being used to maintain a room at 70°F by removing heat from ground water and discharging heat to the room. For the cycle shown in Fig. P7-12,

(a) What is the quality of the fluid leaving the evaporator?

(b) What is the maximum quantity of heat that can be delivered to the room per Btu removed from the groundwater?

(c) If the room is to receive 5750 Btu/hr, what is the minimum horsepower rating of the motor driving the compressor?

For any cycle,

(d) What is the minimum energy that would have to be supplied to pump 5750 Btu into the room using the groundwater as a source?

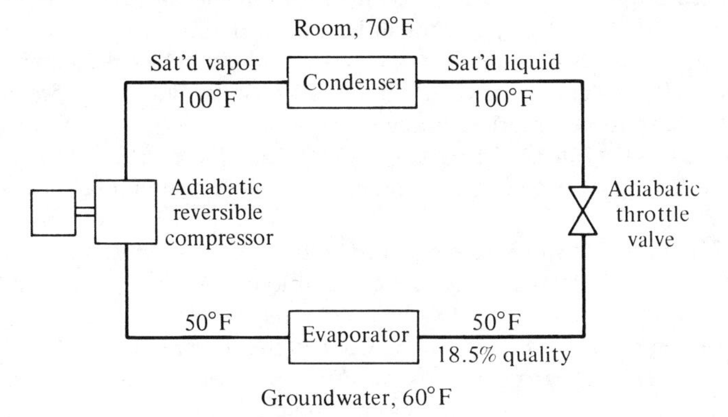

FIGURE P7-12

The properties of the fluid are as follows:

T, °F	P_{sat}, psia	$\underline{S}_{sat\ liq}$ (Btu/lb$_m$ °R)	$\Delta\underline{S}_{vap}$ (Btu/lb$_m$ °R)	$\underline{S}_{sat\ vap}$ (Btu/lb$_m$ °R)
50	81.4	0.0413	0.1266	0.1679
98	128.0	0.0623	0.1036	0.1659
100	131.6	0.0632	0.1026	0.1658
102	135.3	0.0640	0.1018	0.1658

7-13 In a preliminary cooling step of an air liquefaction plant a stream of air at atmospheric pressure and 5°F enters a well-insulated compressor. The compressor discharges the air at 125 psia to a heat exchanger cooled by the ambient air surroundings. The heat exchanger operates with negligible pressure drop so that the compressed air leaves at 125 psia and 105°F. The cooled compressed air is then passed to a well-insulated turbo expander which discharges at atmospheric pressure. The expanded air goes through a second heat exchanger (also isobaric), where it picks up heat from a low-temperature stream in the liquefaction process. The expanded air leaving the second heat exchanger is now at 5°F, in which condition it goes into the compressor to go through the cycle again. The compressor and expander are 80 per cent efficient (based on reversible machines operating over the same pressure ranges). Air may be assumed to behave as an ideal gas with a constant-pressure heat capacity of 6.97 Btu/lb-mol °R. If the work derived from the turbo expander is used to help drive the compressor, determine the net work for the above cycle per unit of heat picked up in the low-temperature heat exchanger. Sketch this cycle on a T–$\underline{S}$ diagram and compare it with a completely reversible cycle performing the same job.

7-14 A conventional refrigerator uses Freon 12, dichlorodifluoromethane, as its working fluid. In the cycle of operations saturated vapor at −10°F enters an insulated compressor whose compression ratio is 9:1 (that is, the outlet pressure is 9 times the inlet pressure). The compressor is 70 per cent efficient based on an adiabatic reversible compression over the same range of pressure. The compressed gas from the compressor is cooled and condensed isobaricly to a saturated liquid. The saturated liquid is passed through an insulated throttle valve whose downstream pressure corresponds to a saturation temperature of −10°F. The mixture from the throttle valve goes to the evaporator, where it absorbs just enough heat from the ice trays and the interior of the refrigerator box to become saturated vapor, which enters the compressor and repeats the cycle. Under summertime conditions (greatest load) it is expected that 110 Btu/min will have to be absorbed in the evaporator.

(a) What is the rate of circulation of Freon 12 in lb_m/min?

(b) How large a motor (rated in terms of its output) should be specified for the compressor?

(c) If the throttle valve were replaced with a turbo expander whose efficiency is 60 per cent (based on adiabatic reversible expansion over same pressure range), and the work from this expander used to help drive the compressor, how large a motor would be needed to supply the remaining work to the compressor?

(d) If a perfect heat pump was operating between the same evaporating and condensing temperatures, how large a motor would it require?

7-15 The refrigeration cycle shown at the left in Fig. P7-15 has been proposed to replace the more conventional cycle at the right. If the circulating fluid is CO_2 and the temperatures entering and leaving the cold box are to be the same for both systems,

(a) Calculate the heat removed from the cold box per pound of CO_2 circulated through the proposed cycle (conventional cycle not operating).

(b) What is the total work input (Btu per pound of CO_2 circulated) to the proposed cycle?

(c) What is the work input (Btu per pound of CO_2 circulated) to the conventional cycle?

(d) Sketch the shape of each cycle on a T–$\underline{S}$ diagram.

(e) Why is one cycle more efficient than the other?

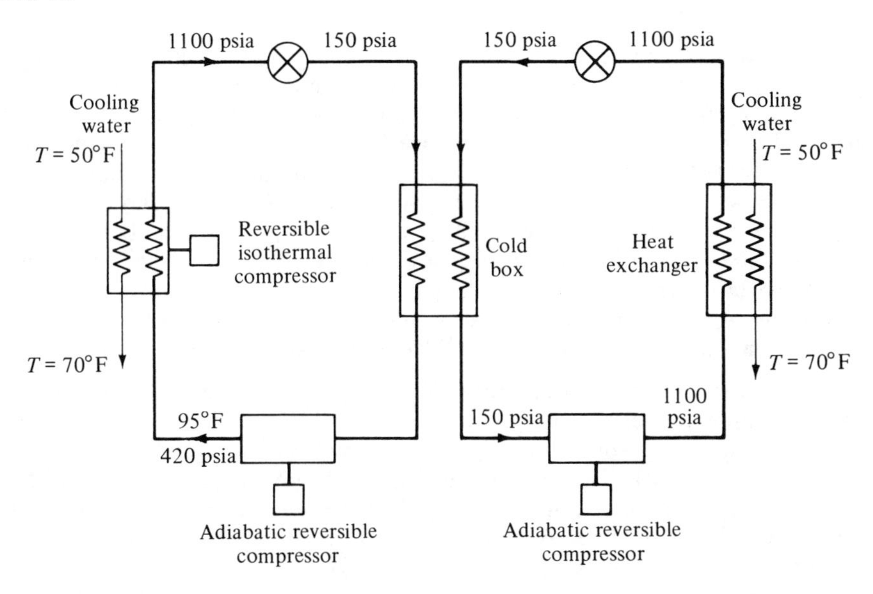

FIGURE P7-15

(f) What is the absolute minimum work of refrigeration for the given conditions entering and leaving the cold box and the given inlet and outlet cooling water temperatures (Btu per pound of CO_2 circulated)?

7-16 For the refrigeration flow diagram of Fig. P7-16, using ammonia, compute the values indicated using

(a) A chart of ammonia properties.

(b) The generalized charts. The compressor may be assumed to be adiabatic and reversible, although an efficiency of 70 to 80 per cent would be reasonably expected for real operating equipment.

$$C_P^* = 6.70 + .0035T(°R) \text{ Btu/lb-mol °R}$$

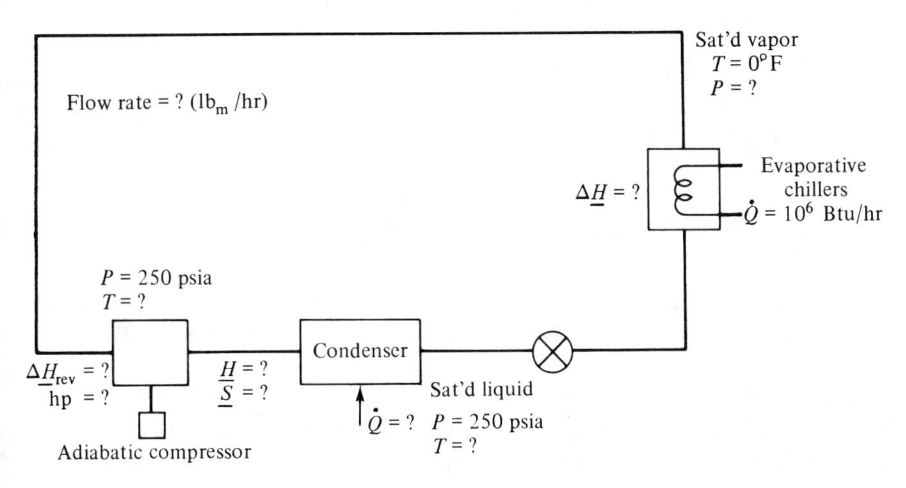

FIGURE P7-16

7-17 Dry ice is to be manufactured by freezing liquid CO_2 (available from another process) at atmospheric pressure. The required cooling is to be provided by a vapor-compression refrigeration cycle that uses Freon 22 as the working fluid. It has been suggested that a two-stage refrigeration cycle (with the condenser in the cold stage giving its heat to the evaporator in the warm stage) will be more efficient than a single-stage cycle. Assuming that heat must be removed from the freezing CO_2 at -130°F and may be rejected at 80°F, compare the coefficients of performance (Btu absorbed from cold source/Btu of work supplied to refrigerator) for a two-stage and a one-stage refrigeration cycle.

(*Note:* Assume isenthalpic expansions for both cycles. You may base your two-stage calculations on a temperature of -20°F in the coupled evaporator–condenser.)

7-18 A refrigerator operates with carbon dioxide as the working fluid. Saturated vapor from the evaporator coils is compressed adiabatically and reversibly from 150 to 800 psia. From the compressor the vapor passes to a condenser, where it becomes a saturated liquid at 800 psia. The liquid is then expanded through a throttle valve to 150 psia, in which condition it enters the evaporator. Use the information below to calculate the work required per Btu of heat picked up at the low temperature in the evaporator.

1. For CO_2,

$$P = \frac{RT}{\underline{V} - b} + \frac{A_2 + B_2T + C_2\exp(-kT)}{(\underline{V} - b)^2} + \frac{A_3 + B_3T + C_3\exp(-kT)}{(\underline{V} - b)^3} + \frac{A_4}{(\underline{V} - b)^4} + \frac{A_5 + B_5T + C_5\exp(-kT)}{(\underline{V} - b)^5}$$

where for P in psia, T in °R, and ρ in lb_m/ft^3, the constants are

$R = 0.24381$ $\quad$ $C_3 = 4.705805$
$b = 0.007495$ $\quad$ $A_4 = -2.112459 \times 10^{-3}$
$A_2 = -8.9273631$ $\quad$ $A_5 = 7.017835 \times 10^{-6}$
$B_2 = 5.262476 \times 10^{-3}$ $\quad$ $B_5 = 1.023511 \times 10^{-8}$
$C_2 = -150.97587$ $\quad$ $C_5 = -4.55437 \times 10^{-4}$
$A_3 = 0.17948621$ $\quad$ $k = 0.01$
$B_3 = -5.770542 \times 10^{-5}$

2.

$$C_V^* = \alpha + \frac{\beta}{T} + \frac{\gamma}{T^2}$$

where for C_V^* in Btu/lb-mol °R and T in °R,

$$\alpha = 14.214$$
$$\beta = -6.53 \times 10^3$$
$$\gamma = 1.41 \times 10^6$$

3. The vapor pressure and saturated liquid volume data may be taken from R.H. Perry et al., *Chemical Engineers' Handbook* (McGraw-Hill Book Company, Inc., New York, 4th ed., 1963).

Check your answer by using the tables of thermodynamic properties for CO_2.

7-19 Carbon dioxide is available at 60°F and 1200 psia. It is proposed to make solid carbon dioxide from it by flash evaporation to 15 psia. If the CO_2 is passed through a valve and the gas allowed to escape freely into the surroundings, and if the solid particles are caught in a bag,

(a) What will be the temperature of the solid?

(b) What fraction of CO_2 will be recoverable as solid?

A $P-\underline{H}$ diagram for CO_2 is available in Appendix D.

7-20 Suppose the Linde process shown in Fig. P7-20 is used in Problem 7-19.

(a) What is the advantage, if any, of circulating the gas from the snow chamber over the coil?

(b) What fraction of the liquid entering the coil shows up as a solid? Assume no extraneous heat losses.

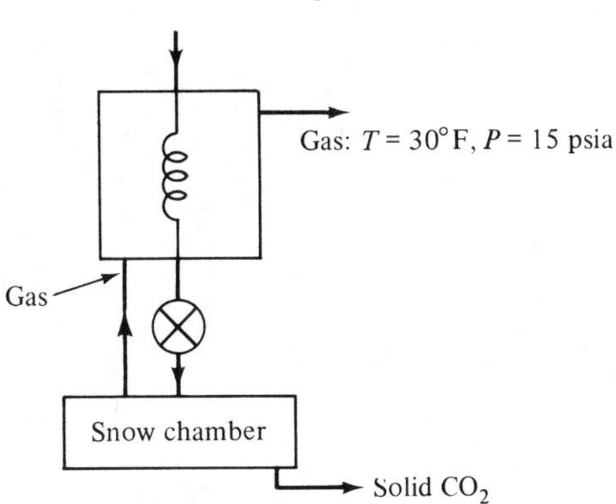

FIGURE P7-20

7-21 A reversible heat pump is used to extract heat from a cool block and transfer it to a warm block. Initially the two blocks are both at the same temperature, 0°C. At the end of the process, the cool block has a temperature of −200°C.

(a) What is the final temperature of the hot block?

(b) How much work has been supplied to the heat pump?

(c) If the heat pump is replaced by a reversible heat engine, how much work can be extracted by letting the engine transfer heat from the hot block to the cool block until the temperatures are equal?

(d) What is the final temperature in (c)?

(e) What is the change in entropy of the universe for each process?

Assume that MC_P for each block = 1.0; all engines, pumps, and heat transfer is reversible.

Thermodynamics of Fluid Flow

8

8.1 The Mechanical Energy Balance

In Chapter 6 we showed how the energy and entropy balances could be combined in such a way as to eliminate all path variables, and thereby we derived the property relations. In this chapter we shall discuss a slightly different combination of the energy and entropy balances which produces the *mechanical energy balance*. This energy balance will then be applied to the study of fluid flow problems, including some interesting and useful applications for the study of compressible flows through pipelines and nozzles.

Derivation of the Mechanical Energy Balance

Let us examine the flow of a fluid through a piece of process equipment such as a pipeline, pump, or expander, as shown in Fig. 8-1. If we choose the piece of equipment and its contents at any instant as a system, the first-law energy balance can be written as

$$dE_{\text{syst}} = \delta Q - \delta W - \Delta[(\underline{H} + \underline{\text{KE}} + \underline{\text{PE}})\cdot \delta M] \tag{8-1}$$

The entropy balance for the system may be written as

$$dS_{\text{syst}} = \int_{\text{volume}} \frac{\delta lw}{T} dV + \int_{\text{surface}} \frac{\delta q}{T} dA - \Delta(\underline{S}\, \delta M) \tag{8-2}$$

For a steady-state process $\Delta(\delta M) = dE_{\text{syst}} = dS_{\text{syst}} = 0$, so equations (8-1)

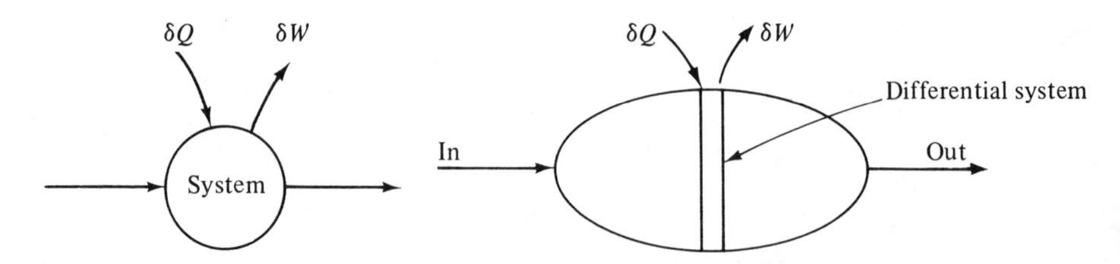

FIG. 8-1. System.

FIG. 8-2. Differential system.

and (8-2) yield, respectively,

$$Q - W - \Delta\left(\underline{H} + \frac{u^2}{2g_c} + \frac{gZ}{g_c}\right)\delta M = 0 \tag{8-3}$$

$$\int_{\text{volume}} \frac{\delta lw}{T} dV + \int_{\text{area}} \frac{dq}{T} dA - (\Delta \underline{S})\,\delta M = 0 \tag{8-4}$$

If we now turn our attention to an infinitesimal portion of the system under investigation as shown in Fig. 8-2, the changes in $\underline{H}$, $\underline{S}$, u, and Z may be represented as differential quantities rather than as differences. (Later we shall integrate the differential equations to obtain the desired difference forms.) The differential forms of equations (8-3) and (8-4) are given by

$$\delta Q - \delta W - d\left(\underline{H} + \frac{u^2}{2g_c} + \frac{gZ}{g_c}\right)\delta M = 0 \tag{8-5}$$

$$\frac{\delta lw}{T} dV + \frac{\delta q}{T} dA = (d\underline{S})\,\delta M \tag{8-6}$$

Since the temperature and δM are constant across the differential system, we may multiply equation (8-6) by $T/\delta M$ to obtain

$$\frac{\delta lw\, dV}{\delta M} + \frac{\delta q\, dA}{\delta M} = T\, d\underline{S} \tag{8-6a}$$

or

$$\underline{lw}\, dV + \underline{q}\, dA = T\, d\underline{S} \tag{8-7}$$

where

$$\underline{lw} = \delta lw/\delta M$$

$$\underline{q} = \delta q/\delta M$$

However, from the definitions of $\underline{lw}$, $\underline{LW}$, $\underline{q}$, and $\underline{Q}$ we see

$$\underline{lw}\, dV = \delta \underline{LW}$$

$$\underline{q}\, dA = \delta \underline{Q}$$

so that equation (8-7) reduces to

$$\delta \underline{LW} + \delta \underline{Q} = T\, d\underline{S} \tag{8-7a}$$

The heat term, $\delta \underline{Q}$, may now be eliminated between equations (8-5) and (8-7a) to give

$$-\delta \underline{LW} - \delta \underline{W} + T\, d\underline{S} - d\left(\underline{H} + \frac{u^2}{2g_c} + \frac{gZ}{g_c}\right) = 0 \tag{8-8}$$

or

$$T\,d\underline{S} - d\underline{H} - \delta\underline{LW} - \delta\underline{W} - d\left(\frac{u^2}{2g_c} + \frac{gZ}{g_c}\right) = 0 \qquad \text{(8-8a)}$$

However, the enthalpy property relation states that

$$d\underline{H} = T\,d\underline{S} + \underline{V}\,dP$$

Therefore,

$$T\,d\underline{S} - d\underline{H} = -\underline{V}\,dP \qquad \text{(8-9)}$$

Equation (8-9) may then be substituted into equation (8-8a) to yield

$$-\delta\underline{LW} - \delta\underline{W} - \underline{V}\,dP - d\left(\frac{u^2}{2g_c} - \frac{gZ}{g_c}\right) = 0 \qquad \text{(8-10)}$$

Equation (8-10) is then integrated across the whole process under consideration (by summing the changes that occur across all the differential elements which comprise the original system) to give

$$-\underline{LW} - \underline{W} - \int_{P_1}^{P_2} \underline{V}\,dP - \Delta\left(\frac{u^2}{2g_c} + \frac{gZ}{g_c}\right) = 0 \qquad \text{(8-11)}$$

Equation (8-11) is the *mechanical energy balance*. The only assumption made in deriving this equation is that the process under consideration is at steady state! (We have also implicitly assumed that the process under consideration occurs through a series of equilibrium states so that the property relation may be used to eliminate $T\,d\underline{S} - d\underline{H}$.)

The work term in the mechanical energy balance is simply the sum of all the works transferred to the differential portions of the system, and, therefore, is the total work transferred from the system to the surroundings. The quantities $\int \underline{V}\,dP$ and $\underline{LW}$, on the other hand, are path functions and must be evaluated over the path followed through the actual process. Thus application of the mechanical energy balance is a very difficult task unless the state of the fluid is known at all times throughout the system.

Applications of the Mechanical Energy Balance

Let us now examine some of the simplifications that are commonly applied to reduce the mechanical energy balance to a more usable form.

Incompressible Fluid Flow: For an incompressible fluid, the specific volume $\underline{V}$ is constant, independent of T and P, so that it may be removed from under the integral sign. That is,

$$\int_{P_1}^{P_2} \underline{V}\,dP = \underline{V}\int_{P_1}^{P_2} dP = \underline{V}\,\Delta P \qquad \text{(8-12)}$$

Since $\underline{V} = 1/\rho$, equation (8-12) becomes

$$\int_{P_1}^{P_2} \underline{V}\,dP = \frac{\Delta P}{\rho} \qquad \text{(8-13)}$$

Equation (8-13) may be substituted into (8-11) to give

$$-\underline{LW} - \delta\underline{W} - \Delta\left(\frac{P}{\rho} + \frac{u^2}{2g_c} + \frac{gZ}{g_c}\right) = 0 \qquad \text{(8-14)}$$

which is *Bernoulli's equation* for one-dimensional fluid flow with work, and lost work. If $\underline{W} = \underline{LW} = 0$, then equation (8-14) reduces to the more familiar form of Bernoulli's equation:

$$\Delta\left(\frac{P}{\rho} + \frac{u^2}{2g_c} + \frac{gZ}{g_c}\right) = 0 \tag{8-15}$$

or

$$\frac{P}{\rho} + \frac{u^2}{2g_c} + \frac{gZ}{g_c} = \text{const} \tag{8-16}$$

SAMPLE PROBLEM 8-1. A pump is taking 60 gal/min of water at 60°F from an open tank and delivering it to nozzles at the top of a spray tower that is 60 ft high (see Fig. S8-1). The pump motor produces 6 hp, and the pump has a

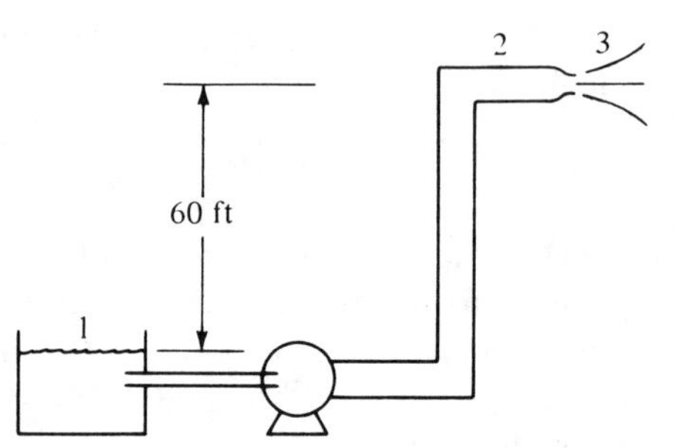

FIGURE S8-1

mechanical efficiency of 75 per cent. The water flows through a 3-in.-i.d. pipe, and losses due to friction in the pipe are estimated to be 10 ft-lb_f/lb_m. If a pressure gauge was installed at the inlet of the nozzle, what would be the reading on the gauge? If there was a frictional loss across the nozzle of 5 ft-lb_f/lb_m of water entering the nozzle, what would be the velocity of the water at the nozzle outlet?

Solution: For the first question, we take as a system the tank, the pump, and all the piping up to, but not including, the nozzle. That is everything between points 1 and 2 on the diagram. A mechanical energy balance may be written around this system as follows:

$$-\int_{P_1}^{P_2} \underline{V}\, dP - \left(\Delta\frac{u^2}{2g_c} + \frac{g}{g_c}\Delta Z\right) - \underline{LW} - \underline{W} = 0$$

In the problem statement we are told that the pump motor delivers 6 hp to the pump. Therefore,

$$\dot{W} = -6 \text{ hp} \times 550 \frac{\text{ft-lb}_f}{\text{sec hp}} = -3300 \text{ ft-lb}_f/\text{sec}$$

is removed from the system. But

$$\underline{W} = \frac{\dot{W}}{\dot{M}}$$

where

$$\dot{M} = \text{mass flow rate} = 60\frac{\text{gal}}{\text{min}} \times 8.35\frac{\text{lb}_m}{\text{gal}} \times \frac{1 \text{ min}}{60 \text{ sec}}$$
$$= 8.35 \text{ lb}_m/\text{sec}$$

Therefore,

$$\underline{W} = \frac{-3300}{8.35} = -396 \text{ ft-lb}_f/\text{lb}_m$$

However, of this, 25 per cent, or 99 $\text{ft-lb}_f/\text{lb}_m$, is degraded to lost work in the pump. Another 10 $\text{ft-lb}_f/\text{lb}_m$ is lost in the piping. Therefore,

$$\underline{LW} = 10 + 99 = 109 \text{ ft-lb}_f/\text{lb}_m$$

At point 1,

$$P = \text{atmospheric pressure} = 0 \text{ psig}$$

$$u = 0$$

$$Z = \text{reference height} = 0$$

at point 2,

$$P = \text{some pressure } P_2\text{, psig}$$

$$Z = 60 \text{ ft above point 1}$$

$$u = \frac{\text{volumetric flow rate}}{\text{area of 3-in.-i.d. pipe}}$$

or

$$u = 60 \frac{\text{gal}}{\text{min}} \times \frac{1 \text{ min}}{60 \text{ sec}} \times \frac{\text{ft}^3}{7.48 \text{ gal}} \times \frac{1}{(3/12)^2 \cdot \pi/4}$$

$$= \frac{64}{7.48 \times \pi} \text{ft/sec} = 2.72 \text{ ft/sec}$$

and the mechanical energy balance reduces to

$$\int_{P_1}^{P_2} \underline{V}\, dP = -\left[\frac{u_2^2 - u_1^2}{2g_c} + \frac{g}{g_c}(Z_2 - Z_1)\right] - \underline{LW} - \underline{W}$$

$$= -\left[\frac{(2.72)^2}{2g_c} + \frac{g}{g_c}(60 - 0)\right] - 109 \text{ ft-lb}_f/\text{lb}_m + 596 \text{ ft-lb}_f/\text{lb}_m$$

or, neglecting u_2^2,

$$\int_{P_1}^{P_2} \underline{V}\, dP = -60 - 109 + 596 \text{ ft-lb}_f/\text{lb}_m = 227 \text{ ft-lb}_f/\text{lb}_m$$

But liquid water is essentially incompressible. Therefore, $\underline{V}$ = const and

$$\int_{P_1}^{P_2} \underline{V}\, dP = \underline{V} \int_{P_1}^{P_2} dP = \underline{V}(P_2 - P_1)$$

But $P_1 = 0$ psig. Therefore,

$$\underline{V}P_2 = 227 \text{ ft-lb}_f/\text{lb}_m$$

$$\underline{V} = 62.4 \text{ ft}^3/\text{lb}_m$$

Therefore, $P_2 = 227(62.4) \text{ lb}_f/\text{ft}^2 = 1.415 \times 10^4 \text{ lb}_f/\text{ft}^2 = 98.2$ psig.

Now choose the nozzle as the system:

$$\underline{W} = 0$$

$$\underline{LW} = 5 \text{ lb}_f\text{-ft}/\text{lb}_m$$

$$\int_{P_2}^{P_3} \underline{V}\, dP = \underline{V}(P_3 - P_2)$$

But $P_3 = 0$ psig. Therefore,

$$\int_{P_2}^{P_3} \underline{V}\, dP = 227 \text{ ft-lb}_f/\text{lb}_m$$

Application of the mechanical energy balance around the nozzle then gives

$$\frac{\Delta u^2}{2g_c} = -\frac{\Delta P}{\rho} - \underline{LW} = [-(-227) - 5] \text{ ft-lb}_f/\text{lb}_m$$
$$= +222 \text{ ft-lb}_f/\text{lb}_m$$

Therefore,

$$\Delta u^2 = u_3^2 - u_2^2 = 1.43 \times 10^4 \text{ ft}^2/\text{sec}^2$$

But $u_2^2 = 7.4 \text{ ft}^2/\text{sec}^2$. Therefore,

$$u_3 = 1.2 \times 10^2 \text{ ft/sec}$$

which is the velocity of the fluid leaving the nozzle tip.

Evaluation of the Lost Work for Flow Through Pipes and Fittings

For horizontal flow of an incompressible liquid through a constant area conduit, the mechanical energy balance reduces to

$$\underline{LW} = \frac{-\Delta P}{\rho} \tag{8-17}$$

That is, the whole pressure drop is associated with the irreversibilities of friction. For flow through circular cylinders, the frictional pressure drop may be expressed in terms of the Fanning friction factor:

$$f = \frac{1}{2}\frac{Dg_c(-\Delta P)}{\rho u^2 L} \tag{8-18}$$

or

$$\frac{-\Delta P}{\rho} = \frac{2u^2 f L}{D g_c} \tag{8-18a}$$

Thus the lost work is expressible in terms of the friction factor, fluid velocity, and the L/D ratio of the pipe:

$$\underline{LW} = -\frac{\Delta P}{\rho} = \frac{2u^2 fL}{Dg_c} \tag{8-19}$$

By means of the technique of dimensional analysis,[1] it may be shown that the friction factor is a function of the Reynolds number ($N_{Re} = Du\rho/\mu$) of the flow, the ratio of pipe roughness to pipe diameter (ϵ/D), and pipe L/D ratio. That is,

$$f = \Phi\left[N_{Re}, \frac{\epsilon}{D}, \frac{L}{D}\right] \tag{8-20}$$

For most problems of engineering significance, the L/D ratio is not an

[1] R. B. Bird, W. E. Stewart, and E. N. Lightfoot, *Transport Phenomena*, John Wiley & Sons, Inc., New York, 1960.

important parameter (that is, the flow is fully developed), and the functional dependence of equation (8-20) reduces to

$$f = \Phi\left[N_{\text{Re}}, \frac{\epsilon}{D}\right] \tag{8-20a}$$

In Fig. 8-3 the Fanning friction factor is given as a function of the Reynolds number and ϵ/D ratio. Because of the difficulty in accurately describing the irreversibilities in a flowing fluid, the information in Fig. 8-3 has been determined experimentally.

Values of ϵ for several commonly used piping materials are listed in Table 8-1.

TABLE 8-1 Roughness of Commonly Used Piping Materials†

Material	Roughness, ft	Material	Roughness, ft
Riveted steel	0.003–0.03	Concrete	0.001–0.01
Cast iron	0.00085	Galvanized iron	0.0005
Commerical steel	0.00015	Wrought iron	0.00015
Drawn tubing	0.000005		

† From S. Whitaker, *Introduction to Fluid Mechanics*, Prentice-Hall, Inc., Englewood Cliffs, N. J., 1968, p. 299.

For noncircular ducts of constant cross section, equation (8-19) is still commonly used to relate the lost work to the friction factor. The pipe diameter, D, is replaced by the "hydraulic diameter" D_h:

$$D_h = \frac{4\,(\text{conduit area})}{\text{wetted perimeter}} \tag{8-21}$$

The friction factor is evaluated from Fig. 8-3 using a Reynolds number based on the hydraulic diameter. (Note that for a conduit of circular cross section the hydraulic and actual diameters are identical.)

Equation (8-19) is useful for relating the lost work to the fluid velocity, pipe L/D ratio, and the Fanning friction factor for flows where the fluid velocity remains constant. If the fluid velocity is changing with position down the pipe (either because the pipe area is changing or because of compressibility effects), we must use a differential form of equation (8-19):

$$\delta \underline{LW} = \frac{2u^2 f}{D g_c}\, dL \tag{8-22}$$

and integrate over the length of the pipeline:

$$\underline{LW} = \int_0^L \frac{2u^2 f}{D g_c}\, dL \tag{8-22a}$$

If the fluid velocity is changing gradually as the fluid moves down the pipeline, we can assume that the local velocity profiles are fully developed, so

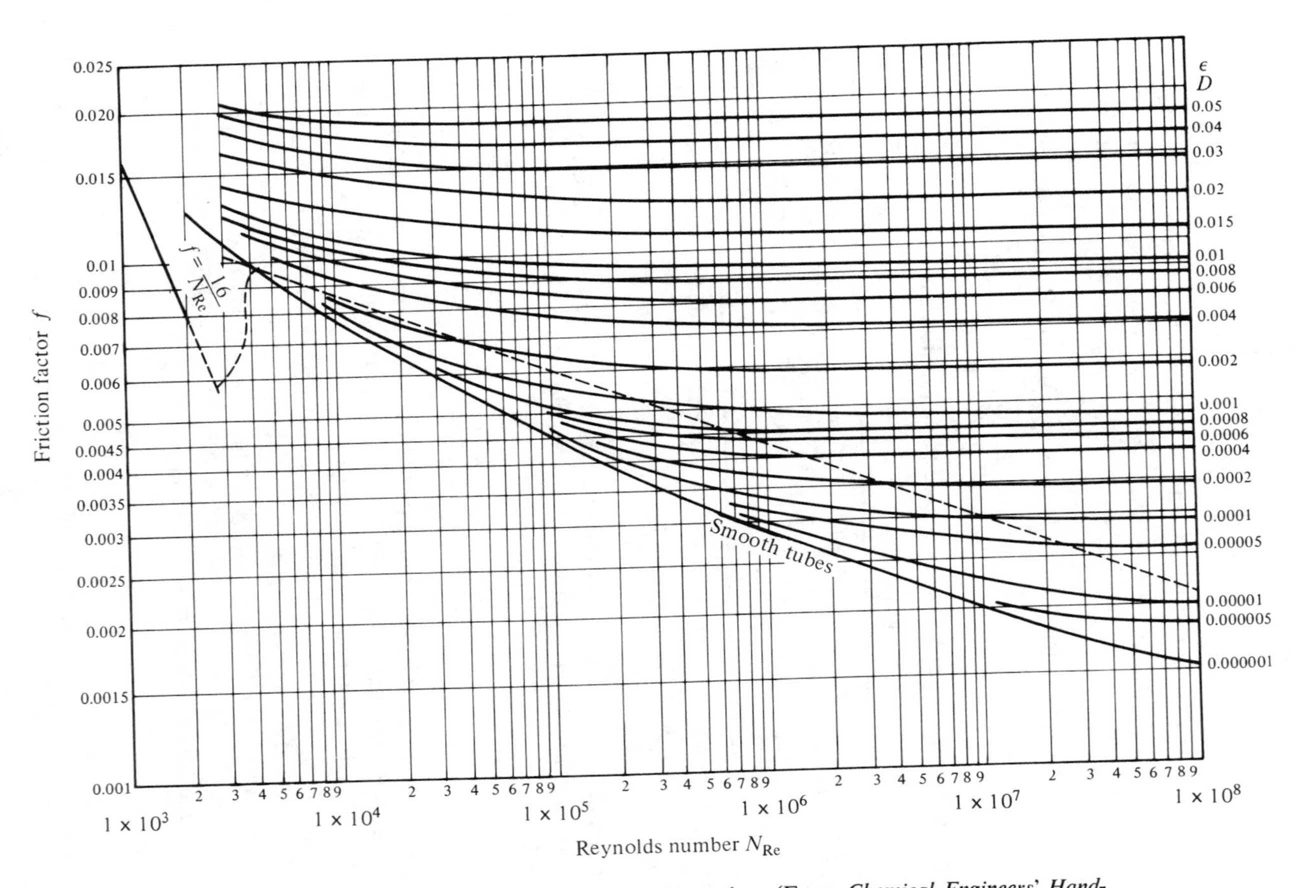

FIG. 8-3. Fanning Friction factor vs. Reynolds number. (From *Chemical Engineers' Handbook*, R. H. Perry et al. eds. McGraw-Hill, Inc., New York, 1963, pg. 20.)

the friction factor–Reynolds number relation portrayed in Fig. 8-3 may be used to express the variation of f with the fluid velocity and fluid properties. For many problems of practical interest, the Reynolds numbers are quite high, say greater than 50,000. In this region Fig. 8-3 shows that the friction factor is reasonably insensitive to changes in the Reynolds number—particularly for commercial pipes with the larger roughness. Therefore, we may often remove the friction factor from under the integral sign in equation (8-22a) and simplify the treatment significantly.

The lost work associated with fittings and valves is usually expressed in terms of the "loss coefficient," k, of the valve or fitting:

$$\underline{LW} = \frac{ku^2}{2g_c} = \left(\frac{-\Delta P}{\rho}\right)_{\text{valve or fitting}} \tag{8-23}$$

For many flow situations of interest (that is, Reynolds numbers greater than 20,000) it is observed that the loss coefficient defined by equation (8-23) is reasonably constant, independent of both diameter and Reynolds number. A list of commonly used loss coefficients for valves and fittings is presented as Table 8-2.

TABLE 8-2 Loss Coefficients for Various Valves and Fittings†

Type of Fitting or Valve	Loss Coefficient, k	Type of Fitting or Valve	Loss Coefficient, k
45° ell		Union	0.04
standard	0.35	Gate valve	
long radius	0.20	open	0.20
90° ell		$\frac{3}{4}$ open	0.90
standard	0.75	$\frac{1}{2}$ open	4.5
long radius	0.45	$\frac{1}{4}$ open	24.0
square or miter	1.3	Diaphragm valve	
180° bend, close return	1.5	open	2.3
Tee, standard		$\frac{3}{4}$ open	2.6
along run, branch blanked off	0.4	$\frac{1}{2}$ open	4.3
used as ell, entering run	1.3	$\frac{1}{4}$ open	21.0
used as ell, entering branch	1.5	Globe valve	
branching flow	~1	open	6.4
Coupling	0.04	$\frac{1}{2}$ open	9.5

† From R. H. Perry et al., *Chemical Engineers' Handbook*, McGraw-Hill, Inc., New York, 4th ed., 1963, Sec. 5. p. 33.

It is interesting to note that equation (8-19), which describes the frictional losses in straight pipe, may also be arranged in terms of a "pseudo loss coefficient":

$$\underline{LW} = \frac{4fL}{D}\frac{u^2}{2g_c} \tag{8-19}$$

where the pseudo loss coefficient is

$$k_{\text{std pipe}} = \frac{4fL}{D} \tag{8-24}$$

In the preceding development we have tacitly assumed that frictional effects were the sole causes of any pressure changes that may have occurred. In this way we were able to express the lost work as $\Delta P/\rho$. The pressure drop was then expressed in terms of the Fanning friction factor or loss coefficient. In situations where pressure changes due to other than frictional effects are present, we shall assume that the lost-work term can still be evaluated as if no other effects were present. That is, we assume that equations (8-19) and (8-23) still apply and that the values of friction factors, or loss coefficients, are unchanged.

SAMPLE PROBLEM 8-2. A large chemical plant requires cooling water at a rate of 1500 gal/min for use in its reactors and condensers. The cooling water will be taken from the municipal water supply and is available at 35 psig. Upon entering the plant, the water is pumped to high pressure and then passed through the plant pipeline. At the end of the pipeline the water is fed to the boiler house. The boiler house requires that the water be available at a pressure greater than 10 psia (to avoid cavitation in the boiler feed pumps). The pipeline comprises two segments, each 2500 ft long, as shown in Fig. S8-2a. In addition to the

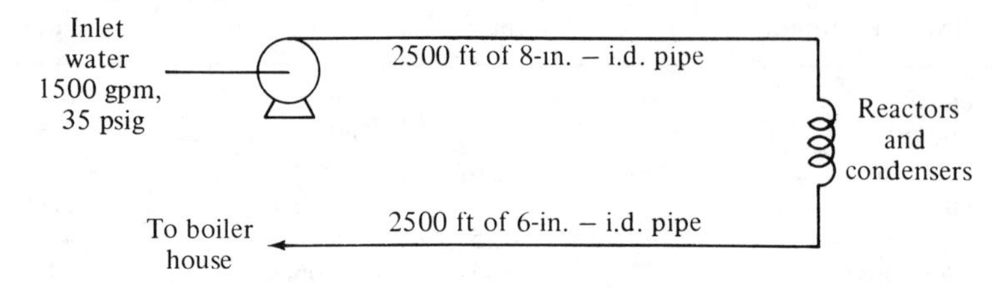

FIGURE S8-2a

straight section, the following valves and fitting are present: 10 6-in. and 10 8-in. elbows (90°); 5 6-in. and 5 8-in. gate valves; and 2 6-in. and 1 8-in. globe valves.

(a) Find the minimum-sized motor that can run the pump (hp).

(b) If pump and motor have an overall efficiency of 70 percent, what is the actual motor size (hp)? If 440-V electric service is available, what current will be drawn by the motor?

(c) What is the pressure at the outlet of the pump? At the reactors and condensers?

Solution: Let us begin by calculating the velocities and Reynolds numbers for both sections of pipe:

$$Q = \text{volumetric flow rate} = 1500 \text{ gal/min} = 200 \text{ ft}^3/\text{min}$$
$$= 3.33 \text{ ft}^3/\text{sec}$$

But

$$\dot{M} = Q\rho = (3.33 \text{ ft}^3/\text{sec})(62.4 \text{ lb}_m/\text{ft}^3) = 206 \text{ lb}_m/\text{sec}$$

For the 8-in. line:

$$A = \frac{\pi D^2}{4} = \frac{\pi}{4}\left(\frac{8}{12}\right)^2 = 0.35 \text{ ft}^2$$

but

$$u = \frac{Q}{A} = \frac{3.33 \text{ ft}^3/\text{sec}}{0.35 \text{ ft}^2} = 9.5 \text{ ft/sec}$$

The Reynolds number is defined by

$$N_{\text{Re}} = \frac{Du\rho}{\mu} = \frac{Du}{\nu}$$

But $\nu = 8.0 \times 10^{-5}$ ft²/sec for H_2O at 100°F. Thus

$$N_{\text{Re}} = \frac{(8/12)(9.5)(\text{ft})(\text{ft/sec})}{8.0 \times 10^{-5}\text{ft}^2/\text{sec}} = 7.8 \times 10^5$$

If we assume that the pipeline is made of commercial steel, Table 8-1 tells us that $\epsilon =$ 0.00015 ft, so

$$\frac{\epsilon}{D} = \frac{(12)(0.00015)}{8} = 0.0002$$

From ϵ/D and N_{Re}, Fig. 8-3 gives us the Fanning friction factor:

$$f_{8\text{-in. line}} = 0.00375$$

Finally, the term $u^2/2g_c$ has the value

$$\frac{u^2}{2g_c} = \frac{9.5}{(2)(32.17)} \frac{(\text{ft/sec})^2}{(\text{lb}_m/\text{lb}_f)(\text{ft/sec}^2)} = 1.40 \text{ ft-lb}_f/\text{lb}_m$$

For the 6-in. line:

$$A = \frac{\pi D^2}{4} = \frac{\pi(\frac{1}{2})^2}{4} = 0.196 \text{ ft}^2$$

$$u = \frac{3.33}{0.196} \text{ ft}^2 = 17.0 \text{ ft/sec}$$

$$N_{\text{Re}} = \frac{Du}{\nu} = \frac{(\frac{1}{2})(17.0)}{8.0 \times 10^{-5}} \frac{(\text{ft})(\text{ft/sec})}{(\text{ft}^2/\text{sec})} = 1.06 \times 10^6$$

$$\frac{\epsilon}{D} = 0.0002$$

$$f = 0.00375$$

$$\frac{u^2}{2g_c} = \frac{(17.0)^2}{(2)(32.17)} \frac{\text{ft/sec}^2}{(\text{lb}_m/\text{lb}_f)(\text{ft/sec}^2)} = 4.5 \text{ ft-lb}_f/\text{lb}_m$$

Now let us add up the total lost work associated with the 8- and 6-in. sections of pipe.

$$\underline{LW} = \frac{u^2}{2g_c} \sum k$$

where k = loss coefficient for various elements in pipeline.

For the 8-in. line:

For the straight pipe $k = 4fL/D$.

$$k_{8\text{-in. pipe}} = \frac{(4)(0.000375)(2500)}{8/12} = 56.4$$

k's for the valves and fittings:

$$\begin{aligned} 10\ 90^\circ \text{ elbows} &= (10)(0.75) &= 7.5 \\ 5 \text{ gate valves} &= (5)(0.2) &= 1.0 \\ 1 \text{ globe valve} & &= 6.4 \\ \text{total } k \text{ for 8-in. pipeline} & &= 71.3 \end{aligned}$$

and the lost work in the 8-in. line is

$$\underline{LW} = \frac{u^2}{2g_c}k = (1.40)(71.3)\ \text{ft-lb}_f/\text{lb}_m$$

or

$$\underline{LW} = 100\frac{\text{ft-lb}_f}{\text{lb}_m}$$

For the 6-in. line:

$$k_{\text{pipe}} = \frac{(4)(0.00375)(2500)}{0.5} = 75.0$$

Valves and fittings:

$$\begin{aligned} 10\ 90^\circ \text{ elbows} &= (10)(0.75) &= 7.5 \\ 5 \text{ gate valves} &= (5)(0.2) &= 1.0 \\ 2 \text{ globe valves} &= (2)(6.4) &= 12.8 \\ \text{total } k & &= 96.3 \end{aligned}$$

and the lost work in this section is

$$\underline{LW} = (96.3)(4.5)\ \text{ft-lb}_f/\text{lb}_m = 433\ \text{ft-lb}_f/\text{lb}_m$$

The total lost work in the circulation system is the sum of the lost work in the 8- and 6-in. sections:

$$\underline{LW}_{\text{total}} = (100 + 433)\ \text{ft-lb}_f/\text{lb}_m = 533\ \text{ft-lb}_f/\text{lb}_m$$

(a) We may now determine the work requirements of the pump by applying the mechanical energy balance around the whole pipeline (Fig. S8-2b):

$$-\underline{LW} - \underline{W} - \Delta\left(\frac{P}{\rho} + \frac{u^2}{2g_c} + \frac{gZ}{g_c}\right) = 0$$

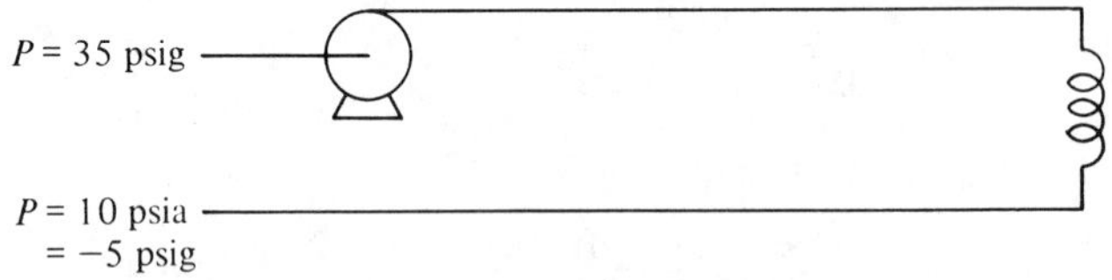

FIGURE S8-2b

Assume that $\Delta gZ/g_c = 0$; we have seen that $\Delta u^2/2g_c$ is quite small, so neglect it. Thus the mechanical energy balance reduces to

$$\underline{W} = -\underline{LW} - \Delta\left(\frac{P}{\rho}\right)$$

But

$$\underline{LW} = 533\ \text{ft-lb}_f/\text{lb}_m$$

$$\frac{\Delta P}{\rho} = \frac{P_2 - P_1}{\rho} = \frac{-5 - 35}{\rho}\ \text{psi} = -\frac{40}{\rho}\ \text{psi}$$

Since $\rho = 62.4\ \text{lb}_m/\text{ft}^3$,

$$\frac{\Delta P}{\rho} = -\frac{(40)(144)}{62.4}\ \frac{\text{lb}_f/\text{ft}^2}{\text{lb}_m/\text{ft}^3} = -92\ \text{ft-lb}_f/\text{lb}_m$$

(with the minus sign indicating that work must be supplied to the pump!)

The mass flow rate $\dot{M} = 206\ \text{lb}_m/\text{sec}$, so the rate of work supplied is

$$\dot{W} = \underline{W}(\dot{M}) = (-441)(206)\ \text{ft-lb}_f/\text{sec}$$
$$= 9.1 \times 10^4\ \text{ft-lb}_f/\text{sec} = 165\ \text{hp}$$

for an ideal pump.

(b) If the efficiency of the pump is 70 per cent, then the actual work supplied must be

$$\dot{W}_{\text{act}} = \frac{\dot{W}_{\text{ideal}}}{\text{eff}} = \frac{165}{0.7} = 235\ \text{hp} = 174\ \text{kW}$$

so the electrical consumption will be 174 kW. If 440-V electrical service is available, the amperage is

$$\text{current} = \frac{\dot{W}}{V} = \frac{174{,}000}{440} = 395\ \text{A!}$$

(c) We may determine the pressure at the pump outlet by applying the mechanical energy balance around the pump. We use the ideal work requirement, neglecting any lost work in the pump. Also assume that $\Delta gZ/g_c = \Delta u^2/2g_c = 0$, so the mechanical energy balance reduces to

$$\underline{W} = -\frac{\Delta P}{\rho} = -441\ \text{lb}_f\text{-ft}/\text{lb}_m$$

Thus

$$\Delta P = (62.4)(441)\ \text{lb}_f/\text{ft}^2 = 2.75 \times 10^4\ \text{lb}_f/\text{ft}^2$$

or

$$\Delta P = 191\ \text{psi}$$

But $P_{\text{in}} = 35$ psig, so $P_{\text{outlet}} = 226$ psig.

Compressible Flows: In the study of compressible flows a relation between $\underline{V}$ and P is necessary before $\int \underline{V}\,dP$ can be evaluated. An equation of state for the fluid provides a relation among pressure, temperature, and volume. Thus if the temperature can be related to the pressure, the desired relationship between P and $\underline{V}$ can be developed and the integral evaluated. For the major portion of the work in this book the ideal gas equation of state will be used, although from time to time the effects of nonideal behavior will be discussed.

Two limiting examples of compressible flows will be studied. Most real flows fall some place between the two, but the limiting conditions serve to set extremes between which all real flows fall.

In certain limited cases, it is possible to have a compressible flow in which the temperature remains constant. For example, consider the flow of natural gas in an underwater section of pipeline where the surrounding water keeps the fluid at a more-or-less constant temperature. For such constant temperature flows $\underline{V}$ may be evaluated directly as a function of P from the equation of state. The velocity, u, may then be directly related to the density through the equation of conservation of mass:

$$\rho u A = \dot{M} = \text{const} \tag{8-25}$$

where A is the flow area, ρ the density of the flowing fluid, and $\dot{M}$ the mass flow rate.

In this manner, the velocity can be expressed as a function of pressure and/or area, so equation (8-10) can be integrated for many problems of interest.

SAMPLE PROBLEM 8-3. The differential form of the mechanical energy balance for the flow of a compressible fluid through a pipeline may be written as

$$\frac{u\,du}{g_c} + \frac{dP}{\rho} + \delta \underline{LW} = 0$$

The lost-work term is evaluated from the equation (8-22):

$$\delta \underline{LW} = \frac{2f}{D}\frac{u^2}{g_c}\,dL$$

where

ρ = fluid density
D = pipe diameter
dL = differential length of pipe
P = pressure
f = friction factor, assumed independent of N_{Re}
u = mean fluid velocity

(a) By proper manipulation and integration of the mechanical energy balance, derive the relation

$$\int_{P_1}^{P_2} \rho\,dP + \frac{G^2}{g_c}\ln\frac{\rho_1}{\rho_2} + \frac{2f}{g_c}\frac{L}{D}G^2 = 0$$

where $G = \rho u$ [=] $\mathrm{lb_m/ft^2\,hr}$.

(b) Ethylene is to be pumped along a 6-in.-i.d. pipe a distance of 5 miles at a mass flow rate of 2 $\mathrm{lb_m/sec}$. The pressure at the end of the pipeline is to be 2 atm, and the flow may be assumed to be isothermal with $T = 60°\mathrm{F}$. If the friction factor may be assumed constant at $f = 0.003$ and ethylene behaves as an ideal gas, determine the required inlet pressure.

Solution: (a) Substitution of the friction-factor equation for the lost-work term in the differential mechanical energy balance yields the relation

$$\frac{dP}{\rho} + \frac{u\,du}{g_c} + \frac{2fu^2}{Dg_c}\,dL = 0$$

Since the velocity, u, changes with position in the pipeline, evaluation of the integral of $(2fu^2/D)\,dL$ requires prior knowledge of the relation between u and L. Therefore, it is to our advantage to eliminate u^2 from the $(2fu^2/D)\,dL$ term before attempting the integration. This is accomplished by dividing the whole equation by u^2 to form

$$\frac{dP}{\rho u^2} + \frac{du}{ug_c} + \frac{2f}{Dg_c}\,dL = 0$$

We may relate u to ρ through the equation of continuity:

$$\rho u = G = \text{const}$$

Therefore, $u = G/\rho$ and

$$du = d\frac{G}{\rho} = G\left(d\frac{1}{\rho}\right) = \frac{-G}{\rho^2}\, d\rho$$

which is substituted above. Multiplication by G^2 and integration over the pipeline then yields the desired relation:

$$\int_{P_1}^{P_2} \rho\, dP + \frac{G^2}{g_c} \ln \frac{\rho_1}{\rho_2} + \frac{2fG^2L}{Dg_c} = 0$$

which is true for all gases and thermal boundary conditions.

(b) If the gas flowing through the pipeline is an ideal gas, the pressure, density, and temperature are related by the ideal gas equation of state:

$$\frac{P}{\rho} = \frac{RT}{M}$$

or

$$\rho = \frac{PM}{RT}$$

where R = ideal gas constant
M = molecular weight (M has been included here in anticipation of expressing ρ in terms of mass rather than moles)

If the flow through the pipeline is also isothermal, then $\rho_1/\rho_2 = P_1/P_2$ and $\int_{P_1}^{P_2} \rho\, dP$ may be integrated to yield

$$\int_{P_1}^{P_2} \rho\, dP = \int_{P_1}^{P_2} \frac{PM}{RT}\, dP = \frac{M}{2RT}(P_2^2 - P_1^2)$$

These relations may be substituted into the previously derived equation to give

$$\frac{g_c}{G^2} \frac{0.5M}{RT}(P_2^2 - P_1^2) + \ln \frac{P_1}{P_2} + \frac{2fL}{D} = 0$$

from which it is possible to relate P to L for any isothermal pipeline flow of an ideal gas. From the problem statement, we are told that

$$P_2 = 2 \text{ atm} = 4240 \text{ lb}_f/\text{ft}^2$$

$$G = \frac{2 \text{ lb}_m/\text{sec}}{\text{pipe area}} = \frac{2 \text{ lb}_m/\text{sec}}{\pi D^2/4} = 10.2 \frac{\text{lb}_m}{\text{ft}^2 \text{ sec}}$$

$$f = 0.003$$

$$L = 5 \text{ miles} = (5)(5280) \text{ ft}$$

$$D = 6 \text{ in.} = 0.5 \text{ ft}$$

$$T = 60°\text{F} = 520°\text{R}$$

The ideal gas constant is

$$R' = \frac{R}{M} = \frac{1545}{M} \frac{\text{lb}_f\text{-ft}}{\text{lb}_m °\text{R}} = \frac{1545}{28} \frac{\text{lb}_f\text{-ft}}{\text{lb}_m °\text{R}}$$

Therefore,

$$\frac{g_c 0.5}{G^2 R'T} = \frac{(0.5)(32.17)(\text{lb}_m/\text{lb}_f)(\text{ft}/\text{sec}^2)}{(10.2)^2(1545/28)(520)(\text{lb}_m/\text{ft}^2 \text{ sec})^2(\text{lb}_f\text{-ft } °\text{R}/\text{lb}_m °\text{R})}$$

$$= 5.6 \times 10^{-7} \text{ ft}^4/(\text{lb}_f)^2$$

Thus the energy balance becomes

$$5.6 \times 10^{-7} \frac{\text{ft}^4}{(\text{lb}_f)^2}\left[\left(4.24 \times 10^3 \frac{\text{lb}_f}{\text{ft}^2}\right)^2 - P_1^2\right] + \ln\left(\frac{P_1}{4.24 \times 10^3 \text{lb}_f}\frac{\text{ft}^2}{}\right)$$
$$+ \frac{(2)(0.003)(5)(5280)}{0.50} = 0$$

or, solving for P_1^2,

$$P_1^2 = 5.84 \times 10^8 \frac{(\text{lb}_f)^2}{\text{ft}^4} + 1.78 \times 10^6 \frac{(\text{lb}_f)^2}{\text{ft}^4} \ln\left(\frac{P_1}{4.24 \times 10^3 \text{lb}_f}\frac{\text{ft}^2}{}\right)$$

Solution by trial and error for P_1 yields

$$P_1 = 24{,}250 \text{ lb}_f/\text{ft}^2 = 11.4 \text{ atm}$$

A careful examination of the equation relating G, P_2, and L for isothermal flow of an ideal gas,

$$\frac{1}{2}\frac{g_c}{G^2 R' T}(P_2^2 - P_1^2) - \ln\frac{P_2}{P_1} + \frac{2fL}{D} = 0 \tag{8-26}$$

derived in Sample Problem 8-3 can lead us to some rather startling conclusions. Rearrange equation (8-26) and solve for G, the mass flow rate per unit area of duct, to obtain

$$G^2 = \frac{g_c}{2R'T}\left[\frac{P_2^2 - P_1^2}{\ln(P_2/P_1) - (2fL/D)}\right] \tag{8-27}$$

or

$$G^2 = \frac{P_1^2 g_c}{2R'T}\left[\frac{1 - (P_2/P_1)^2}{(2fL/D) - \ln(P_2/P_1)}\right] \tag{8-27a}$$

Now suppose we fix the $2fL/D$ term in equation (8-27a) and examine the variation of G^2 with P_2. As P_2 approaches P_1, the $P_1^2 - P_2^2$ term goes to zero, and so does G. This is not suprising—no pressure difference, no flow. But as P_2 approaches zero, we find that the $\ln(P_1/P_2)$ term in the denominator becomes infinite and G_2 approaches zero again. Thus according to equation (8-27a), the G–P_2 curve should appear as shown in Fig. 8-4. The pressure corresponding to the maximum is termed the critical pressure, P_c. (Here and elsewhere in this chapter, P_c is to be distinguished from the pressure used in the evaluation of reduced properties.)

Our physical intuition tells us clearly something is amiss. Decreasing P_2 cannot lead to a decrease in the mass flow rate. As we will soon show (at least for the case of compressible flows through nozzles), the critical pressure is the minimum pressure that can exist *within* the pipeline. If the external pressure is less than P_c an expansion wave will develop at the exit of the pipeline. In the expansion wave the pressure drops almost discontinuously from P_c to the external pressure. Thus the actual curve of mass flow rate, G, versus pressure ratio, P_2/P_1, is as shown in Fig. 8-5.

We may determine the critical pressure ratio (P_c/P_1) by setting the derivative of G (or G^2) with respect to P_2/P_1 equal to zero:

$$\frac{d(G^2)}{d(P_2/P_1)} = \frac{d}{d(P_2/P_1)}\left[\left(\frac{g_c P_1^2}{2R'T}\right)\frac{1 - (P_2/P_1)^2}{(2fL/D) - \ln(P_2/P_1)}\right] = 0 \tag{8-28}$$

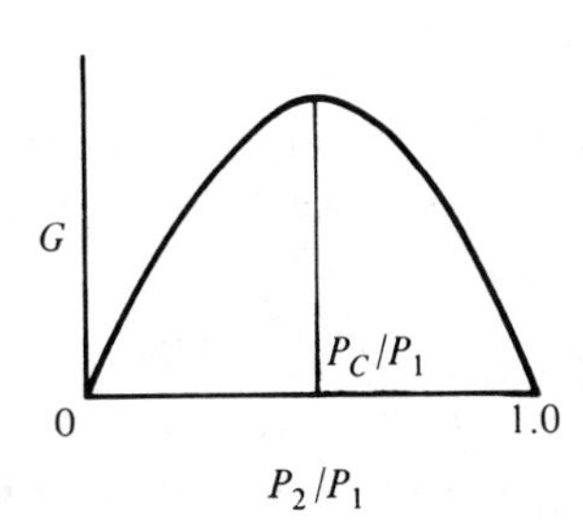

FIG. 8-4. Mass flow rate vs. pressure ratio from equation (8-27a).

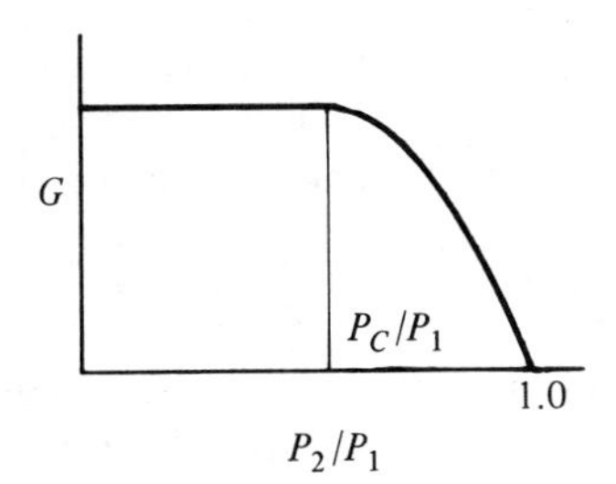

FIG. 8-5. Mass flow rate vs. pressure ratio for isothermal duct flow.

Performing the differentiation for a fixed value of P_1 yields

$$-\frac{2(P_c/P_1)}{(2fL/D) - \ln(P_c/P_1)} + \frac{1-(P_c/P_1)^2}{(P_c/P_1)[2fL/D) - \ln(P_c/P_1)]^2} = 0 \qquad (8\text{-}29)$$

Simplification and collection of terms then yields

$$\frac{2fL}{D} + 0.5 = \ln\left(\frac{P_c}{P_1}\right) + 0.5\left(\frac{P_1}{P_c}\right)^2 \qquad (8\text{-}30)$$

For a particular pipeline and flow conditions the maximum flow rate may be determined as follows:

1. Evaluate $2fL/D$.
2. The critical pressure ratio P_c/P_1 is determined from equation (8-30).
3. The critical pressure ratio is substituted into equation (8-27a) to determine the maximum flow rate.

If it is necessary to pass a specified mass flow rate through the pipeline, the pipeline length must be less than L_{max}, where L_{max} is obtained as follows:

1. We assume that $P_2 = P_c$ in equation (8-27a) and use equation (8-30) to eliminate $2fL/D$ with the result

$$P_c^2 = \frac{G^2R'T}{g_c} \qquad (8\text{-}31)$$

2. P_c is then substituted into equation (8-30) to determine L_{max}, the length of pipeline that corresponds to critical conditions at the end of the pipeline. If the pipeline is shorter than L_{max}, P_2 will be greater than P_c (assuming G is held constant). If the pipeline is of length L_{max}, the outlet pressure must be the critical pressure, if the desired mass flow rate is to be obtained. If the pipeline is longer than L_{max}, there is no outlet pressure which satisfies the equations for the specified mass flow rate. That is, the specified mass flow rate cannot be obtained, and the actual flow rate will be less than the desired one. This phenomenon, in which the desired mass flow rate cannot be obtained no matter what the outlet pressure, is termed *choking* and is a phenomenon we shall encounter frequently throughout the remainder of this chapter.

Adiabatic Flows:

In adiabatic flows it is not possible to establish a priori the relationship between pressure and density, because the temperature is not known as a function of pressure or density. Thus the relationship between P and ρ must be determined before the mechanical energy balance can be integrated. We begin by examining the energy balance around a differential segment of pipeline:

$$d\left(\underline{H} + \frac{u^2}{2g_c} + \frac{gZ}{g_c}\right)\delta M + \delta Q - \delta W = d\left[M\left(\underline{U} + \frac{u^2}{2g_c} + \frac{gZ}{g_c}\right)\right]_{\text{syst}} \tag{8-32}$$

Since the pipeline will be assumed to operate at steady state, the right-hand side of equation (8-32) equals zero. For adiabatic flows, $\delta Q = 0$, and by assumption $\delta W = 0$ and $g\ dZ/g_c = 0$. Thus the energy balance reduces to

$$d\left(\underline{H} + \frac{u^2}{2g_c}\right) = 0 \tag{8-33}$$

or

$$d\underline{H} = -\frac{u\ du}{g_c} \tag{8-34}$$

The enthalpy may in turn be expressed in terms of P, $\underline{V}$, T, and C_P as

$$d\underline{H} = C_P\,dT + \left[\underline{V} - T\left(\frac{\partial \underline{V}}{\partial T}\right)_P\right]dP \tag{8-35}$$

so the energy balance reduces to

$$C_P\,dT + \left[\underline{V} - T\left(\frac{\partial \underline{V}}{\partial T}\right)_P\right]dP = -\frac{u\ du}{g_c} \tag{8-36}$$

The velocity term may be eliminated by application of the continuity equation:

$$\rho u = G = \frac{\dot{M}}{A} = \text{const} \tag{8-37}$$

or

$$-u\ du = \frac{G^2}{\rho^3}\,d\rho \tag{8-38}$$

so

$$C_P\,dT + \left[\underline{V} - T\left(\frac{\partial \underline{V}}{\partial T}\right)_P\right]dP = \frac{G^2}{\rho^3 g_c}\,d\rho \tag{8-39}$$

Equation (8-39) and the equation of state for the fluid, $f(P, \underline{V}, T) = 0$, then allow determination of the relationship between P and ρ (or $\underline{V}$). As we will soon show, this relationship can be expressed in a simple closed form for a gas that obeys the ideal gas equation of state. For this case it is possible to integrate the mechanical energy balance exactly to obtain pressure, temperature, density, and velocity as functions of position within the pipeline. For nonideal equations of state, the relationship between pressure and density becomes extremely complex, and it is usually necessary to resort to numerical integration of the mechanical energy balance.

SAMPLE PROBLEM 8-4. Derive expressions that relate the pressure, temperature, density, and velocity to length during the adiabatic flow of an ideal gas through a pipeline of uniform area. Assume that the inlet conditions are known and that f and C_P are constants.

Solution: We begin with the form of the mechanical energy balance derived in Sample Problem 8-3:

$$\int_1^2 \rho\, dP + \frac{G^2}{g_c} \ln \frac{\rho_1}{\rho_2} + \frac{2fLG^2}{Dg_c} = 0$$

(Point 1 is the entrance of the pipeline; point 2 is any general point within the pipeline.) The relation between P and ρ that is needed to evaluate $\int \rho\, dP$ is obtained by substituting the ideal gas equation of state into the modified energy balance, equation (8-39):

$$C_P\, dT + \left[\underline{V} - T\left(\frac{\partial \underline{V}}{\partial T}\right)_P\right] dP = \frac{G^2}{\rho^3 g_c}\, d\rho$$

For a gas that obeys the ideal gas equation of state,

$$\underline{V} = \frac{RT}{P}$$

the energy balance reduces to

$$C_P\, dT = \frac{G^2}{\rho^3 g_c}\, d\rho$$

This is integrated between the inlet and any general point to give

$$C_P(T - T_1) = -\frac{G^2}{2g_c}\left(\frac{1}{\rho^2} - \frac{1}{\rho_1^2}\right)$$

The temperature may be eliminated by substitution of the equation of state: $T = P/R\rho$ to give

$$C_P\left(\frac{P}{R\rho} - \frac{P_1}{R\rho_1}\right) = -\frac{G^2}{2g_c}\left(\frac{1}{\rho^2} - \frac{1}{\rho_1^2}\right)$$

which may be solved for P:

$$P = \frac{P_1\rho}{\rho_1} - \frac{G^2 R}{2C_P g_c}\left(\frac{1}{\rho} - \frac{\rho}{\rho_1^2}\right)$$

For convenience, define $\lambda = G^2R/2C_Pg_c$, so the relation between P and ρ becomes

$$P = \frac{P_1\rho}{\rho_1} - \lambda\left(\frac{1}{\rho} - \frac{\rho}{\rho_1^2}\right)$$

Since pressure is known directly as a function of density, let us eliminate dP from the integral in the mechanical energy balance. The differential dP is obtained by differentiating the above expression for P:

$$dP = \left[\frac{P_1}{\rho_1} + \lambda\left(\frac{1}{\rho^2} + \frac{1}{\rho_1^2}\right)\right] d\rho$$

(A quick examination of this expression shows that the derivative $d\rho/dP$ is always greater than zero, and bounded.) The integral $\int \rho\, dP$ thus becomes

$$\int_{P_1}^{P} \rho\, dP = \int_{\rho_1}^{\rho} \rho\left[\frac{P_1}{\rho_1} + \lambda\left(\frac{1}{\rho^2} + \frac{1}{\rho_1^2}\right)\right] d\rho$$

$$= \left[\frac{P_1\rho^2}{2\rho_1} + \lambda \ln \rho + \lambda\frac{\rho^2}{2\rho_1^2}\right]_{\rho_1}^{\rho}$$

Substitution of the limits and simplification then gives

$$\int_{P_1}^{P} \rho\, dP = \left[\frac{\rho^2 - \rho_1^2}{2\rho_1^2}\right][P_1\rho_1 + \lambda] + \lambda \ln\left(\frac{\rho}{\rho_1}\right)$$

which is substitued into the mechanical energy balance to give

$$0.5\left[\left(\frac{\rho}{\rho_1}\right)^2 - 1\right](P_1\rho_1 + \lambda) + \lambda \ln\frac{\rho}{\rho_1} - \frac{G^2}{g_c}\ln\frac{\rho}{\rho_1} + \frac{2fL}{D}\frac{G^2}{g_c} = 0$$

or

$$0.5\left[\left(\frac{\rho}{\rho_1}\right)^2 - 1\right](P_1\rho_1 + \lambda) + \left(\lambda - \frac{G^2}{g_c}\right)\ln\frac{\rho}{\rho_1} + \frac{2fL}{D}\frac{G^2}{g_c} = 0$$

Substitution for λ then gives the relation among density, mass flow rate, and length:

$$0.5\left[\left(\frac{\rho}{\rho_1}\right)^2 - 1\right]\left(P_1\rho_1 + \frac{G^2R}{2C_Pg_c}\right) + \left(\frac{G^2R}{2C_Pg_c} - \frac{G^2}{g_c}\right)\ln\frac{\rho}{\rho_1} + \frac{2fL}{D}\frac{G^2}{g_c} = 0$$

Thus for a given mass flow rate, the density as a function of position is as given above. Once the density at any point is known, the pressure, temperature, and velocity are determined from

$$P = \frac{P_1\rho}{\rho_1} - \frac{G^2R}{2C_Pg_c}\left(\frac{1}{\rho} - \frac{\rho}{\rho_1^2}\right)$$

$$T = T_1 - \frac{G^2}{2C_Pg_c}\left(\frac{1}{\rho^2} - \frac{1}{\rho_1^2}\right)$$

$$u = \frac{G}{\rho}$$

For an adiabatic pipeline of known length the expression developed in Sample Problem 8-4 among P, G, and L may be solved for G as a function of the outlet density to give

$$G^2 = \frac{0.5[1 - (\rho/\rho_1)^2]}{\frac{0.25R}{C_Pg_c}\left(\frac{\rho^2}{\rho_1^2} - 1\right) + \frac{R - 2C_P}{2C_Pg_c}\ln\left(\frac{\rho}{\rho_1}\right) + \frac{2fL}{Dg_c}} \tag{8-40}$$

As in the case of isothermal flow, we see that as ρ approaches either ρ_1 ($P_{out} \rightarrow P_1$) or infinity ($P_{out} \rightarrow 0$), G^2 tends to zero. This is clearly incorrect and we may reason, as before, that only the portion of the G–P curve between P_1 and the maximum in G has any physical significance. The conditions at the maximum mass flow rate can be obtained by differentiating G^2 with respect to ρ at constant length and setting the derivative equal to zero. From the critical density the critical pressure, temperature, and velocity may also be determined. (The algebra associated with these calculations will not be presented here.)

If it is desired to pass a specified mass flow rate through an adiabatic pipeline, the pipeline must be shorter than L_{max}, where L_{max} is evaluated in much the same way as for the isothermal pipeline. If the pipeline is longer than L_{max}, no outlet pressure or density will satisfy the governing equations at the specified mass flow rate. Thus the mass flow rate will drop below its specified level, and the pipeline will be "choked."

SAMPLE PROBLEM 8-5. Methane is to be pumped in a high-pressure underground pipeline from Washington, D.C., to New York City, a distance of 250 miles. The pressure in the pipeline at Washington is 3000 psia, and the temperature is 65°F. The pipeline has an inside diameter of 12 in., and it is desired to pump 70 lb_m/sec of methane through the pipeline. Since the pipeline is buried beneath the ground (a poor conductor of heat), you may assume that the flow is adiabatic. At the flow velocities involved, the friction factor may also be assumed constant at $f = 0.003$ independent of velocity.

Methane may be assumed to obey the Beattie–Bridgeman equation of state:

$$P = \frac{RT}{\underline{V}} + \frac{\beta(T)}{\underline{V}^2} + \frac{\gamma(T)}{\underline{V}^3} + \frac{\delta(T)}{\underline{V}^4}$$

where

$$\beta(T) = RB_0T - \frac{RC}{T^2} - A_0$$

$$\gamma(T) = A_0a - RB_0bT - \frac{RB_0C}{T^2}$$

$$\delta(T) = \frac{RB_0bC}{T}$$

The zero-pressure heat capacity of methane is given by

$$C_P^* = (4.52 + 0.0133T) \text{ cal/g-mol °K}$$

when $T =$ °K. Determine the pressure, temperature, velocity, and density as functions of length down the pipeline for these conditions. What happens if the flow rate is increased to 100 lb_m/sec?

Solution: We began by examining the integral mechanical energy balance derived in Sample Problem 8-3:

$$\frac{2fLG^2}{Dg_c} + \int_1^2 \rho \, dP - \frac{G^2}{g_c} \ln \frac{\rho_2}{\rho_1} = 0$$

Let us apply this equation to a small section of pipeline ΔL. Also, apply the mean-value theorem to remove ρ from under the integral sign:

$$\frac{2fG^2}{Dg_c} \Delta L + \rho_{av} \Delta P - \frac{G^2}{g_c} \ln \frac{\rho_2}{\rho_1} = 0$$

If the increments of pipeline are short, we may quite reasonably assume that ρ is relatively constant over the length increment. Thus let us replace ρ_{av} by ρ_1 [for better accuracy we might use $(\rho_1 + \rho_2)/2$, but ρ_1 is actually good enough for our purposes]. If we also replace ρ with $1/\underline{V}$ the mechanical energy balance becomes

$$\frac{2fG^2}{Dg_c} \Delta L + \frac{P_2 - P_1}{\underline{V}_{av}} - \frac{G^2}{g_c} \ln \frac{\underline{V}_1}{\underline{V}_2} = 0$$

a single equation that contains two unknowns. To get the extra equations that are needed to form a complete system of equations we must add the energy equation, continuity equation, and the equation of state. For flow through an increment of pipeline the energy equation gives

$$(\underline{H}_2 - \underline{H}_1) + \frac{u_2^2 - u_1^2}{2g_c} = 0$$

Velocities may be eliminated by means of the continuity equation, $u = G/\rho = G\underline{V}$, so

$$(\underline{H}_2 - \underline{H}_1) + \frac{G^2}{2g_c}(\underline{V}_2^2 - \underline{V}_1^2) = 0$$

Now we must express $\underline{H}_2 - \underline{H}_1$ in terms of the equation of state. Since the equation of state is pressure explicit, we find it somewhat more convenient to work in terms of internal energies rather than enthalpies; thus let

$$\underline{H}_2 - \underline{H}_1 = \underline{U}_2 - \underline{U}_1 + (P\underline{V})_2 - (P\underline{V})_1$$

But

$$d\underline{U} = C_V\, dT + \left(\frac{\partial \underline{U}}{\partial \underline{V}}\right)_T d\underline{V}$$

Let us integrate $d\underline{U}$ over a simple two-step path between points 1 and 2 as shown in Fig. S8-5. Since we are assuming that only small changes occur in any increment of

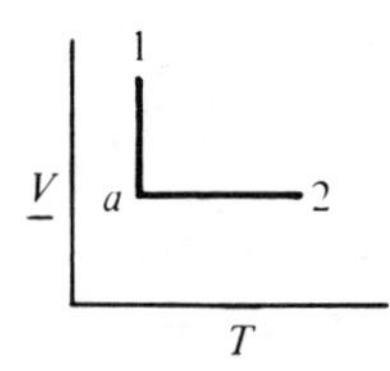

FIGURE S8-5

volume, we shall assume that neither C_V nor $(\partial \underline{U}/\partial \underline{V})_T$ change greatly in value over any length increment but retain their values as calculated at the inlet to the increment. Thus

$$\underline{U}_2 - \underline{U}_1 = C_V(T_2 - T_1) + \left(\frac{\partial \underline{U}}{\partial \underline{V}}\right)_T (\underline{V}_2 - \underline{V}_1)$$

But

$$\left(\frac{\partial \underline{U}}{\partial \underline{V}}\right)_T = \left[T\left(\frac{\partial P}{\partial T}\right)_{\underline{V}} - P\right]$$

which may be evaluated in terms of the equation of state to yield

$$\left(\frac{\partial \underline{U}}{\partial \underline{V}}\right)_T = \left[T\left(\frac{\partial P}{\partial T}\right)_{\underline{V}} - P\right] = \frac{3RC}{T^2\underline{V}^2}\left[1 + \frac{A_0T^2}{3RC}\left(1 - \frac{a}{\underline{V}}\right) + \frac{B_0}{\underline{V}}\left(1 - \frac{b}{\underline{V}}\right)\right]$$

The heat capacity at finite volume may be found from integrating

$$\frac{\partial C_V}{\partial \underline{V}} = T\left(\frac{\partial^2 P}{\partial T^2}\right)_{\underline{V}}$$

with respect to volume from $\underline{V} = \infty$ to $\underline{V}$:

$$C_V(\underline{V}, T) - C_V(\infty, T) = \int_\infty^{\underline{V}} T\left(\frac{\partial^2 P}{\partial T^2}\right)_{\underline{V}} d\underline{V}$$

But from the equation of state,

$$\left(\frac{\partial^2 P}{\partial T^2}\right)_{\underline{V}} = \frac{6RC}{\underline{V}^3T^3}\left(\frac{B_0 b}{\underline{V}^2} - \frac{B_0}{\underline{V}} - 1\right)$$

Substitution and integration then yields $C_V(\underline{V}, T)$:

$$C_V(\underline{V}, T) = C_V^*(T) + \frac{6RC}{T^3}\left(\frac{1}{\underline{V}} + \frac{B_0}{2\underline{V}^2} - \frac{B_0 b}{3\underline{V}^3}\right)$$

Thus the enthalpy difference becomes

$$\underline{H}_2 - \underline{H}_1 = C_V(\underline{V}_1, T_1)(T_2 - T_1) + \left(\frac{\partial \underline{U}}{\partial \underline{V}}\right)_T (\underline{V}_2 - \underline{V}_1) + P_2\underline{V}_2 - P_1\underline{V}_1$$

where C_V and $\partial \underline{U}/\partial \underline{V}$ are evaluated from the equation of state as indicated above. Substitution of the enthalpy difference and the equation of state (for $P_2\underline{V}_2$) in the energy balance, and P_2 in the mechanical energy balance, then gives two equations in two unknowns, which may be solved for T_2 and $\underline{V}_2$ at the end of every pipe increment:

$$C_V(\underline{V}_1, T_1)(T_2 - T_1) + \left(\frac{\partial \underline{U}}{\partial \underline{V}}\right)_T (\underline{V}_2 - \underline{V}_1) + \left[RT_2 + \frac{\beta(T_2)}{\underline{V}_2} + \frac{\gamma(T_2)}{\underline{V}_2^2} + \frac{\delta(T_2)}{\underline{V}_2^3} - P_1\underline{V}_1\right] + \frac{G^2}{2g_c}(\underline{V}_2^2 - \underline{V}_1^2) = 0 \qquad \text{(energy balance)} \tag{a}$$

$$\frac{2fG^2}{Dg_c}\Delta L + \frac{1}{\underline{V}_1}\left[\frac{RT_2}{\underline{V}_2} + \frac{\beta(T_2)}{\underline{V}_2^2} + \frac{\gamma(T_2)}{\underline{V}_2^3} + \frac{\delta(T_2)}{\underline{V}_2^4} - P_1\right] + \frac{G^2}{g_c}\ln\frac{\underline{V}_2}{\underline{V}_1} = 0 \qquad \text{(mechanical energy balance)} \tag{b}$$

Since these two equations are highly nonlinear, a direct solution will not be attempted. Rather we shall use the Newton–Raphson technique. Let us write equations (a) and (b) as

$$E(T_2, \underline{V}_2) = 0 \tag{a$'$}$$

$$F(T_2, \underline{V}_2) = 0 \tag{b$'$}$$

respectively. If we have an approximate solution T_2^k, $\underline{V}_2^k$ to the equations, a (hopefully) better approximation T_2^{k+1}, $\underline{V}_2^{k+1}$ may be obtained from

$$T_2^{k+1} = T_2^k + \frac{E(T_2^k, \underline{V}_2^k)\dfrac{\partial F(T_2^k, \underline{V}_2^k)}{\partial \underline{V}} - F(T_2^k, \underline{V}_2^k)\dfrac{\partial E(T_2^k, \underline{V}_2^k)}{\partial \underline{V}}}{D}$$

$$\underline{V}_2^{k+1} = \underline{V}_2^k + \frac{F(T_2^k, \underline{V}_2^k)\dfrac{\partial E(T_2^k, \underline{V}_2^k)}{\partial T} - E(T_2^k, \underline{V}_2^k)\dfrac{\partial F(T_2^k, \underline{V}_2^k)}{\partial T}}{D}$$

where

$$D = \frac{\partial E(T_2^k, \underline{V}_2^k)}{\partial \underline{V}}\frac{\partial F(T_1^k, \underline{V}_2^k)}{\partial T} - \frac{\partial E(T_2^k, \underline{V}_2^k)}{\partial T}\frac{\partial F(T_2^k, \underline{V}_2^k)}{\partial \underline{V}}$$

See Sample Problem 4-7 for a detailed discussion of the Newton–Raphson iteration procedure which has been used here. The partials of E and F may be readily shown to be

$$\frac{\partial F(T_2, \underline{V}_2)}{\partial \underline{V}_2} = \frac{1}{\underline{V}_1}\left[-\frac{RT_2}{\underline{V}_2^2} - \frac{2\beta(T_2)}{\underline{V}_2^3} - \frac{3\gamma(T_2)}{\underline{V}_2^4} - \frac{4\delta(T_2)}{\underline{V}_2^5}\right] + \frac{G^2}{\underline{V}_2 g_c}$$

$$\frac{\partial F(T_2, \underline{V}_2)}{\partial T_2} = \frac{1}{\underline{V}_1}\left[\frac{R}{\underline{V}_2} + \frac{\beta'(T_2)}{\underline{V}_2^2} + \frac{\gamma'(T_2)}{\underline{V}_2^3} + \frac{\delta'(T_2)}{\underline{V}_2^4}\right]$$

(where the prime represents d/dT) and

$$\frac{\partial E(T_2, \underline{V}_2)}{\partial \underline{V}_2} = \frac{\partial \underline{U}(T_1, \underline{V}_1)}{\partial \underline{V}} + \left[-\frac{\beta(T_2)}{\underline{V}_2^2} - \frac{2\gamma(T_2)}{\underline{V}_2^3} - \frac{3\delta(T_2)}{\underline{V}_2^4}\right] + \frac{G^2\underline{V}_2}{g_c}$$

$$\frac{\partial E(T_2, \underline{V}_2)}{\partial T_2} = C_V(\underline{V}_1, T_1) + \left[R + \frac{\beta'(T_2)}{\underline{V}_2} + \frac{\gamma'(T_2)}{\underline{V}_2^2} + \frac{\delta'(T_2)}{\underline{V}_2^3}\right]$$

As initial guesses to the temperature and volume at the end of the length increment we simply use the values of T and $\underline{V}$ at the entrance of the increment. Once we have calculated the values of T and $\underline{V}$ at the exits of the length increment, we calculate the pressure from the equation of state and the velocity from the continuity equation. We are then ready to begin the next length increment, which uses as its inlet conditions the outlet conditions from the last increment. In this fashion we may march down the whole length of the pipeline, calculating P, T, ρ, and u as we proceed.

As in our previous calculations, we shall find it useful to express all quantities in a consistent set of units. In our case we shall use ft, sec, lb_m, lb_f, and °R, In these units the appropriate input data for the problem are

$$\dot{M} = 100\ lb_m/sec$$
$$P_{inlet} = 3000\ psia = 4.32 \times 10^5\ lb_f/ft^2$$
$$T_{inlet} = 65°F = 525°R$$
$$\text{diam} = 12\ in. = 1\ ft$$
$$\Delta L = 5\ miles = 2.64 \times 10^4\ ft$$
$$L = 250\ miles = 1.32 \times 10^6\ ft$$
$$f = 0.003$$
$$\text{mol wt} = 16.0$$
$$R = 1545\ lb_f\text{-}ft/lb\text{-}mol\ °R = 96.5\ lb_f\text{-}ft/lb_m\ °R$$
$$a = 0.01857\ ft^3/lb_m$$
$$A_0 = 4.8297 \times 10^3\ lb_f\text{-}ft^3/lb_m^2$$
$$b = -0.01589\ ft^3/lb_m$$
$$B_0 = 0.05593\ ft^3/lb_m$$
$$C = 7.476 \times 10^5\ ft\ °R^3/lb_m$$

A program to perform the required calculations is given below. Note that the program begins by inverting the equation of state to find $\underline{V}_{inlet}$ from the given P_{in} and T_{in}. Note the choking at 147 miles for $\dot{M} = 100\ lb_m/sec$.

```
C   ---  PROCRAM TO CALCULATE THE PRESSURE, TEMPERATURE, DENSITY,
C        AND VELOCITY PROFILES FOR THE ADIABATIC FLOW OF A
C        BEATTIE-BRIDGEMAN GAS IN A PIPELINE WITH FRICTIONAL
C        LOSSES.- - -
         INTEGER PFREQ, PMIN
         REAL MFLOW, MVEL, LENGTH
C   ---  FUNCTION STATEMENTS- - -
         BET(T) = R*BO*T - AO - R*C/(T*T)
         GAM(T) = AO*A - R*B*BO*T - R*BO*C/(T*T)
         DEL(T) = R*B*BO*C/(T*T)
         BPRIME(T) = R*BO + 2.0*R*C/(T**3)
         GPRIME(T) = 2.0*R*BO*C/(T**3) - R*B*BO
         DPRIME(T) = -2.0*R*B*BO*C/(T**3)
         DUDV(T,V) = (3.0*R*C)/(T*T*C*C)*(1.0+AO*T*T/(3.0*R*C)*(1.0
       1              -A/V)+BO/V*(1.0-B/V))
         PV(T,V) = R*T+BET(T)/V+GAM(T)/(V*V)+DEL(T)/(V**3)
         CV(T,V) = CPA+CPB*T-R+6.0*R*C/(T**3)*(1./V+.5*BO/(V*V)-B*
       1            BO/(3.*(V**3)))
```

```
        PRES(T,V) = PV(T,V)/V
C  ---  READ INPUT DATA---
   10   READ(105,11)) R,A0,B0,A,B,C
   11   FORMAT(6E10.4)
   20   READ(105,11) CPA,CPB,PSTART,TSTART,DIA,GC
   30   READ(105,31) LENGTH,F,MFLOW,EPSV,EPST
   31   FORMAT(5E10.4)
   40   READ(105,41) NL
   41   FORMAT(I6)
   50   READ(105,51) PFREQ,  PMIN
   51   FORMAT(2I5)
   52   WRITE(108,11) R,A0,B0,A,B,C
   53   WRITE(108,11) CPA,CPB,PSTART,TSTART,DIA,GC
   54   WIRTE(108,31) LENGTH,MFLOW,EPSV,EPST
   55   WRITE(108,41) NL
   56   WRITE(108,51) PFREQ,PMIN
C  ---  BEGIN CALCULATIONS---
   60   MVEL = MFLOW*4.0/(3.1416*DIA*DIA)
C  ---  DETERMINE INITIAL SPECIFIC VOLUME---
   70   T = TSTART
   71   P = PSTART
   72   V = 0.9*R*TSTART/PSTART
   80   BT = BET(T)
   81   GAT = GAM(T)
   82   DET = DEL(T)
   85   I = 0
   90   FOFV = -P + R*T/V + BT/(V*V) + GAT/(V**3) + DET/(V**4)
  100   DFDV = -(R*T/(V*V)+2.*BT/(V**3)+3.*GAT/(V**4)+4.*DET/(V**5))
  110   DV = -FOFV/DFDV
  120   V = V + DV
  130   I = I + 1
  140   IF (ABS(DV/V) .LE. EPSV) GO TO 180
  150   IF (I .LE. 15) GO TO 90
  160   WRITE(108,161)
  161   FORMAT(1H, 36H UNABLE TO DETERMINE INITIAL VOLUME )
  162   WRITE(108, 163), V
  163   FORMAT(1H  , 24H LAST VOLUME ESTIMATE = ,E10.4)
  170   GO TO 10
  180   DELL = LENGTH/NL
  190   LCOUNT = 0
C  ---  BEGIN MOVING DOWN PIPELINE---
  198   WRITE(108,199)
  199   FORMAT(1H  , 22H INLET CONDITIONS ARE )
  200   WRITE(108,201),P,T,MVEL*V,1.0/V
  201   FORMAT(1H  , 5H P = ,E10.4, 5H T = ,E10.4, 7H VEL = ,E10.4,
       1 7H DEN = ,E10.4, //)
  210   SPHT = CV(T, V)
  220   BT = BET(T)
  230   GAT = GAM(T)
  240   DET = DEL(T)
```

```
  250   DUDVT = DUDV(T, V)
  260   BTP = BPRIME(T)
  270   DETP = DPRIME(T)
  280   GATP = GPRIME(T)
  300   P1 = PRES(T, V)
  310   T1 = T
  320   V1 = V
  330   ITER = 0
  340   FRICT = 4.0*F*MVEL*MVEL*DELL/(GC*DIA)
  350   EOFVT = FRICT + 1.0/V1*(PRES(T,V)-P1) + MVEL*MVEL/GC*ALOG(V/V1)
  360   FOFVT = SPHT*(T-T1)+DUDVT*(V-V1)+PV(T,V)-P1*V1 + (MVEL**2)/
     1          (2.*GC)*(V*V-(V1**2))
  370   DEDV = -1./V1*(R*T/(V*V)+2.*BET(T)/(V**3)+
     1         3.*GAM(T)/(V**4)+4.*DEL(T)/(V**5))+MVEL*MVEL/(GC*V)
  380   DEDT = 1./V1*(R/V+BPRIME(T)/(V*V)+GPRIME(T)/(V**3)+
     1         DPRIME(T)/(V**4))
  390   DFDV = DUDVT - (BET(T)/(V*V) + 2.*GAM(T)/(V**3)+
     1         3.*DEL(T)/(V**4)) + MVEL*MVEL*V/GC
  400   DFDT = SPHT + R + BPRIME(T)/V + GPRIME(T)/(V*V) + DPRIME(T)/
     1         (V**3)
  410   DET = DEDV*DFDT - DEDT*DFDV
  420   DV = (FOFVT*DEDT - EOFVT*DFDT)/DET
  430   DT = (EOFVT*DFDV - FOFVT*DEDV)/DET
  440   T = T + DT
  450   V = V + DV
  460   ITER = ITER + 1
  470   IF (ITER.LE. 20) GO TO 500
  480   WRITE(108,481)
  481   FORMAT(1H , 34H UNABLE TO ADVANCE PIPE INCREMENT )
  482   WRITE(108,483), T,V
  483   FORMAT(1H , 41H FINAL ESTIMATES OF T AND V ARE --- TEMP = ,
       1E10.4, 10H VOLUME = ,E10.4)
  490   GO TO 10
  500   IF (EPST .LT. ABS(DT/T)) GO TO 350
  510   IF (EPSV .LT. ABS(DV/V)) GO TO 350
  520   LCOUNT = LCOUNT + 1
C  ---  PIPE INCREMENT COMPLETED---
  530   IF (NL .LE. LCOUNT) GO TO 600
  540   IF (LCOUNT .LE. PMIN) GO TO 570
  550   IF ((LCOUNT/PFREQ)*PFREQ - LCOUNT .NE. 0) GO TO 210
  570   WRITE(108,571), LCOUNT
  571   FORMAT(1H , 7H AFTER, I5, 51H LENGTH INCREMENTS, CONDITIONS IN
       1THE PIPELINE ARE)
  572   WRITE(108,201), PRES(T,V),T,MVEL*V,1./V
  580   GO TO 210
  600   WRITE(108,601)
  601   FORMAT(1H ,39H FINAL CONDITIONS AT PIPELINE EXIT ARE)
  602   WRITE(108,201), PRES(T,V),T,MVEL*V,1./V
  610   GO TO 10
  620   CALL EXIT
```

Output-Case 1

```
.9650E 02 .4840E 04 .5587E-01 .1855E-01-.1587E-01 .7470E 06
.2200E 03 .3580E 00 .4320E 06 .5250E 03 .1000E 01 .3217E 02
.1320E 07 .1500E-02 .7000E 02 .1000E-03 .1000E-03
  100
  10    1
INLET CONDITIONS ARE
P =  .4320E 06 T =  .5250E 03 VEL =  .8224E 01 DEN =  .1084E 02
AFTER   1 LENGTH INCREMENTS, CONDITIONS IN THE PIPELINE ARE
P =  .4302E 06 T =  .5250E 03 VEL =  .8257E 01 DEN =  .1079E 02
AFTER  10 LENGTH INCREMENTS, CONDITIONS IN THE PIPELINE ARE
P =  .4136E 06 T =  .5252E 03 VEL =  .8576E 01 DEN =  .1039E 02
AFTER  20 LENGTH INCREMENTS, CONDITIONS IN THE PIPELINE ARE
P =  .3944E 06 T =  .5253E 03 VEL =  .8987E 01 DEN =  .9917E 01
AFTER  30 LENGTH INCREMENTS, CONDITIONS IN THE PIPELINE ARE
P =  .3742E 06 T =  .5252E 03 VEL =  .9473E 01 DEN =  .9409E 01
AFTER  40 LENGTH INCREMENTS, CONDITIONS IN THE PIPLEINE ARE
P =  .3529E 06 T =  .5251E 03 VEL =  .1006E 02 DEN =  .8861E 01
AFTER  50 LENGTH INCREMENTS, CONDITIONS IN THE PIPELINE ARE
P =  .3301E 06 T =  .5248E 03 VEL =  .1078E 02 DEN =  .8267E 01
AFTER  60 LENGTH INCREMENTS, CONDITIONS IN THE PIPELINE ARE
P =  .3056E 06 T =  .5243E 03 VEL =  .1170E 02 DEN =  .7618E 01
AFTER  70 LENGTH INCREMENTS, CONDITIONS IN THE PIPELINE ARE
P =  .2788E 06 T =  .5234E 03 VEL =  .1291E 02 DEN =  .6901E 01
AFTER  80 LENGTH INCREMENTS, CONDITIONS IN THE PIPELINE ARE
P =  .2489E 06 T =  .5221E 03 VEL =  .1462E 02 DEN =  .6096E 01
AFTER  90 LENGTH INCREMENTS, CONDITIONS IN THE PIPELINE ARE
P =  .2144E 06 T =  .5202E 03 VEL =  .1723E 02 DEN =  .5173E 01
FINAL CONDITIONS AT PIPELINE EXIT ARE
P =  .1726E 06 T =  .5172E 03 VEL =  .2191E 02 DEN =  .4068E 01
```

Output-Case 2

```
.9650E 02 .4840E 04 .5587E-01 .1855E-01-.1587E-01 .7470E 06
.2200E 03 .3580E 00 .4320E 06 .5250E 03 .1000E 01 .3217E 02
.1320E 07 .1500E-02 .1000E 03 .1000E-03 .1000E-03
  100
  10    1
INLET CONDITIONS ARE
P =  .4320E 06 T =  .5250E 03 VEL =  .1175E 02 DEN =  .1084E 02
AFTER   1 LENGTH INCREMENTS, CONDITIONS IN THE PIPELINE ARE
P =  .4283E 06 T =  .5250E 03 VEL =  .1184E 02 DEN =  .1075E 02
AFTER  10 LENGTH INCREMENTS, CONDITIONS IN THE PIPELINE ARE
P =  .3937E 06 T =  .5253E 03 VEL =  .1286E 02 DEN =  .9899E 01
AFTER  20 LENGTH INCREMENTS, CONDITIONS IN THE PIPELINE ARE
P =  .3513E 06 T =  .5251E 03 VEL =  .1444E 02 DEN =  .8820E 01
AFTER  30 LENGTH INCREMENTS, CONDITIONS IN THE PIPELINE ARE
P =  .3029E 06 T =  .5242E 03 VEL =  .1687E 02 DEN =  .7547E 01
AFTER  40 LENGTH INCREMENTS, CONDITIONS IN THE PIPELINE ARE
P =  .2446E 06 T =  .5219E 03 VEL =  .2129E 02 DEN =  .5981E 01
AFTER  50 LENGTH INCREMENTS, CONDITIONS IN THE PIPELINE ARE
```

```
P =   .1652E 06 T =  .5166E 03 VEL =    .3285E 02 DEN =    .3876E 01
AFTER  60 LENGTH INCREMENTS, CONDITIONS IN THE PIPELINE ARE
P =  -.9672E 07 T =  .5069E 12 VEL =  -.6439E 09 DEN =  -.1977E-06
```

8.2 The Velocity of Sound

Before we enter the study of compressible flows in nozzles or diffusers, we shall derive an expression for the speed of sound, or the speed at which an infinitesimal pressure disturbance propagates through a nonmoving medium. This velocity will be of major importance to us later.

Assume that a small pressure disturbance (sound wave) is moving down a tube as shown in Fig. 8-6. Assume that before the wave passes through a

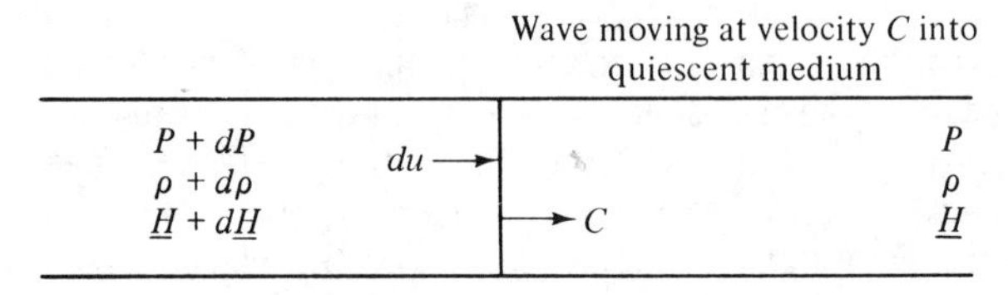

FIG. 8-6. Motion of a small pressure wave through a still media.

region the medium is not moving and that after the wave passes the properties of the substance have changed by only an infinitesimal amount, and the gas is moving toward the wave front with a velocity du.

Let us now examine the same process from the point of view of an observer riding on the wave front, as shown in Fig. 8-7. If we consider the stationary

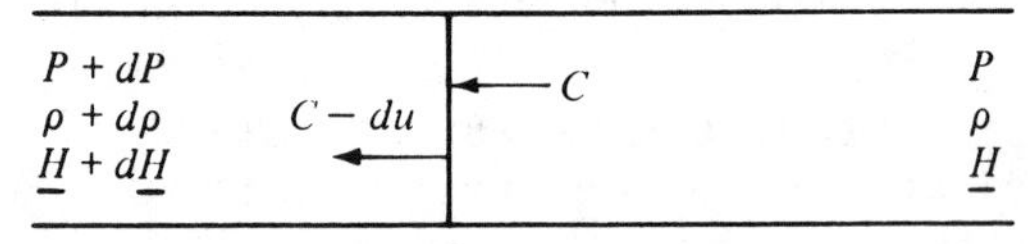

FIG. 8-7. Motion of a small pressure wave as seen by an observer on the wave front.

wave front as a steady state, open system, we may write the energy balance for the system as

$$[(\underline{H} + d\underline{H}) - \underline{H}] + \frac{(C - du)^2 - C^2}{2g_c} = 0 \tag{8-41}$$

or, neglecting second-order differentials,

$$d\underline{H} - \frac{C\,du}{g_c} = 0 \tag{8-42}$$

The continuity equation may be written as

$$(\rho + d\rho)(C - du) = \rho C \tag{8-43}$$

Again if we neglect second-order differentials,

$$C\,d\rho - \rho\,du = 0 \tag{8-44}$$

The $d\underline{H}$ property relation is substituted into equation (8-42) to give

$$T\,d\underline{S} + \frac{dP}{\rho} = \frac{C\,du}{g_c} \tag{8-45}$$

Since the driving force for the sound wave is an infinitesimal pressure change, it seems reasonable to assume that the motion of the wave is reversible. Also, the great speed at which sound waves move does not allow time for much heat transfer, and hence to a good approximation the sound wave moves adiabatically. However, if the wave moves adiabatically and reversibly, it will proceed isentropically. (This conclusion has, in fact, been confirmed by experimental observations.) For isentropic flows $d\underline{S} = 0$, and equation (8-45) reduces to

$$\frac{dP}{\rho} = \frac{C\,du}{g_c} \tag{8-46}$$

Elimination of du between equation (8-44) and (8-46) yields

$$\frac{dP}{d\rho} = \frac{C^2}{g_c} \tag{8-47}$$

where the derivative must be evaluated at constant entropy. Thus equation (8-47) may be more properly written as

$$\frac{C^2}{g_c} = \left(\frac{\partial P}{\partial \rho}\right)_{\underline{S}} \tag{8-47a}$$

For an ideal gas, $(\partial P/\partial \rho)_{\underline{S}}$ may be found from the P–$\underline{V}$ relation of an isentropic process (see Chapter 6) and is given by

$$\left(\frac{\partial P}{\partial \rho}\right)_{\underline{S}} = kRT = k\frac{P}{\rho} \tag{8-48}$$

Therefore, for an ideal gas,

$$\frac{C^2}{g_c} = \frac{kP}{\rho} \tag{8-49}$$

or

$$C = \sqrt{\frac{kPg_c}{\rho}} = \sqrt{kg_cRT} \tag{8-49a}$$

8.3 Flow of Compressible Fluids Through Nozzles and Diffusers

Let us now examine the applications of the principles we have just developed to the study of compressible flow within nozzles and diffusers. A nozzle is a device whose function is to increase the velocity of a fluid by decreasing its pressure; a diffuser is a device that increases the pressure within a fluid by decreasing its kinetic energy. Through the years the term "nozzle" has come to be used for both nozzles and diffusers and will be so used in the following pages.

Since most flows through nozzles occur rapidly, there is little time for heat to be transferred into or out of the working fluid. Therefore, in many practical cases it can be assumed that the flow is adiabatic. In practice, it is found that nozzles may be designed to operate in an almost reversible fashion. That is, practically all the pressure gradient within the nozzle is used to increase the kinetic energy of the working fluid, whereas only a small portion is needed to overcome viscous effects. Therefore, we shall assume that all nozzles studied herein operate in a fully reversible fashion—that is, $\underline{LW} = 0$.

Since the flow in the nozzle is adiabatic, and reversible, it is isentropic; that is, the entropy is constant. Thus the pressure, temperature, density, and enthalpy of the fluid at any point within the nozzle are related by the isentropic conditions. For an ideal gas these relations may be expressed in relatively simple closed form, whereas for nonideal gases these relations are most simply obtained from a suitable property chart, such as a Mollier diagram. (If a suitable property chart is not available, the relationships must be established using the techniques developed in Chapter 6.)

Let us now apply the energy balance and continuity equations around a portion of the simple converging–diverging nozzle shown in Fig. 8-8. If the nozzle is operating at steady state, with no heat, no work, and negligible potential energy effects, the energy balance around the indicated section reduces to

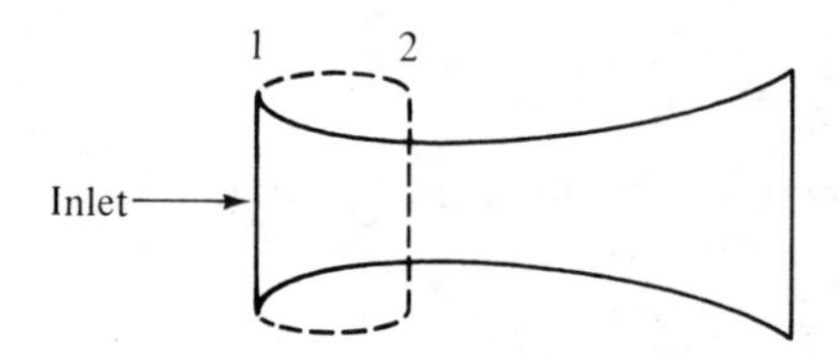

FIG. 8-8. Converging-diverging nozzle.

$$(\underline{H}_2 - \underline{H}_1) + \frac{u_2^2 - u_1^2}{2g_c} = 0 \tag{8-50}$$

(We shall work with the energy balance directly rather than the mechanical energy balance, because the path integral $\int \rho \, dP$ would be extremely difficult to evaluate for isentropic process involving nonideal gases.) If the inlet cross-sectional area of the nozzle is large compared to the cross-sectional areas elsewhere in the nozzle, then $u_1 < u_2$, so $u_1^2 \ll u_2^2$. In this case we may neglect u_1^2 and reduce the energy balance to

$$u_2^2 = 2g_c(\underline{H}_1 - \underline{H}_2) \tag{8-51}$$

(If the inlet velocity is not "negligible," as in the case of many diffusers, then the u_1^2 term must be retained, but the analysis will proceed almost exactly as follows.)

If the pressure at point 2 is known, the enthalpy and temperature, $\underline{H}_2$ and T_2, may be found from a suitable property chart and the inlet conditions,

P_1 and T_1. From equation (8-51) the velocity at point 2 may then be calculated. The density at point 2 may then be calculated from the equation of state and the known pressure and temperature. In addition, if the mass flow rate through the nozzle is specified, the area at point 2 may be determined from the continuity equation:

$$\dot{M} = \rho_2 u_2 A_2 = \text{const} \tag{8-52}$$

or

$$A_2 = \frac{\dot{M}}{\rho_2 u_2} \tag{8-53}$$

SAMPLE PROBLEM 8-6. Steam at 730 psia and 780°F is to be expanded in a converging-diverging nozzle to obtain a high-speed flow. Assume that the incoming steam has negligible velocity and that the flow is isentropic. For a mass flow rate of 10 lb_m/sec determine the velocity, area, and temperatures as functions of pressure throughout the nozzle.

Solution: We may picture the nozzle as shown in Fig. S8-6a. The energy balance may be written about any portion of the nozzle. In particular, let us write the balance

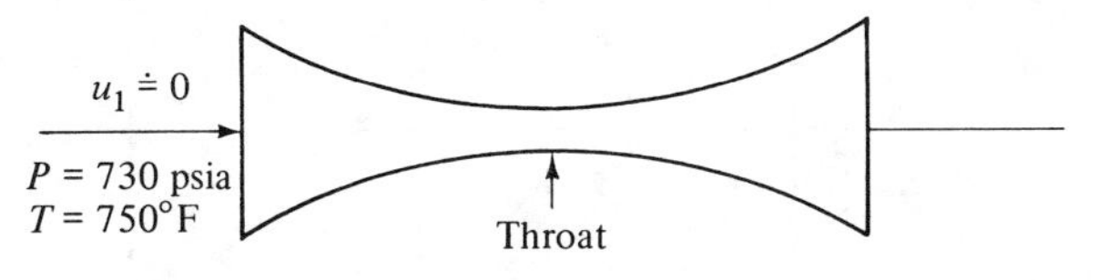

FIGURE S8-6a

around the entrance to the nozzle, and any other point within the nozzle. Assume that $\delta Q = \delta W = g\Delta Z/g_c = 0$ and that u_{in}^2 is negligible. For steady state $\delta M_{\text{in}} = \delta M_{\text{out}}$, and the energy balance reduces to equation (8-51),

$$u^2 = -2g_c(\underline{H} - \underline{H}_1)$$

or

$$u^2 = (\underline{H} - \underline{H}_1) \times 2.0 \times 32.17 \times 778 \frac{\text{lb}_\text{m}}{\text{lb}_\text{f}} \frac{\text{ft}}{\text{sec}^2} \frac{\text{lb}_\text{f}\text{-ft}}{\text{Btu}}$$

$$= (\underline{H} - \underline{H}_1)(5.0 \times 10^4)\ \text{ft}^2\text{-lb}_\text{m}/\text{sec}^2\ \text{Btu}$$

where $\underline{H}$ is the enthalpy at the point under consideration.

We may now proceed to determine the pressure–area profile within the nozzle as follows:

1. Choose a pressure P.
2. Using the initial conditions and the isentropic condition, determine the temperature and enthalpy corresponding to P from the Mollier diagram (Appendix A).
3. Using the temperature and pressure, determine the density (specific volume) from the steam tables (Appendix B).
4. Using the energy balance and the enthalpy, determine the velocity at the point under consideration.
5. From the known mass flow rate, the density, and the velocity, determine the area corresponding to P ($A = \dot{M}/\rho u = \dot{M}\underline{V}/u$).

The calculations are summarized in Table S8-6.

TABLE S8-6 Property Profiles for Converging–Diverging Nozzle

P, psia	T, °R	$\underline{V}$, ft^3/lb_m	$\underline{H}$, Btu/lb_m	$\underline{H} - \underline{H}_1$, Btu/lb_m	u^2, ft^2/sec^2	u, ft/sec	A, ft^2
730	780	0.9472	1390	0	0	0	∞
700	770	0.9806	1385	5	25×10^4	500	1.96×10^{-2}
650	750	1.039	1376	14	70×10^4	835	1.245×10^{-2}
600	725	1.099	1366	24	120×10^4	1090	1.00×10^{-2}
550	700	1.178	1355	35	175×10^4	1320	0.89×10^{-2}
500	680	1.276	1344	46	230×10^4	1512	0.845×10^{-2}
450	650	1.380	1332	58	280×10^4	1670	0.816×10^{-2}
400	620	1.513	1318	72	360×10^4	1895	0.799×10^{-2}
350	588	1.678	1304	86	430×10^4	2070	0.810×10^{-2}
300	550	1.890	1288	102	510×10^4	2258	0.838×10^{-2}
250	510	2.18	1268	122	610×10^4	2465	0.884×10^{-2}
200	460	2.59	1246	144	720×10^4	2680	0.967×10^{-2}
150	400	3.22	1220	170	850×10^4	2910	1.11×10^{-2}

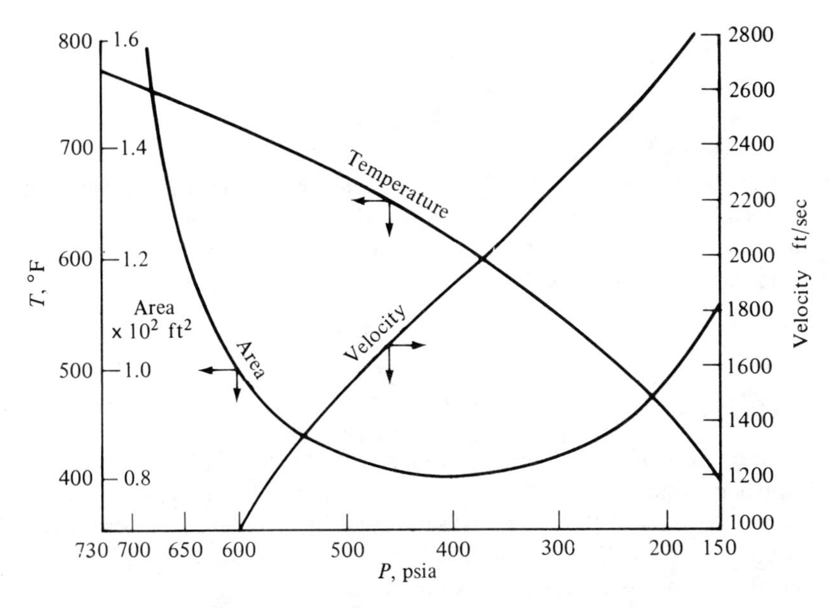

FIGURE S8-6b

We may now plot the velocity, area, and temperature, versus pressure as shown in Fig. S8-6b. An examination of the area–pressure curve indicates that the area of the nozzle passes through a minimum at $P = 395$ psia. As we will show later, this minimum area point, or "throat," is of major importance in determining the operating characteristics of a nozzle. The temperature and velocity at the throat also possess special significance; for the problem at hand they are

$$u_{\text{throat}} = 1860 \text{ ft/sec}$$

$$T_{\text{throat}} = 623°\text{F}$$

For a gas that obeys the ideal gas equation of state the pressure, temperature, and density may be simply related during an isentropic process. Therefore, it is possible to develop closed-form expressions for the relationships among P, T, ρ, u, $\dot{M}$, and A during the isentropic flow of an ideal gas through a nozzle. We shall now establish these relationships and study some of the phenomena associated with them.

We begin, as in the nonideal gas studies, by examining the energy balance written around some general portion of the nozzle as shown in Fig. 8-8:

$$(\underline{H}_2 - \underline{H}_1) + \frac{u_2^2 - u_1^2}{2g_c} = 0 \tag{8-54}$$

or, if we drop the subscript 2 for convenience,

$$\underline{H} - \underline{H}_1 = \frac{u^2 - u_1^2}{2g_c} \tag{8-54a}$$

For a gas that obeys the ideal gas equation of state (and has C_P independent of T),

$$\underline{H}_1 - \underline{H} = C_P(T_1 - T) \tag{8-55}$$

so equation (8-54a) becomes

$$-C_P(T - T_1) = \frac{u^2 - u_1^2}{2g_c} \tag{8-56}$$

Equation (8-56) relates the velocity at any point within the nozzle to the temperature at that point. Most nozzles are designed so that u_1 is a small quantity—especially in relation to the velocities that exist in other portions of the nozzle. Thus for many important cases we may neglect u_1^2 in relation to u^2, and equation (8-56) reduces to

$$u^2 = -2g_cC_P(T - T_1) \tag{8-57}$$

or

$$u^2 = -2g_cC_PT_1\left(\frac{T}{T_1} - 1\right) \tag{8-57a}$$

(Again, retention of the u_1^2 term causes essentially no extra problems where u_1 is not negligible.)

Since the flow in the nozzle is isentropic, the pressure, temperature, and density are related by the equations derived in Chapter 6 for isentropic processes of gases that obey the ideal gas equation of state:

$$\frac{T}{T_1} = \left(\frac{P}{P_1}\right)^{(k-1)/k} \tag{8-58}$$

and

$$\frac{P}{P_1} = \left(\frac{\rho}{\rho_1}\right)^k = \left(\frac{\underline{V}_1}{\underline{V}}\right)^k \tag{8-59}$$

Equation (8-58) is substituted into (8-57a) and yields

$$u^2 = -2g_cC_PT_1\left[\left(\frac{P}{P_1}\right)^{(k-1)/k} - 1\right] \tag{8-60}$$

But

$$C_PT_1 = \frac{k}{k-1}\frac{P_1}{\rho_1} \tag{8-61}$$

and thus the velocity, u, at any point can be expressed solely as a function of the inlet conditions, the ratio of heat capacities, and the pressure:

$$u^2 = -2g_c\frac{k}{k-1}\frac{P_1}{\rho_1}\left[\left(\frac{P}{P_1}\right)^{(k-1)/k} - 1\right] \tag{8-62}$$

SAMPLE PROBLEM 8-7. Develop equation (8-62) directly from the mechanical energy balance.

Solution: The mechanical energy balance around any portion of the nozzle is written as

$$\underline{W} + \underline{LW} + \frac{\Delta u^2}{2g_c} + \frac{g\Delta Z}{g_c} + \int \underline{V}\,dP = 0$$

But for the nozzle we are discussing,

$$\underline{W} = \underline{LW} = \frac{g\Delta Z}{g_c} = 0$$

As previously discussed, if point 1 is taken as the nozzle entrance, u_1^2 is commonly negligible in comparison to u_2^2 and hence is neglected. Thus the mechanical energy balance reduces to

$$\frac{u_2^2}{2g_c} + \int_{P_1}^{P_2} V\,dP = 0$$

But the process between any two points within the nozzle is an isentropic process, and therefore

$$P\underline{V}^k = P_1\underline{V}_1^k = \frac{P_1}{\rho_1^k}$$

Therefore,

$$\underline{V} = \left(\frac{P_1}{P\rho_1^k}\right)^{1/k}$$

so

$$\int_{P_1}^{P_2} \underline{V}\,dP = \int_{P_1}^{P_2} \left(\frac{P_1}{P\rho_1^k}\right)^{1/k} dP = \frac{(P_1)^{1/k}}{\rho_1} \int_{P_1}^{P_2} \frac{dP}{P^{1/k}}$$

Therefore,

$$\begin{aligned}
\frac{u_2^2}{2g_c} = -\frac{(P_1)^{1/k}}{\rho_1}\int_{P_1}^{P_2}\frac{dP}{P^{1/k}} &= -\frac{(P_1)^{1/k}}{\rho_1}\left[\frac{k}{k-1}\left((P_2)^{1-1/k} - (P_1)^{1-1/k}\right)\right] \\
&= -\frac{(P_1)^{1/k}}{\rho_1}\left[\frac{k}{1-k}\left((P_2)^{(k-1)/k} - (P_1)^{(k-1)/k}\right)\right] \\
&= -\frac{(P_1)^{1/k}(P_1)^{1-1/k}}{\rho_1}\frac{k}{1-k}\left[\left(\frac{P_2}{P_1}\right)^{(k-1)/k} - 1\right] \\
&= -\frac{P_1}{\rho_1}\frac{k}{k-1}\left[\left(\frac{P_2}{P_1}\right)^{(k-1)/k} - 1\right]
\end{aligned}$$

Multiplication by $2g_c$ and elimination of the subscript 2 then gives equation (8-62), as desired:

$$u^2 = -2g_c\frac{k}{k-1}\frac{P_1}{\rho_1}\left[\left(\frac{P}{P_1}\right)^{(k-1)/k} - 1\right]$$

We may now relate P to the cross-sectional area of the nozzle, A, by means of the continuity equation:

$$\dot{M} = \rho\, uA = \text{const} \tag{8-63}$$

(where $\dot{M}$ is the flow rate through the nozzle). Equation (8-63) is rearranged to the form

$$A = \frac{\dot{M}}{\rho u} \tag{8-64}$$

Equation (8-62) is now substituted for u in equation (8-64) to form

$$A = \frac{\dot{M}}{\rho\sqrt{2g_c\dfrac{k}{k-1}\dfrac{P_1}{\rho_1}\left[1 - \left(\dfrac{P}{P_1}\right)^{(k-1)/k}\right]}} \tag{8-65}$$

The ρ in equation (8-65) may be expressed in terms of the initial conditions and P by means of equation (8-59):

$$\rho = \rho_1\left(\frac{P}{P_1}\right)^{1/k} \tag{8-59}$$

Therefore,

$$A = \frac{\dot{M}}{\sqrt{\frac{2kg_c}{k-1}P_1\rho_1\left[\left(\frac{P}{P_1}\right)^{2/k} - \left(\frac{P}{P_1}\right)^{(k+1)/k}\right]}} \tag{8-66}$$

Thus we have derived the desired relation between the pressure at any point within the nozzle and the area at that point. Note, however, that the pressure within the nozzle is a function of both the area at that point and the rate of mass flow through the nozzle.

Some very interesting behavior may be observed if we examine equation (8-66). Since the denominator of this equation is a function only of the initial conditions (P_1 and ρ_1) and the pressure within the nozzle, for any given initial conditions equation (8-66) may be rewritten in the form

$$A = \frac{\dot{M}}{\phi(P)} \tag{8-67}$$

where

$$\phi(P) = \sqrt{\frac{2kg_c}{k-1}P_1\rho_1\left[\left(\frac{P}{P_1}\right)^{2/k} - \left(\frac{P}{P_1}\right)^{(k+1)/k}\right]} \tag{8-68}$$

Examination of $\phi(P)$ indicates that $\phi(P) = 0$ for $P = 0$, or $P = P_1$ and is > 0 for $1 > P/P_1 > 0$. Thus it is apparent that $\phi(P)$ must have the form shown in Fig. 8-9. That is, $\phi(P)$ passes through a maximum at some $0 < P < P_1$. We define the pressure at the point where $\phi(P)$ is a maximum as the critical pressure, P_c, of the nozzle. It is a function only of the specific heat ratio and the initial conditions of the fluid entering the nozzle. We may now plot the relation between the area and the pressure within a nozzle, for some specified mass flow rate, $\dot{M}$, as shown in Fig. 8-10. Since $\phi(P)$ goes through a maximum at $P = P_c$, the

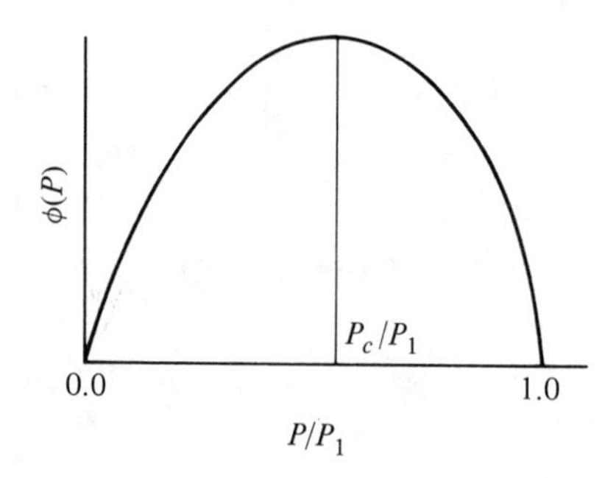

FIG. 8-9. Function $\phi(P)$.

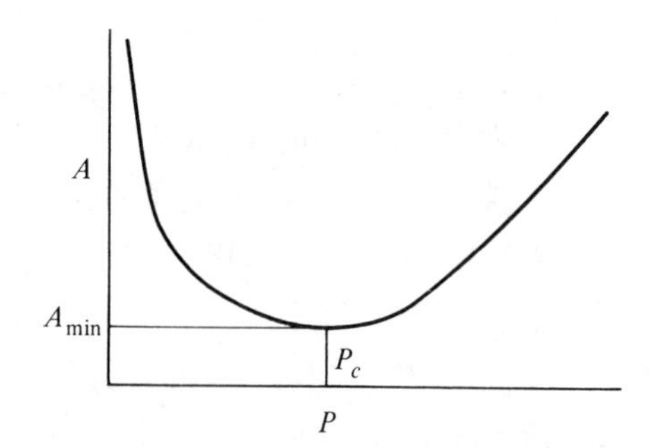

FIG. 8-10. Nozzle area as a function of pressure.

nozzle area must go through a minimum at this pressure, as shown in the figure. Thus we see that if we wish to expand a gas to a pressure less than P_c, we must have a converging–diverging nozzle, that is, a nozzle in which the area first decreases and then increases with distance from the nozzle entrance. The point at which the nozzle area goes through a minimum is known as the *throat* of the nozzle, and the area at this point is appropriately called the *throat area.* We shall now examine the operating characteristics of two different types of nozzles.

Simple Converging Nozzles

A simple converying nozzle is a nozzle whose area continually decreases with increasing distance from the nozzle entrance (Fig. 8-11). The exit of the nozzle is the throat.

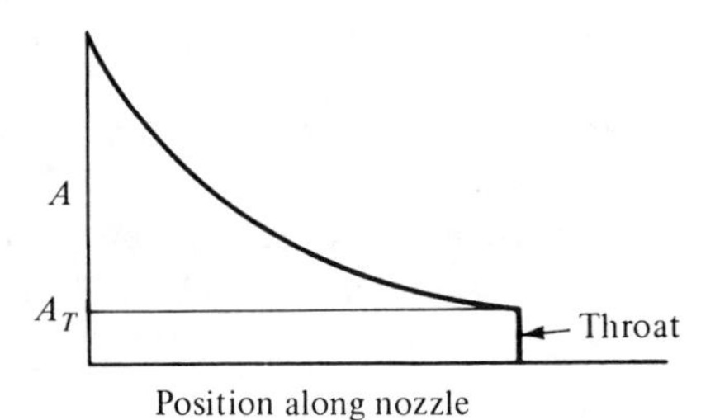

FIG. 8-11. Converging nozzle.

Let us examine the behavior of a simple converging nozzle as the exit pressure is dropped below the inlet pressure. Equation (8-67) is rearranged to the form

$$\dot{M} = A\phi(P) \tag{8-69}$$

and equation (8-69) is applied to the point just before the exit of the nozzle. Therefore,

$$\dot{M} = A_{\text{min}}\phi(P_2) \tag{8-70}$$

where P_2 = the exit pressure.

Examination of Fig. 8-9 shows that for $P_2 = P_1$, $\phi(P_2) = 0$, and no flow occurs. This is equivalent to saying: No pressure drop, no flow! However, as P_2 is reduced, $\phi(P_2)$ increases and, therefore, $\dot{M}$ increases. As P_2 approaches P_c, $\phi(P_2)$ approaches its maximum and so does $\dot{M}$. As P_2 is further decreased, equations (8-67) and (8-68) predict that $\phi(P_2)$ and $\dot{M}$ should decrease again. However, Fig. 8-10 shows that for *any* specific mass flow rate, the lowest pressure that can be attained *within* a simple converging nozzle (where the flow area continually decreases) is the critical pressure P_c. Therefore, for external pressures less than P_c, the pressure just within the nozzle must still be P_c, and both $\phi(P_2)$ and $\dot{M}$ will have their maximum levels. Thus for a particular nozzle we may plot the mass flow rate as a function of the external pressure, as shown in Fig. 8-12. [The dashed line in Fig. 8-12 is the portion of the $\phi(P_2)$ curve that can never be attained in a simple converging nozzle.]

For P_2 less than P_c we find that a sharp discontinuity in the pressure must occur at the nozzle exit, since P_c is the lowest pressure that can exist just within the nozzle; just outside the nozzle, P_2 is less than P_c. This sharp pressure discontinuity at the nozzle throat leads to the formation of a standing rarefaction wave at the tip of the nozzle. The presence of this standing wave may be observed

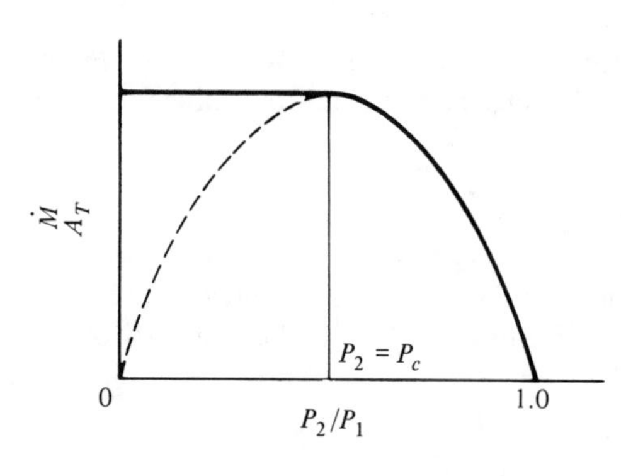

FIG. 8-12. Mass flow rate vs. pressure ratio in a converging nozzle.

from the noise it generates, or by any of a number of techniques for shock- or expansion-wave visualization.

We may now draw the pressure profiles (that is, the pressure–position plots) that exist in a converging nozzle such as illustrated in Fig. 8-11 for three different exit pressures, $P_2 > P_c$, $P_2 = P_c$, and $P_2 < P_c$, as shown in Fig. 8-13.

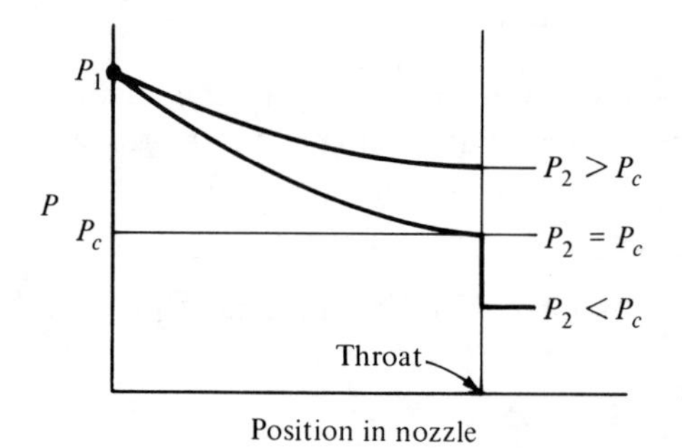

FIG. 8-13. Pressure plots for a converging nozzle.

Let us now determine the maximum mass flow rate that may pass through a simple converging nozzle. For a given value of the throat area, the maximum flow rate will occur where $\phi(P_2)$ is a maximum, that is, at $P_2 = P_c$. The maximum in the $\phi(P_2)$ may be determined by setting the derivative of $\phi(P_2)$ with respect to $P_2 = 0$. Thus at P_c,

$$\frac{d\phi(P_2)}{dP_2} = 0 \tag{8-71}$$

or, because P_1 is fixed,

$$\frac{d\phi(P_2)}{d(P_2/P_1)} = 0 \tag{8-72}$$

But

$$\phi(P_2) = K\left[\left(\frac{P_2}{P_1}\right)^{2/k} - \left(\frac{P_2}{P_1}\right)^{(k+1)/k}\right]^{1/2} \tag{8-68}$$

where

$$K = \sqrt{\frac{2k}{k-1}P_1\rho_1 g_c}$$

Therefore, if $\phi(P_2)$ is to be maximized,

$$\phi^1\left(\frac{P_2}{P_1}\right) = \left[\left(\frac{P_2}{P_1}\right)^{2/k} - \left(\frac{P_2}{P_1}\right)^{(k+1)/k}\right] \tag{8-73}$$

must also be maximized. Equation (8-72) may then be replaced by

$$\frac{d\phi^1}{d(P_2/P_1)} = \frac{d}{d(P_2/P_1)}\left[\left(\frac{P_2}{P_1}\right)^{2/k} - \left(\frac{P_2}{P_1}\right)^{(k+1)/k}\right] = 0 \tag{8-74}$$

Equation (8-74) may be differentiated to yield

$$\frac{2}{k}\left(\frac{P_2}{P_1}\right)^{(2-k)/k} - \frac{k+1}{k}\left(\frac{P_2}{P_1}\right)^{1/k} = 0 \tag{8-75}$$

Equation (8-75) is then solved for P_2/P_1:

$$\frac{P_2}{P_1} = \left(\frac{2}{k+1}\right)^{k/(k-1)} \tag{8-76}$$

But P_2 at the maximum of $\phi(P)$ is the critical pressure; therefore,

$$P_c = P_1\left(\frac{2}{k+1}\right)^{k/(k-1)} \tag{8-77}$$

so P_c depends only on the inlet pressure and the heat capacity ratio.

The critical throat velocity may be calculated by substituting equation (8-77) into (8-62) to yield

$$u_c = \sqrt{\frac{2k}{k+1}\frac{P_1}{\rho_1}g_c} \tag{8-78}$$

We may express the critical throat velocity in terms of the throat conditions by substituting the pressure–density relation

$$\frac{P_1}{\rho_1^k} = \frac{P_c}{\rho_c^k} \tag{8-59}$$

into equation (8-77):

$$\frac{P_1}{\rho_1} = \frac{P_c}{\rho_c}\frac{k+1}{2} \tag{8-79}$$

Equation (8-79) is then substituted into equation (8-78) to give

$$u_c = \sqrt{k\frac{P_c}{\rho_c}} = \sqrt{kRT_c g_c} \tag{8-80}$$

However, we have shown earlier that the speed of sound is given by

$$C = \sqrt{g_c kRT} \tag{8-49a}$$

Therefore, the maximum velocity of an ideal gas in the throat of a simple converging nozzle is identical to the speed of sound at the throat conditions.

We may express the maximum mass flow rate that will pass through a converging nozzle by substituting equation (8-77) for P_c into equation (8-66) to give

$$\dot{M}_c = A_T\left(\frac{2}{k+1}\right)^{1/(k-1)}\sqrt{\frac{2kg_c}{k+1}P_1\rho_1} \tag{8-81}$$

Thus the critical mass flow rate can be calculated if we know the inlet conditions and the throat area. For exit pressures higher than P_c, the mass flow rate will be less than the critical value; for external pressures less than P_c, the throat pressure will be the critical pressure, and the mass flow rate will equal $\dot{M}_c$. On the other hand, equation (8-81) may be used to determine the minimum throat area that will pass a specified mass flow rate. If the area is greater than the minimum, P_2 will be greater than the critical pressure. If the area is less than the minimum, the flow will "choke" and less than the specified mass flow rate will be passed.

SAMPLE PROBLEM 8-8. Air at 300 psia and 100°F enters a converging nozzle with negligible initial velocity. The nozzle exit has a cross-sectional area of 1.0 in². Assuming the flow to be isentropic and air to be an ideal gas at these conditions, calculate the mass flow rate (lb_m/hr) and the linear velocity (ft/sec) at the discharge end

(a) If the discharge pressure is 200 psia.
(b) If the discharge pressure is 100 psia.
For air,

$$C_P = 7.0 \text{ Btu/(lb-mol °R)}$$
$$C_V = 5.0 \text{ Btu/(lb-mol °R)}$$
$$M = 29.0 \text{ lb/lb-mol}$$

Solution: Before we can proceed we must determine the critical pressure for the nozzle:

$$P_c = P_1 \left(\frac{2}{k+1}\right)^{k/(k-1)}$$

But we are told that for air $C_P = 7.0$ and $C_V = 5.0$. Therefore,

$$k = \frac{7.0}{5.0} = 1.40$$

The critical pressure is then calculated from

$$P_c = 300 \left(\frac{2}{2.4}\right)^{1.4/0.4} = 160 \text{ psia}$$

For part (a), the exit pressure is greater than P_c; the flow is subsonic and equations (8-66) and (8-62) may be used directly to determine $\dot{M}$ and u. For part (b), the exit pressure is less than the critical value, so the flow is sonic at the throat and equations (8-78) and (8-81) must be used to calculate u and $\dot{M}$.

(a)
$$\dot{M} = A_T \sqrt{\frac{2kg_c}{k-1} P_1 \rho_1 \left[\left(\frac{P_2}{P_1}\right)^{2/k} - \left(\frac{P_2}{P_1}\right)^{(k+1)/k}\right]}$$

But $P_2 = 200$ psia and $P_1 = 300$ psia; therefore,

$$\frac{P_2}{P_1} = \frac{2}{3}$$

We now use the ideal gas equation to determine ρ_1:

$$\rho_1 = \frac{MP_1}{RT_1} = \frac{(29.0)(300)(\text{lb}_m/\text{lb-mol})\text{ psi}}{(10.731)(560)(\text{psi ft}^3\text{ °R})/(\text{lb-mol °R})} = 1.45 \text{ lb}_m/\text{ft}^3$$

Therefore,

$$P_1\rho_1 = (300)(144)(1.45)\,(\text{lb}_f/\text{ft}^2)\,(\text{lb}_m/\text{ft}^3)$$
$$= 6.27 \times 10^4\ \text{lb}_f/\text{lb}_m/\text{ft}^5$$
$$A_T = 1\ \text{in.}^2 = \tfrac{1}{144}\ \text{ft}^2$$

Therefore, $\dot{M}$ is calculated as

$$\dot{M} = \frac{\text{ft}^2}{144}\sqrt{\frac{(2.0)(1.4)}{0.4}(6.27 \times 10^4)\left[\left(\frac{2}{3}\right)^{2.0/1.4} - \left(\frac{2}{3}\right)^{2.4/1.4}\right](32.17)\frac{\text{lb}_m\text{-ft lb}_f\ \text{lb}_m}{\text{lb}_f\ \text{sec}^2\ \text{ft}^5}}$$

Therefore,

$$\dot{M} = \frac{1}{144}\sqrt{8.5 \times 10^5\ \text{lb}_m^2/\text{sec}^2}$$

or

$$\dot{M} = 6.40\ \text{lb}_m/\text{sec}$$

We may now calculate u from the relation

$$u = \sqrt{\frac{2g_c k}{k-1}\frac{P_1}{\rho_1}\left[1 - \left(\frac{P_2}{P_1}\right)^{(k-1)/k}\right]}$$

Therefore,

$$u = \sqrt{\frac{(2)(1.4)}{0.4}\frac{(32.17)(300)(144)}{1.45}\left[1 - \left(\frac{2}{3}\right)^{0.4/1.4}\right]\frac{\text{lb}_m\text{-ft}}{\text{lb}_f\ \text{sec}^2}\frac{\text{lb}_f\text{-ft}^3}{\text{ft}^2\text{-lb}_m}}$$

or

$$u = \sqrt{6.73 \times 10\,(0.133)\ \text{ft}^2/\text{sec}^2} = 945\ \text{ft/sec}$$

(b) Since the exit pressure is below the critical pressure, we must now use the formulas derived for critical flows. Therefore,

$$\dot{M}_c = A\left(\frac{2}{k+1}\right)^{1/(k-1)}\sqrt{\frac{2kg_c}{k+1}P_1\rho_1}$$

or

$$\dot{M}_c = \frac{\text{ft}^2}{144}\left(\frac{2}{2.4}\right)^{1/0.4}\sqrt{\frac{2.8}{2.4}(6.27 \times 10^4)(32.17)\frac{\text{lb}_m\text{-ft lb}_m\ \text{lb}_f}{\text{lb}_f\ \text{sec}^2\ \text{ft}^5}}$$

Therefore,

$$\dot{M}_c = 4.37 \times 10^{-3}\sqrt{2.36 \times 10^6\ \text{lb}_m^2/\text{sec}^2} = 6.72\ \text{lb}_m/\text{sec}$$

and is the maximum amount of gas this nozzle can pass at an inlet pressure of 300 psia. The throat velocity is sonic, and is evaluated from

$$u_c = \sqrt{\frac{2kg_c}{k+1}\frac{P_1}{\rho_1}} = \sqrt{\frac{(2)(1.4)}{2.4}\frac{(32.17)(300)(144)}{1.45}\frac{\text{lb}_m\text{-ft lb}_f}{\text{lb}_f\ \text{sec}^2\ \text{ft}^2}\frac{\text{ft}^3}{\text{lb}_m}}$$

Therefore,

$$u_c = \sqrt{1.12 \times 10^6\ \text{ft}^2/\text{sec}^2} = 1060\ \text{ft/sec}$$

SAMPLE PROBLEM 8-9. An ideal gas is passed through a nozzle that has been specially designed so that the working fluid remains at the inlet temperature (that is, the flow is isothermal). Is there a maximum (critical) flow rate that may be passed through the nozzle? If so, express the critical mass flow rate and the critical exit velocity in terms of the throat area and the inlet conditions.

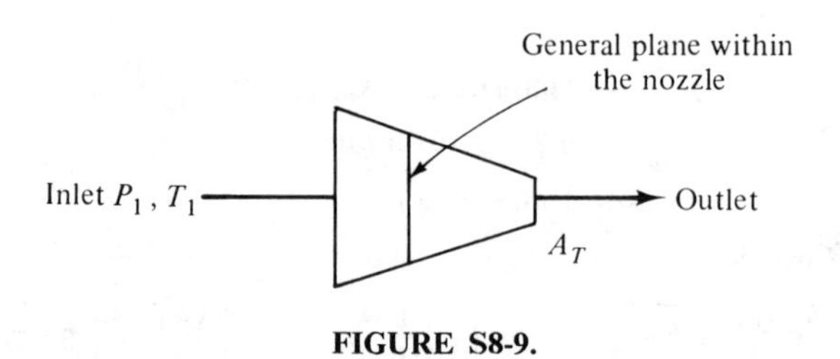

FIGURE S8-9.

Solution: We may picture the nozzle as shown in Fig. S8-9.

Let us now write the mechanical energy balance around a portion of the nozzle which includes everything up to the general plane shown:

$$\underline{LW} + \underline{W} + \int_{P_1}^{P} \underline{V}\, dP + \Delta \frac{u^2}{2g_c} + \frac{g\Delta Z}{g_c} = 0$$

But

$$\underline{LW} = \underline{W} = g\Delta Z = 0$$

Also if the inlet area is large, u_1^2 will normally be quite small in comparison to u^2, and therefore is neglected. Thus the mechanical energy balance reduces to

$$\frac{u^2}{2g_c} = -\int_{P_1}^{P} \underline{V}\, dP$$

or

$$u^2 = -2g_c \int_{P_1}^{P} \underline{V}\, dP$$

But for an ideal gas, $\underline{V} = RT/P$. Therefore, for isothermal flows, we may evaluate the integral as

$$u^2 = -2g_c \int_{P_1}^{P} \frac{RT}{P}\, dP = -2g_c RT \ln \frac{P}{P_1}$$

or

$$u = \sqrt{-2g_c RT \ln \frac{P}{P_1}}$$

The equation of conservation of mass may now be written:

$$\dot{M} = \rho u A = \text{const}$$

Therefore,

$$\frac{\dot{M}}{A} = \rho u = \rho \sqrt{2g_c RT \ln \frac{P_1}{P}}$$

But, for an ideal gas

$$\rho = \frac{P}{RT}$$

Therefore,

$$\frac{\dot{M}}{A} = \frac{P}{RT} \sqrt{2g_c RT \ln \frac{P_1}{P}} = P \sqrt{\frac{2g_c}{RT} \ln \frac{P_1}{P}}$$

We may now apply this equation to the throat of the nozzle, so $\dot{M}$ is expressed as

$$\dot{M} = A_T P_2 \sqrt{\frac{2g_c}{RT} \ln \frac{P_1}{P_2}}$$

Examination of the function $\phi(P_2) = P_2\sqrt{\ln (P_1/P_2)}$ indicates $\phi(P_2) = 0$ for

$P_2 = 0$, or $P_2 = P_1$, and $\phi(P_2) > 0$ for $0 < P_2 < P_1$. Therefore, $\phi(P_2)$ must pass through a maximum for $0 < P_2 < P_1$. As in the case of adiabatic flow, this pressure is called the critical pressure, P_c. Since $\dot{M} = A_T(\text{const})\,\phi(P_2)$, $\dot{M}$ must also pass through a maximum at the critical pressure. (In a manner identical to that used for the adiabatic nozzles, we can show that P_c is the lowest pressure attainable in a converging nozzle. Attainment of lower pressures requires a converging–diverging nozzle. For exit pressures less than P_c, the pressure at the throat is P_c, and a rarefaction wave develops just outside the nozzle throat.) To find the maximum flow rate, we must set $d\phi(P_2)/dP_2 = 0$. Therefore,

$$\frac{d[P_2\sqrt{\ln (P_1/P_2)}]}{d(P_2)} = 0$$

Since P_1 is constant, this also implies that

$$\frac{d[P_2/P_1\sqrt{-\ln (P_2/P_1)}]}{d(P_2/P_1)} = 0$$

Therefore,

$$\sqrt{-\ln\frac{P_2}{P_1}} - \frac{\frac{1}{2}(P_2/P_1)[-\ln (P_2/P_1)]^{-1/2}}{P_2/P_1} = 0$$

or

$$\sqrt{-\ln\frac{P_2}{P_1}} - \frac{1}{2\sqrt{-\ln (P_2/P_1)}} = 0$$

simplifying then yields

$$-\ln\frac{P_2}{P_1} - \frac{1}{2} = 0$$

Thus

$$\ln\frac{P_2}{P_1} = -\frac{1}{2}$$

or

$$\frac{P_2}{P_1} = e^{-1/2} = 0.606$$

Therefore, $P_c = P_1(0.606)$. Thus the critical mass flow rate is given by

$$\dot{M}_c = A_T(0.606P_1)\sqrt{\frac{-2g_c}{RT}\ln 0.606}$$

$$= 0.428A_TP_1\sqrt{\frac{2g_c}{RT}}$$

The critical velocity is given by

$$u_c = \sqrt{-2g_cRT\ln 0.606} = \sqrt{g_cRT}$$

and can be shown to be the velocity at which a reversible, isothermal (rather than adiabatic) pressure disturbance would propagate through a quiescent medium.

8.4 The Converging–Diverging Nozzle

We have shown in Section 8.3 that the maximum exhaust velocity that may be achieved with a simple converging nozzle is the velocity of sound at the throat conditions. However, many nozzle applications, such as thrust nozzles for jet and rocket engines, require exhaust velocities many times the speed of sound. Therefore, the pressures within the nozzles at their exits must be considerably lower than the critical pressure, and the exit areas of these nozzles must be greater than the throat area. That is, the nozzle must contain a converging section that ends at the throat, followed by a diverging section as shown in Fig. 8-14. This type of nozzle is termed a converging–diverging nozzle.

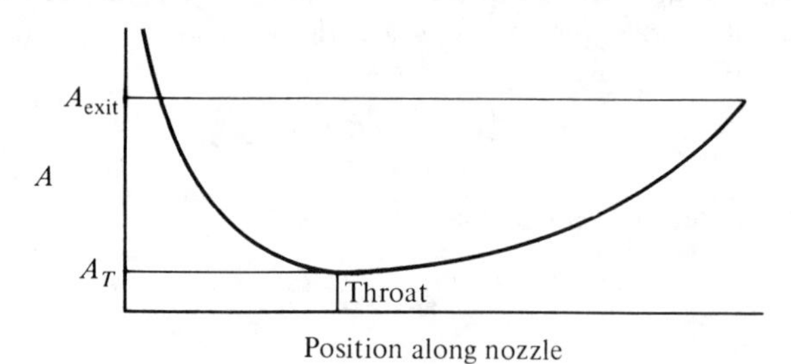

FIG. 8-14. Converging-diverging nozzle.

We may obtain some valuable information about the flows that exist in each of the portions of the converging–diverging nozzle by examining the rate at which P (and thus u) varies with area in the nozzle:

$$A = \frac{\dot{M}}{\rho\sqrt{[2g_c k/(k-1)](P_1/\rho_1)[1-(P/P_1)^{(k-1)/k}]}} \tag{8-66}$$

Equation (8-66) has been derived as an expression between P and A. Let us differentiate this with respect to P at constant mass flow rate and entropy (remember, the expansion within the nozzle is isentropic):

$$\begin{aligned}\left(\frac{\partial A}{\partial P}\right)_S &= \frac{\rho^{-2}(\partial\rho/\partial P)_S\,\dot{M}}{\sqrt{2[kg_c/(k-1)](P_1/\rho_1)[1-(P/P_1)^{(k-1)/k}]}}\\ &\quad+\frac{\dot{M}}{\rho\sqrt{[2kg_c/(k-1)](P_1/\rho_1)}}\,\frac{-\frac{1}{2}}{\sqrt{[1-(P/P_1)^{(k-1)/k}]^3}}\\ &\quad\times\frac{-[(k-1)/k]\,P^{[((k-1)/k)-1]}}{(P_1)^{(k-1)/k}}\end{aligned} \tag{8-82}$$

or

$$\begin{aligned}\left(\frac{\partial A}{\partial P}\right)_S &= \frac{-(\partial\rho/\partial P)_S\,\dot{M}}{\rho^2\sqrt{2[kg_c/(k-1)](P_1/\rho_1)[1-(P/P_1)^{(k-1)/k}]}}\\ &\quad+\frac{\dot{M}P^{-1/k}P_1^{(1-k)/k}[(k-1)/k]}{2\rho\sqrt{[2kg_c/(k-1)](P_1/\rho_1)[1-(P/P_1)^{(k-1)/k}]^3}}\end{aligned} \tag{8-82a}$$

Substitution of equation (8-66) into (8-82a) yields

$$\left(\frac{\partial A}{\partial P}\right)_S = -\frac{A}{\rho}\left(\frac{\partial\rho}{\partial P}\right)_S - A\left[\frac{(P/P_1)^{-1/k}(P_1)^{-1}}{2[kg_c/(k-1)][1-(P/P_1)^{(k-1)/k}]}\right] \tag{8-83}$$

But

$$u^2 = -2\frac{kg_c}{k-1}\frac{P_1}{\rho_1}\left[\left(\frac{P}{P_1}\right)^{(k-1)/k}-1\right] \tag{8-62}$$

Therefore, equation (8-83) becomes

$$\left(\frac{\partial A}{\partial P}\right)_S = -\frac{A}{\rho}\left(\frac{\partial \rho}{\partial P}\right)_S - \frac{A\rho_1}{u^2}\left(\frac{P}{P_1}\right)^{-1/k} \tag{8-84}$$

But we have shown that

$$\frac{\rho_1}{\rho} = \left(\frac{P_1}{P}\right)^{1/k} \tag{8-59}$$

Therefore equation (8-84) simplifies to

$$\left(\frac{\partial A}{\partial P}\right)_S = \frac{A}{\rho}\left[\frac{1}{u^2} - \left(\frac{\partial \rho}{\partial P}\right)_S\right] \tag{8-85}$$

But $(\partial\rho/\partial P)_S = 1/C^2$, where C is the speed of sound. Thus equation (8-85) becomes

$$\left(\frac{\partial A}{\partial P}\right)_S = \frac{A}{\rho u^2}\left(1 - \frac{u^2}{C^2}\right) = \frac{A}{\rho u^2}(1 - M^2) \tag{8-86}$$

where the Mach number, M, has been defined as

$$M = \frac{u}{C} \tag{8-87}$$

Since A, ρ_1, and u^2 are always positive, we may now make some observations about the way pressure will change with area. In particular, we note that a decrease in nozzle area may cause either an increase or a decrease in the nozzle pressure (and hence a decrease or increase in the velocity), depending on the Mach number, M. The effect of area changes on P and u, for constant $\dot{M}$, may be summarized as shown in Table 8-3. For example, if a gas enters a

TABLE 8-3 Effect of Area Changes on *P* and *u*

Flow Type	Mach No.	Change of P with A	Change of u with A
Subsonic	$M < 1$	$\partial P/\partial A > 0$	$\partial u/\partial A < 0$
Supersonic	$M > 1$	$\partial P/\partial A < 0$	$\partial u/\partial A > 0$
Transonic	$M = 1$	$\partial P/\partial A = \infty$	$\partial u/\partial A = 0$

converging nozzle at subsonic speed, the maximum velocity this gas can ever achieve is sonic, because if the gas at any point reached a speed slightly above the sonic velocity, continued flow in the converging nozzle would cause a decrease in the velocity and an increase in the pressure. In this manner, we confirm the result, derived in Section 8.3, that the maximum velocity attainable in a converging nozzle is the speed of sound determined at the throat conditions.

We may now qualitatively describe the pressure and velocity versus area relation of a gas as it passes through a converging–diverging nozzle as follows. The gas enters the converging section of the nozzle at high pressure and low velocity, so $M < 1$. Since the flow is subsonic, the pressure decreases and the velocity increases as the area decreases. This continues until the throat is reached.

At the throat, the flow may be sonic or subsonic (but never supersonic), depending on the exit pressure. If the flow is subsonic, the pressure will increase, and the velocity will decrease in the diverging portion of the nozzle as the area increases. However, if the flow is sonic at the throat, two distinct flow regimes are possible in the diverging portion of the nozzle. If the exit pressure is low enough, the pressure will continue to drop along the nozzle and the flow will become supersonic, with increasing velocity. However, if the exit pressure is above a certain value, the pressure within the nozzle will have to increase in the diverging section of the nozzle, and the flow will be subsonic.

Let us now quantitatively examine the behavior of the converging–diverging nozzle. It is assumed that the throat and exit areas, A_T and A_2, respectively, of the nozzle under consideration are known (or are to be determined). Since the converging section of the converging–diverging nozzle behaves in a manner exactly analogous to a simple converging nozzle, there is a maximum flow rate that may pass through the throat. This maximum occurs when the throat pressure is P_c and the throat velocity is sonic.

Once the maximum flow rate of the nozzle is determined, one may determine the exit pressure P_2 such that $\dot{M}_c/A_2$ satisfies equation (8-66):

$$\frac{\dot{M}}{A} = \sqrt{\frac{2kg_c}{k-1} P_1 \rho_1 \left[\left(\frac{P}{P_1}\right)^{2/k} - \left(\frac{P}{P_1}\right)^{(k+1)/k}\right]} \tag{8-66}$$

It is found that two different pressures, known as the *design pressures* of the nozzle, may satisfy equation (8-66). At the upper design pressure it is found that $P_2 > P_c$; at the lower design pressure it is found that $P_2 < P_c$. The flow conditions in the diverging portion of the nozzle for the two design pressures may be summarized as follows:

1. Upper design pressure: Since $P_2 > P_c$, the pressure in the nozzle must increase with the increasing area. Thus the velocity decreases and the flow is subsonic.

2. Lower design pressure: Since $P_c > P_2$, the pressure within the nozzle must decrease with increasing distance from the throat. Therefore, the velocity continues to increase beyond sonic velocity, and the flow is supersonic.

The pressure profiles (pressure–position plots) for cases 1 and 2 are presented as curves *A* and *B* respectively, in Fig. 8-15.

If the exit pressure, P_2, is greater than the upper design pressure but still less than P_1, it will not be possible for sonic flow to occur in throat, and subsonic flow will be encountered throughout the nozzle. This case is quite similar to flow through a converging nozzle with $P_2 > P_c$. The discharge rate of the nozzle will be less than $\dot{M}_c$ and can be calculated directly from equation (8-66). The pressure profile for the case of P_2 greater than the upper design pressure is illustrated by curve *C* in Fig. 8-15.

If the external pressure, P_2, is less than the lower design pressure, a situation develops which is analogous to that for a simple converging nozzle operating with $P_2 < P_c$. We find that critical flow develops in the throat. Since the pressure decreases as the area increases, supersonic flow develops in the diverg-

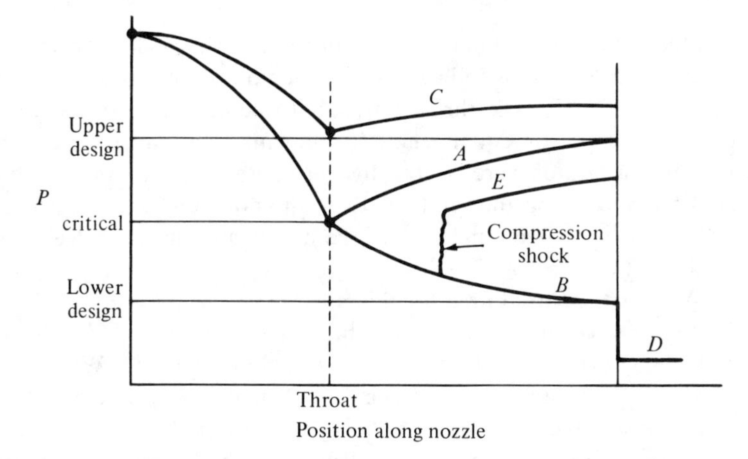

FIG. 8-15. Pressure profiles for a converging–diverging nozzle.

ing portion of the nozzle. However, the lower design pressure is the lowest pressure that may be attained within the nozzle; the surroundings are at a still lower pressure. Thus we find that a pressure discontinuity (an expansion wave) develops at the nozzle exit. This case is illustrated by curve *D* of Fig. 8-15.

If the exit pressure is between the upper and lower design pressure, sonic flow develops at the throat. Since the external pressure is less than the upper design pressure, the flow in the diverging portion of the nozzle begins to expand and become supersonic, as if the exit pressure was at the lower design pressure. However, as the exit of the nozzle is approached, the gas within the nozzle begins to feel the pressure of the surroundings, which is greater than the lower design pressure. Since the gas cannot leave the nozzle at a pressure less than that of the surroundings, there will be a sharp pressure increase some place within the nozzle. This pressure increase manifests itself in the form of a "normal compression shock wave." As the gas passes through the compression shock wave, it is decelerated from supersonic to subsonic flow and experiences a sharp pressure increase. Since the flow leaving the shock is subsonic, its velocity continues to fall while the pressure rises in the remainder of the nozzle. The pressure profile corresponding to operation with an exit pressure between the two design pressures is represented in curve *E* of Fig. 8-15.

When a nozzle is operated so that the exit pressure is between the design pressures, the working fluid is said to be overexpanded. This condition is usually to be avoided in any nozzle designed for long-term operation, because the violent forces developed in the compression shock wave may be strong enough to damage the nozzle. The simplest solution to overexpansion is usually to decrease the exit area until the exit pressure equals one of the design pressures.

SAMPLE PROBLEM 8-10. Liquid hydrogen is to be burned in the combustion chamber of a liquid-fuel rocket. Liquid oxygen, at the stoichiometric rate, will be used as the oxidizing medium. The combustion is designed to operate at 600 psig, and initial calculations indicate that the combustion gases will enter the thrust nozzle at 2000°R. The thrust nozzle is to be designed so that there are equal pressure drops for equal length increments, and must be capable of exhausting to the atmosphere. If the hydrogen flow rate to the combustion chamber is 50 lb_m/sec and the total length of the nozzle is expected to be 6 ft,

(a) Plot the cross-sectional area of the nozzle and the steam velocity as a function of position.

(b) What is the throat area for this nozzle?

(c) It has been suggested that extra thrust may be obtained from the rocket in outer space by expanding the exhaust gases below 14.7 psia. What must the exit area be for an exit pressure of 1 psia? What is the exit velocity? Does this seem to be a reasonable method of obtaining extra thrust?

You may assume that steam is an ideal gas with $C_V = 6.64$ Btu/lb-mol °F under the conditions expected in the nozzle.

Solution: (a) We may picture the combustion chamber and thrust nozzle as shown in Fig. S8-10a. The pressure within the nozzle will vary linearly with position from 614.7 psia at the thrust chamber to 14.7 psia, as shown in Fig. S8-10b. The hydrogen flow

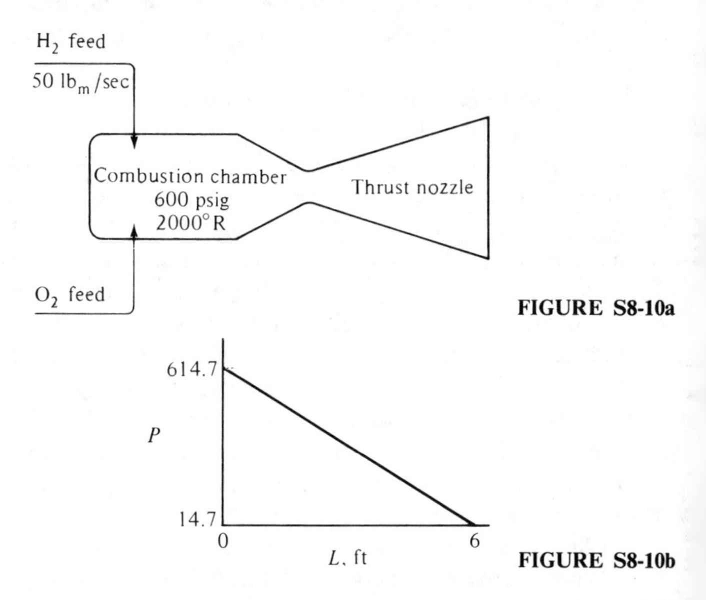

FIGURE S8-10a

FIGURE S8-10b

rate to the combustion chamber is 50 lb_m/sec = 25 lb-mol/sec. Therefore, the steam flow rate into the thrust nozzle is 25 lb-mol/sec = $\dot{M}$.

The area at any point within the nozzle is related to the pressure through the relation

$$A = \frac{\dot{M}}{\sqrt{\frac{2kg_c}{k-1}P_1\rho_1\left[\left(\frac{P}{P_1}\right)^{2/k} - \left(\frac{P}{P_1}\right)^{(k+1)/k}\right]}}$$

where

$$P_1 = 614.7 \text{ psia} = 8.85 \times 10^4 \text{ lb}_f/\text{ft}^2$$

$$\rho_1 = \frac{P_1}{RT_1} = \frac{614.7 \text{ psia}}{(10.75)(2000)} \frac{\text{lb-mol}}{\text{psia ft}^3}$$
$$= 2.86 \times 10^{-2} \text{ lb-mol/ft}^3$$

$$k = \frac{8.63}{6.64} = 1.300; \quad \frac{2}{k} = 1.54; \quad \frac{k+1}{k} = 1.77; \quad \frac{k-1}{k} = 0.231$$

$$\dot{M} = 25 \text{ lb-mol/sec}$$

but

$$\sqrt{\frac{2kg_cP_1\rho_1}{k-1}} = \sqrt{\frac{2(1.3)32.17(8.85 \times 10^4)(2.86 \times 10^{-2})}{0.3(18) \text{ lb}_m/\text{lb-mol}} \frac{\text{lb}_m\text{-ft}}{\text{lb}_f \text{ sec}^2} \frac{\text{lb}_f}{\text{ft}^2} \frac{\text{lb-mol}}{\text{ft}^3}}$$
$$= \sqrt{3.92 \times 10^4 \text{ lb-mol}^2/\text{sec}^2 \text{ ft}^4} = 198 \text{ lb-mol/sec ft}^2$$

Therefore,

$$A = \frac{25 \text{ lb-mol/sec}}{198 \text{ (lb-mol/sec ft}^2) \sqrt{(P/P_1)^{2/k} - (P/P_1)^{(k+1)/k}}}$$
$$= \frac{0.126 \text{ ft}^2}{\sqrt{(P/P_1)^{2/k} - (P/P_1)^{(k+1)/k}}}$$

The velocity of the steam within the nozzle may be calculated as a function of the pressure from the relation

$$u = \sqrt{\frac{2kg_c}{k-1} \frac{P_1}{\rho_1} \left[1 - \left(\frac{P}{P_1}\right)^{(k-1)/k}\right]}$$

but

$$\sqrt{\frac{2kg_cP_1}{(k-1)\rho_1}} = \sqrt{\frac{2(1.3)(32.17)(8.85 \times 10^4)}{(0.3)(2.86 \times 10^{-2})(18)} \frac{(\text{lb}_m\text{-ft/lb}_f \text{ sec}^2)}{(\text{lb-mol/ft}^3)} \frac{(\text{lb}_f)/\text{ft}^2}{(\text{lb}_m/\text{lb-mol})}}$$

Therefore,

$$\sqrt{\frac{2kg_cP_1}{(k-1)\rho_1}} = \sqrt{4.80 \times 10^7 \text{ ft}^2/\text{sec}^2} = 6940 \text{ ft/sec}$$

or

$$u = 6940 \sqrt{1 - \left(\frac{P}{P_1}\right)^{(k-1)/k}} \text{ ft/sec}$$

We may now evaluate both A and u as functions of pressure and position as shown in Table S8-10.

We may also plot the cross-sectional area and steam velocity as a function of position as shown in Fig. S8-10c.

(b) From the plot of A versus X, we may read the throat area as

$$A_T = 0.545 \text{ ft}^2$$

(c) For an exit pressure of 1.0 psia, $P_2/P_1 = 1.63 \times 10^{-3}$. Therefore,

$$A = \frac{0.26 \text{ ft}^2}{\sqrt{(0.00163)^{1.54} - (0.00163)^{1.74}}}$$
$$= 5.05 \text{ ft}^2$$

$$u = 6940 \sqrt{1 - (0.00163)^{0.231}} \text{ ft/sec}$$

TABLE S8-10 Values of Pressure, Area, and Velocity as a Function of Position

X	P, psia	$\frac{P}{P_1}$	$\left(\frac{P}{P_1}\right)^{2/k}$	$\left(\frac{P}{P_1}\right)^{(k+1)/k}$	$\left(\frac{P}{P_1}\right)^{2/k} - \left(\frac{P}{P_1}\right)^{(k+1)/k}$	A, ft^2	$\left(\frac{P}{P_1}\right)^{(k-1)/k}$	$-\left(\frac{P}{P_1}\right)^{(k-1)/k}$	u, ft/sec
0	614.7	1	1	1	0	∞	1	0	0
1	514.7	0.838	0.763	0.731	0.032	0.705	0.96	0.04	1390
2	414.7	0.674	0.545	0.497	0.048	0.575	0.913	0.087	2050
3	314.7	0.512	0.358	0.305	0.053	0.548	0.853	0.147	2680
4	214.7	0.349	0.197	0.155	0.042	0.614	0.784	0.216	3210
5	114.7	0.189	0.077	0.052	0.025	0.798	0.680	0.320	3920
6	14.7	0.0240	0.0032	0.0014	0.0018	2.96	0.422	0.578	5280

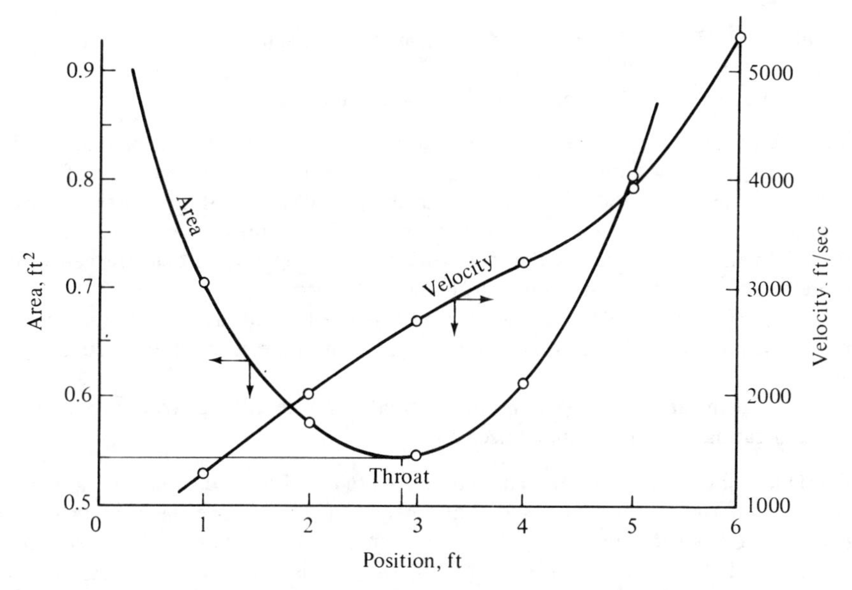

FIGURE S8-10c

Therefore,

$$u = 6100 \text{ ft/sec}$$

Thus we find that we must increase the size, and hence weight, of the nozzle tremendously in order to expand the gas from 14.7 to 1.0 psia, while the exit velocity increases only slightly. This is *not* a practical method of obtaining more thrust from a nozzle, because the added weight of the nozzle will more than likely balance the slight increase in exit velocity.

Problems

8-1 A deep-well pump takes water from 200 ft below the ground and delivers it to a closed storage tank at a mean elevation of 10 ft above the ground. The tank has a volume of 5000 gal and initially contains only air at atmospheric pressure. The air and the water are at 70°F and may be assumed to remain at that temperature at all times. The pump has an efficiency of 75 per cent, and it is driven by an electric motor whose efficiency is 85 per cent. The friction loss in the pipe for the usual pumping velocities is 0.2 ft-$lb_f/(lb_m)$/(ft of pipe). The presence of water vapor in the air in the tank may be neglected. How much electrical energy will have been drawn from the power lines when the tank becomes $\frac{7}{8}$ full?

8-2 Natural gas (assumed to be all methane) is being pumped through a 1-ft-diameter horizontal pipe that is 35 miles long. The upstream pressure is 750 psia and the downstream pressure is 400 psia. A Fanning friction factor of 0.0035 may be assumed. If methane obeys the equation of state

$$\frac{P\underline{V}}{RT} = 1 - \frac{0.4P_r}{T_r}$$

calculate the flow rate and heat transfer through the pipe for the following conditions:

(a) Isothermal flow at 60°F.

(b) Adiabatic flow with an inlet temperature of 60°F.

8-3 The turbine in a hydroelectric generating station receives water from an artifical lake through a 3-ft-diameter concrete pipe 0.5 mile long. The pipe extends from the surface of the lake to the river bed 250 ft below. At all practical flow rates the flow is sufficiently turbulent that a constant Fanning friction factor of $f = 0.005$ may be assumed. In an effort to increase the total work produced by the hydroelectric generators, more water will be allowed to flow through the turbines.

(a) Will an increase in the water flow rate (ft^3/min) always lead to an increased power production? Neglect all losses except friction in the concrete pipe. Justify your answer.

(b) If the answer to (a) is *no*, then at what flow rate will the power production of the generating station be maximized?

8-4 One hundred million standard cubic feet (60°F, 1 atm) per day of radioactive waste gas at 1000°F must be released at a height of 400 ft above the ground to avoid contamination of the surrounding area. A circular stack of uniform diameter is to be used. A draft at the base of the stack of 1 in. of water is available. (The pressure inside the stack base is 1 in. of water less than barometric pressure.) The barometric pressure at the base of the stack is 740 mm Hg and the ambient temperature 60°F. The gas has a molecular weight of 32 and may be considered an ideal gas. Assume that the gas passes through the stack isothermally. What diameter will be required? You can use a Fanning friction factor, $f = 0.016$, and you may assume that u^2 is constant throughout the tower.

8-5 A high-pressure reactor is vented to the surroundings by means of a 3-in.-i.d. pipe that is 4 ft long, as shown in Fig. P8-5. During normal operation a rupture disc

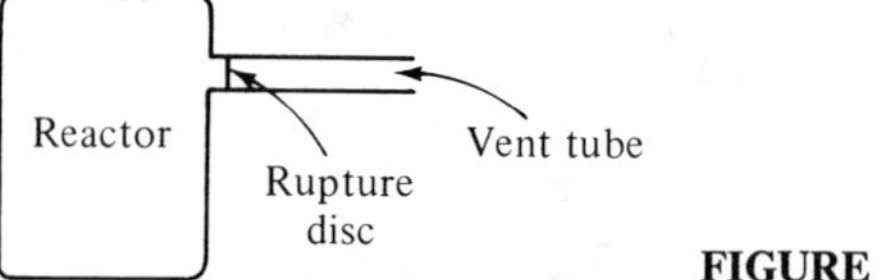

FIGURE P8-5

seals the vent tube. However, if an upset occurs so that the reactor pressure exceeds 100 psig, the rupture disc disintegrates and the reactor is emptied of its contents.

(a) During a test of the vent system the reactor is filled with air at 100 psig when the rupture disc fails. If the flow in the vent line is assumed to be isothermal and of very high Reynolds number, so the friction factor, f, is constant, derive the relation between the rate (lb_m/sec) at which air leaves the reactor (assuming the reactor pressure is 100 psig) and the exit pressure at the vent outlet. (You need *not* substitute numbers.) Use a Fanning friction factor $f = 0.003$.

(b) The vent line would normally be vented to the atmosphere. However, it has been suggested that placing vacuum pumps on the end of the vent line (to decrease the pressure at the end of the vent line) can be used to increase the initial rate at which the reactor empties. If the reactor is filled with air at 100 psig, would these vacuum pumps cause an increase, decrease, or no change at all in the initial rate at which the reactor is emptied? *Justify your answer carefully!*

(c) What is the initial rate at which air at 100 psig would exit from the vent line if the vent leads to the atmosphere? (Assume that the flow is fully developed in the pipeline, so flow field development can be neglected.)

8-6 A high-pressure air line contains a small bleed line whose diameter is $\frac{1}{4}$ in. i.d. and has an effective length of 3 ft. The air in the line is at 800 psia and 80°F. If the valve on the bleed line is accidentally opened, what is the maximum mass flow rate (express in lb_m/hr) of air through the bleed line? What is the exit pressure for the maximum flow rate? What do you suspect will happen if the atmospheric pressure is less than the critical pressure? Calculate the outlet velocity at maximum flow rate, and compare this with sonic velocity for isothermal flow. You may assume that air is an ideal gas at these conditions, the flow in the bleed line may be assumed to be isothermal, and a Fanning friction factor $f = 0.002$ may be assumed.

8-7 An ideal gas whose constant-volume heat capacity is 9.0 Btu/lb-mol °R is passed through two different converging nozzles. In each case the diameter of the discharge throat of the nozzle is 0.4 in., and in each case the gas enters the nozzle at 150 psia and 80°F and is discharged at 120 psia. The nozzles operate differently, however, according to the following description:

(a) This nozzle is well insulated and the friction effect or turbulence is negligible.

(b) This nozzle operates isothermally and the friction effect is negligible.

Show by numerical calculations which nozzle will pass the greatest amount of gas per unit time.

8-8 Gas enters an adiabatic isentropic nozzle at 40 psia and 600°F and with negligible velocity, and is continuously expanded to 1 atm.

(a) At what velocity (ft/sec) will the gas leave the nozzle?

(b) For a gas flow rate of 10 lb_m/hr, what cross-sectional area (ft^2) will be required at the nozzle exit?

The equation of state is $P(\underline{V} - b) = RT$, where $b = 0.5$ ft^3/lb-mol; C_P at 1 atm $= 7.0$ Btu/lb-mol °F; mol wt of gas $= 30$.

8-9 A special research project requires that 0.02 lb-mol of air be fed to a reaction vessel (A), which is initially evacuated (Fig. P8-9). The entire quantity of air is to be

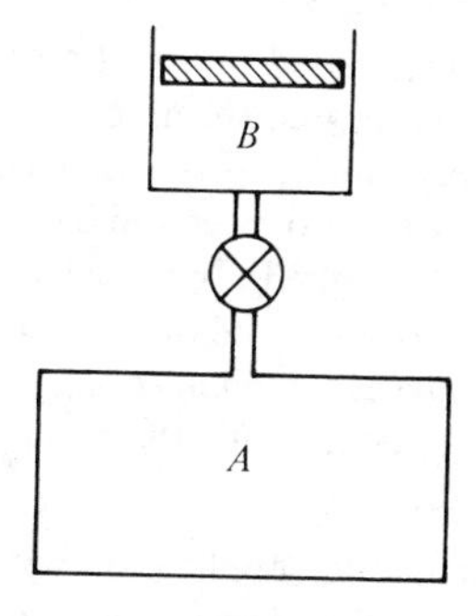

FIGURE P8-9

fed over a 60-sec interval, and the rate of supply must be constant over that interval. It is proposed to accomplish this by feeding the air from tank B through a quick-opening valve and a converging isentropic critical flow nozzle, tank B is to be maintained at constant pressure by a freely floating (assume frictionless) piston. The nozzle will have a cross-sectional area of 0.000027 ft^2 at its throat, and the tanks, line, valve, and nozzle will be well insulated. What is the minimum permissible volume for tank A if the air in B is initially at 100°F and air behaves ideally at all times?

$$C_P = 7 \text{ Btu/lb-mol °R}$$

$$C_V = 5 \text{ Btu/lb-mol °R}$$

$$\frac{P_c}{P_1} = 0.5275 \text{ for air}$$

8-10 A "critical flow prover" is commonly used to measure flow rates accurately in order to calibrate other devices for flow measurement. The instrument consists of a carefully machined orifice stamped with the "air time," defined as the number of seconds required to pass 1 ft³ of air measured at the inlet (upstream) pressure and an inlet temperature of 60°F. The definition thus automatically provides for changes in inlet pressure, while the existence of critical flow ensures that the air time will be independent of discharge pressure. Assuming isentropic flow and ideal gas behavior,

(a) Calculate the air flow (lb_m/hr) through an orifice having an air time of 11.4 sec/ft³ when the inlet temperature is 60°F, the inlet pressure is 40 psia, and the discharge pressure is 15 psia.

(b) Calculate the cross-sectional area (ft²) of an orifice having an air time of 8.5 sec.

(c) Calculate the air flow through the orifice of (a) if the inlet temperature is 80°F, all other conditions remaining the same as in (a).

$$k = 1.4$$

$$\text{mol wt} = 29$$

$$\frac{P_c}{P_1} = 0.5275$$

8-11 Develop the expression for flow through an adiabatic, frictionless nozzle for a gas that has the following properties:

$$\frac{P\underline{V}}{RT} = 1 - 0.4\frac{P_r}{T_r} \qquad C_P = \text{const with } T$$

Find the critical pressure ratio.

8-12 A rocket exhaust gas composed of equal amounts of H_2O, CO_2, CO, and N_2 is to be ejected at a rate of 50 lb_m/sec from an ICBM. The combustion chamber will operate at 2800°K and 200 psia and the gases may be assumed to have a heat-capacity ratio, k, equal to 1.2. *Estimate* the cross-sectional throat area of a nozzle to achieve this performance. If an efficiently designed nozzle has a length approximately 10 times its throat diameter, could this nozzle be used on a rocket that was equipped to use nozzles up to 2 ft in length? If not, suggest an engineering solution to the problem which permits the use of the existing rocket with the same mass ejection rate. Indicate any assumptions made in your solution.

8-13 A nozzle is being designed for application with its exit pressure either very near the inlet pressure or much below it. The exit area of the nozzle is 15 times the throat area, and the nozzle will use air as the working fluid. The air will enter the nozzle at 500 psia and 100°F. Determine the range of pressures over which it is *not safe* to operate this nozzle.

8-14 The air intake of a jet airliner is a diffuser that has the opposite purpose of a nozzle: It increases the pressure of the air flowing through it by performing an adiabatic, reversible deceleration of the air. An intake is to be designed for a plane that is expected to fly at 600 mph at 30,000 ft, where the pressure is 4.6 psia and the temperature is $-50°F$. The diffuser must compress the air to 7.5 psia before the air enters a centrifugal compressor for further compression.

(a) What will be the velocity of the air at the exit from the diffuser?

(b) How large must the cross-sectional area of the inlet and discharge of the diffuser be if it is to compress air at the rate of 15 lb_m/sec?

Multicomponent Systems 9

Introduction

Up to this point our thermodynamic relationships have been limited in application to systems with uniform chemical composition and have dealt only briefly with equilibrium between materials in two or more distinct phases. For instance, we have discussed the properties of pure materials (such as water) and of mixtures with constant composition (such as air). But if we add alcohol or salt to the water or hydrogen to the air, thus changing the composition, our present relationships are inadequate to describe the change in the properties of the system as a function of composition.

We actually want to go well beyond these simple questions to cope with such problems as vapor–liquid equilibrium between petroleum fractions in which a dozen or more components may be present. On each tray of a distillation column, an equilibrium is reached between vapor and liquid mixtures of these components. The compositions are different in each phase and vary markedly from top to bottom of the column. How are these compositions related? We shall see that there are specific methods for treating such problems.

The material in this chapter will extend our thermodynamic relationships so that we can describe systems of two or more chemical components (multicomponent systems) with variable composition. In Chapter 10 we shall discuss multicomponent systems with two or more homogeneous phases present (multiphase systems). Chemical reactions are specifically excluded in this and the next chapter but will be considered in Chapters 11, 12, and 13.

To describe multicomponent systems we must first expand our thermodynamic relations to include the effects of composition. This task requires careful definition and development of several novel concepts.

9.1 Partial Molar Properties

In order that we understand the meaning of thermodynamic terms for a multicomponent system, we shall begin with very basic ideas, recalling first our conclusions for pure components and extending these in a rational manner to multicomponent systems.

We have seen that the total internal energy, U, of a one-phase, pure-component system is a function of two state variables and the size of the system (or number of moles n). For instance,

$$U = U(P, T, n) \tag{9-1}$$

or

$$dU = \left(\frac{\partial U}{\partial P}\right)_{T,n} dP + \left(\frac{\partial U}{\partial T}\right)_{P,n} dT + \left(\frac{\partial U}{\partial n}\right)_{P,T} dn \tag{9-2}$$

The internal energy of a one-phase system of C components must then be given by

$$U = U(P, T, n_1, n_2, n_3, \ldots, n_C) \tag{9-1a}$$

etc.; that is, it is a function of pressure, temperature, and the number of moles of each component present. For any change of state of the system, the change in internal energy will be given by the total differential,

$$dU = \left(\frac{\partial U}{\partial P}\right)_{T,n_i} dP + \left(\frac{\partial U}{\partial T}\right)_{P,n_i} dT + \sum_{i=1}^{C}\left(\frac{\partial U}{\partial n_i}\right)_{P,T,n_j} dn_i \tag{9-3}$$

The subscript n_i in the first two terms denotes that all compositions are held unchanged; the subscript n_j in the third term denotes that the amounts of all species except the ith component are held fixed.

In equation (9-3) we identify the quantity

$$\bar{U}_i \equiv \left(\frac{\partial U}{\partial n_i}\right)_{P,T,n_j} \tag{9-4}$$

as the *partial molar internal energy*. The quantity

$$\bar{U}_i\, dn_i = \left(\frac{\partial U}{\partial n_i}\right)_{P,T,n_j} dn_i \tag{9-5}$$

represents the change in internal energy of the system brought about by adding an increment, dn_i, of component i to the system while holding the pressure, temperature, and all other component amounts constant. The partial molar internal energies, $\bar{U}_i$, of each component in a mixture will depend on the pressure, temperature, and composition of the system but will be independent of the total size of the system. We will use a superscript bar to represent partial molar properties.

Special attention should be paid to the conditions placed on the differentiation to obtain a partial property in that the procedure will be applied to all extensive properties and will be used frequently in the remaining chapters. A partial molar quantity is *always* obtained by differentiation of an extensive property with respect to the number of moles of one of the components *under conditions of constant pressure, temperature, and number of moles of all other components.*

A useful relationship between the partial molar internal energy, $\bar{U}_i$, and the total internal energy, U, can be found in the following manner. Suppose a system of n moles with internal energy U is increased in size by adding to it dn moles of additional material of the same composition. If the total system volume is allowed to increase proportionally, this may be done without changing the temperature and pressure. The change in internal energy of the system is found from equation (9-3):

$$dU = \sum_{i=1}^{C} \bar{U}_i \, dn_i \tag{9-6}$$

Since U is an extensive property and we have not changed the temperature, pressure, or relative mole numbers but only the total system size, the increase in internal energy must be proportional to the change in total system size,

$$dU = \frac{U}{n} dn \tag{9-7}$$

The increase in each of the mole numbers is proportional to the relative amounts of each component originally present,

$$dn_i = \frac{n_i}{n} dn \tag{9-8}$$

Substitution of equations (9-7) and (9-8) in (9-6) and elimination of the factor dn/n on both sides yields the general result

$$U = \sum_{i=1}^{C} \bar{U}_i n_i \tag{9-9}$$

In short, the total internal energy is just the weighted sum of the partial molar internal energies. This result also suggests why the adjective "partial" is employed.

Similar results can be obtained for all other extensive properties of the mixture and their corresponding partial molar properties. For instance,

$$dH = \left(\frac{\partial H}{\partial P}\right)_{T,n_i} dP + \left(\frac{\partial H}{\partial T}\right)_{P,n_i} dT + \sum_{i=1}^{C} \bar{H}_i \, dn_i$$

$$\bar{H}_i = \left(\frac{\partial H}{\partial n_i}\right)_{P,T,n_j}, \quad \text{and} \quad H = \sum_{i=1}^{C} \bar{H}_i n_i \tag{9-10}$$

Exactly equivalent results can be written for S, V, A, and G! Moreover, the relations between these propeties,

$$\begin{aligned} H &= U + PV \\ A &= U - TS \\ G &= H - TS \end{aligned} \tag{9-11}$$

can be differentiated with respect to n_i at constant temperature, pressure, and n_j to produce corresponding relations for the partial molar properties:

$$\begin{aligned} \bar{H}_i &= \bar{U}_i + P\bar{V}_i \\ \bar{A}_i &= \bar{U}_i - T\bar{S}_i \\ \bar{G}_i &= \bar{H}_i - T\bar{S}_i \end{aligned} \tag{9-12}$$

We shall frequently want to refer to the change on mixing of an extensive property such as volume. The total volume of the pure components before mixing is given by

$$V_{\text{components}} = \sum_{i=1}^{C} n_i \underline{V}_i \tag{9-13}$$

Hence the volume change upon mixing is

$$\begin{aligned} \Delta V_{\text{mixing}} &= V_{\text{mixture}} - V_{\text{components}} \\ &= \sum_{i=1}^{C} n_i \bar{V}_i - \sum_{i=1}^{C} n_i \underline{V}_i \\ &= \sum_{i=1}^{C} n_i (\bar{V}_i - \underline{V}_i) \end{aligned} \tag{9-14}$$

Similar results are obtained for all other extensive properties.

In the ensuing discussion of mixtures we shall often find it necessary to refer to the property of a pure component at the same pressure, temperature, and phase as the mixture. This will be done by using the symbol for the pure-component property with a subscript denoting the component. Thus $\underline{V}_i$ denotes the specific volume of pure component i at the same temperature, pressure, and phase as the mixture.

SAMPLE PROBLEM 9-1. Using Table S9-1a,

TABLE 9-1a Specific Gravity of Mixtures of Ethyl Alcohol and Water by Volume and by Weight†, ‡

Specific Gravity	Per cent Alcohol by Volume	Per cent Alcohol by Weight	Grams Alcohol per 100 cm³
1.00000	0.00	0.00	0.00
0.98391	12.40	10.00	9.84
0.97149	24.50	20.01	19.44
0.95745	36.20	30.00	28.73
0.94008	47.00	39.67	37.29
0.93990	47.10	39.76	37.37
0.93971	47.20	39.85	37.45
0.93953	47.30	39.95	37.53
0.93934	47.40	40.04	37.61
0.93916	47.50	40.13	37.69
0.93898	47.60	40.22	37.77
0.93879	47.70	40.32	37.85
0.93861	47.80	40.41	37.93
0.8773	75.00	—	—
0.7939	100.00	100.00	—

† *Handbook of Chemistry and Physics*, Chemical Rubber Publishing Company, Cleveland, 36th ed., 1954–1955, pp. 1932–1938.

‡ Giving the specific gravity at 15.56°C, referred to water at the same temperature. To reduce to specific gravity referred to water at 4°C, multiply by 0.99908.

(a) Calculate $\bar{V}_{CH_3CH_2OH}$ for a solution consisting of 40.04 per cent ethyl alcohol by weight.

(b) Calculate $\bar{V}_{CH_3CH_2OH}$ for a solution consisting of 100 per cent ethyl alcohol by weight.

(c) Calculate $\underline{V}$ for the solution consisting of 40.04 per cent ethyl alcohol by weight.

(d) Calculate $\underline{V}$ for the solution consisting of 100 per cent ethyl alcohol by weight.

Solution: (a) Several techniques may be used to solve this part of the problem. A few of these methods are discussed below:

1. The brute-force technique: By definition,

$$\bar{V}_i = \left(\frac{\partial V}{\partial n_i}\right)_{P,T,n_j} = \lim_{\Delta n_i \to 0} \left(\frac{\Delta V}{\Delta n_i}\right)_{P,T,n_j}$$

That is, $\bar{V}_i$ may be obtained by plotting $(\Delta V/\Delta n_i)_{P,T,n_j}$ versus Δn_i and extrapolating to $\Delta n_i \longrightarrow 0$. Thus $\bar{V}_{alc}$ in a 40.04 per cent (by weight) alcohol–water solution may be found as follows.

We begin by considering 100 cm³ of solution with 40.04 wt. per cent alcohol, 59.96 wt. per cent water. From the data given in the problem, this solution has a density $\rho =$ (specific gravity) (0.99908) $= 0.93848$ g/cm³. Thus the 100 cm³ of solution weighs 93.848 g. Of this, 40.04 per cent, or 37.576 g, is alcohol, and 56.272 g is water. To this original solution, let us add $\Delta n = X$ grams of alcohol, so the solution concentration increases to 40.13 per cent (the next entry on the data table). The density of the new solution is $\rho = (0.93916)(0.99908) = 0.93829$ g/cm³. The total weight of the new solution is

$$M = (93.848 + X)$$

The amount of water in this solution is $(0.5987)(93.848 + X)$ and is equal to the amount in the original 40.04 per cent solution. Thus

$$(0.5987)(93.848 + X) = 56.272$$

which may be solved for X to give

$$X = 0.1423 \text{ g of alcohol added}$$

The volume of the new solution is then $V = M/\rho = 93.990/0.93829 = 100.172$ cm³, so $\Delta V = 0.172$ cm³ and

$$\frac{\Delta V}{\Delta n} = \frac{0.172}{0.143} = 1.20 \text{ cm}^3/\text{g}$$

We perform the same calculations for the next few entries in Table S9-1b. Thus we see

TABLE S9-1b

x_{alc}	Specific Gravity	ρ	x_{H_2O}	$56.272/x_{H_2O}$	$X = [(\ ) - 93.848]$	$V = \frac{93.848 + X}{\rho}$	ΔV	$\Delta V/\Delta n$
40.04	0.93934	0.93848	0.5996	93.848	0	100.00		
40.13	0.93916	0.93829	0.5987	93.990	0.1423	100.172	0.172	1.20
40.22	0.93898	0.93812	0.5978	94.132	0.2838	100.341	0.341	1.20
40.32	0.93879	0.93793	0.5968	94.290	0.4415	100.529	0.529	1.20

that for all intents and purposes the finite difference $\Delta V/\Delta n$ approximates the required derivative and $\bar{V}_{alc} = 1.20$ cm^3/g.

2. A second technique to obtain $\bar{V}$ would be as follows. Take 100 g of water as a base. Then determine the number of grams of alcohol that must be added to this base to form a 10 per cent, 20 per cent, . . . alcohol (by weight) solution. Determine volume of these solutions as functions of concentration. Then plot volume of solution/100 g of water as a function of the number of grams of alcohol (Fig. S9-1). The slope of this curve at any concentration is

$$\text{slope} = \left(\frac{\partial V}{\partial n_{alc}}\right)_{P,T,n_{water}}$$

and hence is the partial molar volume of alcohol at the concentration in question. Thus all one needs to do is determine the slope of this curve at $x_{alc} = 40.04$ per cent.

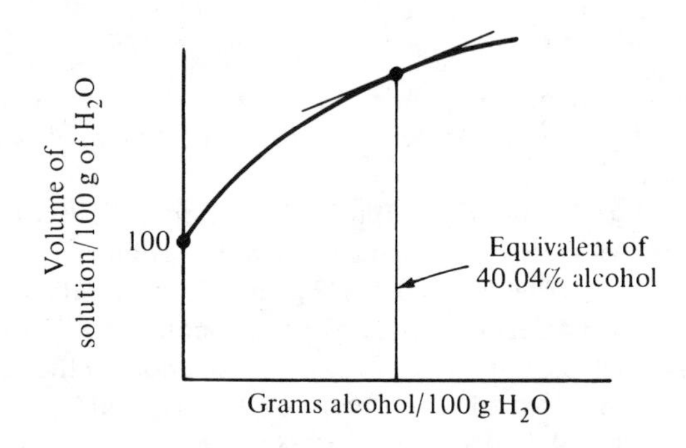

FIGURE S9-1

3. Method of intercepts based on the Gibbs–Duhem equation: This is by far the simplest technique. However, it requires use of the Gibbs–Duhem relation, which is not developed until the next section. The graphical approach is developed in connection with enthalpy–concentration diagrams in Section 9.3 but could be applied to this problem in an exactly analogous fashion.

(b) $\bar{V}_{alc}$ for the pure alcohol is simply $\underline{V}_{alc}$ and is given by

$$\bar{V}_{alc} = \frac{1}{\rho_{alc}} = \frac{1}{(0.7939)(0.99908)} = 1.261 \text{ cm}^3/\text{g}$$

(c) $\underline{V}_{mix} = 1/\rho_{mix} = 1/(0.93934)(0.99908) = 1.0656$ cm^3/g.
(d) Same as (b): $\underline{V}_{alc} = 1.261$ cm^3/g.

A brief comment on the units obtained in the foregoing problem is in order. The derivations should have led the careful reader to expect that the proper units for $\bar{V}_i$ are cm^3/g-mol, not cm^3/g. Conversion requires only division by the molecular weight, but this operation was not carried out because the given units are quite satisfactory. The case is much the same as that in earlier chapters where specific volume $\underline{V}$ was given as often on a mass basis as on a molar basis. The reader can demonstrate for himself that, with suitable redefinition, all results of this and ensuing sections can be interpreted on either a mass or molar basis. Since a great deal of engineering data is given on a mass basis, those will often be the more convenient units. Not until Chapter 11, where we deal with chemical reactions in which mass is conserved (but not always moles), will it be necessary to separate the bases. This apparent ambiguity, then, reduces to but another example of the importance of clearly specifying the units of all numerical answers.

9.2 Partial Molar Gibbs Free Energy—The Chemical Potential

One of the partial molar properties, the partial molar Gibbs free energy, is particularly useful in evaluating the behavior of mixtures. It will be seen later that this quantity plays a central role in the criteria for equilibrium, and hence it is appropriate here to examine in detail the mathematical connection between this and the other thermodynamic properties of mixtures.

We have already seen that

$$dG = \left(\frac{\partial G}{\partial P}\right)_{T,n} dP + \left(\frac{\partial G}{\partial T}\right)_{P,n} dT + \sum_{i=1}^{C} \bar{G}_i \, dn_i \tag{9-15}$$

Hereafter the notation has been modified slightly; the index i is still implied in the first two partial derivatives but is omitted to prevent undue clutter where no ambiguity will result. When there is no change in composition, the above relationship must reduce to the one we have used previously for pure components or mixtures at constant composition:

$$dG = V\,dP - S\,dT \tag{9-16}$$

In other words, we want to preserve the relations established previously,

$$\left(\frac{\partial G}{\partial P}\right)_{T,n} = V \qquad \text{and} \qquad \left(\frac{\partial G}{\partial T}\right)_{P,n} = -S \tag{9-17}$$

Thus we now write equation (9-15) as

$$dG = V\,dP - S\,dT + \sum_{i=1}^{C} \bar{G}_i \, dn_i \tag{9-18}$$

We recall that the Maxwell relationship,

$$\left(\frac{\partial V}{\partial T}\right)_{P,n} = -\left(\frac{\partial S}{\partial P}\right)_{T,n} \tag{9-19}$$

was obtained by equating the derivatives,

$$\frac{\partial^2 G}{\partial P\,\partial T} = \frac{\partial^2 G}{\partial T\,\partial P} \tag{9-20}$$

In like manner, by equating the derivatives obtained from cross differentiation of two other terms of equation (9-18), we obtain

$$\left(\frac{\partial \bar{G}_i}{\partial P}\right)_{T,n} = \left(\frac{\partial V}{\partial n_i}\right)_{P,T,n_j} \tag{9-21}$$

so

$$\left(\frac{\partial \bar{G}_i}{\partial P}\right)_{T,n} = \bar{V}_i \tag{9-22}$$

where the subscript n stands for *all components* and the subscript n_j for *all components but i*. Similarly,

$$\left(\frac{\partial \bar{G}_i}{\partial T}\right)_{P,n} = -\left(\frac{\partial S}{\partial n_i}\right)_{P,T,n_j} = -\bar{S}_i \tag{9-23}$$

Even more useful than the latter is the derivative with respect to temperature of the ratio $\bar{G}_i/T$. Making use of equations (9-23) and (9-12) we find that

$$\begin{aligned}\left(\frac{\partial(\bar{G}_i/T)}{\partial T}\right)_{P,n} &= -\frac{\bar{G}_i}{T^2} + \frac{1}{T}\left(\frac{\partial \bar{G}_i}{\partial T}\right)_{P,n} \\ &= -\frac{1}{T^2}(\bar{H}_i - T\bar{S}_i) + \frac{1}{T}(-\bar{S}_i) \\ &= -\frac{\bar{H}_i}{T^2}\end{aligned} \tag{9-24}$$

Although the central role of the partial molar Gibbs free energy, $\bar{G}_i$, in the description of physical equilibrium has yet to unfold, equations (9-22) and (9-24) already suggest that the partial molar volume, $\bar{V}_i$, and the partial molar enthalpy, $\bar{H}_i$, have important roles in conveying information about the pressure and temperature dependence of $\bar{G}_i$. Although the temperature dependence of $\bar{G}_i$ can also be learned from the partial molar entropy, $\bar{S}_i$, by equation (9-23), it is generally easier to obtain values for $\bar{H}_i$ than for $\bar{S}_i$. Thus we shall use equation (9-24) more frequently than equation (9-23).

It was shown in equation (9-10) that the total Gibbs free energy, G, is related to the partial molar Gibbs free energies by

$$G = \sum_{i=1}^{C} \bar{G}_i n_i \tag{9-25}$$

From this it follows that for any small change,

$$dG = \sum_{i=1}^{C} n_i d\bar{G}_i + \sum_{i=1}^{C} \bar{G}_i\, dn_i \tag{9-26}$$

By comparison to equation (9-18), it is clear that

$$\sum_{i=1}^{C} n_i\, d\bar{G}_i = V\,dP - S\,dT \tag{9-27}$$

and that for any change *at constant temperature and pressure*,

$$\sum_{i=1}^{C} n_i\, d\bar{G}_i = 0 \tag{9-28}$$

This result, known as the *Gibbs–Duhem equation*, will be used frequently in the ensuing material. Exactly similar results can be obtained for the other partial molar extensive properties by repeating the procedure of equations (9-25) to (9-28) for any other property. Thus for any change *at constant temperature and pressure* we may write $\sum_{i=1}^{C} n_i\, d\bar{H}_i = 0$, $\sum_{i=1}^{C} n_i\, d\bar{S}_i = 0$, and so on.

In many situations where the partial molar Gibbs free energy is employed, it is given another name, the chemical potential, and another symbol, μ_i. Indeed we may write

$$\mu_i \equiv \bar{G}_i \tag{9-29}$$

and, if we chose to do so, all the above equations could be rewritten inserting μ_i wherever $\bar{G}_i$ appears. Some authors prefer one symbol, some the other, and the student should be prepared to recognize either symbol.

In some relationships it is more convenient to use the symbol μ_i as seen in the following. If we remember that the internal energy and the Gibbs free energy are related by the equation

$$U = G + TS - PV \tag{9-30}$$

it follows from equation (9-18) that

$$dU = -P\,dV + T\,dS + \sum \mu_i\,dn_i \tag{9-31}$$

On the other hand, we note that $U = U(V, S, n_i)$ and hence that

$$dU = \left(\frac{\partial U}{\partial V}\right)_{S,n} dV + \left(\frac{\partial U}{\partial S}\right)_{V,n} dS + \sum_{i=1}^{C} \left(\frac{\partial U}{dn_i}\right)_{S,V,n_j} dn_i \tag{9-32}$$

In our discussion of relationships for materials of constant composition, we have found that

$$\left(\frac{\partial U}{\partial V}\right)_{S,n} = -P \qquad \text{and} \qquad \left(\frac{\partial U}{\partial S}\right)_{V,n} = T \tag{9-33}$$

and hence

$$dU = -P\,dV + T\,dS + \sum_{i=1}^{C} \left(\frac{\partial U}{\partial n_i}\right)_{S,V,n_j} dn_i \tag{9-34}$$

Since both equations (9-31) and (9-34) must hold for any change in composition, it must follow that

$$\mu_i = \left(\frac{\partial U}{\partial n_i}\right)_{S,V,n_j} \tag{9-35}$$

Similar arguments for H and A yield the results

$$dH = V\,dP + T\,dS + \sum_{i=1}^{C} \mu_i\,dn_i \tag{9-36}$$

$$dA = -P\,dV - S\,dT + \sum_{i=1}^{C} \mu_i\,dn_i \tag{9-37}$$

and

$$\mu_i = \left(\frac{\partial H}{\partial n_i}\right)_{P,S,n_j} = \left(\frac{\partial A}{\partial n_i}\right)_{V,T,n_j} \tag{9-38}$$

Two observations about the chemical potential are needed here. The first is that care should be taken to distinguish between similar-looking partial derivatives such as

$$\left(\frac{\partial H}{\partial n_i}\right)_{P,T,n_j} \qquad \text{and} \qquad \left(\frac{\partial H}{\partial n_i}\right)_{P,S,n_j}$$

The first is the partial molar enthalpy, $\bar{H}_i$, and the second is the chemical potential, μ_i. The two are not equal, as shown in equation (9-12):

$$\mu_i = \bar{G}_i = \bar{H}_i - T\bar{S}_i \tag{9-12}$$

The second observation has to do with the name "chemical potential." It was introduced into the thermodynamic literature because μ_i (or $\bar{G}_i$) is a measure of the chemical energy potential of a component. Thus in equation (9-31), for example, one might say that pressure is a potential that determines the internal energy change associated with a volume change, and temperature a potential that determines the internal energy change associated with an entropy change. By the same token, μ_i (or $\bar{G}_i$) is a potential that determines the internal energy change for a change in composition. Although such an explanation does not justify the use of the quantity in thermodynamic relationships, it does serve to give some appreciation for the nature of this quantity.

9.3 Tabulation and Use of Mixture Property Data

Up to this point we have concentrated on developing a set of mathematical relationships that we can use in describing the thermodynamic behavior of mixtures. Although this task is not completed, it seems appropriate in an engineering textbook to continually relate theory to applications. Thus we pause in this section to consider simple mixing processes, the type of data required to analyze them and methods of tabulating this data. In so doing we use the enthalpy–concentration diagram as an example. However, the discussion and findings relating thereto apply in general to other specific property–composition plots that may be constructed. This is particularly true of the graphical procedures that are developed to evaluate partial enthalpies from $\underline{H}$–x plots.

Enthalpy–Concentration Diagrams

One of the more useful property charts to the engineer concerned with mixing or fractionation processes is the enthalpy–concentration diagram, an example of which is presented in Fig. 9-1 for the NH_3—H_2O system. The reference states (zero-component enthalpies) for this plot have been selected as pure liquid H_2O at 0°C and pure liquid NH_3 at −77°C. It should be noted that such a choice is completely arbitrary and the student will find a variety of choices used in various tabulations throughout the literature. However, for any system it is essential to specify a reference state for each component. The reference states need not be pure components. The concept of the infinitely dilute reference state will be developed later and discussed as one possible alternative to a pure-component reference state.

Data for constructing enthalpy–concentration charts are generally obtained by calorimetric procedures. Suppose, for example, that NH_3 and H_2O are mixed together (as shown in Fig. 9-2) in a flow calorimeter where the pressure is maintained constant and heat is supplied or removed so that the temperature of the mixture is the same as the initial temperature of the pure components. This heat is called the heat of mixing $\underline{Q}_m$ and, if plotted as a function of composition for unit mass of the resulting mixture, produces a curve such as the lower one in Fig. 9-3. Negative values of $\underline{Q}_m$ as shown there imply liberation

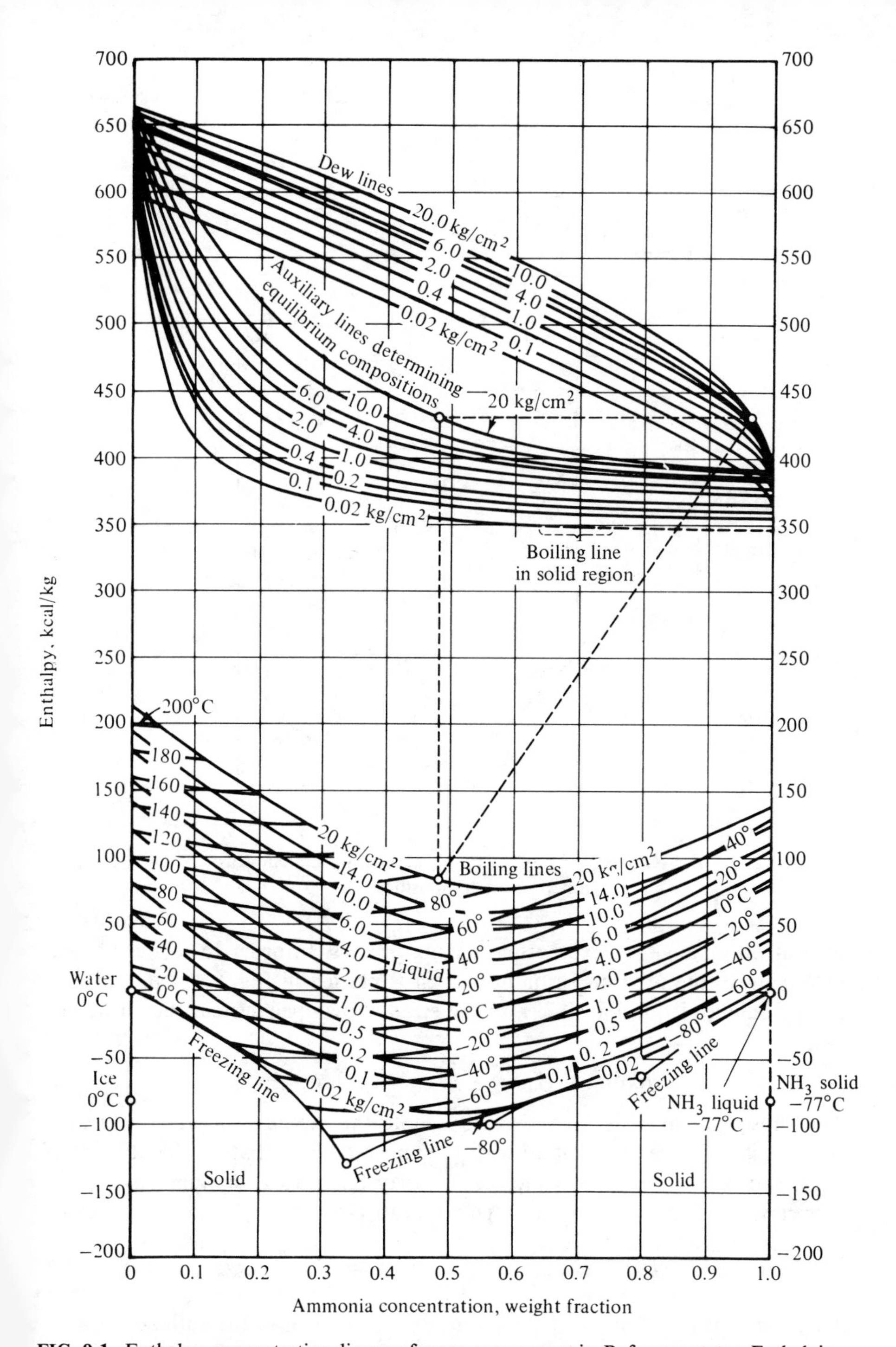

FIG. 9-1. Enthalpy–concentration diagram for aqueous ammonia. Reference states: Enthalpies of liquid water at 0°C and liquid ammonia at −77°C are zero. *Note:* To determine equilibrium compositions, a vertical may be erected from any liquid composition on any boiling line and its intersection with the appropriate auxiliary line determined. A horizontal from this intersection will establish the equilibrium vapor composition on the appropriate dew line: At 48 per cent ammonia and 20 kg/cm² is indicated. [From R. H. Perry et al., *Chemical Engineers Handbook*, McGraw-Hill, Inc., New York, 4th ed., (1963) pp. 3–154.]

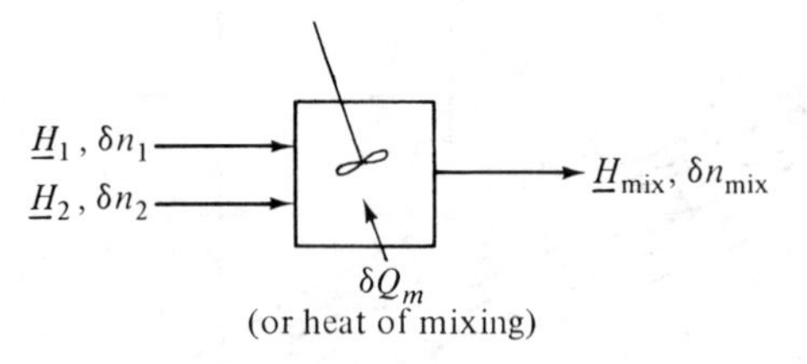

FIG. 9-2. The flow calorimeter.

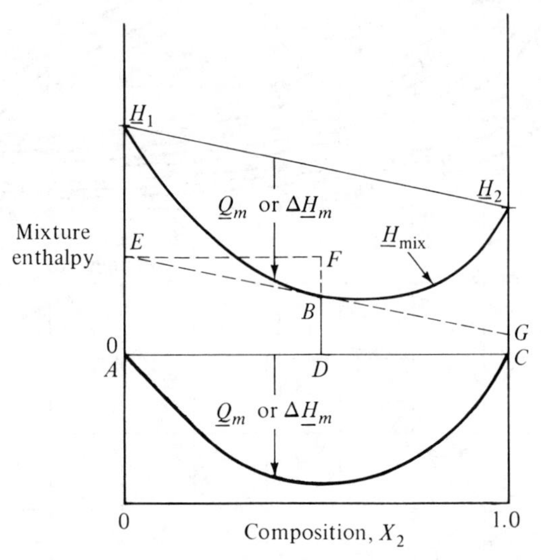

FIG. 9-3. Heat of mixing and mixture enthalpy as functions of composition.

of heat on mixing—an exothermic mixing process. In order to relate the heat of mixing to the mixture enthalpy, which is plotted in Fig. 9-1, we apply the energy balance [equation (9-39)] to the steady-flow open system shown in Fig. 9-2:

$$\underline{H}_1 \delta n_1 + \underline{H}_2 \delta n_2 - \underline{H}_{mix} \delta n_{mix} + \delta Q_m = 0 \tag{9-39}$$

The subscript "mix" is used here to denote the mixture properties. In later sections it is frequently omitted with the understanding that, in the absence of a specific component subscript, the property refers to a mixture or solution property. As seen by reference to equation (9-14),

$$\Delta \underline{H}_m = \underline{H}_{mix} - x_1 \underline{H}_1 - x_2 \underline{H}_2 = \frac{\delta Q_m}{\delta n_{mix}} = Q_m \tag{9-40}$$

Thus the heat measured in the calorimeter is just equal to the enthalpy change upon mixing. In fact $\Delta \underline{H}_m$ is frequently called the heat of mixing. To obtain the upper curve in Fig. 9-3 or the curves shown in Fig. 9-1 ($\underline{H}_{mix}$ versus x) we need to select reference states for the pure components so that their pure-component enthalpies, $\underline{H}_1$ and $\underline{H}_2$, at the temperature of the mixture, can be determined. The mixture enthalpy is then found by rearranging equation (9-40).

Graphical Evaluation of Partial Properties

It is possible to compute the partial molar enthalpies of both components from either of the curves shown in Fig. 9-3. To illustrate the procedure we rewrite equation (9-10) for specific mixture enthalpy:

$$\underline{H}_{\text{mix}} = x_1\bar{H}_1 + x_2\bar{H}_2 \tag{9-41}$$

Since we have two unknown quantities, $\bar{H}_1$ and $\bar{H}_2$, we need another equation. This may be obtained by taking the derivative of equation (9-41) with respect to one mole fraction, say x_2:

$$\frac{d\underline{H}_{\text{mix}}}{dx_2} = \frac{dx_1}{dx_2}\bar{H}_1 + \bar{H}_2 + x_1\frac{d\bar{H}_1}{dx_2} + x_2\frac{d\bar{H}_2}{dx_2} \tag{9-42}$$

Inasmuch as $x_1 = 1 - x_2$, $dx_1/dx_2 = -1$. Moreover, we may eliminate the last two terms by application of the enthalpy form of the Gibbs–Duhem equation:

$$x_1(d\bar{H}_1)_{P,T} + x_2(d\bar{H}_2)_{P,T} = 0 \tag{9-43}$$

Equation (9-42) can now be written

$$\frac{d\underline{H}_{\text{mix}}}{dx_2} = -\bar{H}_1 + \bar{H}_2 \tag{9-44}$$

Equations (9-41) and (9-44) are the two equations from which we can determine $\bar{H}_1$ and $\bar{H}_2$.

A graphical solution technique for this system of equations can be illustrated by combining equations (9-41) and (9-44) to eliminate $\bar{H}_2$,

$$\bar{H}_1 = \underline{H}_{\text{mix}} - x_2\frac{d\underline{H}_{\text{mix}}}{dx_2} \tag{9-45}$$

If we construct the tangent *EBG* to the $\underline{H}_{\text{mix}}$–$x_2$ curve at the composition marked *D* and substitute the line segments for the quantities involved in equation (9-45), noting that $AD = EF$ and $AE = DF$, we find that

$$\begin{aligned}\bar{H}_1 &= DB - AD\frac{-BF}{EF}\\ &= DB + BF = DF\\ &= AE\end{aligned} \tag{9-46}$$

In like manner,

$$\bar{H}_2 = CG \tag{9-46a}$$

If short, the values of the partial molar enthalpies are given by the intercepts of the line *EG* (the tangent to the $\underline{H}_{\text{mix}}$–composition curve at the composition in question) with the two pure-component axes.

If we had chosen to work with the heat of mixing versus composition curve directly we would have found that the tangent to the ΔH_m curve intercepts the pure-component axes at $\bar{H}_1 - \underline{H}_1$ and $\bar{H}_2 - \underline{H}_2$. Thus we must still determine the enthalpies of the pure components, $\underline{H}_1$ and $\underline{H}_2$, before values can be assigned to $\bar{H}_1$ and $\bar{H}_2$.

A similar procedure could be used for obtaining partial molar volumes from data on volume change upon mixing, as was suggested in Sample Problem 9-1. This procedure may require slight alteration when the data are given as densities. However, the technique is still straightforward. The graphical procedure can be used only for binary mixtures, but the preceding analysis can be used for mixtures of any number of components, provided that enough data are at hand to evaluate the required derivatives of the mixture enthalpies. In Chapter 10 these methods will be applied to $\underline{G}$–x plots to obtain partial molar Gibbs free energies, $\bar{G}_i$.

SAMPLE PROBLEM 9-2. It has been suggested that heat-of-solution data may be based on any two arbitrarily selected reference states. Given integral, isothermal heats of solution for ammonia and water at 32°F and the reference states $\underline{H} = 0$ and $\bar{H}_{NH_3} = 0$ for a 20 weight per cent ammonia solution at 32°F, prove that the following quantities are uniquely determined at 32°F:

(a) $\bar{H}_{H_2O}$ in the 20 per cent solution.
(b) $\underline{H}_{H_2O}$.
(c) $\underline{H}_{NH_3}$.
(d) $\bar{H}_{NH_3}$ in an infinitely dilute solution (water).
(e) $\bar{H}_{H_2O}$ in an infinitely dilute solution (water).
(f) $\bar{H}_{H_2O}$ in an infinitely dilute solution (ammonia).
(g) $\underline{H}$ for a 50 per cent solution.

Isothermal heat-of-solution data for ammonia–water solutions have been reported as given in Table S9-2a.

TABLE S9-2a

Pounds of H_2O Added to 1 lb_m of Liquid NH_3 at 32°F	NH_3, wt %	Heat Evolved ($-Q$) Btu/lb_m of NH_3 in Solution
0	100	0
0.111	90	35.5
0.25	80	75.8
0.429	70	121.1
0.667	60	169.7
1.00	50	218.8
1.50	40	270.0
2.33	30	308.2
4.00	20	328.5
9.00	10	343.8
∞	0	358.0

Solution: (a) We are told that the reference states for the NH_3 and water are set by the requirement

$$\underline{H}_{soln} = 0$$
$$\bar{H}_{NH_3} = 0$$

for a 20 wt per cent NH_3 solution at 32°F. We may determine $\bar{H}_{H_2O}$ in the 20 per cent solution by the relation

$$\underline{H} = x_{H_2O}\bar{H}_{H_2O} + x_{NH_3}\bar{H}_{NH_3}$$

which may be solved for $\bar{H}_{H_2O}$ (since the weight fractions, $\bar{H}_{NH_3}$ and $\underline{H}$, are all known) to give

$$\bar{H}_{H_2O} = 0 \text{ at } 32°\text{F in a } 20\% \text{ solution}$$

Thus the two reference enthalpies are now known.

(b) To get $\underline{H}_{H_2O}$ we need simply to determine the quantity $[(\bar{H}_{H_2O})_{20\%\,soln} - \underline{H}_{H_2O}]$, because we already know $(\bar{H}_{H_2O})_{20\%\,soln}$. However, the quantity in brackets is given by the intercept of the line that is tangent to the $(\Delta \underline{H}_m)_{32°F}$ versus x_{NH_3} curve at 20 per cent NH_3 with the pure-water axis. The $\Delta \underline{H}_m$ curve is obtained by dividing the heat absorption on solution by the total weight of solution. These results are listed in Table S9-2b. The data are plotted in Fig. 9-2b. The tangent to the curve at 20 per cent NH_3 is seen to intersect the pure-water axis at $\bar{H}_{H_2O} - \underline{H}_{H_2O} = -6$ Btu/lb$_m$. Thus $\underline{H}_{H_2O} = 6$ Btu/lb$_m$.

TABLE S9-2b

lb_{H_2O} added/lb_{NH_3}	lb_{soln}/lb_{NH_3}	NH_3, wt %	$-Q_{soln}$, Btu/lb_{NH_3}	$-\Delta H_m$ Btu/lb_{soln}
0	1.00	100	0	0
0.111	1.111	90	35.5	32.0
0.25	1.250	80	75.8	60.6
0.429	1.429	70	121.1	84.7
0.667	1.667	60	169.7	101.8
1.00	2.00	50	218.8	109.4
1.500	2.500	40	270.0	108.0
2.333	3.330	30	308.2	92.5
4.00	5.00	20	328.5	65.7
9.00	10.00	10	343.8	34.4
		0	358.0	0

(c) $\underline{H}_{NH_3}$ may be found by extrapolating the tangent to the NH_3 axis as done with the water. However, this is obviously an inaccurate procedure and should be used only when no other approach is possible. Let us take an energy balance around a mixing chamber to which 4 lb of H_2O and 1 lb of NH_3 are isothermally mixed at 32°F (see Fig. S9-2a):

$$4(\underline{H}_{H_2O}) + 1(\underline{H}_{NH_3}) - 5(\underline{H}_{soln}) + Q = 0$$

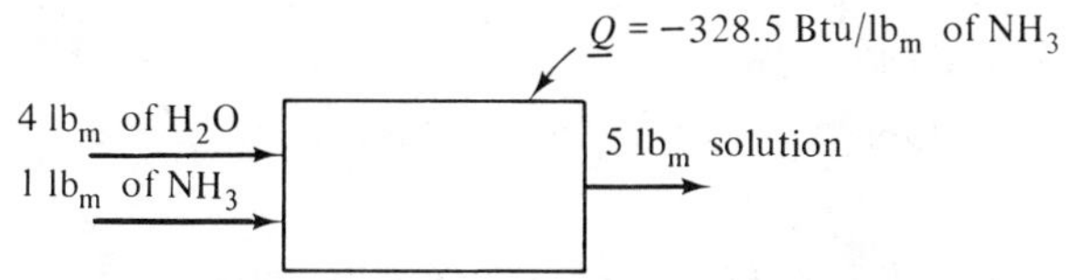

$\underline{Q} = -328.5$ Btu/lb$_m$ of NH_3

4 lb$_m$ of H_2O

1 lb$_m$ of NH_3

5 lb$_m$ solution

FIGURE S9-2a

but $\underline{H}_{H_2O} = 6$ Btu/lb$_m$, $Q = -328.5$ Btu/lb$_m$, and $\underline{H}_{soln} = 0$ Btu/lb$_m$. Thus

$$\underline{H}_{NH_3} = 304 \text{ Btu/lb}_m$$

(d) To get $(\bar{H}_{NH_3})_{\infty \text{ dilute } H_2O}$ we must extrapolate the tangent to the $\Delta \underline{H}_m$ versus x_{NH_3} curve from point $x_{NH_3} = 0$ to the pure NH_3 axis (Fig. S9-2b). This extrapolation is difficult, but we have no choice. The intercept on the NH_3 axis is -360 Btu/lb$_m$ and equals $[(\bar{H}_{NH_3})_{soln} - (\bar{H}_{NH_3})_{pure}]$. But $(\bar{H}_{NH_3})_{pure} = 304$ Btu/lb$_m$, so $(\bar{H}_{NH_3})_{\infty \text{ dilute } H_2O}$ $= 56$ Btu/lb$_m$.

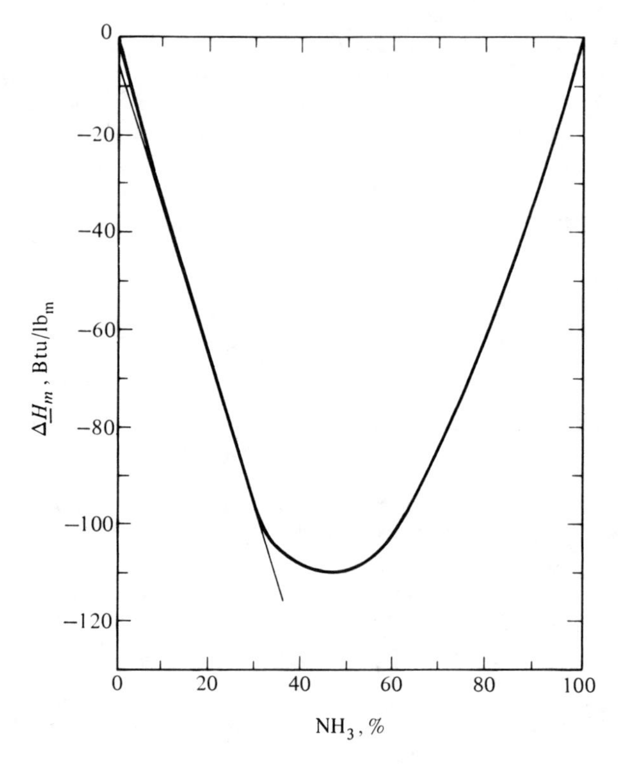

FIGURE S9-2b

(e) $(\bar{H}_{H_2O})_{\infty \text{ dilute } H_2O}$ is pure component enthalpy for water and is simply $\underline{H}_{H_2O}$ $= 6$ Btu/lb$_m$.

(f) $(\bar{H}_{H_2O})_{\infty \text{ dilute } NH_3}$ is obtained by extrapolating the tangent to the $\Delta \underline{H}_m$–x_{NH_3} curve at $x_{NH_3} = 1.0$ to $x_{NH_3} = 0$. Again extrapolation is difficult but gives approximately -360 Btu/lb$_m$, which is equal to $[(\bar{H}_{H_2O})_{\infty \text{ dilute } NH_3} - \underline{H}_{H_2O}]$. Since $\underline{H}_{H_2O} = 6$ Btu/lb$_m$. $(\bar{H}_{H_2O})_{\infty \text{ dilute } NH_3} = -354$ Btu/lb$_m$.

(g) $\underline{H}$ for a 50 per cent solution is found from an energy balance around the mixing cell [as indicated in part (c)]:

$$(1.0)(\underline{H}_{H_2O}) + 1.0(\underline{H}_{NH_3}) - 2.0(\underline{H}) + Q = 0$$

or

$$\underline{H} = (0.5)(6) + (0.5)(304) - (0.5)(218.8)$$
$$= 43.6 \text{ Btu/lb}_m$$

9.4 Fugacity

Although we might use a method similar to that discussed earlier for $\bar{V}_i$ and $\bar{H}_i$ to determine the partial molar Gibbs free energy (or chemical potential), it is a cumbersome approach and would require that we have data on the entropy change on mixing as well as the heat of mixing. As we shall see, it is possible to circumvent this difficulty by making use of the fugacity concept introduced in Chapter 6. In Chapter 6 fugacity was defined as an intensive thermodynamic property which under isothermal conditions was related to differential changes in the Gibbs free energy by equation (6-94). The fugacity of a pure gaseous component, denoted by $f(P, T)$, satisfies the pair of equations

$$RT\left(\frac{\partial \ln f}{\partial P}\right)_T = \underline{V}$$

$$\lim_{P \to 0} \frac{f(P, T)}{P} = 1 \tag{9-47}$$

which constitute the definition of fugacity.

Fugacity is an artificial property in that it is not directly measurable, but the same can be said of enthalpy and the free energies. All are introduced out of convenience and are justified because they help us to express certain thermodynamic relationships. For mixtures we introduce another such property, the partial fugacity of a component i. It will be denoted by the symbol $\bar{f}_i(P, T)$ and is defined by the pair of equations

$$RT\left(\frac{\partial \ln \bar{f}_i}{\partial P}\right)_{T,n} = \bar{V}_i$$

$$\lim_{P \to 0} \frac{\bar{f}_i(P, T)}{y_i P} = 1 \tag{9-48}$$

where y_i is the mole fraction of i in the mixture. Note that partial fugacity is an intensive property and thus is not a "partial property" in the same sense as discussed for the extensive properties earlier. Rather, as we shall see shortly, it is more like a partial pressure.

The convenience of this property in describing the behavior of mixtures will be discussed in later sections, but first we shall summarize the behavior of partial fugacity with changes in pressure and temperature at constant composition. This information will be needed in our later discussions of mixture properties.

The defining equations (9-47) and (9-48) give the dependence of the fugacities on pressure at constant temperature and composition. The dependence of fugacity on temperature may be determined as follows. In light of equation (9-22) we can write

$$RT\left(\frac{\partial \ln f}{\partial P}\right)_T = \underline{V} = \left(\frac{\partial \underline{G}}{\partial P}\right)_T \tag{9-49}$$

$$RT\left(\frac{\partial \ln \bar{f}_i}{\partial P}\right)_{T,n} = \bar{V}_i = \left(\frac{\partial \bar{G}_i}{\partial P}\right)_{T,n} \tag{9-50}$$

These equations may be integrated at constant temperature over the range P^* to P, where the asterisk denotes a reference state at a pressure low enough that $f^* = P^*$ and $\bar{f}^* = y_i P^*$.[1] (The use of an asterisk here has a meaning very similar to its use with specific heats, where C_P^* denoted the specific heat at a pressure low enough that the gas could be considered ideal.) The results of integrating equations (9-49) and (9-50) between these limits are

$$\ln \frac{f}{P^*} = \frac{\underline{G} - \underline{G}^*}{RT} \tag{9-51}$$

$$\ln \frac{\bar{f}_i}{y_i P^*} = \frac{\bar{G}_i - \bar{G}_i^*}{RT} \tag{9-52}$$

Equations (9-51) and (9-52) can in turn be differentiated with respect to temperature at constant pressure (and composition) to give the temperature dependence of the fugacities. As we perform the differentiation we note that, without loss of generality, we may consider P^* to be independent of temperature. The results are

$$\left(\frac{\partial \ln f}{\partial T}\right)_P = \frac{\partial}{\partial T}\left(\frac{\underline{G} - \underline{G}^*}{RT}\right)_P = -\frac{\underline{H} - \underline{H}^*}{RT^2} \tag{9-53}$$

$$\left(\frac{\partial \ln \bar{f}_i}{\partial T}\right)_{P,y} = \frac{\partial}{\partial T}\left(\frac{\bar{G}_i - \bar{G}_i^*}{RT}\right)_{P,y} = -\frac{\bar{H}_i - \bar{H}_i^*}{RT^2} \tag{9-54}$$

Equations (9-51) through (9-54) show us why fugacity is such a useful "stand in" for Gibbs free energy. Equations (9-51) and (9-52) can be used to calculate differences in $\underline{G}$ or $\bar{G}_i$ at constant temperature and composition in terms of changes in fugacity or partial fugacity. Equations (9-53) and (9-54) can be integrated to express differences in $\underline{G}$ or $\bar{G}_i$ at constant pressure and composition in terms of changes in fugacity or partial fugacity at the appropriate temperature and composition conditions. (It should be noted that these equations permit one to evaluate the *change* in $\underline{G}$ or $\bar{G}_i$ between two states, *not the absolute values* of either quantity.)

The treatment of fugacity in Chapter 6 for pure-component systems provides sufficient means to evaluate pure-component fugacities, so equations (9-51) and (9-53) can be used directly in the evaluation $\Delta\underline{G}_T$ or $\Delta\underline{G}_P$.

Use of equations (9-52) and (9-54) require further discussion of the partial fugacity and its evaluation and will be considered in the following sections.

Evaluation of Partial Fugacities from Equations of State

It will be helpful in the development of later sections if we have at hand the integral forms of equations (9-47) and (9-48):

[1] Frequently the pressure P^* is assumed equal to unity in whatever pressure units are in use (for example, 1 atm or 1 mm Hg) and is then dropped from the equations. The symbol P^* is retained here to maintain dimensional consistency and to remind the student of this reference state.

$$\ln \frac{f}{P^*} = \int_{P^*}^{P} \frac{\underline{V}}{RT}\, dP \tag{9-55}$$

$$\ln \frac{\bar{f}_i}{y_i P^*} = \int_{P^*}^{P} \frac{\bar{V}_i}{RT}\, dP \tag{9-56}$$

To eliminate computational problems in the limit $P^* \longrightarrow 0$, we subtract the equation

$$\ln \frac{P}{P^*} = \int_{P^*}^{P} \frac{1}{P}\, dP \tag{9-57}$$

from equations (9-55) and (9-56) to obtain

$$\ln \frac{f}{P} = \int_{P^*}^{P} \left(\frac{\underline{V}}{RT} - \frac{1}{P}\right) dP \tag{9-58}$$

$$\ln \frac{\bar{f}_i}{y_i P} = \int_{P^*}^{P} \left(\frac{\bar{V}_i}{RT} - \frac{1}{P}\right) dP \tag{9-59}$$

We anticipate no further problems arising in the limit of $P^* \longrightarrow 0$, because the integrands on the right-hand side remain bounded as P becomes small. For volume-explicit equations of state the evaluation of $\bar{V}_i$ is straightforward and equation (9-59) can be integrated to obtain $\bar{f}_i$ at given values of P and T.

To facilitate computation of partial fugacities using an equation of state for mixtures which is of the form $P = P(V, T, n_1, n_2, \ldots)$ we need an expression analogous to that obtained in Section 6.9 for pressure-explicit equations. There it was found that the derivative, $(\partial \ln f/\partial \underline{V})_T$, could be expressed in a form that made computation easy for pressure-explicit equations of state.

Before beginning a similar derivation for partial fugacities, let us obtain two needed identities, similar to the Maxwell relations. From equations (9-17), (9-37), and (9-38) it can be seen that

$$\left(\frac{\partial A}{\partial V}\right)_{T,n} = -P \tag{9-60}$$

$$\left(\frac{\partial A}{\partial n_i}\right)_{T,V,n_j} = \mu_i = \bar{G}_i \tag{9-61}$$

$$\left(\frac{\partial G}{\partial P}\right)_{T,n} = V \tag{9-62}$$

$$\left(\frac{\partial G}{\partial n_i}\right)_{P,T,n_j} = \mu_i = \bar{G}_i \tag{9-63}$$

By cross differentiating of equations (9-60) and (9-61) we obtain

$$\left(\frac{\partial P}{\partial n_i}\right)_{T,V,n_j} = -\left(\frac{\partial \mu_i}{\partial V}\right)_{T,n} \tag{9-64}$$

and from equations (9-62) and (9-63),

$$\left(\frac{\partial \mu_i}{\partial P}\right)_{T,n} = \left(\frac{\partial V}{\partial n_i}\right)_{P,T,n_j} = \bar{V}_i \tag{9-64a}$$

Thus the defining relation for partial fugacity [equation (9-48)] may be rearranged as follows:

$$RT\left(\frac{\partial \ln \bar{f}_i}{\partial P}\right)_{T,n} = \bar{V}_i = \left(\frac{\partial \mu_i}{\partial P}\right)_{T,n}$$
$$= \left(\frac{\partial \mu_i}{\partial V}\right)_{T,n}\left(\frac{\partial V}{\partial P}\right)_{T,n}$$
$$= -\left(\frac{\partial P}{\partial n_i}\right)_{T,V,n_j}\left(\frac{\partial V}{\partial P}\right)_{T,n} \tag{9-65}$$

For pressure-explicit equations of state we seek a way of integrating $(\partial \ln \bar{f}_i/\partial V)$ with respect to volume. Thus let us write:

$$RT\left(\frac{\partial \ln \bar{f}_i}{\partial V}\right)_{T,n} = RT\left(\frac{\partial \ln \bar{f}_i}{\partial P}\right)_{T,n}\left(\frac{\partial P}{\partial V}\right)_{T,n} \tag{9-66}$$

which may be combined with equation (9-65) to give

$$RT\left(\frac{\partial \ln \bar{f}_i}{\partial V}\right)_{T,n} = -\left(\frac{\partial P}{\partial n_i}\right)_{T,V,n_j} \tag{9-67}$$

Since the primary terms in the equation of state at large volume ($P \longrightarrow 0$) are

$$P = \frac{nRT}{V} + \text{terms of order } V^{-2}$$

the primary term in the derivative of equation (9-67) will be

$$\left(\frac{\partial P}{\partial n_i}\right)_{T,V,n_j} = \frac{RT}{V} + \text{terms of order } V^{-2}$$

To avoid the problems associated with the ∞ limit in $\int RT\,dV/V$ term, let us add the equation

$$-RT\left(\frac{\partial \ln P}{\partial V}\right)_{T,n} = \frac{RT}{V} - RT\left(\frac{\partial \ln PV}{\partial V}\right)_{T,n}$$

to equation (9-67) and integrate from ∞ to V. In evaluating the resulting expressions we can utilize the fact that

$$\lim_{V\to\infty} \frac{\bar{f}_i}{y_i P} = 1$$

$$\lim_{V\to\infty} \frac{PV}{nRT} = 1$$

Consequently,

$$\int_{\infty}^{V} RT\left(\frac{\partial \ln (\bar{f}_i/P)}{\partial V}\right)_{T,n} dV = \int_{\infty}^{V}\left[\frac{RT}{V} - \left(\frac{\partial P}{\partial n_i}\right)_{T,V,n_j} - RT\left(\frac{\partial \ln PV}{\partial V}\right)_{T,n}\right] dV \tag{9-68}$$

or

$$RT \ln \frac{\bar{f}_i}{P} - RT \ln y_i = \int_{V}^{\infty}\left[\left(\frac{\partial P}{\partial n_i}\right)_{T,V,n_j} - \frac{RT}{V}\right] dV$$
$$- RT \ln PV + RT \ln nRT \tag{9-68a}$$

which is simplified to

$$RT \ln \frac{\bar{f}_i}{Py_i} = \int_V^\infty \left[\left(\frac{\partial P}{\partial n_i} \right)_{T,V,n_j} - \frac{RT}{V} \right] dV - RT \ln \frac{PV}{nRT} \tag{9-68b}$$

Equation (9-68b) can be used to evaluate partial fugacities for components in systems for which a suitable pressure-explicit equation of state is available. The availability of high-speed computing equipment and the continued development of more accurate equations of state make the use of equations (9-59) and (9-68b) ever more feasible and reliable.

Sample Problem 9-3 illustrates the procedure one would use to evaluate partial fugacities if values of $\bar{V}_i$ are available over the range of pressure, temperature, and composition of interest. Integration of equation (9-59) can be carried out either analytically or graphically, depending on the form of the $\bar{V}_i$–P data. The calculated values of $\bar{f}_i$ can then be used to compute changes in $\bar{G}_i$ between two specified pressures by integrating equation (9-50):

$$\int_{\bar{f}_{i_1}}^{\bar{f}_{i_2}} RT(d \ln \bar{f}_i)_{T,y} = (d\bar{G}_i)_{T,y} = (\bar{G}_{i_2} - \bar{G}_{i_1})_{T,y} \tag{9-69}$$

SAMPLE PROBLEM 9-3. The feed material for the direct oxidation of ethylene to ethylene oxide contains a 2 : 1 (stoichiometric) molar ratio of ethylene to oxygen. Its estimated properties and those of pure oxygen and ethylene at 0°C are given in Table S9-3a. Use this information to calculate the partial fugacities of oxygen and ethylene in the mixture at 0°C and 25 atm.

TABLE S9-3a

Pressure, atm	$\underline{V}_{\text{mixture}}$, ft³/lb-mole	$\bar{V}_{C_2H_4}$	$\underline{V}_{C_2H_4}$	$\underline{V}_{O_2}$
1	357.24	336.51	356.16	358.74
5	70.08	69.39	68.93	71.52
10	34.18	33.50	32.96	35.62
15	22.21	21.54	20.94	23.65
20	16.23	15.57	14.90	17.66
25	12.64	11.99	11.24	14.07

Solution: Examination of the equations relating partial fugacity to partial molar volume indicate that at least three equivalent calculations might be made, using either equation (9-56) or (9-59) or yet another equation, obtained by subtracting equation (9-55) from (9-56) to eliminate P^* on the left-hand side. The data here indicate that the difference ($\bar{V}_{C_2H_4} - \underline{V}_{C_2H_4}$) is small over the pressure range of interest; thus we favor the last-mentioned integral:

$$\ln \frac{\bar{f}_i}{y_i f_i} = \frac{1}{RT} \int_{P^*}^{P} (\bar{V}_i - \underline{V}_i)\, dP$$

The integration can be carried out by linear interpolation over the intervals given.

We shall assume that at 1 atm, the materials are nearly ideal (that is, $f_i = P$ and $\bar{f}_i = y_i P$) and perform the integration over the pressure range 1 to 25 atm. Any error so introduced will be small. We now must find a value of f_i before carrying out the integration. We also have a choice of obtaining f_i from equation (9-55) or (9-58) or from the correction factors f/P given in Chapter 6. The latter are, of course, obtained by carrying out the integration indicated at equation (9-58) using the generalized compressibility factor. For ethylene ($T_c = 282°\text{K}$, $P_c = 50.0$ atm) we find that at 0°C and 25 atm ($T_r = 0.97$, $P_r = 0.5$), $f/P = 0.825$ or $f = 20.6$ atm. Thus, for ethylene our calculation becomes

$$\ln \frac{\bar{f}_{C_2H_4}}{(0.667)20.6 \text{ atm}} = \frac{1}{RT} \int_1^{25} (\bar{V}_{C_2H_4} - \underline{V}_{C_2H_4})\, dP$$

$$\ln \frac{\bar{f}_{C_2H_4}}{13.75 \text{ atm}} = \frac{0.522 \text{ atm ft}^3/\text{lb-mol}}{(0.7302)(492) \text{ atm ft}^3/(\text{lb-mol °R})°\text{R}} = 0.0381$$

$$\bar{f}_{C_2H_4} = 13.75 \text{ atm} \exp(0.0381) = 14.28 \text{ atm}$$

Since we were not given partial molar volume data for oxygen, we must first obtain it by making use of the relation

$$\underline{V} = y_1\bar{V}_1 + y_2\bar{V}_2$$

The results are listed in Table S9-3b. Integration over the range 1 to 25 atm is again required but must be done over several intervals. Simple linear interpolation is used

TABLE S9-3b

P, atm	$\bar{V}_{O_2}$	$\bar{V}_{O_2} - \underline{V}_{O_2}$
1	358.70	−0.04
5	71.46	−0.06
10	35.54	−0.08
15	23.55	−0.10
20	17.55	−0.11
25	13.94	−0.13

here. We also note that at 25 atm and 0°C, for oxygen, $T_r = 1.76$, $P_r = 0.5$, and $f/P = 0.98$:

$$\ln \frac{\bar{f}_{O_2}}{(0.33)(0.98)(25) \text{ atm}} = \frac{1}{RT} \int_1^{25} (\bar{V}_{O_2} - \underline{V}_{O_2})\, dP$$

$$\ln \frac{\bar{f}_{O_2}}{8.17 \text{ atm}} = \frac{-2.125 \text{ atm ft}^3/\text{lb-mol}}{(0.7302)(492)°\text{R atm ft}^3/(\text{lb-mol °R})} = 0.0059$$

$$\bar{f}_{O_2} = 8.17 \text{ atm} \exp(-0.0059) = 8.12 \text{ atm}$$

As we can see from Sample Problem 9-3, calculation of partial fugacities from partial volume data is a rather laborious task. The process would have been even more demanding had we chosen to graphically integrate the data or, as we shall see later in this chapter, had we chosen to use an equation of state suitable for gaseous mixtures. Inasmuch as partial fugacities play such a major role in equilibrium calculations in subsequent chapters, it behooves us to develop a method for evaluating $\bar{f}_i$ that overcomes these difficulties. Since many mixtures continue to behave in an "ideal" manner at pressures above P^* (where $\bar{f}_i^* =$

$y_i P^*$), the concept of an ideal solution is suggested in which $\bar{f}_i$ retains the same relationship to composition and fugacity (pressure) as was true at very low pressures. Such behavior is discussed in the next section where the Lewis–Randall rule is considered.

9.5 The Lewis–Randall Rule—Ideal Solutions

A very specific meaning is attached to the term "ideal solution." This is a class of mixtures whose properties can be deduced from a knowledge of the properties of its pure components and the mixture composition. *We choose to define an ideal solution as one that obeys the Lewis and Randall rule,*

$$\bar{f}_i = y_i f_i \tag{9-70}$$

As before, f_i is the fugacity of the pure component in the same phase and at the same temperature and pressure as the mixture. Thus partial fugacities can be evaluated for mixtures obeying the Lewis–Randall rule using the procedures developed in Chapter 6 for pure-component fugacities if the mixture composition is known. Since gaseous or vapor mixtures at low to moderate pressures typically obey the Lewis–Randall rule, equation (9-70) is normally used to evaluate partial fugacities in such systems. Evaluation of $\bar{f}_i$ in liquid and solid solutions that do not obey equation (9-70) is considerably more difficult and is frequently accomplished indirectly by determining partial fugacities in an equilibrium vapor phase. Discussion of such determinations follows in this chapter and again in Chapter 10.

The simplicity of equation (9-70) can be seen by observing that

$$\left(\frac{\partial \ln \bar{f}_i}{\partial P}\right)_{T,y} = \left(\frac{\partial \ln f_i}{\partial P}\right)_T \quad \text{and} \quad \left(\frac{\partial \ln \bar{f}_i}{\partial T}\right)_{P,y} = \left(\frac{\partial \ln f_i}{\partial T}\right)_P \tag{9-71}$$

Substitution of equations (9-70) and (9-71) into (9-49) to (9-54) indicates the following behavior for ideal solutions:

$$\begin{aligned} \bar{V}_i - \underline{V}_i &= 0 \\ \bar{H}_i - \underline{H}_i &= \bar{H}_i^* - \underline{H}_i^* \\ \bar{G}_i - \underline{G}_i &= \bar{G}_i^* - \underline{G}_i^* \\ \bar{S}_i - \underline{S}_i &= \bar{S}_i^* - \underline{S}_i^* \end{aligned} \tag{9-72}$$

The partial molar volume of a component in an ideal solution is seen to be the same as its specific volume at the same temperature and pressure. Thus there is no volume change upon mixing of the components of an ideal solution,

$$\Delta V_m = \sum_{i=1}^{C} n_i(\bar{V}_i - \underline{V}_i) \tag{9-14}$$

The difference between the partial molar and specific enthalpies, entropies, and Gibbs free energies can be seen to be independent of pressure, because the difference is the same at the reference state (right-hand side) as at the arbitrary pressure (left-hand side).

Examination of the behavior of ideal solutions at the reference condition, P^*, permits even further simplification of these expressions. We recognize that P^* may be arbitrarily small and that at a sufficiently low pressure all gases exhibit ideal gas behavior. Hence we shall devote the next section to development of that special class of ideal solutions, the ideal gas mixtures. With this information we are able to more precisely express the quantities on the right-hand side of equation (9-72).

9.6 An Ideal Gas Mixture

A mixture whose components are ideal gases and which itself behaves as an ideal gas is indeed quite simple to describe. If the mixture is an ideal gas, its total volume obeys the ideal gas equation of state:

$$V = \frac{nRT}{P} = (n_1 + n_2 + \cdots)\frac{RT}{P} \tag{9-73}$$

The partial molar volumes are given by

$$\bar{V}_i = \left(\frac{\partial V}{\partial n_i}\right)_{P,T,n_j} = \frac{RT}{P} = \underline{V}_i \tag{9-74}$$

and are equal to the pure-component specific volumes at the same temperature and pressure. Substitution of these results into equations (9-58) and (9-59) confirms that the ideal gas mixture is an ideal solution in that it satisfies the Lewis–Randall rule, equation (9-70). Moreover, the fugacities are equal to the pressure and partial pressures, respectively:

$$f = P \quad \text{and} \quad \bar{f}_i = y_i P = P_i \tag{9-75}$$

The absence of intermolecular forces in any gaseous system that exhibits ideal gas behavior permits us to conclude that there would be no heat effect in the isothermal mixing of ideal gases. Therefore, the internal energy of such mixtures would be equal to the sum of the internal energies of the individual components for isothermal mixing. Thus the partial molar internal energy must equal the specific internal energy at the same temperature and pressure:

$$\bar{U}_i = \underline{U}_i \tag{9-76}$$

Since $\bar{U}_i = \underline{U}_i$ and $\bar{V}_i = \underline{V}_i$ for ideal gas mixtures, we conclude that

$$\bar{H}_i = \bar{U}_i + P\bar{V}_i = \underline{U}_i + P\underline{V}_i$$

or

$$\bar{H}_i = \underline{H}_i \tag{9-77}$$

and thus, upon mixing, there is no change in either internal energy or enthalpy.

However, the mixing of two ideal gases to form an ideal gas mixture is clearly an irreversible process, in that the separation of the mixture into the two pure components would never occur spontaneously. Thus we would expect an entropy increase to result from such a mixing process. To compute this entropy change, let us first recall that one characteristic of an ideal gas mixture is that each component in such a mixture behaves as if it occupied the same total volume as the mixture at the temperature of the mixture, but at the partial pressure of the component. Thus the mixing process is identical to that in which given amounts of pure crystalline solids are mixed at constant temperature and pressure to form a mixed crystal whose volume is equal to the sum of the volumes of the original crystals.

In the mixing of both the ideal gases and ideal crystals, the entropy change on mixing is due to the additional number of particle arrangements, or microstates, that arise from the additional volume elements available to the individual molecules. We showed in Sample Problem 3-4 that the entropy change that occurs during such a process is given by:

$$\Delta S_m = -R\left[\sum_{i=1}^{C} n_i \ln\left(\frac{n_i}{\sum_{i=1}^{C} n_i}\right)\right] = \sum_{i=1}^{C} n_i(\bar{S}_i - \underline{S}_i) \tag{9-78}$$

or

$$\sum_{i=1}^{C} n_i(\bar{S}_i - \underline{S}_i) = -R\left[\sum_{i=1}^{C} n_i \ln n_i + \sum_{i=1}^{C} n_i \ln\left(\sum_{i=1}^{C} n_i\right)\right] \tag{9-78a}$$

Differentiation with respect to n_k at constant pressure, temperature, and n_j gives:

$$(\bar{S}_k - \underline{S}_k) + \sum_{i=1}^{C} n_i\, d\bar{S}_i = -R\left[\ln n_k + \frac{n_k}{n_k} - \sum_{i=1}^{C}\left(\frac{n_i}{\sum_{i=1}^{C} n_i}\right) - \ln \sum_{i=1}^{C} n_i\right] \tag{9-79}$$

Elimination of the second term in the left-hand side by means of the Gibbs–Duhem equation and simplification yields:

$$\bar{S}_k - \underline{S}_k = -R \ln \frac{n_k}{\sum_{i=1}^{C} n_i} \tag{9-80}$$

Since $n_k/\sum_{i=1}^{C} n_i$ is simply y_k, the mole fraction of species k, equation (9-80) reduces to

$$\bar{S}_k - \underline{S}_k = -R \ln y_k \tag{9-81}$$

We can now summarize the mixing behavior for ideal gas systems, using the asterisk to remind us that the earlier reference states were all specified to be at pressures low enough to assure ideal gas behavior:

$$\bar{V}_i^* - \underline{V}_i^* = 0 \tag{9-82}$$

$$\bar{U}_i^* - \underline{U}_i^* = 0 \tag{9-83}$$

$$\bar{H}_i^* - \underline{H}_i^* = 0 \tag{9-84}$$

$$\bar{S}_i^* - \underline{S}_i^* = -R \ln y_i \tag{9-85}$$

$$\bar{G}_i^* - \underline{G}_i^* = \bar{H}^* - \underline{H}^* - T(\bar{S}^* - \underline{S}^*) = RT \ln y_i \tag{9-86}$$

$$\bar{A}_i^* - \underline{A}_i^* = (\bar{U}^* - \underline{U}^*) - T(\bar{S}^* - \underline{S}^*) = RT \ln y_i \tag{9-87}$$

Thus the properties of an ideal gas mixture can be completely described if the properties of each pure component are known.

9.7 Solution Behavior of Real Gases, Liquids, and Solids

The preceding treatment of ideal gas mixtures has provided the relationships with which to simplify equation (9-72). Substituting equations (9-82) through (9-87), we can now write, for *all ideal solutions* (liquid, solid, or gas),

$$\bar{V}_i = \underline{V}_i \qquad \text{and} \qquad \Delta\underline{V}_m = 0 \tag{9-88}$$

$$\bar{U}_i = \underline{U}_i \qquad \text{and} \qquad \Delta\underline{U}_m = 0 \tag{9-89}$$

$$\bar{H}_i = \underline{H}_i \qquad \text{and} \qquad \Delta\underline{H}_m = 0 \tag{9-90}$$

$$\bar{S}_i = \underline{S}_i - R \ln y_i \qquad \text{and} \qquad \Delta\underline{S}_m = -\sum_{i=1}^{C} y_i R \ln y_i \tag{9-91}$$

$$\bar{G}_i = \underline{G}_i + RT \ln y_i \qquad \text{and} \qquad \Delta\underline{G}_m = \sum_{i=1}^{C} y_i RT \ln y_i \tag{9-92}$$

$$\bar{A}_i = \underline{A}_i + RT \ln y_i \qquad \text{and} \qquad \Delta\underline{A}_m = \sum_{i=1}^{C} y_i RT \ln y_i \tag{9-93}$$

Although these results were developed for an ideal gas mixture, it should be noted that the components of an ideal solution may be gases that do not obey the ideal gas equation of state—or even liquid and solid components—provided they combine to form a mixture that obeys the Lewis–Randall rule. In summary, an ideal gas mixture will always behave as an ideal solution, but a system need not conform to the ideal gas law to behave as an ideal solution.

Ideal solution behavior is generally exhibited by gaseous systems, particularly if none of the components are near their critical point. It it also the simplest assumption that can be made if no information is available on the behavior of a mixture, but is frequently a poor assumption for liquid or solid mixtures because of the effects of intermolecular forces in condensed phases.

From equation (9-88) it can be seen that all ideal solutions obey *Amagat's law* of additive volumes:

$$V = \sum_{i=1}^{C} n_i \underline{V}_i \tag{9-94}$$

which states that the volume of a mixture is the weighted sum of the pure component volumes, or that there is no volume change upon mixing. One brief note of caution should be interjected here; a system exhibiting no volume change on mixing need not conform to all the equations for ideal solution behavior. Although the Lewis–Randall rule guarantees that there will be no volume or enthalpy change upon mixing, the reverse is not necessarily true.

In addition to their importance in describing an ideal solution, the results of the previous section can be used to modify the fugacity relationships of Section 9.4 for any mixture (liquid, solid, or gas) because, in the limit, the reference state (*) becomes the ideal gas mixture. Thus, if we relate pure component fugacity, f_i, to partial fugacity, $\bar{f}_i$, by eliminating P^* between equations (9-51) and (9-52), the result is

$$RT \ln \frac{\bar{f}_i}{y_i f_i} = (\bar{G}_i - \underline{G}_i) - (\bar{G}_i^* - \underline{G}_i^*) \tag{9-95}$$

This can now be simplified because we recognize that

$$\bar{G}_i^* - \underline{G}_i^* = RT \ln y_i \tag{9-96}$$

and hence, for *any mixture* (ideal or nonideal),

$$RT \ln \frac{\bar{f}_i}{f_i} = \bar{G}_i - \underline{G}_i \tag{9-97}$$

It can also be shown by combining equations (9-47) and (9-48) that

$$\left(\frac{\partial \ln (\bar{f}_i/f_i)}{\partial P}\right)_{T,y} = \frac{1}{RT}(\bar{V}_i - \underline{V}_i) \tag{9-98}$$

Recalling that $\bar{H}_i^* - \underline{H}_i^* = 0$ from equation (9-84), equations (9-53) and (9-54) may be combined to give

$$\left(\frac{\partial \ln (\bar{f}_i/f_i)}{\partial T}\right)_{P,y} = -\frac{1}{RT^2}(\bar{H}_i - \underline{H}_i) \tag{9-99}$$

It is important to realize that these three equations are applicable to *any* mixture.

9.8 Activity and Activity Coefficient

Inasmuch as many real mixtures are not ideal solutions, we need some additional means for dealing with them. To facilitate later calculations, it proves useful to introduce two new thermodynamic variables, the *activity*, a_i, and the *activity coefficient*, γ_i°. The activity of a component i is defined to be

$$a_i = \frac{\bar{f}_i}{f_i^\circ} \tag{9-100}$$

where f_i° is the fugacity of component i in its standard (or reference) state, denoted by the superscript $^\circ$. The activity coefficient is defined to be

$$\gamma_i^\circ = \frac{a_i}{x_i} = \frac{\bar{f}_i}{x_i f_i^\circ} \tag{9-101}$$

or

$$\bar{f}_i = x_i \gamma_i^\circ f_i^\circ \tag{9-101a}$$

(Note that y is used to denote vapor-phase mole fractions and x liquid and solid mole fractions. Since deviations from ideal solution behavior are more likely in liquid and solid solutions than in gases, x will be used for much the remaining development in this chapter unless the material under discussion relates to gases.) *Both activity and activity coefficient will always be positive quantities whose values will depend on the standard state specified.* Several standard states are commonly used, but they must always be at the same temperature as the mixture.

Standard States

1. The most common and generally useful standard state is the *pure component* (gas, liquid, or solid) *at the same temperature, pressure, and phase as the mixture*. Examining this choice of reference state, we observe that for the activity

$$a_i = \frac{\bar{f}_i}{f_i^\circ} = \frac{\bar{f}_i}{f_i} \tag{9-102}$$

The activity coefficient for this choice of standard state is given by

$$\gamma_i = \frac{a_i}{x_i} = \frac{\bar{f}_i}{x_i f_i} \tag{9-103}$$

where we have dropped the superscript $^\circ$ from γ_i to indicate this particular choice of standard state in the evaluation of γ_i. Thus for an ideal solution and this choice of standard state, $a_i = x_i$ and $\gamma_i = 1$. The deviation of γ_i from unity then provides a measure of the nonideality of the mixture, as seen in Fig. 9-4. It should be noted that even if the mixture is nonideal, γ_i approaches unity in the limit as the mixture becomes pure i.

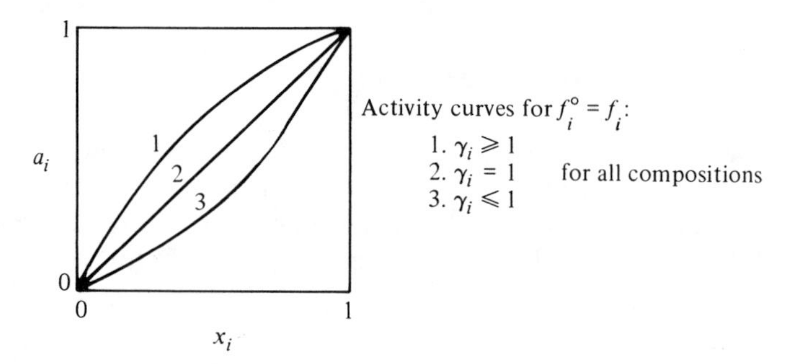

FIG. 9-4. Plot of activity vs. concentration.

2. A frequently encountered choice of standard state for gases is the *pure component at a pressure of* 1 *atm and the temperature of the mixture*. For this choice of standard state the activity is given by

$$a_i = \frac{\bar{f}_i}{f^\circ} = \frac{\bar{f}_i}{(f/P)\cdot 1\text{ atm}} \tag{9-104}$$

where $f^\circ = (f/P)\cdot 1$ atm. Since values of the fugacity coefficient, f/P, are generally unity at a pressure of 1 atm this standard state has the advantage that $f^\circ = 1$ atm and $a_i = \bar{f}_i\, 1\text{ atm}^{-1}$. For ideal gas and ideal solution behavior equation (9-104) reduces to

$$a_i = P_i\, 1\text{ atm}^{-1} \tag{9-105}$$

where P_i is the partial pressure of component i in the gaseous mixture (expressed in atmospheres to maintain consistency of units). When the standard state is chosen to be the pure component at 1 atm, the activity coefficient becomes

$$\gamma_i^\circ = \frac{a_i}{x_i} = \frac{\bar{f}_i}{x_i} 1 \text{ atm}^{-1} \tag{9-106}$$

which for the ideal solution reduces to

$$\gamma_i^\circ = \frac{x_i f_i}{x_i \text{ atm}} = f_i \, 1 \text{ atm}^{-1} = \left(\frac{f}{P}\right) P \, 1 \text{ atm}^{-1} \tag{9-107}$$

where P represents the total pressure of the mixture (in atmospheres) and f/P is evaluated for component i at the temperature and pressure of the solution.

3. For some systems it is not convenient to use the pure component as a reference condition. For instance, if some of the components are gases or solids at the temperature and pressure in question (for example, aqueous solutions of carbon dioxide or of sodium hydroxide), it is often convenient to define the standard state as the component in an *infinitely* dilute solution at the mixture temperature. The meaning of such a choice can be better understood by considering Fig. 9-5, in which the partial fugacity of a component in a mixture is

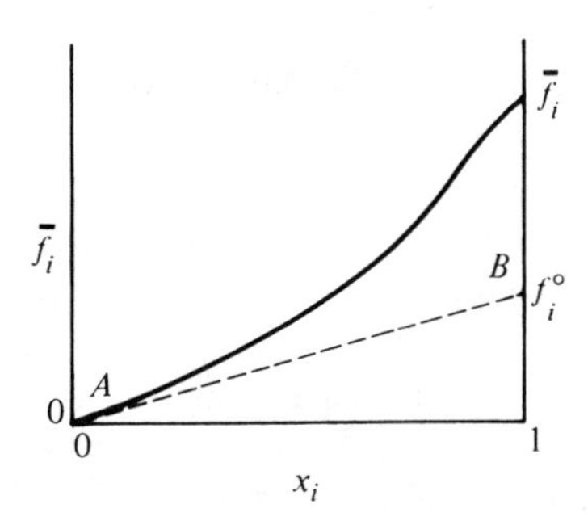

FIG. 9-5. Partial fugacity vs. concentration plot illustrating infinitely dilute standard state.

plotted versus its concentration. If the limiting slope of the curve as $x_i \to 0$ is used to draw a line (A-B) from the origin ($x_i = 0$, $\bar{f}_i = 0$) across the entire composition range, its intersection with the $x_i = 1$ line defines a hypothetical fugacity at point B, which is used as f_i°, the standard-state fugacity.

A plot of a_i–x_i using the infinitely dilute standard state is obtained by dividing each value of $\bar{f}_i$ along the curve by the value for f_i° and replotting it versus x_i as shown in Fig. 9-6.

This choice of standard state is particularly advantageous when one is interested in systems where solutes at low concentrations are involved, because in the dilute region $a_i = x_i$. Many metallurgical and electrochemical systems involve low solute concentrations in alloys, and thus this standard state is often employed. As Fig. 9-6 demonstrates, at high concentrations of i the activity may take on values greater than unity as a result of the standard state chosen. Also shown in Fig. 9-7 is the variation in γ_i° with concentration for the system illustrated in Fig. 9-6 when the infinitely dilute standard state is used. Note that γ_i° has a value of unity in the dilute region and increases at higher values of x_i.

Curves 2 in Figs. 9-6 and 9-7 are included to illustrate a_i versus x_i and γ_i versus x_i if pure component i at the pressure, temperature, and phase of the

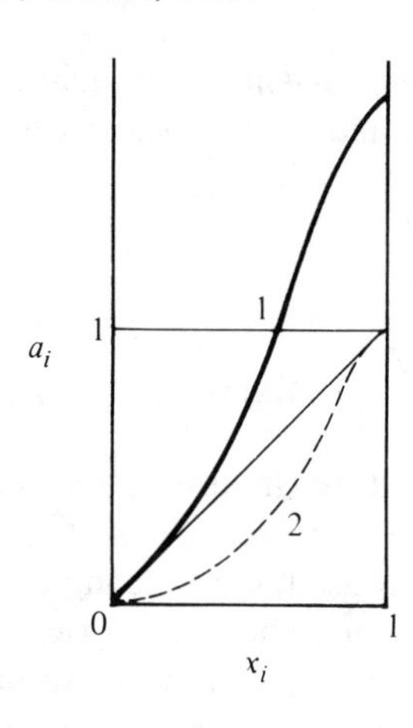

FIG. 9-6. Effect of standard state on activities.

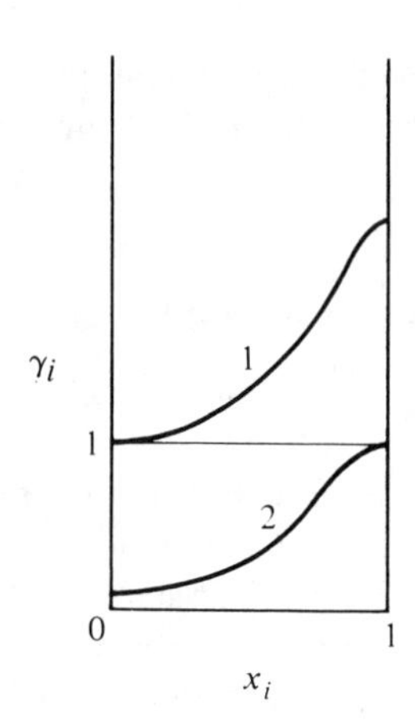

FIG. 9-7. Effect of standard state upon activity coefficients. Curve 1, infinitely dilute standard states; curve 2, Pure i at P and T of mixture as standard state.

mixture is used as the standard state. Values for activity then range between 0 and 1 and $\gamma < 1$.

It is left as an exercise to show that the behavior illustrated by curve 1 in Fig. 9-4 would produce values for the activity and activity coefficient less than unity over the entire composition range if the infinitely dilute standard state is used.

4. Other reference states are occasionally encountered, such as the pure component at the temperature of the mixture but at a low pressure P^*, or the case of infinite dilution with molarity, m_i, substituted for x_i in equation (9-101). The implications of each can be handled systematically by requiring that $\gamma_i^\circ \longrightarrow 1$ at the reference state.

Since the values of activity and activity coefficient of a component depend on the reference state chosen, the reader should always be alert to ascertain what reference or standard state is to be associated with each component of a mixture. For gaseous systems, reference conditions 1 and 2 are generally used; for liquids and solids, condition 1 is the most common, although 3 and 4 are occasionally used.

SAMPLE PROBLEM 9-4. Estimate the partial fugacities of ethylene and oxygen in a 2 : 1 molar mixture at 25 atm and 0°C, assuming that the mixture is an ideal solution. Compare the result with that of Sample Problem 9-3 and evaluate the activity and activity coefficient for the mixture described there.

Solution: We have obtained in Sample Problem 9-3 the fugacities of pure ethylene and pure oxygen at 25 atm and 0°C. Thus, using the Lewis–Randall rule we obtain

$$\bar{f}_{C_2H_4} = (0.667)(20.6) \text{ atm} = 13.75 \text{ atm}$$

$$\bar{f}_{O_2} = (0.333)(24.5) \text{ atm} = 8.16 \text{ atm}$$

which are, respectively, smaller and larger than the values obtained in Sample Problem 9-3. The activities and the activity coefficients for the partial fugacities of Sample Problem 9-3 are obtained with equations (9-102) and (9-103), if we choose the standard states to be the pure component gases at 25 atm and °C.

$$a_{C_2H_4} = \frac{\bar{f}_{C_2H_4}}{f_{C_2H_4}} = \frac{14.28 \text{ atm}}{20.6 \text{ atm}} = 0.693$$

$$a_{O_2} = \frac{\bar{f}_{O_2}}{f_{O_2}} = \frac{8.12 \text{ atm}}{24.5 \text{ atm}} = 0.331$$

$$\gamma_{C_2H_4} = \frac{a_{C_2H_4}}{y_{C_2H_4}} = \frac{0.693}{0.667} = 1.04$$

$$\gamma_{O_2} = \frac{a_{O_2}}{y_{O_2}} = \frac{0.331}{0.333} = 0.994$$

Comparison of the computations in Sample Problems 9-3 and 9-4 should make it quite clear why we prefer to invoke the ideal solution assumption whenever it is justified. Our justification may, in practice, weigh the probable error against the additional labor when mixture data are available.

9.9 Variation of Activity Coefficient with Temperature and Composition

The relationship between the activity coefficient and other mixture parameters for the case of the pure-component standard state at the pressure and temperature of the mixture is given by equations (9-108) through (9-110). It should be noted that equation (9-108) results from integration of the relationship $d\bar{G}_i = RT\, d \ln \bar{f}_i$, which is valid only at constant temperature—thus the necessity of choosing the standard state at the mixture temperature:

$$RT \ln \frac{\bar{f}_i}{f_i} = RT \ln (\gamma_i x_i) = \bar{G}_i - \underline{G}_i \tag{9-108}$$

Differentiation with respect to pressure and temperature gives

$$\left(\frac{\partial \ln (\bar{f}_i/f_i)}{\partial P}\right)_{T,x} = \left(\frac{\partial \ln \gamma_i}{\partial P}\right)_{T,x} = \frac{\bar{V}_i - \underline{V}_i}{RT} \tag{9-109}$$

$$\left(\frac{\partial \ln (\bar{f}_i/f_i)}{\partial T}\right)_{P,x} = \left(\frac{\partial \ln \gamma_i}{\partial T}\right)_{P,x} = -\frac{\bar{H}_i - \underline{H}_i}{RT^2} \tag{9-110}$$

Information regarding the concentration dependence of the activity coefficient for a component in a solution can be obtained by further development of the Gibbs–Duhem equation [(9-27) or (9-28)]:

$$\sum_{i=1}^{C} x_i (d\bar{G}_i)_{P,T} = 0 \tag{9-28}$$

Modification of this equation can be achieved by noting that differentiation of equation (9-108) at constant T and P yields

$$(d\bar{G}_i)_{P,T} = RT(d \ln \bar{f}_i)_{P,T} \tag{9-111}$$

Substitution of equation (9-111) into equation (9-28) yields

$$RT \sum_{i=1}^{C} x_i \, d \ln \bar{f}_i = 0 \tag{9-112}$$

or

$$\sum_{i=1}^{C} x_i \, d \ln \bar{f}_i = 0 \tag{9-112a}$$

This equation relates the partial fugacities of components in a solution at constant temperature and pressure as the composition is varied. Integration of equation (9-112a) for a binary system, *A-B*, between any two compositions 1 and 2 yields

$$\ln \bar{f}_{A_2} - \ln \bar{f}_{A_1} = -\int_1^2 \frac{x_B}{x_A} \, d \ln \bar{f}_B \tag{9-113}$$

In most cases one knows $\bar{f}_B$ at specific compositions corresponding to datum points rather than the functional relationship between $\bar{f}_B$ and composition. Thus graphical integration of equation (9-113) is frequently required. Such a procedure is illustrated by Fig. 9-8. As can be seen from the plot, $\ln \bar{f}_B$ approaches

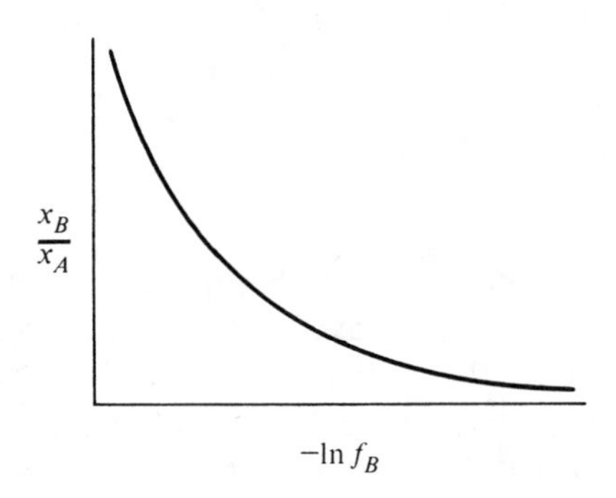

FIG. 9-8. Graphical integration of the Gibbs–Duhem equation.

minus infinity as $x_B \longrightarrow 0$. Similarly, as $x_B \longrightarrow 1$, the ratio $x_B/x_A \longrightarrow \infty$. Thus graphical integration of the right-hand side of equation (9-113) produces an indeterminate area if integrated from either $x_B = 0$ or $x_B = 1$ to an intermediate composition. Since pure-component fugacities are frequently known, these are the points from which one would normally choose to initiate the integration were it not for the above difficulties. If one knows values for $\bar{f}_A$ and $\bar{f}_B$ at an intermediate composition, these difficulties can be avoided and one could use this procedure to evaluate $\bar{f}_A$ at other compositions, assuming that $\bar{f}_B$ is known at a sufficient number of compositions to generate the curve in Fig. 9-8 over the composition range of interest.

Use of equation (9-100) permits replacement of $\bar{f}_i$ in equation (9-112) with the following:

$$\bar{f}_i = a_i f_i^\circ \tag{9-100a}$$

Since f_i° is a constant chosen at a given *P*, *T*, and composition, equation (9-112a) can be rewritten in terms of activities as follows:

$$\sum_{i=1}^{C} x_i \, d \ln a_i = 0 \tag{9-114}$$

It is left as an exercise to demonstrate that difficulties similar to those experienced with equation (9-113) are encountered with this form of the Gibbs–Duhem equation also.

The above difficulties can generally be avoided if equation (9-101) is used to replace $\bar{f}_i$ in terms of the activity coefficient,

$$\bar{f}_i = x_i \gamma_i^\circ f_i^\circ \tag{9-101a}$$

yielding

$$\sum_{i=1}^{C} [x_i (d \ln \gamma_i^\circ)_{P,T} + x_i \, d \ln x_i] = 0 \tag{9-115}$$

Again we have set $d \ln f_i^\circ = 0$, since f_i° is a constant. In addition, $d \ln x_i = dx_i/x_i$, so the second term in equation (9-115) becomes

$$\sum_{i=1}^{C} x_i \, d \ln x_i = \sum_{i=1}^{C} x_i \frac{dx_i}{x_i} = \sum_{i=1}^{C} dx_i = 0$$

and equation (9-115) reduces to the most commonly used form of the Gibbs–Duhem equation:

$$\sum_{i=1}^{C} x_i \, d \ln \gamma_i^\circ = 0 \tag{9-116}$$

This equation when integrated for a binary mixture between compositions 1 and 2 yields

$$\ln \gamma_{A_2}^\circ - \ln \gamma_{A_1}^\circ = -\int_1^2 \frac{x_B}{x_A} d \ln \gamma_B^\circ \tag{9-117}$$

Again graphical integration is frequently necessary, as shown in Fig. 9-9. In this case, as $x_B \longrightarrow 0$, γ_B° approaches a finite constant. Thus, if one chooses to

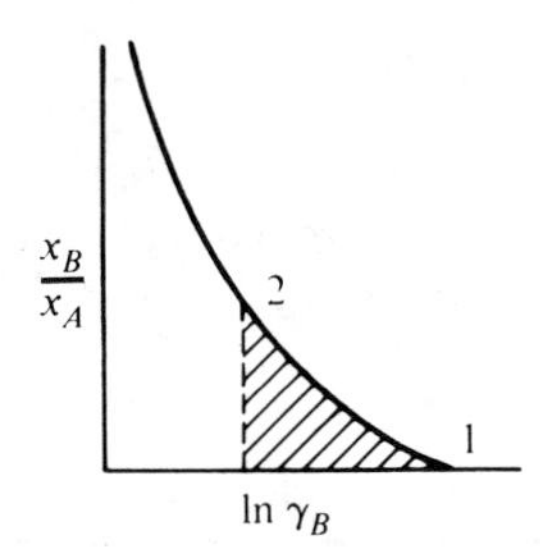

FIG. 9-9. Graphical integration of the Gibbs–Duhem equation.

integrate from pure A (composition 1) to any composition of interest (composition 2), equation (9-117) becomes

$$\ln (\gamma_A^\circ)_2 - \ln (\gamma_A^\circ)_1 = -\int_1^2 \frac{x_B}{x_A} d \ln \gamma_B^\circ = -(\text{area shaded in Fig. 9-9})$$

SAMPLE PROBLEM 9-5. Using Table S9-5a, predict the activity coefficients for water as a function of composition, and compare with the values listed in Perry (same place) for this system. Standard states have been chosen to be pure liquids at 1 atm and the temperature of the system.

TABLE S9-5a Liquid Solutions of Water and *n*-Propyl Alcohol at 1 atm†

Mole fraction *n*-propanol	Activity Coefficient *n*-propanol
0.0	12.0
0.05	8.0
0.10	4.9
0.20	2.8
0.30	1.9
0.40	1.5
0.50	1.3
0.60	1.2
0.70	1.15
0.80	1.10
0.90	1.05
1.00	1.00

† R. H. Perry et al., *Chemical Engineers' Handbook*, McGraw-Hill, Inc., New York, 4th ed., 1963, Fig. 13-1.

Solution: Let us assume that temperature corrections to the Gibbs–Duhem equation are negligible since the temperature range of the data is fairly small. (The normal boiling point of *n*-propyl alcohol is 101.6°C.) Since the standard states have been chosen to be the pure component at the pressure and temperature of the system, we may drop the ° superscripts in equation (9-117) to obtain:

$$\ln(\gamma_{H_2O})_2 - \ln(\gamma_{H_2O})_1 = -\int_1^2 \frac{x_{alc}}{x_{H_2O}}\, d\ln\gamma_{alc}$$

If we choose state 1 as pure water, $(\gamma_{H_2O})_1 = 1.0$ and we get

$$\ln(\gamma_{H_2O})_2 = -\int_{x_{H_2O}=1.0}^{x_{H_2O}} \frac{x_{alc}}{x_{H_2O}}\, d\ln\gamma_{alc}$$

Thus let us prepare a plot of x_{alc}/x_{H_2O} vs. $\ln\gamma_{alc}$ as shown in Fig. S9-5. The required data are given in Table S9-5b. From this plot we may graphically evaluate the required integral. (Actually the trapezoid rule was used for numerical values.) The results are

TABLE S9-5b

x_{alc}	x_{H_2O}	γ_{alc}	$\ln\gamma_{alc}$	x_{alc}/x_{H_2O}
0.0	1.00	12.0	2.48	0.00
0.05	0.95	8.0	2.08	0.0525
0.10	0.90	4.9	1.59	0.111
0.20	0.80	2.8	1.03	0.250
0.30	0.70	1.9	0.64	0.428
0.40	0.60	1.5	0.41	0.667
0.50	0.50	1.3	0.26	1.000
0.60	0.40	1.2	0.18	1.500
0.70	0.30	1.15	0.14	2.333
0.80	0.20	1.10	0.095	4.000
0.90	0.10	1.05	0.049	9.000
1.00	0.00	1.00	0.00	

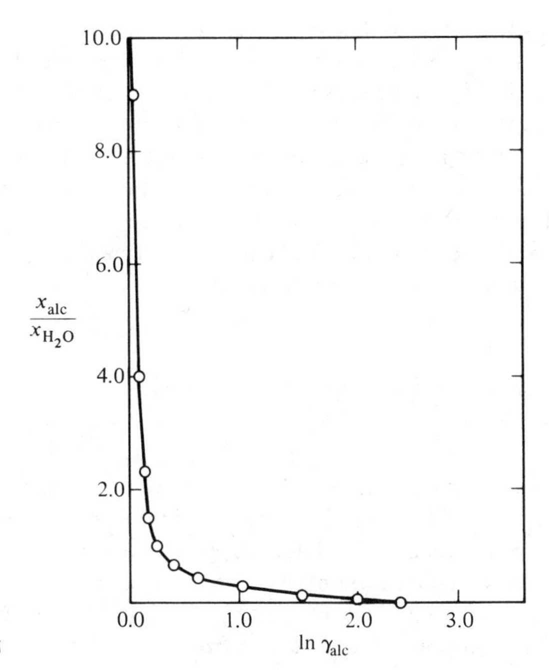

FIGURE S9-5

TABLE S9-5c

k	$x^k_{H_2O}$	$\ln \gamma^k_{H_2O}$	$-\int_{x^k_{H_2O}}^{x^{k+1}_{H_2O}} \frac{x_{alc}}{x_{H_2O}}\, d \ln \gamma_{alc}$	$\ln \gamma^{k+1}_{H_2O}$	$\gamma^k_{H_2O}$	$(\gamma^k_{H_2O})_{Perry}$
0	1.0	0.00	0.0105	0.0105	1.00	1.00
1	0.95	0.0105	0.0401	0.0506	1.01	1.05
2	0.90	0.0506	0.1011	0.152	1.05	1.10
3	0.80	0.152	0.166	0.318	1.16	1.20
4	0.70	0.318	0.126	0.444	1.37	1.30
5	0.60	0.444	0.125	0.569	1.56	1.50
6	0.50	0.569	0.100	0.669	1.77	1.80
7	0.40	0.669	0.077	0.746	1.95	1.95
8	0.30	0.746	0.142	0.888	2.11	2.25
9	0.20	0.888	0.299	1.187	2.43	2.50
10	0.10	1.187			3.28	2.85
11	0.00					

given in Table S9-5c and compared with values presented in Perry. It is observed that the predictions for all points except the last are well within our ability to read the charts in Perry. The discrepancy in the last point is probably due to errors in evaluating the small difference in the alcohol activity over this concentration range.

Equation (9-116) may also be written in terms of partial derivatives, as indicated here for a binary mixture:

$$x_A\left(\frac{\partial \ln \gamma_A^\circ}{\partial x_A}\right)_{P,T} + x_B\left(\frac{\partial \ln \gamma_B^\circ}{\partial x_A}\right)_{P,T} = 0 \tag{9-118}$$

Application of the Gibbs–Duhem equation to data obtained in vapor–liquid equilibrium studies is frequently useful, both in checking the consistency of the data and in calculating activity coefficients of a nonvolatile species in the liquid phase. Such data are normally obtained with either pressure or temperature varying. Thus, strictly speaking, the several forms of the Gibbs–Duhem equation that we have just developed for conditions of constant T and P are not applicable. In Chapter 10 this matter will be discussed further and the corrections necessary to utilize the Gibbs–Duhem equation where P and T changes affect the results significantly will be introduced.

9.10 Prediction of P–$\underline{V}$–T Properties for Real Gas Mixtures

Knowledge of the variation of the P–$\underline{V}$–T behavior of a mixture with composition permits evaluation of $\underline{U}$, $\underline{H}$, $\underline{S}$, $\underline{G}$, and $\underline{A}$ relative to any arbitrarily selected reference states from the relationships developed in Chapters 6 and 9. Although the P–$\underline{V}$–T properties of many pure gases are tabulated in the literature in handbooks such as Perry's[2] and the properties of many others can be predicted from the generalized compressibility charts or from equations of state, the added degree of freedom for mixtures of gases (due to variation in composition) makes it unrealistic to attempt to tabulate P–$\underline{V}$–T properties for gas mixtures. There are a few exceptions for mixtures encountered quite frequently. Air is one such example, and the 3 : 1 molar mixture of hydrogen and nitrogen encountered in ammonia synthesis is another. Consequently, if the P–$\underline{V}$–T properties of a gas mixture are required, it is usually necessary either to make experimental measurements or, preferably, to use one of the available rules for predicting these properties from information on the pure components.

Amagat's law of additive volumes is one such rule,

$$V = \sum_{i=1}^{C} n_i \underline{V}_i \tag{9-119}$$

It states that the total volume of a mixture should equal the sum of the volumes of the pure components at the same temperature and pressure as the mixture. Since in computations it is frequently convenient to use compressibility factors, each volume in equation (9-119) is expressed in terms of its compressibility factor from which a compressibility factor for the mixture can be determined:

$$\begin{aligned} Z_m &= \frac{PV}{nRT} = \frac{P}{nRT} \sum_{i=1}^{C} n_i \underline{V}_i \\ &= \frac{P}{nRT} \sum_{i=1}^{C} \frac{n_i Z_i RT}{P} = \sum_{i=1}^{C} \frac{n_i}{n} Z_i \\ &= \sum_{i=1}^{C} y_i Z_i \end{aligned} \tag{9-120}$$

[2] R. H. Perry et al., *Chemical Engineers' Handbook*, McGraw-Hill, Inc., New York, 4th ed., 1963, p. 3-101 ff.

Thus the compressibility factor for the mixture is the sum of the individual compressibility factors weighted by its mole fraction. Each compressibility factor is determined at the pressure and temperature of the mixture. As was shown previously, all ideal solutions obey Amagat's rule.

Another useful rule is Dalton's law of additive pressures,

$$P = \sum_{i=1}^{C} \mathscr{P}_i \tag{9-121}$$

This rule states that the pressure of a mixture is the sum of the pure-component pressures, $\mathscr{P}_i$. Each $\mathscr{P}_i$ is the pressure that would be associated with n_i moles of pure i present in the mixture if it occupied the entire volume of the mixture at the temperature of the mixture. The script letter is used in $\mathscr{P}_i$ to distinguish this quantity from P_i, the partial pressure, because the two are not always equal. Again, let us examine how we might find the compressibility factor for a mixture that obeys this rule:

$$\begin{aligned} Z_m &= \frac{PV}{nRT} = \frac{V}{nRT}\sum_{i=1}^{C} \mathscr{P}_i \\ &= \sum \frac{n_i}{n}\frac{\mathscr{P}_i V}{n_i RT} = \sum_{i=1}^{C} y_i \mathscr{Z}_i \end{aligned} \tag{9-122}$$

This appears to be similar to the result from Amagat's law, but there is an important distinction. The compressibility factor,

$$\mathscr{Z}_i = \frac{\mathscr{P}_i V}{n_i RT} \tag{9-123}$$

applies for n_i moles of component i occupying the entire volume of the mixture at the same temperature as the mixture. It is, therefore, not the same factor determined for Amagat's law and is given a different symbol here to emphasize the distinction. Dalton's law is empirical and may predict nonideal mixture behavior.

Another empirical rule that provides a convenient means for computing the compressibility factor for a mixture is the pseudo critical law. This rule states that "pseudo critical" properties for a mixture are computed from the critical properties of the components present in the mixture as follows:

$$\begin{aligned} P_{c,p} &= \sum_{i=1}^{C} y_i P_{c,i} \\ T_{c,p} &= \sum_{i=1}^{C} y_i T_{c,i} \end{aligned} \tag{9-124}$$

where the subscript p indicates a pseudo critical property. These pseudo critical properties may then be used to compute a pseudo reduced temperature and pressure for the mixture:

$$\begin{aligned} T_{r,p} &= \frac{T}{T_{c,p}} \\ P_{r,p} &= \frac{P}{P_{c,p}} \end{aligned} \tag{9-125}$$

The assertion is then made that the mixture behaves as if it were a single specie at these reduced conditions. The compressibility factor for the mixture is determined from the generalized compressibility charts in the usual fashion. Since this rule is simply an empirical scheme that seems to give reasonable results, no physical significance should be attached to these pseudo critical properties. Indeed, a gas mixture may not have a single well-defined critical point.

Just as the computer has created increased interest in the more complex equations of state as a means of evaluating pure-component behavior, such is also the case for mixtures. Equations of state also have applications in predicting mixture properties when a procedure for computing mixture coefficients from the coefficients of the pure materials is known. Some mixture rules for the Redlich–Kwong, Beattie–Bridgeman, and Benedict–Webb–Rubin equations of state (see Section 3.8) are given below.

The mixing rules proposed by Redlich and Kwong[3] are

$$a_m = \left(\sum_{i=1}^{C} y_i a_i^{1/2}\right)^2$$
$$b_m = \sum_{i=1}^{C} y_i b_i \tag{9-126}$$

Prausnitz and Chueh[4] have proposed an alternative rule for a_m which they find to give superior accuracy in some instances.

The mixing rules proposed by Beattie[5] for the Beattie–Bridgeman equation of state are

$$\begin{aligned} (A_{0m})^{1/2} &= \sum y_i (A_{0i})^{1/2} \\ a_m &= \sum y_i a_i \\ B_{0m} &= \sum y_i B_{0i} \\ b_m &= \sum y_i b_i \\ c_m &= \sum y_i c_i \end{aligned} \tag{9-127}$$

The normal mixing rules for the Benedict–Webb–Rubin equation of state are

$$\begin{aligned} A_{0m} &= (\sum y_i A_{0i}^{1/2})^2 \\ B_{0m} &= \sum y_i B_{0i} \\ C_{0m} &= (\sum y_i C_{0i}^{1/2})^2 \\ a_m &= (\sum y_i a_i^{1/3})^3 \\ b_m &= (\sum y_i b_i^{1/3})^3 \\ c_m &= (\sum y_i c_i^{1/3})^3 \\ \alpha_m &= (\sum y_i \alpha_i^{1/3})^3 \\ \gamma_m &= (\sum y_i \gamma_i^{1/2})^2 \end{aligned} \tag{9-128}$$

Alternative rules, particularly for A_{0m} and B_{0m} are often used.[6]

[3] O. Redlich and J. N. S. Kwong, *Chem. Rev.*, *44*, 233 (1949).

[4] J. M. Prausnitz and P. L. Chueh, *Computer Calculations for High-Pressure Vapor-Liquid Equilibria*, Prentice-Hall, Inc., Englewood Cliffs, N. J., 1968.

[5] J. A. Beattie, *J. Amer. Chem. Soc.*, *51*, 19 (1924).

[6] R. V. Orye, *Ind. Eng. Chem. Process Design Develop.*, *8*, 579–588 (1969).

In general, the prediction of mixture properties from equations incorporating these mixture coefficients will be no more successful than the prediction of pure-component properties with the same equations. The mixture rules cited above are empirical and the subject of continuing discussion in the literature, as indicated above. For guidance in selecting alternative mixture rules some additional help can be found in the references cited. Although these mixture rules have proved to be reasonably reliable, confirmation by comparison with experimental observation should be considered whenever feasible.

The mixture rules above do not account for interactions between unlike pairs of molecules if they are greatly different from interactions between like pairs. Some workers[7] favor the virial equation of state for mixtures because its mixture rules have the same theoretical basis as the equation itself and the rules explicitly identify an interaction parameter. For instance, the mixing rule for the second virial coefficient is

$$B_m = \sum_{i=1}^{C} \sum_{j=1}^{C} y_i y_j B_{ij} \tag{9-129}$$

where the B_{ii} are the second virial coefficients based on the interaction of like pairs of molecules and the B_{ij} $(i \neq j)$ are coefficients based on interaction of unlike pairs of molecules. In principle the B_{ij} can be evaluated by procedures analogous to those used to obtain the B_{ii}.

SAMPLE PROBLEM 9-6. A stream containing a 3 : 1 molar ratio of hydrogen to nitrogen is used as the feed gas for an ammonia synthesis plant. The stream is compressed to 400 atm and heated to 300°C before entering the catalytic converter. The specific volume of the mixture at these conditions is needed to size the reactor. Perry's *Handbook*[8] gives information from which it can be found that $Z_m = 1.155$. Compare this value with that predicted by each of the following procedures:

(a) Ideal gas.
(b) Dalton's law and the generalized Z chart.
(c) Amagat's law and the generalized Z chart.
(d) Pseudo critical method.

Solution: (a) The ideal gas computation for the mixture is particularly straightforward. We first need to establish the values to be used for R and T:

$$R = 0.7302 \text{ atm ft}^3/\text{lb-mol °R}$$

$$T = (300 + 273)(1.8)\text{°R} = 1032\text{°R}$$

Thus

$$\underline{V} = \frac{RT}{P} = \frac{(0.7302)(1032)}{400} \frac{\text{atm ft}^3 \text{ °R}}{\text{lb-mol atm}} = 1.884 \text{ ft}^3/\text{lb-mol}$$

$$Z_m = 1.0$$

[7] J. M. Prausnitz, C. A. Eckert, R. V. Orye, and J. P. O'Connell, *Computer Calculations for Multicomponent Vapor-Liquid Equilibria*, Prentice-Hall, Inc., Englewood Cliffs, N. J., 1967.

[8] R. H. Perry et al., *Chemical Engineers' Handbook*, McGraw-Hill, Inc., New York, 4th ed., 1963, Table 3-162, p. 3-104.

(b) The application of Dalton's Law to this problem requires knowledge of the pure-component pressures, $\mathscr{P}_{N_2}$ and $\mathscr{P}_{H_2}$. Since these are not known in advance, we must set up an iterative computation (trial and error) in which we first assume values for these pressures and then check our assumptions. The best initial guess of the pure-component pressures is to assume they are equal to the partial pressures.

Trial 1:

$$\mathscr{P}_{H_2} = P_{H_2} = (0.75)(400) = 300 \text{ atm}$$

$$\mathscr{P}_{N_2} = P_{N_2} = (0.25)(400) = 100 \text{ atm}$$

The critical properties of nitrogen and hydrogen are as follows:

$$\text{Hydrogen:} \quad P_c = 12.8 \text{ atm}, \quad T_c = 59.9°\text{R}$$

$$\text{Nitrogen:} \quad P_c = 33.5 \text{ atm}, \quad T_c = 227°\text{R}$$

Thus, based on the assumed pure-component pressures and the known temperature, we may compute the reduced properties for each component:

$$\text{Hydrogen:} \quad P_r = \frac{300}{12.8} = 23.4, \quad T_r = \frac{1032}{59.9} = 17.2$$

$$\text{Nitrogen:} \quad P_r = \frac{100}{35.5} = 2.99, \quad T_r = \frac{1032}{227} = 4.55$$

The compressibility factors obtained from the generalized compressibility charts for the components are

$$\mathscr{Z}_{H_2} = 1.13 \qquad \mathscr{Z}_{N_2} = 0.99$$

and the mixture compressibility and specific volume are

$$Z_m = y_{H_2}\mathscr{Z}_{H_2} + y_{N_2}\mathscr{Z}_{N_2}$$
$$= (0.75)(1.13) + (0.25)(0.99) = 1.095$$

$$\underline{V} = \frac{ZRT}{P} = \frac{(1.095)(0.7302)(1032)}{400} = 2.063 \text{ ft}^3/\text{lb-mol}$$

It is necessary to check assumed values of pure-component pressures. This may be done by solving equation (9-123) for $\mathscr{P}_{H_2}$ and substituting the necessary values from the above computations:

$$\mathscr{P}_{H_2} = \frac{\mathscr{Z}_{H_2} n_{H_2} RT}{V} = \frac{\mathscr{Z}_{H_2} y_{H_2} RT}{\underline{V}}$$

$$= \frac{(1.13)(0.75)(0.7302)(1032)}{2.063} \frac{\text{lb-mol atm ft}^3\ °\text{R}}{\text{lb-mol ft}^3\ °\text{R}} = 310 \text{ atm}$$

$$\mathscr{P}_{N_2} = \frac{(0.99)(0.25)(0.7302)(1032)}{2.063} \frac{\text{lb-mol atm ft}^3\ °\text{R}}{\text{lb-mol ft}^3\ °\text{R}} = 90 \text{ atm}$$

Trial 2: Now we must repeat the above computations until we obtain no further correction for the pure-component pressures.

$$\text{Hydrogen:} \quad P_r = \frac{310}{12.8} = 24.2, \quad \mathscr{Z}_{H_2} = 1.13$$

$$\text{Nitrogen:} \quad P_r = \frac{90}{33.5} = 2.70, \quad \mathscr{Z}_{N_2} = 0.99$$

These are the same compressibilities that we found in trial 1, so the remainder of the above computations are not repeated here and the mixture volume is the same as was found in trial 1.

(c) The component compressibilities to be used here are based on the total pressure. Thus we may write immediately:

$$\text{Hydrogen:}\quad P_r = \frac{400}{12.8} = 31.25,$$

$$T_r = 17.2$$

$$Z_{H_2} = 1.16$$

$$\text{Nitrogen:}\quad P_r = \frac{400}{33.5} = 11.94$$

$$T_r = 4.55$$

$$Z_{N_2} = 1.17$$

The mixture compressibility and the mixture volume then follow:

$$\begin{aligned} Z_m &= Z_{H_2}y_{H_2} + Z_{N_2}y_{N_2} \\ &= (0.75)(1.16) + (0.25)(1.17) \\ &= 1.16 \end{aligned}$$

$$\underline{V} = \frac{Z_m RT}{P} = \frac{(1.16)(0.7302)(1832)}{(400)} \frac{\text{ft}^3\ \text{atm}\ ^\circ\text{R}}{\text{atm lb-mol}\ ^\circ\text{R}} = 2.185\ \text{ft}^3/\text{lb-mol}$$

(d) The first step here is to compute the pseudo critical properties of this mixture, following equation (9-124):

$$\begin{aligned} P_{c,p} &= y_{H_2}P_{c,H_2} + y_{N_2}P_{c,N_2} \\ &= (0.75)(12.8)\ \text{atm} + (0.25)(33.5)\ \text{atm} = 17.98\ \text{atm} \end{aligned}$$

$$\begin{aligned} T_{c,p} &= y_{H_2}T_{c,H_2} + y_{N_2}T_{c,N_2} \\ &= (0.75)(59.9)^\circ\text{K} + (0.25)(227)^\circ\text{K} = 101.9^\circ\text{K} \end{aligned}$$

The pseudo reduced properties are

$$P_{r,p} = \frac{400\ \text{atm}}{17.98\ \text{atm}} = 22.25$$

$$T_{r,p} = \frac{1032^\circ\text{K}}{101.9^\circ\text{K}} = 10.1$$

The compressibility corresponding to these reduced properties is 1.21 and the corresponding volume is

$$\underline{V} = \frac{ZRT}{P} = \frac{(1.21)(0.7302)(1032)}{400} \frac{\text{atm ft}^3\ ^\circ\text{R}}{\text{atm lb-mol}\ ^\circ\text{R}} = 2.280\ \text{ft}^3/\text{lb-mol}$$

The results are summarized in Table S9-6. It can be seen that all the rules of the pre-

TABLE S9-6

Method	Compressibility	Volume, ft³/lb-mol	Error, %
Perry's *Handbook*	1.155	2.176	—
Ideal gas	1.0	1.884	−13.4
Dalton's law	1.095	2.063	−5.2
Amagat's law	1.16	2.186	0.43
Pseudo critical	1.21	2.280	4.8

vious section give better results than the ideal gas law. It should be observed that neither Dalton's law nor the pseudo critical method predict ideal solution behavior except when the compressibility is nearly unity (ideal gas).

9.11 Prediction of Mixture Properties for Liquid and Solid Systems

Whereas the most common method of predicting deviations from ideal solution behavior for gaseous mixtures is to select an appropriate mixing rule for use with an equation of state, for liquid and solid systems it has generally been more fruitful to attempt to express the activity coefficient as a function of composition. Although physical meaning is frequently assigned to the coefficients appearing in such expressions, evaluation of the coefficients without recourse to experimental data is generally difficult and subject to error. The coefficients can, however, be treated as empirical (or semiempirical) constants and their values determined from experimental values of the activity coefficient or from mixing properties. Quite accurate predictions of nonideality often result from these empirical fitting procedures.

Since many of the solution data for liquid and solid systems are determined indirectly by measuring the properties of the equilibrium vapor phase, much of our discussion will be postponed until multicomponent phase equilibrium is treated in Chapter 10.

The common starting point for development of either empirical or theoretical expressions for activity coefficients is the excess mixture properties. Once the excess properties of the solution are known as a function of composition, it is relatively straightfoward to determine the activity coefficients—also as functions of composition. In what follows we shall define the excess mixture properties and show their significance in describing nonidealities. Thereafter we discuss various expressions that have been found useful in correlating activity coefficients. Finally we shall discuss the categories of solution theories that might be used to make predictions with limited mixture data or even in the absence of data. Of these theories, most attention is given to the regular solution theory, which, as we shall demonstrate, represents a slight relaxation of the ideal solution mixing rules. Regular solution theory does, however, permit modeling of systems in which molecular interactions that lead to nonzero enthalpies of mixing occur. However, the resulting model is fairly restrictive and should be considered as only a first-order refinement to the ideal solution model. Since it is not our intent to present an exhaustive treatment of solution theory in this development, the interested reader is referred to any of several excellent reference works[9,10,11] for detailed discussion of more advanced concepts.

[9] J. M. Prausnitz, *Molecular Thermodynamics of Fluid-Phase Equilibrium*, Prentice-Hall, Inc., Englewood Cliffs, N.J., 1969.

[10] J. H. Hildebrand and R. L. Scott, *Reqular Solutions*, Prentice-Hall, Inc., Englewood Cliffs, N.J., 1962.

[11] E. A. Guggenheim, *Mixtures*, Oxford University Press, New York, 1952.

9.12 Excess Properties

We define an excess property to be the amount by which the actual change of an extensive property upon mixing differs from the change that would occur if the solution were ideal. Thus, taking the total Gibbs free energy as an example, we write

$$G_{\text{mix}} = \sum_{i=1}^{C} n_i \bar{G}_i$$

$$G_{\text{components}} = \sum_{i=1}^{C} n_i \underline{G}_i$$

$$\Delta G_m^I = \sum_{i=1}^{C} n_i (\bar{G}_i - \underline{G}_i)^I = \sum_{i=1}^{C} n_i (RT \ln x_i) \tag{9-130}$$

where the superscript I indicates ideal mixing behavior. The excess Gibbs free energy of the mixture is defined as

$$\Delta G^E \equiv \Delta G_m - \Delta G_m^I \tag{9-131}$$

or

$$\Delta G^E \equiv G_{\text{mix}} - (G_{\text{components}} + \Delta G_m^I)$$

$$= \sum_{i=1}^{C} n_i (\bar{G}_i - \underline{G}_i - RT \ln x_i) \tag{9-131a}$$

Excess properties for the other extensive properties are defined just as for the Gibbs free energy in equation (9-131a):

$$\Delta H^E = \Delta H_m - \Delta H_m^I = \Delta H_m \tag{9-132}$$

$$\Delta V^E = \Delta V_m - \Delta V_m^I = \Delta V_m \tag{9-133}$$

$$\Delta U^E = \Delta U_m - \Delta U_m^I = \Delta U_m \tag{9-134}$$

$$\Delta S^E = \Delta S_m - \Delta S_m^I = \Delta S_m + R \sum_{i=1}^{C} x_i \ln x_i \tag{9-135}$$

$$\Delta A^E = \Delta A_m - \Delta A_m^I = \Delta A_m - R \sum_{i=1}^{C} x_i \ln x_i \tag{9-136}$$

Excess properties are interrelated in precisely the same way that total properties are:

$$\Delta H^E = \Delta U^E + P \, \Delta V^E \tag{9-137}$$

$$\Delta G^E = \Delta H^E - T \, \Delta S^E \tag{9-138}$$

$$\Delta A^E = \Delta U^E - T \, \Delta S^E \tag{9-139}$$

Similarly, excess properties can be differentiated in a manner analogous to the total properties to produce a whole family of equations similar to those developed in Chapter 6 for pure-component systems and earlier in this chapter for mixtures.

Equation (9-131) can be related to the activity coefficient (for the case of pure-components at the pressure and temperature of the system as the reference states) by substituting for the quantity $\bar{G}_i - \underline{G}_i$,

$$\bar{G}_i - \underline{G}_i = RT \ln (\gamma_i x_i) \tag{9-108}$$

to obtain

$$\Delta G^E = \sum_{i=1}^{C} n_i RT \ln \gamma_i \tag{9-140}$$

or, for 1 mole of solution,

$$\Delta \underline{G}^E = \sum_{i=1}^{C} x_i RT \ln \gamma_i \tag{9-141}$$

The derivative of equation (9-140) with respect to the number of moles of specie j, n_j, at constant T, P, and n_k (where n_k represents all mole amounts held constant except the jth component) yields the partial excess property, $\bar{G}_j^E$, as shown in equations (9-142) through (9-144).

$$\left(\frac{\partial \Delta G^E}{\partial n_j}\right)_{P,T,n_k} = \sum_{i=1}^{C} RT \ln \gamma_i \left(\frac{\partial n_i}{\partial n_j}\right)_{P,T,n_k} + \sum_{i=1}^{C} n_i RT \left(\frac{\partial \ln \gamma_i}{\partial n_j}\right)_{P,T,n_k} \tag{9-142}$$

Equation (9-142) may be simplified since we note that $(\partial n_i/\partial n_j)_{P,T,n_k} = 0$ unless $i = j$, and that

$$\sum_{i=1}^{C} n_i RT \left(\frac{\partial \ln \gamma_j}{\partial n_j}\right)_{P,T,n_k} = RTn \sum_{i=1}^{C} x_i \left(\frac{\partial \ln \gamma_i}{\partial n_j}\right)_{P,T,n_k} = 0 \tag{9-143}$$

by virtue of the Gibbs–Duhem equation (9-118). Dropping the last sum in equation (9-142) then yields the desired result

$$\left(\frac{\partial \Delta G^E}{\partial n_j}\right)_{P,T,n_k} = \Delta \bar{G}_j^E = RT \ln \gamma_j \tag{9-144}$$

which indicates that we only need to predict the excess Gibbs free energy of a mixture as a function of composition to be able to determine the activity coefficients of all its components.

Should we wish to use a different reference state (than the pure component at the pressure and temperature of the system) for component j, we would obtain the similar result

$$\left(\frac{\partial \Delta G^E}{\partial n_j}\right)_{P,T,n_k} = RT \ln \gamma_j^\circ + RT \ln \frac{f_i^\circ}{f_i} = RT \ln\left(\gamma_j^\circ \frac{f_i^\circ}{f_i}\right) \tag{9-145}$$

Since the ratio f_i°/f_i is a constant, independent of composition, we discover that the new activity coefficient differs from the former by a constant multiplying factor f_i°/f_i, which was already apparent in equation (9-101).

Equation (9-141) can be rearranged as follows:

$$\frac{\Delta \underline{G}^E}{RT} = \sum_{i=1}^{C} x_i \ln \gamma_i \tag{9-146}$$

This equation proves to be a convenient departure point for the development of expressions used to correlate and predict activity coefficients as a function of composition.

We may determine some of the characteristics that any functional relationship between the excess Gibbs free energy and composition must satisfy as follows. If the pure-component standard state (pressure and temperature of the system) is used for a solution that exists across the entire composition range, then $\gamma_i \longrightarrow 1$ as $x_i \longrightarrow 1$. Since as $x_i \longrightarrow 1$ all other compositions approach zero, it is seen that equation (9-146) predicts that $\Delta \underline{G}^E = 0$ for pure components. Recalling our definition of an excess property, this is precisely what would be expected. Thus we observe that any functional relationship between $\Delta \underline{G}^E/RT$ and composition must reduce to zero in the limit of any pure component. That is,

$$\frac{\Delta \underline{G}^E}{RT} = \phi(x_1, x_2 \cdots x_i \cdots x_C) \tag{9-147}$$

where

$$\frac{\Delta \underline{G}^E}{RT} = 0 \qquad \text{for any } x_i = 1 \tag{9-148}$$

Although we choose to consider only binary systems in the development that follows, this is done merely for convenience. In general, the techniques are equally applicable to multicomponent systems as well, although the algebraic complexities mount as we increase the number of components.

The simplest functional relationship for a binary system that satisfies the conditions in equations (9-147) and (9-148) is

$$\frac{\Delta \underline{G}^E}{RT} = x_1 x_2 A \tag{9-149}$$

where A is an adjustable parameter that can be varied to provide the best fit with the available data for a system. This simple expression exhibits symmetry in the excess Gibbs free energy about the composition $x_1 = x_2 = 0.5$ and may have either positive or negative values, depending on the sign of the parameter A. Systems that exhibit positive values of $\Delta \underline{G}^E$ are termed *positive deviators* from ideal solution behavior. Activity coefficients in such systems (for pure-component standard states) are greater than unity. Systems with negative values of ΔG^E are termed *negative deviators* and the activity coefficients range between 0 and 1. Although the symmetrical behavior predicted by equation (9-149) is not typical of most systems, this simple expression is frequently applicable to systems in which the components do not differ appreciably in size, shape, and chemical behavior. Regular solutions to be discussed later in this chapter will be shown to conform to this type of behavior.

One can easily conceive of other more complex relationships which satisfy equation (9-148) and which would permit skewed behavior about the composition $x_1 = x_2 = 0.5$. Redlich and Kister[12] proposed one such equation in which the constant parameter A is replaced with a series expansion in $(x_1 - x_2)$ and additional adjustable parameters:

$$\frac{\Delta \underline{G}^E}{RT} = x_1 x_2 [A + B(x_1 - x_2) + C(x_1 - x_2)^2 + \cdots] \tag{9-150}$$

[12] O. Redlich and A. T. Kister, *Ind. Eng. Chem.*, *40*, 345 (1948).

The large number of adjustable parameters permits improved correlation and prediction of thermodynamic data.

Wohl,[13] recognizing the importance of component volumes in the mixing process, proposed an expression that accounts directly for molecular volumes rather than mole fractions:

$$\frac{\Delta \underline{G}^E}{RT\sum_i q_i x_i} = \sum_{ik} z_i z_k a_{ik} + \sum_{ikj} z_i z_k z_j a_{ikj} + \cdots \tag{9-151}$$

where q_i represents the molar volume of component i, z_i represents the volumetric fraction of component i, and the $a_i \cdots$ are empirical constants. As written, the equation is applicable to multicomponent systems and takes into account two- and three-particle interactions between similar and dissimilar molecules. Additional terms can be added to treat interactions of four or more particles, but the data required to evaluate the parameters seldom justify it.

If Wohl's three-suffix equation is applied to a binary system it can be written as

$$\frac{\Delta \underline{G}^E}{RT(q_1 x_1 + q_2 x_2)} = z_1 z_2(2a_{12}) + z_1^2 z_2(3a_{112}) + z_1 z_2^2(3a_{122}) \tag{9-152}$$

where a_{12} is an adjustable parameter relating to the interaction between one molecule of each component and a_{112} and a_{122} are adjustable parameters relating to the interaction of two molecules of one component with one of the other. It should be noted that the coefficients with identical subscripts, a_{11}, a_{22}, etc., must be zero to satisfy the conditions of equation (9-148).

Equation (9-152) can be used to derive expressions for activity coefficients in terms of either volume fractions, z_i, or mole fractions, x_i. The connection is simply

$$z_i = \frac{x_i q_i}{\sum_{j=1}^{C} x_j q_j} \tag{9-153}$$

Equations (9-154) and (9-155) are examples of the former; they include three adjustable parameters A', B', and q_1/q_2:

$$\ln \gamma_1 = z_2^2\left[A' + 2\left(B'\frac{q_1}{q_2} - A'\right)z_1\right] \tag{9-154}$$

$$\ln \gamma_2 = z_1^2\left[B' + 2\left(A'\frac{q_2}{q_1} - B'\right)z_2\right] \tag{9-155}$$

For symmetric systems where $q_1 = q_2$, equations (9-154) and (9-155) become equivalent to the three-suffix Margules equations.[14] In terms of mole fractions, the Margules equations can be written as

$$\ln \gamma_1 = \frac{A}{2}x_2^2 + \frac{B}{3}x_2^3 \tag{9-156}$$

$$\ln \gamma_2 = \frac{A + B}{2}x_2 - \frac{B}{3}x_2^3 \tag{9-157}$$

[13] K. Wohl, *Trans. Amer. Inst Chem. Eng.*, *42*, 215 (1946).

[14] M. Margules, *Sitzber. Akad. Wiss. Wien., Math.-Naturw. Kl.*, (II) *104*, 1243 (1895).

If $q_1/q_2 = A'/B'$, equations (9-154) and (9-155) reduce to the van Laar equations[15]:

$$\ln \gamma_1 = z_1^2 A' \tag{9-158}$$

$$\ln \gamma_2 = z_1^2 B' \tag{9-159}$$

Expressed in terms of mole fractions, a generally more useful form, the van Laar equations, are

$$\ln \gamma_1 = \frac{a}{[1 + (ax_1/bx_2)]^2} \tag{9-160}$$

$$\ln \gamma_2 = \frac{b}{[1 + (bx_2/ax_1)]^2} \tag{9-161}$$

where a and b are adjustable parameters. Evaluation of the parameters requires experimentally determined values of the activity coefficients. a and b can be expressed in terms of activity coefficients and composition from equations (9-160) and (9-161) as follows:

$$a = \ln \gamma_1\left(1 + \frac{x_2 \ln \gamma_2}{x_1 \ln \gamma_1}\right)^2 \tag{9-162}$$

$$b = \ln \gamma_2\left(1 + \frac{x_1 \ln \gamma_1}{x_2 \ln \gamma_2}\right)^2 \tag{9-163}$$

It should be noted that at $x_1 = 0$, $\ln \gamma_1 = a$ and $\ln \gamma_2 = 0$, while at $x_2 = 0$, $\ln \gamma_2 = b$ and $\ln \gamma_1 = 0$. Thus, if binary behavior in the region of low concentration is known, the empirical constants a and b can be evaluated. As we shall see in Chapter 10, these constants are frequently evaluated from vapor–liquid equilibrium data for systems of interest.

The Margules and van Laar equations are the most extensively used of the simple equations for correlation of activity coefficients. They are mathematically simple and require only two adjustable parameters. The merits of the former are obvious; the merit of the latter is that the parameters can, in principle, be evaluated from one datum point. With the advent of computers, several other slightly more complex equations (although still with only two adjustable parameters) have found increasing application. The most strongly touted of these is the Wilson equation[16]:

$$\begin{aligned} \ln \gamma_1 &= -\ln(x_1 + Ax_2) + x_2\left(\frac{A}{x_1 + Ax_2} - \frac{B}{Bx_1 + x_2}\right) \\ \ln \gamma_2 &= -\ln(x_2 + Bx_1) - x_1\left(\frac{A}{x_1 + Ax_2} - \frac{B}{Bx_1 + x_2}\right) \end{aligned} \tag{9-164}$$

Orye and Prausnitz[17] have extensively studied the application of this equation to miscible systems and found its representation to be generally comparable and sometimes superior to either the Margules or the van Laar equations.

[15] J. J. Van Laar, *Z. Phys. Chem.*, *72*, 723 (1910).
[16] G. M. Wilson, *J. Amer. Chem. Soc.*, *86*, 127 (1964).
[17] R. V. Orye and J. M. Prausnitz, *Ind. Eng. Chem.*, *57*, 19 (1965).

Scatchard and Hamer[18] have proposed a two-parameter equation which, in addition, contains the ratio of the molar specific volumes of the two pure components. This offers some additional flexibility which may be useful in fitting data. It can be obtained as a special case of the Wohl equation and provides correlations that are generally intermediate between those of Margules and van Laar.

For systems that are strongly nonideal or of limited miscibility, the equations of Renon and Prausnitz[19] may be useful. In these areas they have definite advantages over the Wilson equation (which cannot predict limited miscibility).

The construction of theoretical or empirical expressions for the excess Gibbs free energy of mixtures has been the subject of much research, which is continuing. As yet none of the expressions postulated are so generally superior to others that we can cite any one as being the best. The selection of an expression for the excess Gibbs free energy (and its implied equations for activity coefficients) for a given application depends on such considerations as (1) the need for simplicity of equations, (2) the availability of data with which to evaluate the parameters, (3) the need to make predictions without extensive data, (4) the desire for accurate correlations, (5) the desire to ascribe physical meaning to the parameters, or (6) unusual nonidealities, including immiscibility of components in some concentration ranges.

The expressions for excess Gibbs free energy set forth by Wohl, Wilson, and Renon are all capable of producing correlations for application to multicomponent systems. The Wohl equation is mathematically the most straightforward and has the apparent virtue of allowing the user to retain terms to any polynomial degree in concentration, introducing additional adjustable parameters at each increase. Quite accurate data are needed, however, before expressions more complex than two adjustable parameters (for a binary system) are justified. The parameters used for fitting binary behavior also appear in the Wohl correlations for activity coefficients in ternary (three-component) mixtures. A separate ternary coefficient is also included. The latter can often be neglected or a small number of ternary data used to select a value so as to refine the correlation.

The Wilson and Renon equations, although slightly more complex than the Wohl equation, have been tested extensively and have produced correlations of binary data that are the equal of the Wohl equation (in either its Margules or van Laar forms). Although the Wilson and Renon equations are limited by theoretical considerations to only two adjustable parameters for fitting binary data, their application to ternary or higher-order systems does not introduce any parameters not found in the equations for each binary pair. This is indeed a virtue, for it obviates the need for data on the more complex mixtures. Their testing in correlations of ternary data seems to have been successful in many systems.

[18] G. Scatchard and W. J. Hamer, *J. Amer. Chem. Soc.*, *57*, 1805 (1935).

[19] H. Renon and J. M. Prausnitz, *Amer. Ind. Chem. Engrs. J.*, *14*, 135 (1968).

Many workers in regular solution theory have advanced models by which to predict the excess energy change upon mixing. Among them are van Laar, Scatchard, and Hildebrand.[20] Their analyses, however, are beyond the scope of this work.

9.13 Regular Solutions

Let us now turn to the classification of liquid theories that lead to the excess free-energy functions, such as those mentioned above. A brief digression into the common rationale of these theories will help in that classification. We may think of an ideal binary solution (in the Lewis and Randall sense) as one in which the molecules of *A* and *B* interact in nearly the same fashion when they occur in *A*-*B* pairs as when they occur in *A*-*A* and *B*-*B* pairs. For such mixtures the excess Gibbs free energy is, of course, zero. Thus it is argued that nonzero excess Gibbs free energies are attributable to *A*-*B* interactions, which are quite different from *A*-*A* or *B*-*B* interactions. We can, moreover, speculate that these nonideal interactions are attributable to either physical or chemical intermolecular forces. Among the theories based on physical interactions between molecules are the regular solution and the athermal solution theories. Among those based on chemical interaction are the association and solvation theories. Only the regular solution theory will be discussed in this development. For purposes of discussion here we shall assume a binary regular solution to be one that meets the following conditions:

$$\left.\begin{aligned} \Delta \underline{S}^E &= 0 \\ \Delta \underline{G}^E &= \Delta \underline{H}_m = x_A x_B W \end{aligned}\right\} \text{ definition of regular solution} \tag{9-165}$$

where W is a parameter that relates to the molecular interaction of dissimilar molecules in the solution. Since many systems exhibit behavior where $\Delta \underline{H}_m \neq 0$, this model is often useful for systems that clearly do not conform to ideal solution behavior but yet do not experience sufficiently strong attractive or repulsive forces that the assumption of $\Delta \underline{S}^E = 0$ is an unreasonable approximation. In general, this is most likely to be valid for mixtures whose molecules, although chemically dissimilar, are of a similar size.

For n moles of a binary solution that conforms to regular solution behavior, the mixing properties can be expressed as follows:

$$\Delta S_m = -nR(x_A \ln x_A + x_B \ln x_B) = -R(n_A \ln x_A + n_B \ln x_B) \tag{9-166}$$

$$\Delta H_m = n x_A x_B W = (n_A + n_B) x_A x_B W \tag{9-167}$$

$$\Delta G_m = (n_A + n_B) x_A x_B W + RT(n_A \ln x_A + n_B \ln x_B) \tag{9-168}$$

From equation (9-144) we can express the activity coefficient as

$$\ln \gamma_A = \frac{1}{RT}\left(\frac{\partial \Delta G^E}{\partial n_A}\right)_{P,T,n_B} \tag{9-144}$$

[20] J. H. Hildebrand, *J. Amer. Chem. Soc.*, *51*, 66 (1929).

But for a regular solution ΔG^E can be written as

$$\Delta G^E = \Delta H^E - T\,\Delta S^E = (n_A + n_B)x_A x_B W \tag{9-169}$$

Differentiation of (9-169) with respect to n_A at constant P, T, n_B yields

$$\left(\frac{\partial \Delta G^E}{\partial n_A}\right)_{P,T,n_B} = \left(\frac{\partial[(n_A n_B W)/(n_A + n_B)]}{\partial n_A}\right)_{P,T,n_B}$$

$$= \frac{n_B^2 W}{(n_A + n_B)^2} = x_B^2 W \tag{9-170}$$

Substituting (9-170) in (9-144) yields

$$\ln \gamma_A = -\frac{x_B^2 W}{RT} \tag{9-171}$$

$$\ln \gamma_B = -\frac{x_A^2 W}{RT} \tag{9-172}$$

Equations (9-167), (9-168), (9-171), and (9-172) can be used to calculate ΔH_m, ΔG_m, γ_A, and γ_B if the interaction parameter W is known. W is generally determined either from heat of mixing data or determination of γ_A or γ_B at a given temperature. If W is permitted to assume either positive or negative values, the regular solution model can be applied to systems that exhibit both positive and negative deviations from ideal solution behavior. The special case for which $W = 0$ is seen to reduce the ideal solution model. The regular solution model predicts symmetrical behavior for the ΔH_m, ΔG_m and activity coefficients as shown in Figs. 9-10, 9-11, and 9-12.

Before concluding our discussion of regular solutions, it should be empha-

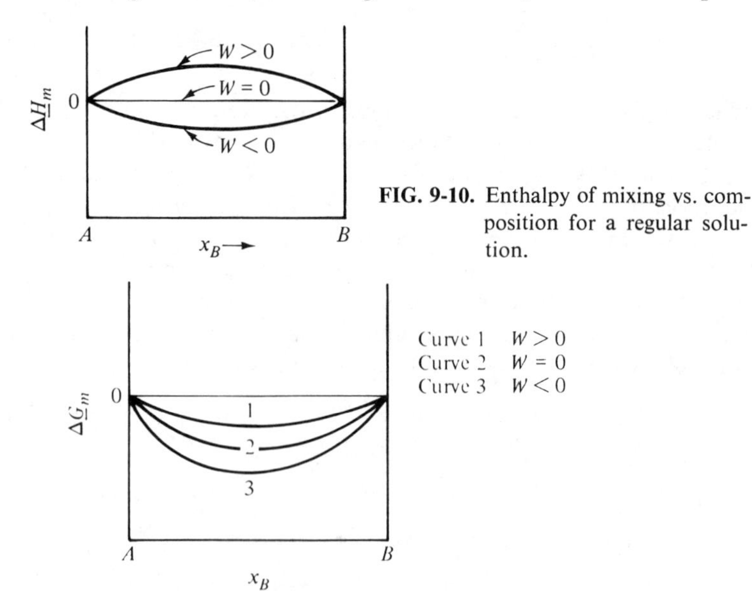

FIG. 9-10. Enthalpy of mixing vs. composition for a regular solution.

FIG. 9-11. Free energy of mixing vs. composition for regular solution.

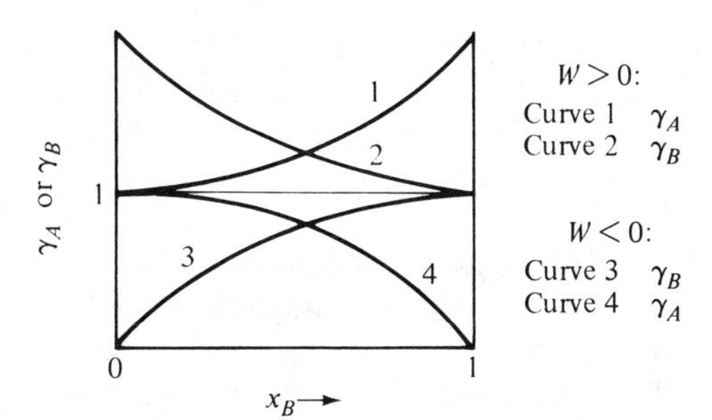

FIG. 9-12. Activity coefficients vs. composition for regular solution.

sized that the greater the absolute value of the interaction energy, W, the greater the likelihood that ΔS^E will be nonzero. As W increases in the positive direction, an indication of repulsion between the components A and B, the tendency toward immiscibility increases. As W becomes more negative, the attractive forces between A and B become large relative to the forces between like components. If such forces become sufficiently large, compounds can form in the system. These considerations will be considered in some detail in Chapter 10 when phase equilibria are discussed.

9.14 Other Solution Theories

Under our earlier classification of liquid theories we identified the regular and athermal solution theories as being based on the premise that the nonideal molecular interactions were physical in nature. Other theories have been based on nonideal interactions that are chemical in nature. These theories are described here superficially to indicate the kinds of mixtures for which they might be either theoretically or empirically useful.

Athermal Solutions

The name *athermal* derives from the postulate, common to all athermal theories, that there be no heat of mixing ($\Delta H^E = 0$). The modeling effort is then placed on obtaining expressions for the excess entropy of mixing. Such behavior is often a reasonable approximation for mixtures of molecules that are chemically similar, although perhaps of quite different size. Among the athermal theories that have proved useful in predicting nonideal behavior is the equation of Flory and Huggins[21] for polymer solutes in chemically similar solvents. The Wilson equation (9-164) also has its theoretical origin as an athermal solution theory.

Strictly speaking, athermal solution theories are valid only for mixtures that exhibit no heat of mixing, so their application in predicting nonidealities should generally be limited to mixtures with small heats of mixing. The some-

[21] P. J. Flory, *J. Chem. Phys.*, *9*, 660 (1941); *10*, 51 (1942); and M. L. Huggins, *J. Phys. Chem.*, *9*, 440 (1941); *Ann. N.Y. Acad. Sci,. 43*, 1 (1942).

what-more-general success of the Wilson equation as an empirical correlation should not be allowed to obscure this fact.

Chemical-Interaction Theories

If the molecular species in a mixture interact chemically, significant non-idealities may occur. Among the theories that deal with such interaction are the associated solution theories and the solvated solution theories. In the former, allowances are made for the fact that molecules may associate by hydrogen bonding or other interaction. The larger associated molecules that result will effectively reduce the number of moles present in the system. At the simplest level, non-ideality can be explained by discerning the true, rather than the apparent, composition of the solution. In solvated solution theories, the components are presumed to form more strongly bonded complexes, which again will lead to a discrepancy between the true and apparent composition of the solution. These theories are flexible and capable of predicting both positive and negative deviations from ideality. The difficulty of ascertaining what "true" chemical species are present and the equilibrium between species limit their utility.

Problems

9-1 Isothermal heat of solution data for ammonia–water solutions have been reported as follows:

Pounds of H_2O Added to 1 lb_m of Liquid NH_3 at 32°F	NH_3, wt %	Heat Evolved ($-Q$) Btu/lb_m NH_3 in Solution
0	100	0
0.111	90	35.5
0.25	80	75.8
0.429	70	121.1
0.667	60	169.7
1.00	50	218.8
1.50	40	270.0
2.33	30	308.2
4.00	20	328.5
9.00	10	343.8
$\sim\infty$	~ 0	358.0

The C_P of liquid NH_3 = 1.1 Btu/lb_m °F.

(a) Calculate $\underline{H}$ for a 20 per cent ammonia solution, referred to pure liquid water at 32°F and pure liquid ammonia at −40°F.

(b) Calculate Q for the process of absorbing 30 lb_m of water vapor (saturated) at 60°F in 10 lb_m of 80 per cent ammonia solution at 32°F, if the final product is to be at 32°F.

(c) Plot the 32°F isotherm on an enthalpy–concentration curve referred to pure liquid water at 32°F and pure liquid ammonia at 32°F.

(d) Using the plot from (c), determine $\underline{H}$ for an ammonia solution obtained by adiabatically mixing 25 lb_m of pure liquid water at 32°F with 75 lb_m of a 60 per cent ammonia solution at 32°F.

(e) A 10 per cent ammonia solution at 32°F is to be concentrated to form an 85 per cent solution at 32°F by adding pure liquid ammonia adiabatically. At what temperature should the pure ammonia be supplied to the mixer?

(f) Determine Q for the process of mixing 20 lb_m of pure liquid water, 30 lb_m of pure liquid ammonia, and 50 lb_m of a 20 per cent ammonia solution, if the mixing is done isothermally at 32°F.

(g) If the process in (f) is to be carried out by first mixing the ammonia and the 20 per cent solution, then removing the necessary heat, and finally adding the water adiabatically, determine $\underline{H}$ for the intermediate mixture after the necessary heat has been removed.

9-2 The partial molar enthalpies for aqueous solutions of sulfuric acid at 25°C are as follows:

Moles of H_2O/mole of H_2SO_4	$\bar{H}_{H_2O}$, cal/g-mol	$\bar{H}_{H_2SO_4}$, cal/g-mol
∞	0	0
1,600	−0.48	4,811
200	−2.16	5,842
10	−233	9,632
2	−2,315	18,216
1	−4,731	21,451
0	—	23,540

(a) What are the reference states for the above data?

(b) Sketch the 25°C isotherm on an enthalpy–concentration diagram for the above reference states.

(c) Sketch the 25°C isotherm using as reference states pure H_2O and H_2SO_4 at 25°C. Indicate how the points on the curve would actually be calculated.

(d) Calculate Q (Btu/lb_m 90 per cent solution) for diluting 90 per cent (by weight) H_2SO_4 with H_2O to produce 50 per cent (by weight) H_2SO_4.

9-3 One pound of liquid water and $1\frac{1}{2}$ lb_m of H_2SO_4, both at 32°F, are mixed adiabatically. The temperature of the mixture rises to 250°F. The specific heat of the resulting mixture is 0.635. From this information. Sketch the enthalpy–concentration isotherm (in Btu per pound of solution) for water–sulfuric acid mixtures at 32°F relative to water and sulfuric acid at the same temperature, showing the coordinates of the three known points.

9-4 (a) Determine the heat capacity of a 30 wt per cent caustic solution at 150°F. (An enthalpy–concentration diagram for this system is given in Appendix D.)

(b) If steam at 400°F and 25 psia were bubbled through a 60 wt per cent caustic solution in a well-insulated container in such a manner that the solution remained at 340°F and the water vapor leaving the solution is in thermal equilibrium with the solution at 20 psia, would the concentration of the solution increase or decrease? Explain.

(c) One hundred pounds of 50 wt per cent NaOH solution at 150°F is diluted with water at 130°F to give a solution containing 20 wt per cent NaOH. (1) If the final solution is to be at 140°F, how much heat must be removed? (2) If the mixing is adiabatic, what will be the final temperature of the solution after mixing?

(d) What is the latent heat of evaporation of 1 lb_m of water from a 50 wt per cent NaOH solution at 250°F?

9-5 In a water–ethyl alcohol mixture with a water mole fraction of 0.4, the partial volume of ethyl alcohol is observed to be 57.5 cm^3. If the mixture density is 0.8494, calculate the partial volume of water.

9-6 Sulfuric acid is concentrated by spraying it into the top of a brick-lined tower in which it flows down against an ascending stream of hot gas. The dilute acid is 73 weight per cent H_2SO_4 (corresponding to $H_2SO_4 + 2H_2O$) and enters the tower at 18°C. The concentrated acid is 91.5 weight per cent H_2SO_4 (corresponding to $2H_2SO_4 + H_2O$) and is withdrawn from the bottom of the tower at 200°C. The hot gas enters the bottom of the tower at 1000°C and leaves the top at 200°C. Calculate the number of gram-moles of gas entering the bottom per gram of dilute acid fed at the top of the tower.

$$(C_P)_{\text{gas}} = 6.5 + 0.001T(°\text{K}) \text{ cal/g-mol °K}$$

$$(C_P)_{91.5\%\ H_2SO_4} = 0.40 \text{ cal/g °C}$$

The heat liberated when water and H_2SO_4 are mixed isothermally at 18°C is as follows:

g-mols of H_2O added to 1 g-mol H_2SO_4	0.5	1.0	2.0	5.0	10.0
Heat liberated, cal	3,750	6,710	9,760	13,750	15,880

9-7 From the following data prepare an enthalpy–concentration diagram for solutions of NaOH in water at 68°F and 180°F. Give enthalpies as Btu/lb_m solution, and cover the concentration range from 0 to 40 mass per cent NaOH. Select the reference states as pure water at 32°F and pure NaOH at 68°F.

Calculate the latent heat of vaporization for 1 lb_m of water evaporating from a 30 per cent NaOH solution at 180°F. If 100 lb_m of a 10 per cent solution is mixed with 50 lb_m of a 35 per cent solution, how much heat must be removed if the temperature is

C_p of NaOH Solutions, Btu/lb_m °F

	Mass % NaOH			
T, °F	10	20	30	40
60	0.897	0.859	0.837	0.815
100	0.911	0.875	0.855	0.826
140	0.918	0.884	0.866	0.831
180	0.922	0.886	0.869	0.832

Heat of Solution of NaOH in water at 68°F

lb_m of NaOH Added to 1 lb_m of H_2O	Heat Evolved, Btu/lb_m of NaOH
0	455
0.0417	457.04
0.0870	462.15
0.1363	464.13
0.220	460.0
0.316	445.53
0.429	417.7
0.666	340.8

maintained at 68°F from beginning to end? If the initial solutions are at 68°F and the process is adiabatic, what will be the final temprature?

9-8 Eighty pounds of 10 wt per cent NH_3 at 60°F is mixed with 120 lb_m of 85 wt per cent NH_3 at 60°F.

(a) What is the temperature rise if mixing takes place adiabatically?

(b) How many pounds of cooling water would be needed to hold the temperature at 60°F while mixing is taking place if cooling water is available at 40°F and undergoes a 10°F temperature rise?

9-9 Refer to the H_2O–H_2SO_4 enthalpy–concentration diagram in the following:

(a) Estimate the heat capacity of 40 wt per cent H_2SO_4 solution at 100°F.

(b) What reference states were used in constructing the diagram?

(c) If water at 72°F is used to dilute an 80 per cent H_2SO_4 solution at 32°F to 60 per cent, calculate the heat that must be removed per pound of solution if the solution temperature is to remain at 32°F.

(d) Sketch the 32°F isotherm on the diagram corresponding to the following reference states: $\bar{H}_{H_2O}$ in pure $H_2O = 0$ and $\underline{H}_{60\%} = 0$, both at 32°F.

9-10 One pound of a 40 per cent NaOH solution at 110°F is to be mixed adiabatically with a more dilute solution to obtain 3 lb_m of a 20 per cent solution at 190°F.

(a) Specify the temperature and composition of the dilute solution.

(b) How much heat will be evolved when 1 lb_m of the 20 per cent solution at 190°F is isothermally diluted with an infinite quantity of water?

(c) What is the enthalpy of the 40 per cent solution at 110°F referred to water at 64°F instead of 32°F, the value used in the chart supplied with this book.

9-11 An ammonia-absorption refrigeration cycle is pictured in Fig. P9-11. The solution in the absorber and stripper is ammonia and water. An enthalpy–concentration diagram for ammonia and water is given in Fig. 9-1.

(a) Steam heating is used to remove the absorbed ammonia from its water solution in the stripper. If the stripper operates at 180°C and a pressure of 20 kg/cm^2, what is the composition of the dilute liquid leaving the stripper?

(b) Eight pounds of dilute solution enter the absorber for every pound of ammonia vapor. The dilute solution enters at 40°C and the ammonia arrives as saturated vapor from the evaporator, which operates at 1 kg/cm^2. The absorber produces concentrated solution at 20°C. How much heat is removed from the absorber per pound of concentrated solution?

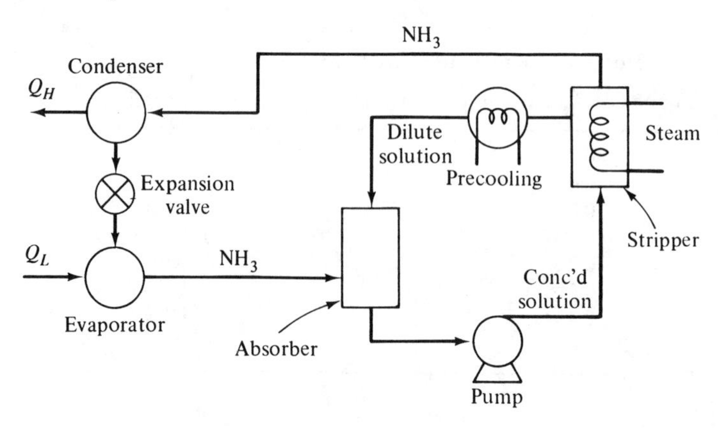

FIGURE P9-11

(c) Could all the ammonia vapor be absorbed adiabatically? *If so*, what would be the temperature of the concentrated solution leaving the absorber? *If not*, what is the minimum amount of heat that must be removed?

Pressure, kg/cm^2	Enthalpy of Saturated Ammonia Vapor, kcal/kg
0.1	357
1.0	374
10	390
20	392

9-12 It is often desirable to have empirical expressions for activity coefficients as functions of composition. The following forms of activity coefficient functions have been proposed for an isothermal binary solution where the standard states are the pure components at essentially the same pressure and temperature as the solution:

(a) $\gamma_A = Ax_A$; $\gamma_B = Bx_B$.

(b) $\gamma_A = 1 + Ax_B$; $\gamma_B = 1 + Bx_A$.

(c) $\ln \gamma_A = Ax_B$; $\ln \gamma_B = Bx_A$.

(d) $\ln \gamma_A = Ax_B^n$; $\ln \gamma_B = Bx_B^n$.

(e) $\ln \gamma_A = \dfrac{AB^2x_B^2}{(Ax_A + Bx_B)^2}$; $\ln \gamma_B = \dfrac{A^2Bx_A^2}{(Ax_A + Bx_B)^2}$.

Criticize the acceptability of each of the above suggestions, supporting your conclusions with quantitative developments.

Multicomponent Phase Equilibrium 10

Introduction

In Chapter 9 we developed methods for evaluating the properties of single-phase multicomponent mixtures. The various mixture rules permit computation of these properties for a considerable variety of systems, both ideal solutions and certain classes of nonideal solutions. Little was said, however, about the application of these rules because their application in thermodynamics is generally encountered in connection with phase and chemical equilibrium. This is particularly true in chemical engineering, where we are interested in such processes as transferring a solute from one liquid to another by taking advantage of relative solubilities (extraction), or in altering the composition of a liquid by taking advantage of the relative volatilities of its constituents (fractional distillation).

Accordingly, in this chapter, we shall reopen the question of phase equilibrium, this time for multicomponent systems. We shall make use of the criteria for equilibrium developed in Chapter 6 as well as the expressions of Chapter 9 for dealing with multicomponent mixtures.

10.1 Criteria for Equilibrium

Discussions of equilibrium and the criteria relating thereto have occurred several times in earlier chapters and will arise again in Chapters 11, 12, and 13, where chemically reacting systems are considered. One can choose to think of equilibrium as represented by a minimization of internal energy, U; the enthalpy, H, the Helmholtz free energy, A; the Gibbs free energy, G; or the maximization of entropy, S, depending on the constraints placed on the system. In developing the entropy concept we demonstrated that entropy was maximized at equilibrium for systems with fixed U and V. It can also be readily shown that if S and V are fixed, U will be minimized at equilibrium; if V and T are fixed, A is minimized, and if P and T are fixed, G is minimized. Since P and T are easily measured and controlled parameters in actual processes, the latter criterion has proved most useful in engineering applications.

In Chapters 6 and 11 we used the relationship $-W_{\text{rev}} = \Delta G_T$ to relate work production and the Gibbs free energy to our concept of equilibrium. At this point we choose (merely for brevity) to begin with minimization of the Gibbs free energy at constant temperature and pressure as the criterion from which we develop the working relationships we need for multicomponent systems. It should be observed that a development paralleling that in Chapter 6 for single-component phase equilibria could be included. The interested reader can refer to this treatment for additional justification for the condition $\Delta G = 0$ at equilibrium.

The development that follows considers two phases I and II, but generalization to any number of phases is easily shown. If we consider a two-phase system where mass is being transferred from one phase to the other, we observe that the free-energy change for the system is the sum of the changes occurring in each phase:

$$dG = dG^{\text{I}} + dG^{\text{II}} \tag{10-1}$$

For a multicomponent system at constant T and P, the free-energy change for each phase is given by

$$dG = \sum_{i=1}^{C} \bar{G}_i \, dn_i \tag{10-2}$$

Substitution of equation (10-2) into (10-1) yields

$$dG = \sum_{i=1}^{C} \bar{G}_i^{\text{I}} \, dn_i^{\text{I}} + \sum_{i=1}^{C} \bar{G}_i^{\text{II}} \, dn_i^{\text{II}} \tag{10-3}$$

Since the summations cover the same range and, by conservation of mass, $dn_i^{\text{II}} = -dn_i^{\text{I}}$ for each component present, we can combine the summations:

$$dG = \sum_{i=1}^{C} (\bar{G}_i^{\text{II}} - \bar{G}_i^{\text{I}}) \, dn_i^{\text{II}} \tag{10-4}$$

Our criterion for equilibrium should be generally valid; that is, it should not depend on any particular choice of the dn_i. Thus the total Gibbs free energy will be a minimum ($dG = 0$) for an arbitrary choice of the dn_i only if

$$\bar{G}_i^{\text{II}} = \bar{G}_i^{\text{I}} \tag{10-5}$$

for all C components. Thus, at equilibrium the partial molar Gibbs free energy (or chemical potential) of each component must be the same in each phase present in the system. Equation (10-5) can be expressed in terms of partial fugacities:

$$\bar{f}_i^{\text{I}} = \bar{f}_i^{\text{II}} = \bar{f}_i^{\text{III}} = \cdots \tag{10-6}$$

As we shall see in the ensuing discussion, this relationship is the starting point for all multicomponent phase equilibrium computations.

Although the discussion above has assumed that both phases are at the same temperature and pressure (typically the case for phase equilibrium), it can be shown that the equilibrium criterion [equation (10-5)] is also valid even if the two phases are at different pressures, provided only that these pressures are constant. The development of the equilibrium requirement $dG = 0$ in Chapter 6 required only that the temperatures of the two phases be identical. Thus even if $P^{\text{I}} \neq P^{\text{II}}$, equation (10-5) can be obtained as long as $dP^{\text{I}} = dP^{\text{II}} = 0$, and $T^{\text{I}} = T^{\text{II}}$. Let us consider for example the diffusion of water across the semipermeable membrane of a reverse osmosis cell as shown in Fig. 10-1. If P^{I}

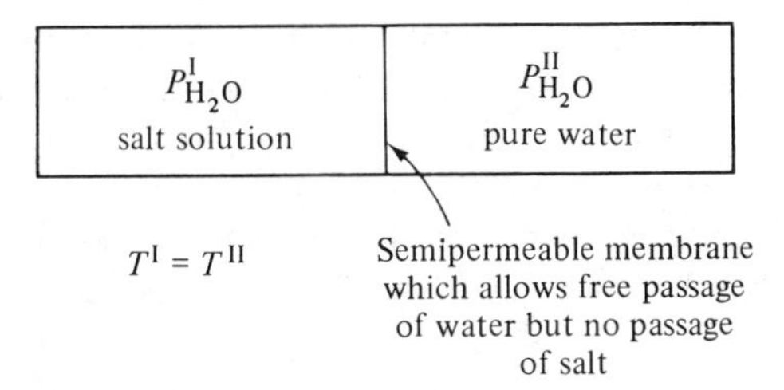

FIG. 10-1. The reverse osmosis cell.

is sufficiently greater than P^{II}, then water will diffuse from the salt solution into the pure water. If we adjust the pressure such that the two phases are in equilibrium, what conditions must be met? For the equilibrium conditions the Gibbs free energy must be at a minimum so that the dG of equation (10-4) vanishes. Thus

$$\sum_{i=1}^{2} (\bar{G}_i^{\text{II}} - \bar{G}_i^{\text{I}})\, dn_i = 0 \tag{10-4a}$$

However, the semipermeable membrane prevents the salt from diffusing, and thus $dn_{\text{salt}} = 0$. The equilibrium requirement thus reduces to equation (10-5) applied only to the water component. As we will show in our later discussion, the concentration difference between phases I and II will require $P^{\text{I}} > P^{\text{II}}$ if $\bar{G}^{\text{I}}_{H_2O} = \bar{G}^{\text{II}}_{H_2O}$. If we remove the semipermeable membrane, then salt is also permitted to diffuse and the new equilibrium conditions would require equality of the partial molar Gibbs free energies of the salt as well as the water. The only way in which both the salt and water partial molar Gibbs free energies can be constant across an open boundary would be for the salt concentration and the pressure to be uniform. This example illustrates that the conditions for equilibrium are greatly dependent on the constraints placed upon the process.

10.2 Computation of Partial Fugacities

Section 9.5 discussed the evaluation of partial fugacities using equations of state, experimental determinations of $\bar{V}_i$, and the Lewis–Randall rule. Although these approaches are generally adequate for vapor phases, they frequently prove inadequate for liquid or solid phases. Few reliable equations of state exist for condensed phase mixtures and their relatively incompressible behavior makes it impractical to obtain the data needed to graphically integrate equations (9-50) or (9-59). Frequently, nonidealities in the behavior of condensed phases rule out the use of the Lewis–Randall rule as a reliable means of calculating partial fugacities for liquid or solid solutions.

The relative ease of determining vapor-phase fugacities coupled with the result of equation (10-6) provides one practical way to determine condensed phase-mixture fugacities. It will be recalled from equation (9-101a) that the partial fugacity for a component can be expressed as

$$\bar{f}_i = x_i \gamma_i^\circ f_i^\circ \tag{9-101a}$$

where γ_i°, the activity coefficient, can generally be assumed to equal unity for gases (when the pure gas at the temperature and pressure of the system is chosen as the standard state). However, γ_i° is often not equal to unity for liquids and solids. Substitution of (9-101a) into (10-6) for liquid–vapor equilibrium (with the pure component at the temperature and pressure of the system taken as the standard states so that $f_i^\circ = f_i$) produces

$$y_i^V \gamma_i^V f_i^V = x_i^L \gamma_i^L f_i^L \tag{10-7}$$

Since γ_i^V can generally be assumed to equal unity, and y_i and f_i^V can be directly determined, equation (10-7) is useful for determining either x_i^L or γ_i^L, provided the other is known.

One important point should be made regarding f_i^L (the fugacity of the pure liquid i at the temperature and pressure of the mixture): In Section 6.12 we observed that it is frequently safe to approximate f_i^L by the fugacity of the pure liquid at its vapor pressure at the temperature in question [equation (6-139)]. This equivalence is assumed in almost all vapor–liquid equilibrium work and will be a tacit assumption in much of what follows.

SAMPLE PROBLEM 10-1. In the evaporative desalinization of sea water, the salt water is first heated to its boiling point and then partially vaporized. The vapor, which is essentially pure water, is then condensed and collected to give the product stream. If the entering sea water is initially 1.5 mol percent NaCl, estimate the boiling temperature of the solution at 1 atm. $\Delta \underline{H}_V = 970$ Btu/lb$_m$ and may be assumed constant. The vapor may be assumed to be ideal and to have negligible density with respect to the liquid. Carefully list any assumptions you make; you should not need any data other than those supplied.

Solution: We may picture the boiling as shown in Fig. S10-1. Water will spontaneously leave the liquid phase (that is, the solution will boil) when

$$\bar{f}^L_{H_2O} > \bar{f}^V_{H_2O}$$

At $P = 1$ atm and approximately 212°F, $(f/P)^V = 1$, so $\bar{f}^V_{H_2O} = 1$ atm. Thus the solution will boil when $\bar{f}^L_{H_2O} > 1$ atm. Since the mole fraction of water in the solution is very close to unity, we may assume that the solution is ideal—at least with regard

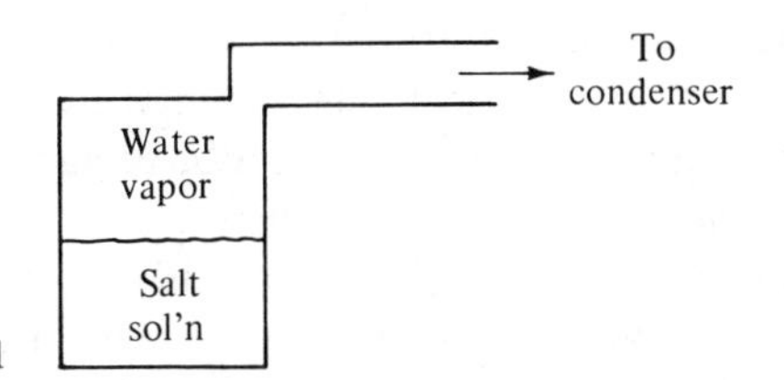

FIGURE S10-1

to water. Thus we may use the Lewis–Randall rule for $\bar{f}^L_{H_2O}$:

$$\bar{f}^L_{H_2O} = x_{H_2O} f_{H_2O}$$

At this pressure and temperature $f_{H_2O} = P'_{H_2O}$ (the vapor pressure at the temperature of the solution). Thus we find that

$$\bar{f}^L_{H_2O} = x_{H_2O} P'_{H_2O} > 1 \text{ atm}$$

Since $x_{H_2O} = 0.985$, we find that

$$P'_{H_2O} > \frac{1.0 \text{ atm}}{0.985} = 1.015 \text{ atm}$$

Thus the minimum solution boiling temperature will be the temperature at which the vapor pressure of water is 1.015 atm. We may determine the variation of the vapor pressure of water by means of the Clausius–Clapeyron equation. However, at the conditions involved, the assumptions of the simplified form of this equation are justifiable, and we shall use it:

$$\frac{d(\ln P'_{H_2O})}{d(1/T)} = -\frac{\Delta \underline{H}_V}{R}$$

which may be integrated to give

$$\ln \frac{P'_2}{P'_1} = -\frac{\Delta \underline{H}_V}{R}\left(\frac{1}{T_2} - \frac{1}{T_1}\right) = \frac{\Delta \underline{H}_V}{R}\left(\frac{T_2 - T_1}{T_2 T_1}\right)$$

Let us take condition 1 as the normal boiling point of water:

$$P'_1 = 1 \text{ atm} \qquad T_1 = 212°\text{F} = 672°\text{R}$$

Condition 2 is the unknown temperature for which

$$P'_2 = 1.015 \text{ atm} \qquad \Delta \underline{H}_V = 970 \text{ Btu/lb}_m$$

Thus T_2 is found from

$$\ln \frac{1.015}{1.000} = \frac{(970)(18)\,(\text{Btu/lb}_m)\,(\text{lb}_m\text{/lb-mol})}{(1.987)\text{ Btu/lb-mol °R}} \frac{T_2 - 672}{T_2(672)} \text{ °R}^{-1}$$

which is solved for T_2 to give

$$T_2 = 672.8°\text{R} = 212.8°\text{F}$$

for a boiling-point elevation of 0.8°F.

Equation (10-7) is particularly useful in determining activity coefficients in liquid (or solid) solutions of known composition where equilibrium vapor-phase measurements can be made. Values of γ_i^L (or γ_i^S) at known compositions can then be used to evaluate the empirical constants in one of the correlating expressions or models discussed in Chapter 9. For systems where accurate vapor-phase measurements cannot be made, such as in metallic systems exhibiting very low pressures, electrochemical measurements are often used to determine γ versus x values, from which empirical constants can then be evaluated. Although not discussed explicitly in Chapter 13, the necessary relationships for making such determinations will be developed there.

Once γ_i^L is known over the composition range of interest, equation (10-7) can be used to determine the liquid-phase composition for a given vapor mixture or vice versa. Thus in a distillation operation where one is separating a binary liquid mixture into two streams, each enriched in one of the two components, repeated application of equation (10-7) to the separation process permits one to determine the number of equilibrations required to perform the separation. Further discussion of such applications occurs in the later sections and sample problems of this chapter.

In applying equation (10-7) to a two-phase system, it should be remembered that the equation can be written for each component that is present in the two phases. For a C-component system, C such equations will result, each equation containing two composition unknowns. The fugacities of the pure components are functions only of the pressure and temperature of the system. Recognizing that $\sum_{i=1}^{C} x_i = 1$ and $\sum_{i=1}^{C} y_i = 1$, the total number of independent unknown compositions equals $2(C - 1)$. If T and P are added, the total number of unknowns possible in a C-component, two-phase equilibrium system are $2(C - 1) + 2$. Activity coefficients, although they may not be known, are functions of the same independent unknowns and thus would not add to the $2(C - 1) + 2$ unknowns identified earlier. The degrees of freedom, or the number of unknowns that must be specified to uniquely fix a system, are equal to the difference between the number of unknowns and the number of equations available. Thus for this case

$$2(C - 1) + 2 - C = f \tag{10-8}$$

where f of the unknowns must be fixed. For a binary system $C = 2$; for a ternary, $C = 3$; and so on. Equation (10-8) is a special case of the Gibbs phase rule

$$P + f = C + 2 \tag{10-9}$$

where P = phases
f = degrees of freedom
C = components

which will be developed completely in Chapter 12.

For a binary, two-phase system specification of any two of the four independent parameters T, P, x, or y will fix each of the others. Since evaluation of f_i^L and f_i^V would require a knowledge of the temperature and pressure to evaluate the fugacity coefficients, f/P, and the vapor pressure, failure to specify either of these complicates the solution of the series of simultaneous equations because of the complex functional dependence of f/P and the vapor pressure on pressure and temperature. This difficulty is illustrated in Sample Problem 10-2.

SAMPLE PROBLEM 10-2. A mixture of gas containing 10 percent *n*-butane and 90 percent *n*-pentane is originally at 1 atm and 140°C. The mixture is compressed isothermally to a point where condensation occurs (see Table S10-2a).

(a) What is the pressure when the first drop condenses? What is the composition of that drop?

(b) What are the pressure and compositions when the last drop condenses?

TABLE S10-2a

	Vapor Pressure at 140°C, atm	T_c, °K	P_c, atm
n-butane	29.756	425.8	36
n-pentane	13.86	470.2	33

Solution: (a) At the pressure when the first drop of liquid condensate forms, the system will consist of a vapor of known composition in equilibrium with a liquid of unknown composition. The temperature will be 140°C, but the pressure is unknown. Our starting point is equation (10-7).

If we denote *n*-butane with a subscript 4 and *n*-pentane with a subscript 5, we can write these equations as

$$\gamma_4^L x_4 f_4^L = \gamma_4^V y_4 f_4^V$$
$$\gamma_5^L x_5 f_5^L = \gamma_5^V y_5 f_5^V$$
$$x_4 + x_5 = 1.0$$

Since butane and pentane are similar molecules and since we have no better information, we shall assume that both vapor and liquid phases form ideal solutions,

$$\gamma_4^L = \gamma_4^V = \gamma_5^L = \gamma_5^V = 1.0$$

with respect to the pure-component standard state at the pressure and temperature of the mixture. Next we evaluate the liquid fugacities, assuming they are equal to those at the respective vapor pressures, which we denote as P'_4 and P'_5. These fugacities may be obtained from the generalized charts:

$$P'_{4r} = \frac{29.756}{36} = 0.827 \qquad T_{4r} = \frac{413.2}{425.8} = 0.970$$

$$\frac{f_4}{P'_4} = 0.68 \text{ or } f_4^L = (0.68)(29.756) = 20.2 \text{ atm}$$

and

$$P'_{5r} = \frac{13.86}{33} = 0.420 \qquad T_{5r} = \frac{413.2}{470.2} = 0.879$$

$$\frac{f_5}{P'_5} = 0.79 \text{ or } f_5^L = (0.79)(13.86) = 10.94 \text{ atm}$$

If we rewrite the original three equations, incorporating our assumptions to this point,

$$x_4 = \frac{y_4 f_4^V}{f_4^L} = \frac{(0.1)(f/P)_4 P}{20.2 \text{ atm}}$$

$$x_5 = \frac{y_5 f_5^V}{f_5^L} = \frac{(0.9)(f/P)_5 P}{10.94 \text{ atm}}$$

$$x_4 + x_5 = 1.0$$

We see that if we assume a value for the total pressure, P, then we may evaluate the fugacity coefficient, f/P, for each component and calculate a value for x_4 and x_5. The assumed value of pressure will be correct if the liquid mole fractions total 1.0; otherwise a new value will have to be assumed and checked.

It seems reasonable to expect that the condensation will occur between the highest and lowest vapor pressure of the components involved. Moreover, since the mixture here is largely pentane, the condensation should begin at a pressure near that end of the pressure range. As a first guess, let us try 15 atm. The computations are given in Table S10-2b.

TABLE S10-2b

Trial	P, atm	P_{4r}	$(f/P)_4$	x_4	P_{5r}	$(f/P)_5$	x_5	$x_4 + x_5$
1	15	0.416	0.85	0.063	0.455	0.77	0.954	1.017
2	14.75	0.410	0.85	0.062	0.447	0.77	0.938	1.000

Since the first trial produced values of x_4 and x_5 that were too large, the pressure was reduced for the second trial. The trend of the sum $(x_4 + x_5)$ with pressure is found by observing that each mole fraction is proportional to the product, $(f/P)P$, but the ratio f/P is not strongly dependent on pressure. Thus $x_4 + x_5$ is almost directly proportional to P. The second P is thus chosen as 15 atm/1.017 or 14.75 atm. The second trial is clearly sufficient.

(b) When the last drop condenses, the liquid will have a composition of 90 per cent pentane, 10 per cent butane. The remaining vapor will be lean in pentane, which condenses more readily than butane. The computation here is analogous to that before:

$$y_4 = \frac{x_4 f_4^L}{f_4^V}$$

$$y_5 = \frac{x_5 f_5^L}{f_5^V}$$

$$y_4 + y_5 = 1.0$$

After a few trials it is found that at $P = 15.35$ atm, $(f/P)_4 = 0.84$, $(f/P)_5 = 0.76$, $y_4 = 0.157$, and $y_5 = 0.843$.

Note that for a mixture the pressure increases during condensation at constant temperature.

The terms "dew point" and "bubble point" are usually assigned to the conditions just computed in Sample Problem 10-2. At the dew point (depending on the direction in which it is approached) the first drop of condensate, or dew, appears or the last drop of liquid evaporates. At the bubble point, either the first bubble of vapor appears in a liquid or the last bit of vapor condenses. Between these conditions there are varying amounts of liquid and vapor in equilibrium.

10.3 Description of Vapor–Liquid Equilibrium

Before undertaking further quantitative description of vapor–liquid equilibrium, let us examine some figures that will help our qualitative understanding of the subject. It is not practical to attempt to show all equilibrium data (for even a binary mixture) on a single chart, for to do so would require a multidimensional plot. For a binary, we typically hold one parameter constant. For example, we may fix either the vapor or liquid composition and show the pressure-temperature behavior of the dew point or bubble point, respectively. The result would look like the line *HBJDK* in Fig. 10-2. An isothermal compression, beginning

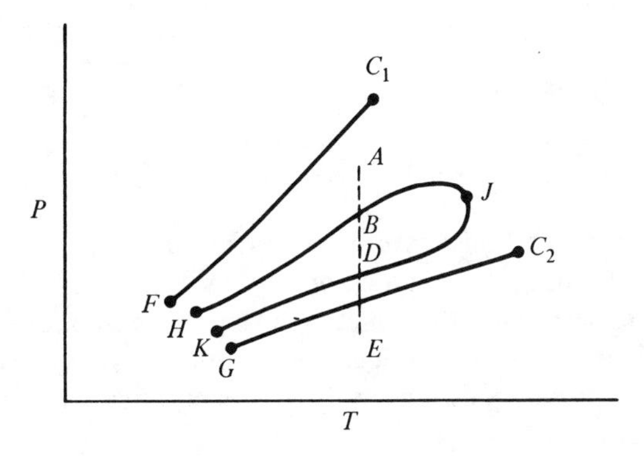

FIG. 10-2. Pressure-temperature diagram for binary mixtures.

with vapor at E, passes through the dew point, D, the bubble point, B, and ends with compressed liquid A. The pure-component vapor-pressure curves, F-C_1 and G-C_2, terminating at the critical points C_1 and C_2, are shown for comparison. Pure-component, vapor–liquid equilibrium is represented by points on these lines; mixture vapor–liquid equilibrium is represented by the entire region between the dew-point line *KDJ* and the bubble-point line *HBJ*, which meet at point J. To complete the description of the binary mixture, we could draw similar envelopes for other compositions. The figure would soon become quite filled with criss-crossing lines. Hence, when we want information about a number of compositions, we normally employ other graphical representations.

Many practical applications occur at either constant temperature or constant pressure and can be adequately described by diagrams such as those shown in Fig. 10-3. Points A, B, C, and D represent the behavior of pure components; horizontal lines between the equilibrium vapor and liquid lines connect compositions that occur at equilibrium for given values of temperature and pressure. An additional plot of x versus y at constant temperature or pressure might be constructed from these. It would typically appear as in Fig. 10-4, where the dashed line represents the line $x = y$. With this qualitative introduction, we resume our quantitative treatment. A comprehensive discussion of vapor–liquid equilibrium is not feasible here because of the variety of combinations possible.[1] We shall treat several classes of mixtures which are useful in many practical problems. These include (1) low-pressure, ideal solutions; (2) low-pressure, nonideal

[1] The reader wishing to explore these topics more comprehensively can find elaboration on nearly every aspect of this chapter in the excellent monograph by J. M. Prausnitz, *Molecular Thermodynamics of Fluid-Phase Equilibria*, Prentice-Hall, Inc., Englewood Cliffs, N.J., 1969.

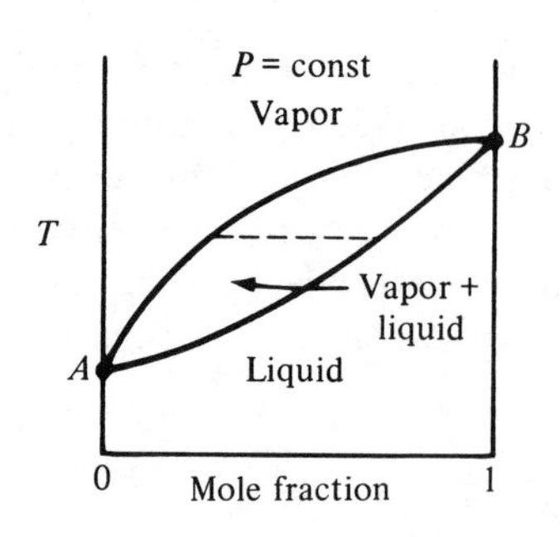

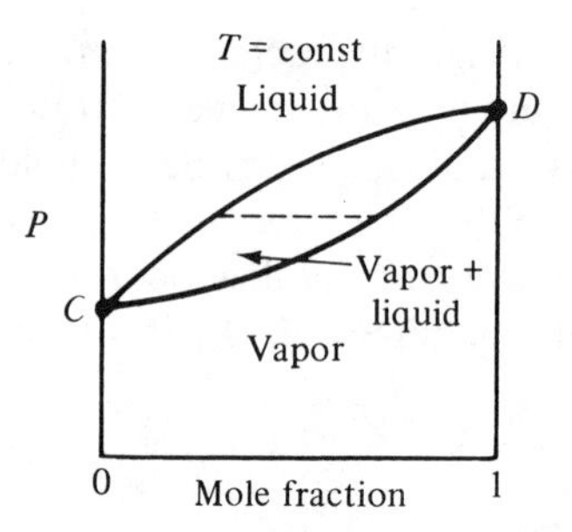

FIG. 10-3. Temperature composition and pressure composition diagrams for binary mixtures.

liquids; and (3) hydrocarbon mixtures. The procedures developed here should give guidance in treating other types of systems.

At pressures less than 10 atm it is normally reasonable to assume that gases form ideal solutions even though the components may not be ideal gases under the conditions in question. With this simplification, our criterion for equilibrium becomes

$$\gamma_i^L x_i f_i^L = y_i f_i^V \tag{10-10}$$

where the standard state is chosen as the pure component at the pressure and temperature of the system.

Moreover, the pressure is often sufficiently low that we may equate pressures and fugacities to obtain the equation

$$\gamma_i^L x_i P_i' = y_i P \tag{10-11}$$

In practice the behavior of any solution can be fitted with this equation by letting γ_i^L assume whatever value is required. If the foregoing assumptions are not met, however, the value of γ_i^L may include contributions from the fugacity coefficients as well. If it is further assumed that the liquid is an ideal solution, the result is *Raoult's law*,

$$x_i P_i' = y_i P = P_i \tag{10-12}$$

This is the simplest quantitative rule for phase equilibrium but is reliable for

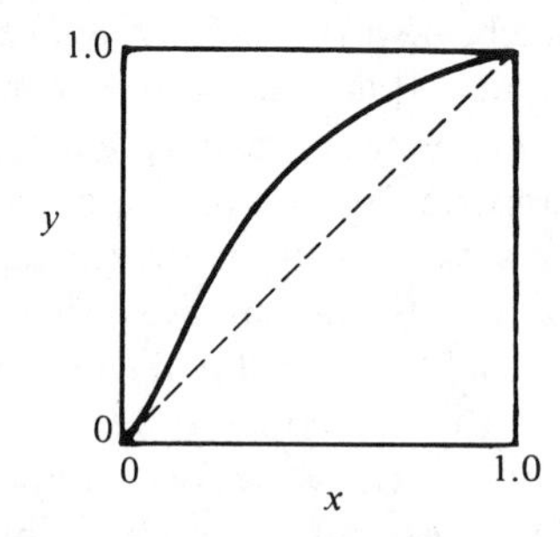

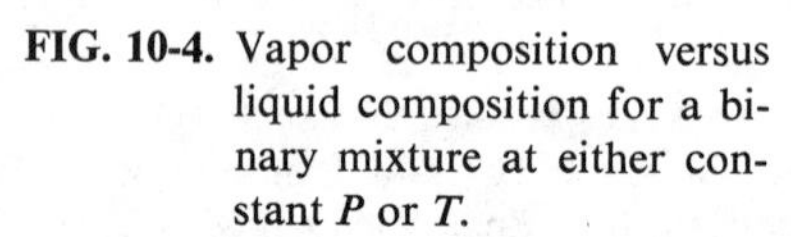
FIG. 10-4. Vapor composition versus liquid composition for a binary mixture at either constant P or T.

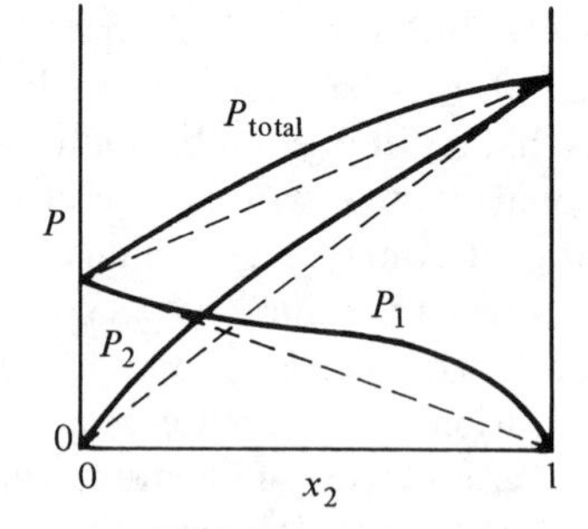

FIG. 10-5. Comparison of Raoult's law with a typical nonideal binary mixture.

only a limited number of real mixtures. However, as indicated in Fig. 10-5 it does provide a useful standard against which to compare vapor–liquid equilibrium. The dashed lines represent Raoult's-law predictions and the solid line a typical nonideal solution. Both components show a positive deviation from Raoult's law, which [see equation (10-11)] implies that the liquid activity coefficients are both greater than unity.

Another linear rule frequently used for nonideal solutions is *Henry's law*, which is usually written

$$x_i H_i = P_i = y_i P \tag{10-13}$$

where H_i is the Henry's-law constant for component i.

Although neither Henry's nor Raoult's law is generally applicable over the entire composition range, they are frequently used to describe dilute nonideal solutions. In particular Henry's law is often used to describe the solute, while Raoult's law is applied to the solvent. Although the equations are valid in the limit of infinite dilution, the concentration range of applicability may be quite small.

In view of our discussion of reference or standard states in Section 9.8, it is pertinent to pause here in our discussion to ask what reference states are commonly presumed when Henry's law is used. The general expression for activity coefficient, given at equation (9-103), can be inserted in equation (10-6) to give

$$\gamma_i^{\circ L} x_i f_i^{\circ L} = \gamma_i^{\circ V} y_i f_i^{\circ V} \tag{10-14}$$

The vapor standard state is the usual one: pure gas at the temperature and pressure of the mixture. From this choice, if the gas is ideal, we can deduce that

$$\gamma_i^{\circ V} y_i f_i^{\circ V} = y_i P = P_i \tag{10-15}$$

We can select for the liquid reference state either (1) the pure liquid or (2) the solute at infinite dilution, both at the temperature and pressure of the mixture. Under the first assumption and the premise that $f_i^L = P_i'$, we obtain

$$\begin{aligned} x_i \gamma_i^{\circ L} f_i^{\circ L} &= x_i \gamma_i^L P_i' = P_i = x_i H_i \\ H_i &= \gamma_i^L P_i' \end{aligned} \tag{10-16}$$

We see that with this choice of standard state, Henry's law implies a constant activity coefficient, although not necessarily unity. It reduces to Raoult's law if the activity coefficient is unity.

For the infinite dilution standard state we obtain

$$\gamma_i^{\circ L} f_i^{\circ L} = H_i \tag{10-17}$$

If we recall our requirement that the activity coefficient approach the value of unity in the standard state and note that Henry's-law coefficients are experimentally determined by fitting data in the limit of infinite dilution, we conclude that $\gamma_i^{\circ L} = 1.0$ and that $f_i^{\circ} = H_i$ for the infinite dilution standard state. We normally consider Henry's law to be applicable over whatever concentration range the partial fugacity can be predicted by linear extrapolation of dilute solution behavior. The slope of the partial fugacity versus the concentration curve in the dilute diagram equals H_i or f_i°. Beyond this range $\gamma_i^{\circ L}$ will differ from unity and Henry's law will cease to be applicable.

SAMPLE PROBLEM 10-3. It is necessary to estimate the solubility of gaseous nitrogen in liquid carbon tetrachloride at 25°C and a partial pressure of nitrogen of one atm. A hypothetical nitrogen vapor pressure at 25°C may be taken as 1000 atm, and the fugacity coefficient correction is assumed to be unimportant.

(a) Calculate the mole fraction of nitrogen present in the liquid CCl_4 at equilibrium if the two species form an ideal solution.

(b) From regular solution theory it is estimated that

$$\ln \gamma_{N_2}^L = 0.526(1 - x_{N_2})^2$$

where pure liquid N_2 at the pressure and temperature of system is taken as the reference state. What is the equilibrium mole fraction of nitrogen in CCl_4 under these circumstances?

(c) What is the Henry's-law constant for this system?

Solution: (a) We may calculate the liquid composition from the equilibrium condition:

$$\bar{f}_{N_2}^L = \bar{f}_{N_2}^V$$

Since the vapor is essentially pure nitrogen at 1 atm pressure, $\bar{f}_{N_2}^V = 1$ atm. If the Lewis–Randall rule is used to describe the liquid N_2–CCl_4 solution,

$$\bar{f}_{N_2}^L = x_{N_2} f_{N_2}^L = 1 \text{ atm}$$

If the hypothetical nitrogen vapor pressure is taken as 1000 atm and the fugacity coefficient and the effect of pressure on liquid fugacity are neglected, then $f_{N_2}^L = 1000$ atm and x_{N_2} is obtained from

$$x_{N_2} = \frac{1 \text{ atm}}{f_{N_2}^L} = \frac{1 \text{ atm}}{10^3 \text{ atm}} = 10^{-3} \text{ mole fraction}$$

(b) For a nonideal solution we may write

$$\bar{f}_{N_2}^L = x_{N_2} \gamma_{N_2}^{\circ L} f_{N_2}^{\circ L} = 1 \text{ atm}$$

However, the reference state is still pure liquid, so $f_{N_2}^{\circ L} = f_{N_2}^L = 1000$ atm. The liquid-phase activity coefficient for this reference state is given by

$$\ln \gamma_{N_2}^L = 0.526(1 - x_{N_2})^2$$

or

$$\gamma_{N_2}^L = \exp [0.526(1 - x_{N_2})^2]$$

so the equilibrium condition becomes

$$x_{N_2} \exp [0.526(1 - x_{N_2})^2] (1000 \text{ atm}) = 1 \text{ atm}$$

which must be solved for x_{N_2} by trial and error. Clearly x_{N_2} will still be quite small, so the iteration scheme should be arranged as shown below and in Table S10-3, with $x_{N_2} = 10^{-3}$ as a reasonable first guess:

$$x_{N_2}^{k+1} = \frac{10^{-3}}{\exp [0.526(1 - x_{N_2}^k)^2]}$$

TABLE S10-3

k	$x^k_{N_2}$	$(1 - x^k_{N_2})^2$	$\exp[0.526(1 - x^k_{N_2})^2]$	$x^{k+1}_{N_2}$
1	0.001	0.998	1.69	5.92×10^{-4}
2	0.000592	0.999	1.69	5.92×10^{-4}

Thus $x_{N_2} = 5.92 \times 10^{-4}$ mole fraction.

(c) The Henry's-law constant is given by

$$H_{N_2} = \lim_{x_{N_2} \to 0} \frac{\bar{f}^L_{N_2}}{x_{N_2}}$$

But

$$\bar{f}^L_{N_2} = \gamma^L_{N_2} x_{N_2} f^L_{N_2}$$

so

$$H_{N^2} = \lim_{x_{N_2} \to 0} \frac{\gamma^L_{N_2} x_{N_2} f^L_{N_2}}{x_{N_2}} = \lim_{x_{N_2} \to 0} \gamma^L_{N_2} f^L_{N_2}$$

But

$$\gamma^L_{N_2} = \exp[0.526(1 - x_{N_2})^2]$$

so

$$H_{N_2} = \lim_{x_{N_2} \to 0} \exp[0.526(1 - x_{N_2})^2](1000 \text{ atm})$$

or

$$H_{N_2} = 1700 \text{ atm}$$

Most liquids form solutions whose behavior is somewhat nonideal. However, unless we have data on the specific system, we must frequently resort to the assumption of ideal liquid solution and hope that it is not in serious error. Although means exist for predicting nonideality without specific data, as indicated in Chapter 9, these methods generally involve extensive computations based on such properties as the free volume or cohesive energy density of the pure components and have not been included within the scope of this book. For many systems of industrial significance—particularly hydrocarbon mixtures—it has proved convenient to describe nonideal solution behavior in terms of the equilibrium vaporization ratio, or K factor:

$$K_i = \frac{y_i}{x_i} \tag{10-18}$$

For an equilibrium mixture, the ratio y_i/x_i is given by

$$\frac{y_i}{x_i} = \frac{\gamma_i^{\circ L} f_i^{\circ L}}{\gamma_i^{\circ V} f_i^{\circ V}} \tag{10-7a}$$

so the K factor becomes

$$K_i = \frac{\gamma_i^{\circ L} f_i^{\circ L}}{\gamma_i^{\circ V} f_i^{\circ V}} \tag{10-19}$$

If we use the conventional standard states of pure gases and liquids at the pressure and temperature of the system, equation (10-19) becomes

$$K_i = \frac{\gamma_i^L (f/P)_{P_i',T} P_i'}{\gamma_i^V (f/P)_{P,T}\, P} \tag{10-20}$$

For an ideal solution with ideal gas behavior, $\gamma_i = f/P = 1$ and equation (10-20) becomes

$$K_i = \frac{P_i'}{P} \tag{10-21}$$

a constant, independent of composition or the constituents of the mixture. On the other hand, if we deal with nonideal mixtures, we should expect the K factor in general to be a function not only of temperature and pressure, but of composition and chemical makeup as well.

For systems of chemically similar compounds, especially the light hydrocarbons, it is experimentally observed that the K factors are only weakly sensitive to composition and thus may be assumed to be independent of these variables with only minor error. The resulting K factors, although not rigorously correct, do yield good engineering results, and are particularly useful for preliminary or approximate designs. Figures 10-6 and 10-7 present one set of correlations for the K factors of light hydrocarbons as functions of temperature and pressure. Other, more extensive, correlations are also available when more precise results are required.[2]

SAMPLE PROBLEM 10-4. A compressed liquid mixture of 42 mol per cent propane and 58 mol per cent ethane is to be fed to a flash tank where the pressure is reduced and a liquid and vapor stream separated. The flashing operation will be carried out at 90°F and is to vaporize 40 mol per cent of the liquid feed. At what pressure should the tank be operated? What will be the compositions of the vapor and liquid streams?

The vapor pressures of ethane and propane at 90°F are 670 and 180 psia, respectively. Equilibrium ratios can be obtained from Figs. 10-6 and 10-7.

Solution: We may picture the flash tank as shown in Fig. S10-4. The component material balances yield three independent equations:

$$x_2 L + y_2 V = x_{2F} F$$
$$x_3 L + y_3 V = x_{3F} F$$
$$x_2 + x_3 = 1$$
$$y_2 + y_3 = 1$$

The equilibrium ratios, $K_i = y_i/x_i$, provide two additional equations. Once we specify that $L/F = 0.6$ and $V/F = 0.4$ and the feed compositions $x_{2F} = 0.58$ and $x_{3F} = 0.42$, there remain only five unknown quantities: two x's, two y's, and P. We do not have explicit relations for the K's in terms of pressure, but the charts in Figs. 10-6 and

[2] See, for example, the *Engineering Data Handbook* of the Natural Gasoline Supply Men's Association, 421 Kennedy Bldg., Tulsa, Oklahoma.

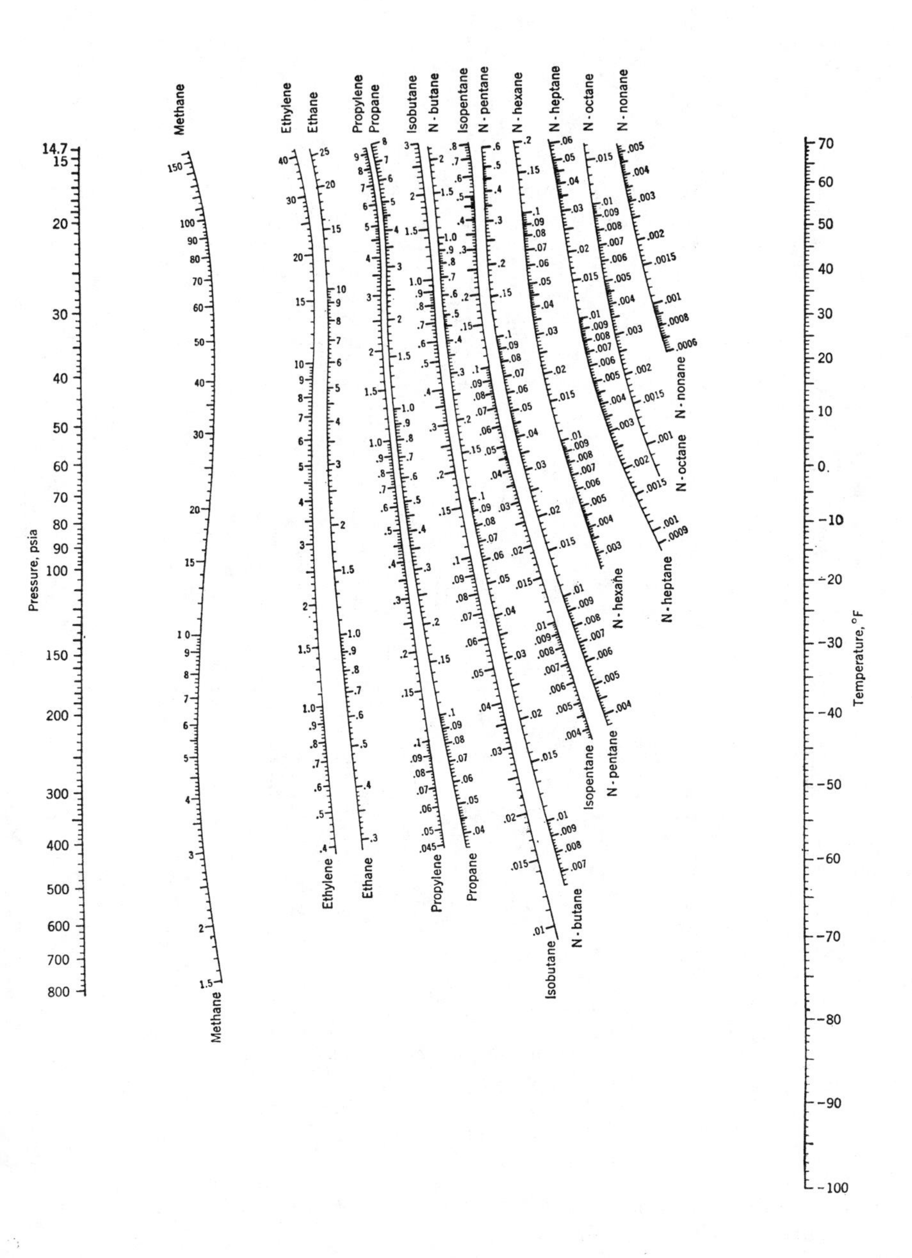

FIG. 10-6. Equilibrium constants in light-hydrocarbon systems, low-temperature range. [Reproduced by permission from C. L. DePriester, *Chem. Eng. Progr. Symp.* Ser. 7, **49** (1953).]

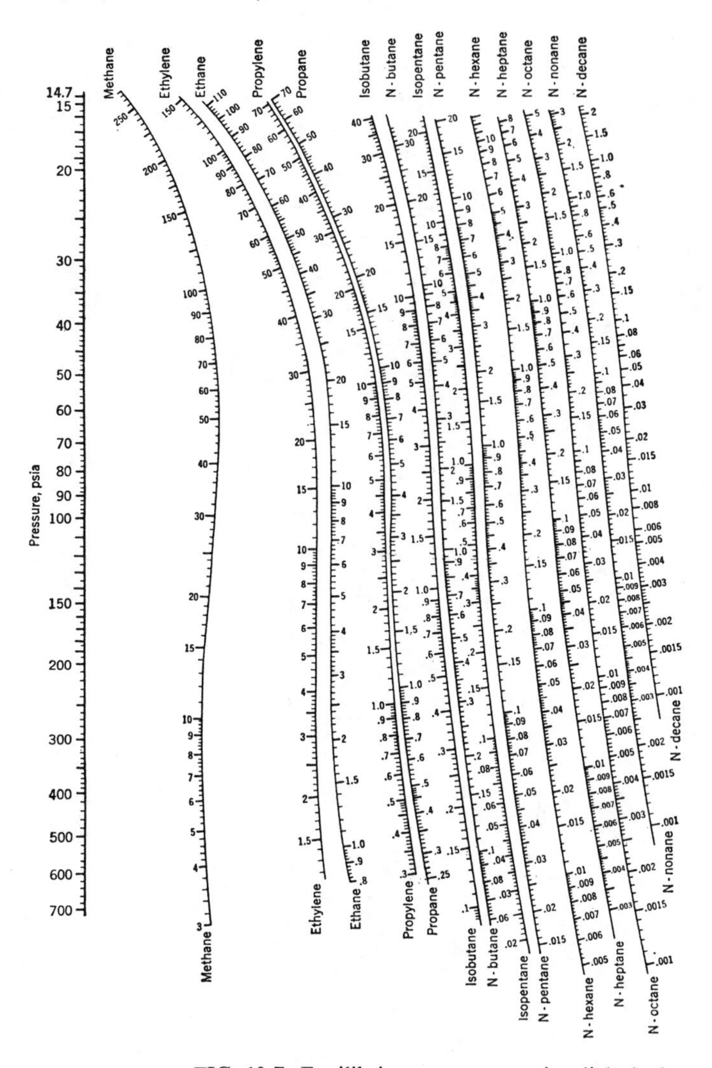

FIG. 10-7. Equilibrium constants in light-hydrocarbon systems, high-temperature range. [Reproduced by permission from C. L. De Priester, *Chem. Eng. Progr. Symp.* Ser. 7, **49** (1953).]

10-7 enable us to determine the K's once a value of P is selected. Thus, if we choose a pressure, say 200 psia, we find that

$$K_2 = 2.65 \qquad \text{and} \qquad K_3 = 0.90$$

These can be used to compute the liquid-phase compositions from

$$x_i = \frac{x_{iF}}{L/F + K_i(V/F)}$$

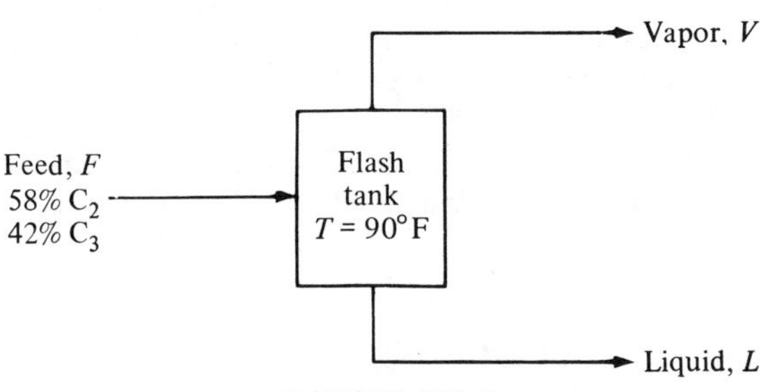

FIGURE S10-4

Thus

$$x_2 = \frac{0.58}{0.6 + (2.65)(0.4)} = 0.349$$

$$x_3 = \frac{0.42}{0.6 + (0.90)(0.4)} = 0.437$$

$$x_2 + x_3 = 0.786$$

Since the liquid mole fractions do not add to unity, we must select another pressure and repeat the process until we find a pressure at which $x_2 + x_3 = 1$. Examination of the charts in Figs. 10-6 and 10-7 indicate that the K's decrease with increasing pressure at constant temperature. Consequently, the pressure must be greater than 200 psia. At 400 psia we find that

$$K_2 = 1.50 \qquad K_3 = 0.53$$

$$x_2 = 0.483 \qquad x_3 = 0.517$$

$$y_2 = 0.726 \qquad y_3 = 0.274$$

We observe that at this pressure $\sum_{i=1}^{C} x_i = \sum_{i=1}^{C} y_i = 1.0$ and we have found the correct solution.

SAMPLE PROBLEM 10-5. It is desired to condense 50 mol per cent of a vapor containing 25 mol per cent ethyl alcohol in 75 per cent water. The condensation is to be performed at 78.15°C. At what pressure must the condenser be operated? At 78.15°C the vapor pressures of pure water and alcohol are: water, 329 mm Hg; alcohol, 755 mm Hg. Although the gaseous solutions may be assumed to be ideal, the liquid solutions clearly are not. The liquid activity coefficients may, however, be adequately described by the van Laar equations:

$$\ln \gamma_1 = \frac{A}{[1 + (Ax_1/Bx_2)]^2}; \qquad \ln \gamma_2 = \frac{B}{[1 + (Bx_2/Ax_1)]^2}$$

where

$$A = 1.00$$

$$B = 0.845$$

and the standard states are pure liquids at the temperature of the system. The subscript 1 represents water, and subscript 2 ethanol.

Solution: The component material balances of this system reduce to the following equations, since $L = V = F/2$. Only one of these equations is independent, since the sum of the mole fractions must be unity. For a basis of 2 moles of feed:

$$\begin{aligned} y_1 + x_1 &= 2x_{1F} \\ y_2 + x_2 &= 2x_{2F} \end{aligned} \qquad \text{(a)}$$

The two unknown compositions, say x_1 and y_1, must be further connected by the equilibrium ratio, K_1. The problem statement has suggested that the gaseous solutions are ideal, but the liquid solutions are nonideal. To further simplify the problem, it seems reasonable to assume that, at the low pressures in question, the fugacity coefficients are all unity. For pure component standard states we may write

$$\frac{y_i}{x_i} = K_i = \frac{\gamma_i P'_i}{P} \tag{b}$$

Combining equations (a) and (b) and solving for the pressure gives two equations of the form

$$P = \frac{\gamma_i x_i P'_i}{2x_{iF} - x_i} \tag{c}$$

Since the activity coefficients are known only as functions of composition from the van Laar expressions, these equations cannot be solved directly. An iterative procedure can, however, be formulated. If we first guess a composition, say x_1, we can calculate both activity coefficients. If one of equations (c) is used to calculate the pressure, the other can be used to check the assumed value of x_1. Another checking procedure would be to use both of equations (c) to calculate P until both yield the same value.

Since water is less volatile than ethanol, its liquid-phase composition should be greater than its entering gas-phase composition. From this argument we make a first guess of $x_1 = 0.80$. The van Laar equations then give

$$\gamma_1 = \exp\left\{\frac{1.0}{[1 + 0.8/(0.2)(0.845)]^2}\right\} = 1.031$$

$$\gamma_2 = \exp\left\{\frac{0.845}{[1 + (0.2)(0.845)/0.8]^2}\right\} = 1.777$$

And from equation (c) we calculate the pressure

$$P = \frac{\gamma_2 x_2 P'_2}{2x_{2F} - x_2} = \frac{(1.777)(0.2)(755)}{(2)(0.25) - 0.2} = 894 \text{ mm Hg}$$

And check the assumed value of x_1,

$$x_1 = \frac{2x_{1F}}{1 + (\gamma_1 P'_1/P)} = \frac{(2)(0.75)}{1 + [(1.031)(329)/894]} = 1.087$$

This is, of course, implausibly high and we cannot simply use it as our second guess. It does indicate that x_1 should be larger than our original guess, and further computations produce the following solution:

$$x_1 = 0.88 \qquad y_1 = 0.62$$
$$P = 470 \text{ mm Hg}$$

10.4 Azeotropic Behavior

Figures 10-8 and 10-9 illustrate the T– and P–composition behavior for a special class of systems, which we term *azeotropes*. Figure 10-8 differs from 10-5 in that the total vapor pressure passes through a maximum at an intermediate composition. The T–composition behavior for the same system at the pressure of

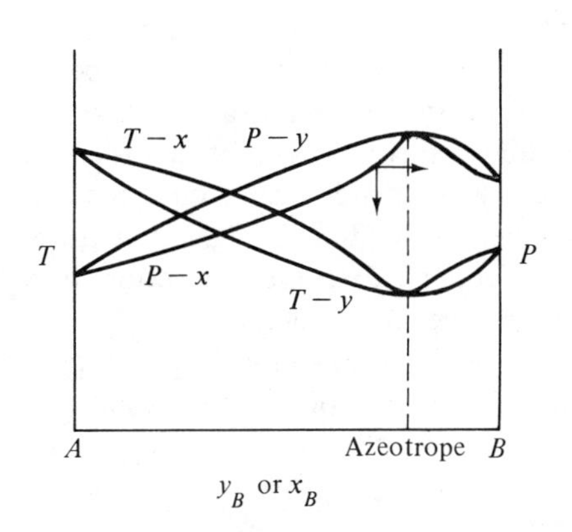

FIG. 10-8. Binary system exhibiting a minimum boiling temperature azeotrope.

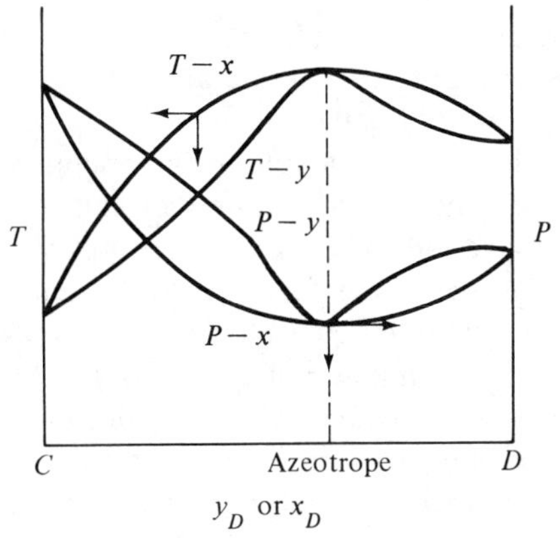

FIG. 10-9. Binary system exhibiting a maximum boiling temperature azeotrope.

the azeotrope is represented on the same plot. It will be noted that this system exhibits a minimum in its boiling point at precisely the same composition that produced a maximum in the vapor-pressure curve. (Since the azeotropic composition is a function of temperature and pressure, the P–composition curve must be for the azeotrope temperature indicated by the T–composition curve and vice versa.) Such a system is classified as a *minimum-boiling azeotrope* and will be shown to exhibit positive deviations from ideal solution behavior.

The behavior of the system illustrated in Fig. 10-9 is the opposite of that shown in Fig. 10-8; the vapor pressure of the solution at constant temperature passes through a minimum at precisely the same composition that the T–x (or boiling-point) curve exhibits a maximum. This type of system is referred to as a *maximum-boiling azeotrope*. Such a system always produces negative deviations from ideal solution behavior.

The two types of azeotropic behavior have a common feature in that the equilibrium composition of liquid and vapor in each type of system is identical at the azeotrope. Figure 10-10 illustrates the same characteristic behavior in

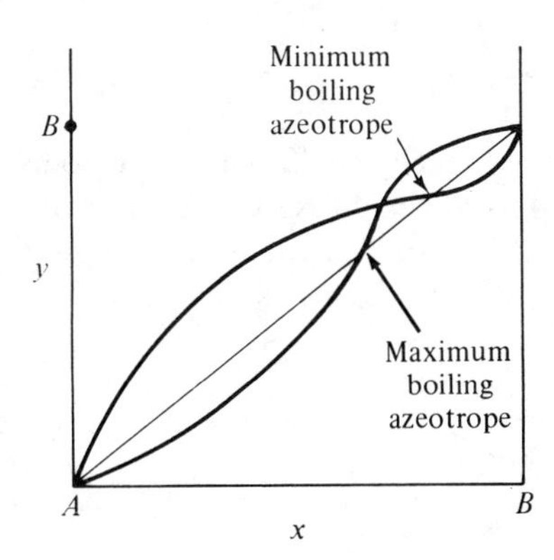

FIG. 10-10. Liquid-vapor equilibrium for azeotropic mixtures.

terms of a y–x plot at constant temperature. The minimum-boiling behavior is encountered considerably more often than the maximum-boiling type in systems normally of interest to engineers. The isopropyl ether–isopropyl alcohol system is an example of the minimum-boiling azeotrope. Such behavior arises from the repulsive intermolecular forces, which cause each component to behave somewhat as if it were a pure component rather than being in a mixture. However, these forces are not quite strong enough to cause separation into two separate liquid phases—a type of behavior yet to be discussed.

Maximum-boiling systems, such as the acetone–chloroform system, generally involve solvation effects. During the mixing of these components, heat is evolved because of the energy liberation on solution. The union of the species leaves fewer molecules of each type free (than the concentration would suggest) to contribute to the total system pressure.

An azeotrope behaves much the same as a pure component in that its transition from the liquid to vapor phase (or vice versa) occurs at a constant pressure and temperature and without any change in the composition of either phase during the vaporization process. Thus a separations process that relies on a difference in the equilibrium composition of the two phases cannot effect the separation of an azeotrope into pure components. As we shall observe shortly, there is some similarity between the behavior of an azeotrope and a eutectic mixture in this regard (eutectics are solid–liquid analogs of azeotropes).

For systems in which the composition of the azeotrope is known at a given temperature and pressure, it is a simple procedure to evaluate the activity coefficients at the azeotropic composition. Equation (10-7) can be written for each component, A and B. If pure components at the pressure and temperature of the azeotrope are specified as the standard state for each phase, then

$$x_A \gamma_A^L f_A^L = y_A \gamma_A^V f_A^V$$
$$x_B \gamma_B^L f_B^L = y_B \gamma_B^V f_B^V \tag{10-22}$$

Since $x_A = y_A$ and $x_B = y_B$ at the azeotrope and, $\gamma_A^V = \gamma_B^V = 1$ is a reasonable assumption, γ_A^L and γ_B^L can be expressed as follows:

$$\gamma_A^L = \frac{f_A^V}{f_A^L}$$
$$\gamma_B^L = \frac{f_B^V}{f_B^L} \tag{10-23}$$

Using the fugacity coefficient charts to evaluate the fugacities of the pure liquid and vapors at the pressure and temperature of the azeotrope (with the assumption that the vapor pressure of each pure component is not affected by the total pressure of the system), equations (10-23) can be expressed as

$$\gamma_A^L = \frac{(f/P)_{P,T}P}{(f/P)_{P'_A T}P'_A}$$
$$\gamma_B^L = \frac{(f/P)_{P,T}P}{(f/P)_{P'_B T}P'_B} \tag{10-24}$$

If the fugacity coefficients are approximately equal for the pure liquid and vapor, further simplification is possible and leads to the results

$$\gamma_A^L = \frac{P}{P'_A}$$

$$\gamma_B^L = \frac{P}{P'_B} \tag{10-25}$$

Since the vapor pressures are functions of temperature, a knowledge of only the temperature and pressure at which an azeotrope occurs is sufficient to evaluate γ_A^L and γ_B^L. However, it should be remembered that γ is generally a function of composition, and thus a knowledge of γ without knowing the composition to which it applies is of limited value. If the composition of the azeotrope is known, one can use this information to evaluate the empirical constants in the van Laar or Margules equations (or other correlations discussed in Chapter 9), and obtain some estimate of the concentration dependence of γ at the given values of pressure and temperature. Since γ is not strongly dependent on pressure, such an expression would provide a good approximation to liquid-phase behavior at other pressures as well. Equation (9-110) permits one to calculate changes in the activity coefficient with temperature at constant P and x if $\bar{H}_i$ is known as a function of temperature.

SAMPLE PROBLEM 10-6. A solution containing 10 mol per cent carbon disulfide and 90 mol per cent acetone is to be separated in a reversible, continuous-flow, isothermal, isobaric separator into pure acetone and the azeotrope that forms at 39.25°C under atmospheric pressure. The azeotrope contains 61 mol per cent carbon disulfide and 39 mol per cent acetone.

It is desired to estimate the work needed to perform the above separation if the feed solution is all liquid at 39.25°C and the products are withdrawn as liquids at the same temperature.

At 39.25°C the vapor pressure of CS_2 is 604 mm Hg; the vapor pressure of $(CH_3)_2CO$ at the same temperature is 400 mm Hg. At atmospheric pressure the vapor phase may be assumed to behave as an ideal gas. The liquid solution, however, does not behave as an ideal solution, but the activity coefficients may be described by the van Laar equations.

Solution: We may picture the separator as shown in Fig. S10-6. The streams are numbered as labeled. The isothermal, reversible work of separation is given by

$$-W_{\text{rev}} = \Delta G_T = G_{\text{outlet}} - G_{\text{inlet}}$$

If we designate the molal flows of the individual streams as n's, then the work is given by

$$W_{\text{rev}} = n_1\underline{G}_1 - (n_2\underline{G}_2 + n_3\underline{G}_3)$$

but

$$\underline{G}_1 = x_{1,\text{ace}}\bar{G}_{1,\text{ace}} + x_{1,\text{cs}_2}\bar{G}_{1,\text{cs}_2}$$
$$\underline{G}_2 = x_{2,\text{ace}}\bar{G}_{2,\text{ace}} + x_{2,\text{cs}_2}\bar{G}_{2,\text{cs}_2}$$
$$\underline{G}_3 = x_{3,\text{ace}}\bar{G}_{3,\text{ace}} + x_{3,\text{cs}_2}\bar{G}_{3,\text{cs}_2}$$

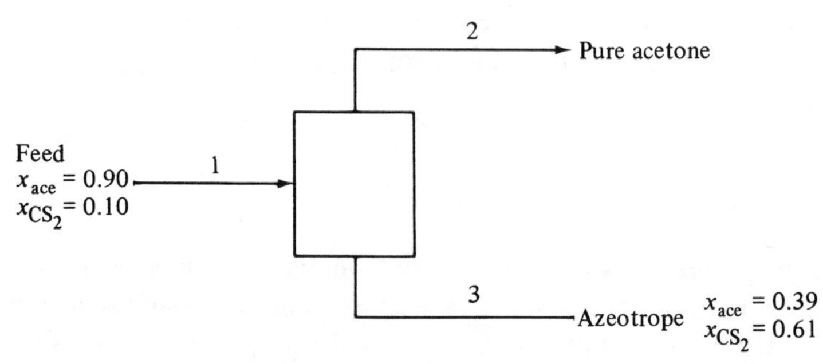

FIGURE S10-6

Since the compositions are fixed, our problem reduces to determining the molal flows and the partial molar Gibbs free energies. We may obtain the molal flows by a series of mole balances around the separator. Let us choose as a basis of calculation 1 lb-mol of solution entering the separator, $n_1 = 1.0$ mol. The entering solution thus contains 0.1 mol of CS_2. All of this appears in stream 3, which is 0.61 mole fraction CS_2. Thus

$$n_3 = \frac{0.10}{0.61} = 0.164 \text{ mol}$$

n_2 is found by a total mole balance:

$$n_2 = n_1 - n_3 = 0.836 \text{ mol}$$

The mole fractions are as given in Table S10-6a.

TABLE S10-6a

Stream No.	x_{ace}	x_{CS_2}	n
1	0.90	0.10	1
2	1.00	0.0	0.836
3	0.39	0.61	0.164

The partial molar Gibbs free energies may be evaluated from

$$\bar{G}_i = \underline{G}_i^\circ + RT \ln \frac{\bar{f}_i}{f_i^\circ}$$

where we shall choose the standard states to be pure liquids at the pressure and temperature of system (for all streams). The reference-state Gibbs free energies are then arbitrarily chosen equal to zero. (The actual value will not matter, because this number appears in both the inlet and outlet terms and will always cancel if there is no chemical reaction.) Thus

$$\bar{G}_i = RT \ln \frac{\bar{f}_i}{f_i}$$

The partial fugacity may now be expressed in terms of the activity coefficients as

$$\bar{f}_i = \gamma_i x_i f_i$$

so

$$\bar{G}_i = RT \ln \gamma_i x_i = (1.987)(560.5) \ln \gamma_i x_i \text{ Btu/lb-mol}$$

We must now determine the activity coefficients as functions of composition. We are told that the van Laar equations,

$$\ln \gamma_a = \frac{A}{[1 + (Ax_a/Bx_b)]^2}, \qquad \ln \gamma_b = \frac{B}{[1 + (Bx_b/Ax_a)]^2}$$

are an adequate representation of the liquid activity coefficients. Thus we must obtain values for the van Laar constants A and B. Since there are two constants, we need two pieces of information with which to fit the constants. These pieces of information are simply the activity coefficients at the azeotrope. We showed in equation (10-25) that the activity coefficient at an azeotrope is given by

$$\gamma_i = \frac{P}{P'_i}$$

(The superscript L has been dropped from γ_i for purposes of these calculations.) Thus at the azeotrope,

$$x_{\text{ace}} = 0.39 \qquad \gamma_{\text{ace}} = \frac{760}{400} = 1.900$$

$$x_{CS_2} = 0.61 \qquad \gamma_{CS_2} = \frac{760}{604} = 1.258$$

Solution of the van Laar equations for A and B in terms of the known points gives

$$A = \ln \gamma_{\text{ace}}\left(1 + \frac{x_{CS_2} \ln \gamma_{CS_2}}{x_{\text{ace}} \ln \gamma_{\text{ace}}}\right)^2$$

$$B = \ln \gamma_{CS_2}\left(1 + \frac{x_{\text{ace}} \ln \gamma_{\text{ace}}}{x_{CS_2} \ln \gamma_{CS_2}}\right)^2$$

Substituting the azeotrope point data for the γ's and x's gives

$$A = 1.558 \qquad B = 1.787$$

so the activity coefficients are given by

$$\ln \gamma_{\text{ace}} = \frac{1.558}{[1 + (1.558x_{\text{ace}}/1.787x_{CS_2})]^2} \text{ and } \ln \gamma_{CS_2} = \frac{1.787}{[1 + (1.787x_{CS_2}/1.558x_{\text{ace}})]^2}$$

We may tabulate the mole fractions, molal flows, activity coefficients, and partial Gibbs free energies of the various streams then as shown in Table S10-6b. The total stream Gibbs free energies are then evaluated (Table S10-6c).

TABLE S10-6b

Stream	x_{ace}	x_{CS_2}	γ_{ace}	γ_{CS_2}	n	$x_{\text{ace}}\gamma_{\text{ace}}$	$x_{CS_2}\gamma_{CS_2}$	$\bar{G}^*_{\text{ace}}$	$\bar{G}^*_{CS_2}$
1	0.90	0.10	1.020	4.08	1.0	0.920	0.408	−93.	−1000
2	1.00	0.00	1.00	—	0.836	1.00	—	0	—
3	0.39	0.61	1.900	1.258	0.164	0.74	0.76	−335	−297

* where $\bar{G}$ = Btu/lb-mol

TABLE S10-6c

Stream	$x_{ace}\bar{G}_{ace}$	$x_{CS_2}\bar{G}_{CS_2}$	$\underline{G} = \sum_{i=1}^{C} x_i\bar{G}_i$	$n\underline{G}$
1	-0.835×10^2	-1.0×10^2	-1.83×10^2	-183
2	0	0	0	0
3	-1.30×10^2	-1.81×10^2	-3.11×10^2	-51

The total work requirement is then

$$W_{rev} = +n_1\underline{G}_1 - (n_2\underline{G}_2 + n_3\underline{G}_3)$$
$$= -183 - (0 - 51.0) \text{ Btu/lb-mol}$$
$$W_{rev} = -132 \text{ Btu/lb-mol}$$

10.5 Phase Equilibrium Involving Other Than Vapor–Liquid Systems

The engineer often encounters separation problems involving other than the vapor–liquid systems just discussed. For example, liquid–liquid extraction between two immiscible phases, adsorption on a solid surface from either a liquid or vapor phase, and leaching or crystallization involving liquid–solid phase equilibria are all examples familiar to the chemical engineer. The same criteria discussed earlier for phase equilibria apply to these applications as well:

$$\bar{f}_i^L = \bar{f}_i^S, \; \bar{f}_i^V = \bar{f}_i^S \quad \text{or} \quad \bar{f}_i^{L_1} = \bar{f}_i^{L_2} \tag{10-26}$$

Although the subject of heterogeneous chemical reactions is discussed in later chapters, it is appropriate to point out at this point that an intermediate step in many reactions involves adsorption on a catalyst. Such catalysts are frequently in the solid form. Although they do not actually take part in the reaction, the catalysts serve as a promoter by providing active sites on their surface at which the reacting species can combine with considerably less difficulty than in the fluid phase itself. The study of the kinetics of catalyzed reactions necessitates attention to the behavior of components at, and in the region of, the catalyst surface. Often it is assumed, for lack of better information, that the reacting component, either reactant or product, in the fluid phase is in equilibrium with the same constituent in the adsorbed state at the catalyst surface. Such a condition implies the equality of partial fugacities, as stated in equations (10-26). The validity of such an assumption clearly depends on the rates at which reacting species can get to and from the active sites relative to the rate at which they react at the site.

Such considerations fall outside the realm of equilibrium thermodynamics. Although we allude to such rate phenomena in Chapter 14, which deals with irreversible (nonequilibrium) processes, these processes are typically covered in courses that treat mass transfer and kinetics. Mention is made here simply to emphasize this important consideration, which often is either overlooked or assumed to be an inseparable part of the reaction mechanism when chemical kinetics are studied. The importance of this intermediate step in catalytic processes is demonstrated by the dependence that reaction rates occasionally display with regard to agitation or mixing effects.

Similarly, in leaching studies, where a component moves from a solid to a fluid phase, it is important to recognize that equilibrium (although frequently assumed to exist across the fluid–solid interface) does not exist with respect to the partial fugacities of the components in the bulk phases. Molecular movement across the interface is ordinarily assumed to occur at a much faster rate than the diffusion of the component from the bulk to the interface, particularly in the solid phase. Nevertheless, it is important in thermodynamics to identify clearly the criterion of equal partial fugacity between phases as the condition that defines equilibrium for such processes. It is equally important to recognize in the rate analysis of such a process that it is the magnitude of the partial fugacity difference which serves as the driving force for component transfer. These same rate considerations apply to phase equilibrium between two fluid phases. However, rate effects can often be minimized in fluid systems by thoroughly dispersing the two phases in each other.

Phase behavior in liquid–solid systems is normally mapped on a T–x diagram, much as described earlier for vapor–liquid systems. Whereas vapor–liquid systems show strong pressure dependence, the same is not true for liquid–solid regions of the phase diagram. Thus Fig. 10-11, which illustrates a system

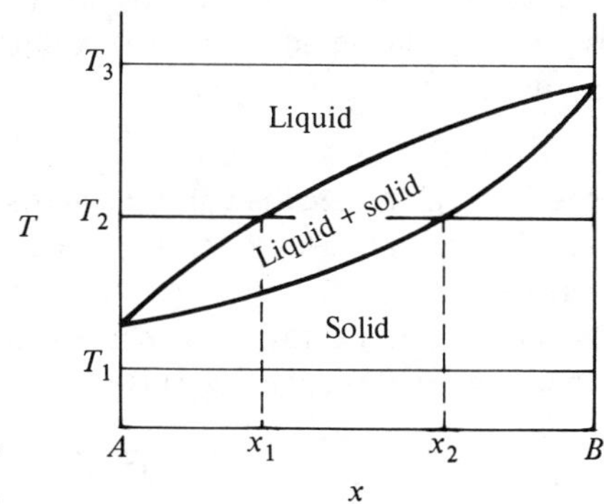

FIG. 10-11. Phase diagram for mutually soluble solid components.

in which both components are soluble across the entire composition range for both phases, is an adequate representation of the phase behavior for a large range of pressures.

A plot of the activity for component B in the above system at the three temperatures indicated is shown in Fig. 10-12. The conventional standard states are used at each temperature and the behavior is assumed to conform to the ideal solution theory. At temperature T_2 where B exists in both the liquid and solid phases, curves 2 and 3 are based on pure solid and liquid standard states respectively. Deviations from ideality typically occur for condensed phases, and the curves could be adjusted easily to account for either positive or negative deviations.

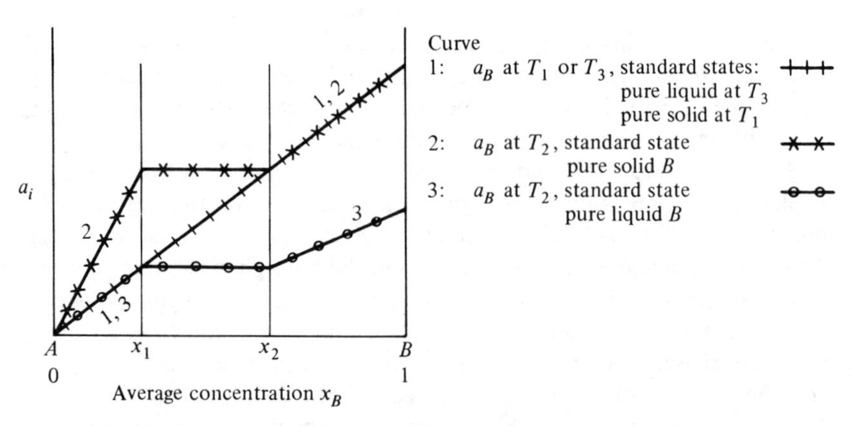

FIG. 10-12. Activity concentration diagrams for the components of Fig. 10-10.

Since the pure solid and pure liquid (either A or B) are not in equilibrium at any of the three temperatures, the pure component liquid and solid fugacities are different. Thus the activities shown in curves 2 and 3 are different. The ratio of activities between curves 2 and 3 is constant, and given by the ratio of liquid to solid pure component fugacities.

$$\frac{(a_B)_2}{(a_B)_3} = \frac{\bar{f}_B/f_B^{\circ S}}{\bar{f}_B/f_B^{\circ L}} = \frac{f_B^{\circ L}}{f_B^{\circ S}} > 1 \tag{10-27}$$

Although the pure solid and liquid phases (either pure A, or pure B) are not in equilibrium at T_2, the liquid mixture at composition x_1 is in equilibrium with the solid mixture at x_2. Thus

$$(\bar{f}_B^L)_{x_1} = (\bar{f}_B^S)_{x_2} \tag{10-28}$$

and the activity of specie B is constant across the region $x_1 < x_B < x_2$ for any specific choice of standard state. The behavior of component A is similar to that described for B.

The Gibbs free energy change between pure liquid B and pure solid B at T_2 (where liquid B is subcooled) is obtained from

$$\underline{G}_B^S - \underline{G}_B^L = RT \ln \frac{f_B^S}{f_B^L} \tag{10-29}$$

or substituting equation (10-27) for the fugacity ratio

$$\underline{G}_B^S - \underline{G}_B^L = RT \ln \frac{(a_B)_3}{(a_B)_2} < 1 \tag{10-30}$$

Liquid–solid systems exhibit a wide variety of behavior other than the simple solution behavior shown in Fig. 10-11. The eutectic system, shown in Fig. 10-13, is somewhat analogous to azeotropic behavior in that the phase transition takes place at constant temperature when the mixture is at the eutectic composition. It has the distinct difference in that the solid actually consists of two phases of compositions 2 and 3, in proportions such that the average composition is 1. (This two-phase nature of the solidified eutectic may be observed by microscopic examination only. To the naked eye, the solidified eutectic

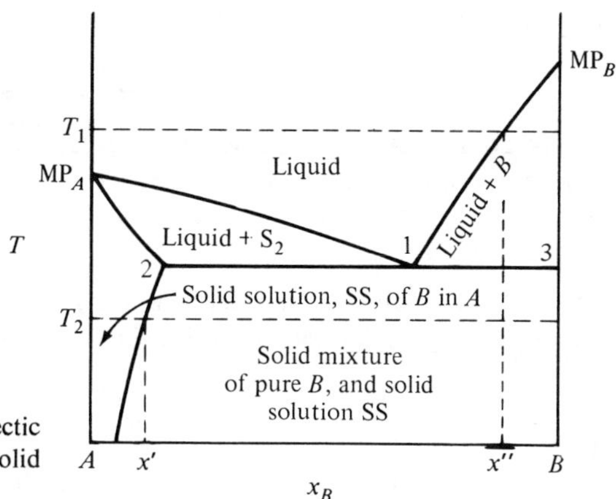

FIG. 10-13. Phase diagram for a eutectic system with limited solid solubility.

would appear as a single phase.) Note also that Fig. 10-13 indicates that solid A dissolves some B in the solid phase, whereas A is essentially insoluble in solid B. The two components appear mutually soluble in the liquid region. The activities of species A and B as functions of composition for the phase diagram shown in Fig. 10-13 are illustrated in Fig. 10-14 for the temperatures T_1 and T_2. It should be noted that at temperature T_1 the activity remains constant from x'' to pure B, where liquid solution is in equilibrium with pure solid B. At T_2 the activity remains constant between x' and x'', where solid solution is in equilibrium with pure solid B.

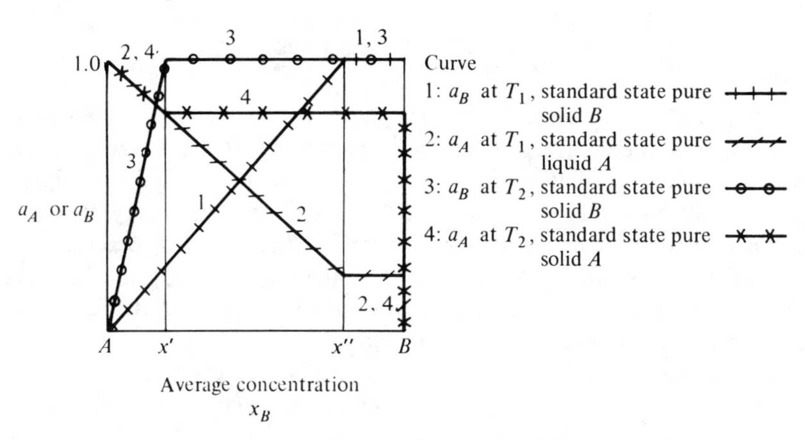

FIG. 10-14. Activity concentration diagram for the system of Fig. 10-13.

10.6 Free Energy–Composition Diagrams

A plot of the Gibbs free energy with respect to composition can be constructed for a multicomponent system as suggested by Fig. 9-11 for a regular solution. The Gibbs free energy change on mixing is given by

$$\Delta \underline{G}_m = x_A(\bar{G}_A - \underline{G}_A) + x_B(\bar{G}_B - \underline{G}_B) \tag{10-31}$$

However,

$$x_A\bar{G}_A + x_B\bar{G}_B = \underline{G}_{\text{soln}} \tag{10-32}$$

If pure components are chosen as the reference states so that $\underline{G}_A = \underline{G}_B = 0$, then equation (10-31) reduces to

$$\Delta\underline{G}_m = \underline{G}_{\text{soln}} \tag{10-31a}$$

Thus for pure-component reference states the $\Delta\underline{G}_m$–x plot is equivalent to a $\underline{G}_{\text{soln}}$–$x$ plot. Consider such a plot, as shown in Fig. 10-15, and recall the devel-

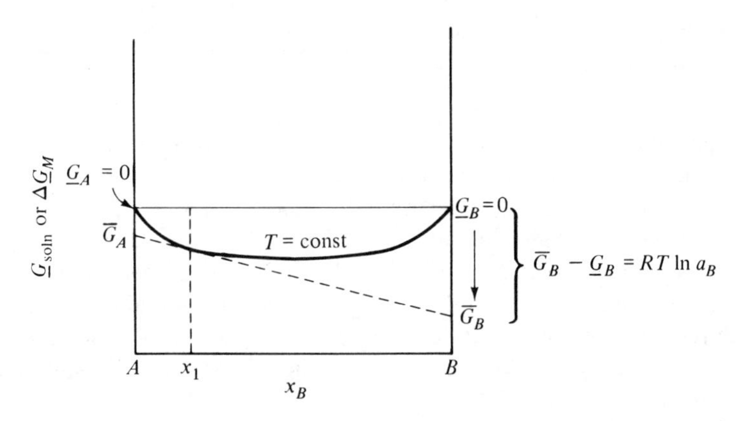

FIG. 10-15. Gibbs free energy–concentration diagram for totally soluble materials.

opment for the enthalpy–concentration ($\underline{H}$–x) plots in Section 9.3. We showed at that time that the intersections of the tangent to the $\underline{H}$–x curve with the pure-component axes occur at points corresponding to the partial enthalpies. A similar derivation would show that the tangent to the free-energy curve intersects the pure-component axes at values corresponding to $\bar{G}_A$ and $\bar{G}_B$, as indicated in Fig. 10-15. As shown in Fig. 10-15, the quantity $\bar{G}_B - \underline{G}_B$ can be directly related to activity; a similar result is obtained for component A. From such a diagram one could construct an activity–concentration plot.

Figure 10-15 represents the $\underline{G}$–x diagram for a system with a single phase existing across the entire composition range. If one contrasts the behavior of this system with that shown in Fig. 10-16, it can be seen that over a portion of the composition range ($x_2 - x_3$), the system's free energy is minimized if two phases of compositions x_2 and x_3 are present instead of a single phase of any intermediate composition, such as x_1. This type of behavior leads to immiscibility, or phase separation, and would correspond to a phase diagram as shown in Fig. 10-17, where T_1 represents the temperature to which Fig. 10-16 might correspond.

We may determine the compositions of the two phases that will result from the separation of the mixture of overall composition x_1 on Fig. 10-16

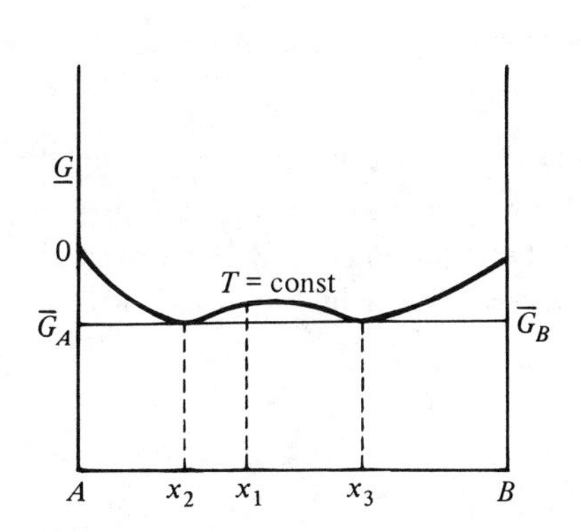

FIG. 10-16. Gibbs free energy–concentration diagram for partially soluble materials.

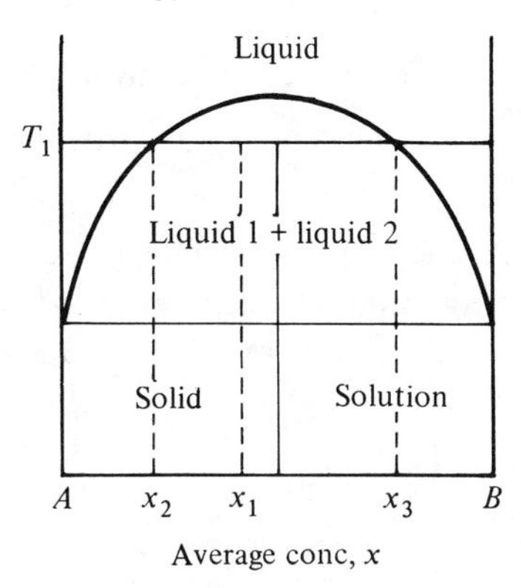

FIG. 10-17. Phase diagram for partially soluble materials.

as follows. We recall that the condition for equilibrium in a system at constant T and P is minimization of the total Gibbs free energy of the mixture. We showed, in addition, that this was equivalent to requiring that the partial molar Gibbs free energy of each specie be constant across the two-phase region. However, we also showed in Chapter 9 that the tangent curve of any mixture property versus composition (in a binary system) intersected the pure-component axes at the values of the partial molar properties for the composition where the tangent is drawn. Thus in Fig. 10-16 the points x_2 and x_3 must be such that the tangents to the $\underline{G}$–x curve at those points must both intersect the $x_A = 0$ and $x_A = 1$ axes at the same values of $\bar{G}_A$ and $\bar{G}_B$. The only way that this is possible is for the same line to be tangent to the $\underline{G}$–x curve at both x_2 and x_3. Thus x_2 and x_3 lie along the line that is tangent to both lobes of the $\underline{G}$–x curve, and are at the two points of tangency of this line with the $\underline{G}$–x curve. In general any mixture of composition $x_2 < x_B < x_3$ will separate into two phases with compositions x_2 and x_3 as shown in Fig. 10-18. The relative amounts of each phase will be determined by the bulk composition of the original mixture.

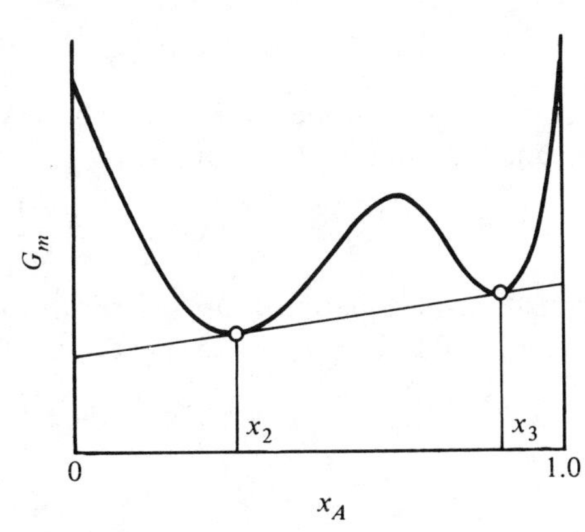

FIG. 10-18. Gibbs free energy–concentration diagram for partially soluble materials.

For a regular solution with the pure component reference states:

$$\underline{G}_{\text{soln}} = \Delta \underline{G}_m = \Delta \underline{H}_m - T \Delta \underline{S}_m$$
$$= x_A x_B W + RT(x_A \ln x_A + x_B \ln x_B) \tag{10-33}$$

If W is assumed to have a constant value characteristic of the A-B interaction energy and T is permitted to vary so that the parameter, W/RT, takes on the values 0, 1, 2, and 3, the free-energy curves shown in Fig. 10-19 will result.

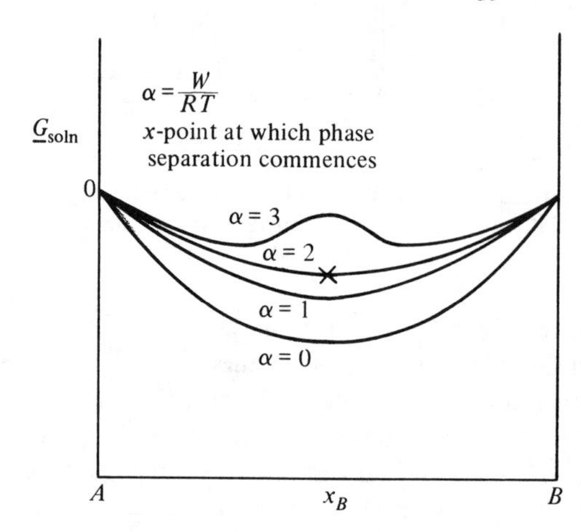

FIG. 10-19. Gibbs free energy–concentration diagram for regular solutions.

The nature of regular solution behavior guarantees symmetry of the curve about the composition $x = 0.5$. Thus the $\underline{G}_{\text{soln}}$–$x$ curve must pass through an extrema at this point and

$$\frac{\partial \underline{G}_{\text{soln}}}{\partial x_B} = 0 \tag{10-34}$$

must always hold at $x_B = 0.5$. *The temperature at which an inflection point first appears in the curve is termed the critical mixing temperature*, indicated by the symbol T_{CM}, and is the temperature at which immiscibility first appears in the system. This temperature may be determined by looking for the first appearance of an inflection point in the $\underline{G}_{\text{soln}}$–$x$ curve, that is, the temperature at which

$$\frac{\partial^2 \underline{G}_{\text{soln}}}{\partial x_B^2} = 0 \tag{10-34a}$$

is first satisfied for some x_B. Differentiating equation (10-33) twice with respect to x_B and equating to 0 produces

$$\frac{\partial^2 \underline{G}_{\text{soln}}}{\partial x_B^2} = -2W + RT\left(\frac{1}{x_B} + \frac{1}{1 - x_B}\right) = 0 \tag{10-35}$$

Examination of equation (10-35) indicates that an inflection point will occur first at $x_B = 0.5$. For $x_B = 0.5$, equation (10-35) becomes

$$-2W + 4RT = 0 \tag{10-36}$$

or

$$\frac{W}{RT} = 2 \tag{10-36a}$$

Thus, for values of $W/RT \leq 2$, a single phase exists across the entire composition range; for $W/RT > 2$, the $\underline{G}_{\text{soln}}$–$x$ curve will have two inflection points, such as those represented by the $\alpha = 3$ curve in Fig. 10-19. The location of the two minima in the $\underline{G}_{\text{soln}}$–$x$ curve depends on the value of W/RT; the greater the value of W/RT the farther apart the minima occur. At values of W/RT just greater than 2, the minima are very close together and the immiscible phases are similar in composition. Thus, as the temperature decreases far below T_{CM}, the value of W/RT increases and the compositions of the two equilibrium liquid phases become more widely separated.

Throughout the development W has been assumed to be independent of temperature and composition. The composition independence is required by regular solution theory. Independence of W with respect to temperature, although assumed in the development of the critical mixing temperature, is only an approximation. One would expect some temperature variation in W, even in systems that obey regular solution theory.

In more general solution behavior one might encounter the first inflection point (which appears at the critical mixing temperature) at some composition other than $x_B = 0.5$, as shown in Fig. 10-20. Critical mixing would again occur at that composition for which $\partial^2\underline{G}/\partial x_B^2$ first equals zero. For temperatures

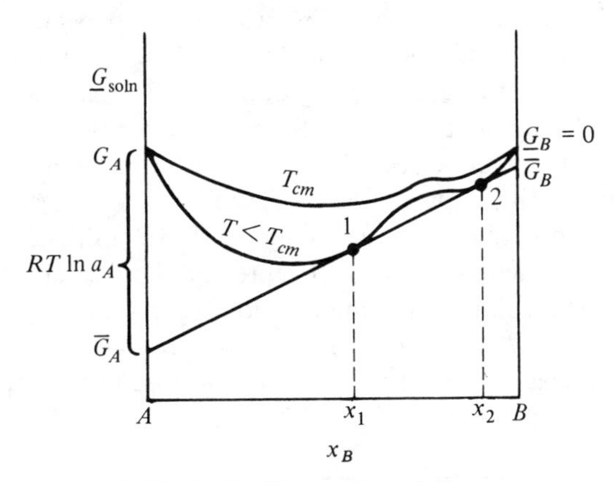

FIG. 10-20. Nonregular solutions.

less than T_{CM}, the $\underline{G}_{\text{soln}}$–$x$ curve might resemble the lower curve in Fig. 10-20. As previously indicated, two equilibrium phases would exist for any overall composition between points 1 and 2. These phases would have compositions given by x_1 and x_2. The activity of components A and B will remain constant for overall mixture compositions between x_1 and x_2 and can be evaluated graphically, as shown in Fig. 10-20.

Large positive values of W are indicative of a strong preference for A-A and B-B bonding as opposed to A-B. Consequently, at large values of W the $\Delta\underline{H}_M$ term dominates and the free-energy curve may exhibit convex upward behavior over the entire composition range, indicating negligible solubility of either component in the other. Such behavior would be exhibited in highly immiscible liquid or solid systems.

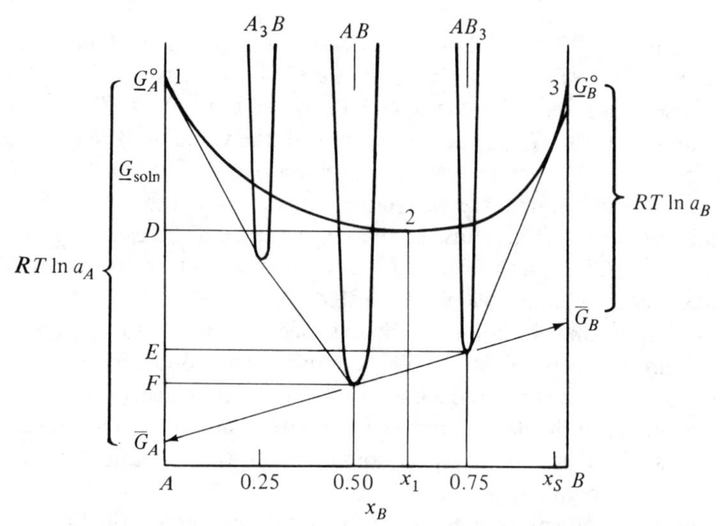

FIG. 10-21. Gibbs free energy–composition diagram for systems exhibiting solid compounds.

In many systems molecules will combine in certain ratios to produce particularly stable groupings termed *compounds*. A free energy–composition diagram for such a system is shown in Fig. 10-21. Here the groupings, A_3B, AB, and AB_3 form particularly stable groupings with free energies below that exhibited by a solution of the two components across the composition range, such as curve 1-2-3. It can be seen that mixtures of A and B at composition x_2 would experience a much lower free energy if the molecules arranged themselves with an approximately equal number of AB and AB_3 clusters, since $0.5E + 0.5\mathrm{F} \ll D$, where E, F, and D represent the free energy of AB_3, AB, and a hypothetical solution of composition x_2 respectively. Thus between $x_B = 0.5$ and $x_B = 0.75$ the two compounds AB and AB_3 would exist in equilibrium and the activity of A and B between those two compositions would remain constant, as indicated in Fig. 10-21.

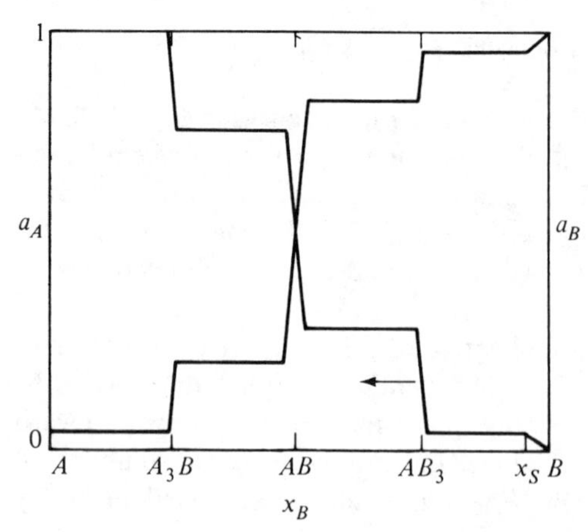

FIG. 10-22. Activities as a function of composition for systems with solid compounds.

Many variations of the type of behavior cited above will be found in actual systems, as will be seen by examining any compilation of phase diagrams. However, the stable phases will always be those that produce the lowest free energy possible for the system. A plot of activity versus composition for the system of Fig. 10-21 is shown in Fig. 10-22. The composition x_S indicates the maximum solubility of A in B at this temperature. Activities for both A and B vary with composition for values of $x_B > x_S$. The Gibbs–Duhem relationship developed in Chapter 9 must apply to such systems and can be used to determine the activity of one component from data on the other component if the phase behavior is known.

10.7 Application of the Gibbs–Duhem Equation

We have shown that the partial fugacities of each component in an equilibrium mixture must be constant in all phases. Since partial fugacities may be measured with relative ease in the vapor phase, most solid or liquid partial fugacities are determined by measurements on a vapor that is in equilibrium with the solid or liquid under investigation. As indicated earlier, such procedures are applicable only to volatile constituents. Frequently one or more of the components is present to only a small degree in the vapor, and accurate measurements are difficult to make. In such instances the Gibbs–Duhem equation (9-116),

$$\sum_{i=1}^{C} x_i \, d(\ln \gamma_i)_{P,T} = 0 \qquad (9\text{-}116)$$

can be used to calculate the activity coefficient of a nonvolatile component if the activities of the other species are known. For a binary system it will be recalled that both pressure and temperature cannot be held constant, and composition of the liquid and vapor varied without violating the phase rule. Thus vapor–liquid or vapor–solid data may be either isothermal or isobaric, but not both. Nevertheless, equation (9-116) or equivalent forms of the Gibbs–Duhem equation are frequently applied to systems where temperature and/or pressure varies. The errors in such an application have been investigated[3] and found to be small for most cases, although there is greater likelihood of error in isobaric (nonisothermal) applications than in isothermal (nonisobaric) situations.

SAMPLE PROBLEM 10-7. The vapor pressure of water over aqueous sodium chloride solutions has been determined experimentally at 100°C. The results may be represented by the equation

$$\text{vapor pressure} = 760 - (1560)\,(\text{moles of NaCl/moles of } H_2O)\ \text{mm Hg}$$

A saturated solution of sodium chloride in water at 100°C contains 0.121 moles

[3] B. F. Dodge and N. V. Ibl, *Chem. Eng. Sci.*, **2**, 120 (1953), and J. M. Prausnitz, *Molecular Thermodynamics of Fluid-Phase Equilibria*, Prentice-Hall, Inc., Englewood Cliffs, N.J., 1969, p. 215.

of NaCl/mole of H_2O. A new device has been proposed to separate solid sodium chloride from a 5 mol per cent salt solution. The products will be pure sodium chloride and pure water, both at 100°C. The device will be fed continuously with the 5 per cent solution at 100°C. Determine the minimum work per 100 lb-mol of feed solution processed in the proposed equipment.

Solution: Let us picture the separation as shown in Fig. S10-7. All calculations will be based on 100 lb-mol of solution entering the separator. Choose pure components at

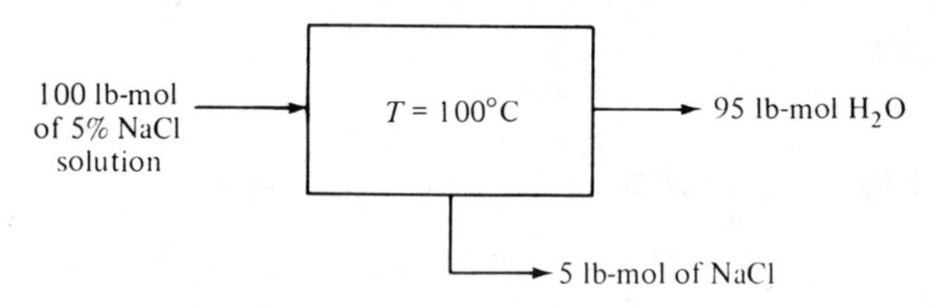

FIGURE S10-7

the pressure and temperature of the system as the standard states: pure liquid water; pure solid salt. The minimum work requirements will be those associated with a reversible process. Under isothermal conditions,

$$
\begin{aligned}
-W_r &= G_{\text{prod}} - G_{\text{react}} \\
&= (G_{\text{prod}} - G^\circ_{\text{prod}}) - (G_{\text{react}} - G^\circ_{\text{react}}) + (G^\circ_{\text{prod}} - G^\circ_{\text{react}})
\end{aligned}
$$

where "prod" and "react" refer to outlet and inlet streams, respectively. Since we have chosen G° to be the same for products and reactants (and there is no chemical reaction), then $(G^\circ_{\text{prod}} - G^\circ_{\text{react}})$ equals zero, and W_r is given by

$$-W_r = (G_{\text{prod}} - G^\circ_{\text{prod}}) - (G_{\text{react}} - G^\circ_{\text{react}})$$

Splitting the product and reactant Gibbs free energies into their component parts then gives

$$
\begin{aligned}
-W_r = &\{5[\bar{G}_{\text{NaCl}} - G^\circ_{\text{NaCl}}] + 95[\bar{G}_{\text{H}_2\text{O}} - G^\circ_{\text{H}_2\text{O}}]\}_{\text{prod}} \\
&- \{5[\bar{G}_{\text{NaCl}} - G^\circ_{\text{NaCl}}] + 95[\bar{G}_{\text{H}_2\text{O}} - G^\circ_{\text{H}_2\text{O}}]\}_{\text{react}}
\end{aligned}
$$

or

$$
\begin{aligned}
-\frac{W_r}{RT} = &\left[5\ln\left(\frac{\bar{f}}{f^\circ}\right)_{\text{NaCl}} + 95\ln\left(\frac{\bar{f}}{f^\circ}\right)_{\text{H}_2\text{O}}\right]_{\text{prod}} \\
&- \left[5\ln\left(\frac{\bar{f}}{f^\circ}\right)_{\text{NaCl}} + 95\ln\left(\frac{\bar{f}}{f^\circ}\right)_{\text{H}_2\text{O}}\right]_{5\%\ \text{soln}}
\end{aligned}
$$

But the water and salt are in their standard states in the product stream, so these two terms drop out and W_r/RT reduces to

$$\frac{W_r}{RT} = \left[5\ln\left(\frac{\bar{f}}{f^\circ}\right)_{\text{NaCl}} + 95\ln\left(\frac{\bar{f}}{f^\circ}\right)_{\text{H}_2\text{O}}\right]_{5\%\ \text{soln}}$$

For water in the inlet stream,

$$
\begin{aligned}
f^\circ_{\text{H}_2\text{O}} &= 760 \text{ mm Hg} = \text{pure-component fugacity} \\
\bar{f}_{\text{H}_2\text{O}} &= P_{\text{H}_2\text{O}} = [760 - 1560(5/95)] = 676 \text{ mm Hg}
\end{aligned}
$$

At these low pressures

$$\frac{\bar{f}_{H_2O}}{f^\circ_{H_2O}} = 0.89$$

and

$$\ln \frac{\bar{f}_{H_2O}}{f^\circ_{H_2O}} = -0.116$$

For salt we need to use the Gibbs–Duhem equation, because we do not have salt fugacities directly but do have water fugacities over a range of compositions:

$$x_1\, d \ln \bar{f}_1 + x_2\, d \ln \bar{f}_2 = 0$$

Let component 1 be salt:

$$d \ln \bar{f}_1 = -\frac{1 - x_1}{x_1}\, d \ln \bar{f}_2$$

or, integrating between solid salt and 5 per cent solution remembering that $\bar{f}^S_{1\,\text{solid}} = \bar{f}^L_{1\,\text{sat}}$

$$\int_{\text{sat}}^{5\%\ \text{soln}} d \ln \bar{f}_1 = -\int_{\text{sat}}^{5\%\ \text{soln}} \frac{1 - x_1}{x_1}\, d \ln \bar{f}_2$$

Integrating the left-hand side gives the desired quantity:

$$\int_{\text{sat}}^{5\%\ \text{soln}} d \ln \bar{f}_1 = \ln \frac{(\bar{f}_1)_{5\%\ \text{soln}}}{(\bar{f}_1)_{\text{sat}}} = \ln\left(\frac{\bar{f}}{f^\circ}\right)_{\text{NaCl}}$$

The right-hand side is now integrated as follows. Observe that

$$\bar{f}_2 = P_2 = 760 - 1560\frac{x_{\text{NaCl}}}{x_{H_2O}}$$

or

$$\bar{f}_2 = 760 - 1560\frac{x_1}{1 - x_1}$$

so the integrated Gibbs–Duhem equation becomes

$$\ln\left(\frac{\bar{f}}{f^\circ}\right)_{\text{NaCl}} = -\int_{\text{sat}}^{5\%} \frac{1 - x_1}{x_1}\, d \ln\left(760 - 1560\frac{x_1}{1 - x_1}\right)$$

or letting $y = x_1/(1 - x_1)$, we can rewrite this equation as

$$\ln\left(\frac{\bar{f}}{f^\circ}\right)_{\text{NaCl}} = -\int_{\text{sat}}^{5\%} \frac{-1560\, dy}{y(760 - 1560y)} = 1560 \int_{\text{sat}}^{5\%} \frac{dy}{760y - 1560y^2}$$

Evaluation of the integral on the right-hand side (which can be shown to be in a standard form) yields

$$\ln\left(\frac{\bar{f}}{f^\circ}\right)_{\text{NaCl}} = -2.38$$

so the work expression becomes

$$\frac{W_r}{RT} = (5)(-2.38) + (95)(-0.116)$$

$$= -22.9$$

$$W_r = -(1.987)(672)(22.9) = -3.09 \times 10^4 \text{ Btu/100 lb-mol}$$

with the minus sign indicating that work must be supplied.

In instances where the errors associated with pressure and/or temperature variations are not negligible, or an assessment of their magnitude is required, the following procedures may be used. The pressure dependence of the activity coefficient for constant T and x is expressed as

$$\left(\frac{\partial \ln \gamma_i}{\partial P}\right)_{T,x} = \frac{\bar{V}_i - \underline{V}_i}{RT} \tag{10-37}$$

while the temperature dependence of the activity coefficient at constant P and x is expressed as

$$\left(\frac{\partial \ln \gamma_i}{\partial T}\right)_{P,x} = -\frac{\bar{H}_i - \underline{H}_i}{RT^2} \tag{10-38}$$

For situations where the temperature and pressure are not constant, the Gibbs–Duhem equation may be expanded to the form

$$\sum_{i=1}^{C} x_i\, d\ln \gamma_i = \sum_{i=1}^{C} x_i \left(\frac{\partial \ln \gamma_i}{\partial P}\right)_{T,x} dP + \sum_{i=1}^{C} x_i \left(\frac{\partial \ln \gamma_i}{\partial T}\right)_{P,x} dT \tag{10-39}$$

(The interested reader can derive this result by combining the multicomponent dG property relationship with the relation $\underline{G} = \sum_{i=1}^{C} \bar{G}_i x_i$. Finally the definition of the activity coefficient must also be introduced.)

Equations (10-37) and (10-38) are now substituted into equation (10-39) to give

$$\sum_{i=1}^{C} x_i\, d\ln \gamma_i = \sum_{i=1}^{C} x_i \left(\frac{\bar{V}_i - \underline{V}_i}{RT}\right) dP - \sum_{i=1}^{C} x_i \left(\frac{\bar{H}_i - \underline{H}_i}{RT^2}\right) dT \tag{10-40}$$

$$= \left(\frac{\Delta \underline{V}_m}{RT}\right) dP - \left(\frac{\Delta \underline{H}_m}{RT^2}\right) dT \tag{10-40a}$$

Application of equation (10-40a) to a set of isobaric binary data for which γ_A and $\Delta\underline{H}_m$ have been measured at various x_A and T yields for γ_B,

$$d\ln \gamma_B = -\frac{x_A}{x_B}\, d\ln \gamma_A - \frac{\Delta \underline{H}_m}{RT^2}\, dT \tag{10-41}$$

Integration of equation (10-41) is complicated by the fact that the temperature-correction term includes $\Delta\underline{H}_m$, which is a function of composition. This difficulty is illustrated in Fig. 10-23.

Since the activity coefficient is a state function, the path along which equation (10-41) is integrated is unimportant except that both of the terms on the right-hand side should in theory be integrated along the *same* path. Since γ_A is measured at values of T and x_A that correspond to the actual path the system follows in going from state 1 to state 2, shown by the solid line in Fig. 10-23, a graphical integration of that term along the system path can be achieved. However, the value of the temperature-correction term depends on the concentration at which it is evaluated. If evaluated at x_{A_1}, the effective path followed is 1-3, a path quite divergent from 1-2. To minimize such an error one can proceed in a stepwise manner to integrate equation (10-41). Thus one can incorporate an approximate value for $\Delta\underline{H}_m$ over each concentration increment.

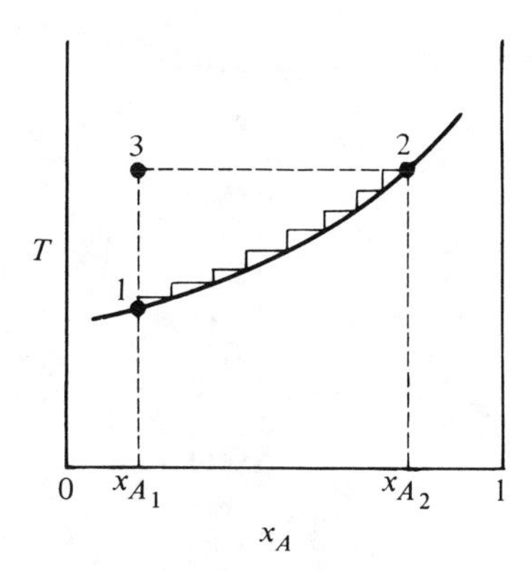

FIG. 10-23. Correction of the Gibbs–Duhem equation for changes in temperature.

As previously indicated, such corrections are frequently ignored completely if temperature changes or $\Delta \underline{H}_m$ are not appreciable. Since $\Delta \underline{V}_m/RT$ is generally small in magnitude, pressure corrections may frequently be safely neglected. However, such corrections, if necessary, could be handled in a manner analagous to that used for the temperature changes.

Another application of the Gibbs–Duhem equation for binary systems at low pressures and essentially constant temperature, where the f/P corrections can be ignored, is obtained by substituting the relation for the liquid-phase activity coefficient

$$\gamma_i^L = \frac{y_i P}{x_i P'_i} \tag{10-42}$$

into equation (9-118). After much manipulation, the following result is obtained:

$$\frac{d \ln P}{dx_1} = \frac{y_1 - x_1}{y_1(1 - y_1)} \frac{dy_1}{dx_1} \tag{10-43}$$

or

$$\frac{dy_1}{dx_1} = \frac{y_1(1 - y_1)}{y_1 - x_1} \frac{d \ln P}{dx_1} \tag{10-43a}$$

Equation (10-43a) can be used to generate or interpolate vapor-phase compositions from data on P versus x. To start the computation we need one piece of y versus x data. If the starting point for these computations is the point $x_1 = y_1 = 0$, we need in addition the limit of the above equation under those conditions. Application of l'Hôspital's rule yields that limiting form,

$$\lim_{x_1 \to 0} \frac{d \ln P}{dx_1} = \lim_{x_1 \to 0} \left(\frac{dy_1}{dx_1} - 1\right) \tag{10-44}$$

SAMPLE PROBLEM 10-8. Given in Table S10-8a are liquid composition versus total pressure data for the system toluene (1) and acetic acid (2) at 70°C.

(a) Using the Gibbs–Duhem equation, calculate the composition of the vapor phase as a function of liquid-phase composition.

(b) Calculate the activity coefficients for toluene and acetic acid in the liquid phase as a function of composition.

TABLE S10-8a

x_1	P, mm Hg
0.0000	136.0
0.125	175.3
0.231	195.6
0.3121	204.9
0.4019	213.5
0.4860	218.9
0.5349	221.3
0.5912	223.4
0.6620	225.0
0.7597	225.1
0.8289	222.7
0.9058	216.6
0.9565	210.7
1.000	202.0

Solution: Before attempting any computations, we note one important feature of the x–P data—that the pressure of the system takes on a maximum value at $x_1 = 0.7597$. Consequently, we anticipate that the mixture will have an azeotrope in this vicinity. This fact will play a role in our interpretation of the results.

(a) The relations necessary for calculation of the vapor-phase composition have been derived and are given at equations (10-43a) and (10-44). Let us rewrite the first in the form

$$\frac{dy_1}{dx_1} = \frac{y_1(1 - y_1)}{y_1 - x_1} \frac{d \ln P}{dx_1} \tag{a}$$

from which we see that, if $d \ln P/dx_1$ is known at one value of x_1, y_1, the value of the derivative dy_1/dx_1 can be calculated and we can estimate the value of y_1 at a nearby point, $x_1 + \Delta x_1$, from the expression

$$(y_1)_{x_1 + \Delta x_1} = (y_1)_{x_1} + \left(\frac{dy_1}{dx_1}\right)_{x_1} \Delta x_1 \tag{b}$$

With this new value of y_1 and the value of the derivative $d \ln P/dx_1$ at the new point, a new value of dy_1/dx_1 is calculated from equation (a). Another step can then be taken using equation (b). In this stepwise fashion, values of y_1 can be calculated at all values of x_1 from 0 to 1. The error associated with the stepping procedure can be made smaller by choosing smaller Δx_1 if necessary. To begin the calculations we need to know the value of y_1 at some value of x_1. It is known, of course, that $y_1 = 0$ at $x_1 = 0$, $y_1 = 1$ at $x_1 = 1$, and that $x_1 = y_1$ at the azeotrope. Any of these points could provide a starting point for the calculations. Evaluation of x and y at the azeotrope requires careful graphical interpolation of the given data and will not be used here, because it is not applicable to other systems which do not exhibit azeotropes. We choose, instead, to start at the point $x_1 = y_1 = 0$. Since equation (a) is undefined in this limit, we need to use equation (10-44) for the first step. It can be written

$$\frac{dy_1}{dx_1} = 1 + \frac{d \ln P}{dx_1} \qquad \text{at } x_1 = 0 \tag{c}$$

which enables us to begin the computations.

In carrying out the numerical procedure, we must also decide how to evaluate the derivative $d \ln P/dx_1$ from the data given. Values of $\ln P$ can be computed and plotted from the given data and tangents drawn to provide the derivatives, but if this graphical step can be avoided, it will simplify our computations. If we recall the definition of a derivative from calculus, it is seen that the following expression is an approximation of the derivative:

$$\left(\frac{d \ln P}{dx_1}\right)_{x_1} \cong \frac{(\ln P)_{x_1+\Delta x_1} - (\ln P)_{x_1}}{x_1 + \Delta x_1 - x_1} \tag{d}$$

If we begin our computations at $x_1 = 0$, we first use equation (d) to find

$$\frac{d \ln P}{dx_1} = \frac{\ln(175.3/136)}{0.125} = 2.031$$

Then, from equation (c),

$$\frac{dy_1}{dx_1} = 1 + 2.031 = 3.031$$

$$y_1 = 0 + (3.031)(0.125) = 0.379 \qquad \text{at } x_1 = 0.125$$

The second step proceeds similarly. At $x_1 = 0.125$,

$$\frac{d \ln P}{dx_1} = \frac{\ln(195.6/175.3)}{0.231 - 0.125} = 1.050$$

Now, and in further steps, we can use equation (a):

$$\frac{dy_1}{dx_1} = \frac{(0.379)(1 - 0.379)(1.050)}{0.379 - 0.125} = 0.953$$

The increment in y_1 is, consequently,

$$\Delta y_1 = (0.953)(0.231 - 0.125) = 0.101$$

and we find that, at $x_1 = 0.231$,

$$y_1 = 0.379 + 0.101 = 0.480$$

The process shown above, if carried out for each increment, using a bit greater precision, yields the results shown in the first two columns of Table S10-8b. Examina-

TABLE S10-8b *x* **versus** *y* **as Calculated from the Given Data**

x_1	$y_{1\,(\text{fwd})}$	$y_{1\,(\text{bkwd})}$	$y_{1\,(\text{comb})}$
0.0000	0.00000		0.000
0.1250	0.37884		0.379
0.2310	0.48042		0.480
0.3121	0.52691		0.527
0.4019	0.57462		0.575
0.4860	0.60997		0.610
0.5349	0.63090		0.631
0.5912	0.65381		0.654
0.6620	0.67961	0.74941	0.680
0.7597	0.68510	0.74401	0.744
0.8289	0.71610	0.78627	0.786
0.9058	0.76616	0.86303	0.863
0.9565	0.80159	0.91433	0.914
1.0000	0.84488	1.00000	1.000

tion of these results shows that an azeotrope is predicted in the vicinity of $x_1 = 0.68$ but that we have failed to reach the known result of $y_1 = 1$ at $x_1 = 1$. In short, we have a method for calculating y–x data (Table S10-8b) from P–x data but we must refine our numerical calculations so as to give more accurate predictions. Examination of equation (a) reveals that at the azeotrope the derivative $d \ln P/dx_1$ changes sign, as does the difference $y_1 - x_1$. If these changes do not occur simultaneously in our numerical computations, then significant errors may be introduced. These and other errors can accumulate to give poor results for higher values of x_1. One simple way to avoid these problems is to begin the computations at $x_1 = 1$. This should at least give better results for the larger values of x_1. With the observation that

$$\frac{dy_1}{dx_1} = 1 - \frac{d \ln P}{dx_1} \qquad \text{at } x_1 = 1 \tag{c'}$$

we can calculate the values of y_1 shown in column 3 of Table S10-8b. This computation was stopped after y_1 was observed to *increase* as x_1 *decreased* from 0.7597 to 0.662. Evidently the accumulated numerical errors again have affected the computations. Since we believe that the values in the second column (forward computations) are best below the azeotrope and the values in the third column are best above the azeotrope (backward computations), then the values shown in the fourth column (combination of the above) represent our best estimates of the y data from these computations.

TABLE S10-8c ***P* versus *x* data Expanded by Graphical Interpolation**

x_1	P, mm Hg
0.0000	136.0
0.0250	145.0
0.0500	153.5
0.0750	161.4
0.1000	168.5
0.1250	175.3
0.1750	186.5
0.2310	195.6
0.2700	200.5
0.3121	204.9
0.3700	209.5
0.4019	213.5
0.4450	216.4
0.4860	218.9
0.5349	221.3
0.5912	223.4
0.6620	225.0
0.7150	225.5
0.7597	225.1
0.8289	222.7
0.8700	220.0
0.9058	216.6
0.9350	213.7
0.9565	210.7
0.9700	208.6
0.9850	205.5
1.0000	202.0

These combined data predict an azeotrope at $x_1 = 0.72$. The idea of solving this problem in two parts by integrating the x versus P data from $x_1 = 0$ to the azeotrope and then from $x_1 = 1$ to the azeotrope has been justified here on an entirely pragmatic basis. The numerical and physical reliability of this method is supported in a more rational way by van Ness[4] who summarizes his findings as follows: "The general conclusion to be drawn . . . is that the numerical integration of the coexistence equation [equation (10–43)] must in all cases proceed in the direction of increasing pressure." Since the pressure takes on a maximum value at the azeotropic composition in this problem, our method is the correct one.

Since the increments in x_1 were relatively large in the previous computations, we might expect that we could further refine the computations by choosing smaller increments. This calls for some interpolation of the P–x data. Again, seeking to minimize the effort, linear interpolation could be used between the data points given in the problem statement. Although this can be done without graphical assistance, it turns out that, for this problem, it does not greatly change the results shown in Table S10-8b. Computational problems are still encountered near the azeotrope composition but can be circumvented as before. Table S10-8c shows the P–x data expanded from 14 to 27 points by graphical (nonlinear) interpolation. The interpolated values occur primarily near the ends of the composition range and near the azeotrope.

Computations with these data, using the same method as before, yield the results shown in the first two columns of Table S10-8d. The results of this calculation are also compared to that done without interpolation in the next two columns of Table S10-8d. It is apparent that our first calculation was adequate to within "engineering" accuracy except perhaps near $x_1 = 0$.

TABLE S10-8d Summary of Calculated Vapor-Phase Compositions and Liquid-Phase Activity Coefficients

	y_1			Activity Coefficients	
x_1	With Interpolation	Without Interpolation	Error	γ_1	γ_2
0.000	0.000	0.000	0.000	2.456	1.000
0.125	0.316	0.379	0.063	2.192	1.008
0.231	0.440	0.480	0.040	1.846	1.047
0.3121	0.498	0.527	0.029	1.618	1.100
0.4019	0.559	0.575	0.016	1.470	1.158
0.486	0.603	0.610	0.007	1.344	1.245
0.5349	0.627	0.631	0.004	1.284	1.306
0.5912	0.654	0.654	0.000	1.223	1.391
0.662	0.688	0.680	−0.008	1.158	1.525
0.7597	0.730	0.744	0.014	1.071	1.860
0.8289	0.780	0.786	0.006	1.037	2.109
0.9058	0.859	0.863	0.004	1.017	2.381
0.9565	0.921	0.914	−0.007	1.004	2.822
1.000	1.000	1.000	0.000	1.000	3.200

(b) Calculations of the activity coefficients are a relatively straightforward matter once y and x data are available. The assumptions that have permitted derivation of equa-

[4] H. C. van Ness, *A. I. Ch. E. Jour.*, *16*, 18–22 (1970).

tions (a) and (c) were that the gaseous phase is ideal and that the fugacity coefficients are unity. Reference states are pure components at the mixture temperature. Consequently,

$$\gamma_1 = \frac{y_1 P}{x_1 P'_1} \quad \text{and} \quad \gamma_2 = \frac{(1 - y_1)P}{(1 - x_1)P'_2} \tag{e}$$

where $P'_1 = 202$ mm Hg and $P'_2 = 136$ mm Hg, the vapor pressures of the pure components at the mixture temperature. Slight problems are encountered with γ_1 at $x_1 = 0$ and with γ_2 at $x_1 = 1$. These limiting activities can be computed with the help of l'Hôspital's rule and equations (c) and (c′):

$$\lim_{x_1 \to 0} \gamma_1 = \frac{P + (dP/dx_1)}{P'_1} \quad \text{and} \quad \lim_{x_1 \to 1} \gamma_2 = \frac{P - (dP/dx_1)}{P'_2} \tag{f}$$

The calculated activity coefficients are shown in Table S10-8d.

The Gibbs–Duhem equation can also be used to produce consistency tests for experimental binary vapor-liquid equilibrium data. We begin with equation (9-118)

$$x_A \frac{d \ln \gamma_A^\circ}{dx_A} + x_B \frac{d \ln \gamma_B^\circ}{dx_A} = 0 \tag{9-118}$$

or

$$\left[\frac{d}{dx_A}(x_A \ln \gamma_A^\circ) - \ln \gamma_A^\circ \frac{dx_A}{dx_A}\right] + \left[\frac{d}{dx_A}(x_B \ln \gamma_B^\circ) - \ln \gamma_B^\circ \frac{dx_B}{dx_A}\right] = 0 \tag{10-45}$$

Combination of terms, noting that $dx_B/dx_A = -1$, yields

$$\frac{d}{dx_A}(x_A \ln \gamma_A^\circ + x_B \ln \gamma_B^\circ) = \ln \gamma_A^\circ - \ln \gamma_B^\circ \tag{10-45a}$$

Integration of (10-45a) from $x_A = 0$ to $x_A = 1$ gives

$$\begin{aligned} \int_0^1 \ln \frac{\gamma_A^\circ}{\gamma_B^\circ} dx_A &= (x_A \ln \gamma_A^\circ + x_B \ln \gamma_B^\circ)_{x_A=1} \\ &\quad -(x_A \ln \gamma_A^\circ + x_B \ln \gamma_B^\circ)_{x_A=0} \\ &= (\ln \gamma_A^\circ)_{x_A=1} - (\ln \gamma_B^\circ)_{x_A=0} \end{aligned} \tag{10-46}$$

To this point we have assumed only that the activity coefficients remain finite, so $x \ln \gamma^\circ \longrightarrow 0$ as $x \longrightarrow 0$. Equation (10-46) can be further simplified, depending on the choice of reference states for the activity coefficients. In the common case of $\gamma_i \longrightarrow 1$ as $x_i \longrightarrow 1$, equation (10-46) becomes

$$\int_0^1 \ln \frac{\gamma_A}{\gamma_B} dx_A = 0 \tag{10-47}$$

The consistency of experimental activity coefficients if they are available over the entire concentration range can be checked by plotting $\ln(\gamma_A/\gamma_B)$ versus concentration as demonstrated in Fig. 10-24. The net area beneath the curve should equal zero, as indicated by equation (10-47). For other choices of standard states an equation comparable to (10-47) can be obtained.

This test is frequently referred to as the *integral test* inasmuch as it requires data over the entire composition range. It represents a necessary, but

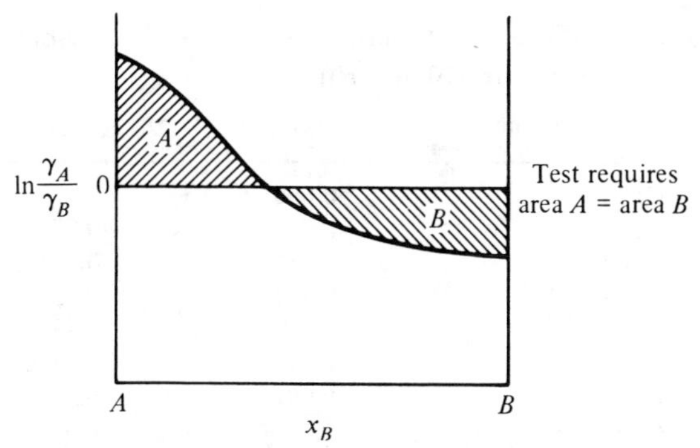

FIG. 10-24. The integral consistency test for vapor–liquid equilibrium data.

not sufficient, condition which experimental data must satisfy. Compensating errors in the data could very well exist such that this criterion is satisfied even though the actual values of activity coefficients are in error.

A second test, referred to as the *differential test*, can also be developed from the Gibbs–Duhem equation, which applies at a specific composition. By rearranging equation (9-118) we can deduce

$$\frac{d \ln \gamma_B^\circ}{dx_B} = -\frac{x_A}{1 - x_A} \frac{d \ln \gamma_A^\circ}{dx_B} \tag{10-48}$$

If $\ln \gamma_A^\circ$ and $\ln \gamma_B^\circ$ are plotted with respect to x_B, as shown in Fig. 10-25, the slopes at any given concentration should satisfy equation (10-48). This is a more rigorous test than the integral test but still cannot guarantee that the data are accurate.

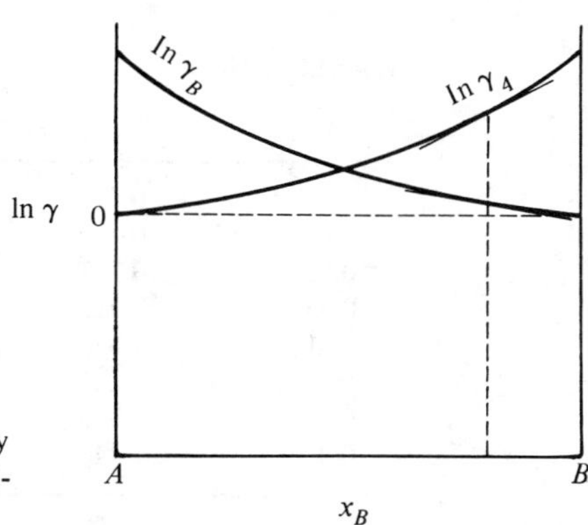

FIG. 10-25. The differential consistency test for vapor–liquid equilibrium data.

SAMPLE PROBLEM 10-9

(a) Test the thermodynamic consistency of the data in Table S10-9a with an appropriate form of the Gibbs–Duhem equation.

(b) Fit the data to the van Laar equation, and recalculate the activity coefficients at acetone mole fractions of 0.163 and 0.883.

TABLE S10-9a Activity Coefficients for Acetone- *cis*-Dichloroethylene System at 760 mm Hg† ‡

Mole Fraction Acetone	γ_{Ace}	$\gamma_{Dichlor}$
0.023	0.608	0.993
0.053	0.711	0.974
0.163	0.764	0.973
0.357	0.854	0.934
0.516	0.917	0.891
0.736	0.988	0.782
0.883	0.987	0.781
0.979	1.00	0.694

† From N. Alpert, D. G. Flom, and P. J. Elving, *Ind. Eng. Chem.*, 1183 (May 1951).
‡ Standard states are pure liquids at the temperature and pressure of the system.

Solution: We shall test the consistency of the data by means of the integral test—equation (10-47):

$$\int_{x_1=0}^{x_1=1} \ln \frac{\gamma_1}{\gamma_2}\, dx_1 = 0$$

This test is strictly applicable only to data taken at constant temperature and pressure. Since these data are not at constant temperature, the test can, at best, give us a qualitative check on the overall consistency.

We may graphically evaluate the integral in the consistency test by plotting $\ln (\gamma_1/\gamma_2)$ versus x_1, as shown in Fig. 10-24. The data for the plot are tabulated in Table S10-9b. We have arbitrarily labeled acetone component 1, the *cis*-dichloroethylene 2. The plot is presented in Fig. S10-9. Within the accuracy of the data, the check appears satisfactory.

TABLE S10-9b

x_1	γ_1/γ_2	$\ln (\gamma_1/\gamma_2)$
0.023	0.612	−0.492
0.053	0.730	−0.315
0.163	0.785	−0.242
0.357	0.915	−0.0887
0.516	1.030	0.0296
0.736	1.263	0.233
0.883	1.266	0.236
0.979	1.440	0.364

(b) We shall fit the van Laar constants by observing that

$$A = \lim_{x_1 \to 0} \ln \gamma_1$$

$$B = \lim_{x_2 \to 0} \ln \gamma_2$$

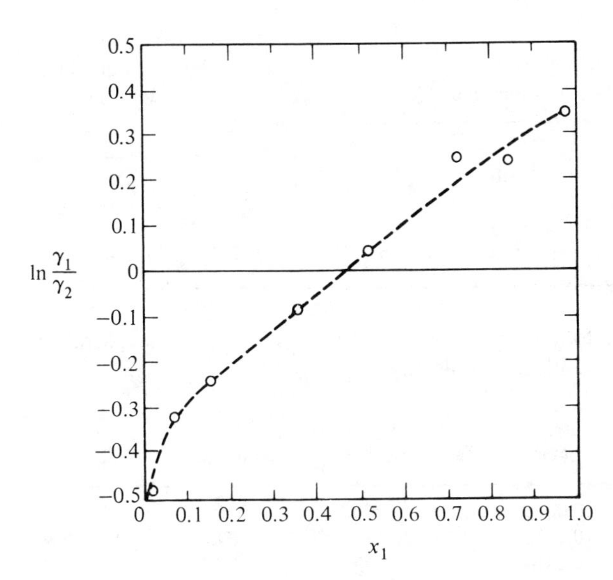

FIGURE S10-9

At $x_1 \longrightarrow 0$, $\gamma_2 \longrightarrow 1$, so $\ln(\gamma_1/\gamma_2) \longrightarrow \ln \gamma_1$. Thus the van Laar constant is the left-hand limit of the $\ln(\gamma_1/\gamma_2)$ versus x_1 curve and has the value

$$A \approx -0.58$$

Similarly, as $x_1 \longrightarrow 1$, $x_2 \longrightarrow 0$, and the right-hand limit is $-B$:

$$B \approx -0.38$$

so the van Laar equation is

$$\ln \gamma_1 = \frac{-0.58}{[1 + (0.58x_1/0.38x_2)]^2} \qquad \ln \gamma_2 = \frac{-0.38}{[1 + (0.38x_2/0.58x_1)]^2}$$

Evaluating γ_1 and γ_2 at $x_1 = 0.163$ and 0.883 yields the results in Table S10-9c and we see that the van Laar equation is apparently a reasonable fit at these points.

TABLE S10-9c

x_1	$\gamma_{1\,(\text{calc})}$	$\gamma_{1\,(\text{obs})}$	$\gamma_{2\,(\text{calc})}$	$\gamma_{2\,(\text{obs})}$
0.163	0.708	0.764	0.980	0.973
0.883	0.998	0.987	0.725	0.781

Problems

10-1 Calculate the activity coefficient of NH_3 in a 10 per cent NH_3 solution using the following data at 32°F for a H_2O–NH_3 system. Use conventional standard states for liquids. Does this system follow Henry's law at 10 per cent?

NH_3 in liquid, mol %	Partial Pressure of NH_3 in Equilibrium Vapor, psia
5	0.26
10	0.52
15	0.90
50	19.36
100	62.29

10-2 For the following data at 69.94°C, assume that the partial pressure of each component is identical with its fugacity.

x_1 (toluene)	x_2 (acetic acid)	P_1 (toluene)	P_2 (acetic acid)
0.0000	1.0000	0	136
0.1250	0.8750	54.8	120.5
0.2310	0.7690	84.8	110.8
0.3121	0.6879	101.9	103.0
0.4019	0.5981	117.8	95.7
0.4860	0.5140	130.7	88.2
0.5349	0.4651	137.6	83.7
0.5912	0.4088	154.2†	78.2
0.6620	0.3380	155.7	69.3
0.7597	0.2403	167.3	57.8
0.8289	0.1711	176.2	46.5
0.9058	0.0942	186.1	30.5
0.9565	0.0435	193.5	17.2
1.0000	0.0000	202	0

Source: International Critical Tables, 1st ed., Vol. III, pp. 217, 223, and 288, McGraw-Hill Book Company, New York, 1928.

† This is evidently a typographical error. The correct figure is near 145.2.

(a) Draw a graph of $\bar{f}_1$ versus x_1. Indicate Raoult's law by a dashed line.

(b) Draw a graph of $\bar{f}_2$ versus x_2. Indicate Raoult's law and Henry's law, each by a dashed line.

(c) Find the constant in Henry's law for acetic acid in toluene solutions by extrapolating a graph of $\bar{f}_2/x_2$ versus x_2 to infinite dilution.

(d) Calculate the activities and activity coefficients of acetic acid on the basis of an f_2° established from Henry's law. Plot these values versus x_2.

(e) Calculate the activities and activity coefficients of acetic acid when the pure liquid is taken as the standard state. Plot these values on the same graph as in (d).

(f) Calculate the activities and activity coefficients of toluene, the solvent in these solutions. What standard states did you use?

10-3 Suppose we have a chamber separated into two halves by means of a membrane that is permeable to water vapor molecules but not to nitrogen molecules. At a given instant of time the temperature, pressures, and concentrations in the two halves are

$T = 932.1°R$	$T = 932.1°R$
$P = 10.93$ atm	$P = 43.71$ atm
$x_{H_2O} = 0.1$	$x_{H_2O} = 0.32503$
$x_{N_2} = 0.9$	$x_{N_2} = 0.97$

only H_2O

FIGURE P10-3

found to be the values in Fig. P10-3. Assuming ideal solutions where necessary, answer the following questions:

(a) What are the partial pressures of water in each half?

(b) Is the system at equilibrium subject to the constraints of the problem?

(c) If the system is not at equilibrium, calculate the driving force for attainment of equilibrium.

(*Note:* At these conditions both nitrogen and water may be assumed to follow the generalized compressibility chart.)

10-4 A vessel containing a sodium chloride solution at 25°C is connected by an overhead tube to one containing pure water. The temperature of the latter vessel is adjusted until no transfer of water takes place through the tube; that is, equilibrium conditions are attained. This temperature is found to be 22°C. Using only the fact that the heat of vaporization of water is 585 cal/g in this temperature range, determine the activity of water in the sodium chloride solution.

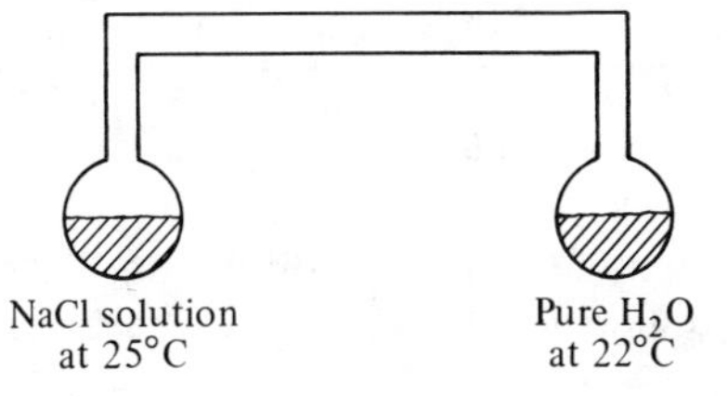

FIGURE P10-4

10-5 A salt solution (brine) containing 4.1 mol per cent salt is slowly heated until boiling occurs. At a pressure of 1.0 atm, it is found that the brine boils at 216.2°F. The evolved vapor is essentially pure water, and may be assumed to be in thermodynamic equilibrium with the boiling brine. Calculate the liquid-phase activity coefficient for the water from the above information.

10-6 The following data were obtained experimentally at 1000 psia and 100°F.

Component	Equilibrium-Phase Compositions, Mole Fraction	
	Vapor	Liquid
Methane	0.657	0.284
Ethylene	0.187	0.1705
Isobutane	0.156	0.5455

Assuming ideal solution behavior in both phases, compute the fugacity of pure methane in the liquid phase. Note that this is one method for obtaining such information above the critical point of a compound.

10-7 Compute and compare vapor–liquid distribution coefficients, $K = y/x$ for the methane–ethylene–isobutane system at 500 psia and 100°F by the following methods. Plot the results as log K versus log P.

(a) Assume ideal gas behavior and ideal solutions (assume that the effect of pressure on liquid fugacity is negligible).

(b) Experimental data.

(c) Compute K for isobutane only, including the effect of pressure on liquid fugacity in the assumption of ideal solutions.

	Phase Compositions, Mole Fraction			
	500 psi, 100°F		1000 psi, 100°F	
	y	x	y	x
Methane	0.3355	0.069	0.657	0.284
Ethylene	0.4815	0.279	0.187	0.1705
Isobutane	0.1830	0.652	0.156	0.5455

Source: M. Benedict, E. Solomon, and L. C. Rubin, *Ind. Eng. Chem.*, *37*, 55 (1945).

10-8 The partial pressures of water and *n*-propyl alcohol over solutions at 25°C are given in the following table:

x_{alc}	0	0.05	0.1	0.2	0.4	0.6	0.8	0.9	0.96	1.0
p_{H_2O}	23.8	23.2	22.7	21.8	21.7	19.9	13.4	8.1	4.2	0
p_{alc}	0	10.8	13.2	13.6	14.2	15.5	17.8	19.4	20.8	21.8

Test these data for their overall consistency. Fit the van Laar equation to these data and determine the composition at which the activity coefficients are the same for the water and the alcohol.

10-9 The following data have been taken on the partial pressures of water and ammonia over their liquid solutions at 20°C:

$\frac{\text{g of } NH_3}{1000 \text{ g of } H_2O}$ in Liquid	Partial Pressure NH_3, mm Hg	Partial Pressure H_2O, mm Hg
0	0	17.5
20	12.0	17.1
25	15.0	17.0
30	18.2	16.9
40	24.9	16.7
50	31.7	16.5
75	50.0	15.9
100	69.6	15.3
150	114	14.1
200	161	12.9
250	227	11.6
300	298	10.5
∞	6430	0

Show by calculations whether these are good data. Compute the minimum work to separate (at 20°C) a 16.67 mass per cent ammonia solution into pure gaseous ammonia at 760 mm Hg pressure and pure liquid water at 760 mm Hg pressure.

10-10 Sketch the y–x equilibrium curve for the system n-butane and n-hexane at a total pressure of 20 atm by three methods.

(a) Assuming Raoult's law and ideal gases.
(b) Using fugacity corrections.
(c) From the following experimental data at 20 atm.
The mole fraction of n-butane is as follows:

x_4	0.200	0.400	0.600	0.800
y_4	0.349	0.619	0.799	0.912

It is believed that five select points should be adequate to represent the curve. As (0, 0) and (1, 1) are two excellent points, only three need be calculated for part (a). As (a) and (c) will provide two curves, the curve for (b) should be drawn from points at 160°C and 180°C only.

10-11 Figure P10-11 refers to the enthalpy relationships of a completely miscible binary system at a constant pressure of 1 atm. The mole fraction of component B is plotted

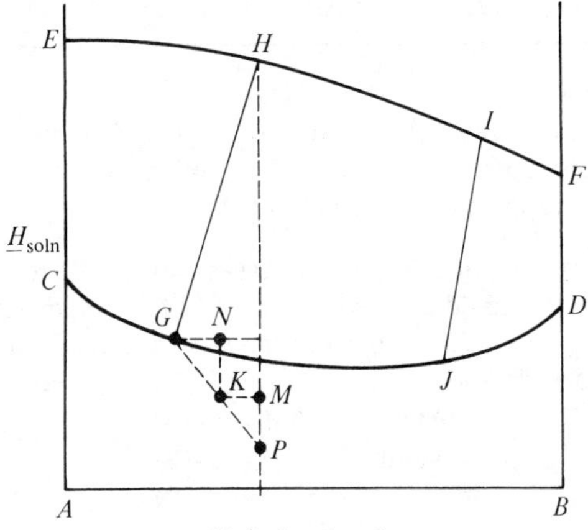

FIGURE P10-11

from left to right and the enthalpy of the mixture, relative to the pure components in chosen reference states, is plotted vertically. Curve CD represents the enthalpy of the liquid at its boiling point as a function of composition and curve EF represents the enthalpy of the vapor above the liquid. GH and IJ are typical tielines relating equilibrium liquid–vapor compositions. A feed liquid, corresponding to point K, is separated by continuous flash distillation into a liquid of composition G and a vapor of composition H. Show that the heat required per mole of the vapor is represented by $\overline{PH}$.

10-12 A refinery by-product gas whose composition is 20 mol per cent methane, 30 mol per cent n-butane, and 50 mol per cent n-pentane is available at 120°C and 10 atm. To what pressure must the mixture be compressed isothermally to liquify 50 mol per cent of it? For the final temperature and pressure obtained, what are the K values of the three components

(a) Computed from the composition of each phase determined above?

(b) Computed on the basis of Raoult's law and Dalton's law?

10-13 A tank containing equimolal quantities of propane, *n*-butane, and *n*-pentane is under an absolute pressure of 15 atm. The temperature in the tank is 80°C. What is the condition of the mixture in the tank?

10-14 A mixture containing 30 mol per cent propane, 50 mol per cent butane, and 20 mol per cent hexane is originally at 1 atm and 120°C. If this mixture is compressed isothermally to a point where condensation begins to occur, what is the pressure of the mixture? What is the composition of the first drop condensing? What is the minimum work per 100 lb mol of mixture handled?

10-15 What is the composition of the vapor in equilibrium with a liquid mixture of 60.6 mol per cent *n*-propane and 39.4 mol per cent *n*-butane at 80°F? Assume the vapor pressures of *n*-propane and *n*-butane at 80°F are not known, but the following data are available at 70°F.

	P' psia	ΔH_V Btu/lb$_m$	T_c, °R	P_c, psia	Mol Wt
n-propane	124	145	665	617	44.09
n-butane	32	156	767	530	58.12

For small ranges in temperature, $\Delta \underline{H}_V$ and the compressibility factor of the gas may be assumed to remain constant. The specific volume of the liquid may be assumed to be very small compared to the specific volume of the gas.

10-16 The pressure on a liquified petroleum gas is maintained at 40 atm. Calculate the temperature to which the liquid must be heated to initiate vaporization, assuming that the total pressure has no effect on the vapor pressure of the pure components. The liquid composition is as follows:

ethane: 15 mol per cent
propane: 40 mol per cent
n-butane: 45 mol per cent

10-17 In a petroleum refining plant, an equimolar mixture of ethane, propane, and normal butane is fed to a flash separator in which the pressure is 200 psia and the

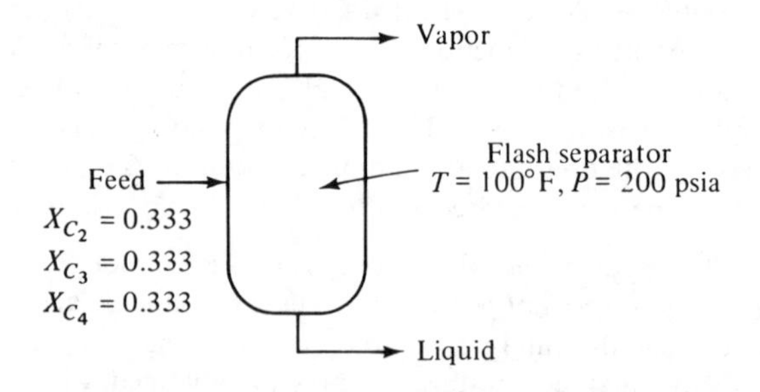

FIGURE P10-17

temperature 100°F, as shown in Fig. P10-17. Assuming ideal solution behavior, calculate the liquid to vapor ratio and the composition of each phase.

10-18 A mixture of gas containing 15 mol per cent n-butane and 85 mol per cent n-pentane is to be employed in a petrochemical process operating at atmospheric pressure. However, this gas mixture is available at 140°C and 14 atm pressure, so an enterprising young engineer suggests passing the gas through a turbine to obtain some work before allowing it to go to the process reactor. The turbine, of course, would be well-insulated. When this idea is suggested to the section chief, the chief wonders whether condensation may take place in the turbine, with resulting erosion of the blades. Show by calculations whether condensation could occur in the turbine.

$$n\text{-}C_4H_{10}: \quad C_P^* = 21.3 + 0.031T(°F) \quad \text{Btu/lb-mol °R}$$
$$n\text{-}C_5H_{12}: \quad C_P^* = 25.1 + 0.045T(°F) \quad \text{Btu/ lb-mol °R}$$

10-19 Total pressure versus composition data for a binary mixture at constant temperature are found to be correlated well by the polynomial

$$P = P_2' + 3(P_1' - P_2')x_1 - 2(P_1' - P_2')x_1^2$$

where P_1' and P_2' are the respective vapor pressures and x_1 is liquid-phase mole fraction.

(a) Determine if the mixture forms an azeotrope and, if so, what is the azeotrope composition?

(b) Compute y_1 versus x_1 from the available information if $P_1' = 2P_2'$.

10-20 Benzene and ethanol form an azeotropic mixture. Consequently, small amounts of benzene are sometimes added to solvent grades of ethanol in order to prevent industrious chemical engineering students from purifying the ethanol solvent by distillation for an after-finals party. Given that at 45°C a liquid phase containing 61.6 mol per cent ethanol (the remainder being benzene) is in equilibrium with a vapor containing 43.4 mol per cent ethanol at 302 mm Hg and that the vapor pressures of pure ethanol and benzene at 45°C are, respectively, 172 and 222 mm Hg, verify that at 45°C an ethanol–benzene azeotrope forms at 37.5 mol per cent ethanol. What is the pressure above the liquid phase when the azeotrope is formed?

10-21 The pure liquids A and B have the following properties at 90°C:

	Liquid A	Liquid B
Vapor pressure, mm Hg	1,000	700
$\Delta\underline{H}_V$, cal/mol	12,000	10,000
Molecular weight	80	20
Viscosity, centipoise	0.4	0.6
Specific gravity	0.8	1.0

The liquid-phase activity coefficients are given by

$$\ln \gamma_A = x_B^2$$
$$\ln \gamma_A = x_A^2$$

and the vapors behave as ideal gases.

(a) What is the total vapor pressure in equilibrium with a 50 mol per cent mixture of A and B at 90°C?

(b) Will an azeotrope be found and, if so, will it be maximum or minimum boiling? Why?

(c) If $\ln \gamma_A = 4x_B^2$, calculate an expression for $\ln \gamma_B$ as a function of composition.

10-22 At atmospheric pressure the system carbon tetrachloride–ethyl alcohol exhibits an azeotrope at 61.3 mol per cent CCl_4 with a boiling point of 64.95°C. The vapor pressures of CCl_4 and C_2H_5OH at this temperature are 530.9 and 448.8 mm Hg, respectively.

(a) Calculate the activity coefficient of each component in the liquid phase for the above azeotrope.

(b) Using the van Laar equation, estimate the activity of CCl_4 in a solution containing 50 mol per cent CCl_4. Assume that the activity coefficients are independent of temperature over this range.

10-23 A dilute solution of sodium chloride at 100°C is being rotated in a centrifuge at 400 rev/sec. After equilibrium conditions have been reached, a sample of the solution is taken from a point 4 cm from the axis of rotation. Analysis shows this sample to be 0.110 m (that is, 0.110 g-mol of NaCl/1000 g of H_2O). From the following data estimate the concentration of sodium chloride in a sample of solution taken from a point 16 cm from the axis of rotation. The vapor pressure of H_2O over NaCl solutions at 100°C = 760 − 1560(mol of NaCl/mol of H_2O). The density of NaCl solutions at 100°C is as follows:

NaCl, wt %	Density, g/cm^3
0	0.9584
1	0.9651
2	0.9719

10-24 (a) As part of an experimental setup a column 300 ft long is filled with methane. The temperature of the gas is 40°F throughout and the pressure at the top of the column is 1 atm. A sample is withdrawn from the top and is analyzed with a mass spectrometer. It is found to contain 1.12 per cent C^{13} methane and the rest C^{12} methane. At equilibrium what will be the concentration of C^{13} methane at the bottom? (Assume that only H^1 is present. The superscripts refer to isotopes.)

(b) Suppose the methane is put into a cylinder 1 ft long and rotated about one end at 6000 rpm. The concentration of C^{13} methane at the axis is 1.12 per cent. What will the concentration of C^{13} methane be at the other end?

10-25 A natural gas well extends down into the earth 9200 ft. When it is first discovered, the pressure at the top of the well is 1900 psia and the composition at the top is determined to be 85 mol per cent methane and 15 mol per cent ethane. If the gas composition is assumed constant throughout the well and the average temperature is 120°F, what is the pressure at the bottom of the well? If one considers that the well has been standing untouched for millions of years and so is at equilibrium, what are the pressure and composition at the bottom, assuming a temperature of 120°F throughout?

10-26 The vapor–liquid equilibrium data at 35°C for the ethanol–chloroform system are as follows:

Mole Fraction Chloroform in Liquid	Mole Fraction Chloroform in Vapor	Total Vapor Pressure, mm Hg
0.0000	0.0000	102.78
0.0100	0.0414	106.20
0.0200	0.0832	109.83
0.0500	0.2000	121.58
0.1000	0.3588	143.23
0.2000	0.5754	190.19
0.3000	0.6844	228.88
0.4000	0.7466	257.17
0.5000	0.7858	276.98

The vapor pressure of pure chloroform = 295.11 mm Hg. Assuming that the vapors behave as ideal gases:

(a) Calculate the activities of chloroform and ethanol as functions of composition.

(b) Does this system exhibit regular solution behavior?

(c) If so, what is the value of W/RT at this temperature?

Equilibrium in Chemically Reacting Systems

11

Introduction

In Chapters 6 and 10 we introduced the concept of equilibrium in nonreacting systems and demonstrated that equilibrium exists in the system if any of the three following statements is true:

1. The entropy of the system is maximized, subject to the constraint of constant internal energy and volume.

2. No potential (P, T, $\bar{G}$, etc.) differences exist within the system.

3. The system is not capable of producing any work when it is isolated from its surroundings.

It was indicated that these three conditions are actually equivalent and can be used to derive many other useful expressions about systems that are in equilibrium. In particular, it was shown that, if no barriers (constraints) are present to limit the transfer of chemical species from one part of the system to another, the following conditions must exist at all points within a system at equilibrium:

$$T, P, \bar{G}_A = \mu_A, \bar{G}_B = \mu_B, \ldots, \bar{G}_C = \mu_C = \text{constants at all points within the system} \quad (11\text{-}1)$$

$$\bar{f}_A, \bar{f}_B, \ldots, \bar{f}_C = \text{constants at all points within the system}$$

That is, neither the temperature (or the pressure) of the system, nor the partial molar Gibbs free energy $\bar{G}_i$ (or equivalently the chemical potential μ_i, or the partial fugacity $\bar{f}_i$) may change from point to point within the system. From these conditions we found that the criteria for equilibrium in nonreacting systems were easily established and used. In this chapter we shall discuss the development and application of similar criteria for equilibrium in chemically reacting systems. Before we can discuss equilibrium for systems in which chemical reactions may occur, however, we must develop a notation capable of unambiguously describing the various, and sometimes subtle, changes that take place when chemical reactions occur. This notation is introduced in the following section.

11.1 Work Production from Chemically Reacting Systems

Let us begin our discussion by choosing a system in which a chemical reaction is occurring. For the purposes of discussion, let us examine the steady-flow isothermal chemical reactor shown in Fig. 11-1, where the components A_1, A_2, ... react to form the products ..., A_{C-1}, A_C according to the reaction

$$-\alpha_1 A_1 - \alpha_2 A_2 - \cdots \longrightarrow \cdots \alpha_{C-1} A_{C-1} + \alpha_C A_C \tag{11-2}$$

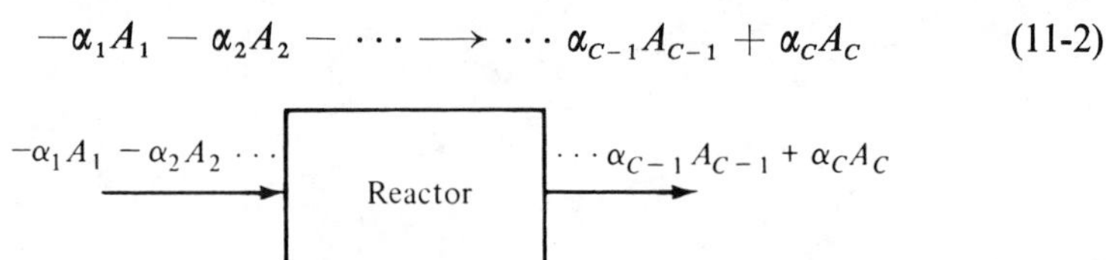

FIG. 11-1. Steady-flow isothermal chemical reactor.

The stoichiometric coefficient of the ith species in the chemical reaction is α_i, and is negative or positive depending on whether the ith species is a reactant (−) or product (+). For example, the reaction between hydrogen and chlorine to form hydrogen chloride would be written

$$H_2 + Cl_2 = 2HCl \tag{11-3}$$

or

$$-\alpha_1 A_1 - \alpha_2 A_2 = \alpha_3 A_3 \tag{11-3a}$$

where

$$\begin{aligned} \alpha_1 &= -1 \qquad & A_1 &= H_2 \\ \alpha_2 &= -1 \qquad & A_2 &= Cl_2 \\ \alpha_3 &= 2 \qquad & A_3 &= HCl \end{aligned} \tag{11-4}$$

Clearly we could label hydrogen component 2 and chlorine 1. Although any order is allowable (as long as we attach a minus sign before the stoichiometric coefficients of the reactants), it is conventional to give the reactants the lowest numbers and the products the highest numbers. (We shall follow this convention throughout the remaining discussion.) Thus the hydrogen chloride of the preceding example would not be labeled component 1 or 2 but would always be component 3.

It will frequently be somewhat more convenient in our later work if we rearrange our chemical reactions so that all species appear on the same side:

$$\alpha_1 A_1 + \alpha_2 A_2 + \cdots + \alpha_{C-1} A_{C-1} + \alpha_C A_C = 0 \tag{11-5}$$

or

$$\sum_{i=1}^{C} \alpha_i A_i = 0 \tag{11-5a}$$

where the α's and A's have exactly the same meaning as before. Once again, a negative α_i indicates a reactant and a positive α_i represents a product. Using this scheme, the hydrogen–chlorine reaction would be written as

$$-H_2 - Cl_2 + 2HCl = 0$$

Although we have written the reaction between hydrogen and chlorine with the coefficients given in equation (11-4), we might just as correctly have written it as

$$-\lambda H_2 - \lambda Cl_2 + 2\lambda HCl = 0 \tag{11-6}$$

where λ is any number other than zero. Thus we find that there is no unique set of α's that describe a particular reaction. For the sake of convenience and clarity we shall arbitrarily choose λ such that $-\alpha_1 = 1$. This means that we may get a different set of α's if we rearrange the order of the chemical species in a particular reaction, and therefore we must always specify which component has been chosen as number 1.

Let us return to the isothermal chemical reactor of Fig. 11-1. It is assumed that reactants, in their stoichiometric ratio, enter the reactor and are completely converted into the stoichiometric amounts of products. Upon completion we shall extend the development that follows to cover circumstances in which either, or both, of these conditions are not satisfied.

It was shown in Chapter 6 that the maximum work which may be extracted from any isothermal flow process is the negative of the Gibbs-free-energy change. Thus for the chemical reactor we may write

$$(-W_R)_T = (\Delta G)_T = G_{\text{prod}} - G_{\text{react}} \tag{11-7}$$

where the work and Gibbs free energies are based on a given quantity of *mass* that enters and leaves the reactor. (The reactor is operating at steady state, so the *mass* flow rate out equals the *mass* flow rate in.) Equation (11-7) is put on a per mole basis by dividing the whole equation by a given number of moles. Although there is no unique value that *must* be chosen, we shall arbitrarily choose the number of moles of species 1, n_1, that enter the reactor. Thus equation (11-7) becomes

$$\left[-\frac{W_R}{(n_1)_{\text{in}}}\right]_T = \left[\frac{\Delta G}{(n_1)_{\text{in}}}\right]_T = \frac{G_{\text{prod}}}{(n_1)_{\text{in}}} - \frac{G_{\text{react}}}{(n_1)_{\text{in}}} \tag{11-8}$$

or

$$(-\underline{\mathscr{W}}_R)_T = (\Delta \underline{\mathscr{G}})_T = \underline{\mathscr{G}}_{\text{prod}} - \underline{\mathscr{G}}_{\text{react}} \tag{11-8a}$$

where the script character with the bar beneath it is meant to read "per mole of species 1 entering the reactor." The distinction between a normal specific property such as $\underline{G}$ and the property $\mathscr{G}$ we have just defined is an extremely important one. A specific property is simply equal to the total property divided by the *total* mass (or moles); the script symbol represents the total property divided by mass (or moles) of a particular component, in this case 1. (This distinction is not critical when specific properties are expressed on a mass basis, because total mass is conserved during any process. However, the number of moles are not necessarily conserved in a chemical reaction, and thus if specific properties are expressed in terms of moles, extreme care must be exercised to insure consistency.)

The term $\mathscr{G}_{\text{prod}} - \mathscr{G}_{\text{react}}$ is now expressed in terms of the partial molar properties of the individual species as follows:

$$\mathscr{G}_{\text{prod}} - \mathscr{G}_{\text{react}} = \sum_{\text{prod}} \frac{(\bar{G}_i n_i)_{\text{out}}}{(n_1)_{\text{in}}} - \sum_{\text{react}} \frac{(\bar{G}_i n_i)_{\text{in}}}{(n_1)_{\text{in}}} \tag{11-9}$$

where $(n_1)_{\text{in}}$ = moles of species 1 which enter the reactor. If we also define

$$\underline{n}_i = \frac{n_i}{(n_1)_{\text{in}}}$$

equation (11-9) reduces to

$$\mathscr{G}_{\text{prod}} - \mathscr{G}_{\text{react}} = \sum_{\text{prod}} (\bar{G}_i \underline{n}_i)_{\text{out}} - \sum_{\text{react}} (\bar{G}_i \underline{n}_i)_{\text{in}} \tag{11-9a}$$

Since it is assumed that all components are present in their stoichiometric ratio, $(\underline{n}_i)_{\text{in}} = -\alpha_i$ for a reactant and $(\underline{n}_i)_{\text{out}} = \alpha_i$ for a product. Thus equation (11-9a) becomes

$$\mathscr{G}_{\text{prod}} - \mathscr{G}_{\text{react}} = \sum_{\text{prod}} (\alpha_i \bar{G}_i)_{\text{out}} + \sum_{\text{react}} (\alpha_i \bar{G}_i)_{\text{in}} \tag{11-10}$$

and equation (11-8a) becomes

$$(-\mathscr{W}_R)_T = \sum_{\text{prod}} (\alpha_i \bar{G}_i)_{\text{out}} + \sum_{\text{react}} (\alpha_i \bar{G}_i)_{\text{in}} \tag{11-11}$$

It is obviously impossible to tabulate partial molar Gibbs free energies for all possible combinations of temperature, pressure, and composition. However, we have developed (in Chapter 9) methods for predicting the variation of $\bar{G}_i$ with pressure, temperature, and composition. Thus if we tabulate values of the Gibbs free energies at a prespecified set of reference conditions, we may calculate the free energies at any other conditions. Thus we now refer all partial molar Gibbs free energies to a set of arbitrary reference conditions. These conditions may be different for each component. The only requirement that we shall place upon the reference state is that it must be at the temperature of the system.

Although a great variety of choices for the composition and pressure of the reference state exist, the reference state is most frequently (but not always) chosen as the pure component, at the temperature of the system and at either 1 atm pressure or at the pressure of the system. We shall see that each of these choices has certain advantages and disadvantages. Choice of the system pressure for the reference state leads to a particularly simple expression for the equilibrium criterion. However, a change in the system pressure then changes the reference states, and hence the equilibrium expression. The choice of 1 atm pressure, on the other hand, leads to a slightly more complicated equilibrium expression, but the reference states do not vary with the system pressure—an obvious advantage when the system pressure must vary or is not known. Since the free energies of formation (that is, the free energies of the individual components) are usually tabulated for pure components at 1 atm pressure, this is the reference state we will most commonly use. Because the reference state (or, as it is frequently called, the standard state) is intimately related to the equilibrium expressions we shall develop,' it is extremely important that we choose, specify, and use the correct standard state in all equilibrium calculations.

The partial molar Gibbs free energy of any component is now expressed in terms of the standard-state free energy, $\underline{G}^\circ$, as shown below:

$$\bar{G}_i = (\bar{G}_i - \underline{G}_i^\circ) + \underline{G}_i^\circ \tag{11-12}$$

Equation (11-12) is then substituted into equation (11-11) for the various components to give

$$(-\underline{\mathscr{W}}_R)_T = \sum_{\text{prod}} \alpha_i[(\bar{G}_i)_{\text{out}} - \underline{G}_i^\circ] + \sum_{\text{react}} \alpha_i[(\underline{G}_i)_{\text{in}} - \underline{G}_i^\circ] + \sum_{i=1}^{C} \alpha_i \underline{G}_i^\circ \tag{11-13}$$

Since the reference state has been chosen at the temperature of the system, we have shown (Chapter 10) that

$$\bar{G}_i - \underline{G}_i^\circ = RT \ln \frac{\bar{f}_i}{f_i^\circ} \tag{11-14}$$

$$\bar{G}_i - \underline{G}_i^\circ = RT \ln a_i \tag{11-14a}$$

where a_i is the activity of the ith component and is defined as $a_i = \bar{f}_i/f_i^\circ$. Substitution of equation (11-14) into equation (11-13) then yields

$$(-\underline{\mathscr{W}}_R)_T = \sum_{\text{prod}} \alpha_i RT \ln a_i)_{\text{out}} + \sum_{\text{react}} \alpha_i RT \ln a_i)_{\text{in}} + \sum_{i=1}^{C} \alpha_i \underline{G}_i^\circ \tag{11-15}$$

The term $\sum_{i=1}^{C} \alpha_i \underline{G}_i^\circ$ is equal to the Gibbs-free-energy change that would occur if the reactants (based on 1 mol of species 1) in their standard states were completely converted to the stoichiometric amount of products in their standard state. This quantity is termed the "standard-state Gibbs-free-energy change of reaction" and is written as

$$\Delta\underline{\mathscr{G}}^\circ = \sum_{i=1}^{C} \alpha_i \underline{G}_i^\circ = \sum_{\text{prod}} \alpha_i \underline{G}_i^\circ + \sum_{\text{react}} \alpha_i \underline{G}_i^\circ \tag{11-16}$$

Thus equation (11-15) reduces to

$$(-\underline{\mathscr{W}}_R)_T = RT \sum_{\text{prod}} \alpha_i \ln(a_i)_{\text{out}} + RT \sum_{\text{react}} \alpha_i \ln(a_i)_{\text{in}} + \Delta\underline{\mathscr{G}}^\circ \tag{11-17}$$

but

$$\alpha_i \ln a_i = \ln a_i^{\alpha_i} \tag{11-18}$$

$$\ln a_i^{\alpha_i} + \ln a_j^{\alpha_j} = \ln(a_i^{\alpha_i} \cdot a_j^{\alpha_j}) \tag{11-18a}$$

so equation (11-17) becomes

$$(-\underline{\mathscr{W}}_R)_T = RT \ln[\prod_{\text{prod}} (a_i)_{\text{out}}^{\alpha_i}] + RT \ln[\prod_{\text{react}} (a_i)_{\text{in}}^{\alpha_i}] + \Delta\underline{\mathscr{G}}^\circ \tag{11-19}$$

or

$$(-\underline{\mathscr{W}}_R)_T = RT \ln \frac{\prod_{\text{prod}} (a_i)_{\text{out}}^{\alpha_i}}{\prod_{\text{react}} (a_i)_{\text{in}}^{-\alpha_i}} + \Delta\underline{\mathscr{G}}^\circ \tag{11-19a}$$

where the symbol $\prod_{\text{prod}} \left(\prod_{\text{react}}\right)$ represents "the repeated product" over all products (reactants). For example,

$$\prod_{i=1}^{4} B_i = B_1 \cdot B_2 \cdot B_3 \cdot B_4 \tag{11-20}$$

We now define the activity ratio, J_a, as

$$J_a = \frac{\prod_{\text{prod}} (a_i)_{\text{out}}^{\alpha_i}}{\prod_{\text{react}} (a_i)_{\text{in}}^{-\alpha_i}} \tag{11-21}$$

and equation (11-19a) becomes

$$(-\underline{\mathscr{W}}_R)_T = RT \ln J_a + \Delta\underline{\mathscr{G}}^\circ \tag{11-22}$$

In the evaluation of the activity ratio, J_a, it must be remembered that *the activities of the various species in the products must be determined from the state of the products, while the activities of the species in the reactants must be evaluated from the state of the reactants.*

Equation (11-22) expresses the maximum (reversible) work that can be obtained from the complete conversion of reactants to products in a steady-flow isothermal chemical reactor. This work is expressed in terms of the standard-state change in Gibbs free energy between the reactants and products and the activities of the reactants and products.

Let us now consider the case in which some material passes through the reactor unchanged. (This material may be unreacted reactants, products that are initially present in the reactant stream, or an inert diluent that takes no part in the reaction.) We begin as before by observing that the reversible work which may be obtained from any isothermal process is simply the change in Gibbs free energy between the entering stream and the leaving stream:

$$(-W_R)_T = \Delta G = G_{\text{out}} - G_{\text{in}} \tag{11-23}$$

Once again this is placed on a per-mole basis by dividing by $(n_1)_{\text{in}}$, the number of moles of species 1 that enter the reactor:

$$\frac{(-W_R)_T}{(n_1)_{\text{in}}} = \frac{\Delta G}{(n_1)_{\text{in}}} = \frac{G_{\text{out}}}{(n_1)_{\text{in}}} - \frac{G_{\text{in}}}{(n_1)_{\text{in}}} \tag{11-24}$$

or

$$(-\underline{\mathscr{W}}_R)_T = \Delta\underline{\mathscr{G}} = \underline{\mathscr{G}}_{\text{out}} - \underline{\mathscr{G}}_{\text{in}} \tag{11-24a}$$

Now, when we attempt to describe $\underline{\mathscr{G}}_{\text{out}}$ and $\underline{\mathscr{G}}_{\text{in}}$, we must include not only that material which reacts, but also that material which goes through unreacted. For example, if an inert material undergoes a change in its concentration between inlet and outlet (because the reaction produces more or less moles than it consumes), then its partial molar Gibbs free energy, or chemical potential, will change. Therefore, its contribution to the total Gibbs free energy of the inlet and outlet streams will be different and the chemically inert species will contribute to the reversible work production. (This effect can also occur in equimolar reactions if nonideal solution behavior, or pressure variations, cause significant changes in the chemical potentials of the nonreacting species.)

The Gibbs free energies of the inlet and outlet streams are expressed in terms of the partial molar Gibbs free energies of the reacting species:

$$-\underline{\mathscr{W}}_R = \underline{\mathscr{G}}_{\text{out}} - \underline{\mathscr{G}}_{\text{in}} = \sum_{i=1}^{C} (\underline{n}_i \bar{G}_i)_{\text{out}} - \sum_{i=1}^{C} (\underline{n}_i \bar{G}_i)_{\text{in}} \tag{11-25}$$

The extent of reaction, X, is now defined as the number of "moles of reaction" that occur—that is, the number of moles of species 1 that react divided by α_1. Since we have arbitrarily set $\alpha_1 = 1$ (at least for our single-reaction systems), the extent of reaction simply reduces to the number of moles of species 1 that react.

The outlet mole numbers, $(n_i)_{\text{out}}$, may then be expressed in terms of the inlet mole numbers, $(n_i)_{\text{in}}$, and the extent of reaction as

$$(n_i)_{\text{out}} = (n_i)_{\text{in}} + \alpha_i X \tag{11-26}$$

Dividing by $(n_i)_{\text{in}}$ then gives

$$(\underline{n}_i)_{\text{out}} = (\underline{n}_i)_{\text{in}} + \alpha_i \underline{X} \tag{11-26a}$$

where $\underline{X} = X/(n_1)_{\text{in}}$, and is easily seen to be the fraction of species 1 which enters the reactor that is reacted during the reaction.

Equation (11-26a) is substituted into (11-25) to give

$$-\underline{\mathscr{W}}_R = \sum_{i=1}^{C} [(\underline{n}_i)_{\text{in}} + \alpha_i \underline{X}](\bar{G}_i)_{\text{out}} - \sum_{i=1}^{C} (\underline{n}_i)_{\text{in}} (\bar{G}_i)_{\text{in}} \tag{11-27}$$

Equation (11-27) is then rearranged to give

$$-\underline{\mathscr{W}}_R = \sum_{i=1}^{C} (\underline{n}_i)_{\text{in}}[(\bar{G}_i)_{\text{out}} - (\bar{G}_i)_{\text{in}}] + \sum_{i=1}^{C} \underline{X}\alpha_i (\bar{G}_i)_{\text{out}} \tag{11-27a}$$

However, the difference $(\bar{G}_i)_{\text{out}} - (\bar{G}_i)_{\text{in}}$ can be expressed as

$$\begin{aligned}(\bar{G}_i)_{\text{out}} - (\bar{G}_i)_{\text{in}} &= R \ln \frac{(\bar{f}_i)_{\text{out}}}{(\bar{f}_i)_{\text{in}}} \\ &= RT \ln \frac{(\bar{f}_i)_{\text{out}}/f_i^\circ}{(\bar{f}_i)_{\text{in}}/f_i^\circ} = RT \ln \frac{(a_i)_{\text{out}}}{(a_i)_{\text{in}}}\end{aligned} \tag{11-28}$$

so equation (11-27) becomes

$$-\underline{\mathscr{W}}_R = RT \sum_{i=1}^{C} (\underline{n}_i)_{\text{in}} \ln \frac{(a_i)_{\text{out}}}{(a_i)_{\text{in}}} + \underline{X} \sum_{i=1}^{C} \alpha_i (\bar{G}_i)_{\text{out}} \tag{11-29}$$

As in our previous discussion, we now refer $(\bar{G}_i)_{\text{out}}$ to the reference-state Gibbs free energies:

$$\begin{aligned}(\bar{G}_i)_{\text{out}} &= [(\bar{G}_i)_{\text{out}} - \underline{G}_i^\circ] + \underline{G}_i^\circ \\ &= RT \ln (a_i)_{\text{out}} + \underline{G}_i^\circ\end{aligned} \tag{11-14}$$

Equation (11-14) is substituted into equation (11-29) and the result rearranged to give

$$-\underline{\mathscr{W}}_R = RT\sum_{i=1}^{C}(\underline{n}_i)_{\text{in}}\ln\frac{(a_i)_{\text{out}}}{(a_i)_{\text{in}}} + \underline{X}\left\{RT\ln\left[\frac{\prod_{\text{prod}}(a_i)_{\text{out}}^{\alpha_i}}{\prod_{\text{react}}(a_i)_{\text{out}}^{-\alpha_i}}\right] + \Delta\underline{\mathscr{G}}^\circ\right\} \tag{11-30}$$

Examination of the term

$$\frac{\prod_{\text{prod}}(a_i)_{\text{out}}^{\alpha_i}}{\prod_{\text{react}}(a_i)_{\text{out}}^{-\alpha_i}}$$

indicates that it is quite similar to J_a except that the reactant activities are evaluated at the outlet conditions. Since both the product and reactant activities are evaluated at the outlet conditions, we call this term the "outlet activity ratio," $(J_a)_{\text{out}}$:

$$(J_a)_{\text{out}} = \frac{\prod_{\text{prod}}(a_i)_{\text{out}}^{\alpha_i}}{\prod_{\text{react}}(a_i)_{\text{out}}^{-\alpha_i}} = \prod_{i=1}^{C}(a_i)_{\text{out}}^{\alpha_i} \tag{11-31}$$

At a later point in our discussion we shall also have occasion to examine the activity ratio based on inlet conditions. Thus, let us also define

$$(J_a)_{\text{in}} = \frac{\prod_{\text{prod}}(a_i)_{\text{in}}^{\alpha_i}}{\prod_{\text{react}}(a_i)_{\text{in}}^{-\alpha_i}} = \prod_{i=1}^{C}(a_i)_{\text{in}}^{\alpha_i} \tag{11-31a}$$

Equation (11-30) is now expressed in terms of the outlet activity ratio:

$$-\underline{\mathscr{W}}_R = RT\sum_{i=1}^{C}(\underline{n}_i)_{\text{in}}\ln\frac{(a_i)_{\text{out}}}{(a_i)_{\text{in}}} + \underline{X}[RT\ln(J_a)_{\text{out}} + \Delta\underline{\mathscr{G}}^\circ] \tag{11-32}$$

Equation (11-32) may then be used to evaluate the maximum amount of work that can be obtained from the partial reaction of a chemically reacting mixture in a reversible, isothermal-flow chemical reactor.

For those cases where the total Gibbs-free-energy change is due mainly to the reaction rather than to changes in the partial Gibbs free energies of the unreacted materials, we may rearrange equation (11-32) into a slightly more convenient form. We begin by observing that the inlet mole number, $(\underline{n}_i)_{\text{in}}$, may be split into two parts: the portion that actually reacts ($-\alpha_i\underline{X}$ for a reactant, zero for products, or inerts) and $\underline{\eta}_i$, the portion that passes through the reactor unreacted:

$$\begin{aligned}(\underline{n}_i)_{\text{in}} &= \underline{\eta}_i - \underline{X}\alpha_i && \text{for a reactant}\\ &= \underline{\eta}_i && \text{for a product or inert}\end{aligned} \tag{11-33}$$

Substitution of equation (11-33) into (11-32) gives

$$\begin{aligned}-\underline{\mathscr{W}}_R = RT\sum_{\text{prod+inert}}(\underline{\eta}_i)\ln\frac{(a_i)_{\text{out}}}{(a_i)_{\text{in}}} + RT\sum_{\text{react}}(\underline{\eta}_i - \underline{X}\alpha_i)\ln\frac{(a_i)_{\text{out}}}{(a_i)_{\text{in}}}\\ + \underline{X}[RT\ln(J_a)_{\text{out}} + \Delta\underline{\mathscr{G}}^\circ]\end{aligned} \tag{11-34}$$

or

$$-\underline{\mathscr{W}}_R = RT\sum_{i=1}^{C}\underline{\eta}_i\ln\frac{(a_i)_{\text{out}}}{(a_i)_{\text{in}}} + \underline{X}\left[RT\ln(J_a)_{\text{out}} - RT\sum_{\text{react}}\alpha_i\ln\frac{(a_i)_{\text{out}}}{(a_i)_{\text{in}}} + \Delta\underline{\mathscr{G}}^\circ\right] \tag{11-34a}$$

but

$$\ln(J_a)_{\text{out}} = \sum_{i=1}^{C} \alpha_i \ln(a_i)_{\text{out}} = \sum_{\text{react+prod}} \alpha_i \ln(a_i)_{\text{out}} \tag{11-31a}$$

Thus

$$RT \ln (J_a)_{\text{out}} - RT \sum_{\text{react}} \alpha_i \ln \frac{(a_i)_{\text{out}}}{(a_i)_{\text{in}}} = RT \sum_{\text{prod}} \alpha_i \ln(a_i)_{\text{out}} + \sum_{\text{react}} \alpha_i \ln(a_i)_{\text{in}}$$
$$= RT \ln J_a$$

so equation (11-34a) reduces to

$$-\mathscr{W}_R = RT \sum_{i=1}^{C} \underline{\eta}_i \ln \frac{(a_i)_{\text{out}}}{(a_i)_{\text{in}}} + \underline{X}(RT \ln J_a + \Delta\underline{\mathscr{G}}^\circ) \tag{11-35}$$

which is the desired form. When the activities of the unreacted species do not change greatly between the inlet and outlet streams, it is frequently a reasonable *approximation* (but only an approximation) to neglect the first term in equation (11-35); see, for example, Sample Problem 11-1.

SAMPLE PROBLEM 11-1. A hydrogen fuel cell that uses air as the oxidizing medium is being designed. The cell will be fed pure hydrogen and pure air. A 20 percent excess of air will be used, and the waste gas will be purged from the top of the cell. Pure liquid water is the main product. The cell will be operated at 25°C and 1 atm pressure. What is the maximum amount of work that can be recovered per mole of hydrogen (H_2) fed to the cell? Assume that all the hydrogen is reacted.

For the reaction $H_2 + \frac{1}{2}O_2 = H_2O$,

$$\Delta\underline{\mathscr{G}}^\circ = -56{,}700 \text{ cal/g-mol}$$

where the standard states are pure liquid water and pure gaseous hydrogen and oxygen at 1 atm and 25°C.

Solution: We may picture the fuel-cell operation as shown in Fig. S11-1. Since it is desired to know the work liberated per mole of H_2, let us call hydrogen species 1,

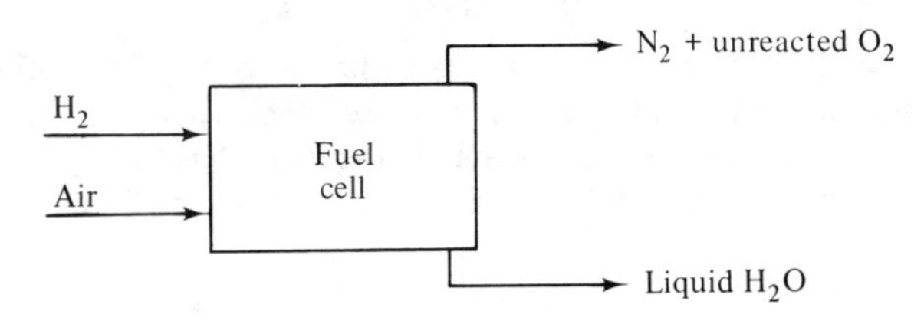

FIGURE S11-1

oxygen species 2, and water species 3. The stoichiometric coefficients for the reaction are then

$$\alpha_1 = -1 \quad \text{(hydrogen)}$$
$$\alpha_2 = -\tfrac{1}{2} \quad \text{(oxygen)}$$
$$\alpha_3 = 1 \quad \text{(water)}$$

and our basis of calculation is 1 mol of H_2 fed to the reactor. Since it is expected that the major portion of the Gibbs free energy change will be due to chemical reaction, we use equation (11-35) to determine the reversible work.

For each mole of hydrogen fed to the fuel cell, we have 0.6 mol of oxygen and 2.4 mol of N_2 (for a total of 3 mol of air). According to the problem statement, all hydrogen is consumed, so "1 mol of reaction" has occurred for each mole of H_2 fed to the cell. Thus $\underline{X} = 1.0$ and $\mathscr{W}_R$ is given by

$$-\mathscr{W}_R = \Delta\mathscr{G}^\circ + RT \ln J_a + RT \sum_{i=1}^{C} \underline{\eta}_i \ln \frac{(a_i)_{\text{out}}}{(a_i)_{\text{in}}}$$

But

$$\Delta\mathscr{G}^\circ = -56{,}700 \text{ cal/g-mol}$$

$$R = 1.987 \text{ cal/g-mol °K}$$

$$T = 25°\text{C} = 298°\text{K}$$

$$J_a = \frac{\prod_{\text{prod}} (a_i)_{\text{out}}^{\alpha_i}}{\prod_{\text{react}} (a_i)_{\text{in}}^{-\alpha_i}}$$

Only O_2 and N_2 go through the reactor without complete conversion; thus $\underline{\eta}_{H_2} = 0$. The unreacted amounts of N_2 and O_2 are given by

$\underline{\eta}_i = (\underline{n}_i)_{\text{in}} + \alpha_i \underline{X}$ (for reactant or inert)

$\underline{\eta}_{O_2} = 0.6$ mol of O_2 entering $-$ 0.5 mol of O_2 reacted $=$ 0.1 mol of O_2 unreacted

$\underline{\eta}_{N_2} = 2.4$ mol of N_2 entering $-$ none reacted $=$ 2.4 mol of N_2 unreacted

Thus

$$-\mathscr{W}_R = \Delta\mathscr{G}^\circ + RT \ln J_a + RT\left[0.1 \ln \frac{(a_{O_2})_{\text{out}}}{(a_{O_2})_{\text{in}}} + 2.4 \ln \frac{(a_{N_2})_{\text{out}}}{(a_{N_2})_{\text{in}}}\right]$$

Let us now evaluate the various activities. Both the inlet hydrogen and outlet water are in their standard states, and thus $(a_{H_2})_{\text{in}} = (a_{H_2O})_{\text{out}} = 1.0$. We are at liberty to choose the standard state for nitrogen as the pure gas at 1 atm and 25°C and evaluate the activities of the oxygen and nitrogen in the inlet and outlet streams from the expression

$$a_i = \frac{\bar{f}_i}{f_i^\circ} = \frac{y_i \gamma_i f_i}{f_i^\circ} = \frac{y_i \gamma_i \nu_i P}{f_i^\circ}$$

where we have used the pure-component-at-the-pressure-and-temperature-of-the-system standard state for the evaluation of the activity coefficients. For both the oxygen and nitrogen $f_i^\circ = 1$ atm. In addition, the pressure in the fuel cell is 1 atm, so $P = 1$ atm. At these low pressures we may reasonably assume that $\nu_i = \gamma_i = 1$ so the activities reduce to

$$a_i = y_i$$

or

$$(a_{O_2})_{\text{in}} = 0.20$$

$$(a_{O_2})_{\text{out}} = \frac{0.1}{2.5} = 0.0400$$

$$(a_{N_2})_{\text{in}} = 0.80$$

$$(a_{N_2})_{\text{out}} = \frac{2.4}{2.5} = 0.96$$

The reversible work production is then given by

$$\underline{\mathscr{W}}_R = -\left\{-56{,}700 + (1.987)(293)\left[\ln\frac{1.0}{(1.0)(0.2)^{1/2}} + 0.1\ln\frac{0.04}{0.20} + 2.4\ln\frac{0.96}{0.80}\right]\right\}\frac{\text{cal}}{\text{g-mol}}$$

or

$$\underline{\mathscr{W}}_R = (56{,}700 - 630)\ \text{cal/g-mol} = 56{,}070\ \text{cal/g-mol}$$

11.2 Development of the Equilibrium Constant

Let us now consider what happens to the total Gibbs free energy of a reacting mixture as the mixture passes through an isothermal, constant-pressure flow reactor as shown in Fig. 11-2.

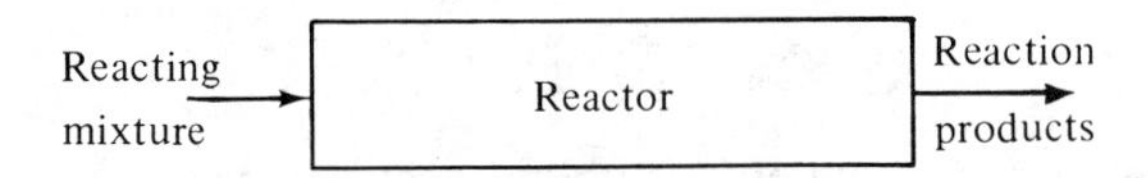

FIG. 11-2. Chemical reactor.

We have shown in Chapter 6 that as long as the total Gibbs free energy of the reacting mixture is decreasing, the reaction will continue to proceed spontaneously—that is, without the surroundings supplying work to drive the reaction forward. However, when the Gibbs free energy of the reacting mixture reaches a minimum, the reaction will proceed no further spontaneously, because the surroundings would have to provide work if the Gibbs free energy is to increase, and this violates our assumption of spontaneity. Similarly, the reaction cannot even proceed (spontaneously) in the reverse direction, because this would also cause an increase in the Gibbs free energy of the reacting mixture and hence require the surroundings to supply work. We may indicate these various situations on a plot of $\underline{\mathscr{G}}_{\text{mix}}$ versus $\underline{X}$ as shown in Fig. 11-3. (Since $(n_1)_{\text{in}}$ is a constant, the minimum in $\underline{\mathscr{G}}_{\text{mix}}$ is also the minimum in G_{mix}.) Once the reaction

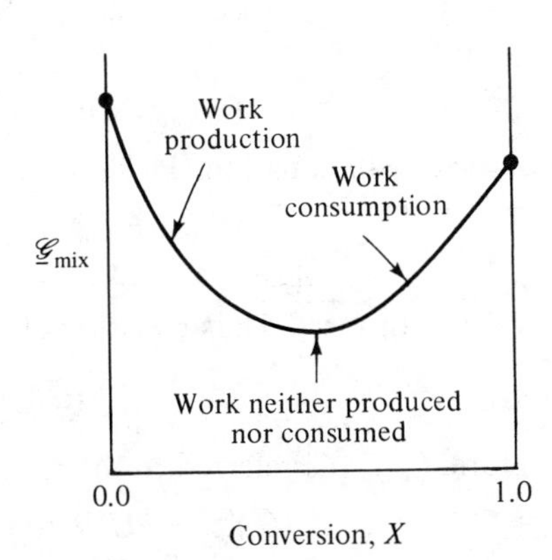

FIG. 11-3. Gibbs free energy as a function of conversion.

has proceeded to the minimum in the $\mathscr{G}$–$\underline{X}$ curve, it may clearly go no further (in either direction) without the surroundings supplying work. Clearly, the reacting mixture at this point is at its equilibrium condition subject to the imposed constraints of constant temperature and pressure. There is no way that this mixture can supply additional work to the surroundings unless we alter the constraints.

We may determine the conditions that correspond to the minimum in the $\mathscr{G}$–$\underline{X}$ curve by setting $d\mathscr{G}/d\underline{X} = 0$. (We know that this is a minimum rather than a maximum from the nature of the $\mathscr{G}$–$\underline{X}$ curve.) Although we may determine the minimum in the $\mathscr{G}$–$\underline{X}$ curve by a straight differentiation of either equation (11-32) or (11-35) with respect to $\underline{X}$, this is a rather tedious procedure, which may be avoided if we remember that the outlet Gibbs free energy may be expressed as

$$\mathscr{G}_{\text{out}} = \sum_{i=1}^{C} (\underline{n}_i)_{\text{out}}(\bar{G}_i)_{\text{out}}$$

Thus the minimum in the outlet $\mathscr{G}$ is obtained by setting

$$\frac{d\mathscr{G}_{\text{out}}}{d\underline{X}} = 0 = \frac{d}{d\underline{X}}\left[\sum_{i=1}^{C} (\underline{n}_i)_{\text{out}}(\bar{G}_i)_{\text{out}}\right] \tag{11-36}$$

or

$$0 = \sum_{i=1}^{C} \frac{d}{d\underline{X}}[(\underline{n}_i)_{\text{out}}(\bar{G}_i)_{\text{out}}] \tag{11-36a}$$

which upon differentiation becomes

$$0 = \sum_{i=1}^{C}\left[(\underline{n}_i)_{\text{out}}\frac{(d\bar{G}_i)_{\text{out}}}{d\underline{X}} + (\bar{G}_i)_{\text{out}}\frac{(d\underline{n}_i)_{\text{out}}}{d\underline{X}}\right] \tag{11-37}$$

However, the first term when summed over all components at constant T and P gives zero via the Gibbs–Duhem equation, so equation (11-37) reduces to

$$0 = \sum_{i=1}^{C} (\bar{G}_i)_{\text{out}}\frac{(d\underline{n}_i)_{\text{out}}}{d\underline{X}} \tag{11-38}$$

As we have shown, the outlet mole numbers may be expressed in terms of the inlet mole numbers and the extent of reaction as

$$(n_i)_{\text{out}} = (n_i)_{\text{in}} + X\alpha_i \tag{11-26}$$

or

$$(\underline{n}_i)_{\text{out}} = (\underline{n}_i)_{\text{in}} + \underline{X}\alpha_i \tag{11-26a}$$

Thus the derivative in equation (11-38) becomes

$$\frac{d(\underline{n}_i)_{\text{out}}}{d\underline{X}} = \alpha_i \tag{11-39}$$

where $\alpha_i = 0$ for an inert. Thus equation (11-38) reduces to

$$0 = \sum_{i=1}^{C} (\bar{G}_i)_{\text{out}}\alpha_i \tag{11-40}$$

Now, as before, $(\bar{G}_i)_{\text{out}}$ is referred to the reference state conditions

$$(\bar{G}_i)_{\text{out}} = RT \ln (a_i)_{\text{out}} + \underline{G}^\circ \tag{11-14a}$$

which is substituted into equation (11-40) to give

$$\sum_{i=1}^{C} \alpha_i \left[G_i^\circ + RT \ln (a_i)_{\text{out}} \right] = 0 \tag{11-41}$$

The first term in the summation is recognized as $\Delta\mathscr{G}^\circ$, the standard-state Gibbs-free-energy change, while the log terms are combined to yield the overall result:

$$\Delta\mathscr{G}^\circ + RT \ln \left[\frac{\prod_{\text{prod}} (a_i)_{\text{out}}^{\alpha_i}}{\prod_{\text{react}} (a_i)_{\text{out}}^{-\alpha_i}} \right] = 0 \tag{11-42}$$

Equation (11-42) thus relates the standard-state Gibbs-free-energy change to the activities of the reacting species which coexist in a mixture that is at chemical equilibrium. The term in brackets is called the *equilibrium constant*, K_a:

$$K_a = \frac{\prod_{\text{prod}} (a_i)_{\text{out}}^{\alpha_i}}{\prod_{\text{react}} (a_i)_{\text{out}}^{-\alpha_i}} = \prod_{i=1}^{C} (a_i)_{\text{out}}^{\alpha_i} \tag{11-43}$$

Equations (11-42) and (11-43) may then be solved for K_a to give

$$K_a = \exp\left(\frac{-\Delta\mathscr{G}^\circ}{RT} \right) \tag{11-44}$$

or

$$-\Delta\mathscr{G}^\circ = RT \ln K_a \tag{11-44a}$$

Thus if we can determine $\Delta\mathscr{G}^\circ$ for a given reaction, equations (11-44) permit us to evaluate the equilibrium constant for the reaction.

Examination of equation (11-43) indicates that the definition of the equilibrium constant greatly resembles our definition of $(J_a)_{\text{out}}$. Indeed, the definitions appear at first glance to be identical. However, there is a subtle, but major, distinction between the two that must be remembered: *The equilibrium constant applies only to the activities in a reacting mixture that is at chemical equilibrium.* The outlet activity ratio, $(J_a)_{\text{out}}$, on the other hand, is simply a number that may always be calculated from the state of the stream exiting from the chemical reactor under study. *In the special case where the exit stream is at chemical equilibrium, $(J_a)_{\text{out}}$ and K_a are equal.* In the more general case, where the exit stream is not in chemical equilibrium, K_a is still calculated from equation (11-44), but $(J_a)_{\text{out}}$ will no longer be equal to the equilibrium constant so calculated.

We may determine much about the state of a chemically reacting mixture by examining the activity ratio calculated from its present compositions. For example, let us consider a stream entering a chemical reactor. If $(J_a)_{\text{in}}$ is equal to K_a, the mixture is at equilibrium, and no reaction will occur. If $(J_a)_{\text{in}}$ is less than K_a, the activities of the products are less than their equilibrium value, while the activities of the reactants are above their equilibrium value. If the reaction is allowed to proceed, the activities of the products will increase, while those of the reactants will decrease. At constant temperature and pressure, the only way to increase the activity of a species is to increase its mole fraction. Thus as the reaction proceeds, the mole fractions of the products will increase, while the mole fractions of the reactants will decrease. Therefore, reactants are consumed and products produced, so the reaction proceeds in the forward direction.

If $(J_a)_{in}$ is greater than K_a, the reverse is true. That is, the activities and mole fractions of the products are too high, and the reaction will proceed in the reverse direction—opposite to that assumed in the calculation of $(J_a)_{in}$ and K_a.

The term "equilibrium constant" is actually a misnomer, because K_a is not a true constant but varies with temperature. However, K_a, does not vary with pressure (unless the standard-state conditions change) or composition of the reacting species, and therefore has become referred to as a "constant."

As we shall soon demonstrate, the equilibrium constant is directly related to the equilibrium compositions in a reacting mixture. Therefore, knowledge of K_a will tell us much about the maximum extent to which any given reaction proceeds. From these observations we may begin to understand the great importance of knowing, or being able to calculate, K_a. The remainder of this chapter will deal with the determination and use of K_a under many different circumstances.

11.3 Evaluation of $\Delta \underline{\mathscr{G}}^\circ$ from Gibbs Free Energies of Formation

The Gibbs free energies of formation of several commonly encountered substances *at 25°C and 1 atm pressure* are given in Table 11-1. The free energies of all pure elements have been arbitrarily chosen as zero.

TABLE 11-1 Standard Gibbs Free Energies of Formation at 25°C (kcal/mol)†

Substance	$\Delta \underline{G}_f^\circ$	Substance	$\Delta \underline{G}_f^\circ$	Substance	$\Delta \underline{G}_f^\circ$
SO_3(g)	−88.52	NO(g)	20.66	CH_4(g)	−12.14
HCl(g)	−22.74	H_2S(g)	− 7.87	C_2H_6(g)	− 7.6
C_2H_4O(g) (ethylene oxide)	− 2.79	SO_2(g)	−71.7	C_3H_8(g)	− 5.61
CO(g)	−32.81	NH_3(g)	− 3.94	C_2H_4(g)	16.34
CO_2(g)	−94.26	H_2O(l)	−56.70	C_2H_2(g)	50.7
Methanol (l)	−40.0				
Benzene (l)	29.06				
Acetic Acid (l)	−94.5				
Ethanol (l)	−40.2				
Acetaldehyde (l)	−31.9				
Ethylene Glycol (l)	−77.12				

The (l) or (g) after each component indicates the equilibrium phase at 1 atm.

† Selected mainly from F. D. Rossini et al., *Selected Values of Properties of Hydrocarbons and Related Compounds*, API Research Project 44, Carnegie Institute of Technology, Pittsburgh, 1953; F. D. Rossini et al., *Selected Values of Chemical Thermodynamic Properties*, National Bureau of Standards Circular 500, 1952.

The Gibbs free energy of formation for many other compounds may be found in standard references, such as the API Research Project 44 report cited. In cases where the Gibbs free energy of formation of a particular compound cannot be found in the standard references, it may be estimated via the "group contribution theory" of Anderson, Beyer, and Watson[1] and Brown.[2] According to this theory, the thermodynamic properties of an ideal gas can be broken down into contributions that are attributable to each of the atomic groups. By examining many compounds whose structure and properties were already known, it was possible to estimate the contributions of each of the various atomic groupings. For a further discussion of this technique, as well as tables of the various group contributions, the interested reader is referred to the textbook of Hougen, Watson, and Ragatz.[3]

The standard-state Gibbs-free-energy change for a specific reaction can be determined from the Gibbs free energies of formation of the reacting species as follows. The reaction equation is written as

$$\sum_{i=1}^{C} \alpha_i A_i = 0 \tag{11-45}$$

The standard Gibbs-free-energy change of reaction is defined as

$$\Delta\mathscr{G}° = \sum_{i=1}^{C} \alpha_i \underline{G}_i^\circ \tag{11-46}$$

The Gibbs free energies of the individual species may then be replaced by their Gibbs free energies of formations:

$$\underline{G}_i^\circ = (\Delta \underline{G}_i^\circ)_f + \sum_{k=1}^{Mi} a_{ki} \underline{G}_k^\circ \tag{11-47}$$

where a_{ki} = number of atoms of atomic species k in molecule i
M_i = number of different atomic species in molecule i
$\underline{G}_k^\circ$ = Gibbs free energy of pure elemental k

Thus

$$\Delta\mathscr{G}° = \sum_{i=1}^{C} \alpha_i (\Delta \underline{G}_i^\circ)_f + \sum_{i=1}^{C} \alpha_i \sum_{k=1}^{M_i} a_{ki} \underline{G}_k^\circ \tag{11-48}$$

Although the Gibbs free energies of the elements, $\underline{G}_k^\circ$, are usually arbitrarily set equal to zero, it is easy to show that whatever values are chosen for the Gibbs free energies of the elements, the last term in equation (11-48) always vanishes. Thus equation (11-48) reduces to

$$\Delta\mathscr{G}° = \sum_{i=1}^{C} \alpha_i (\Delta \underline{G}_i^\circ)_f \tag{11-49}$$

A similar expression can, of course, be derived for the standard-state change of

[1] J. W. Anderson, G. H. Beyer, and K. M. Watson, *Natl. Petrol. News, Tech. Sec., 36,* R476 (July 5, 1944).

[2] J. M. Brown, University of Wisconsin, Department of Chemical Engineering, Special Problems Project Department, June 1953.

[3] O. A. Hougen, K. M. Watson, and R. A. Ragatz, *Chemical Process Principles—Part II, Thermodynamics,* John Wiley & Sons, Inc., New York, 2nd ed., 1959, pp. 1004–1013.

any property on reaction, in terms of the property changes of formation of the reacting species.

In cases where it is desired to use standard states other than those used in Table 11-1, it is usually a simple task to correct the values given in the table to whatever conditions are desired.

SAMPLE PROBLEM 11-2. Determine the free energy of formation of pure water vapor at 25°C and 1 atm.

Solution: From Table 11-1, (ΔG_f) for liquid water is given by

$$\Delta \underline{G}_f^{\text{liq}} = -56.70 \text{ kcal/g-mol} = \underline{G}_{H_2O}^{\text{liq}} - \tfrac{1}{2}\underline{G}_{O_2} - \underline{G}_{H_2}$$

However, at 25°C the vapor pressure of liquid H_2O is only 23.76 mm Hg. Therefore, $f_{H_2O} = 23.76$ mm Hg (assuming negligible change of liquid fugacity with pressure and ideal gas behavior). For water vapor at 1 atm and 25°C, $f_{H_2O}^{\text{vap}} = 760$ mm Hg. The change in $\underline{G}_{H_2O}$ from 23.76 mm Hg to 1 atm is given by

$$\begin{aligned}(\underline{G}_{H_2O}^{\text{vap}}) - (\underline{G}_{H_2O}^{\text{liq}}) &= RT \ln \frac{f^{\text{vap}}}{f^{\text{liq}}} \\ &= (1.987)(298) \text{ cal °K/g-mol °K} \cdot \ln \frac{760}{23.76} \\ &= 2.05 \text{ kcal/g-mol}\end{aligned}$$

Therefore,

$$\begin{aligned}(\Delta \underline{G}_f^{\text{vap}}) &= \underline{G}_{H_2O}^{\text{vap}} - \tfrac{1}{2}\underline{G}_{O_2} - \underline{G}_{H_2} \\ &= \underline{G}_{H_2O}^{\text{liq}} - \tfrac{1}{2}\underline{G}_{O_2} - \underline{G}_{H_2} + (\underline{G}_{H_2O}^{\text{vap}} - \underline{G}_{H_2O}^{\text{liq}}) \\ &= -56.70 \text{ kcal/g-mol} + 2.05 \text{ kcal/g-mol}\end{aligned}$$

or

$$(\Delta \underline{G}_f^{\text{vap}})\ 1 \text{ atm} = -54.65 \text{ kcal/g-mol}$$

which is the desired free energy of formation of water *vapor* at 1 atm and 25°C.

SAMPLE PROBLEM 11-3. Determine the equilibrium constant at 25°C for the reaction

$$CH_4 + H_2O = CO + 3H_2$$

Solution: Since both the standard-state free energies and the equilibrium constants are functions of the standard-state conditions, we must specify these quantities before we can calculate a meaningful value of the equilibrium constant. Since the free energy of formation data we have is for pure components at 1 atm pressure and 25°C, this is the most reasonable choice for standard-state conditions. The standard-state free-energy change may then be calculated from the free energies of formation as shown in equation (11-49):

$$\Delta \underline{\mathscr{G}}^{\circ} = \sum_{i=1}^{4} \alpha_i (\Delta \underline{G}_f^{\circ})_i$$

Therefore,

$$\Delta \underline{\mathscr{G}}^{\circ} = (\Delta \underline{G}_f^{\circ})_{CO} + 3(\Delta \underline{G}_f^{\circ})_{H_2} - (\Delta \underline{G}_f^{\circ})_{CH_4} - (\Delta \underline{G}_f^{\circ})_{H_2O}$$

But

$$(\Delta \underline{G}_f^\circ)_{H_2} = 0$$

Therefore,

$$\Delta\mathscr{G}^\circ = -32.81 \text{ kcal/g-mol} - [-12.14 + (-56.70)] \text{ kcal/g-mol}$$

or

$$\Delta\mathscr{G}^\circ = 36.03 \text{ kcal/g-mol}$$

But, $RT \ln K_a = -\Delta\mathscr{G}^\circ$; therefore,

$$\ln K_a = \frac{-36.03 \text{ kcal/g-mol}}{(298.17)(0.00198) \text{ kcal/g-mol}} = -61.0$$

Therefore,

$$K_a = e^{-61.0} = 10^{-26.5}$$

or

$$K_a = 3.1 \times 10^{-27}$$

As we shall soon show, an equilibrium constant this small indicates that the reaction will not proceed to any appreciable extent at this temperature.

SAMPLE PROBLEM 11-4. The equilibrium constant for the reaction $H_2 + \frac{1}{2}O_2 = H_2O^{liq}$ has been determined to be 1.6×10^{41} at 25°C, where standard states are pure components at 1 atm pressure and 25°C. Calculate the minimum work required to dissociate 1 lb mol of water at 25°C and 1 atm if

(a) Pure and separated hydrogen and oxygen are to be produced.

(b) A stoichiometric *mixture* of oxygen and hydrogen is to be produced.

Solution: (a) As in Sample Problem 11-1 we suspect the major portion of the Gibbs-free-energy change will come from the reaction. Thus we use equation (11-35) to determine the work requirements:

$$-\underline{\mathscr{W}}_R = \underline{X}(\Delta\mathscr{G}^\circ + RT \ln J_a) + RT \sum_{i=1}^{C} \underline{\eta}_i \ln \frac{(a_i)_{out}}{(a_i)_{in}}$$

Since it is desired to completely dissociate the water, $\underline{X} = 1.0$ and $\underline{\eta}_i = 0$. Thus

$$-\underline{\mathscr{W}}_R = \Delta\mathscr{G}^\circ + RT \ln J_a$$

The dissociation reaction is written

$$H_2O = H_2 + \tfrac{1}{2}O$$

so the stoichiometric coefficients are as given in Table S11-4.

TABLE S11-4

i	A_i	α_i
1	H_2O	-1
2	H_2	$+1$
3	O_2	$+\frac{1}{2}$

Thus the minimum work expression becomes

$$-\underline{\mathscr{W}}_R = \Delta\mathscr{G}^\circ + RT \ln \frac{(a_{O_2})_{out}^{1/2}(a_{H_2})_{out}}{(a_{H_2O})_{in}}$$

Since $\mathscr{W}_R$ is the maximum work that can be produced by a reaction, $(-\mathscr{W}_R)$ is also the minimum amount of work that must be supplied to cause the given reaction to occur (whichever is applicable). Thus if we can determine the activities of the various components at the desired reaction conditions and $\Delta\mathscr{G}°$, we can determine $-\mathscr{W}_R$ directly.

We may calculate $\Delta\mathscr{G}°$ from the equilibrium constant using equation (11-44a):

$$\Delta\mathscr{G}° = -RT \ln K_a$$

But $K_a = 1.6 \times 10^{41}$ for the reaction

$$H_2 + \tfrac{1}{2}O_2 - H_2O^{liq} = 0$$

at 25°C. Therefore, K_a for the reverse reaction

$$H_2O^{liq} - H_2 - \tfrac{1}{2}O_2 = 0$$

is $K_a = (1.6 \times 10^{41})^{-1}$ at 25°C. $\Delta\mathscr{G}°$ for the dissociation reaction at 25°C is then given by

$$\begin{aligned} \Delta\mathscr{G}° &= -RT \ln(1.6 \times 10^{41})^{-1} = RT \ln(1.6 \times 10^{41}) \\ &= (1.987)(298)(1.8) \text{ Btu °K °R/(lb-mol °R °K)} \ln (1.6 \times 10^{41}) \\ &= 1.02 \times 10^5 \text{ Btu/lb-mol} \end{aligned}$$

Since all the components of this reaction occur in their standard states (pure components at 1 atm and 25°C), the activities of the water in the inlet, and the hydrogen, and oxygen in the outlet are all unity. Therefore, the work consumption is given by

$$\begin{aligned} -\mathscr{W}_R &= \Delta\mathscr{G}° + RT \ln \frac{(1)^{1/2} \cdot 1}{1} = \Delta\mathscr{G}° \\ &= 1.02 \times 10^5 \text{ Btu/lb-mol} \end{aligned}$$

so 1.02×10^5 Btu of work must be supplied to dissociate 1 lb-mol of liquid water at 1 atm into pure gaseous oxygen and hydrogen.

(b) It is now desired to determine the work necessary to dissociate 1 lb-mol of liquid water into a stoichiometric mixture of H_2 and O_2.

As in part (a), we may express the minimum work from the equation

$$-\mathscr{W}_R = \Delta\mathscr{G}° + RT \ln \frac{(a_{O_2})_{out}^{1/2}(a_{H_2})_{out}}{(a_{H_2O})_{in}}$$

Since $\Delta\mathscr{G}°$ is independent of the reaction conditions (except T), it is the same as in part (a) and is given by

$$\Delta\mathscr{G}° = 1.02 \times 10^5 \text{ Btu/lb-mol}$$

The liquid water is still in its standard state, so $(a_{H_2O})_{in} = 1.0$.

However, the hydrogen and oxygen are no longer pure substances and thus do not have unit activities. We may determine the activities of the hydrogen and oxygen in the outlet stream from the relation

$$a_i = \frac{\bar{f}_i}{f°} = \frac{\gamma_i y_i f_i}{f_i^\circ}$$

Assuming ideal vapor-phase solutions, $\gamma_i = 1.0$. Since the system pressure and temperature equal the standard-state pressure and temperature, $f_i = f_i^\circ$, and the activity reduces to $(a_i)_{out} = (y_i)_{out}$. The stoichiometric mixture of H_2 and O_2 has mole fractions

$$\begin{aligned} (y_{H_2})_{out} &= 0.66 \\ (y_{O_2})_{out} &= 0.33 \end{aligned}$$

Therefore, the work of dissociation is given by

$$-\mathscr{W}_R = 1.02 \times 10^5 \text{ Btu/lb-mol} + (1.987)(2.98)(1.8) \text{ Btu °R °K /(lb-mol °R °K)} \ln \frac{(0.66)(0.33)^{1/2}}{1}$$

$$= (1.02 \times 10^5) \text{ Btu/lb-mol} - (1.03 \times 10^3) \text{ Btu/lb-mol}$$

Therefore,

$$-\mathscr{W}_R = 1.0 \times 10^5 \text{ Btu/lb-mol}$$

11.4 The Equilibrium Constant in Terms of Measurable Properties

Let us now show how the equilibrium constant may be expressed in terms of measurable, and easily calculable, properties. By the definition of K_a and a_i,

$$K_a = \prod_{i=1}^{C} a_i^{\alpha_i} = \prod_{i=1}^{C} \left(\frac{\bar{f}_i}{f_i^\circ}\right)^{\alpha_i} \tag{11-50}$$

Although the subscript "out" has not been written in equation (11-50), we should remember that the equilibrium constant only applies to outlet properties. We have dropped the "out" subscript for the sake of clarity. Equation (11-50) may be rewritten in the form

$$K_a = \frac{\prod_{i=1}^{C} \bar{f}_i^{\alpha_i}}{\prod_{i=1}^{C} f_i^{\circ\,\alpha_i}} \tag{11-51}$$

Therefore,

$$K_a = \frac{K_{\bar{f}}}{K_{f^\circ}} \tag{11-52}$$

where

$$K_{\bar{f}} = \prod_{i=1}^{C} \bar{f}_i^{\alpha_i}$$
$$K_{f^\circ} = \prod_{i=1}^{C} f_i^{\circ\alpha_i} \tag{11-53}$$

For many gas-phase systems, the Lewis–Randall rule can be applied and $\bar{f}_i$ becomes

$$\bar{f}_i = y_i f_i \tag{11-54}$$

where f_i is the fugacity of pure i at the pressure and temperature of the reacting system. For gases the pure component fugacity, f_i, is then calculated from

$$f_i = \left(\frac{f}{P}\right)_i P = \nu_i P \tag{11-55}$$

where $(f/P)_i = \nu_i$, the fugacity coefficient for component i evaluated at the pressure and temperature of the system. For nonideal solutions, the Lewis–Randall rule may be modified by addition of an activity coefficient γ_i:

$$\bar{f}_i = y_i \gamma_i \nu_i P \tag{11-56}$$

In this manner it would possible to correct the development that follows for nonideal behavior, if this should be necessary. The effects of nongaseous components will be discussed in Chapter 12.

Substitution of equations (11-54) and (11-55) into equation (11-53) yields

$$K_{\bar{f}} = \prod_{i=1}^{C} (y_i \nu_i P)^{\alpha_i} \tag{11-57}$$

Thus

$$K_{\bar{f}} = K_y K_\nu K_P \tag{11-58}$$

where

$$K_y = \prod_{i=1}^{C} y_i^{\alpha_i}$$

$$K_\nu = \prod_{i=1}^{C} \nu_i^{\alpha_i}$$

$$K_P = \prod_{i=1}^{C} P^{\alpha_i} = P^{\sum_{i=1}^{C} \alpha_i}$$

Thus for an ideal mixture of gases, K_a can be expressed as

$$K_a = \frac{K_y K_\nu K_P}{K_{f^\circ}} \tag{11-59}$$

For a nonideal mixture, a K_γ will also appear in the numerator.

Under most circumstances we shall know P and T for the reaction under consideration, so K_P and K_ν can be calculated directly. From the standard-state conditions, K_{f° can also be determined in a straightforward manner. Therefore, we see that K_a can then be directly related to the equilibrium compositions, as expressed by K_y (and vice versa). Thus measurement of one set of equilibrium conditions and compositions may be used to determine K_a (from which we can also obtain $\Delta\mathscr{G}^\circ$). Once K_a is determined for one set of conditions, we may then use this value to calculate K_y for many other sets of reaction conditions as long as T is constant. Because of the direct relation between K_y and the equilibrium compositions of a reaction, knowledge of K_a, P, and T along with the reaction feed composition is usually sufficient to completely determine the equilibrium compositions of the reaction. A calculation of this nature will be presented in Sample Problem 11-5.

SAMPLE PROBLEM 11-5. The oxidation of SO_2 to SO_3 occurs according to the reaction

$$SO_2 + \tfrac{1}{2}O_2 = SO_3$$

At 610°C the equilibrium constant for this reaction is

$$K_a = \frac{a_{SO_3}}{a_{SO_2}(a_{O_2})^{1/2}} = 8.5$$

where the reference states are pure gaseous components at 1 atm.

Gas containing 12 mol per cent SO_2, 8 mol per cent O_2, and 80 mol per cent N_2 from sulfur burners is fed to a catalytic reactor at 1 atm for conversion to SO_3. If the reactors are operated at 610°C, what is the equilibrium composition of the gases leaving the reactor?

Solution: For an ideal (Lewis–Randall) mixture of gases, the equilibrium constant K_a has been expressed in equation (11-59) as

$$K_a = \frac{K_y K_P K_\nu}{K_{f^\circ}}$$

Since the standard states for all components have been chosen as the pure gases at 1 atm, the further assumption of ideal gas behavior for all pure components leads to the results

$$f^\circ_{SO_2} = f^\circ_{SO_3} = f^\circ_{O_2} = 1 \text{ atm}$$

and

$$\nu_{SO_2} = \nu_{SO_3} = \nu_{O_2} = 1$$

Consequently

$$K_\nu, = 1 \qquad \text{and} \qquad K_{f^\circ} = 1 \text{ atm}^{-1/2}$$

Furthermore, because the reactor operates at $P = 1$ atm, $K_P = 1 \text{ atm}^{-1/2}$. Thus the equilibrium expression reduces to

$$K_a = K_y = \frac{y_{SO_3}}{y_{SO_2}(y_{O_2})^{1/2}} = 8.5$$

We may now express the *exit* compositions of all components in terms of one variable, say the amount of SO_2 converted to SO_3. These expressions are then substituted into the equilibrium expression and solved for the degree of conversion as follows.

For equilibrium calculations it is generally most convenient to base the calculations on a specified number of moles of total feed rather than on any one component. Thus let us base our calculations on 100 mol of feed; 100 mol of feed contains 0 mol of SO_3, 12 mol of SO_2, 8 mol of O_2, and 80 mol of N_2. Assume that X mol of reaction occurs, so X mol of SO_2 is converted to SO_3.

The outlet mole numbers for all species are obtained from

$$(n_i)_{\text{out}} = (n_i)_{\text{in}} + \alpha_i X$$

Thus the outlet stream contains X mol of SO_3, $12 - X$ mol of SO_2, $8 - 0.5X$ mol of O_2, and 80 mol of N_2, for a total of $100 - 0.5\,X$ mol. The mole fraction of each component in the exit stream is given by

$$y_{SO_3} = \frac{X}{100 - 0.5X}$$

$$y_{SO_2} = \frac{12 - X}{100 - 0.5X}$$

$$y_{O_2} = \frac{8 - 0.5X}{100 - 0.5X}$$

The mole fractions may now be substituted into the equilibrium expression to give

$$\frac{X/(100 - 0.5X)}{\dfrac{12 - X}{100 - 0.5X}\left(\dfrac{8 - 0.5X}{100 - 0.5X}\right)^{1/2}} = 8.5$$

or, upon simplification,

$$\frac{X}{12 - X}\sqrt{\frac{100 - 0.5X}{8 - 0.5X}} = 8.5$$

Solution of this equation for X then allows us to completely determine the equilibrium compositions.

Before attempting solution of equation for X it is wise to examine the equation to determine if it has any particular properties that will allow us to get a good approximate answer without going to the effort of obtaining an exact solution. The first thing to do is to check the range of possible values of X—easily seen to be $0 < X < 12$. Now let us check to see if the solution lies near either of the extremes of this range. As $X \longrightarrow 0$ the left-hand side of the equation approaches zero while the right-hand side is constant. Therefore, the solution will not lie at small X. On the other hand, as $X \longrightarrow 12$, the left-hand side of the equation gets much larger than the right side, so the solution is not in this region either. Thus the actual X must be someplace toward the middle of the range. Examination of the equation then shows that the term $\sqrt{(100 - 0.5X)/(8 - 0.5X)}$ will not vary greatly with moderate changes in X, in the range of X's we expect. Therefore, rearrange the equation to the form

$$\frac{X}{12 - X} = 8.5\sqrt{\frac{8 - 0.5X}{100 - 0.5X}} = C$$

and assume that over any one trial the right-hand side will remain constant. Solving for X we now obtain

$$X = (12 - X)C$$
$$X(1 + C) = 12C$$

or, in terms of an iteration scheme,

$$X^{n+1} = \frac{12C}{1 + C} \qquad \text{where } C = 8.5\sqrt{\frac{8 - 0.5X^n}{100 - 0.5X^n}}$$

Now we use a trial-and-error procedure to solve for X. As a first guess, choose $X = 6.0$ (see Table S11-5). The final solution will be extremely close to $X = 7.65$.

TABLE S11-5

X^n	$0.5X$	$\sqrt{\frac{8 - 0.5X}{100 - 0.5X}}$	C	$12C$	$1 + C$	X^{n+1}
6.0	3.0	0.227	1.93	23.2	2.93	7.95
7.95	3.98	0.205	1.74	20.9	2.74	7.62
7.62	3.81	0.210	1.77	21.3	2.77	7.68

Thus we see that by using a little analysis, we have developed a method of solution that will give a relatively accurate answer to a complex equation with a minimum of effort. (Although this particular equation could have been solved by analytical means—squaring both sides yields a cubic equation whose solution can be found—this is frequently not true for complex reactions, and therefore a trial-and-error solution is often used. We could also have chosen to use the Newton–Raphson iteration for this problem; however, the scheme we have illustrated is considerably quicker for hand solution. For a computer solution, the Newton–Raphson procedure would have been preferred.)

Since $X = 7.65$, $X/12$, or 63.5 per cent, of the incoming SO_2 is converted to SO_3 at equilibirum conditions!

11.5 Effect of Pressure on Equilibrium Conversions

We have shown that the equilibrium constant K_a is directly related to the standard-state free-energy difference between the reactants and the products by the relation

$$-RT \ln K_a = \Delta\mathscr{G}^\circ \tag{11-60}$$

Since it is obviously inconvenient to tabulate K_a as a function of both T and P, we may wonder if it is not possible instead to calculate the changes that occur in K_a when the pressure and temperature of a system are changed. In addition, it is useful to have some feeling for how these variations affect the equilibrium compositions of the reacting mixture.

As shown in equation (11-60), the equilibrium constant K_a is directly related to the change in standard-state Gibbs free energy by

$$\ln K_a = -\frac{\Delta\mathscr{G}^\circ}{RT} \tag{11-60a}$$

Since $\Delta\mathscr{G}^\circ$ is a function only of the choice of standard states for the products and reactants, it is not a function of the pressure of the system, unless of course the standard-state conditions are related to the system pressure. It will be assumed in everything that follows that the standard-state conditions are not related to the system pressure! Since the temperature of the system is also independent of the pressure, the equilibrium constant, K_a, is *independent* of the pressure of the system. However, since $\Delta\mathscr{G}^\circ$ and T both vary with the temperature, K_a will certainly be a function of T.

Although we have shown that the equilibrium constant, K_a, is independent of pressure, this in no way means that the equilibrium *composition* of a reacting mixture is necessarily independent of pressure. Let us now examine the ways in which changing pressure may vary the composition of a reacting mixture.

As shown in equation (11-59), the equilibrium constant for a gas-phase reaction can be split into four separate parts.

$$K_a = \frac{K_f}{K_{f^\circ}} = \frac{K_y K_P K_v}{K_{f^\circ}} \tag{11-59}$$

where the K's have been defined in equations (11-53) and (11-58).

However, K_{f° is a function only of the standard states and, like K_a, is independent of pressure. Thus we find that

$$K_y \cdot K_P \cdot K_v = K_a \cdot K_{f^\circ} = \text{function of temperature} \tag{11-61}$$

However, we are really interested in variations in K_y, since we may determine the equilibrium compositions directly once K_y is known. Qualitatively we may say: The greater the value of K_y, the further a reaction will proceed in the forward direction (reactants → products). From equation (11-59) we may express K_y as

$$K_y = \frac{K_a \cdot K_{f^\circ}}{K_P \cdot K_\nu} \tag{11-62}$$

Thus the greater the quantity $K_\nu \cdot K_P$, the smaller will be the yield of the products.

For many practical applications $K_\nu = 1$, or at least K_ν varies very slowly with pressure, so at constant temperature K_y varies nearly inversely as K_P. However,

$$K_P = P^{\sum_{i=1}^{C} \alpha_i}$$

Therefore, if $\sum_{\text{prod}} \alpha_i > -\sum_{\text{react}} \alpha_i$, K_P will increase with increasing pressure and K_y will decrease, thereby decreasing the equilibrium yield. For example, consider the reaction $\alpha_2 = 2$, $\alpha_1 = -1$, where the number of moles of products > moles of reactants. We may expect the equilibrium yield of A_2 to decrease with increasing P. On the other hand, if $-\sum_{\text{react}} \alpha_i > \sum_{\text{prod}} \alpha_i$, K_P will decrease with increasing P and K_y will increase with P, thereby increasing the yield. For example, the equilibrium yield of NH_3 from the reaction

$$N_2 + 3H_2 = 2NH_3$$

may be increased by increasing P.

If $-\sum_{\text{react}} \alpha_i = \sum_{\text{prod}} \alpha_i$, K_P is independent of pressure, and unless K_ν changes significantly with pressure, K_y and the equilibrium composition will be relatively independent of pressure.

SAMPLE PROBLEM 11-6. It is desired to raise the equilibrium conversion of SO_2 to SO_3 in Sample Problem 11-5 from 63.5 per cent at 610°C to 80 per cent by changing the system pressure. At what pressure must the reactor now be operated? (You may still assume ideal solution and ideal gas behavior.)

Solution: We have shown that the equilibrium constant may be related to the pressure of the system and the equilibrium compositions by the expression

$$K_a = \frac{K_y \cdot K_P \cdot K_\nu}{K_{f^\circ}}$$

For the problem at hand $K_\nu = 1.0$, since all components are assumed to be ideal. In addition, since the standard states for all components have been chosen as pure gases at 1 atm, the standard-state fugacities are all 1 atm, so $K_{f^\circ} = 1 \text{ atm}^{-1/2}$. Therefore, K_a reduces to

$$K_a = \frac{K_y K_P}{1 \text{ atm}^{-1/2}}$$

But, from Problem 11-5, $K_a = 8.5$; therefore,

$$K_y K_P = 8.5(1 \text{ atm})^{-1/2}$$

Determination of K_y will now allow us to solve for K_P, which is related to the reactor pressure by the expression

$$K_P = \frac{P}{PP^{1/2}} = P^{-1/2}$$

Therefore,

$$P^{-1/2} = \frac{8.5 \text{ atm}^{-1/2}}{K_y}$$

But

$$K_y = \frac{y_{SO_3}}{y_{SO_2}(y_{O_2})^{1/2}}$$

Base all future calculations on 100 mol of feed to reactor: Therefore, the feed contains: 12 mol of SO_2, 8 mol of O_2, and 80 mol of N_2. For 80 per cent conversion of SO_2 to SO_3,, the product stream will contain

$$\begin{array}{llll}
SO_3: & (0.8)(12) & = 9.6 \text{ mol of } SO_3 & y_{SO_3} = 0.101 \\
SO_2: & 12 - (0.80)(12) & = 2.4 \text{ mol of } SO_2 & y_{SO_2} = 0.0252 \\
O_2: & 8 - (0.40)(12) & = 3.2 \text{ mol of } O_2 & y_{O_2} = 0.0337 \\
N_2: & & = \underline{80.0} \text{ mol of } N_2 & y_{N_2} = 0.840 \\
 & & 95.2 &
\end{array}$$

Therefore,

$$K_y = \frac{0.101}{(0.0252)(0.0337)^{1/2}} = 21.8$$

The value for K_y is then substituted into the expression for the system pressure, to give

$$P^{-1/2} = \frac{8.5 \text{ atm}^{-1/2}}{21.8} = 0.39 \text{ atm}^{-1/2}$$

or

$$P = \left(\frac{21.8}{8.5}\right)^2 \text{ atm} = 6.6 \text{ atm}$$

Thus, if the reactor is operated at 6.6 atm, an equilibrium conversion of 80 per cent of the SO_2 to SO_3 may be expected.

SAMPLE PROBLEM 11-7. The thermodynamic properties of saturated sodium vapor are given in Table S11-7a. (The atomic weight of sodium is 22.9.)

(a) Calculate the equilibrium constant for the dissociation reaction $Na_2 = 2Na$ at the temperatures for which data are given. (*Note*: This is a vapor-phase reaction.)

(b) In view of the increase of K_a with temperature, how do you explain the increasing molecular weight of the mixture, which indicates that the reaction is being driven toward the left?

TABLE S11-7a

T, °R	Mol Wt of the Vapor	Vapor Pressure, psia
1300	23.791	0.0392
1700	24.922	1.317
2100	26.060	15.39
2700	27.352	150.5

Solution: (a) Before attempting to calculate an equilibrium constant, it is necessary to specify standard states for all components. Let us choose pure-component vapors at 1 psia pressure and the temperature of the system as our standard states. If we now assume ideal gas behavior (probably not a bad assumption except possibly at the highest pressure), then $f_i^\circ = 1$ psia and $f_i = P$. If we also assume ideal gaseous mixtures, $\bar{f}_i = y_i f_i = y_i P$. But

$$a_i = \frac{\bar{f}_i}{f_i^\circ}$$

Therefore,

$$a_i = \frac{y_i P}{1 \text{ psia}} = y_i P \text{ psia}^{-1}$$

From the definition of K_a,

$$K_a = \frac{(a_{Na})^2}{a_{Na_2}} = \frac{(y_{Na})^2 P^2}{y_{Na_2} P} \text{ psia}^{-1}$$

or

$$K_a = \frac{(y_{Na})^2 P}{y_{Na_2}} \text{ psia}^{-1}$$

We may calculate the composition of the vapor from its molecular weight by means of the relation

$$\text{mol wt} = (y_{Na_2} \cdot M_{Na_2}) + (y_{Na} \cdot M_{Na})$$

But $M_{Na_2} = 45.8$, $M_{Na} = 22.9$, and $y_{Na_2} + y_{Na} = 1.0$ or $y_{Na_2} = 1.0 - y_{Na}$. Therefore

$$\text{mol wt} = (1.0 - y_{Na})(45.8) + y_{Na}(22.9)$$

which may be solved for y_{Na} to yield

$$y_{Na} = 2.0 - \frac{\text{mol wt}}{22.9}$$

We may solve now for K_a as shown in Table S11-7b,

TABLE S11-7b

T	Mol Wt	$\frac{\text{Mol Wt}}{22.9}$	y_{Na}	y_{Na_2}	$(y_{Na})^2$	$(y_{Na})^2/y_{Na_2}$	P, psia	K_a
1300	23.791	1.039	0.961	0.039	0.924	23.7	3.92×10^{-2}	0.928
1700	24.922	1.087	0.913	0.087	0.834	9.58	1.317	12.6
2100	26.06	1.139	0.861	0.139	0.742	5.34	15.39	82.0
2700	27.352	1.192	0.808	0.192	0.652	3.40	150.5	512

(b) As may be seen from the equilibrium constants calculated in part (a), the higher temperatures tend to greatly favor the dissociation of Na_2. However, at higher temperatures the vapor pressure of liquid sodium increases greatly. The increasing vapor pressure tends to drive the reaction back toward the formation of more Na_2. This pressure effect is strong enough to overcome the unfavorable shift in the equilibrium constant, and the net result is to obtain more Na_2. This is only one of a series of examples where factors other than the equilibrium constant can play a significant role in determining equilibrium compositions.

11.6 Variation of K_a with Changes in Temperature

Although we have shown that K_a is not a function of the system pressure, it is well known that K_a may vary with system temperature. We shall now develop a means of calculating the variation of K_a with changes in temperature. Since K_a is related to $\Delta\mathscr{G}^\circ$ by the relation

$$R \ln K_a = -\frac{\Delta\mathscr{G}^\circ}{T} \tag{11-63}$$

our problem reduces to one of determining the variation of $\Delta\mathscr{G}^\circ/T$ with respect to temperature. The derivative of K_a with T is given by

$$R\frac{d \ln K_a}{dT} = -\frac{d}{dT}\frac{\Delta\mathscr{G}^\circ}{T} \tag{11-64}$$

Since $\Delta\mathscr{G}^\circ$ refers to a specific pressure (that of the standard state), the derivative in equation (11-64) must be evaluated at constant pressure. This derivative can in turn be expressed as

$$\begin{aligned}\frac{\partial}{\partial T}\left(\frac{\Delta\mathscr{G}^\circ}{T}\right)_P &= \frac{\partial}{\partial T}\left(\frac{1}{T}\sum_{i=1}^{C}\alpha_i \underline{G}_i^\circ\right)_P \\ &= \frac{\partial}{\partial T}\left(\sum_{i=1}^{C}\alpha_i \frac{\underline{G}_i^\circ}{T}\right)_P = \sum_{i=1}^{C}\alpha_i \frac{\partial}{\partial T}\left(\frac{\underline{G}_i^\circ}{T}\right)_P\end{aligned} \tag{11-65}$$

However, we showed in Chapter 10 that

$$\frac{\partial}{\partial T}\left(\frac{\underline{G}_i^\circ}{T}\right)_P = -\frac{\underline{H}_i^\circ}{T^2} \tag{11-66}$$

which may be substituted into equation (11-65) to give

$$\frac{\partial}{\partial T}\left(\frac{\Delta\mathscr{G}_i^\circ}{T}\right)_P = -\sum_{i=1}^{C}\frac{\alpha_i \underline{H}_i^\circ}{T^2} = -\frac{\Delta\mathscr{H}^\circ}{T^2} \tag{11-67}$$

where

$$\Delta\mathscr{H}^\circ = \sum_{i=1}^{C}\alpha_i \underline{H}_i^\circ \tag{11-68}$$

is termed the *standard-state enthalpy change* (or heat) *of reaction*. Equation (11-67) (which is commonly called the *Gibbs–Helmholtz equation*) may then be substituted into equation (11-64) to give the desired relation:

$$R\frac{d}{dT}(\ln K_a) = \frac{\Delta\mathscr{H}^\circ}{T^2} \tag{11-69}$$

or

$$\frac{d}{dT}(\ln K_a) = \frac{\Delta\mathscr{H}^\circ}{RT^2} \tag{11-69a}$$

Equation (11-69a) may then be integrated between temperatures T_1 and T_2 to give

$$\int_{T_1}^{T_2} d \ln K_a = \int_{T_1}^{T_2} \frac{\Delta \mathscr{H}^\circ}{RT^2}\, dT \tag{11-70}$$

or

$$(\ln K_a)_{T_2} - (\ln K_a)_{T_1} = \int_{T_1}^{T_2} \frac{\Delta \mathscr{H}^\circ}{RT^2}\, dT \tag{11-71}$$

which becomes

$$\ln \frac{(K_a)_{T_2}}{(K_a)_{T_1}} = \int_{T_1}^{T_2} \frac{\Delta \mathscr{H}^\circ}{RT^2}\, dT \tag{11-71a}$$

Therefore, if we know K_a at any temperature T_1 and can express $\Delta \mathscr{H}^\circ$ as a function of temperature, then K_a at any other temperature T_2 can be evaluated directly from equation (11-71a).

Unfortunately, $\Delta \mathscr{H}^\circ$ is itself usually a function of temperature, and this function must be determined before equation (11-71a) may be integrated accurately. If we know $\Delta \mathscr{H}^\circ$ at any temperature T^* and the specific heats of the products and reactants, we may evaluate $\Delta \mathscr{H}^\circ$ at any other temperature T. Since $\mathscr{H}^\circ$ is a state property (it is simply the enthalpy at the standard-state pressure), then the enthalpy change $(\mathscr{H}_3^\circ - \mathscr{H}_1^\circ)$ between points 1 and 3 of

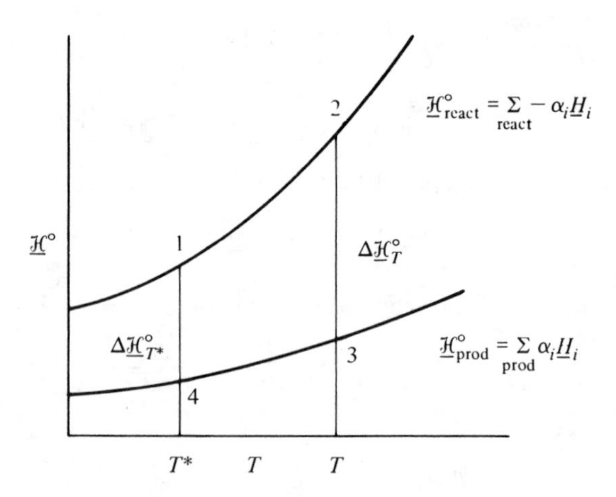

FIG. 11-4. Standard-state enthalpies.

Fig. 11-4 must be independent of the path chosen between states 1 and 3. Therefore,

$$\mathscr{H}_3^\circ - \mathscr{H}_1^\circ = (\mathscr{H}_4^\circ - \mathscr{H}_1^\circ) + (\mathscr{H}_3^\circ - \mathscr{H}_4^\circ) = (\mathscr{H}_2^\circ - \mathscr{H}_1^\circ) + (\mathscr{H}_3^\circ - \mathscr{H}_2^\circ) \tag{11-72}$$

However, $\mathscr{H}_4^\circ - \mathscr{H}_1^\circ = \Delta \mathscr{H}_{T^*}^\circ$, the standard-state heat of reaction at T^*, while $\mathscr{H}_3^\circ - \mathscr{H}_2^\circ$ equals $\Delta \mathscr{H}_T^\circ$, the standard heat of reaction at the temperature T. The enthalpy change between points 1 and 2 is then expressed as

$$\mathscr{H}_2^\circ - \mathscr{H}_1^\circ = \sum_{\text{react}} -\alpha_i(\underline{H}_i^\circ)_{T_2} - \sum_{\text{react}} -\alpha_i(\underline{H}_i^\circ)_{T_1}$$
$$= -\sum_{\text{react}} \alpha_i[(\underline{H}_i^\circ)_{T_2} - (\underline{H}_i^\circ)_{T_1}] \tag{11-73}$$

However, since the process between points 1 and 2 is at constant pressure (P°), the individual enthalpy changes are given by

$$(\underline{H}_i^\circ)_{T_2} - (\underline{H}_i^\circ)_{T_1} = \int_{T^*}^{T} (C_P^\circ)_i \, dT \tag{11-74}$$

which may be substituted into equation (11-73) to give

$$\mathscr{H}_2^\circ - \mathscr{H}_1^\circ = \sum_{\text{react}} -\alpha_i \int_{T^*}^{T} (C_P^\circ)_i \, dT \tag{11-75}$$

Similarly, we may show that

$$\mathscr{H}_3^\circ - \mathscr{H}_4^\circ = \sum_{\text{prod}} \alpha_i \int_{T^*}^{T} (C_P^\circ)_i \, dT \tag{11-76}$$

Equations (11-76) and (11-75), along with the definitions of $\Delta\mathscr{H}_{T^*}^\circ$ and $\Delta\mathscr{H}_T^\circ$, are substituted into equation (11-72) to give

$$\Delta\mathscr{H}_T^\circ = \Delta\mathscr{H}_{T^*}^\circ + \sum_{\text{prod}} \alpha_i \int_{T^*}^{T} (C_P^\circ)_i \, dT - \sum_{\text{react}} -\alpha_i \int_{T^*}^{T} (C_P^\circ)_i \, dT \tag{11-77}$$

or

$$\Delta\mathscr{H}_T^\circ = \Delta\mathscr{H}_{T^*}^\circ + \sum_{i=1}^{C} \alpha_i \int_{T^*}^{T} (C_P^\circ)_i \, dT \tag{11-77a}$$

Defining,

$$\Delta\mathscr{C}_P^\circ = \sum_{i=1}^{C} \alpha_i (C_P^\circ)_i \tag{11-78}$$

equation (11-77a) becomes

$$\Delta\mathscr{H}_T^\circ = \Delta\mathscr{H}_{T^*}^\circ + \int_{T^*}^{T} \Delta\mathscr{C}_P^\circ \, dT \tag{11-79}$$

Equation (11-79) now allows us to calculate $\Delta\mathscr{H}_T^\circ$ (at any temperature) provided only that we know $\Delta\mathscr{H}_{T^*}^\circ$ (at a particular temperature) and the heat capacities of the reacting species.

Substitution of equation (11-79) into (11-71a) then allows direct calculation of K_a at any temperature:

$$\ln\frac{(K_a)_{T_2}}{(K_a)_{T_1}} = \int_{T_1}^{T_2} \frac{\Delta\mathscr{H}_{T^*}^\circ + \int_{T^*}^{T} \Delta\mathscr{C}_P^\circ \, dT}{RT^2} \, dT \tag{11-80}$$

SAMPLE PROBLEM 11-8. A mixture of nitrogen and hydrogen, in the proportions 4 mols of hydrogen to 1 mol of nitrogen, is being fed continuously to a catalytic ammonia converter. The pressure in the converter is 200 atm, and the temperature is held constant at 450°F. Compute the maximum (equilibrium) conversion of nitrogen to ammonia from the reaction

$$N_2 + 3H_2 = 2NH_3$$

$K_a = 14.4 \times 10^{-6}$ at 500°C for this reaction. Standard states are pure components at 1 atm pressure. $\Delta\mathscr{H}^\circ = -22{,}000$ cal/g-mol N_2 at 18°C and 1 atm pressure. The heat capacities in cal/g-mol °K at 1 atm ($T =$ °K) are

$$(C_P^\circ)_{NH_3} = 6.70 + 0.0063T$$
$$(C_P^\circ)_{N_2} = 6.76 + 0.00606T$$
$$(C_P^\circ)_{H_2} = 6.62 + 0.00081T$$

Solution: Since we wish to calculate the equilibrium composition for this reaction at 450°F, our first task will be to determine the equilibrium constant at this temperature. From the equilibrium constant we shall determine K_y, from which the equilibrium conversion can be obtained. Since we do not have a value for K_a at 450°F but do have one at 500°C, we shall use equation (11-71a) to calculate the variation of K_a with T. Before evaluating the integral $\int_{T_1}^{T_2} (\Delta\mathscr{H}^\circ/T^2)\,dT$ we will have to express $\Delta\mathscr{H}^\circ$ as a function of T. We do this by means of equation (11-79):

$$\Delta\mathscr{H}_T^\circ = \Delta\mathscr{H}_{T^*}^\circ + \int_{T^*}^{T} \Delta\mathscr{C}_P^\circ\,dT$$

where

$$\Delta\mathscr{C}_P^\circ = 2(C_P^\circ)_{NH_3} - [(C_P^\circ)_{N_2} + 3(C_P^\circ)_{H_2}] = [(2.0)(6.70 + 0063T)] \text{ cal/g-mol °K}$$
$$-[6.76 + 0.00606T + (3)(6.62 + 0.00081T)] \text{ cal/g-mol °K}$$

or

$$\Delta\mathscr{C}_P^\circ = (-13.22 + 0.0096T) \text{ cal/g-mol °K}$$

where $T =$ °K. Substitution of this expression for $\Delta\mathscr{C}_P^\circ$ and the value of $\Delta\mathscr{H}_{18°C}^\circ$ into the $\Delta\mathscr{H}_T^\circ$ expression yields

$$\Delta\mathscr{H}_T^\circ = -22{,}000 \text{ cal/g-mol} + \int_{291}^{T} (-13.22 + 0.0096T)\,dT \text{ cal/g-mol °K}$$

or

$$\Delta\mathscr{H}_T^\circ = (-18{,}560 - 13.22T + 0.0048T^2) \text{ cal/g-mol}$$

where $T =$ °K. We may now substitute this expression for $\Delta\mathscr{H}_T^\circ$ into equation (11.71a) to yield

$$\ln\frac{(K_a)_{450°F}}{(K_a)_{500°C}} = \int_{500°C}^{450°F} \frac{\Delta\mathscr{H}_T^\circ}{RT^2}\,dT$$

or after converting all temperatures to °K

$$R\ln\frac{(K_a)_{505°K}}{(K_a)_{773°K}} = \int_{773°K}^{505°K} \frac{-18{,}560 - 13.22T + 0.0048T^2}{T^2}\,dT \text{ cal/°K}^2 \text{ g-mol}$$

Therefore,

$$1.987\frac{\text{cal}}{\text{g-mol °K}}\ln\frac{(K_a)_{505°K}}{14.4 \times 10^{-6}} = \left[(18{,}560)\left(\frac{1}{505} - \frac{1}{773}\right) - 13.22\ln\frac{505}{773} + (0.0048)(505 - 773)\right] \text{ cal/g-mol °K}$$

or

$$\ln\frac{(K_a)_{505°K}}{14.4 \times 10^{-6}} = \frac{12.72 + 5.63 - 1.29}{1.987} = 8.85$$

Therefore,

$$(K_a)_{505^\circ K} = 14.4 \times 10^{-6} \cdot e^{8.85} = 0.0768 \text{ at } 450^\circ F\ (505^\circ K)$$

K_a is related to the properties of the reacting mixture through the expression

$$K_a = \frac{K_y K_P K_\nu}{K_{f^\circ}} \quad \text{or} \quad K_y = \frac{K_a K_{f^\circ}}{K_P K_\nu}$$

The standard-state conditions are all pure components at 1 atm. Therefore, $f^\circ = 1$ atm, so $K_{f^\circ} = 1 \text{ atm}^{-2}$. K_P is related to the system pressure by the expression

$$K_P = \frac{P^2}{PP^3} = \frac{1}{P^2}$$

Therefore,

$$K_P = (200 \text{ atm})^{-2} = 2.5 \times 10^{-5} \text{ atm}^{-2}$$

Substitution of K_P and K_{f° into the above expression yields

$$K_y = \frac{K_a K_{f^\circ}}{K_P K_\nu} = \frac{(0.0768)(1 \text{ atm}^{-2})}{2.5 \times 10^{-5} \text{ atm}^{-2} K_\nu} = \frac{3.07 \times 10^3}{K_\nu}$$

Let us now set up a table of property values (Table S11-8a) for the reacting mixture based on 1 mol of nitrogen feed and X mol of N_2 reacting. Calculation of $K_\nu = (\nu_{NH_3})^2/\nu_{N_2}(\nu_{H_2})^3$ now gives

$$K_\nu = \frac{(0.73)^2}{(1.11)(1.12)^3} = 0.342$$

and indicates that ideal behavior is not a good approximation under these conditions. Substitution of K_ν into the K_y expression then gives

$$K_y = \frac{3.07 \times 10^3}{0.342} = 9.0 \times 10^3$$

TABLE S11-8a

	N_2	H_2	NH_3	Total
Initial	1	4	0	5
Formed by reaction	$-X$	$-3X$	$2X$	$-2X$
Final	$1 - X$	$4 - 3X$	$2X$	$5 - 2X$
Final mole fraction	$(1 - X)/(5 - 2X)$	$(4 - 3X)/(5 - 2X)$	$2X/(5 - 2X)$	1.0
Reduced pressure	5.97	15.2	1.79	
Reduced temperature	4.0	15.6	1.24	
Fugacity coefficient	1.11	1.12	0.73	

But

$$K_y = \frac{(y_{NH_3})^2}{y_{N_2}(y_{H_2})^3}$$

Therefore,

$$9000 = \frac{\left(\dfrac{2X}{5 - 2X}\right)^2}{\dfrac{1 - X}{5 - 2X}\left(\dfrac{4 - 3X}{5 - 2X}\right)^3}$$

or

$$9000 = \frac{(4X^2)(5 - 2X)^2}{(1 - X)(4 - 3X)^3}$$

Solution of this equation for X then gives the equilibrium conversions of N_2 to NH_3. Since this is a fourth-order equation, a trial-and-error solution is appropriate. Examination of the range of possible value for X shows that $0 \leq X \leq 1$; a rapid check of the extremes $X \longrightarrow 0$ and $X \longrightarrow 1$ shows that the actual X must be fairly near 1.0. [How could

$$\frac{4X^2 (5 - 2X)^2}{(1 - X)(4 - 3X)^3}$$

be as large as 9000 without $1 - X$ being quite small?] Examination of the equation for X shows that the only term which will vary rapidly with X in the range $X \longrightarrow 1$ is the term $X - 1$. Therefore, as the trial-and-error scheme, rearrange the equation into the form

$$1 - X^{n+1} = \frac{4(X^n)^2(5 - 2X^n)^2}{(9000)(4 - 3X^n)^3}$$

and let $X = 1$ be a first trial variable (see Table S11-8b). Thus the final solution is obtained on the second iteration.

TABLE S11-8b

X^n	$4(X^n)^2$	$(5 - 2X^n)^2$	$(4 - 3X^n)^3$	$1 - X^{n+1}$	X^{n+1}
1.0	4.0	9	1.0	0.004	0.996
0.996	3.97	9.04	1.03	0.004	0.996

The maximum conversion of N_2 to NH_3 under these conditions, with this feed, is then $X/1.0 = 0.996$, or 99.6 per cent conversion.

11.7 Adiabatic Reactions

In the earlier sections we worked with problems of equilibrium in constant-temperature systems and found that we could reduce most of the problems to the solution of a single algebraic (although sometimes complex) equation that could then be solved for the equilibrium compositions and conversions. However, many reactions that are of industrial importance are carried out under conditions which can be more closely represented as adiabatic than isothermal. The reason for this is that many industrial reactions are performed in large-scale plant equipment in which design limitations usually do not allow provision for sufficient heat exchange to provide any significant level of thermal control. Therefore, in many practical problems the temperature in the reactor will not be known before the problem is solved. Without knowledge of the exit temperature we shall not be able to determine K_a (which must be evaluated at T_{exit}, since this is the temperature at which the exit stream leaves the reactor), and therefore we shall not have sufficient information to calculate the exit compositions. Since the exit compositions and temperature are interrelated, they must be determined simultaneously. As may be imagined, this complicates matters significantly. Let us first examine one or two of the simpler cases of adiabatic reactions. In particular, we shall examine problems to which solutions can be obtained with a reasonable amount of effort.

The first simplification we shall discuss is the case in which the equilibrium constant for the reaction under study is so great that the reaction goes almost to completion independent of the final temperature. For example, if hydrogen is burned in a large excess of air so that the outlet temperature is less than 2000°F, the equilibrium constant

$$K_a = \frac{a_{H_2O}}{a_{H_2}(a_{O_2})^{1/2}}$$

for the reaction is greater than 10^7 (based on standard states of pure gases at 1 atm) and at least 99 percent of the H_2 will be reacted. Obviously in this situation even a fairly large change in the outlet temperature will not have a significant effect on the equilibrium conversion of hydrogen and oxygen to water. In cases such as this we may "uncouple" the energy and equilibrium equations and solve for the outlet temperature, assuming essentially complete chemical reaction. We then must use the calculated temperature to check the equilibrium constant and ensure that the reaction does proceed essentially to completion. If not, we must correct our assumption about the degree of reaction and proceed again. Later we shall examine this type of problem in greater detail.

We shall now develop a method of predicting the exit temperature of an adiabatic reaction. We begin by choosing a reaction system. For the case at hand, let us choose an open-system, constant-pressure reactor. If the reactor is operating at steady state with negligible changes in potential and kinetic energy, an energy balance around the reactor (the system) may be written as

$$\Delta\mathscr{H} = \mathscr{Q} - \mathscr{W} \tag{11-81}$$

If neither heat nor work is transferred to the surroundings, $\mathscr{Q} = \mathscr{W} = 0$ and equation (11-81) becomes

$$\Delta\mathscr{H} = 0 \tag{11-82}$$

We shall now attempt to develop an expression for the overall change in enthalpy of a reacting stream as a function of the extent of reaction and outlet temperature. Once this is complete, our task of finding the outlet conditions from the reaction reduces to one of determining which outlet conditions give an enthalpy change that satisfies the energy-balance equation (11-82). (If the reaction is not adiabatic—or work is transferred to, or from, the system—the energy balance must be rewritten to take these effects into account.)

Since thermodynamic properties are not path functions, we may postulate any convenient reaction mechanism to calculate property changes. Thus let us assume that the reactor is divided into two sections as shown in Fig. 11-5. The reactants enter the first section, where they react isothermally to form the

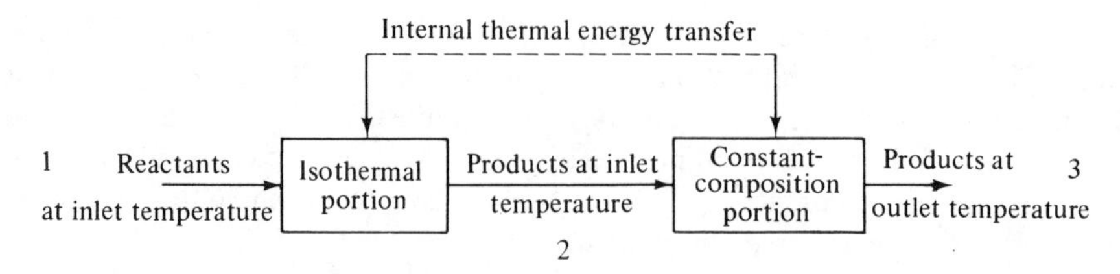

FIG. 11-5. Two-stage adiabatic reactor.

products. In the second section of the reactor the products absorb (or liberate) some fraction of the heat liberated (absorbed) in the first stage but do not change composition. (It is assumed that the material leaving the first stage is at the final product composition.)

The overall enthalpy change across the reactor, $\Delta\mathscr{H}_{1-3}$, may be expressed as the sum of the enthalpy changes across both sections of our model reactor:

$$\Delta\mathscr{H}_{1-3} = \Delta\mathscr{H}_{1-2} + \Delta\mathscr{H}_{2-3} \tag{11-83}$$

where the enthalpy, $\mathscr{H}$, of a stream is given by

$$\mathscr{H} = \sum_{i=1}^{C} \underline{n}_i \bar{H}_i \tag{11-84}$$

If we assume that the reacting mixture is an ideal mixture, then the only enthalpy changes that occur in the isothermal section of our model reactor are those due to the chemical reaction. Thus for every mole of species 1 that reacts, the enthalpy change is, by definition, the enthalpy change on reaction, $\Delta\mathscr{H}_{\text{react}}$, *evaluated at the inlet temperature*, T_1. Thus if each mole of species 1 that entered the reactor reacted, $\Delta\mathscr{H}_{1-2}$ would simply equal $(\Delta\mathscr{H}_{\text{react}})_{T_1}$. If X fraction of species 1 that enters the reaction actually reacts, the total change in enthalpy due to reaction is given by

$$\Delta\mathscr{H}_{1-2} = \underline{X}(\Delta\mathscr{H}_{\text{react}})_{T_1} \tag{11-85}$$

Since the temperature change in section 2 of the model reactor occurs at constant pressure and composition, the enthalpy changes of the individual species are given by the integral of C_P (again assuming ideal solutions):

$$(\bar{H}_i)_3 - (\bar{H}_i)_2 = (\underline{H}_i)_3 - (\underline{H}_i)_2 = \int_{T_2}^{T_3} (C_P)_i \, dT \tag{11-86}$$

but the total enthalpy change is given by

$$\mathscr{H}_3 - \mathscr{H}_2 = \sum_{i=1}^{C} (\underline{n}_i)_{\text{out}}(\bar{H}_i)_3 - (\bar{H}_i)_2 \tag{11-87}$$

so

$$\mathscr{H}_3 - \mathscr{H}_2 = \sum_{i=1}^{C} (\underline{n}_i)_{\text{out}} \int_{T_1=T_2}^{T_3} (C_P)_i \, dT \tag{11-88}$$

or

$$\mathscr{H}_3 - \mathscr{H}_2 = \int_{T_1}^{T_3} (\mathscr{C}_P)_{\text{exit}} \, dT \tag{11-88a}$$

where

$$(\underline{\mathscr{C}}_P)_{\text{exit}} = \sum_{i=1}^{C} (\underline{n}_i)_{\text{out}}(C_P)_i$$

Substitution of equations (11-88a) and (11-85) into (11-83) then yields

$$\Delta\underline{\mathscr{H}}_{1-3} = \underline{X}(\Delta\underline{\mathscr{H}}_{\text{react}})_{T_1} + \int_{T_1}^{T_3} (\underline{\mathscr{C}}_P)_{\text{exit}}\, dT \tag{11-89}$$

Equation (11-89) is then substituted into the overall energy balance to give

$$0 = \underline{X}(\Delta\underline{\mathscr{H}}_{\text{react}})_{T_1} + \int_{T_1}^{T_3} (\underline{\mathscr{C}}_P)_{\text{exit}}\, dT \tag{11-90}$$

Therefore, if we know $\Delta\underline{\mathscr{H}}_{\text{react}}$ at T_1, the fractional conversion, $\underline{X}$, and express $(\underline{\mathscr{C}}_P)_{\text{exit}}$ as a function of T, we may solve equation (11-90) for the outlet temperature T_3.

SAMPLE PROBLEM 11-9. Pure hydrogen is to be burned in a hydrogen torch using air as the oxidizing medium. Both the hydrogen and air enter the torch at 25°C. If the torch is adjusted to give 200 per cent more air than the stoichiometric ratio and the combustion is adiabatic, what is the flame temperature (assume that air = 80 per cent N_2, 20 per cent O_2, molar). The data are given in Table S11-9a, where C_P = cal/g-mol °K, T = °K.

TABLE S11-9a

C_P	O_2	$6.14 + 3.102 \times 10^{-3}T$
C_P	N_2	$6.524 + 1.250 \times 10^{-3}T$
C_P	H_2	$6.947 - 0.120 \times 10^{-3}T$
C_P	H_2O	$7.256 + 2.290 \times 10^{-3}T$

Reaction $\quad H_2 + \frac{1}{2}O_2 = H_2O$ (vapor)

$$(\Delta\underline{\mathscr{H}}_{\text{react}})_{25^\circ\text{C}} = -57.8 \text{ kcal/g-mol } H_2$$

$K_a = 1.0 \times 10^{40}$ at 25°C with pure vapors at 1.0 atm as the standard states for all components (even H_2O).

Solution: The basis of all calculatons is 1 mol of H_2 entering the reaction. One mol of H_2 requires $\frac{1}{2}$ mol of O_2. But we are told that the burner is adjusted for a 200 per cent excess of O_2. Therefore, the actual O_2 rate = 1.5 mol, which includes 6.0 mol of N_2. The inlet and outlet streams are then summarized as given in Table S11-9b. The outlet stream assumes total combustion of H_2.

TABLE S11-9b

	N_2	O_2	H_2	H_2O	Total
Inlet	6.00	1.50	1.0	0.0	8.50
Outlet	6.00	1.0	0.0	1.0	8.00

Since we have $\Delta\mathscr{H}_{\text{react}}$ at the temperature of the entering feed stream, we may substitute it directly into equation (11-90). [If this were not the case, we would simply calculate $\Delta\mathscr{H}_{\text{react}}$ at the feed temperature by means of equation (11-79).]

We now proceed to calculate $(\mathscr{C}_P)_{\text{exit}}$ as follows (basis, 1 mol of H_2 feed):

$$\begin{aligned}(\mathscr{C}_P)_{\text{exit}} &= 6.00(C_P)_{N_2} + 1.0(C_P)_{O_2} + 1.0(C_P)_{H_2O} \\ &= [(6.0)(6.524 + 1.250 \times 10^{-3}T) + (6.148 + 3.102 \times 10^{-3}T) \\ &\quad + (7.256 + 2.30 \times 10^{-3}T)] \text{ cal/g-mol °K} \qquad (T = \text{°K}) \\ &= (53.21 + 12.90 \times 10^{-3}T) \text{ cal/g-mol °K}\end{aligned}$$

Substitution of $(\mathscr{C}_P)_{\text{exit}}$ and $(\Delta\mathscr{H}_{\text{react}})_{T_1}$ into equation (11-90) then yields

$$\int_{298}^{T_2} (53.21 + 12.90 \times 10^{-3}T)\, dT \text{ cal °K/g-mol °K} = 57{,}800 \text{ cal/g-mol}$$

or

$$53.21(T_2 - 298) + 6.45 \times 10^{-3}(T_2^2 - 298^2)] = 57{,}800$$

Collecting terms yields

$$6.45 \times 10^{-3}T_2^2 + 53.21T_2 - 74{,}300 = 0$$

or

$$T_2^2 + 8.25 \times 10^3 T_2 - 1.15 \times 10^7 = 0$$

We may now solve for T_2 by means of the quadratic equation

$$T_2 = \frac{-8.25 \times 10^3 + \sqrt{(8.25 \times 10^3)^2 + (4)(1.15 \times 10^7)}}{2.0}$$

Thus

$$T_2 = 1215\text{°K}$$

or

$$T_2 = 2187\text{°R} = 1727 \text{ °F}$$

We must now check to see if the assumption of complete reaction is reasonable. We shall use the Gibbs–Helmholtz equation. Thus we must calculate $(\Delta\mathscr{H}^\circ)_T$:

$$\begin{aligned}(\Delta\mathscr{H}^\circ)_T &= (\Delta\mathscr{H}^\circ)_{298} + \int_{298}^{T} \Delta\mathscr{C}_P^\circ\, dT \\ &= (-57{,}800) \text{ cal/g-mol} + \int_{298}^{T} [(7.256 + 2.298 \times 10^{-3}T) \\ &\quad -(6.945 - 0.120 \times 10^{-3}T) - 0.5(6.148 + 3.102 \times 10^{-3})]\, dT \text{ cal °K/g-mol °K}\end{aligned}$$

Thus

$$\begin{aligned}(\Delta\mathscr{H}^\circ)_T &= \left[-57{,}800 + \int_{298}^{T} (-2.75 + 0.87 \times 10^{-3}T)\, dT\right] \text{cal/g-mol} \\ &= [-57{,}800 - 2.75(T - 298) + 0.44 \times 10^{-3}(T^2 - 298^2)] \text{ cal/g-mol}\end{aligned}$$

or

$$\Delta\mathscr{H}_T^\circ = (-58{,}600 - 2.75T + 4.4 \times 10^{-4}T^2) \text{ cal/g-mol} \qquad (T = \text{°K})$$

But

$$R \ln \frac{(K_a)_{T_2}}{(K_a)_{T_1}} = \int_{T_1}^{T_2} \frac{(\Delta\mathscr{H}^\circ)_T}{T^2}\, dT$$

Therefore,

$$R \ln \frac{(K_a)_{1185}}{(K_a)_{298}} = \int_{298}^{1215} \left(\frac{-58{,}600}{T^2} - \frac{2.75}{T} + 4.4 \times 10^{-4} \right) dT \text{ cal/g-mol °K}^2$$

$$R \ln \frac{(K_a)_{1215}}{(K_a)_{298}} = \left[58{,}600 \left(\frac{1}{1215} - \frac{1}{298} \right) - 2.75 \ln \frac{1215}{298} \right.$$

$$\left. + 4.4 \times 10^{-4}(1215 - 298) \right] \text{cal/g-mol °K}$$

or

$$(2.303)(1.987) \text{ cal/g-mol °K} \log_{10} \frac{(K_a)_{1215}}{(K_a)_{298}}$$

$$= (-148.4 - 386 + 0.4) \text{ cal/g-mol °K}$$

so

$$\log_{10} \frac{(K_a)_{1215}}{1.0 \times 10^{40}} = -33.2$$

which yields

$$(K_a)_{1215} = (1.0 \times 10^{40})(6.3 \times 10^{-34}) = 6.3 \times 10^6$$

Now let us check the approximate equilibrium composition. Amounts of each component at equilibrium, assuming X mol of H_2 have reacted, are summarized in Table S11-9c. But

$$K_a = \frac{a_{H_2O}}{a_{H_2}(a_{O_2})^{1/2}} = \frac{\bar{f}_{H_2O}/f^\circ_{H_2O}}{(\bar{f}_{H_2}/f^\circ_{H_2})(\bar{f}_{O_2}/f^\circ_{O_2})^{1/2}}$$

$$f^\circ_{H_2} = f^\circ_{O_2} = f^\circ_{H_2} = 1.0 \text{ atm}$$

Assuming ideal solutions of ideal components,

$$\bar{f}_i = y_i f_i$$

But $f_i = 1$ atm for all components. Therefore, $\bar{f}_i = y_i$ atm and

$$K_a = \frac{y_{H_2O}}{y_{H_2}(y_{O_2})^{1/2}}$$

TABLE S11-9c

	N_2	H_2	O_2	H_2O	Total
Amounts	6.0	$1 - X$	$1.5 - 0.5X$	X	$8.5 - 0.5X$
Mole fraction	$\frac{6.0}{8.5 - 0.5X}$	$\frac{1 - X}{8.5 - 0.5X}$	$\frac{1.5 - 0.5X}{8.5 - 0.5X}$	$\frac{X}{8.5 - 0.5X}$	1.0

or

$$6.3 \times 10^6 = \frac{y_{H_2O}}{y_{H_2}(y_{O_2})^{1/2}} = \frac{X/(8.5 - 0.5X)}{\frac{1 - X}{8.5 - 0.5X}\left(\frac{1.5 - 0.5X}{8.5 - 0.5X}\right)^{1/2}}$$

Therefore,

$$6.3 \times 10^6 = \frac{X(8.5 - 0.5X)^{1/2}}{(1 - X)(1.5 - 0.5X)^{1/2}}$$

As $X \longrightarrow 1.0$ we recognize that $1 - X$ is the term most sensitive to small change in X; therefore, rearrange the equation in the following manner:

$$1 - X^{n+1} = \frac{X^n(8.5 - 0.5X^n)^{1/2}}{(6.3 \times 10^6)(1.5 - 0.5X^n)^{1/2}}$$

X^n	$0.5X^n$	$1.5 - 0.5X^n$	$8.5 - 0.5X^n$	$1 - X^{n+1}$
1.0	0.5	1.0	8.0	1.3×10^{-6}

Thus the reaction proceeds 99.99987 per cent to completion, and our original assumption of essentially complete reaction is quite justifiable and correct.

The analysis presented in Sample Problem 11-9 is simple and straightforward. However, for some important problems—especially those where the attainment of extremely high temperatures is desired—the assumption of complete reaction may be in serious error. Therefore, in problems of this nature it is necessary to simultaneously solve for both the equilibrium composition and the final temperature. We shall illustrate the complexity of these calculations in Sample Problem 11-10. For simplicity we shall use the data of Sample Problem 11-9, but it should be recognized that in a real calculation at least three terms would be necessary to give an accurate representation of C_P at the extremely high temperatures involved. We shall also neglect the effects of any side reactions, such as formation of the oxides of N_2, because the inclusion of these effects would complicate the problem tremendously.

SAMPLE PROBLEM 11-10. To increase the operating temperature of the hydrogen torch of Sample Problem 11-9, pure oxygen at exactly the stoichiometric rate replaces air as the oxidizing medium. Neglecting all reactions other than the combination of hydrogen and oxygen and using the data of Sample Problem 11-9, determine the adiabatic flame temperature for these operating conditions.

Solution: We may summarize the feed and exit compositions as shown in Table S11-10a. The basis is 1 mol of H_2 feed and X mol of H_2 reacted.

TABLE S11-10a

	Inlet Amount	Outlet Amount	Outlet Composition
O_2	0.5	$(0.5)(1 - X)$	$\frac{1 - X}{3 - X}$
H_2	1	$1 - X$	$\frac{1 - X}{1.5 - 0.5X}$
H_2O	0	X	$\frac{X}{1.5 - 0.5X}$
Total	1.5	$1.5 - 0.5X$	1.0

We may now express the standard-state heat of reaction and the equilibrium constant as shown in Sample Problem 11-9:

$$\Delta\mathscr{H}^\circ_T = (-58{,}600 - 2.75T + 4.4 \times 10^{-4}T^2)\ \text{cal/g-mol}$$

$$R \ln \frac{(K_a)_T}{1.0 \times 10^{40}} = \left[-58{,}600\frac{T-298}{T\cdot 298} - 2.75 \ln \frac{T}{298} + 4.4 \times 10^{-4}(T-298)\right] \text{cal/g-mol °K}$$

where $T =$ °K. The outlet temperature may be related to the equilibrium conversion of H_2 to H_2O by:

$$\int_{298}^{T} (\mathscr{C}_P)_{\text{exit}} = -X(\Delta\mathscr{H}_{\text{react}})_{298}$$

where $(\mathscr{C}_P)_{\text{exit}}$ is also dependent on the equilibrium composition.

We shall assume that the reaction which occurs in the flame occurs at great speed, so the combustion products are always in equilibrium at the temperature of the flame. Therefore, the compositions of the reacting species are related by the equilibrium expression

$$K_a = \frac{a_{H_2O}}{a_{H_2}(a_{O_2})^{1/2}}$$

which, for $P = 1$ atm, standard states at 1 atm, and ideal gases reduces to

$$K_a = \frac{y_{H_2O}}{y_{H_2}(y_{O_2})^{1/2}} = \frac{X/(1.5 - 0.5X)}{\dfrac{1-X}{1.0 - 0.5X}\left(\dfrac{1-X}{3-X}\right)^{1/2}}$$

or

$$K_a = \frac{X(3-X)^{0.5}}{(1-X)^{1.5}} \tag{a}$$

but K_a is related to the outlet temperature through the equation

$$R \ln \frac{(K_a)_T}{10^{40}} = \left[-58{,}600\frac{T-298}{298T} - 2.75 \ln \frac{T}{298} + 4.4 \times 10^{-4}(T-298)\right] \frac{\text{cal}}{\text{g-mol °K}}$$

or

$$\log_{10}(K_a)_T = 40 - 43.0\left(1 - \frac{298}{T}\right) - 0.602 \ln \frac{T}{298} + 9.4 \times 10^{-5}\,(T - 298) \tag{b}$$

where $T =$ °K, and T is related to X through the relation

$$\int_{298}^{T} (\mathscr{C}_P)_{\text{exit}}\, dT = X(\Delta\mathscr{H}_{\text{react}})_{298} = X(57{,}800)\ \text{cal/g-mol}$$

where

$$(\mathscr{C}_P)_{\text{exit}} = 0.5(1-X)(C_P)_{O_2} + (1-X)(C_P)_{H_2} + X(C_P)_{H_2O}$$

or

$$(\mathscr{C}_P)_{\text{exit}} = \frac{1-X}{2}(6.148 + 3.102 \times 10^{-3}T) + (1-X)(6.947 - 1.20 \times 10^{-4}T) + X(7.256 + 2.298 \times 10^{-3}T)\ \text{cal/g-mol °K}$$

Therefore,

$$(\mathscr{C}_P)_{\text{exit}} = [10.02 - 2.77X + T(1.43 \times 10^{-3} + 0.87X)]\ \text{cal/g-mol °K}$$

Substitution of the expression for $(\mathscr{C}_P)_{\text{exit}}$ into the energy-balance expression then gives

$$\int_{298}^{T} [10.02 - 2.77X + T(1.43 \times 10^{-3} + 0.87X)]\, dT \text{ cal °K/g-mol °K}$$
$$= 57{,}800X \text{ cal/g-mol}$$

or

$$57{,}800X = (10.02 + 2.77X)(T - 298) + (7.15 \times 10^{-4} + 4.35 \times 10^{-4}X)(T^2 - 298^2)$$

Division by 298°K and rearrangement then yields the following form

$$194X = (10.02 + 2.77X)\left(\frac{T}{298} - 1\right) + (0.213 + 0.130X)\left[\left(\frac{T}{298}\right)^2 - 1\right] \qquad \text{(c)}$$

Equations (a), (b), and (c) now form a system of three equations in three unknowns (K_a, X, and T) which are completely sufficient to allow us to solve for all unknown quantities. Unfortunately, these equations are highly nonlinear as well as completely coupled. Therefore, solution of these equations is a difficult task and will require a trial-and-error procedure. Examination of the variables involved indicates that the flame temperature, T, seems to be the most suitable trial variable. Once we guess a final T, we may calculate X directly from equation (c) and K_a from equation (b). Using the X just calculated, we may calculate K_a from equation (a). When the two K_a's agree, we are done.

We begin the calculation by estimating the temperature that would be obtained from complete combustion. In this case a first approximation of $T = 3500$°K is obtained, and we use this value to begin our procedure. Rearrangement of equation (c) to solve for X in terms of T gives

$$X\left[194 - 2.77\left(\frac{T}{298} - 1\right) - 0.130\right]\left[\left(\frac{T}{298}\right)^2 - 1\right]$$
$$= \left(\frac{T}{298} - 1\right)\left(10.02 + 0.213\left[\frac{T}{298} + 1\right]\right)$$

or

$$X = \frac{[(T/298) - 1]\{10.02 + 0.213[(T/298) + 1]\}}{194 - [(T/298) - 1]\{2.77 + 0.130[(T/298) + 1]\}}$$

where $T =$ °K. Substitution of $T = 3500$°K then gives

$$X = \frac{(10.75)(10.02 + 2.72)}{194 - (10.75)(2.77 + 1.66)} = \frac{137}{156.5} = 0.875$$

Evaluation of $(K_a)_T$ from T yields

$$\log_{10}(K_a)_T = 40 - 43.0\left(1 - \frac{298}{3500}\right) - 0.602 \ln \frac{3500}{298} + 9.4 \times 10^5(3500 - 298)$$
$$= -0.60$$

or

$$K_a = 0.250$$

Calculation of K_a from equation (a) then yields

$$K_a = \frac{(0.875)(3.0 - 0.875)^{1/2}}{(0.125)^{1.5}} = \frac{(0.875)(1.456)}{0.044} = 29.1$$

Comparison of the K_a's shows that T is too high and the calculations must be continued; we summarize the remaining calculations in Table S11-10b. Since it would be difficult to predict a better guess of T from a comparison of the two K_a's, we shall simply guess a new T that is in the right direction from the old T and use the results of the second and first calculations to suggest a reasonable third guess. This will be continued until the K_a's agree to any desired degree of accuracy. The final temperature will be approximately 2725°K, and flame will consume 62 per cent of the H_2.

TABLE S11-10b

T	X	$(K_a)_T$	$(K_a)_X$
3500	0.875	0.250	29.1
2900	0.712	1.43	12.9
2700	0.615	4.46	4.00
2730	0.620	4.00	4.10

Before leaving the area of adiabatic reactions—in particular combustion reactions—it is wise to note some of the simplifications inherent in the analysis of Sample Problem 11-10 and some of the difficulties that might be expected in a more realistic problem of this nature. We have neglected all reactions other than the simple combination of H_2 and O_2 to form H_2O. However, at the temperatures attainable with this reaction, the dissociation of molecular hydrogen and oxygen to atomic hydrogen and oxygen (or in some cases even more complex side reactions) may be quite significant and can exert a controlling influence on the equilibrium behavior. Calculation of these effects will be considerably more complex than shown in Sample Problem 11-10 and almost without exception would require use of a digital computer to solve the sets of equations that are formed. In addition to the effects of side reactions many problems are complicated by the presence of more than one primany reactant. For example, automobile fuel is a mixture of as many as 20 or 30 separate chemical species, all of which have different equilibrium behavior. Since at true equilibrium each compound must satisfy its own equilibrium equations, we may have as many as 20 or 30 equations to satisfy simultaneously—a problem capable of challenging even the biggest and fastest of today's computing machines. (As may be expected, problems of this nature are not usually solved in an exact fashion. Instead, approximate solutions based on empirical laws and past experience are usually used for most engineering applications.)

Finally, no reaction is ever perfectly adiabatic—in particular, an open flame may not even be close. Conduction and convection usually carry a small portion of the flame's energy away. However, with ultra-high-temperature flames, radiation will remove a great portion of the energy from the flame, thereby invalidating the assumption of an adiabatic reaction. Fortunately, the effects of heat loss from the reaction system may be accounted for (when the heat loss can be determined) simply by including a heat-loss term in the overall energy balance.

11.8 Equilibrium with Competing Reactions

We have written the stoichiometric equation for a single reaction as

$$\sum_{i=1}^{C} \alpha_i A_i = 0 \tag{11-91}$$

When we have several chemical reactions occurring simultaneously, we must write a separate equation for each reaction. These separate reactions may be written in an extremely compact form:

$$\sum_{i=1}^{C} \alpha_{ji} A_i = 0 \qquad j = 1, 2, \cdots, N \tag{11-92}$$

where j = reaction index

A_i = ith species in reacting mixture

α_{ji} = stoichiometric coefficient of the ith species in the jth reaction

N = number of reactions under consideration

Let us consider, for example, the series of reactions among C, O_2, CO, and CO_2:

$$\begin{aligned} -C - \tfrac{1}{2}O_2 + CO \qquad\qquad &= 0 \qquad j = 1 \\ -\tfrac{1}{2}O_2 - CO + CO_2 &= 0 \qquad j = 2 \\ -C \qquad + 2CO - CO_2 &= 0 \qquad j = 3 \end{aligned} \tag{11-93}$$

The coefficient matrix for this series of reactions is then given by the values in Table 11-2.

TABLE 11-2

i	1	2	3	4
j A_i	C	O_2	CO	CO_2
1	-1	$-\frac{1}{2}$	1	0
2	0	$-\frac{1}{2}$	-1	1
3	-1	0	2	-1

Examination of the series of reactions shows that the third reaction is simply the first minus the second. Thus the third reaction yields no useful information that was not contained in reactions (1) and (2). Clearly reactions (1) and (2) are sufficient to completely describe this reacting system. We shall call this minimum set of reactions which fully describe a reacting system a set of "primary reactions." [We could also have chosen reactions (1) and (3), or (2) and (3), for the primary reactions with equal validity, because any two equations will give the third.] Thus we see that if a set of reactions is linearly dependent (that is, each can be expressed as a linear combination of the others), some reactions are redundant and may be excluded from consideration. The minimum number of primary reactions is equal to the rank of the matrix α_{ij}.

We may determine both the minimum number of primary reactions and one set of primary reactions by means of the Gauss elimination algorithm, as indicated below. (For a proof of the validity of this procedure, the reader is referred to *Mathematical Methods in Chemical Engineering*.[4]

Express the stoichiometric coefficients as a matrix:

$$\begin{pmatrix} \alpha_{11} & \alpha_{12} & \alpha_{13} & \cdots & \alpha_{1C} \\ \alpha_{21} & \alpha_{22} & \alpha_{23} & \cdots & \alpha_{2C} \\ \cdot & & & & \\ \cdot & & & & \\ \cdot & & & & \\ \alpha_{N1} & \alpha_{N2} & \alpha_{N3} & \cdots & \alpha_{NC} \end{pmatrix}$$

Divide the first row by α_{11} to give

$$\begin{pmatrix} 1 & \alpha_{12}/\alpha_{11} & \alpha_{13}/\alpha_{11} & \cdots & \alpha_{1C}/\alpha_{11} \\ \alpha_{21} & \alpha_{22} & \alpha_{23} & \cdots & \alpha_{2C} \\ \cdot & & & & \\ \cdot & & & & \\ \cdot & & & & \\ \alpha_{N1} & \alpha_{N2} & \alpha_{N3} & \cdots & \alpha_{NC} \end{pmatrix}$$

Now subtract α_{21}, times the first row from the second, α_{31} times the first row from the third row, and so on, to introduce all zeros in the first column except in the (1,1) position:

$$\begin{pmatrix} 1 & \alpha_{12}/\alpha_{11} & \alpha_{13}/\alpha_{11} & \cdots & \alpha_{1C}/\alpha_{11} \\ 0 & b_{22} & b_{23} & \cdots & b_{2C} \\ \cdot & & & & \\ \cdot & & & & \\ \cdot & & & & \\ 0 & b_{N2} & b_{N3} & \cdots & b_{NC} \end{pmatrix}$$

where

$$b_{ji} = \alpha_{ji} - \alpha_{j1}\frac{\alpha_{1i}}{\alpha_{11}} \qquad \text{for } 2 \leq i \leq C,\ 2 \leq j \leq N$$

If there is not a zero in the (2, 2), proceed as before. Divide the second row by b_{22}, and then introduce zeros into all the elements of the second column beneath the (2, 2) position. Proceed in this fashion until the last row or column that has an element along the main diagonal (where $i = j$) is encountered. If at any point in the calculation a diagonal element of zero is encountered, switch the column whose diagonal element is zero with the next column across. Continue switching columns (any column that has not had a 1 introduced in its

[4] N. R. Amundson, *Mathematical Methods in Chemical Engineering*, Prentice-Hall, Inc., Englewood Cliffs, N.J., 1966.

diagonal element may be used in the switching) until a nonzero number is placed in the diagonal element under consideration. If no nonzero number can be introduced, proceed to the next row and attempt to introduce a nonzero number in its diagonal element if one is not there already. Continue moving down the diagonal until a nonzero diagonal element is located. Then proceed as before. Divide the row by the diagonal element, and introduce zeros into the elements below the diagonal element where the 1 has just been placed. When the last diagonal element has been completed, the matrix should contain zeros in all elements below the diagonal and a set of 1's and 0's along the diagonal. The number of 1's along the diagonal is the number of independent reactions; the rows that contain the 1's represent a set of primary reactions. (This is by no means the only set of primary reactions; we may obtain other equally valid sets of primary reactions by renumbering the reactions and starting over again.)

SAMPLE PROBLEM 11-11. During the catalytic oxidation of ammonia to nitric oxide the following reactions are believed to occur:

$$4NH_3 + 5O_2 = 4NO + 6H_2O \tag{a}$$

$$4NH_3 + 3O_2 = 2N_2 + 6H_2O \tag{b}$$

$$4NH_3 + 6NO = 5N_2 + 6H_2O \tag{c}$$

$$2NO + O_2 = 2NO_2 \tag{d}$$

$$2NO = N_2 + O_2 \tag{e}$$

$$N_2 + 2O_2 = 2NO_2 \tag{f}$$

Determine the minimum number of chemical reactions necessary to completely describe this system. Determine one set of independent reactions.

Solution: Let us number the molecular species as follows: $A_1 = NH_3$, $A_2 = O_2$, $A_3 = NO$, $A_4 = H_2O$, $A_5 = N_2$, and $A_6 = NO_2$. The coefficient matrix is then

$$\begin{pmatrix} -4 & -5 & 4 & 6 & 0 & 0 \\ -4 & -3 & 0 & 6 & 2 & 0 \\ -4 & 0 & -6 & 6 & 5 & 0 \\ 0 & -1 & -2 & 0 & 0 & 2 \\ 0 & 1 & -2 & 0 & 1 & 0 \\ 0 & -2 & 0 & 0 & -1 & 2 \end{pmatrix}$$

Divide the first row by -4 and zero the remaining elements in the first column to give the first transformed matrix:

$$\begin{pmatrix} 1 & \frac{5}{4} & -1 & -1\frac{1}{2} & 0 & 0 \\ 0 & 2 & -4 & 0 & 2 & 0 \\ 0 & 5 & -10 & 0 & 5 & 0 \\ 0 & -1 & -2 & 0 & 0 & 2 \\ 0 & 1 & -2 & 0 & 1 & 0 \\ 0 & -2 & 0 & 0 & -1 & 2 \end{pmatrix}$$

Since we have a nonzero element in the second diagonal element, divide the second row by its diagonal element and introduce zeros into the remaining subdiagonal elements of the second column:

$$\begin{pmatrix} 1 & \frac{5}{4} & -1 & -1\frac{1}{2} & 0 & 0 \\ 0 & 1 & -2 & 0 & 1 & 0 \\ 0 & 0 & 0 & 0 & 0 & 0 \\ 0 & 0 & -4 & 0 & 1 & 2 \\ 0 & 0 & 0 & 0 & 0 & 0 \\ 0 & 0 & -4 & 0 & 1 & 2 \end{pmatrix}$$

We now have a zero in the third diagonal element. Since all the other elements across the third row are also zero, we can get no nonzero element into the third diagonal element. Thus we proceed to the fourth diagonal element. This also has a zero, but switching the fourth and third (or fifth) columns gives a nonzero diagonal element:

$$\begin{pmatrix} 1 & \frac{5}{4} & -1\frac{1}{2} & -1 & 0 & 0 \\ 0 & 1 & 0 & -2 & 1 & 0 \\ 0 & 0 & 0 & 0 & 0 & 0 \\ 0 & 0 & 0 & -4 & 1 & 2 \\ 0 & 0 & 0 & 0 & 0 & 0 \\ 0 & 0 & 0 & -4 & 1 & 2 \end{pmatrix}$$

Now divide the fourth row by -4 and zero the subdiagonal elements to get

$$\begin{pmatrix} 1 & \frac{5}{4} & -1\frac{1}{2} & -1 & 0 & 0 \\ 0 & 1 & 0 & -2 & 1 & 0 \\ 0 & 0 & 0 & 0 & 0 & 0 \\ 0 & 0 & 0 & 1 & -\frac{1}{4} & -\frac{1}{2} \\ 0 & 0 & 0 & 0 & 0 & 0 \\ 0 & 0 & 0 & 0 & 0 & 0 \end{pmatrix}$$

Examination of this matrix shows that all subdiagonal elements are zero and that the diagonal elements are all either zero or unity. Thus the elimination is complete. Since the diagonal has three nonzero elements, we need three primary reactions. One such set will be reactions 1, 2, and 4—the rows which contain the nonzero elements along the diagonal:

$$\begin{aligned} 4NH_3 + 5O_2 &= 4NO + 6H_2O \qquad & j = 1 \\ 4NH_3 + 3O_2 &= 2N_2 + 6H_2O \qquad & j = 2 \\ 2NO + O_2 &= 2NO_2 \qquad & j = 4 \end{aligned}$$

Having established a set of primary reactions, we now turn our attention to the question of determining the equilibrium compositions of a mixture in which several reactions are occurring simultaneously. We shall discuss two separate, but equivalent, techniques for determining the equilibrium compositions.

If the reacting system consists of a small number of reactions, we may describe the equilibrium compositions of the whole system in terms of the primary reactions. As in the case of single reactions, the equilibrium corresponds to the minimum in the Gibbs free energy of the reacting system. The outlet Gibbs free energy of the system is expressed as

$$G_{out} = \sum_{i=1}^{C} (n_i\bar{G}_i)_{out} \tag{11-94}$$

The minimum free energy is obtained by setting the differential of G_{out} to zero:

$$(dG)_{out} = 0 = \sum_{i=1}^{C} (\bar{G}_i)_{out}(dn_i)_{out} + \sum_{i=1}^{C} (n_i)_{out}(d\bar{G}_i)_{out} \tag{11-95}$$

The second sum vanishes through the Gibbs–Duhem equation, so the equilibrium condition reduces to

$$\sum_{i=1}^{C} (\bar{G}_i)_{out}(dn_i)_{out} = 0 \tag{11-95a}$$

Let us now assume that each reaction proceeds independently to an extent X_j. That is, X_j moles of reaction j occur, independent of what occurs in the other reactions. The amount of species i *produced* by chemical reaction j is then simply the stoichiometric coefficient, α_{ji}, multiplied by the extent of reaction, X_j. The total amount of species i formed by all the chemical reaction is obtained, then, by summing the amounts formed by each of the individual reactions. Thus

$$(n_i)_{out} = (n_i)_{in} + \sum_{j=1}^{N} X_j\alpha_{ji} \tag{11-96}$$

The equilibrium condition [equation (11-95a)] then becomes

$$0 = \sum_{i=1}^{C} (\bar{G}_i)_{out} \sum_{j=1}^{N} \alpha_{ji}\, dX_j \tag{11-97}$$

Since $(\bar{G}_i)_{out}$ is independent of j, the $(\bar{G}_i)_{out}$ may be brought inside the j summation and the order of summation reversed to give

$$0 = \sum_{j=1}^{N} \left[\sum_{i=1}^{C} \alpha_{ji}(\bar{G}_i)_{out}\right] dX_j \tag{11-98}$$

The summation must vanish for any dX_j at the minimum in G_{out}, so the coefficients of each dX_j must vanish independently:

$$\sum_{i=1}^{C} \alpha_{ji}(\bar{G}_i)_{out} = 0 \qquad \text{for } 1 \leq j \leq N \tag{11-99}$$

As in the case of single reactions, we refer the outlet partial molar Gibbs free energies to the reference-state conditions and introduce the activities. After some rearrangement, equation (11-99) then reduces to

$$RT\ln(K_a^j) + \Delta\mathscr{G}^{\circ j} = 0 \qquad \text{for } 1 \leq j \leq N \tag{11-100}$$

where

$$\Delta\mathscr{G}^{\circ j} = \sum_{i=1}^{C} \alpha_{ji}\underline{G}_i^\circ$$

$$K_a^j = \prod_{i=1}^{C} (a_i)^{\alpha_{ji}}$$

That is, the equilibrium constant must be satisfied for each separate reaction.

(It is left as an exercise to prove that the equilibrium constants for all reactions will be satisfied if the constants for any primary set of reactions are satisfied.)

If a large number of simultaneous reactions must be considered, it is frequently not computationally attractive to attempt solution of the set of nonlinear simultaneous equilibrium equations for the equilibrium compositions. In these cases we may use one of the standard methods of optimization to determine the set of outlet mole numbers that corresponds to the minimum outlet Gibbs free energy of the reacting mixture (subject to the mass-balance constraints). Since no equilibrium constants are required, direct minimization of the Gibbs free energy does not require knowledge, or specification, of the system of reaction equations—an obvious advantage for large systems, where the number of possible reactions can be staggering. As we shall see, the first technique is best suited for hand calculations on systems with only two, or at most three, competing reactions. The second technique, on the other hand, is best suited for computer calculations involving three or more reactions. We shall now demonstrate, by means of sample problems, calculations of both types.

SAMPLE PROBLEM 11-12. An electric generating station burns coke (essentially pure carbon) in air to provide heat for the main boilers. The gases leaving the combustion chamber are at 1150°F and may be assumed to be in complete equilibrium at a pressure of 1.0 atm. Determine the equilibrium composition of the gases leaving the burner.

The following reactions are known to occur:

$$C + \tfrac{1}{2}O_2 = CO \tag{1}$$

$$CO + \tfrac{1}{2}O_2 = CO_2 \tag{2}$$

$$CO_2 + C = 2CO \tag{3}$$

$$(K_a)^1 = 2.2 \times 10^{11}$$

$$(K_a)^2 = 1.0 \times 10^{12}$$

$$(K_a)^3 = 0.22$$

at $T = 1150°F$. Standard states are pure gaseous O_2, CO, and CO_2, and pure solid carbon at 1 atm pressure. (As we will show in Chapter 12, as long as any unreacted carbon remains, it is always in its standard state. Thus the activity of carbon is equal to unity independent of the amount of carbon left!)

Solution: As we have shown in our earlier discussion, these three reactions are redundant, and we only need consider two of them as the primary reactions. Any two are acceptable. We arbitrarily choose the first two:

$$C + \tfrac{1}{2}O_2 = CO \tag{1}$$

$$CO + \tfrac{1}{2}O_2 = CO_2 \tag{2}$$

The equilibrium constants for these two reactions may be written as

$$(K_a)^1 = \frac{a_{CO}}{a_C(a_{O_2})^{1/2}}$$

$$(K_a)^2 = \frac{a_{CO_2}}{a_{CO}(a_{O_2})^{1/2}}$$

Since both the reaction system and the standard states are at 1 atm pressure, $f_i = f_i^\circ$ for the gaseous components. (As mentioned earlier, $a_C = 1.0$ independent of the mass of carbon.) Assuming that the gaseous reactants form ideal solutions in each other, $\gamma_i = 1.0$ and the activities of the gaseous components are expressed as

$$a_i = y_i$$

Therefore, the equilibrium expression shown above reduces to

$$(K_a)^1 = \frac{y_{CO}}{(y_{O_2})^{1/2}} = 2.2 \times 10^{11}$$

$$(K_a)^2 = \frac{y_{CO_2}}{y_{CO}(y_{O_2})^{1/2}} = 1.0 \times 10^{12}$$

Let us now express the composition of the reacting species in terms of the conversion of carbon to CO and CO_2. The basis of all calculations will be 5 mol of inlet air, containing 1 mol of O_2 and 4 mol of N_2. Assuming that the extents of reaction for the two reactions are X_1 and X_2, we may summarize the inlet and outlet conditions as shown in Table S11-12.

TABLE S11-12

	O_2	N_2	CO	CO_2	Total
Inlet amount	1	4	0.0	0.0	5
Outlet amount	$1 - (0.5X_1 + X_2)$	4	X_1	X_2	$5 + 0.5X_1$
y_i (outlet)	$\frac{1 - (X_2 + 0.5X_1)}{5 + 0.5X_1}$		$\frac{X_1}{5 + 0.5X_2}$	$\frac{X_2}{5 + 0.5X_1}$	

We may now substitute the values of y_i into the equilibrium expressions to obtain two equations in the two unknowns, X_1 and X_2:

$$2.2 \times 10^{11} = \frac{X_1/(5.0 + 0.5X_1)}{\left[\frac{1 - (X_2 + 0.5X_1)}{5 + 0.5X_1}\right]^{1/2}}$$

$$1.0 \times 10^{12} = \frac{X_2/(5.0 + 0.5X_1)}{\frac{X_1}{5.0 + 0.5X_1}\left[\frac{1 - X_2 - 0.5X_1}{5.0 + 0.5X_1}\right]^{1/2}}$$

The solution to these two equations is then sufficient to completely satisfy all equilibrium conditions. We may simplify the two equilibrium expressions to the following form:

$$2.2 \times 10^{11} = \frac{X_1}{\sqrt{[1 - (X_2 + 0.5X_1)](5.0 + 0.5X_1)}} \tag{d}$$

$$1.0 \times 10^{12} = \frac{X_2}{X_1}\sqrt{\frac{5.0 + 0.5X_1}{1 - X_2 - 0.5X_1}} \tag{e}$$

Multiplication of expressions (d) and (e) yields

$$2.2 \times 10^{23} = \frac{X_2}{1 - X_2 - 0.5X_1} \tag{f}$$

or

$$2.2 \times 10^{23} - X_2(1 + 2.2 \times 10^{23}) - 1.1 \times 10^{23}\, X_1 = 0$$

Neglecting 1.0 in comparison to 2.2×10^{23} and solving for X_2 in terms of X_1 then gives

$$X_2 = 1.0 - 0.5X_1$$

Next we eliminate $1 - X_2 - 0.5X_1$ between equations (d) and (e) by division to yield

$$0.22 = \frac{X_1^2}{X_2(5.0 + 0.5X_1)}$$

or, substituting the expression for X_2,

$$0.22 = \frac{X_1^2}{(1.0 - 0.5X_1)(5.0 + 0.5X_1)}$$

which now can be rearranged to

$$(0.22)(5.0 - 2.5X_1 - 0.25X_1^2) = X_1^2$$

or

$$1.055X_1^2 + 0.55X_1 - 1.10 = 0$$

Solution using the quadratic formula then yields

$$X_1 = \frac{-0.55 \pm \sqrt{(0.55)^2 + (4)(1.055)(1.10)}}{2(1.055)}$$
$$= -0.261 \pm 1.060$$

Neglecting the negative root this gives

$$X_1 = 0.799$$
$$X_2 = 1 - 0.5X_1 = 0.600$$

We may estimate the O_2 composition, $[1 - (X_2 + 0.5X_1)]/(5 + 0.5X_1)$, by rearranging expression (f):

$$1 - (X_2 + 0.5X_1) = \frac{X_2}{2.2 \times 10^{23}} = \frac{0.600}{2.2 \times 10^{23}} = 2.7 \times 10^{-24}$$

But

$$y_{O_2} = \frac{1 - X_2 - 0.5X_1}{5 + 0.5X_1} = \frac{2.7 \times 10^{-24}}{5 + (0.5)(0.799)} = 5.0 \times 10^{-25}$$

$$y_{CO} = \frac{X_1}{5.0 + 0.5X_1} = 0.148$$

$$y_{CO_2} = \frac{X_2}{5.0 + 0.5X_1} = 0.111$$

Thus we can see that the oxygen has reacted just about completely and is more or less evenly split between the CO and CO_2 that formed. If the temperature of combustion is changed, we find that the ratio of CO/CO_2 will change, because $(K_a)^1$ and $(K_a)^2$ are affected quite differently by changes in temperature. Therefore, it is possible to produce almost any desired CO/CO_2 ratio by changing the temperature. However, even at very high temperatures the greatest portion of oxygen will be consumed, since $(K_a)^1$ does not decrease rapidly with T. At 10,000°F $(K_a)^1$ is still greater than 10^4 and assures us that little oxygen can coexist with carbon at any reasonable temperature.

SAMPLE PROBLEM 11-13. Ethane is to be steam cracked to form hydrogen, which will be recovered at a later stage in the process. Ethane and steam are mixed over a cracking catalyst at 1000°K and 1 atm. The feed contains 4 moles of H_2O per mole of C_2H_6. Assuming that no carbon is deposited, determine the equilibrium composition of the effluent mixture. Assume that only the following compounds will be present in the outlet stream: CH_4, C_2H_4, C_2H_2, CO_2, CO, O_2, H_2, H_2O, and C_2H_6. At 1000°K the free energies of formation of the various compounds are as given in Table S11-13a, where the pressure is 1 atm.

TABLE S11-13a

	$(\Delta \underline{G}_f^\circ)$, kcal/g-mol		$(\Delta \underline{G}_f^\circ)$, kcal/g-mol
O_2	0.0	CH_4	4.61
H_2	0.0	C_2H_6	26.13
H_2O	−46.030	C_2H_4	28.249
CO	−47.942	C_2H_2	40.604
CO_2	−94.610		

Solution: Because of the extremely complex nature of this process, we shall not attempt to describe all the various competing reactions, but will use a scheme first described by White, Johnson, and Dantzig[5] to determine the mixture composition which corresponds to the minimum total Gibbs free energy of the reacting mixture.

The total Gibbs free energy of a system of C compounds containing n_i moles of component i is written as

$$G = \sum_{i=1}^{C} n_i \bar{G}_i = \sum_{i=1}^{C} n_i[(\bar{G}_i - \underline{G}_i^\circ) + \underline{G}_i^\circ]$$

But

$$\bar{G}_i - \underline{G}_i^\circ = RT \ln \frac{\bar{f}_i}{f_i^\circ}$$

so the total Gibbs free energy is given by

$$G = \sum_{i=1}^{C} n_i\left(\underline{G}_i^\circ + RT \ln \frac{\bar{f}_i}{f_i^\circ}\right)$$

For the problem at hand, all components are gases. The pressure of the reaction is also the standard-state pressure (1 atm). Thus the activities, $\bar{f}_i/f_i^\circ$, of the reacting components are given by

$$\frac{\bar{f}_i}{f_i^\circ} = y_i \gamma_i = \frac{n_i}{n}\gamma_i$$

where n is the total number of moles in the reacting mixture, including any nonreacting species.

At the low pressures and high temperatures involved, it is quite reasonable to assume ideal gas-phase solutions, $\gamma_i = 1$, so

$$\frac{\bar{f}_i}{f_i^\circ} = \frac{n_i}{n}$$

Thus the total Gibbs free energy becomes

$$G = \sum_{i=1}^{C} n_i\left(\underline{G}_i^\circ + RT \ln \frac{n_i}{n}\right)$$

[5] W. B. White, S. M. Johnson, and G. B. Dantzig, *J. Chem. Phys.*, *28*, no. 5, 751 (1958).

Since RT is a constant, independent of composition, the minimum in G occurs at the same compositions as the minimum in G/RT:

$$\frac{G}{RT} = \sum_{i=1}^{C} n_i\left(\frac{G_i^\circ}{RT} + \ln \frac{n_i}{n}\right)$$

Our problem is reduced to finding the component amounts, n_i, which correspond to the minimum in G/RT. The problem is, however, complicated by the fact that not every combination of n_i is acceptable. The n_i are constrained by the conservation of atomic species. Thus the n_i must satisfy the following relations:

$$\sum_{i=1}^{C} a_{ji}n_i = b_j \qquad \text{for } 1 \leq j \leq M$$

where a_{ji} are the number of gram-atoms of (atomic) element j in a mole of molecule i, M is the number of elements present in the reacting mixture (three in our case: C, H, O), and b_j is the total number of gram-atoms of element j in the reacting mixture (determined by the ethane and steam fed in this case). We also assume that if the Gauss–Jordan elimination procedure were performed on the a_{ji} matrix, we would obtain M 1's along the main diagonal. (For the rare case when less than M 1's are obtained, the reader is referred to Amundson's book for a discussion of what must be done with the coefficient matrix before proceeding.)

When inert materials are present, the sum indicated above is extended to include all components present. Each inert material is treated as a new (atomic) element which does not appear in any of the other chemical species. For example, if we had nitrogen in our problem, we would call nitrogen chemical species $i = 10$, and element $j = 4$. Since the new element does not appear in any other reacting species, $a_{4j} = 0$ except for $i = 10$, where $a_{4,10} = 1.0$. Similarly, the inert chemical species (nitrogen) contains no other element except itself, so $a_{j,10} = 0$, except $j = 4$ where $a_{4,10} = 1.0$.

We shall now develop the *method of steepest descent*, with which we shall determine the minimum in $F = G/RT$. The method is iterative in that it uses one approximate solution for the minimum to find, hopefully, a better approximation. The iteration is continued until some desired degree of convergence is attained.

We begin by assuming that we have some approximate solution (after k iterations) for the mole numbers $n_1^k, n_2^k, \ldots, n_C^k$ which satisfies the conservation-of-atoms equations. The total Gibbs-free-energy function G^k/RT is expressed as

$$\frac{G^k}{RT} = F^k = \sum_{i=1}^{C} n_i^k\left(\frac{G_i^\circ}{RT} + \ln \frac{n_i^k}{n^k}\right)$$

Now expand F in a Taylor's-series expansion around $n_1^k, n_2^k, \ldots, n_C^k$:

$$F^{k+1} = F^k + \sum_{i=1}^{C} (n_i^{k+1} - n_i^k)\frac{\partial F^k}{\partial n_i} + \sum_{i=1}^{C}\sum_{l=1}^{C} \frac{1}{2}(n_i^{k+1} - n_i^k)(n_l^{k+1} - n_l^k)\frac{\partial^2 F^k}{\partial n_i \, \partial n_l}$$
$$+ \text{ higher-order terms}$$

Neglect the higher-order terms and perform the required differentiation to obtain, after some simplication,

$$\frac{\partial F^k}{\partial n_i} = \frac{G_i^\circ}{RT} + \ln \frac{n_i}{n}$$

$$\frac{\partial^2 F^k}{\partial n_i^2} = \frac{1}{n_i} - \frac{1}{n}$$

$$\frac{\partial^2 F^k}{\partial n_i \, \partial n_l} = -\frac{1}{n}$$

so the approximation for F^{k+1} becomes

$$F^{k+1} = F^k + \sum_{i=1}^{C} \left(\frac{G_i^\circ}{RT} + \ln \frac{n_i^k}{n^k}\right)(n_i^{k+1} - n_i^k) + \frac{1}{2} \sum_{i=1}^{C} n_i^k \left(\frac{n_i^{k+1} - n_i^k}{n_i^k} - \frac{n^{k+1} - n^k}{n^k}\right)^2$$

We shall now attempt to find the values of n_i^{k+1} which minimize F^{k+1} but still satisfy the mass-balance equations. To that end, let

$$\phi^{k+1} = F^{k+1} + \sum_{j=1}^{M} \pi_j\left(-\sum_{i=1}^{C} a_{ji} n_i^{k+1} + b_j\right)$$

where the π_j are termed the *Lagrangian multipliers*. Notice that as long as the mass-balance equations are satisfied, $\phi^{k+1} = F^{k+1}$. Since we shall require that the mass-balance equations be satisfied, $\phi = F$, and we may reasonably minimize ϕ rather than F. From our calculus we know that the minimum (or maximum) in a function occurs when all the partial derivatives of the function vanish. Thus the minimum in ϕ^{k+1} occurs whenever

$$\frac{\partial \phi^{k+1}}{\partial n_i^{k+1}} = 0 \qquad \text{for } 1 \leq i \leq C$$

(We know this must be a minimum since G increases monotonically as we move away from the equilibrium point.) Differentiating ϕ^{k+1} gives

$$\frac{\partial \phi^{k+1}}{\partial n_i^{k+1}} = \left(\frac{G_i^\circ}{RT} + \ln \frac{n_i^k}{n^k}\right) + \left(\frac{n_i^{k+1}}{n_i^k} - \frac{n^{k+1}}{n^k}\right) - \sum_{j=1}^{M} \pi_j a_{ji} = 0$$

which may be written for all i, yielding C equations in $C + M + 1$ (the C values of n_i^{k+1}, M values of π_j, and n^{k+1}) unknowns. The other $M + 1$ equations are the M mass-balance equations,

$$\sum_{i=1}^{C} a_{ji} n_i^{k+1} = b_j \qquad j = 1, 2, \ldots, M$$

and the definition of n^{k+1},

$$n^{k+1} = \sum_{i=1}^{C} n_i^{k+1}$$

We may now in principle solve for all the values at the end of the iteration. It is possible, however, to reduce the size of the system of equations that must be solved, if a number of the variables are hand eliminated as shown below. Solve for n_i^{k+1} in the $\partial \phi / \partial n_i = 0$ equations:

$$n_i^{k+1} = n^{k+1} \frac{n_i^k}{n^k} + n_i^k \left(\sum_{j=1}^{M} \pi_j a_{ji}\right) - n_i^k \left(\frac{G_i^\circ}{RT} + \ln \frac{n_i^k}{n^k}\right)$$

Now sum over all i:

$$n^{k+1} = n^{k+1} \frac{n^k}{n^k} + \sum_{i=1}^{C} n_i^k \sum_{j=1}^{M} \pi_j a_{ji} - \sum_{i=1}^{C} n_i^k \left(\frac{G_i^\circ}{RT} + \ln \frac{n_i^k}{n^k}\right)$$

which reduces to

$$\sum_{i=1}^{C} n_i^k \sum_{j=1}^{M} \pi_j a_{ji} = \sum_{i=1}^{C} n_i^k \left(\frac{G_i^\circ}{RT} + \ln \frac{n_i^k}{n^k}\right)$$

In the left-hand term reverse the order of summation to give

$$\sum_{i=1}^{C} n_i^k \sum_{j=1}^{M} \pi_j a_{ji} = \sum_{j=1}^{M} \pi_j \sum_{i=1}^{C} a_{ji} n_i^k = \sum_{j=1}^{M} \pi_j b_j$$

so we get

$$\sum_{j=1}^{M} \pi_j b_j = \sum_{i=1}^{C} n_i^k \left(\frac{G_i^\circ}{RT} + \ln \frac{n_i^k}{n^k} \right)$$

Substitute the equations for n_i^{k+1} into the mass-balance equations to give

$$\sum_{i=1}^{C} a_{ji} \left[n^{k+1} \frac{n_i^k}{n^k} + n_i^k \sum_{l=1}^{M} \pi_l a_{li} - n_i^k \left(\frac{G_i^\circ}{RT} + \ln \frac{n_i^k}{n^k} \right) \right]_{j=1,2\cdots M}$$
$$= b_j$$

But since n_i^k satisfies the mass-balance equations, $\sum_{i=1}^{C} a_{ji}\, n_i^k = b_j$, the first term reduces to $b_j(n^{k+1}/n^k)$, and the whole expression becomes

$$\sum_{i=1}^{C} \left(a_{ji} \sum_{l=1}^{M} \pi_l a_{li} \right) n_i^k + b_j \left(\frac{n^{k+1}}{n^k} - 1 \right) = \sum_{i=1}^{C} {}_{j=1,2\cdots M}\; a_{ji} n_i^k \left(\frac{G_i^\circ}{RT} + \ln \frac{n_i^k}{n^k} \right)$$

The a_{ji} in the first term of the left-hand side may be brought into the second summation and the order of summations reversed to finally yield

$$\sum_{l=1}^{M} \pi_l \sum_{i=1}^{C} a_{ji} a_{li} n_i^k + b_j \left(\frac{n^{k+1}}{n^k} - 1 \right) = \sum_{i=1}^{C} {}_{j=1,2\cdots M}\; a_{ji} n_i^k \left(\frac{G_i^\circ}{RT} + \ln \frac{n_i^k}{n^k} \right)$$

which is a system of M equations (one for each j) in $M + 1$ unknowns (M values of π_l and n^{k+1}). The last equation is simply

$$\sum_{l=1}^{M} \pi_l b_l = \sum_{i=1}^{C} n_i^k \left(\frac{G_i^\circ}{RT} + \ln \frac{n_i^k}{n^k} \right)$$

We now write this system of equations in standard form. If we let $r_{jl} = \sum_{i=1}^{C} a_{ji} a_{li} n_i^k$ and $u = [(n^{k+1}/n^k) - 1]$ we get

$$\pi_1 r_{11} + \pi_2 r_{12} + \cdots + \pi_M r_{1M} + b_1 u = \sum_{i=1}^{C} a_{1i} n_i^k \left(\frac{G_i^\circ}{RT} + \ln \frac{n_i^k}{n^k} \right)$$

$$\pi_1 r_{21} + \pi_2 r_{22} + \cdots + \pi_M r_{2M} + b_2 u = \sum_{i=1}^{C} a_{2i} n_i^k \left(\frac{G_i^\circ}{RT} + \ln \frac{n_i^k}{n^k} \right)$$

$$\vdots$$

$$\pi_1 r_{M1} + \pi_2 r_{M2} + \cdots + \pi_M r_{MM} + b_M u = \sum_{i=1}^{C} a_{Mi} n_i^k \left(\frac{G_i^\circ}{RT} + \ln \frac{n_i^k}{n^k} \right)$$

$$\pi_1 b_1 + \pi_2 b_2 + \cdots + \pi_M b_M = \sum_{i=1}^{C} n_i^k \left(\frac{G_i^\circ}{RT} + \ln \frac{n_i^k}{n^k} \right)$$

The linear system of $M + 1$ equations in $M + 1$ unknowns is solved for the π_j and u. From these, a new set of n_i^{k+1} are now calculated. These values of n_i^{k+1}, if all positive, could be used as the starting values for the next iteration. However, a small extra step prevents the possibility of negative n_i's and guarantees convergence of the iteration scheme.

Let us consider the computed changes $\Delta_i = n_i^{k+1} - n_i^k$ to be direction numbers which indicate the "best" direction of travel but not necessarily the "best" length of travel. Rather, let us restrict the distance traveled to some fraction λ of the calculated travel. The value of λ will be chosen as the largest value that satisfies the two conditions (1) n_i^{k+1} are positive, and (2) the derivative $d(G/RT)/d\lambda$ does not become positive (that is, the minimum is not passed). The derivative is easily computed, because

$$\frac{d(G/RT)}{d\lambda} = \sum_{i=1}^{C} \Delta_i \left(\frac{G_i^\circ}{RT} + \ln \frac{n_i^k + \lambda \Delta_i}{n^k + \lambda \Delta} \right) \qquad \text{where } \Delta = n^{k+1} - n^k$$

(Prove this last expression as an exercise.)

The procedure for finding the best λ need not be terribly precise, and only a minor part of the computation time should be spent in this phase of the procedure. When the solution is distant, it is more important to complete the next iteration to find an improved direction of travel. When the solution is near, λ will approach unity anyway. Note that the mass-balance constraints are satisfied for all values of λ. (We shall examine the problem of finding λ during the actual programming steps.)

Let us now return to our original problem. The feed composition to the cracker is 1 mol of C_2H_6 for each 4 mol of H_2O. The a_{ji} matrix is written as shown in Table S11-13b. The b_j are then calculated from

$$b_j = \sum_{i=1}^{C} a_{ji} n_i$$

$$b_1 = 2 \text{ atoms of C}$$

$$b_2 = 14 \text{ atoms of H}$$

$$b_3 = 4 \text{ atoms of O}$$

TABLE S11-13b

			1	2	3	4	5	6	7	8	9
j	Atom	Comp	CH_4	C_2H_4	C_2H_2	CO_2	CO	O_2	H_2	H_2O	C_2H_2
1	C		1	2	2	1	1	0	0	0	2
2	H		4	4	2	0	0	0	2	2	6
3	O		0	0	0	2	1	2	0	1	0
		n_i°	0	0	0	0	0	0	0	4	1

Now let us make an initial guess of the equilibrium composition (Table S11-13c). It must of course satisfy the mass-balance equations. (*Note:* Trace amounts of all compounds are specified in the initial guess, to avoid problems with the logarithm terms.)

TABLE S11-13c

i	Component	Initial Guess
1	CH_4	0.001
2	C_2H_4	0.001
3	C_2H_2	0.001
4	CO_2	0.993
5	CO	1.000
6	O_2	0.007
7	H_2	5.992
8	H_2O	1.000
9	C_2H_6	0.001

If at any point in the calculations the amount of any species gets so low that the logarithm term does cause trouble, this component should be dropped from the reaction mixture (except insofar as an initial reactant effects the b_j's) and the calculation repeated without the offending component. The results of this smaller calculation will adequately represent the equilibrium compositions and mole numbers of the species present in significant amounts. We may then *estimate* the amounts of the significant components by making one more iteration, in which we include these components. Also, since we know that the total number of moles or the mole fractions of the individual species in the mixture should not change greatly, let us simplify our problem by assuming that the Lagrangian multipliers will remain constant over this final iteration. (n_i, however, will be allowed to change.) Thus express the total Gibbs free energy of the mixture as

$$\frac{G}{RT} = F = \sum_{i=1}^{C} n_i\left(\frac{G_i^\circ}{RT} + \ln\frac{n_i}{n}\right) + \sum_{j=1}^{M} \pi_j\left(-\sum_{i=1}^{C} a_{ji}n_i + b_j\right)$$

where the summations now include *all* components, even the insignificant ones. Assuming that the π_j are constants, take $(\partial F/\partial n_i)_{n_{j\neq i}}$ and set it equal to zero, because this is the equilibrium condition:

$$0 = -\frac{\partial}{\partial n_i}\left[\sum_{k=1}^{C} n_k\left(\frac{G_k^\circ}{RT} + \ln\frac{n_k}{n}\right) + \sum_{j=1}^{M} \pi_j\left(-\sum_{k=1}^{C} a_{jk}n_k + b_j\right)\right]$$

or

$$\left(\frac{G_i^\circ}{RT} + \ln\frac{n_i}{n}\right) + n_i\frac{\partial}{\partial n_i}\ln n_i + \sum_{k=1}^{C} n_k\left(\frac{-\partial \ln n}{\partial n_i}\right) - \sum_{j=1}^{M} \pi_j a_{ji}$$

which reduces to

$$0 = \frac{G_i^\circ}{RT} + \ln\frac{n_i}{n} - \sum_{j=1}^{M} \pi_j a_{ji}$$

Solving for n_i then gives an approximate value of the equilibrium concentration of the trace components:

$$n_i = n\exp\left(\sum_{j=1}^{M} \pi_j a_{ji} - \frac{G_i^\circ}{RT}\right)$$

A program for computing the equilibrium composition by the method of steepest descent follows.

```
C    ---   PROGRAM TO APPLY THE METHOD OF STEEPEST DESCENT TO REACTION
C    ---   PROBLEMS GOAL IS TO DETERMINE EQUILIBRIUM COMPOSITIONS FOR
C    ---   COMPLEX REACTION ---
           INTEGER C,M,A(3,9)
           REAL N(9),NN(9),LAM,NT,NNT,LAMH,MF(9),LL
           DIMENSION PI(3),B(3),AA(11,11),BB(11),CC(11),GRT(9),DEL(9),RHO(9),
          1DFDL(11),DELH(9),PN(9)
           LOGICAL CHECK
           DELTA=1.0E-3
           ITMAX = 20
C    ---   READ INPUT DATA ---
  10       READ ( 105, 20 ) C,M
  20       FORMAT ( 2I10)
C    ---   READ THE REACTION COEFFICIENT MATRIX A ROW AT A TIME ---
  30       READ (105,40) (( A(J,I), I=1,C), J=1,M)
  40       FORMAT(16I5)
C    ---   READ THE INLET AMOUNTS OF THE REACTANTS ---
  50       READ( 105, 60 ) (NN(I), I=1,C)
```

```
   60     FORMAT(9F4.1)
C  ---    READ G/RT FOR EACH COMPONENT ---
   71     READ ( 105, 72) ( GRT(I), I=1,C)
   72     FORMAT(8(F8.4,1X),F8.4)
C  ---    READ THE INITIAL GUESS OF THE EXIT COMPOSITION ---
   70     READ(105,73) (N(I),I=1,C)
   73     FORMAT(9(F5.3,1X))
          WRITE(108,20) C,M
          WRITE (108, 74)
   74     FORMAT (26H THE COEFFICIENT MATRIX IS)
          DO 75 J=1, M
   75     WRITE(108,40) (A(J,I),I=1,C)
          WRITE(108,77) (NN(I) GRT(I), N(I), I=1,C)
   77     FORMAT (36H THE NN(I), GRT(I), N(I) VECTORS ARE / (3(10X, F10.4))
C  ---    BEGIN CALCULATIONS—CALCULATE THE B(J) VECTOR AND CHECK
C  ---    MASS BALANCE ---
   80     DO 100 J= 1,M
   81     IF(IT.LT.10) IFREQ=1
   85     B(J) = 0
   90     DO 100 I=1,C
  100     B(J) = B(J) + A(J,I)*NN(I)
  110     CHECK = .FALSE.
  120     DO 160 J = 1,M
  130     BCHECK = 0.0
  140     DO 150 I=1,C
  150     BCHECK = BCHECK + A(J,I)*N(I)
  160     IF( ABS( BCHECK- B(J))  .GT. DELTA ) CHECK = .TRUE.
  170     IF( .NOT. CHECK ) GO TO 220
  180     WRITE( 108, 190)
  190     FORMAT( 1H0, 42H BAD INITIAL GUESS B VECTORS DO NOT CHECK )
  200     GO TO 70
C  ---    BEGIN EVALUATION OF THE NEW COMPOSITIONS ---
  220     IT = 1
C  ---    DETERMINE TOTAL MOLE NUMBER---
  225     NT = 0.
  226     DO 227 I=1,C
  227     NT = NT + N(I)
C  ---    LOAD THE COEFFICIENT MATRICES ---
  230     DO 290  J=1, M
  235     CC(J) = 0.
  240     DO 270  I=1,M
  250     AA(J,I) = 0.0
  260     DO 270  L=1,C
  270     AA(J,I) = AA(J,I) + A(J,L)*A(I,L)*N(L)
  279     DO 280 I=1,C
  280     CC(J) = CC(J) + A(J,I)*N(I)*(GRT(I) + ALOG(N(I)/NT))
  290     AA(J,M+1) = B(J)
  310     DO 320  I=1,M
  320     AA( M+1,I) = B(I)
  330     AA(M+1,M+1) = 0.0
```

```
  340   CC(M+1) = 0.0
  350   DO 360 I=1,C
  360   CC(M+1)= CC(M+1) + N(I)*(GRT(I) + ALOG(N(I)/NT))
C  ---  SOLVE FOR THE LAGRANGE MULTIPLIERS AND U ---
  370   CALL LINEQ( AA, CC, BB, M+1, 11, IFLAG)
  380   DO 390  J=1, M
  390   PI(J)=BB(J)
  400   U=BB(M+1)
C  ---  EVALUATE THE NN(I) AND NNT FROM THE SOLUTION ---
  410   NNT = NT*(U+1.0)
  420   DO 460 I=1,C
  430   NN(I) = NNT*N(I)/NT - N(I)*(GRT(I) + ALOG(N(I)/NT))
  440   DO 450  J=1,M
  450   NN(I) = NN(I) + N(I)*PI(J)*A(J,I)
  460   DEL(I) = NN(I) - N(I)
  465   DELN = NNT - NT
C  ---  EVALUATE MAXIMUM LAMBDA WHICH SATISFIES CONDITION 1 ---
  473   LAM = 1.0
  480   DO 520 I=1,C
  490   IF( NN(I) .GT.0.0) GO TO 520
  500   LL=.99999*N(I)/(N(I)-NN(I))
  510   IF(LL.LT.LAM) LAM=LL
  520   CONTINUE
  514   DO 515 I=1,C
  515   RHO(I)=N(I)+DEL(I)*LAM
C  ---  LAM NOW CONTAINS MAXIMUM LAMBDA FOR CONDITION 1 ---
C  ---  NOW FIND LAMBDA FOR SECOND CONDITION- DECREASE LAMBDA IN
C  ---  10 PER CENT DECREMENTS AND CHECK AFTER EACH DECREASE ---
  550   DO 610 L=0,10
  560   LL=LAM*(1.0-L/10.)
  570   DFDL(L+1)=0.0
  580   DO 590 I=1,C
  590   DFDL(L+1)=DFDL(L+1)+DEL(I)*(GRT(I)+ALOG((N(I)+LL*DEL(I))/(NT+LL*
       1DELN)))
  610   CONTINUE
  607   L=1
  600   IF(DFDL(L).LE.0.0) GO TO 650
  599   L=L+1
  598   IF (L.EQ 11) GO TO 640
  597   GO TO 600
C  ---  REVERT TO PREVIOUS ITERATION AND USE DECREASING LAMBDA
C  ---  VALUES UNTIL THE SECOND CONDITION IS SATISFIED---
  640   LLL= LLL+1
  641   DO 642 I=1,C
  642   N(I) = PN(I) - LLL*LAMH*DELH(I)/10.
  643   NT = PNT - LLL*LAMH*DELNH/10.
  644   IF (LLL.NE.11) GO TO 230
  620   WRITE ( 108, 630)
  630   FORMAT( 40H NO LAMBDA FOUND, TRY NEW CALCULATION )
  645   GO TO 10
```

```
C   ---  LAMBDA FOUND, UPDATE N S AND RETURN FOR NEW ITERATION ---
   650   LAM=LAM*(1.-(L-1.)/10.)
   651   LLL = 0
   660   DO 670  I=1,C
   670   N(I) = N(I) + LAM*DEL(I)
   680   NT = NT + LAM*DELN
   690   IT=IT+1
   700   WRITE(108,710) ( IT-1), LAM, ( N(I),I=1,C)
   710   FORMAT ( 1H0, 7H AFTER, I5, 17H ITERATIONS LAM =, F6.3, 17H, THE
       1 NEW N'S ARE/(50X,  E15.8))
C   ---  NOW CHECK FOR CONVERGENCE ---
   720   CHECK = .TRUE.
   730   DO 740  I=1,C
   740   IF (ABS((N(I)-PN(I))/N(I)) .GT. DELTA) CHECK = .FALSE.
   750   IF ( CHECK) GO TO 800
   751   LAMH=LAM
   752   DO 753 I=1,C
   755   PN(I) = N(I)
   753   DELH(I) =DEL(I)
   754   DELNH=DELN
   756   PNT = NT
   760   IF( IT .LT. ITMAX) GO TO 230
   770   WRITE( 108, 780)
   780   FORMAT( 40H  ITERATION FAILED TO CONVERGE BAIL OUT)
   790   GO TO 10
   800   WRITE( 108, 810) (N(I), I=1,C)
   810   FORMAT( 46H  CONVERGENCE OBTAINED, FINAL MOLE NUMBERS ARE / /
       1 ( 50X, E15.8))
   814   DO 815 I = 1, C
   815   MF(I) = N(I)/NT
   816   WRITE(108,817) (MF(I),I=1,C)
   817   FORMAT(1H0,26H FINAL MOLE FRACTIONS ARE // (30X, E15.8))
   818   GO TO 10
   820   CALL EXIT
   830   END
```

```
METHOD OF STEEPEST DESCENT FOR COMPLEX REACTIONS
                                  9        3
THE COEFFICIENT MATRIX IS
                             1    2    2    1    1    0    0    0    2
                             4    4    2    0    0    0    2    2    6
                             0    0    0    2    1    2    0    1    0
THE NN(I), GRT(I), N(I) VECTORS ARE
          0.0000       2.3200     0.0010
          0.0000      14.2169     0.0010
          0.0000      20.4348     0.0010
          0.0000     -47.6145     0.9930
          0.0000     -24.1278     1.0000
```

```
       0.0000      0.0000    0.0070
       0.0000      0.0000    5.9920
       4.0000    -23.1656    1.0000
       1.0000     13.2104    0.0010
AFTER     1 ITERATIONS LAM = 1.000, THE NEW N'S ARE
                                     .10938976E-02
                                     .75732405E-03
                                     .61796804E-03
                                     .98675599E 00
                                     .10078603E 01
                                     .69999970E-07
                                     .59747432E 01
                                     .10186276E 01
                                     .76960835E-03
AFTER     2 ITERATIONS LAM = 1.000, THE NEW N'S ARE
                                     .12519209E-02
                                     .54364557E-03
                                     .33050375E-03
                                     .97168496E 00
                                     .10241839E 01
                                     .69999995E-12
                                     .59619357E 01
                                     .10324462E 01
                                     .56547021E-03
AFTER     3 ITERATIONS LAM = 1.000, THE NEW N'S ARE
                                     .15325138E-02
                                     .30302582E-03
                                     .91654536E-04
                                     .94809951E 00
                                     .10489189E 01
                                     .70000027E-17
                                     .59403656E 01
                                     .10548821E 01
                                     .32985888E-03
AFTER     4 ITERATIONS LAM = 1.000, THE NEW N'S ARE
                                     .20243466E-02
                                     .11174462E-03
                                     .91654506E-09
                                     .91409395E 00
                                     .10833915E 01
                                     .29215280E-17
                                     .59069072E 01
                                     .10884206E 01
                                     .13334811E-03
AFTER     5 ITERATIONS LAM = 1.000, THE NEW N'S ARE
                                     .31062442E-02
```

```
                                        .11174470E-08
                                        .80409509E-09
                                        .85750212E 00
                                        .11393782E 01
                                        .22509365E-18
                                        .58481499E 01
                                        .11456175E 01
                                        .66988199E-05
AFTER     6 ITERATIONS LAM = 1.000, THE NEW N'S ARE
                                        .57588897E-02
                                        .25099837E-08
                                        .64613616E-09
                                        .76924967E 00
                                        .12249906E 01
                                        .22509373E-23
                                        .57519710E 01
                                        .12365100E 01
                                        .40183474E-06
AFTER     7 ITERATIONS LAM = 1.000, THE NEW N'S ARE
                                        .20492855E-01
                                        .12101550E-07
                                        .29269667E-09
                                        .52938755E 00
                                        .14501194E 01
                                        .19691715E-22
                                        .54679084E 01
                                        .14911055E 01
                                        .10431349E-06
AFTER     8 ITERATIONS LAM = 1.000, THE NEW N'S ARE
                                        .45719592E-01
                                        .38144542E-07
                                        .33829415E-09
                                        .54184775E 00
                                        .14124322E 01
                                        .12955502E-21
                                        .54046880E 01
                                        .15038722E 01
                                        .15696846E-06
AFTER     9 ITERATIONS LAM = 1.000, THE NEW N'S ARE
                                        .63274425E-01
                                        .73600451E-07
                                        .31863897E-09
                                        .54447695E 00
                                        .13922481E 01
                                        .61275978E-21
                                        .53546526E 01
                                        .15187979E 01
```

```
                    .15952549E-06
AFTER    10 ITERATIONS LAM = 0.090, THE NEW N'S ARE
                    .66482713E-01
                    .92619775E-07
                    .31511799E-09
                    .54492056E 00
                    .13885962E 01
                    .19499157E-20
                    .53454713E 01
                    .15215626E 01
                    .15717592E-06
AFTER    11 ITERATIONS LAM = 0.000, THE NEW N'S ARE
                    .66513144E-01
                    .94719113E-07
                    .31509097E-09
                    .54492443E 00
                    .13885619E 01
                    .36073643E-20
                    .53453838E 01
                    .15215892E 01
                    .15716868E-06
AFTER    12 ITERATIONS LAM = 0.100, THE NEW N'S ARE
                    .66422293E-01
                    .93692643E-07
                    .31518904E-09
                    .54491195E 00
                    .13886652E 01
                    .32452700E-20
                    .53456439E 01
                    .15215108E 01
                    .15723135E-06
AFTER    13 ITERATIONS LAM = 0.000, THE NEW N'S ARE
                    .66435369E-01
                    .93849470E-07
                    .31517494E-09
                    .54491375E 00
                    .13886503E 01
                    .34125131E-20
                    .53456065E 01
                    .15215221E 01
                    .15722235E-06
AFTER    14 ITERATIONS LAM = 0.000, THE NEW N'S ARE
                    .66455772E-01
                    .94094473E-07
                    .31515294E-09
                    .54491655E 00
                    .13886271E 01
```

```
                                  .36850613E-20
                                  .53455481E 01
                                  .15215397E 01
                                  .15720832E-06
AFTER    15 ITERATIONS LAM = 0.002, THE NEW N'S ARE
                                  .66447989E-01
                                  .94001106E-07
                                  .31516134E-09
                                  .54491548E 00
                                  .13886360E 01
                                  .35848265E-20
                                  .53455703E 01
                                  .15215330E 01
                                  .15721367E-06
AFTER    16 ITERATIONS LAM = 0.000, THE NEW N'S ARE
                                  .66456398E-01
                                  .94102264E-07
                                  .31515227E-09
                                  .54491663E 00
                                  .13886264E-01
                                  .37041382E-20
                                  .53455463E 01
                                  .15215402E 01
                                  .15720789E-06
AFTER    17 ITERATIONS LAW = 0.000, THE NEW N'S ARE
                                  .66456398E-01
                                  .94102264E-07
                                  .31515227E-09
                                  .54491663E 00
                                  .13886264E 01
                                  .37041382E-20
                                  .53455463E 01
                                  .15215402E 01
                                  .15720789E-06
CONVERGENCE OBTAINED, FINAL MOLE NUMBERS ARE
                                  .66456398E-01
                                  .94102264E-07
                                  .31515227E-09
                                  .54491663E 00
                                  .13886264E 01
                                  .37041382E-20
                                  .53455463E 01
                                  .15215402E 01
                                  .15720789E-06

FINAL MOLE FRACTIONS ARE
                                  .74947313E-02
                                  .10612540E-07
                                  .35541824E-10
                                  .61453884E-01
```

```
.15660467E 00
.41774038E-21
.60285291E 00
.17159424E 00
.17729383E-07
```

Table S11-13d summarzes a set of six primary reactions for the species in this problem. In principle, any six such reactions could be selected as long as they are linearly independent. To check the adequacy of these primary reactions and the consistency of the results of this problem, which did not rely explicitly on any set of

TABLE S11-13d

	Reactions	$\Delta\mathscr{G}°$(Kcal/g-mol)	$K_a = \left(\exp - \frac{\Delta\mathscr{G}°}{RT}\right)$	$K_a = \prod_{i=1}^{C} y_i^{\alpha_{ji}}$
1	$H_2O + CO \longrightarrow CO_2 + H_2$	-0.638	1.38	1.37
2	$C_2H_6 \longrightarrow C_2H_4 + H_2$	2.12	0.344	0.361
3	$C_2H_4 \longrightarrow C_2H_2 + H_2$	12.36	2.0×10^{-3}	2.01×10^{-3}
4	$H_2 + \frac{1}{2}O_2 \longrightarrow H_2O$	-46.03	1.0×10^{10}	1.4×10^{10}
5	$CH_4 + H_2O \longrightarrow CO + 3H_2$	-6.52	2.66×10	2.68×10
6	$C_2H_2 + 3H_2 \longrightarrow 2CH_4$	-31.40	7.1×10^6	7.2×10^6

reactions, the equilibrium constants K_a have been computed, based first on $\Delta\mathscr{G}°$ and then on the mole fractions computed by this program. Agreement is good, although some disagreement for reaction 4 is evident. The latter is probably attributable to the very low concentration of O_2 (less than 10^{-21} mole fraction).

Problems

11-1 Hydrogen is to be formed by the steam cracking of methane according to the reaction

$$CH_4 + 2H_2O = CO_2 + 4H_2$$

The reaction will be performed at 600°C, where the equilibrium constant K_a is

$$\log_{10} K_a = 0.75$$

The standard states are pure gases at 1 atm. If the reaction pressure is 1 atm and a 50 per cent excess of steam is used, what is the fractional conversion of CH_4 to H_2? What is the composition of the exhaust stream?

11-2 Methanol is manufactured commercially by reacting carbon monoxide and hydrogen according to the reaction

$$CO + 2H_2 = CH_3OH$$

Determine the equilibrium constant for this reaction at 25°C. What are your standard states? If a 20 per cent excess of CO is used and the reaction is run at 15 atm at 25°C, what is the conversion of H_2 to alcohol? How much heat is liberated (or consumed) per mole of CO that reacts? (Values of the free energy and enthalpy of formation are given in Appendix E.)

11-3 The equilibrium constant for the gas-phase reaction $A = B + C$ at 400°K is $K_a = 1.0$ (standard states: pure gases at 1 atm and $T = 400$°K). It is proposed to obtain a conversion of A to B and C by passing pure A into a catalytic reactor operating at 400°K and with sufficiently long residence time that equilibrium is achieved.

(a) At what pressure should the reactor be operated to obtain 90 per cent conversion of A to B and C?

(b) Instead of feeding pure A into the reactor, an equimolar mixture of A and an inert diluent I is used as the feed. If the pressure is maintained at 1 atm, what is the equilibrium conversion of A to B and C?

(c) Instead of using a constant-pressure flow reactor, the reaction will be run in an isothermal constant-volume reaction vessel. Initially (that is, before any reaction has occurred), the vessel is filled with pure A at 1 atm and 400°K. Find the equilibrium conversion and final pressure in the system.

11-4 Isomerization of a paraffin hydrocarbon is the conversion of a straight carbon chain to a branched chain. A petroleum plant is isomerizing normal butane to isobutane in a gas-phase reaction at 15 atm and 300°F. For this reaction, at 300°F the standard free-energy change, $\Delta\mathscr{G}^\circ = 97.5$ cal/g-mol $n\text{C}_4$. If the feed to the reactor is 93 mol percent n-butane, 5 mol per cent isobutane, and 2 mol per cent HCl (catalyst), what is the maximum conversion to isobutane?

	T_c, °F	P_c, psia
$n\text{-}C_4H_{10}$	305.6	550.7
$i\text{-}C_4H_{10}$	275.0	529.1
HCl	124.5	1200

11-5 Methanol can be synthesized in accordance with the reaction

$$CO + 2H_2 = CH_3OH(g)$$

K_a for this reaction is 1.377 at a temperature of 127°C.

(a) Calculate the equilibrium yield of methanol when 1 lb-mol of CO and 2 lb-moles of hydrogen react at 127°C and a pressure of 5 atm. (Assume that methanol is in the vapor phase.)

(b) How many lb-moles of CO should be contacted with 2 lb-moles of hydrogen if it is desired to obtain the maximum possible concentration of methanol in the equilibrium gas mixture at 127°C and 5 atm pressure?

11-6 Assume that 0.01 lb-mol of nitrogen is added to a constant-volume bomb that already contains 0.03 lb-mol of hydrogen. The bomb, which has a volume of 1.95 ft³, is then placed in a constant-temperature bath at 910°R. When equilibrium has been established, the pressure in the bomb is 10 atm.

(a) Calculate $\Delta\mathscr{G}_T^\circ$ (Btu/lb-mol) for the reaction

$$N_2 + 3H_2 \rightleftharpoons 2NH_3$$

at 910°R and 10 atm. Refer to the usual standard states.

(b) How many lb-moles of ammonia would be present at equilibrium if 2 lb-moles of nitrogen are added to 3 lb-moles of hydrogen at a constant temperature of 910°R and a constant pressure of 10 atm? What is the concentration of the ammonia in the gas mixture?

(c) Repeat (b) for the addition of 1 lb-mol of nitrogen and 1 lb-mol of argon to 3 lb-moles of hydrogen.

It may be assumed that all three gases behave ideally at 910°R and 10 atm pressure.

11-7 Stoichiometric proportions of SO_2 and O_2 are placed in a constant-pressure bomb at 1 atm and allowed to come to equilibrium isothermally according to the reaction

$$SO_2 + \tfrac{1}{2}O_2 = SO_3$$

The resulting equilibrium gas analyses at two temperatures are as follows:

	1000°K	1200°K
SO_2	42.4	60.4
O_2	21.2	30.2
SO_3	36.4	9.4
	100.0	100.0

(a) If 2 g-moles of air and 1 g-mol of SO_2 were to be placed in the bomb and allowed to come to equilibrium isothermally at 1000°K, how many moles of SO_3 would be present at equilibrium?

(b) Calculate the isothermal enthalpy change on reaction (cal/g-mol of SO_3 formed), assuming it to be independent of temperature in the range 1000 to 1200°K.

(c) What is the change in entropy [cal/(°K)(g-mol of SO_3 formed)] accompanying the original reaction at a constant temperature of 1000°K and a constant pressure of 1 atm?

11-8 Gaseous hydrogen is to be manufactured by steam cracking of methane in a catalytic reactor according to the reaction

$$CH_4 + 2H_2O = CO_2 + 4H_2$$

at 900°F and 1 atm. Equilibrium data, K_a versus $1/T$, are provided in Fig. P11-8. The standard states for all gaseous components are pure components at 1 atm.

(a) If 5 moles of steam are fed to the reactor for every mole of methane, what is the equilibrium composition of methane in the product stream?

(b) Is this reaction exothermic or endothermic? How much heat must be added or removed from the reactor per mole of methane fed?

(c) Will an increase in pressure result in an increased or decreased yield?

(d) It has been suggested that the equilibrium yield of this reaction can be increased by diluting the feed stream with an inert gas. Do you agree? Justify your answer clearly. You may assume ideal gaseous mixtures.

$$\text{Yield} = \frac{\text{moles of } CH_4 \text{ reacted}}{\text{moles of } CH_4 \text{ fed}}$$

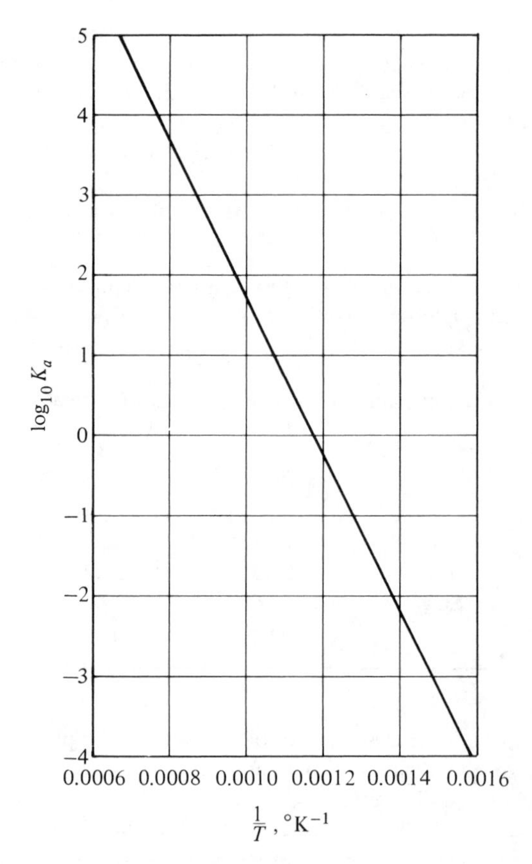

FIGURE P11-8

11-9 It is proposed to make hydrogen cyanide by the reaction $C_2H_2 + N_2 = 2HCN$. The data available on the reactant and product compounds are as follows:

Compound	C_2H_2	N_2	HCN
$\underline{G}°$ at 25°C, Referred to Constituent Elements at 25°C, cal/g-mol	50,000	0	28,700
$\underline{H}°$ at 25°C, Referred to Constituent Elements at 25°C, cal/g-mol	54,000	0	31,200
$\underline{S}°$ at 25°C, Referred to Pure Compound at 0°K and 1 atm, cal/g-mol °K	47.997	45.767	48.230
T_c, °K	308.7	126.0	456.7
P_c, atm	61.6	33.5	50.0
Heat Capacity $C_P^* = a + bT + cT^2$, cal/g-mol °K			
a	7.331	6.449	5.974
$b \times 10^3$	12.622	1.413	10.208
$c \times 10^6$	−3.889	−0.0807	−4.317

The standard states are pure compounds as gases at 1 atm fugacity. Suppose stoichiometric proportions of the reactants at atmospheric pressure are fed to an isothermal reactor at 25°C. If the reaction goes to equilibrium, how much acetylene is converted? Is any work done on or by the reactor? How do you justify this in the light of your value of $\Delta\mathscr{G}^\circ$ for this reaction? If the reactor operates at 300°C, will the conversion be improved? According to Le Châtelier's principle, increasing the pressure should not help, because there is no change in moles of products and reactants. Calculate the conversion at 25°C and 200 atm using the Lewis–Randall approximation. How do you explain your answer?

11-10 Methanol, CH_3OH, is decomposed by a catalyst into CO and H_2. At equilibrium conditions at 190°C and 1 atm pressure CH_3OH is 98 per cent decomposed into CO and H_2. At 298°K the "enthalpy change of formation" ($\Delta \underline{H}_f$) of gaseous H_2O from the elements with each at 1 atm pressure is $-68{,}313$ cal/g-mol. For the formation of CO_2 from CO and $\frac{1}{2}O_2$ under the same conditions $\Delta \underline{H}_f$ is $-67{,}623$ cal. The $\Delta\mathscr{H}$ at 298°K for the following reaction is $-182{,}580$ cal/g-mol.

$$CH_3OH(g) + \tfrac{3}{2}O_2(g) = CO_2(g) + 2H_2O(g)$$

The molal heat capacities (C_P) in calories/g-mol °K at 1 atm pressure are

$$\begin{aligned} \text{gaseous } H_2&: \quad C_P = 6.5 + 0.0009T(°K) \\ \text{gaseous } CO&: \quad C_P = 8.5 + 0.001T(°K) \\ \text{gaseous } CH_3OH&: \quad C_P = 2 + 0.03T(°K) \end{aligned}$$

From these data derive an expression for $\ln K_a$ as a function of temperature for the formation of CH_3OH from CO and H_2.

11-11 The vapor-phase hydration of ethylene to ethanol is represented by the reaction

$$C_2H_4 + H_2O = C_2H_5OH$$

Experimental measurements have determined the equilibrium constant at two temperatures as

$$K_a = 6.8 \times 10^{-2} \text{ at } 145°C$$
$$K_a = 1.9 \times 10^{-3} \text{ at } 320°C$$

where the standard states are the usual ones-pure gases at 1 atm fugacity. The heat capacities of the reactants and products in the vapor phase in cal/g-mol °K are

$$\begin{aligned} C_2H_4&: \quad C_P = 6.0 + 0.015T(°K) \\ H_2O&: \quad C_P = 7.0 + 0.00277T(°K) \\ C_2H_5OH&: \quad C_P = 4.5 + 0.038T(°K) \end{aligned}$$

If ethylene and water vapor are passed to a reactor at 400°C and 10 atm pressure, calculate the maximum percentage conversion of ethylene to ethanol if stoichiometric proportions of the reactants are used.

11-12 Nitric acid can be made by air oxidation of nitric oxide (NO) to NO_2 and absorption in water. It has been suggested that nitric oxide be made from air by heating the air to a very high temperature at atmospheric pressure. To make this process economically interesting, it is necessary to obtain 1 per cent nitric oxide so that the recovery of the NO by absorption will not be too expensive. What is the minimum temperature that must be used to obtain this minimum concentration assuming the gases reach equilibrium at this temperature?

$$(\Delta \underline{G}_f^\circ)_{298} = 20.72 \text{ kcal/mol NO}$$
$$(C_P^*)_{O_2} = 8.27 + 2.58 \times 10^{-4}T$$
$$(C_P^*)_{N_2} = 6.50 + 10^{-3}T$$
$$(C_P^*)_{NO} = 8.05 + 2.33 \times 10^{-4}T$$
$$(\Delta \underline{H}_f^\circ)_{298} = 21.6 \text{ kcal/mol NO}$$
$$(\Delta \underline{S}_f^\circ)_{298} = 50.34 \text{ kcal/mol NO °K}$$

where $T =$ °K and the standard states are pure gases at 1 atm.

11-13 It is proposed to synthesize methane by the reaction

$$2CO(g) + 2H_2(g) = CH_4(g) + CO_2(g)$$

(a) If equal volumes of CO and H_2 are placed in a reaction vessel at 1000°K and 50 atm pressure, what will be the concentration of CH_4 in the vessel when equilibrium is established?

(b) What will be the equilibrium concentration of CH_4 if 10 per cent excess CO is used?

	C_P at 1 atm, cal/g-mol °K (T in °K)	$\Delta \underline{H}_f$, kcal/g-mol, at 25°C and 1 atm	$\Delta \underline{G}_f$, kcal/g-mol, at 25°C and 1 atm
CO	$6.60 + 0.0012T$	−26.416	−32.808
H_2	$6.62 + 0.00081T$	—	—
CH_4	$5.34 + 0.0115T$	−17.889	−12.140
CO_2	$10.34 + 0.00274T - \frac{195,500}{T^2}$	−94.052	−94.260

(*Note:* Reduced properties of H_2 are generally calculated from the equations

$$T_r = \frac{T}{T_c + 8°\text{K}} \quad \text{and} \quad P_r = \frac{P}{P_c + 8 \text{ atm}}$$

11-14 Ethyl acetate may be produced by the catalytic dehydrogenation of ethyl alcohol according to the reaction

$$2C_2H_5OH = CH_3COOC_2H_5 + 2H_2$$

Equilibrium constants for this reaction have been measured at 181 and 201.5°C as 1.075 and 1.705, respectively, with the standard states being pure gases at 1 atm pressure. Heat capacities of the several components are given in cal/g-mol°K (with T in °K) as

$$H_2: \quad C_P^* = 6.744 + 2.7 \times 10^{-4}T + 1.956 \times 10^{-7}T^2$$
$$C_2H_5OH: \quad C_P^* = 5.56 + 4.522 \times 10^{-2}T - 1.639 \times 10^{-5}T^2$$
$$CH_3COOC_2H_5: \quad C_P^* = 2.27 + 0.88T - 3.086 \times 10^{-5}T^2$$

It is proposed to design a reactor to be fed continuously with ethanol vapor at 150°C and 1 atm. The product stream is to be at 201.5°C and 1 atm. If equilibrium is attained in the product stream, how much heat must be supplied to the reactor per gram-mole of ethanol fed?

11-15 For the gaseous hydration of ethylene to ethanol,

$$C_2H_4 + H_2O = C_2H_5OH$$

Dodge[6] reports:

$$\Delta\mathscr{G}^\circ = (28.60T - 9740) \text{ cal/g-mol}$$

($T = $ °K) where the standard states are pure gases at 1 atm. If the hydration is performed at 1 atm and 300°C,

(a) Is this reaction endothermic or exothermic?
(b) How much heat is liberated (or consumed) during the formation of 1 g-mol of ethanol? How does this vary from $\Delta\mathscr{H}^\circ$?

11-16 Natural gas (essentially pure methane) is burned as a heating fuel in a 25 per cent excess of air. Determine the flame temperature (assume adiabatic combustion) and the exit composition of the combustion products.

$$CH_4 + 2O_2 = CO_2 + 2H_2O$$

$$(C_P)_{O_2} = 6.148 + 3.10 \times 10^{-3}T - 9.23 \times 10^{-7}T^2$$
$$(C_P)_{CH_4} = 3.381 + 1.80 \times 10^{-2}T - 4.3 \times 10^{-6}T^2$$
$$(C_P)_{CO_2} = 6.214 + 1.04 \times 10^{-2}T - 3.55 \times 10^{-6}T^2$$
$$(C_P)_{H_2O} = 7.256 + 2.30 \times 10^{-3}T + 2.83 \times 10^{-7}T^2$$
$$C_P = \text{cal/g-mol °K} \qquad T = \text{°K}$$

11-17 Carbon monoxide is burned in a stoichiometric amount of air. Air and CO are fed to the burner at 25°C. If heat losses from the burner are estimated to be 47,500 Btu/lb-mol of CO feed, determine the composition and temperature of the gas leaving the burner.

$$(C_P)_{CO} = 6.42 + 1.0 \times 10^{-3}T$$
$$(C_P)_{CO_2} = 6.21 + 7 \times 10^{-3}T$$
$$(C_P)_{N_2} = 6.524 + 1.25 \times 10^{-3}T$$
$$(C_P)_{O_2} = 6.148 + 2.0 \times 10^{-3}T$$
$$C_P = \text{cal/g-mol °K} \qquad T = \text{°K}$$

11-18 Rework Problem 11-17 assuming that only 3500 Btu/lb-mol is lost.

11-19 Ammonia has been proposed as a fuel for an experimental rocket motor (Fig. P11-19). Gaseous NH_3 and air (to supply oxygen) in the stoichiometric ratio are inject-

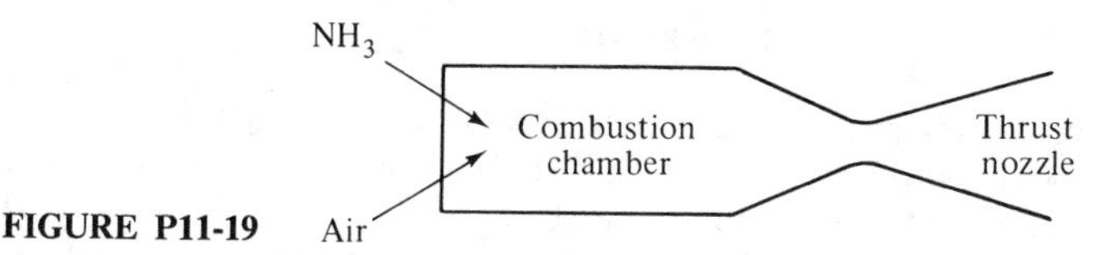

FIGURE P11-19

[6] B. F. Dodge, *Chemical Engineering Thermodynamics*, McGraw-Hill Book Company, Inc., New York, 1944.

ed into the rocket motor at 150°F and 300 psia, where they react to form nitrogen and water according to the reaction

$$2NH_3 + \tfrac{3}{2}O_2 = N_2 + 3H_2O$$

If the combustion chamber is insulated so that the reaction is essentially adiabatic, determine the temperature and approximate composition of the combustion gases as they enter the thrust nozzle. You may use the following average specific heats:

$$(C_P)_{N_2} = 8 \text{ Btu/lb-mol °R}$$
$$(C_P)_{H_2O} = 10.0 \text{ Btu/lb-mol °R}$$
$$(C_P)_{O_2} = 8.2 \text{ Btu/lb-mol °R}$$
$$(C_P)_{NH_3} = 13 \text{ Btu/lb-mol °R}$$

11-20 Gases from the burner in a contact sulfuric acid plant have the following composition (molar): 7.8 per cent SO_2, 10.8 per cent O_2, and 81.4 per cent N_2. The gas mixture at 450°C and 1 atm is passed through a catalytic converter, where the SO_2 is oxidized to SO_3. If the converter operates adiabatically, what is the maximum per cent of SO_2 converted to SO_3? How does this compare with the maximum conversion if the converter operates isothermally?

C_P in cal/g-mol °K (with T in °K)

$$SO_2 = 8.43 + 0.0024T$$
$$O_2 \text{ and } N_2 = 6.84 + 0.00034T$$
$$SO_3 = 10.6 + 0.002T$$

$\Delta\mathscr{H}°$ at 0°C for the above reaction is −22,000 cal/g-mol of SO_3 formed. The equilibrium constant K_a for the above reaction is

$$\log_{10}K_a = \frac{4950}{T(°K)} - 4.7$$

where the standard states are pure gases at atmospheric pressure.

11-21 A stationary gas-turbine plant is constructed according to the scheme shown in Fig. P11-21. Air at 70°F enters the compressor, where its pressure is raised to 6 atm.

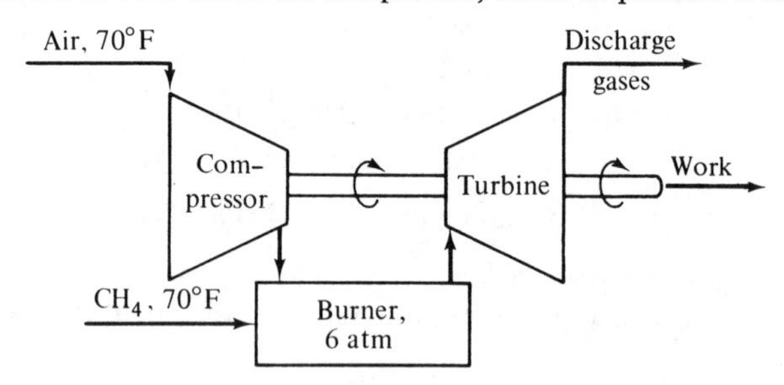

FIGURE P11-21

From the compressor the air passes to a burner, where methane is added at 70°F. The resulting hot gases pass to a turbine, which discharges to atmospheric pressure. The barometer is normal at 14.7 psia, and air may be considered to be 21 per cent O_2 and 79 per cent N_2 on a mol basis. In order to keep the turbine inlet temperature down, 300 per cent excess air will be used to burn the methane in the burner. Using the following data, calculate the work output of this plant per lb-mol of methane fed.

The compressor and turbine are both 80 per cent efficient on an adiabatic reversible basis. There is negligible pressure drop in any line and negligible heat losses to the surroundings. All gases may be assumed to behave as ideal gases. The isothermal enthalpy change of combustion of methane at 70°F is −345,000 Btu/lb-mol. The average heat capacities of the various compounds in Btu/lb-mol °F are tabulated below:

Compound	C_P
CO_2	10.3
H_2	8.7
N_2	7.2
O_2	7.5

11-22 A plant produces methanol from hydrogen and carbon monoxide. In the process, stoichiometric proportions of CO and H_2 are compressed isothermally and reversibly at 25°C from 1 to 310 atm. The compressor discharges the gas continuously to a well-insulated catalytic reactor (Fig. P11-22), where the following reaction takes place:

$$CO(g) + 2H_2(g) \overset{\text{cat.}}{=} CH_3OH(g)$$

The methanol leaving the reactor is in equilibirum with the CO and H_2 in the stream. There is negligible pressure drop across the reactor.

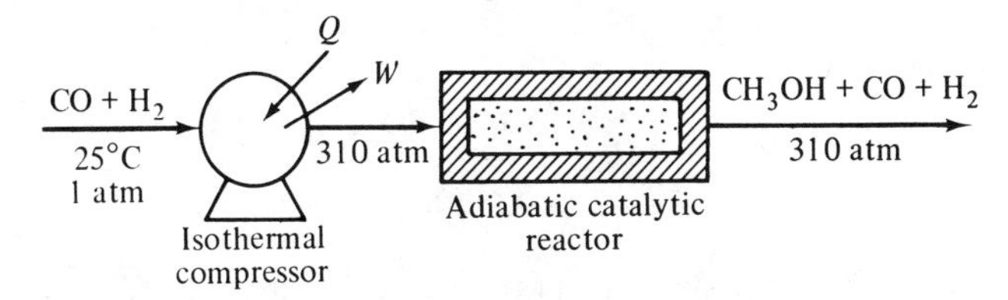

FIGURE P11-22

(a) Calculate the work done by the compressor in calories per g-mol of carbon monoxide fed. Also determine on the same basis the amount of heat transferred to the surroundings from the compressor.

(b) What is the conversion of carbon monoxide to methanol?

(c) What is the temperature of the exit stream from the reactor?

The C_p's at 1 atm (cal/g-mole °K) are

$$CO(g): \quad 6.60 + 0.0012T(°K)$$
$$H_2(g): \quad 6.62 + 0.00081T(°K)$$
$$CH_3OH(g): \quad 3.53 + 0.0256T(°K)$$

Critical Constants	T_c, °K	P_c, atm
CO	134	35
H_2	33	12.8
CH_3OH	513	78.7

11-23 SO_2 is to be oxidized with a 50 per cent excess of oxygen (from air) in a catalytic reactor (Fig. P11-23) according to the reaction

$$SO_2 + \tfrac{1}{2}O_2 = SO_3$$

For optimum yield from the reaction it is desirable to keep the reactor isothermal at a temperature of 425°C. This will be accomplished by preheating the reactants (available at 100°F) as shown in Fig. P11-23. The reactants are first passed through tubes in

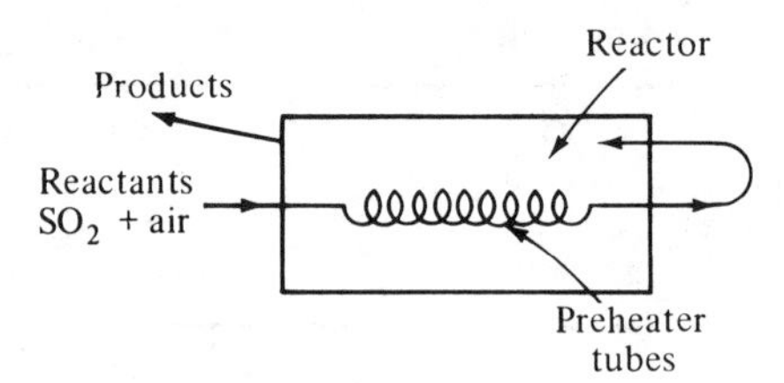

FIGURE P11-23

the catalyst bed, where the reactants are heated to a temperature of 425°C. The reactants are then passed through the catalyst bed, where they react at constant temperature. If the heat absorbed by the reactants does not equal that liberated during the reaction, steam or cooling water will be used to provide the necessary additional heating or cooling.

(a) If equilibrium is attained within the reactor, what is the exit composition?

(b) How much heat must be supplied or removed from the reactor per lb-mol of SO_2 fed to maintain isothermal operation? Must heat be added or removed from the reactor?

For the reaction

$$SO_2 + \tfrac{1}{2}O_2 = SO_3$$

$$\log_{10} K_a = \frac{5186.5}{T} + 0.611 \log_{10} T - 6.7497$$

(T = °K) where the standard states are pure gases at 1 atm pressure.[7] The heat capacities of the reactants may be taken as

$$(C_P)_{N_2} = 7.0 \text{ Btu/lb-mol °R}$$
$$(C_P)_{O_2} = 7.0 \text{ Btu/lb-mol °R}$$
$$(C_P)_{SO_2} = 12.0 \text{ Btu/lb-mol °R}$$

11-24 Hydrogen peroxide (H_2O_2) is used as a fuel for the small thruster rockets by which our astronauts turn and roll their space capsule. The hydrogen peroxide is stored as a pure liquid at 25°C and 1 atm pressure aboard the space capsule. When thrusting is needed, the peroxide is passed over a catalyst, where it decomposes according to the reaction

$$H_2O_2(l) = H_2O(g) + \tfrac{1}{2}O_2$$

The decomposition products are then fed to the thrust nozzles where they expand, and provide the desired force.

(a) Assuming that the decomposition goes to completion, what is the temperature of the steam and oxygen that enter the thruster?

(b) Obtain a reasonable estimate of the equilibrium constant for the reaction at this temperature. Is the assumption of part (a) reasonable?

[7] M. Bodenstein and W. Pohl, *Z. Electrochem.*, *11*, 373 (1905).

At 25°C,

$$(\Delta G_f^\circ)_{H_2O_2(l)} = -28.2 \text{ kcal/g-mol}$$
$$(\Delta \underline{H}_f^\circ)_{H_2O_2(l)} = -44.8 \text{ kcal/g-mol}$$
$$(C_P)_{O_2} = 8.0 \text{ Btu/lb-mol °R}$$
$$(C_P)_{H_2O} = 10.0 \text{ Btu/lb-mol °R}$$

The standard states are pure liquid H_2O_2 at 1 atm; pure vapors, H_2O, O_2 at 1 atm.

11-25 Carbon monoxide from a water gas plant is being burned with air in an adiabatic reactor. Both the carbon monoxide and air are fed to the reactor at 70°F and atmospheric pressure. Ten per cent more air is used than theoretically required by the stoichoimetry of the reaction. The constant-pressure heat capacities in Btu/lb-mol °R are as follows:

CO: $7.0 + 0.0005T$
N_2: $7.0 + 0.0005T$
CO_2: $9.0 + 0.0014T$
H_2O: $8.0 + 0.0012T$
O_2: $7.1 + 0.0006T$

where $T =$ °F. The ideal gas law may be assumed.

(a) If the products from the reactor are at atmospheric pressure, what is the maximum temperature they may attain?

(b) If the exact stoichoimetric amount of air is used, show quantitatively what will happen to the temperature of the products.

11-26 Although the oxidation of carbon monoxide to carbon dioxide is highly favorable according to free-energy considerations, the kinetics are quite unfavorable at low temperatures and low CO concentrations. (This is the reason carbon monoxide is a major pollutant of the air.) In an effort to reduce the CO effluent from a chemical process, a catalytic converter is being designed as shown in Fig. P11-26. The plant effluent

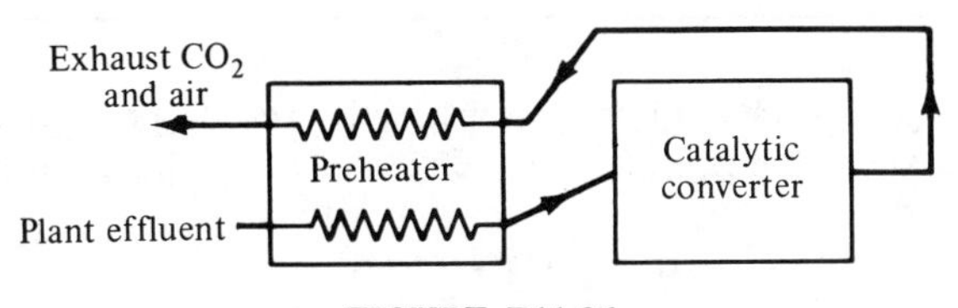

FIGURE P11-26

(25°C and 1 atm) contains 3 per cent CO and 97 per cent air (mole basis). The effluent is first passed through a preheater, where it is heated by the converter product stream. After the preheater the effluent is passed into the catalytic converter, which produces an exit stream that is essentially in equilibrium with itself. (We assume that no chemical reaction occurs in the preheater because no catalyst is present.) The exit stream from the converter is then used to preheat the entering gas in a countercurrent preheater. Assume that the preheater is of sufficient size that the plant effluent is preheated to a temperature that is 500°F cooler than the temperature of the gas leaving the catalytic converter.

(a) Determine the temperature of the streams entering and leaving the catalytic converter.

(b) Determine the composition of the gas leaving the converter.

You may assume that $(C_P)_{O_2} = (C_P)_{N_2} = 7.0$ Btu/lb-mol °R, while $(C_P)_{CO} = 8.0$ Btu/lb-mol °R, and $(C_P)_{CO_2} = 13.5$ Btu/lb-mol °R over these temperatures.

11-27 A stream of monatomic hydrogen is available from an arc at 1000°K. These atoms are made to recombine on a suitable catalyst surface. What is the maximum temperature that may be generated by this reaction? (Remember to account for dissociation of H_2.) For $H_2 = 2H$,

$$\Delta\mathscr{G}^\circ = 81{,}000 - 3.5T \ln T + 4.5 \times 10^{-4}T^2 + 1.17T$$

where T is in °K, $\Delta\mathscr{G}^\circ$ is in cal/g-mol, standard states are pure gases at 1 atm. Assume as an approximation that

$$(C_P)_{H_2} = 7 \text{ cal/g-mol °K}$$
$$(C_P)_{H} = 5 \text{ cal/g-mol °K}$$

11-28 At 400°K and 1 atm n-pentane isomerizes according to the following reactions:

$$C_5H_{12_{norm}} = C_5H_{12_{iso}}$$
$$C_5H_{12_{norm}} = C_5H_{12_{neo}}$$

The equilibrium mixture contains 9.5 per cent n-pentane, 64.2 per cent isopentane, and 26.3 per cent neopentane (all molar per cent).

(a) Calculate $\Delta\mathscr{G}^\circ$ for the reaction $C_5H_{12_{iso}} = C_5H_{12_{neo}}$ at 400°K assuming the following standard states: (1) pure components at 1 atm, and (2) pure components at 5 atm.

(b) If the pressure on the system is increased until a liquid phase appears, calculate its initial composition. Assume that the temperature remains constant at 400°K.

	T_c, °C	P_c, atm	Vapor Pressure, psia at 400°K
n-Pentane	197.2	33	120
Isopentane	187.8	32.8	170
Neopentane	159	33	250

11-29 Synthesis gas for use in Fischer–Tropsch plants can be produced from methane by partial oxidation with oxygen (assume 100 per cent). The major reactions involved are

$$CH_4 + \tfrac{1}{2}O_2 = CO + 2H_2 \tag{a}$$
$$CH_4 + H_2O = CO + 3H_2 \tag{b}$$
$$H_2 + CO_2 = H_2O + CO \tag{c}$$
$$C + CO_2 = 2CO \tag{d}$$

Reaction (a) can be neglected because the unreacted oxygen is essentially zero. Further, reaction (d) can be treated separately (that is, for the present assume that no carbon is formed). Thus only reactions (b) and (c) need be considered. It is desired to obtain a final adiabatic equilibrium temperature of 2200°F, using reactants preheated to 1000°F. The operating pressure is to be 20 atm. Determine

(a) The ratio of O_2/CH_4 in the reactants.
(b) The final H_2/CO ratio, at equilibrium.
(c) Mole fraction of CH_4 in the reacting mixture at equilibrium.
(d) The oxygen concentration at equilibrium according to reaction (a).

(e) Whether carbon deposition is thermodynamically possible according to equation (d).

The K_a values at 2200°F are

$$(K_a)^a = 1.3 \times 10^{11}$$
$$(K_a)^b = 1.7837 \times 10^5$$
$$(K_a)^c = 2.6058$$
$$(K_a)^d = 1.3293$$

The standard states are at 1 atm and all gases behave ideally. The enthalpies of ideal gases for the above elements at absolute zero temperature (Btu/lb-mol) are as follows:

	CH_4	H_2O	CO_2	CO	H_2	O_2
1000°F	−13,492	−90,546	−154,958	−38,528	10,100	10,690
2200°F	8,427	−78,213	−139,009	−28,837	18,927	20,831

Heterogeneous Equilibrium and the Gibbs Phase Rule 12

Introduction

In Chapter 11 we considered chemical equilibrium between the various components of a homogeneous, gas-phase reacting mixture. In this chapter we shall discuss chemical equilibrium in multiphase reacting mixtures. Since each separate phase may be thought of as a separate system capable of undergoing changes in mass and composition, equilibrium between all components in every phase must exist if true equilibrium is to be established.

12.1 The Gibbs Phase Rule

Before beginning our discussion of equilibrium in heterogeneous (multiphase) reacting systems, we shall find it valuable to derive a relation that will tell us the number of state properties (such as P, T, or composition) of a heterogeneous system that we must specify before all other state properties of the phases in the system are fixed. For example, we indicated in Chapter 3 that specification of the pressure, temperature, and number of moles of each component completely specified the state of a single-phase system.

We define f, the number of *degrees of freedom* of the system, to be the number of properties of a heterogeneous system that we must specify before fixing the values of the other state properties of the individual phases (*not* the mixture properties of the system as a whole). The degrees of freedom of the system are related to the number of phases present and the number of components in these phases by means of the *Gibbs phase rule*, which was first derived by J. Willard Gibbs in 1875. It ranks as one of the truly great contributions to the physical sciences.

We begin our derivation by attempting to describe the minimum number of independent properties that will completely characterize a general system of P phases containing C components. We shall then enumerate the relations between these unknowns imposed by the fundamental requirements of equilibrium. Subtraction of the number of equations from the total number of unknowns will then give us the number of properties of the system of phases which must be specified before all others are fixed, that is, the number of degrees of freedom.

We have seen in Chapter 3 that the state of a single-phase is determined by specifying $C + 1$ variables P, T, and the $(C - 1)$ independent mole fractions of the C components. If, in addition, the total size of the system (not a state property) were desired, we would require one additional variable n, the total number of moles of all components. Since some confusion may arise in the mind of the reader who feels that he must specify P, T, and the mole numbers of all species, we caution him that in specifying these $C + 2$ variables he has set the state and the size of the system. Only $C + 1$ need be specified to determine the state of each phase. Consequently we have $P(C + 1)$ unknowns which describe the properties of P coexisting phases in a C component system.

If we presume that there are no barriers to mass or energy transfer between the phases then the following equations must be satisfied:

$(P - 1)$ equations of the form

$$P^1 = P^2 = P^3 = \cdots = P^P \tag{12-1}$$

$(P - 1)$ equations of the form

$$T^1 = T^2 = T^3 = \cdots = T^P \tag{12-2}$$

C sets of $(P - 1)$ equations of the form

$$\begin{aligned} \bar{G}_1^1 &= \bar{G}_1^2 = \bar{G}_1^3 = \cdots \bar{G}_1^P \\ \bar{G}_2^1 &= \bar{G}_2^2 = \bar{G}_2^3 = \cdots \bar{G}_2^P \\ &\vdots \\ \bar{G}_C^1 &= \bar{G}_C^2 = \bar{G}_C^3 = \cdots \bar{G}_C^P \end{aligned} \tag{12-3}$$

for a total of $(C + 2)(P - 1)$ equations in $CP + P$ unknowns. The number of degrees of freedom is given by the difference between the number of unknowns and the equations that relate these unknowns:

$$f = (CP + P) - (CP + 2P - C - 2) \tag{12-4}$$

or

$$f = C + 2 - P \tag{12-5}$$

which is the Gibbs phase rule as it applies to nonreacting systems (in which P, T, and $x_1 \cdots x_{C-1}$ are the only variables needed to describe the state of a single-phase system). If surface (or electric, or magnetic) effects are also important, then these properties, in addition to P, T, $x_1 \cdots x_{C-1}$, must be specified to fix the state of a single-phase system, and the constant in equation (12-5) must be increased by the additional number of new external variables such as surface area, magnetic field strength, and so on:

$$F = C - P + 2 + NV \tag{12-6}$$

where NV = number of new external variables.

The derivation of the phase rule for systems with competing chemical reactions follows exactly the same logic as that for nonreacting systems, except that we have one additional equilibrium expression which relates the activities (or partial molar Gibbs free energies) of the reacting species for each independent reaction. Therefore, if we have N independent competing chemical reactions, the phase rule becomes

$$f = C - N - P + 2 \tag{12-7}$$

which, of course, may be used for systems with no reactions simply by setting $N = 0$.

SAMPLE PROBLEM 12-1. How many degrees of freedom exist for a single-component, single-phase system?

Solution: For a single-component, single-phase system, $C = P = 1$ and $N = 0$. Therefore, the phase rule tell us that

$$f = 1 - 0 - 1 + 2 = 2$$

Therefore, two properties of the system must be specified before all other properties of the system are fixed. Thus, if we specify P and T, then $\underline{H}$, $\underline{U}$, $\underline{S}$, $\underline{V}$, and so on, are all fixed quantities. Indeed, it is just this behavior that permits us to generate and use property tables and charts without having to question the validity of assuming that any one property is an exact function of only two others. We have also made extensive use of this fact in our discussions of the mathematics of properties, where we frequently made statements of the form $\underline{U} = \underline{U}(T, P)$, $\underline{S} = \underline{S}(T, \underline{V})$, and so on.

SAMPLE PROBLEM 12-2. It is desired that a mixture of salt and water have a specific vapor pressure. In how many ways may this vapor pressure be obtained, and what are the variables that would most likely be adjusted to obtain the desired vapor pressure?

Solution: There are to be two components and no reactions for the problem at hand. Since it is desired to control the vapor pressure above a liquid, we must assume that two phases are present (the vapor and the solution). Therefore, the degrees of freedom for this system are given by

$$f = C - N - P + 2 = 2 - 0 - 2 + 2 = 2$$

That is, there will be two degrees of freedom. However, the problem statement indicates that a certain system pressure is desired. Therefore, one degree of freedom is specified, so there is only one degree of freedom left. Thus, if we choose a salt concentration in the mixture, the temperature and all other properties of the mixture must be fixed.

Conversely, if we choose a temperature, the salt composition (if a satisfactory one exists) and all other properties are fixed.

SAMPLE PROBLEM 12-3. Zinc oxide (ZnO) in the solid phase is to be reduced by solid carbon according to the reaction

$$ZnO + C = CO + Zn \tag{a}$$

However, it is also known that the following side reactions occur to some extent:

$$ZnO + CO = CO_2 + Zn \tag{b}$$

$$CO_2 + C = 2CO \tag{c}$$

$$CO = C + \tfrac{1}{2}O_2 \tag{d}$$

$$CO_2 = CO + \tfrac{1}{2}O_2 \tag{e}$$

$$Zn + \tfrac{1}{2}O_2 = ZnO \tag{f}$$

If the zinc produced during these reactions is to be collected as a liquid at the pressure and temperature of the reaction, how many degrees of freedom does this system contain?

Solution: We must now apply the phase rule as it applies to reacting systems:

$$f = C - N - P + 2$$

The reaction system will contain six components (ZnO, C, CO, Zn, CO_2, O_2) and four phases (solid ZnO; solid C; gas containing CO, CO_2, O_2 and Zn; and liquid Zn). Therefore,

$$f = 6 - N - 4 + 2 = 4 - N$$

Six possible reactions are observed in the reacting system. However, of these six, the Gaussian elimination developed in Chapter 11 may be used to show that only *three* of the reactions are independent. Reaction (c) is the difference between (a) and (b); reaction (e) is the sum of reactions (c) and (d); and reaction (f) is the sum of reactions (a) and (d), so $N = 3$ and

$$f = 4 - 3 = \text{one degree of freedom}$$

Thus, if we fix *either* the temperature *or* the pressure of the system, all other properties are then set.

12.2 Equilibrium in Heterogeneous Reactions

In Chapter 11 we showed that the Gibbs free-energy change on reaction was given by

$$\Delta\underline{\mathscr{G}} = -\underline{\mathscr{W}}_R = \underline{X}(\Delta\underline{\mathscr{G}}^\circ + RT \ln J_a) + RT \sum_{i=1}^{C} \underline{\eta}_i \ln \frac{(a_i)_{\text{out}}}{(a_i)_{\text{in}}} \tag{12-8}$$

where

$$J_a = \frac{\prod_{\text{prod}} (a_i)_{\text{out}}^{\alpha_i}}{\prod_{\text{react}} (a_i)_{\text{in}}^{-\alpha_i}}$$

from which we could also determine the reversible work that is available from any particular reaction. We then noted that if the reaction was allowed to proceed spontaneously until the outlet steam was in equilibrium with itself, the maximum extent of reaction was that which corresponded to the minimum in the $\mathscr{G}_{\text{out}}$–$\underline{X}$ curve. By setting the derivative of $\mathscr{G}_{\text{out}}$ with respect to $\underline{X}$ equal to zero, we showed that the equilibrium conditions were described by

$$RT \ln K_a = -\Delta\underline{\mathscr{G}}^\circ \tag{12-9}$$

where

$$K_a = \prod_{i=1}^{C} (a_i)_{\text{out}}^{\alpha_i}$$

We then proceeded to study gas-phase reacting mixtures and demonstrated how activities could be related to pressures and compositions once the standard-state conditions were chosen. In this manner we were able to relate compositions that could be expected from a reaction mixture to the equilibrium compositions. In this section we shall develop a similar line of reasoning which will be used to perform the same types of calculations for reactions in heterogeneous (multiphase) systems as were performed for reactions in homogeneous systems.

The only assumptions we made during the derivation of equation (12-8) was that the reaction take place under isothermal conditions. We did *not* make, nor do we require, any assumption about the phase behavior of either the products or the reactants. Therefore, with the proper choice of standard-state conditions (for calculating $\Delta\underline{\mathscr{G}}^\circ$ and J_a), equation (12-8) applies for heterogenous as well as homogeneous reactions. If a reaction is allowed to proceed until the outlet stream is in equilibrium with itself, the Gibbs free energy of the outlet mixture must be at its minimum (subject to restraints of constant temperature and pressure) independent of whether the reacting mixture is homogeneous or heterogeneous. Thus, at equilibrium, $(d\mathscr{G})_{\text{out}}/d\underline{X} = 0$ for any reaction. Since the expression for $\mathscr{G}_{\text{out}}$ is identical for heterogeneous and homogeneous reactions, the condition for $(d\mathscr{G})_{\text{out}}/d\underline{X} = 0$ will also be the same for heterogeneous and homogeneous reactions. That is,

$$RT \ln K_a = -\Delta\underline{\mathscr{G}}^\circ \tag{12-9}$$

applies for all equilibrium mixtures, irrespective of the phase behavior of the products or reactants. Thus our problem of describing equilibrium in heterogeneous reactions simply reduces to a problem of choosing and specifying the proper standard states. As in the case of homogeneous reactions, the choices of standard states for heterogeneous reactions are completely arbitrary (except for the temperature—which must always be the temperature of the system). Therefore, we may choose different standard states for the different components. In particular, we shall see that vastly different standard states are frequently chosen for gaseous, liquid, or solid components. For example, let us consider the water–gas reaction

$$C_{\text{coke}} + H_2O_{\text{steam}} = CO + H_2$$

The equilibrium activity ratio, K_a, is written as

$$K_a = \frac{a_{CO}a_{H_2}}{a_C a_{H_2O}} = \frac{(\bar{f}/f^\circ)_{CO}(\bar{f}/f^\circ)_{H_2}}{(\bar{f}/f^\circ)_C(\bar{f}/f^\circ)_{H_2O}} \tag{12-10}$$

Normally the standard states for the three gaseous components would be chosen as the pure gases at either 1 atm pressure or at the pressure of the system. The choice of a standard state for the carbon (coke) is still left. We could choose pure gaseous carbon at some specific pressure as the standard state. However, carbon at most normal temperatures is a solid with an extremely low, and hence difficult to measure, vapor pressure. Therefore, pure carbon in any gaseous state would be a rather inconvenient choice for the reference conditions. On the other hand, we may choose pure *solid* carbon at some pressure, usually 1 atm, and the temperature of the system as the standard state. When this choice is made, we find that the activity of the solid carbon becomes extremely simple to calculate. As long as any solid carbon remains (assuming negligible solubility of the gases in the solid carbon), this carbon is essentially in its pure form. Thus the partial fugacity of carbon is equal to its pure-component value independent of the extent of reaction. The variation of the pure-component fugacity from the standard-state pressure to the pressure of the reaction is given by

$$\ln \frac{f_i}{f_i^\circ} = \int_{P^\circ}^{P} \frac{\underline{V}_i}{RT}\, dP \tag{12-11}$$

However, $\underline{V}_i$ for most solids is *extremely* small with respect to RT, so f_i is essentially independent of pressure, except over extremely large pressure ranges. (Remember that we found this same result when considering the change in pure-component fugacity of a liquid with pressure.) Thus we find that the pure-component fugacity and partial fugacity of the solid carbon are essentially equal to the standard-state fugacity, except at very high pressures. Since the partial fugacities and standard-state fugacities are equal, the activity of the solid carbon is unity independent of the extent of reaction, provided only that some solid carbon is still present. Thus we see that a very convenient choice of standard states for solid components is the pure solid component at some pressure and the temperature of the system. (You may assume that this choice of standard states has been made for *all* solid components for the remainder of this text, unless you are specifically told otherwise.)

Although examples of solid reactants appearing in their standard states are by far more numerous than examples of any other pure phases, they are by no means the only examples of pure phases that we may encounter. For example, we have already discussed the dissociation of pure liquid water into hydrogen and oxygen gas. If the standard state for water is chosen as pure liquid water at the temperature of the dissociation and some arbitrary pressure, then the liquid water is in its standard state and will have an activity of unity (provided, of course, that f° does not vary greatly with pressure). Thus we see that any time a pure phase is encountered, we should always consider the possibility that the material may be in its standard state and hence have an activity of unity.

SAMPLE PROBLEM 12-4. A 2-liter constant-volume bomb is evacuated and then filled with 0.10 g-mol of methane, after which the temperature of the bomb and its contents is raised to 1000°C. At this temperature the equilibrium pressure is measured to be 7.02 atm. Assuming that methane dissociates according to the reaction

$$CH_4(g) = C(s) + 2H_2(g)$$

and that ideal gas and solution behavior prevails, what is K_a for this reaction at 1000°C?

Solution: Before even considering calculation of the equilibrium constant, we must specify our choices of standard states! For the problem at hand we choose the following standard states:

CH_4 and H_2: pure gases at 1 atm and 1000°C
carbon: pure solid carbon at 1 atm and 1000°C

With these standard states, $f^\circ_{CH_4} = f^\circ_{H_2} = 1$ atm and the activity of carbon is unity independent of the amount of carbon present. The equilibrium constant, K_a, is defined as

$$K_a = \frac{a^2_{H_2} a_C}{a_{CH_4}} = \frac{(\bar{f}/f^\circ)^2_{H_2} \cdot 1.0}{(\bar{f}/f^\circ)_{CH_4}}$$

or

$$K_a = \frac{\bar{f}^2_{H_2}}{\bar{f}_{CH_4}} \cdot 1 \text{ atm}^{-1}$$

We may now express the partial fugacities of H_2 and CH_4 in terms of the mole fractions and total pressures as

$$\bar{f}_i = \gamma_i y_i f_i$$

But $\gamma_i = 1$ and $f_i = P(f/P) = P = 7.02$ atm. Therefore,

$$\bar{f}_i = y_i(7.02 \text{ atm})$$

and K_a reduces to

$$K_a = \frac{(y_{H_2})^2}{y_{CH_4}}(7.02)$$

We may calculate the mole fractions of the H_2 and CH_4 as follows. The total number of moles of gas in the flask is determined from the ideal gas law (assuming that the solid carbon has negligible volume):

$$n = \frac{PV}{RT} = \frac{(7.02)(2.0) \text{ atm liters}}{(0.0821)(1273) \text{ atm liters °K/g-mol °K}}$$

or

$$n = 0.134 \text{ g-mol}$$

Now assume that x moles of CH_4 is reacted. Then at equilibrium we have the following amount of each gaseous component:

$$\begin{aligned} CH_4 &= 0.10 - x \\ H_2 &= 2x \\ \hline & \quad 0.1 + x \end{aligned}$$

But the total number of moles = 0.134. Therefore,

$$x = 0.034$$

and the mole fractions of CH_4 and H_2 are given by

$$y_{CH_4} = \frac{0.066}{0.134} = 0.493$$

$$y_{H_2} = \frac{0.068}{0.134} = 0.507$$

The equilibrium constant, K_a, is then given by

$$K_a = 7.02 \frac{(0.507)^2}{0.493} = 3.66$$

SAMPLE PROBLEM 12-5. The oxide of tungsten may be reduced with hydrogen according to the reaction

$$WO_2(s) + 2H_2 \rightleftharpoons W(s) + 2H_2O$$

The reaction has been studied at atmospheric pressure, and the equilibrium ratio of the partial pressure of water to that of hydrogen has been measured at a number of temperatures and found to be as given in Table S12-5a. From these data, determine the heat of reaction at 900°C.

TABLE S12-5a

T, °C	P_{H_2O}/P_{H_2}
734	0.45
800	0.53
828	0.65
865	0.75
900	0.82
941	1.00
1000	1.16
1036	1.29

Solution: We may determine the heat of reaction for a given reaction, if we know the variation of K_a with T, from the Gibbs–Helmholtz equation:

$$R\frac{d \ln K_a}{dT} = \frac{\Delta\mathscr{H}^\circ}{T^2}$$

or

$$\frac{d(\ln K_a)}{d(1/T)} = -\frac{\Delta\mathscr{H}^\circ}{R}$$

Thus, if we plot ln K_a versus $1/T$, the slope of the ensuing curve will be $-(\Delta\mathscr{H}^\circ/R)$. Since the reaction is oçcurring at low pressure, we may assume that $(\partial\underline{H}/\partial P)_T = 0$, so the heat of reaction is equal to the standard-state enthalpy change. Therefore,

$$\frac{d(\ln K_a)}{d(1/T)} = -\frac{\Delta\underline{H}^\circ}{R}$$

We may now determine the equilibrium constant from the partial pressure ratio as follows:

$$K_a = \frac{a_{W}(a_{H_2O})^2}{a_{WO_2}(a_{H_2})^2}$$

Choosing the standard states as close to the reaction conditions as possible, we choose

W and WO_2: pure solids at 1 atm and T of system
H_2O and H_2: pure gases at 1 atm and T of system

With this choice of standard states $a_W = a_{WO_2} = 1.0$ and $f^\circ_{H_2O} = f^\circ_{H_2} = 1.0$ atm. The equilibrium expression thus reduces to

$$K_a = \left(\frac{\bar{f}_{H_2O}}{\bar{f}_{H_2}}\right)^2$$

At 1 atm we may assume ideal solutions of ideal gases, so the partial fugacities are equal to the partial pressures. Therefore,

$$K_a = \left(\frac{P_{H_2O}}{P_{H_2}}\right)^2$$

and we may relate K_a and T as shown in Table S12-5b on the following page.

We may now prepare a plot of ln K_a versus $1/T$ as shown in Fig. S12-5. Although a fair amount of scatter is present in the data, we estimate the slope of the line to be about

$$\frac{-\Delta\mathscr{H}^\circ}{R} = -1.19 \times 10^4 \ ^\circ\text{K}$$

Therefore,

$$\Delta\mathscr{H}^\circ = 1.19 \times 10^4 \times 1.987 \text{ (cal/g-mol)}(^\circ\text{K}/^\circ\text{K})$$

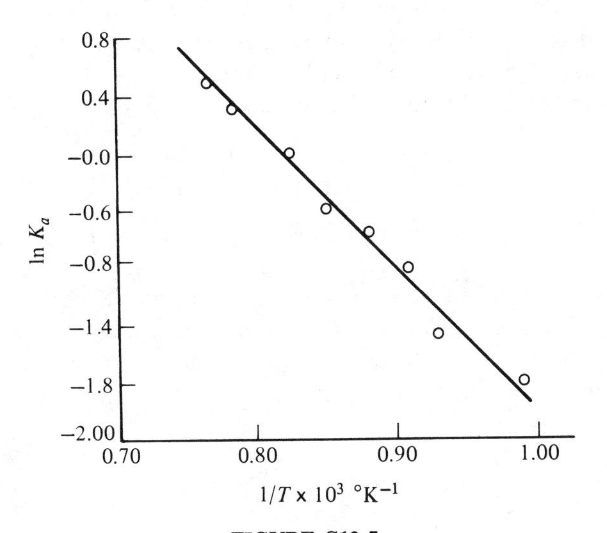

FIGURE S12-5

TABLE S12-5b

T, °C	T, °K	$1/T$, °K ($\times 10^{-3}$)	P_{H_2O}/P_{H_2}	K_a	$\ln K_a$
734	1007	0.99	0.45	0.202	−1.600
800	1073	0.931	0.53	0.281	−1.268
828	1101	0.909	0.65	0.422	−0.815
865	1138	0.880	0.75	0.563	−0.574
900	1173	0.852	0.82	0.672	−0.397
941	1214	0.824	1.00	1.00	0.0
1000	1273	0.785	1.16	1.35	0.300
1036	1309	0.765	1.29	1.66	0.505

or

$$\Delta \underline{H}^\circ = \Delta \mathscr{H}^\circ = 23.6 \text{ kcal/g-mol}$$

so 23.6 kcal of heat must be added to isothermally reduce 1 g-mol of WO_2 with hydrogen at 900°C.

12.3 Effect of Solid Phases Upon Equilibrium

As we have seen in Sample Problems 12-4 and 12-5, the actual amount of a reacting solid phase does not affect the equilibrium composition, as long as there is enough of the solid component present to form a pure phase. This comes about because the activity of the material in the solid phase is unity independent of the amount. In contrast, the activity of a component in a gas phase is quite dependent on the amount of the component present. Thus we can expect that the equilibrium behavior of a reacting mixture with solid phases may be quite different from that exhibited by a gas-phase (or liquid-solution) reaction. This difference can best be exhibited by looking at the respective equilibrium expressions for a simple gas-phase and a solid–gas reaction.

For purposes of illustration, let us examine the following reactions:

$$C(s) + \tfrac{1}{2}O_2 = CO \qquad \text{(Reaction 1)} \tag{12-12}$$

$$CO + \tfrac{1}{2}O_2 = CO_2 \qquad \text{(Reaction 2)} \tag{12-13}$$

For simplicity assume that the reactions are occurring at 1 atm pressure, so all gas components may be assumed to behave in an ideal manner. Then if the standard states are taken as being pure components (either gases or solids) at the pressure and temperature of the system, the activity of the solid carbon is unity, and the activities of the gaseous components are equal to their mole fractions. Thus the equilibrium expressions for the two reactions may be written as

$$(K_a)^1 = \frac{y_{CO}}{(y_{O_2})^{1/2}} \tag{12-14}$$

$$(K_a)^2 = \frac{y_{CO_2}}{y_{CO}(y_{O_2})^{1/2}} \tag{12-15}$$

Now let us examine the following situation: Suppose carbon is mixed with a large excess of oxygen and reaction (12-12) is allowed to proceed. Since K_a for reaction (12-12) is normally a very large number, y_{CO} must be much greater than y_{O_2} at equilibrium. However, we are told that the reaction initially had a large excess of oxygen over carbon, and therefore a carbon balance will indicate that there is not sufficient carbon present to reduce the O_2 content to a point where equilibrium can exist. That is, the outlet activity ratio, $(J_a)_{out}$, will be less than K_a when all the carbon is *exhausted.* Therefore, this reaction cannot reach equilibrium until enough carbon is added so that the outlet activity ratio equals the equilibrium constant. Thus we see that it is possible to *completely consume* a solid-phase reactant (up to the point where a single crystal of solid remains) without reaching equilibrium conditions. In this manner the reaction may be driven to completion with respect to the solid component. Although this phenomenon is not overly important in the oxidation of carbon, it does form the cornerstone of many metallurgical operations, and hence is extremely valuable.

That this behavior is impossible to obtain in a gas-phase reaction can be shown as follows. Suppose we wish to burn CO in an excess of O_2. As the reaction proceeds and the amount of CO unreacted decreases, the mole fraction of CO also decreases, and the outlet activity ratio, $(J_a)_{out}$, increases. By reducing the CO content far enough, it is always possible to have the outlet activity ratio equal the equilibrium constant so that equation (12-15) is satisfied. Therefore, equilibrium can always be attained in a gas-phase reaction. By the same token we see that a gas-phase reaction can never be driven to completion (although if K_a is extremely high, the reaction may go *almost* to completion) in the way a reaction involving a solid phase can be.

As we have demonstrated in Chapter 11, the direction a given reaction will proceed may be determined by comparing the inlet activity ratio with the equilibrium constant. If

$(J_a)_{in} < K_a$	reaction will proceed in direction written
$(J_a)_{in} = K_a$	reactants already in equilibrium with themselves; no reaction
$(J_a)_{in} > K_a$	reaction will proceed in reverse direction from that written

However, as we have seen, the tendency of a reaction to proceed toward equilibrium may be overridden by mass-balance considerations when solid (or pure liquid) phases are present.

SAMPLE PROBLEM 12-6. In the reduction of roasted zinc oxide concentrates the following reaction takes place:

$$ZnO(s) + CO(g) = Zn(g) + CO_2(g)$$

The purified zinc is removed as a vapor. Following the reduction, the gaseous zinc, CO_2, and CO mixture is cooled so that the zinc condenses to form a liquid and is withdrawn. However, at the temperatures encountered during condensation, the equilibrium shifts and favors the reformation of ZnO and CO, unless the CO/CO_2 ratio is great enough. If the zinc condensation is to be performed at 1000°K, what is the minimum ratio of CO to ZnO in the feed stream that will prevent the liquid zinc from reoxidizing in the CO_2–CO mixture that is exhausted from the reduction chamber?

Assume that all gases and solutions behave ideally. The equilibrium constant for the reduction reaction is given as a function of temperature by

$$R(\ln K_a) = -\frac{47{,}290}{T} - 4.46 \ln T + 0.00156T + 2.455 \times 10^{-7}T^2 + 60.305$$

where $T = {}^\circ K$, $R = 1.987$, and K_a is the equilibrium constant. The following standard states were used in the determination of K_a:

CO and CO_2:	pure gases at 1 atm and temperature of the system
ZnO:	pure solid at 1 atm and temperature of the system
Zn:	pure liquid zinc under 1 atm pressure and temperature of the system

The vapor pressure of pure liquid zinc is given by

$$\log_{10} P' = 8.108 - \frac{6163.1}{T}$$

where P' = vapor pressure in mm Hg
T = temperature in °K

Solution: If it is desired that the reverse reaction,

$$Zn + CO_2 = ZnO + CO$$

not occur, all we need to do is guarantee that $(J_a)_{out}$ for the forward reaction is less than K_a at 1000°K, so the reaction is still tending toward the formation of Zn and CO_2. We may express $(J_a)_{out}$ from the activities of the individual species as

$$(J_a)_{out} = \left(\frac{a_{Zn}a_{CO_2}}{a_{ZnO}a_{CO}}\right)_{out}$$

Since ZnO is in its standard state, $a_{ZnO} = 1.0$. (We must assume that just enough ZnO forms to establish a pure phase, and that the activity of the ZnO is unaffected by pressure.)

The standard states for CO and CO_2 are pure gases at 1 atm. Therefore, $f^\circ_{CO} = f^\circ_{CO_2} = 1.0$ atm. Assuming an ideal gaseous solution of CO and CO_2, we can relate $\bar{f}_{CO}$ and $\bar{f}_{CO_2}$ to their mole fractions and the system pressure as

$$\bar{f}_{CO} = y_{CO}P$$
$$\bar{f}_{CO_2} = y_{CO_2}P$$

Therefore, a_{CO_2}/a_{CO} is given by

$$\frac{a_{CO_2}}{a_{CO}} = \frac{y_{CO_2}P}{y_{CO}P} = \frac{y_{CO_2}}{y_{CO}}$$

The standard state for zinc is the pure liquid at 1 atm pressure and the temperature of the system, in this case 1000°K. If $\underline{V}$ for the liquid zinc is low, the fugacity of pure liquid zinc will not vary greatly from the fugacity of the zinc vapor (at its vapor pressure), which is in equilibrium with liquid zinc. By the same reasoning the standard-state fugacity for liquid zinc will be equal to the vapor pressure of the liquid zinc at the temperature of the system. Since the partial and standard-state fugacities for the zinc are equal, the activity of the zinc is unity, and the outlet activity ratio for the reduction reaction simplifies to

$$(J_a)_{out} = \left(\frac{a_{CO_2}}{a_{CO}}\right)_{out} = \left(\frac{y_{CO_2}}{y_{CO}}\right)_{out}$$

If $(J_a)_{out} < K_a$, the reverse reaction cannot occur. Thus determination of K_a at 1000°K fixes the minimum ratio of CO/CO_2 which will prevent reoxidation of Zn to ZnO:

$$R \ln K_a = -\frac{47{,}290}{T} - 4.46 \ln T + 0.00156T + 2.455 \times 10^{-7}T^2 + 60.305$$

At $T = 1000$°K,

$$R \ln K_a = (-47.290 - 30.8 + 1.56 + 0.25 + 60.305) \text{ cal/g-mol °K}$$

or

$$R \ln K_a = -15.98 \text{ cal/g-mol °K}$$

or

$$\ln K_a = -8.04$$

so

$$K_a = 3.21 \times 10^{-4}$$

Therefore, y_{CO_2}/y_{CO} must be less than 3.21×10^{-4} to prevent any significant reoxidation of the zinc. Since 1 mol of ZnO produces 1 mol of CO_2, we must have at least $3.12 \times 10^3 + 1.0$ mol of CO initially per mole of ZnO if we are to prevent any reoxidation of the zinc during condensation.

SAMPLE PROBLEM 12-7. FeO is to be reduced by CO in a small pressurized laboratory converter to form Fe according to the reaction

$$FeO(s) + CO(g) = Fe(s) + CO_2(g) \tag{1}$$

What is the maximum/minimum (state which) pressure at which this reaction may be carried out before the side reaction

$$Fe(s) + CO(g) = FeO(s) + C(s) \tag{2}$$

will occur? The reduction is to be performed at 1800°F. The equilibrium constants for the two reactions at 1800°F are

$$(K_a)^1 = 0.39$$

$$(K_a)^2 = 0.0256$$

where the standard states are pure gases or solids at 1800°F and 1 atm pressure.

Solution: We may write the outlet activity ratios for the two reactions as

$$(J_a)^1_{out} = \left(\frac{a_{Fe}a_{CO_2}}{a_{FeO}a_{CO}}\right)_{out}$$

and

$$(J_a)^2_{out} = \left(\frac{a_{FeO}a_C}{a_{Fe}a_{CO}}\right)_{out}$$

Since the solid components, Fe, FeO, and C, are all in their standard states, we can say that $a_{Fe} = a_{FeO} = a_C = 1.0$. The activity ratios then reduce to

$$(J_a)^1_{out} = \left(\frac{a_{CO_2}}{a_{CO}}\right)_{out}$$

$$(J_a)^2_{out} = \left(\frac{1}{a_{CO}}\right)_{out}$$

Since it is desired to prevent reaction (2) from occurring, $(J_a)^2_{out}$ must be greater than $(K_a)^2$, so any carbon that is produced will react with the FeO to produce Fe and CO. Therefore,

$$(J_a)^2_{out} = \left(\frac{1}{a_{CO}}\right)_{out} > (K_a)^2$$

or

$$(a_{CO})_{out} < \frac{1}{(K_a)^2} = \frac{1}{0.0256} = 39.0$$

We may relate $(a_{CO})_{out}$ to the mole fraction of CO and the total pressure as follows:

$$(a_{CO})_{out} = \left(\frac{\gamma_{CO}y_{CO}f_{CO}}{f^\circ_{CO}}\right)_{out}$$

Assuming ideal solutions of ideal gases, $f_{CO} = P$ (the total system pressure), $f^\circ_{CO} = 1$ atm, and $\gamma_{CO} = 1.0$. Therefore,

$$(a_{CO})_{out} = (y_{CO})_{out}\, P \cdot 1\ \text{atm}^{-1}$$

Substitution of this expression for $(a_{CO})_{out}$ into the activity ratio expression then gives

$$(y_{CO})_{out} P < 39.0\ \text{atm}$$

or

$$P < \frac{39.0}{(y_{co})_{out}}\ \text{atm}$$

We may determine $(y_{CO})_{out}$ from the equilibrium expression for reaction (1). Since reaction (1) is the major reaction, we shall assume that it is at equilibrium up to the point where reaction (2) sets in. Thus $(J_a)^1_{out} = (K_a)^1$, so

$$\left(\frac{a_{CO_2}}{a_{CO}}\right)_{out} = (K_a)^1 = 0.39$$

Again assuming ideal mixtures of ideal gases,

$$(a_{CO_2})_{out} = (y_{CO_2})_{out} P$$

$$(a_{CO})_{out} = (y_{CO})_{out} P$$

Substitution of the activity expressions into the equilibrium constant then yields

$$\left(\frac{y_{CO_2}}{y_{CO}}\right)_{out} = 0.39$$

Assuming that CO_2 and CO are the only gaseous components present in the converter,

$$(y_{CO_2})_{out} + (y_{CO})_{out} = 1.0$$

or

$$(y_{CO_2})_{out} = 1.0 - (y_{CO})_{out}$$

Therefore,

$$\left(\frac{1 - y_{CO}}{y_{CO}}\right)_{out} = 0.39$$

Solution of the last equation for y_{CO} then gives

$$(y_{CO})_{out} = 0.72$$

We may substitute this value for $(y_{CO})_{out}$ into the expression for the system pressure to get

$$P < \frac{39.0 \text{ atm}}{0.72} = 54.2 \text{ atm}$$

Therefore, as long as P is *less* than 54.2 atm, reaction (2) cannot occur!

In summarizing our discussion of equilibrium in multiphase, or heterogeneous, systems it is useful to remember that the basic relation

$$-\Delta\mathscr{G}^\circ = RT \ln K_a = RT \ln \prod_{i=1}^{C} (a_i)_{out}^{\alpha_i} \tag{12-16}$$

holds no matter how many phases are present. If one (or more) of the components exist in several phases, the activity of that component must be the same in all phases if true equilibrium exists. Thus we can choose to evaluate the activity of this component in whatever phase it is most convenient to do so. Since in certain instances the activity of a component in a given phase is known a priori (for example, a pure solid), the task of evaluating J_a can be greatly simplified by the choice of the phase in which to calculate the various activities. The preceding sample problems have served to illustrate the application of this useful strategy.

12.4 Competing Heterogeneous Reactions

In the reduction of iron ore in a blast furnace the following is a partial list of the many reactions that may be expected to occur:

$$\begin{aligned}
FeO + C &= Fe + CO \\
C + \tfrac{1}{2}O_2 &= CO \\
CO + \tfrac{1}{2}O_2 &= CO_2 \\
FeO + CO &= Fe + CO_2 \\
2CO &= C + CO_2 \\
Fe + \tfrac{1}{2}O_2 &= FeO \\
CaCO_3 &= CaO + CO_2 \\
H_2O + C &= CO + H_2 \\
H_2O + CO &= CO_2 + H_2 \\
FeO + H_2 &= Fe + H_2O \\
H_2 + \tfrac{1}{2}O_2 &= H_2O
\end{aligned}$$

Although many other reactions are possible, the list includes most of the important reactions. As in the case of competing homogeneous reactions, we may attack problems of this nature from two separate, but equivalent, paths: (1) We may attempt to write equilibrium expressions for the primary reactions that describe the system, or (2) we may attempt to find the compositions of the reacting species that correspond to the minimum Gibbs free energy (subject, of course, to the conditions of constant T and P and satisfaction of the conservation of atomic species).

The first method contains a new difficulty when heterogeneous systems are considered: Some reactions may not proceed to equilibrium (the second technique handles this problem automatically). Thus our first problem is to decide which, if any, reactions may not proceed to equilibrium. Once this is determined, we proceed as before to solve the remaining equilibrium expressions simultaneously. However, since the equilibrium conditions so determined will determine whether a given side reaction will proceed or not, it is frequently impossible to know beforehand which reactions will attain equilibrium and which will not, so trial and error are frequently needed. Clearly this is an extremely unappealing procedure if more than two or three primary reactions are to be considered. For larger problems of this nature a slight modification of the method of White, Dantzig, and Johnson[1] described in Chapter 11 is much to be preferred (even for hand calculations). We now demonstrate both types of calculations by means of sample calculations.

SAMPLE PROBLEM 12-8. A laboratory-scale continuous blast furnace is operated with pure FeO, C, and O_2 feed streams, in the ratio 1 mol of FeO/3 mol of C/1 mol of O_2. The furnace is operated at 1800°F and 1 atm pressure. Assume cocurrent flow of the solid and gaseous streams. The following are assumed to be the only significant reactions that occur:

$$C + \tfrac{1}{2}O_2 = CO \tag{1}$$

$$CO + \tfrac{1}{2}O_2 = CO_2 \tag{2}$$

$$CO_2 + C = 2CO \tag{3}$$

$$FeO + CO = Fe + CO_2 \tag{4}$$

$$FeO + C = Fe + CO \tag{5}$$

$$Fe + \tfrac{1}{2}O_2 = FeO \tag{6}$$

The equilibrium constants for these reactions at 1800°F are

$$(K_a)^1 = 2.0 \times 10^9$$

$$(K_a)^2 = 1.4 \times 10^7$$

$$(K_a)^3 = 1.4 \times 10^2$$

$$(K_a)^4 = 4.0 \times 10^{-1}$$

$$(K_a)^5 = 5.6 \times 10^1$$

$$(K_a)^6 = 3.6 \times 10^7$$

[1] W. B. White, S. M. Johnson, and G. B. Dantzig, *J. Chem. Phys.*, *28*, no. 5, 751 (1958).

when the standard states are chosen as the pure components (gases or solids) at 1 atm and 1800 °F. Assuming that equilibrium exists in all reactions where possible, determine the composition of the gases leaving the converter and the conversion of FeO to Fe.

Solution: Examination of the six possible reactions by means of Gaussian elimination indicates that there are only three independent reactions. For the purposes of this problem, let us use reactions (1), (2), and (4) as the primary reactions, with equations (3), (5) and (6) being considered the dependent equations. Thus, we shall now work with only reactions (1), (2), and (4). (This is not the only set of reactions that could be chosen.) The basis of all calculations is 1 mol of FeO, 3 mol of C, and 1 mol of O_2. Since we are attempting to convert FeO to Fe, let us *assume* that reaction 4 goes to completion, so there is no FeO left. We can now calculate the equilibrium ratio of CO/CO_2 from reactions (1) and (2) and then check the assumption about reaction (4). Assume that at the end of the reaction there are x moles of CO and y moles of CO_2 in the exhaust stream (that is, that the extents of reaction are $X_1 = x$, $X_2 = y$). The exhaust-gas composition may then be summarized as shown in Table S12-8.

TABLE S12-8

Composition	Amount	Mole Fraction
O_2	$\frac{3}{2} - y - \frac{1}{2}x$	$\frac{3 - 2y - x}{3 + x}$
CO	x	$\frac{2x}{3 + x}$
CO_2	y	$\frac{2y}{3 + x}$
	$\frac{1}{2}(3 + x)$	1

We may now use the equilibrium expressions for reactions (1) and (2) to determine x and y as follows:

$$(K_a)^1 = \frac{a_{CO}}{a_C(a_{O_2})^{1/2}}$$

$$(K_a)^2 = \frac{a_{CO_2}}{a_{CO}(a_{O_2})^{1/2}}$$

If we assume ideal components, $f^\circ_{CO} = f^\circ_{O_2} = f^\circ_{CO_2} = 1.0$. Since the reactions occur at 1 atm, $f_{CO} = f_{O_2} = f_{CO_2} = 1.0$. If we now assume ideal gaseous solutions, the activities of the gaseous components reduce to their respective mole fractions. Since the solid phases of carbon, iron, and iron oxide are all in their standard states, their activities are identically equal to unity. Therefore, the equilibrium expressions reduce to

$$(K_a)^1 = 2.0 \times 10^9 = \frac{y_{CO}}{(y_{O_2})^{1/2}} = \frac{2x/(3 + x)}{\left(\dfrac{3 - 2y - 2x}{3 + x}\right)^{1/2}}$$

$$(K_a)^2 = 1.4 \times 10^7 = \frac{y_{CO_2}}{y_{CO}(y_{O_2})^{1/2}} = \frac{2y/(3 + x)}{\dfrac{2x}{3 + x}\left(\dfrac{3 - 2y - 2x}{3 + x}\right)^{1/2}}$$

which in turn reduce to

$$2.0 \times 19^9 = \frac{2x}{\sqrt{(3 - 2y - x)(3 + x)}}$$

$$1.4 \times 10^7 = \frac{y\sqrt{3 + x}}{x\sqrt{3 - 2y - x}}$$

Multiplication of the two equilibrium expressions gives

$$2.8 \times 10^{16} = \frac{2y}{3 - 2y - x}$$

or

$$\frac{3 - 2y - x}{2y} = 3.5 \times 10^{-17}$$

Therefore, we see that the oxygen concentration is going to be extremely small, so our assumption of complete reduction of FeO to Fe will probably be correct! Since

$$3 - x - 2y = (7.0 \times 10^{-17})y$$

we may say, without significant error, that

$$x + 2y = 3.0$$

Division of the two equilibrium expressions then yields

$$1.43 \times 10^2 = \frac{2.0x^2}{y(3 + x)}$$

But $y = 1.5 - 0.5x$. Therefore,

$$71.5(1.5 - 0.5x)(3 + x) = x^2$$

which may be solved for x to yield

$$x = 2.96$$

Substitution of $x = 2.96$ into the expression

$$y = \frac{2.0x^2}{143(3 + x)}$$

then yields

$$y = 2.0 \times 10^{-2}$$

We have used this equation to determine y rather than the simpler form

$$y = 1.5 - 0.5x$$

because the simpler form leads to the subtraction of two almost equal numbers and a chance for loss in accuracy.

Thus the exit gases from the furnace contain, $(0.5)(5.96) = 2.9$ mol and have a composition

$$y_{O_2} = 0.02\left(\frac{3.5 \times 10^{-17}}{2.98}\right) = 2.4 \times 10^{-17}$$

$$y_{CO} = \frac{2x}{5.96} = 0.993$$

$$y_{CO_2} = \frac{2y}{5.96} = 6.71 \times 10^{-3}$$

Using these values for the exit composition we must check our assumption that no FeO remains unreduced. Since we have chosen equation (4) as our primary equation, we shall calculate $(J_a)^4_{out}$. If $(J_a)^4_{out} < (K_a)^4$, then the reaction will proceed to completion:

$$(J_a)^4_{out} = \left(\frac{a_{Fe}a_{CO_2}}{a_{FeO}a_{CO}}\right)_{out}$$

but $a_{Fe} = a_{FeO} = 1.0$ and $a_{CO_2} = y_{CO_2}$; $a_{CO} = y_{CO}$. Therefore, $(J_a)^4_{out}$ reduces to

$$(J_a)^4_{out} = \left(\frac{y_{CO_2}}{y_{CO}}\right)_{out} = \frac{6.68 \times 10^{-3}}{0.993} = 6.74 \times 10^{-3}$$

But $(K_a)^4 = 4.0 \times 10^{-1}$. Since $(J_a)^4_{out}$ is less than $(K_a)^4$, the reduction of FeO to Fe must go to completion as assumed, and the exit-gas composition we calculated on the basis of this assumption is therefore correct!

SAMPLE PROBLEM 12-9. A blast furnace for the continuous reduction of ferric oxide (FeO) is fed the following feed mixture: 1 mol of FeO (solid), 2.0 mol of coke (solid carbon), and 1.5 mol of an equimolar mixture of H_2 and CO, and 0.5 mol of oxygen. Determine the composition of the off-gas and the fractional conversion of FeO to Fe. Will any excess carbon remain? (Assume that all solid phases are immiscible at these temperatures; it is not true, but it is a place to begin.) The reaction section of the blast furnace will operate at 2000°F, and it will be assumed that thermodynamic equilibrium is established. The standard-state Gibbs free energies for the reacting species at 2000°F are given in Table S12-9.

TABLE S12-9

Compound	Name	$G°$, kcal/g-mol
1	Fe	0.0
2	FeO	−41.7
3	C	0.0
4	CO	−55.3
5	CO_2	−94.7
6	H_2	0.0
7	O_2	0.0
8	H_2O	−41.0

Solution: This reacting system is too complex for us to attempt to describe by means of a series of primary reactions. Rather, we shall describe the equilibrium conditions by the minimum in the total Gibbs free energy. Thus we may (and will) use an extention of the method developed in Chapter 11 for determining equilibrium compositions in a complex reacting mixture.

We shall derive the method of steepest descent for heterogeneous reacting systems in much the same way as we did for homogeneous systems. Slight differences in the derivation and results will become apparent as we proceed.

We begin by assuming that we have some approximate solution $(n_1^k, n_2^k, \ldots, n_c^k)$ for the equilibrium mole numbers of the reacting mixture. We shall assume that S

solid components (mutually insoluble) may be present and that they will be numbered chemical components 1, 2, . . . , S. The remaining chemical components are numbered $S+1$, $S+2, \ldots, C$ and will be assumed to be gaseous.

The total Gibbs free energy for the kth iteration reaction mixture is given by

$$F^k = \frac{G^k}{RT} = \sum_{i=1}^{S} n_i^k \frac{G_i^\circ}{RT} + \sum_{i=S+1}^{C} n_i^k \left(\frac{G_i^\circ}{RT} + \ln \frac{n_i^k}{n^k}\right)$$

where n = sum of mole numbers of all gases:

$$n = \sum_{i=S+1}^{C} n_i$$

The mass-balance (conservation of atomic species) constraints are still given by

$$\sum_{i=1}^{C} a_{ji} n_i = b_j \qquad \text{for } 1 \le j \le M$$

where M = number of atomic species present
a_{ji} = number of j atoms in chemical compound i

Let us expand the total Gibbs free energy about the values after the kth iteration:

$$F^{k+1} = F^k + \sum_{i=1}^{C} (n_i^{k+1} - n_i^k) \frac{\partial F^k}{\partial n_i} + \frac{1}{2} \sum_{i=1}^{C} \sum_{l=1}^{C} (n_i^{k+1} - n_i^k)(n_l^{k+1} - n_l^k) \frac{\partial^2 F^k}{\partial n_i \, \partial n_l}$$

where F^k is as given above.

The values of the partials of F^k are given by

$1 \le i \le S$	$S+1 \le i \le C$
$\dfrac{\partial F^k}{\partial n_i} = \dfrac{G_i^\circ}{RT}$	$\dfrac{\partial F^k}{\partial n_i} = \dfrac{G_i^\circ}{RT} + \ln \dfrac{n_i^k}{n^k}$
$\dfrac{\partial^2 F^k}{\partial n_i \, \partial n_l} = 0$	$\dfrac{\partial^2 F^k}{\partial n_i^2} = \dfrac{1}{n_i^k} - \dfrac{1}{n^k}$
	$\dfrac{\partial^2 F^k}{\partial n_i \, \partial n_l} = \begin{cases} 0 & \text{for } \quad 1 < l \le S \\ -1/n^k & \text{for } S+1 \le l \le C \end{cases}$

These are substituted into the Taylor's series to give:

$$\begin{aligned} F^{k+1} = {} & \sum_{i=1}^{S} n_i^k \frac{G_i^\circ}{RT} + \sum_{i=S+1}^{C} n_i^k \left(\frac{G_i^\circ}{RT} + \ln \frac{n_i^k}{n^k}\right) + \sum_{i=1}^{S} (n_i^{k+1} - n_i^k) \frac{G_i^\circ}{RT} \\ & + \sum_{i=S+1}^{C} (n_i^{k+1} - n_i^k) \left(\frac{G_i^\circ}{RT} + \ln \frac{n_i^k}{n^k}\right) \\ & + \frac{1}{2} \sum_{i=S+1}^{C} n_i^k \left[\left(\frac{n_i^{k+1} - n_i^k}{n_i^k}\right) - \left(\frac{n^{k+1} - n^k}{n^k}\right)\right] \end{aligned}$$

As in the case of homogeneous reactions, we introduce the mass-balance equations by means of the Lagrangian multipliers:

$$\Phi^{k+1} = F^{k+1} + \sum_{j=1}^{M} \pi_j \left(-\sum_{i=1}^{C} a_{ji} n_i^{k+1} + b_j\right)$$

As long as the mass-balance equations are satisfied, $\Phi = F$, and thus the minimum in Φ is also the minimum in F.

The condition for minimizing Φ^{k+1} is the vanishing of all the partials of Φ^{k+1} with respect to n_i^{k+1}. Thus let us set

$$\frac{\partial \Phi^{k+1}}{\partial n_i^{k+1}} = 0 \qquad \text{for} \quad 1 \le i \le C$$

Differentiation of the expression for Φ^{k+1} then yields

$$\frac{\partial \Phi^{k+1}}{\partial n_i^{k+1}} = \begin{cases} \left(\frac{G_i^\circ}{RT} - \sum_{j=1}^{M} \pi_j a_{ji}\right) = 0 & 1 \le i \le S \\ \left(\frac{G_i^\circ}{RT} + \ln \frac{n_i^k}{n^k}\right) + \left(\frac{n_i^{k+1}}{n_i^k} - \frac{n^{k+1}}{n^k}\right) - \sum_{j=1}^{M} \pi_j a_{ji} = 0 & S+1 \le i \le C \end{cases}$$

These equations (C of them) contain $C + M + 1$ unknowns. The other $M + 1$ equations are the M mass-balance relations

$$\sum_{i=1}^{C} a_{ji} n_i^{k+1} = b_j \qquad 1 \le j \le M$$

and the definition of n^{k+1}

$$n^{k+1} = \sum_{i=S+1}^{C} n_i^{k+1}$$

The size of the system of equations that must be solved is reduced by hand eliminating some of the unknowns. To this end, we proceed as follows. Solve for n_i^{k+1} in the $\partial\Phi^{k+1}/\partial n_i^{k+1} = 0$ equations ($S + 1 \le i \le C$):

$$n_i^{k+1} = n_i^k \left[\frac{n^{k+1}}{n^k} + \sum_{j=1}^{M} \pi_j a_{ji} - \left(\frac{G_i^\circ}{RT} + \ln \frac{n_i^k}{n^k}\right)\right] \qquad S+1 \le i \le C$$

Sum these over all $S + 1 \le i \le C$, remembering that $n = \sum_{i=S+1}^{C} n_i$:

$$n^{k+1} = n^k \frac{n^{k+1}}{n^k} + \sum_{i=S+1}^{C} n_i^k \sum_{j=1}^{M} \pi_j a_{ji} - \sum_{i=S+1}^{C} n_i^k \left(\frac{G_i^\circ}{RT} + \ln \frac{n_i^k}{n^k}\right)$$

Simplifying yields

$$\sum_{i=S+1}^{C} n_i^k \sum_{j=1}^{M} \pi_j a_{ji} = \sum_{i=S+1}^{C} n_i^k \left(\frac{G_i^\circ}{RT} + \ln \frac{n_i^k}{n^k}\right)$$

The remaining $\partial\Phi^{k+1}/\partial n_i^{k+1}$ equations are

$$\sum_{j=1}^{M} \pi_j a_{ji} = \frac{G_i^\circ}{RT} \qquad 1 \le i \le S$$

Multiply these by n_i^k and sum on $1 \le i \le S$ and add to the sum of the other $\partial\Phi^{k+1}/\partial n_i^{k+1}$ equations to give

$$\sum_{i=1}^{C} n_i^k \sum_{j=1}^{M} \pi_j a_{ji} = \sum_{i=1}^{S} n_i^k \frac{G_i^\circ}{RT} + \sum_{i=S+1}^{C} n_i^k \left(\frac{G_i^\circ}{RT} + \ln \frac{n_i^k}{n^k}\right)$$

Reversing the summations of the left-hand side gives

$$\sum_{i=1}^{C} n_i^k \sum_{j=1}^{M} \pi_j a_{ji} = \sum_{j=1}^{M} \pi_j \sum_{i=1}^{C} n_i^k a_{ji}$$

Since n_i^k satisfies the mass-balance equations, we then obtain

$$\sum_{j=1}^{M} \pi_j b_j = \sum_{i=1}^{S} n_i^k \frac{G_i^\circ}{RT} + \sum_{i=S+1}^{C} n_i^k \left(\frac{G_i^\circ}{RT} + \ln \frac{n_i^k}{n^k}\right)$$

The n_i^{k+1} must also satisfy the mass-balance equations. Thus for $S + 1 \le i \le C$ substitute the values of n_i^{k+1} found from the $\partial\Phi^{k+1}/\partial n_i^{k+1}$ equations into the mass balances. For $1 \le i \le S$, leave n_i^{k+1}:

$$\sum_{i=1}^{S} a_{ji} n_i^{k+1} + \sum_{i=S+1}^{C} a_{ji} \left[n^{k+1} \frac{n_i^k}{n^k} + n_i^k \sum_{l=1}^{M} \pi_l a_{li} - n_i^k \left(\frac{G_i^\circ}{RT} + \ln \frac{n_i^k}{n^k}\right)\right] = b_j \qquad 1 \le j \le M$$

The known terms are cleared to the right-hand side to give

$$\sum_{i=S+1}^{C} a_{ji}n_i^k \sum_{l=1}^{M} a_{li}\pi_l + \sum_{i=1}^{S} a_{ji}n_i^{k+1} + n^{k+1} \sum_{i=S+1}^{C} a_{ji}\left(\frac{n_i^k}{n^k}\right) = b_j + \sum_{i=S+1}^{C} a_{ji}n_i^k\left(\frac{G_i^\circ}{RT} + \ln\frac{n_i^k}{n^k}\right)$$

Reversal of the summation order and introduction of the matrix $r_{jl} = \sum_{i=S+1}^{C} a_{ji}a_{li}n_i^k$ then puts this equation into a form similar to that derived for the homogeneous reaction case:

$$\sum_{l=1}^{M} r_{jl}\pi_l + \sum_{i=1}^{S} a_{ji}n_i^{k+1} + n^{k+1}\left(\sum_{i=S+1}^{C} a_{ji}\frac{n_i^k}{n^k}\right) = b_j + \sum_{i=S+1}^{C} a_{ji}n_i^k\left(\frac{G_i^\circ}{RT} + \ln\frac{n_i^k}{n^k}\right)$$

$$1 \le j \le M$$

These M equations, along with the previously derived equations

$$\sum_{j=1}^{M} \pi_j a_{ji} = \frac{G_i^\circ}{RT} \qquad 1 \le i \le S$$

and

$$\sum_{j=1}^{M} \pi_j b_j = \sum_{i=1}^{S} n_i^k \frac{G_i^\circ}{RT} + \sum_{i=S+1}^{C} n_i^k\left(\frac{G_i^\circ}{RT} + \ln\frac{n_i^k}{n^k}\right)$$

then give us the reduced set of $M + 1 + S$ equations which must be solved for the $M + 1 + S$ unknowns (π_j, n^{k+1}, and $n_1^{k+1}, n_2^{k+1}, \ldots, n_S^{k+1}$). The n_i^{k+1} for the gas-phase components are then determined directly in terms of n^{k+1}.

A program to perform the required calculations is presented below. The concentrations of any trace gaseous components may be estimated by an extension of the method presented with the homogeneous reaction problem.

```
      INTEGER C, M, A, S
      REAL N, NN, LAM, NT, NNT, LL
      LOGICAL CHECK
      DIMENSION A(25,15), PI(25), B(25), AA(16,16), BB(16), CC((16), N(25)
      DIMENSION NN(25), GRT( 25 ), DEL(25), RHO(25), DFDL(11), R(20,20)
      DIMENSION X(25)
C        READ INPUT DATA AND BEGIN SETTING UP
   10 READ (5, 20) C,M,S, ITMAX
   20 FORMAT ( 4I5 )
      DELTA = 0.0001
   25 FORMAT ( " C = ", I3, ", M = ", I3, ", S = ", I3 )
C        READ COEFFICIENT MATRICES A ROW AT A TIME SOLIDS FIRST
      DO 29 J=1,M
   29 READ (5,30) (A(J,I),I=1,C)
   30 FORMAT (13I5)
C        READ INLET AMOUNTS
      READ (5,40) (NN(I), I=1,C)
   40 FORMAT ( 6F10.5 )
      NT =0
      DO 45 I=1,C
   45 X(I) = 1.0
C        READ G/RT FOR EACH COMPONENT
      READ ( 5,60) ( GRT(I), I=1,C)
   60 FORMAT ( 6F10.5 )
   63 FORMAT ( 10X, F10.5)
```

```
C         READ THE INITIAL COMPOSITION GUESSES
   75     READ( 5,40 ) ( N(I), I=1,C)
          WRITE (6,25) C,M,S
          WRITE ( 6,33 )
   33     FORMAT ( / " THE COEFFICIENT MATRIX A(J,I) IS GIVEN BY "/)
          DO 50 I=S+1,C
   50     NT = NT+ N(I)
          DO 34 J=1,M
   34     WRITE (6,30) ( A(J,I),I=1,C)
          WRITE (6,70)
   70     FORMAT (/ " THE GRT(I), NN(I), AND N(I) VECTORS ARE "/)
          WRITE (6,67) ( GRT(I), NN(I), N(I), I=1,C)
   67     FORMAT ( E10.4, 5X, E10.4, 5X, E10.4)
C         BEGIN CALCULATIONS- -CALCULATE B VECTORS AND CHECK MASS
     1    BAL
          DO 100 J=1,M
          IFREQ = 1
          B(J) = 0.0
          DO 100 I=1,C
  100     B(J) = B(J) + A(J,I) * NN(I)
          CHECK = .FALSE.
          DO 160 J=1,M
          BCHECK = 0.0
          DO 150 I=1,C
  150     BCHECK = BCHECK+A(J,I) * N(I)
  160     IF( ABS( BCHECK-B(J)) .GT. (DELTA * B(J))) CHECK =.TRUE.
          IF ( .NOT. CHECK ) GO TO 220
          WRITE ( 6,190)
  190     FORMAT ( " BAD INITIAL GUESS - B(J) VECTORS DO NOT CHECK ")
          GO TO 75
C         BEGIN EVALUATION OF THE NEW COMPOSITIONS
  220     IT = 1
C         LOAD THE COEFFICIENT MATRIX
  235     DO 230 J=1,M
          DO 230 L=1,M
          AA(J,L) =0.0
          DO 230 I = S+1,C
  230     AA(J,L) = AA(J,L) + A(J,I)*A(L,I)*N(I)
          DO 270 J=1,M
          CC(J) = B(J)
          DO 260 I=1,S
  260     AA(J,I+M) = A(J,I)
          AA( J, M+S+1) = 0.0
          DO 270 I=S+1,C
          CC(J) = CC(J) + A(J,I)*N(I)*( GRT(I) + ALOG( N(I)/NT))
  270     AA(J,M+S+1) = AA(J,M+S+1) + A(J,I)*N(I)/NT
          DO 280 L= M+1, M+S
          DO 275 K=1,M
  275     AA(L,K) = A(K,L - M)
```

```
          DO 278 K=M+1, M+S+1
    278   AA(L,K) = 0.0
    280   CC(L) = GRT( L-M)
          DO 290 K=1, M
    290   AA( M+S+1,K ) = B(K)
          DO 300 K = M+1, M+S+1
    300   AA(M+S+1,K) = 0.0
          L=M+S+1
          CC(L) = 0.0
          DO 310 K=1,S
    310   CC(L) = CC(L) + N(K)*GRT(K)
          DO 320 K=S+1,C
    320   CC(L) = CC(L) + N(K)*( GRT(K) + ALOG( N(K)/NT ) )
C           INVERT THE MATRIX AND SOLVE FOR PI'S, NT, AND SOLID N(I)S
          CALL LINEQ ( AA, CC, BB, L, 16, IFLAG )
          DO 340 K=1,M
    340   PI(K) = BB(K)
          DO 360 K=M+1, M+S
    360   NN(K-M) = BB(K)
          NNT = BB( M+S+1 )
C           DETERMINE NEW GASEOUS MOL NUMBERS --
          DO 380 I = S+1,C
          SUM = 0.0
          DO 370 J=1,M
    370   SUM = SUM + PI(J)*A(J,I)
    380   NN(I) = N(I)*( NNT/NT + SUM -GRT(I) -ALOG( N(I)/NT ) )
C           EVALUATE MAXIMUM LAMBDA WHICH SATISFIES CRITERION 1
          DO 400 I=1, C
    400   DEL(I) = NN(I) - N(I)
          DELT = NNT-NT
          LAM = 1.0
          DO 420 I=1,C
          IF ( NN(I) .GT. 0.0 ) GO TO 420
          LL = ( N(I)*0.99999/ ( N(I) - NN(I)))
    420   IF ( LL .LT. LAM ) LAM = LL
          DO 430 I=1,C
    430   RHO(I) = N(I) + LAM*DEL(I)
C           LAM NOW CONTAINS MAXIMUM LAMBDA FOR CONDITION 1
C           FIND LAMBDA FOR SECOND CONDITION- DECREASE IN 10%
C           DECREMENTS
          DO 520 L=1,11
          LL = LAM * ( 1.0 -(L-1)/10. )
          DFDL(L) = 0.0
          DO 510 I=1,S
    510   DFDL(L) = DFDL(L) + DEL(I)*GRT(I)
          DO 520 I=S+1,C
    520   DFDL(L) = DFDL(L) + DEL(I)*( GRT(I) + ALOG( (N(I) + LL*
         1DEL(I))/(NT + DELT*LL)))
          DO 530 L=1,11
          L1 =L
```

```
 530    IF ( DFDL(L) .LE. 0.0 ) GO TO 550
 540    WRITE ( 6, 545)
 545    FORMAT ( " NO LAMBDA FOUND, TRY NEW CALCULATION ")
        GO TO 10
C         LAMBDA FOUND, UPDATE N(I)S, AND RETURN FOR NEW ITERATION
 550    LAM = LAM* ( 1.0 - (L1-1)/10.0
        DO 570 I=1,C
 570    N(I) = N(I) + LAM*DEL(I)
        NT = NT + LAM*DELT
C         PRINT RESULTS OF THE ITERATION - -
        WRITE ( 6,590 ) IT, LAM, NNT
 590    FORMAT ( " AFTER ", I4, " ITERATIONS LAM = ", F5.3,
     1  " AND NNT = ", E 15.8)
        WRITE (6,600) ( N(I), I=1,C)
 600    FORMAT ( 15X, E15.8 )
        IT = IT +1
C         CHECK FOR CONVERGENCE
        CHECK = .TRUE.
        DELTA2 = DELTA
        DO 620 I=1,C
 620    IF ( ABS(DEL(I)/N(I)) .GT. DELTA2) CHECK = .FALSE.
        IF ( CHECK ) GO TO 650
        IF ( IT .LT. ITMAX ) GO TO 235
        WRITE ( 6,630)
 630    FORMAT ( " THE ITERATION FAILED- -BAIL OUT ")
        GO TO 10
 650    WRITE (6,670)
 670    FORMAT ("CONVERGENCE OBTAINED, FINAL MOL NUMBERS ARE" )
        WRITE ( 6,490) ( N(I),I=1,C)
        GO TO 10
 490    FORMAT (/( I0X, EI5,8))
        END
```

The results of the computations are shown below:

```
                C = 8, M = 4, S = 3
THE COEFFICIENT MATRIX A(J,I) IS GIVEN BY
      1    1    0    0    0    0    0    0
      0    1    0    1    2    0    2    1
      0    0    1    1    1    0    0    0
      0    0    0    0    0    2    0    2
THE GRT(I), NN(I), AND N(I) VECTORS ARE
 0.               0.               .9500E 00
 -.8530E 01       .1000E 01        .5000E-01
 0.               .2000E 01        .1000E 01
 -.1130E 02       .7500E 00        .1520E 01
 -.1940E 02       0.               .2300E 00
 0.               .7500E 00        .5000E-01
 0.               .5000E 00        .1000E-01
 -.8390E 01       0.               .7000E 00
```

```
AFTER 1 ITERATIONS LAM = 0.002 AND NNT = .22407055E 02
                .99999950E 00
                .50000017E-06
                .95598321E 00
                .15575884E 01
                .23642842E 00
                .50286383E-01
                .99203398E-02
                .69971362E 00
AFTER 2 ITERATIONS LAM = 0.000 AND NNT = .23279819E 02
                .10000000E 01
                .50000004E-11
                .95598277E 00
                .15575887E 01
                .23642848E 00
                .50286386E-01
                .99203391E-02
                .69971361E 00
AFTER 3 ITERATIONS LAM = 0.000 AND NNT = .23279827E 02
                .10000000E 01
                .50000011E-16
                .95598277E 00
                .15575887E 01
                .23642848E 00
                .50286386E-01
                .99203391E-02
                .69971361E 00
AFTER 4 ITERATIONS LAM = 0.000 AND NNT = .23279827E 02
                .10000000E 01
                .50000009E-21
                .95598277E 00
                .15575887E 01
                .23642848E 00
                .50286386E-01
                .99203391E-02
                .69971361E 00
```

At first glance it would appear that the computation has converged. However a closer examination indicates some problems. The first sign of difficulty is tha LAM=0.0—an indication that the computation is not proceeding. Also the sum o the gaseous mol numbers does not equal NNT. To see what has gone wrong, we mus examine the gaseous and solid mol numbers predicted by the matrix inversion routin before the lambda corrections are applied. The results of these computations for th fifth iteration are shown below.

```
THE NEW AND PARTLY CORRECTED NS ARE
     Predicted by          Corrected for
     matrix inversion      LAM = 0.0
```

```
 .24596692E 02     .10000000E 01
-.23596692E 02     .50000017E-31
-.19803576E 02     .95598277E 00
 .19282584E 02     .15575887E 01
 .32709915E 01     .23642848E 00
 .18037804E 00     .50286386E-01
-.23748565E-01     .99203391E-02
 .56962196E 00     .69971361E 00
```

Notice that the mole numbers for FeO and C are highly negative in the first column (uncorrected). By choosing LAM sufficiently small, the corrected values in the second column are all made positive. However, since the activity of FeO is not affected by its amount (in the gas phase this is not true, of course), the activity of FeO has not been reduced during the iteration. Since the reaction

$$\text{FeO} + \text{C} \longrightarrow \text{Fe} + \text{CO}$$

still has a strong tendency to occur ($J_a \ll K_a$), the computations will attempt to form more Fe and CO—so much so that the quantities of Fe and C remaining will be negative again. Once again a small lambda will have to be chosen to correct these negative values. Thus as long as FeO (or for that matter, any solid component which is not present in the equilibrium mixture) is carried in the computation, the amount of FeO present in the uncorrected mole numbers will go negative, and the calculation will cease to proceed in a useless manner. We overcome this problem by simply removing FeO from the computation scheme. The atom balances are corrected for the lack of FeO in the input stream by adding 1 mole Fe and $\frac{1}{2}$ mole O_2 to the inlet component amounts. In this fashion, the correct atomic abundances (b_j's) are retained. At this point we cannot determine whether carbon will have to be removed from the computation scheme, but this will be ascertained as the calculations proceed. The calculations with FeO removed are shown below.

```
            C = 7, M = 4, S = 2
    THE COEFFICIENT MATRIX A(J,I) IS GIVEN BY
          1   0   0   0   0   0   0
          0   0   1   2   0   2   1
          0   1   1   1   0   0   0
          0   0   0   0   2   0   2
    THE GRT(I), NN(I), AND N(I) VECTORS ARE
     0.               .1000E 01     .1000E 01
     0.               .2000E 01     .1000E 01
    -.1130E 02        .7500E 00     .1520E 01
    -.1940E 02       0.             .2300E 00
     0.               .7500E 00     .5000E-01
     0.               .1000E 01     .3500E-01
    -.8390E 01       0.             .7000E 00
AFTER 1 ITERATIONS LAM = 0.057 AND NNT = .34364278E 01
                 .10000000E 01
                 .91394320E 00
                 .16078523E 01
                 .22820450E 00
```

```
                        .64261996E-01
                        .34999994E-06
                        .68573800E 00
AFTER 2 ITERATIONS LAM = 0.143 AND NNT = .31314964E 01
                        .10000000E 01
                        .83591122E 00
                        .17178045E 01
                        .19628430E 00
                        .11037307E 00
                        .34999972E-11
                        .63962693E 00
AFTER 3 ITERATIONS LAM = 1.000 AND NNT = .32811111E 01
                        .10000000E 01
                        .21888886E 00
                        .24918504E 01
                        .39260711E-01
                        .57037185E 00
                        .19425290E-10
                        .17962815E 00
AFTER 4 ITERATIONS LAM = 0.521 AND NNT = .35264620E 01
                        .10000000E 01
                        .91009465E-01
                        .26039081E 01
                        .55082398E-01
                        .71407293E 00
                        .47775312E-10
                        .35927067E-01
AFTER 5 ITERATIONS LAM = 1.000 AND NNT = .33956981E 01
                        .10000000E 01
                        .10430186E 00
                        .25705221E 01
                        .75176014E-01
                        .72087416E 00
                        .13522224E-09
                        .29125844E-01
AFTER 6 ITERATIONS LAM = 1.000 AND NNT = .33915117E 01
                        .10000000E 01
                        .10848828E 00
                        .25626705E 01
                        .78841267E-01
                        .72035299E 00
                        .24136278E-09
                        .29647014E-01
AFTER 7 ITERATIONS LAM = 0.900 AND NNT = .33914267E 01
                        .10000000E 01
                        .10856483E 00
                        .25625203E 01
                        .78914889E-01
                        .72035006E 00
```

```
                                 .28598954E-09
                                 .29649936E-01
          AFTER 8 ITERATIONS LAM = 1.000 AND NNT = .33914266E 01
                                 .10000000E 01
                                 .10857337E 00
                                 .25625035E 01
                                 .78923107E-01
                                 .72034974E 00
                                 .29622358E-09
                                 .29650260E-01
          NO LAMBDA FOUND, TRY NEW CALCULATION
```

Although the computation has not converged to the level specified in the program, examination of the results of iterations 7 and 8 indicates that with the exception of oxygen, all other species are at their equilibrium conditions. The oxygen is causing a minor problem because its relative amount is so low. We could strike oxygen from the calculation, but clearly little would be gained. The results of iteration 8 are a satisfactory solution. Note that solid carbon is present in the final equilibrium mixture.

12.5 Effects of Reaction Rates

Although we have carefully avoided the subject of reaction rates up to this point in our discussion of chemical equilibrium, it is now desirable to briefly discuss this most vital area.

Since no reaction occurs at an infinite rate (although admittedly some reactions do occur quite rapidly), all reactions require a finite amount of time to reach equilibrium. Since the reactor in which a specific reaction is to be performed must be large enough to allow the reactants to reach the desired degree of conversion while still within the reactor, it is apparent that the rate at which the reaction progresses is as important as the maximum, or equilibrium, conversion for the reaction. A reaction that takes years to approach equilibrium is obviously of little economic value, even if the equilibrium that is finally reached is quite favorable. On the other hand, a reaction that approaches equilibrium rapidly may be economically attractive even if the equilibrium yield is quite small.

Since both reaction rates and equilibrium conversions are temperature dependent, it is frequently possible to adjust the temperature of the reaction so as to reach an economic optimum in the design of the reactor. Reaction rates, for the vast majority of reactions, increase with temperature, while equilibrium conversions may increase, or decrease, with temperature depending on the sign of $(\Delta\mathscr{H}^\circ)_{\text{react}}$. In the majority of the industrially important reactions $(\Delta\mathscr{H}^\circ)_{\text{react}}$ is negative (that is, the reactions are exothermic), so the equilibrium yield decreases with increasing temperature. For example, in the catalytic conversion of $N_2 + 3H_2 \rightleftharpoons 2NH_3$, the equilibrium conversions are quite high at low temperatures. However, at these temperatures the reaction rates are so low as to make the reaction totally unattractive economically. At higher temperatures the reaction rates become more reasonable but the equilibrium yield drops off. Obviously some compromise between rate and equilibrium conversion must be struck before an operating plant can be designed and built. In the NH_3 synthesis the reaction rates are so low at even moderate temperatures (where the equilibrium conversions are only mediocre) that this reaction is normally run at fairly high temperatures, where the equilibrium conversions are very low. Since the NH_3 reaction is pressure sensitive, the synthesis is performed at extremely high pressures in an attempt to increase the equilibrium conversions. However, even at the elevated pressure used, the maximum conversion normally anticipated in an ammonia converter is only about 10 percent—well below the 99+ percent maximum conversions that are indicated by equilibrium calculations at low temperatures.

Reaction rates can also play a significant role in determining actual conversions in reactors where competing reactions may occur. Since the reaction rates for different reactions may vary considerably with temperature, it is often possible to exhibit some degree of control over the competing reactions by choosing the reaction temperature at a value where the desired reaction, or reactions, have the highest rates. For example, in discussing the reduction of ZnO with carbon, Lewis, Radasch, and Lewis make the following remarks[2]:

> In the reduction of zinc oxide with carbon in a conventional retort at atmospheric pressures three main reactions are involved. At temperatures of 600 to 700°C and above, the reaction $ZnO + CO = Zn + CO_2$ is rapid. At 900 to 1100°C ZnO reacts with carbon to produce Zn and CO. At 1100 to 1300°C, the range of temperature in which the bulk of the distillation is normally conducted, the rate of reduction of CO_2 by carbon is appreciable. At no stage in the operation is it desirable to have any significant amount of CO_2 in the exit gases, since as the gases are cooled, the reaction $ZnO + CO = Zn + CO_2$ reverses and a portion of the Zn vapors is re-oxidized. To suppress formation of CO_2 the usual practice is to provide a large excess of carbon in the original charge, so that the gas phase in the retort is in contact mainly with carbon, and to heat the charge as quickly as possible to temperatures at which the reaction $CO_2 + C = 2CO$ is rapid.

Problems

12-1 According to the phase rule, when H_2O liquid and vapor coexist alone in a system, that system possesses one degree of freedom. How do you explain the fact that a lake and water vapor can coexist at a given pressure for different temperatures?

[2] W. K. Lewis, A. H. Radasch, and H. C. Lewis, *Industrial Stoichiometry*, McGraw-Hill Book Company, Inc., New York, 1954, p. 356.

12-2 The reaction

$$H_2O + C(s) = CO + H_2$$

is frequently referred to as the *water–gas reaction.*

(a) From the data below calculate $\Delta\mathscr{G}°$ for this reaction proceeding to the right at 298°K and 1 atm; also at 1000°K and 1 atm. Specify the standard states used.

(b) Would a mixture of H_2O and C if brought to equilibrium at either of these temperatures and 1 atm contain CO or H_2? Your answer must be justified.

(c) Express K_a in terms of (1) activities and (2) concentrations. Indicate any necessary assumptions.

$\Delta \underline{G}_f$ in kcal/g-atom of oxygen, at 1 atm:

T, °K	$H_2O(g)$	$CO(g)$
298	−54.6	−32.8
1000	−46.0	−47.9

12-3 In the manufacture of producer gas, steam is passed over coke. The following reactions are known to occur:

$$C + H_2O = CO + H_2 \qquad (a)$$

$$CO + H_2O = CO_2 + H_2 \qquad (b)$$

$$C + CO_2 = 2CO \qquad (c)$$

(a) How many of these reactions are independent?

(b) If we fix the pressure and temperature of the reaction, how many other degrees of freedom are left?

(c) Steam is slowly passed over a coke bed at 1 atm and 1500°F and allowed to come to equilibrium with respect to the three reactions indicated above. If the equilibrium constants for these reactions are

$$(K_a)^1 = 10.0$$

$$(K_a)^2 = 1.0$$

$$(K_a)^3 = 10.0$$

where the standard states are pure gases and solids at 1 atm and the temperature of the system, what is the equilibrium composition of the gas leaving the coke bed reactor? (You may assume that there is always an excess of coke.)

12-4 The decomposition reaction

$$A = B + C$$

is performed at 100°C in the liquid state. The equilibrium constant for the reaction at 100°C is $K_a = 2.0$ where all standard states are pure liquids at 100°C. The vapor pressures at 100°C of the various components are

$$P'_A = 5 \text{ atm}$$

$$P'_B = 20 \text{ atm}$$

$$P'_C = 2 \text{ atm}$$

Calculate the equilibrium pressure of this reaction at 100°C and the composition of all liquid and vapor phases if the following assumptions are valid: (1) All vapors and their mixtures are ideal. (2) Liquid B is completely immiscible with mixtures of the liquids A and C; liquids A and C, on the other hand, are totally soluble in each other. (3) The A–C mixture is ideal.

12-5 The standard free-energy change for the formation of ethylene from the elements may be represented by the equation

$$\Delta \mathscr{G}^\circ = 5700 + 21.1T \text{ cal/g-mol}$$

where T is in °K and the reaction is given as

$$2C(\text{graphite}) + 2H_2(g) = C_2H_4(g)$$

On the basis of thermodynamic calculations, discuss the feasibility of the industrial preparation of ethylene by the above reaction.

12-6 At 828°F and 200 atm, the mole fraction of ammonia at equilibrium in the reaction

$$2FeN(s) + 3H_2(g) = 4Fe(s) + 2NH_3(g)$$

is 0.9142.

(a) Calculate K_a at the above pressure and temperature.

(b) Ammonia at 300 atm and 828°F is charged in a reactor with iron. The reactor is held at 250 atm and 828°F. A gas stream of hydrogen and ammonia containing 50 percent ammonia on a mole basis is drawn from the reactor at the temperature and pressure of the reactor. Calculate the maximum work extracted (or the minimum work required).

12-7 Hydrated sodium carbonate decomposes according to the reaction

$$Na_2CO_3 \cdot H_2O(s) = Na_2CO_3(s) + H_2O(g)$$

The equilibrium pressure in atmospheres of water vapor during this reaction is given by the relationship

$$\ln P' = 18.3 - \frac{6910}{T}$$

where T is in °K. Assuming that water vapor behaves ideally, derive an expression for $\Delta \mathscr{G}^\circ$ for the above reaction, as a function of temperature.

12-8 Ferrous oxide is reduced to metallic iron by passing a mixture of 20 per cent CO and 80 per cent N_2 over it at a temperature of 1000°C and a pressure of 1 atm:

$$FeO(s) + CO(g) = Fe(s) + CO_2(g)$$

At 1000°C, $\Delta \mathscr{G}^\circ$ for this reaction is 2300 cal/g-mol. How many pounds of metallic iron will be produced if 1000 ft^3 of CO (at 1000°C and 1 atm) is introduced and allowed to come to equilibrium?

12-9 The following partial pressure of CO and CO_2 are in equilibrium with austenite (a solid solution of iron and carbon) of the stated carbon contents at 1000°C according to the equation

$$2CO(g) = C(\text{in austenite}) + CO_2$$

C, wt %	P_{CO}, atm	P_{CO_2}, atm
0.13	0.891	0.109
0.45	0.9660	0.9340
0.96	0.9862	0.0138
1.50	0.9928	0.0072

The solution of 1.50 per cent is saturated with graphite at 1000°C.

(a) Find the activity of carbon in each case, taking pure graphite as the standard state.

(b) Plot the activities of carbon versus carbon concentration up to 2.0 per cent carbon.

(c) What will be the ratio of pressures of CH_4 to H_2 (at 1 atm total pressure) in equilibrium with iron containing 0.6 per cent carbon at 1000°C?

$\Delta \underline{G}_f$ for CH_4 at 1000°C and 1 atm is 11.78 kcal/mol.

12-10 It is desired to explore possibilities for decreasing the energy requirements necessary for decomposing Al_2O_3 to Al and O_2 according to the reaction

$$Al_2O_3(s) = 2Al(l) + \tfrac{3}{2}O_2(g)$$

$\Delta \mathscr{G}^\circ_{2500^\circ K} = 210$ kcal/mol of Al_2O_3. The standard states are pure solid Al_2O_3, pure liquid Al, and pure gaseous O_2 at 1 atm.

(a) If the oxygen pressure can be maintained at 10^{-6} atm, what is the maximum Al activity for which the reaction will proceed to the right?

(b) Can you think of any method by which this activity might realistically be achieved?

(c) Calculate the minimum work requirements to decompose 1 mole of Al_2O_3 to pure Al liquid and pure oxygen at a pressure of 0.1 atm.

12-11 It is known that carbon monoxide does not react with sodium hydroxide at any appreciable rate at room temperature and pressure. However, it is suggested that the reaction should be tried at higher temperatures such as 300°C, in order to produce sodium formate. The expected reaction is

$$NaOH(s) + CO(g) = HCOONa(s)$$

Using the following data and assuming the usual standard states, demonstrate by calculations whether you believe the suggestion to try the reaction at 300°C is worthwhile.

	$\Delta \underline{G}_f$ at 25°C, 1 atm kcal/g-mol	$\Delta \underline{H}_f$ at 25°C, 1 atm kcal/g-mol	C_P^*, cal/g-mol °K
Co(g)	− 32.808	−26.416	$6.34 + 1.84 \times 10^{-3}T$
NaOH(s)	− 90.600	−101.960	12.5
HCOONa(s)	−147.00	−155.030	18.3

12-12 It is proposed to "clean" some tungsten wires (that is, to remove all traces of the oxide WO_2) by heating the wires to 2000°K in a hydrogen atmosphere (1 atm total pressure).

(a) What is the maximum water impurity level that can be tolerated in the hydrogen?

(b) If the temperature of the wire is raised, would higher purity be required? Show your calculations.

For WO_2, $\Delta \underline{G}_f = -131{,}600 + 36.6T$, at 1 atm; for H_2O, $\Delta \underline{G}_f = -58{,}900 + 13.1T$, at 1 atm where $\Delta \underline{G}_f$ is in cal/g-mol of compound and T is in °K.

12-13 Pure HI is introduced into a reaction vessel and allowed to equilibrate at 25°C. If the total pressure in the vessel is 3×10^{-4} mm Hg, what is the composition of the vapor phase?

$$P'_{I_2} \text{ at } 25°\text{C} = 4 \times 10^{-4} \text{ atm}$$

$$\underline{V}_{I_2} \text{ at } 25°\text{C} = 13 \text{ cm}^3/\text{g-atom}$$

$$(\Delta \underline{G}_f)_{HI} = 0.365 \text{ kcal/g-mol at } 25°\text{C, 1 atm.}$$

Also, what is the activity of the iodine in this system?

12-14 Determine the minimum pressure that must be exerted on solid graphite (carbon) to form diamond (also carbon). Assume that the diamonds and graphite form pure solid phases that are incompressible and insoluble in each other. The temperature is 25°C. For the reaction

$$C_{\text{graphite}} = C_{\text{diamond}}$$

$$\Delta \underline{\mathscr{G}}^\circ_{25°\text{C}} = 685 \text{ cal/g-mol}$$

(standard-state pure solids at 1 atm pressure). The density of graphite is 2.25 g/cm³; that of diamond is 3.51 g/cm³.

12-15 The following data were taken on the equilibrium pressure of CO_2 over $CaCO_3$ at various temperatures. In this reaction calcium carbonate decomposes to calcium oxide and carbon dioxide according to the equation $CaCO_3 = CaO + CO_2$. From the table of data estimate the standard heat of decomposition of $CaCO_3$ at 750°C.

T, °C	P, mm Hg
587	1.00
631	4
701	23
743	60
786	134
800	183
852	381
894	716

12-16 The following equilibrium data have been determined for the reaction

$$\text{NiO(s)} + \text{CO(g)} = \text{Ni(s)} + \text{CO}_2\text{(g)}$$

T, °C	663	716	754	793	852
$K_a \times 10^{-3}$	4.535	3.323	2.554	2.037	1.577

(a) Find $\Delta\mathscr{H}^\circ$, K_a, and $\Delta\mathscr{G}^\circ$ at 1000°K by use of graphical techniques.

(b) Would an atmosphere of 15 per cent CO_2, 5 per cent CO, and 80 per cent N_2 oxidize nickel at 1000°K?

12-17 The heat of formation of silver oxide is −7300 cal/g-mol at 25°C. Calculate the approximate temperature at which silver oxide begins to decompose on heating (a) in pure oxygen and (b) in air (both at $P = 1$ atm). Use the following data:

	Entropy, cal/g-mol °K $\underline{S}^\circ_{298°K,\ 1\ atm}$	C_P, cal/g-mol °K
Ag_2O	29.1	15.7
O_2	49.0	7.0
Ag	10.2	6.1

12-18 You have been asked to design a piece of equipment in which gaseous titanium tetrachloride will be reacted isothermally with pure liquid magnesium to form solid titanium and liquid magnesium chloride. As the designer, you should determine the heat effect of the reaction. The following data are available.[3] For the reaction,

$$TiCl_4(g) = Ti(s) + 2Cl_2(g)$$

$$\Delta\mathscr{G}^\circ = 180{,}700 + 1.8T\log T - 34.65T \text{ cal/g-mol } TiCl_4$$

The temperature range is 298 to 1700°K. For the reaction

$$MgCl_2(l) = Mg(l) + Cl_2(g)$$

$$\Delta\mathscr{G}^\circ = 147{,}850 + 13.58T\log T - 72.77 \text{ cal/g-mol}$$

The temperature range is 987 to 1376°K. How much heat will be absorbed or released by the formation of 1 g of titanium at 727°C?

12-19 For the reaction

$$Si(s) + SiO_2(s) = 2SiO(g)$$

$$\Delta\mathscr{G}^\circ = 160{,}000 + 13.8T\log_{10}T - 121.55T \text{ cal/g-mol}$$

where $T =$ °K.

(a) What is the pressure of SiO(g) above pure Si and pure SiO_2 at 1000°K?

(b) What is the $\Delta\mathscr{H}^\circ$ at 1000°K?

(c) What is the standard entropy of SiO(g) at 1000°K?

Substance	S° at 1000°K, cal/g-mol °K
Si	11.36
SiO_2	27.85

[3] O. Kubaschewski and E. Evans, *Metallurgical Thermochemistry*, John Wiley & Sons, Inc., New York, 1956.

12-20 The standard free-energy change for the reduction of chromium oxide by hydrogen is given approximately as follows:

$$Cr_2O_3(s) + 3H_2(g) = 2Cr(s) + 3H_2O(g)$$

$$\Delta\mathscr{G}^\circ = 97{,}650 - 28.6T, \text{ cal g-mol}, \qquad T = {}^\circ K$$

(a) Find the maximum partial pressure of water vapor in otherwise pure hydrogen at 1 atm in which chromium can be heated without oxidation at 1500°K.

(b) Is the oxidation of chromium by water vapor exothermic or endothermic?

(c) What can you conclude from the above data about the heat effect at 1500°K for the oxidation of chromium by pure oxygen?

(d) Would the equilibrium in the equation above be affected by a change in pressure of the hydrogen–water vapor mixture from 1 to 2 atm? To 200 atm?

12-21 In the manufacture of formaldehyde a mixture of air with methanol vapor is brought into reaction on a silver catalyst. In this process the silver slowly loses its metallic luster and partially disintegrates. Use the following data to examine whether this might be due to the formation of silver oxide ($P = 1$ atm, $T = 550$°C). For silver oxide at 25°C, and 1 atm

$$\Delta \underline{G}_f = -2590 \text{ cal/g-mol}$$

$$\Delta \underline{H}_f = -7310 \text{ cal/g-mol}$$

$$(C_P)_{Ag} = 6.4 \text{ cal/g-mol °K}$$

$$(C_P)_{Ag_2O} = 15.7 \text{ cal/g-mol °K}$$

$$(C_P)_{O_2} = 9.5 \text{ cal/g-mol °K}$$

12-22 A dissociates to B and C according to the reaction

$$A(g) = \tfrac{1}{2}B(s) + 2C(g)$$

When C is contacted with B at 400°K and a pressure of 2.5 atm, the resulting equilibrium gas composition is 50 per cent A and 50 per cent C. A and C may be assumed to behave ideally at 2.5 atm.

(a) Calculate $\Delta\mathscr{G}^\circ$ for the above dissociation reaction at 400°K.

(b) Calculate the equilibrium gas composition if 1 g-mol of A dissociates at 500°K and 2.5 atm.

(c) Repeat part (b) for dissociation at 400°K and 100 atm.

Component	C_P†	$\Delta \underline{H}_f$‡	T_c, °K	P_c, atm
A	$11.3 + 0.0086T$	13,400	364	33.3
B	$5.0 + 0.016T$	−18,200	502	53
C	$3.9 + 0.0053T$	11,450	320	20

† cal/g-mol °K; T in °K.
‡ $\Delta \underline{H}$ of formation at 400°K, cal/g-mol.

12-23 An equimolar mixture of CO and steam at 1 atm and 600°C will be passed over a catalyst for conversion to H_2 via the water-gas shift:

$$CO + H_2O = CO_2 + H_2$$

(a) Assuming that the reaction goes to equilibrium, is there any danger of significant carbon deposition via the side reaction

$$2CO = C + CO_2?$$

(b) Does consideration of a second side reaction,

$$C + H_2O = CO + H_2$$

yield any new information?

(c) Examination of the inlet conditions to the reactor suggests that the second reaction might deposit carbon in the inlet sections of the reactor, regardless of what occurs at the end of the reactor. Is it possible to have a stable carbon phase (that is, one that is in equilibrium with the other reacting species at that point) in the inlet sections of the reactor?

Equilibrium data are supplied in Appendix E.

12-24 Limestone dissociates according to the reaction

$$CaCO_3(s) = CaO(s) + CO_2$$

with an equilibrium pressure (that is, the pressure of CO_2 in equilibrium with the mixture of solids) which is given, as a function of temperature, by the relationship

$$\log P = -\frac{11{,}355}{T} - 5.388 \log T + 26.238$$

where the units are atmospheres and °K. If limestone and coke are mixed at elevated temperatures, carbon monoxide is produced by the side reaction

$$C + CO_2 = 2CO$$

At 1000°C the equilibrium constant for the latter reaction is $K_a = 126$, based on standard states of pure solids at 1000°C and pure gases at 1000°C and 1 atm.

(a) Evaluate K_a for the decomposition of limestone at 1000°C.

(b) What will be the composition of the gas phase in equilibrium with an equimolar mixture of limestone and coke?

(c) What will be the pressure of the gas phase in (b)?

(d) If the pressure is decreased from the value found above, and maintained there, what effect will this have on the system?

12-25 The following equilibrium constants were obtained at 700°C, with the Fe, FeO, and C being present as solids and the H_2, CO, and CO_2 being present as gases:

$$Fe + H_2O(g) = FeO + H_2 \qquad K_a = 2.35$$

$$Fe + CO_2 = FeO + CO \qquad K_a = 1.52$$

$$C + H_2O(g) = CO + H_2 \qquad K_a = 1.55$$

(a) Will a mixture consisting of equimolecular proportions of water vapor and hydrogen at 1 atm total pressure be capable of reducing FeO at the given temperature?

(b) What will be the effect of increasing the total pressure?

(c) Will a gaseous mixture of 12 mol per cent CO, 9 per cent CO_2, and 79 per cent inert gas such as N_2 at a total pressure of 1 atm tend to deposit carbon?

(d) At what pressures can the latter reaction occur?

12-26 The reactions listed below are expected to occur simultaneously when equal molar quantities of CO and H_2 are fed to a reaction vessel and allowed to come to equilibrium at 1000°K and a pressure of 500 atm.

(a) Calculate the composition of the equilibrium gas under these conditions.

	$\Delta\mathscr{G}^\circ$ at 1000°K, cal/g-mol
$CO + H_2 = \frac{1}{2}CH_4 + \frac{1}{2}CO_2$	2942
$CO + H_2 = C + H_2O$	1906
$C + H_2O = \frac{1}{2}CH_4 + \frac{1}{2}CO_2$	1036

(b) Could the conversion to CH_4 be increased by increasing the amount of C in the system? Explain.

(c) Would any of the above reactions proceed spontaneously if products and reactants were maintained in their standard states?

(d) Does the extent of conversion for an exothermic reaction increase or decrease with increasing temperature? Assume that the heat of reaction is insensitive to temperature changes.

12-27 It has been suggested that water is an extremely dangerous material to use for extinguishing hot fires where large amounts of iron (or other metals) are present. It is feared that the iron (or other metals that may be present) may decompose the water to form metal oxide and liberate elemental hydrogen. For iron the reaction would be

$$Fe + H_2O \rightleftharpoons FeO + H_2$$

(Equilibrium data for this reaction are in Appendix E.)

(a) If the temperature in a certain fire is 1800°F, determine the ratio y_{H_2}/y_{H_2O} which would be in equilibrium with the iron.

(b) Does it appear that this reaction will proceed to any appreciable extent at 1800°F. (Justify your answer!)

(c) Determine the amount of heat liberated per mole of iron oxidized according to the above reaction.

12-28 At 800°C the equilibrium partial pressure of nitrogen for the reaction

$$Mg_3N_2(s) = 3Mg(l) + N_2$$

is 10^{-15} atm. If a liquid bismuth solution containing 2.5 mol per cent magnesium is blanketed with an argon atmosphere at 2 atm pressure and 800°C, what is the maximum concentration of N_2 that can be tolerated in the gas without forming Mg_3N_2? Assume that N_2 and Mg_3N_2 are insoluble in bismuth and that the activity coefficient of the magnesium in the bismuth solution is $\gamma^\circ_{Mg} = 10^{-3}$, where the standard state of magnesium is pure liquid at 800°C and 1 atm pressure. Also assume that bismuth does not form a stable nitride under these conditions and that the vapor pressures of Mg and Bi are negligible.

12-29 In a process for making concentrated nitric acid, a gaseous mixture of N_2 and NO_2 is charged to a reaction vessel containing very dilute nitric acid. The reaction that takes place is

$$2NO_2(g) + H_2O(l) = HNO_3(aq) + NO(g)$$

(a) Using the data below, calculate the K_a for this reaction at 650°F and 200 atm.

(b) If in a certain operation 5 moles of a 2 mole per cent nitric acid solution (aqueous) and 10 moles of a gas mixture containing 50 mole per cent NO are charged to the vessel at 650°F and 200 atm, calculate the concentration of the aqueous nitric acid solution when equilibrium has been attained.

$$\text{vapor pressure of pure } HNO_3 \text{ at } 650°F = 1950 \text{ psia}$$

$$\text{vapor pressure of pure } H_2O \text{ at } 650°F = 2208 \text{ psia}$$

$$(C_P)_{NO_2} = 7.0 + 7.1 \times 10^{-3}T - 8 \times 10^{-7}T^2 \text{ Btu/lb-mol °F}$$

$$(C_P)_{H_2O} = 18 \text{ Btu/lb-mol °F}$$

$$(C_P)_{NO} = 8.05 + 0.233 \times 10^{-3}T - 1.563 \times 10^{-5}T^2 \text{ Btu/lb-mol °F}$$

$$(C_P)_{HNO_3} = 28.64 - 8.0 \times 10^{-3}T \text{ Btu/lb-mol °F}$$

where T is in °R, and the heat capacities are for the appropriate phases. Assume that the heat capacities of the liquids are not affected by their being mixed. Assume that pressure has no effect on enthalpy of the liquids.

At 25°C, and 1 atm	$\Delta \underline{G}_f$	$\Delta \underline{S}_f$
HNO_3 (aqueous for all concentrations encountered here)	−78.36	−0.14590
NO_2(g)	12.26	−0.01442
NO(g)	20.72	0.00956
H_2O(l)	−56.69	−0.03907

where $\Delta \underline{G}_f$ is given in kcal/g-mol and $\Delta \underline{S}_f$ is given in kcal/g-mol °K. The standard states for gases are pure gases at 650°C and 1 atm; for liquids the standard state is the pure liquid under its own vapor pressure.

	P_c, atm	T_c, °K
NO_2	75.3	367.2
NO	65.0	179.0

Assume that for HNO_3 and H_2O vapors the f/P in the existing state equals the f/P in the standard state. The following equations represent approximations for the partial pressure of HNO_3 and water over solutions within a range of concentration of 10 to 75 mole per cent HNO_3 at 650°F:

$$P_{HNO_3} = \frac{210x}{40 - x} + 0.15\frac{140x^2}{40 - x}$$

$$P_{H_2O} = 2208 - \frac{2940x}{40 - x}$$

where P is in psia and x is concentration of HNO_3 in mole per cent.

12-30 An equimolar mixture of ethylene and water is being fed to a catalytic reactor operating at a uniform temperature and pressure of 200°C and 600 psia, respectively.

The product mixture, consisting of ethylene, water, and ethyl alcohol, is at complete chemical and physical equilibrium when removed under reactor conditions. From the following data and any other information and assumptions you care to supply, estimate the phase compositions of the product and the relative amounts of the phases. How would these change if the reactor were operated at only 200 psia? Show your analysis of both conditions by the phase rule.

	C_2H_4	H_2O	C_2H_5OH
Vapor pressure at 200°C, psia	—	226	429
T_c, °K	282.4	647.3	516.3
P_c, atm	50.0	28.3	63.1
C_P, cal/g-mol °K ($T°$ is in °K)	$6.0 + 0.015T$	$7.0 + 0.00277T$	$4.5 + 0.038T$
$\Delta \underline{G}_f$, ideal gas, 1 atm, 25°C, cal/g-mol	16,282	−54,635	−40,300
$\Delta \underline{H}_f$, ideal gas, 1 atm, 25°C, cal/g-mol	12,496	−57,798	−56,240
$\Delta \underline{H}_V$ at 200°C, cal/g-mol	—	8,343	5,421

Electrochemical Processes 13

Introduction

An electrochemical process is defined as a chemical reaction in which the transfer of electrons through an external circuit occurs. The external circuit may either receive electrical energy from the electrochemical process, as occurs during the discharge of a battery, or the external circuit may supply electrical energy to the process, as in the case of electrolysis or the recharging of a battery.

Although in theory many reactions can be performed as an electrochemical process, only a limited number of reactions have actually yielded usable electrochemical processes. However, since the potential gains to be realized from a practical electrochemical process that yields large amounts of electrical energy are very great, much effort has been, and continues to be, expended in an effort to develop more practical electrochemical reactions.

The advantage of an electrochemical process, as against a standard combustion–heat engine cycle, for converting chemical energy to electrical energy may be seen by examining the maximum efficiency that may be obtained from the two paths.

The maximum work (or in the case of an electrochemical process, electrical energy) that can be obtained from any process is given, as indicated in Chapter 11, by minus the isothermal change in the Gibbs free energy. That is,

$$(-\underline{W}_R)_T = (\Delta \underline{G})_T = (\Delta \underline{H})_T - T(\Delta \underline{S})_T \tag{13-1}$$

Since the maximum amount of heat that could have been recovered from this process is given by the isothermal enthalpy change, the thermal efficiency, or fraction of heat recovered as work, is given by

$$\eta = \frac{(-\underline{W}_R)_T}{(\Delta \underline{H})_T} = \frac{(\Delta \underline{H})_T - T(\Delta \underline{S})_T}{(\Delta \underline{H})_T} \tag{13-2}$$

On the other hand, if the reaction is used to supply heat to a heat engine, the maximum work is obtained from the Carnot limitation:

$$(-\underline{W}_R)_T = (\Delta \underline{H})_T \frac{T_H - T_L}{T_H} \tag{13-3}$$

and the maximum thermal efficiency reduces to

$$\eta = \frac{T_H - T_L}{T_H} \tag{13-4}$$

Since the $T\,\Delta\underline{S}$ term in equation (13-2) is in general a small fraction of $\Delta\underline{H}$ (except at very high temperatures), the thermal efficiency of an electrochemical process may be quite high, particularly at low temperatures. On the other hand, the efficiency of the Carnot-limited heat engine is quite low and will not begin to approach that of the electrochemical process except at very high operating temperatures, frequently temperatures higher than currently available construction materials can withstand for extended periods of time.

The effect of temperature on the maximum thermal efficiency that can be obtained from the oxidation of hydrogen in either an electrochemical process or a Carnot-limited heat engine is shown in Fig. 13-1. The Carnot-limited heat engine is assumed to reject heat at a temperature of 110°F. As seen in Fig. 13-1,

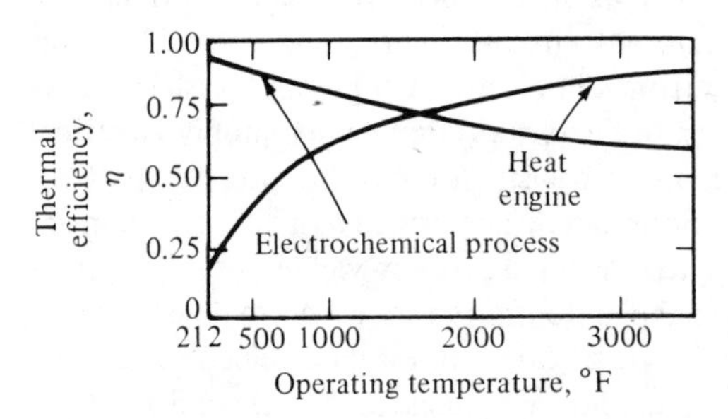

FIG. 13-1. Maximum efficiency for the oxidation of H_2.

the electrochemical oxidation is nearly 100 per cent efficient at low temperatures, and is still more efficient than the Carnot-limited heat engine (which uses hydrogen as a fuel to supply the required heat) up to an operating temperature of approximately 1800°F, well above the maximum temperature attainable in present-day steam boilers.

The unusually high thermal efficiencies attainable with electrochemical processes may be explained by noting that at no point in the electrochemical process is energy transferred in the form of heat. Therefore, the classical Carnot efficiency simply does not apply.

From the preceding discussion we can begin to appreciate the desirability of developing a practical electrochemical process for the oxidation of such commonly used fuels as coal or petroleum derivatives. Unfortunately, at the present time no such scheme exists. However, much effort is being expended in an effort to develop such an electrochemical process, and it seems likely that the future will see an ever-increasing amount of electrical energy being generated by means of electrochemical reactions.

Electrochemical processes may be broadly divided into two categories: those which supply electrical energy to the surroundings, and those which receive electrical energy from the surroundings. The processes that supply electrical energy to the surroundings may be referred to as batteries or fuel cells, depending on whether or not reactants can be continuously fed to the reaction vessel. If the reaction vessel is charged once and not replenished during operation, the unit is termed a *battery*. Examples of batteries encountered in daily life include the lead storage battery common to almost all automobile ignition systems, and the zinc–carbon batteries used to supply electricity for flashlights and many children's toys. In some instances the electrochemical reaction that occurs in a battery may be reversed, as in the recharging of an automobile battery. In other cases, however, the reaction cannot be appreciably reversed, and the battery, once exhausted, must be replaced. Because the amount of electrical energy a battery can supply before replacement or recharging is proportional to the amount of reactants that can be sealed in the reaction vessel, batteries are not likely to offer an economical source of great amounts of electrical energy.

Fuel cells, on the other hand, operate in a continuous fashion, with reactants being continuously fed to the cell while products are continuously withdrawn. Thus, in theory, the only limitation on the electrical energy that a fuel cell can produce is the period of time over which the cell will be operated. Unfortunately, practical considerations such as electrode reaction rates (which severely limit usable current densities) prevent the use of present-day fuel cells in applications where high power consumption is necessary. Current research is aimed at developing electrodes that will reduce these power restrictions.

Electrochemical processes that receive electrical power from the external circuit are frequently termed *electrolysis reactions* (discounting, of course, those processes where a battery is being recharged). Electrolysis reactions are used in many important industrial processes, such as the electrolysis of (1) aluminum oxide to yield metallic aluminum; (2) magnesium chloride to yield metallic magnesium; (3) brine to yield sodium hydroxide, chlorine, and hydrogen; and many others.

Although the various electrochemical reactions, as they occur in batteries, fuel cells, and electrolysis cells, may all appear quite different, the thermodynamic analyses for these reactions are all the same. That this is so will become self-evident as we develop our analysis of electrochemical processes. For the purpose of illustration we shall restrict our discussion to the analysis of simple galvanic cells (a form of battery). However, the analysis and discussion that follow can easily be extended to cover any electrochemical process.

13.1 The Simple Galvanic Cell

The simple galvanic cell is composed of two half-cells, or *electrodes*, that are connected by a semipermeable membrane which allows the passage of ions but does not allow bulk flow of the electrolyte in which the ions exist. At the anodic electrode, or *anode*, electrons are liberated and enter the external electrical circuit; at the *cathode*, electrons are received from the external circuit. A Daniell galvanic cell is illustrated in Fig. 13-2.

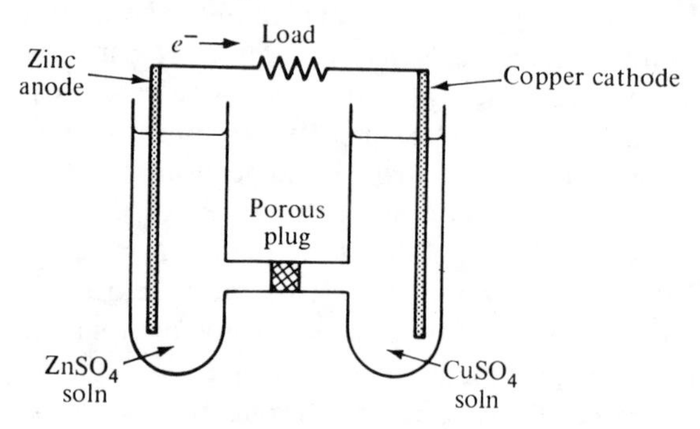

FIG. 13-2. Daniell galvanic cell.

During normal operation of the cell the zinc electrode operates as an anode, liberating electrons according to the reaction

$$Zn = Zn^{2+} + 2e^- \tag{13-5}$$

The copper electrode acts as a cathode, receiving electrons from the external circuit according to the equation

$$Cu^{2+} + 2e^- = Cu \tag{13-5a}$$

The total cell reaction is then

$$Zn + Cu^{2+} = Cu + Zn^{2+} \tag{13-6}$$

As the cell operates the zinc electrode dissolves while copper precipitates on the copper electrode. The sulfate ions that are liberated when the Cu^{2+} ions precipitate must diffuse through the semipermeable membrane to the zinc electrode, where they neutralize the zinc ions that are entering solution. If electrons (current) are removed from the zinc electrode faster than the sulfate ions can migrate across the semipermeable membrane, an electrical imbalance will develop between the two electrodes. When this imbalance occurs, the galvanic cell is said to be *polarized*, and its output voltage drops below the zero-current (or reversible) voltage. This occurrence is familiar to anyone who has attempted to start an automobile on a cold morning. After cranking the starting motor for a few moments the battery appears to be weakening. However, the battery will soon regain its original life if given a rest. This phenomenon can be easily explained in terms of polarization: In cold weather the diffusion rates of the ions are greatly reduced, so the battery is quickly polarized under the great current drain of starting (up to 150 A in a 12-V system and 300 A in a 6-V system). Once the battery is polarized, the voltage drops off and the battery appears to be dying. However, if the load is removed from the battery, the polarizing ions are allowed to diffuse to their proper places. Electrical neutrality is soon restored, and the output voltage of the battery returns to its original value. The battery is then ready to resume its task.

Discharge of a polarized galvanic cell is a nonreversible process and will not be considered in great detail in this book. Instead, we shall limit our discussion to cases in which the currents removed from the cell under consideration are so small that no polarization occurs, and the operation may be treated as a reversible process. The reactions that occur in a galvanic cell may be reversed if the load is replaced by a *back voltage* that is greater than the voltage of the cell. During this reverse reaction, usually called *recharging*, the cell is storing electrical energy in the form of chemical potential differences. This energy may later be recovered (almost completely in a well-designed cell) as the cell is discharged. Thus the common name *storage battery* for the galvanic cell used in automobile ignition systems. It should be noted that during charging of a Daniell cell the copper electrode dissolves liberating electrons to the external circuit and hence is the anode, while zinc precipitates on the zinc electrode, which now acts as the cathode.

Let us now examine the application of thermodynamic principles to the study of galvanic cells (or, by simple extension, any electrochemical process). During the discharging of a galvanic cell, the compositions of the reacting species are always changing. Thus, if we are to analyze the behavior of the cell at any instant, we must examine the differential changes in the cell properties as the discharging occurs. If we assume that the cell is held at constant temperature and pressure, the maximum work that can be obtained is given by minus the change in Gibbs free energy of the reacting mixture:

$$(-\delta W_R)_T = (dG)_T \tag{13-7}$$

In an electrochemical process, work is released in the form of electrical energy and is equal to the electrical charge transferred multiplied by the potential difference:

$$\delta W = E\,\delta q \tag{13-8}$$

where q = charge transferred in coulombs (always positive)
E = potential difference (E will be positive when W is positive; that is, if the cell is supplying work to the surroundings it has a + voltage; if the cell is receiving work from the surroundings its voltage will be negative)

Equation (13-8) may be placed on a per mole basis by dividing by n_1, the number of moles of species 1 that were initially charged to the reactor (cell):

$$\frac{\delta W}{n_1} = \frac{E\,\delta q}{n_1} \tag{13-9}$$

or

$$\delta \underline{W} = E\,\delta \underline{q}, \tag{13-9a}$$

where the letter with an underbar represents "per mole of species 1 charged to the cell."

The differential charge transferred during the reaction, $\delta \underline{q}$, may be expressed as

$$\delta \underline{q} = \mathscr{N} F(d\underline{X}) \tag{13-10}$$

where $\underline{X}$ = extent of the electrochemical reaction—moles of specie 1 reacted per mole of specie 1 charged to the cell

$\mathscr{N}$ = number of moles of electrons liberated per mole of species 1 that reacts

F = Faraday constant: the charge of 1 mole of electrons

F = 96,439 coulombs/equivalent

or

F = 23,068 cal/volt-equivalent

Thus the reversible work of a cell is related to its voltage by the expression

$$\delta \underline{W} = \mathscr{N} F E\, d\underline{X} \tag{13-11}$$

or

$$\frac{\delta \underline{W}}{d\underline{X}} = \mathscr{N} F E = \underline{\mathscr{W}} \tag{13-11a}$$

where $\underline{\mathscr{W}}$ = work liberated per mole of species 1 that reacts.

Substitution of equation (13-7) for the reversible work then gives us an expression for the cell voltage in terms of the Gibbs-free-energy change for reactants to products:

$$-\frac{d\underline{G}}{d\underline{X}} = \mathscr{N} F E = -\Delta \underline{\mathscr{G}} \tag{13-12}$$

where $\Delta \underline{\mathscr{G}}$ = Gibbs-free-energy change per mole of species 1 reacted.

Since $\Delta \underline{\mathscr{G}}$ is a function of the compositions of the reactants and products, we now seek a simple relation between $\Delta \underline{\mathscr{G}}$ and the compositions of the reacting species. As in our studies of simple chemical processes, we now choose a convenient standard state for each component involved in the various electrode reactions. The activities of the reacting species are then expressed in terms of the standard-state conditions. Following a development similar to that in Chapter 11, the Gibbs-free-energy change is expressed as

$$\Delta \underline{\mathscr{G}} = \Delta \underline{\mathscr{G}}^\circ + RT\left[\ln J_a + \sum_{i=1}^{C} \underline{\eta}_i \ln \frac{(a_i)_{\text{out}}}{(a_i)_{\text{in}}}\right] \tag{13-13}$$

where

$$J_a = \frac{\prod_{\text{prod}} (a_i)_{\text{out}}^{\alpha_i}}{\prod_{\text{react}} (a_i)_{\text{in}}^{-\alpha_i}} \qquad \left[\begin{array}{l}\text{(subscripts “out” and “in” refer to product and}\\ \text{reactant electrodes rather than actual outlets and}\\ \text{inlets)}\end{array}\right]$$

η_i = moles of specie i unreacted, per mole of species 1 which react

$\Delta\mathscr{G}^\circ = \sum_{i=1}^{C} \alpha_i \underline{G}_i^\circ$ = the Gibbs-free-energy change per mole reaction when reactants and products are in standard states

For the remainder of our work we shall assume that the change in Gibbs free energy of the nonreactants (per mole of species 1 that reacts) may be neglected. In general this is a rather good approximation—particularly with respect to galvanic cell measurements. For an illustration of the the magnitude of these terms in a “typical” fuel-cell calculation, the reader is referred back to Sample Problem 11-1, where this term has been retained. Neglecting the non-reacting terms in equation (13-13) gives

$$\Delta\mathscr{G} = \Delta\mathscr{G}^\circ + RT \ln J_a \tag{13-13a}$$

which may be substituted into equation (13-12) to give

$$-\mathscr{N}FE = \Delta\mathscr{G}^\circ + RT \ln J_a \tag{13-14}$$

However, $\Delta\mathscr{G}^\circ$ itself is most easily measured by measuring the voltage produced by a galvanic cell when all the products and reactants are in their standard state so that $J_a = 1$. Thus

$$\Delta\mathscr{G}^\circ = -\mathscr{N}FE^\circ \tag{13-15}$$

so

$$\Delta\mathscr{G} = -\mathscr{N}FE = -\mathscr{N}FE^\circ + RT \ln J_a \tag{13-16}$$

or

$$E = E^\circ - \frac{RT}{\mathscr{N}F} \ln J_a \tag{13-16a}$$

Equation (13-16a) is known as the *Nernst equation*, and is used to relate the voltages of a nonstandard state cell to the standard-state voltages and the activities of the reacting species in the non-standard-state cell. (Had we considered a continuous-flow electrochemical reactor—such as a fuel cell—the same result would have been attained. The only difference would be that J_a would have its conventional meaning. The reader should derive this result as an exercise.)

Normally the following standard states are used with electrochemical processes:

1. Pure gases at 1 atm pressure.
2. Pure solid phases (such as the metallic copper and zinc of the Daniell cell).
3. Electrolyte solutions at some prescribed concentration. Frequently, but by no means universally, the slope of the $\bar{f}_i$–m_i curve (where m_i is the molality of species i and is given by m_i = g-mol of i/1000 g of solvent) as $m_i \longrightarrow 0$ is

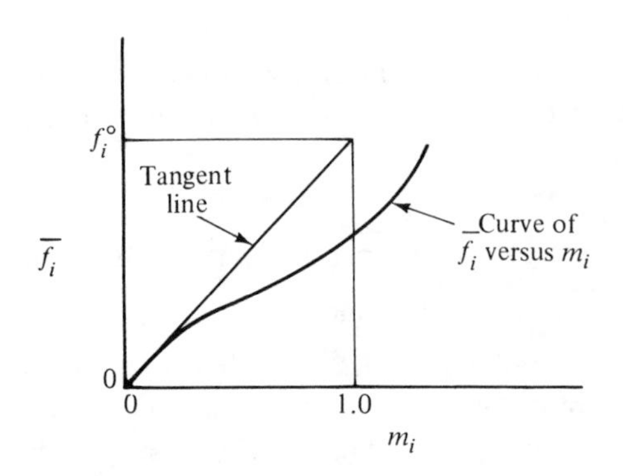

FIG. 13-3. Determination of the standard-state fugacity for an electrolyte in solution.

used. The values of f_i° are given by the intersection of the tangent line at $m_i \longrightarrow 0$ with the line $m_i = 1.0$, as shown in Fig. 13-3. The standard state illustrated in Fig. 13-3 is frequently referred to as infinite-dilution standard state and has the useful property that $\gamma_i^\circ \longrightarrow 1$ as $m_i \longrightarrow 0$ (when γ_i° is defined such that $\bar{f}_i = m_i \gamma_i^\circ f_i^\circ$).

SAMPLE PROBLEM 13-1. A simple Daniell cell is initially charged with 0.1 m solutions of $CuSO_4$ and $ZnSO_4$. At 25°C the cell voltage is measured and found to be $E = 1.100$ V. Assuming that the electrolyte solutions are ideal,

(a) Determine the voltage of the cell (at 25°C) after it has been discharged to the point where 90 per cent of the Cu^{2+} has been deposited on the cathode.

(b) Determine the concentrations of Zn^{2+} and Cu^{2+} when the cell has been discharged to the point where the cell voltage is $E = 0.0$ V ($T = 25$°C).

Solution: We choose the following standard states to use for an analysis: (1) pure solid metallic copper and zinc at 25°C, and (2) 0.1 m solutions of $CuSO_4$ and $ZnSO_4$ at 25°C. Since these are the conditions under which the cell voltage of the problem statement was measured:

$$E^\circ = 1.100 \text{ V}$$

We may determine the voltage at any other concentration by means of the Nernst equation:

$$E = E^\circ - \frac{RT}{\underline{\mathscr{N}}F} \ln J_a$$

The overall reaction for the Daniell cell is written as

$$Zn + Cu^{2+} = Cu + Zn^{2+}$$

Therefore, J_a for this reaction is given by

$$J_a = \frac{(a_{Cu}\, a_{Zn^{2+}})_{out}}{(a_{Zn}\, a_{Cu^{2+}})_{in}}$$

However, the solid zinc and copper electrodes are in their standard states, and hence $a_{Zn} = a_{Cu} = 1.0$ as long as any zinc and copper are left. The expression for J_a then reduces to

$$J_a = \frac{(a_{Zn^{2+}})_{out}}{(a_{Cu^{2+}})_{in}}$$

We may express the activities of the zinc and copper ions at their respective electrodes in terms of their concentrations by means of an expression of the type

$$a_{Zn^{2+}} = \frac{\bar{f}_{Zn^{2+}}}{f^{\circ}_{Zn^{2+}}} = \frac{\gamma^{\circ}_{Zn^{2+}} x_{Zn^{2+}} f^{\circ}_{Zn^{2+}}}{f^{\circ}_{Zn^{2+}}} = x_{Zn^{2+}} \gamma^{\circ}_{Zn^{2+}}$$

Since the activity coefficient is defined so that the standard state activity is unity, $\gamma^{\circ}_{Zn^{2+}}$ in the standard state is given by the expression

$$\gamma^{\circ}_{Zn^{2+}} = \frac{a_{Zn^{2+}}}{x^{\circ}_{Zn^{2+}}} = \frac{1}{x^{\circ}_{Zn^{2+}}}$$

If we assume that the zinc and copper ion solutions are ideal, then the activity coefficients are independent of composition, and the activities of the zinc and copper ion solutions are given by:

$$a_{Zn^{2+}} = \frac{x_{Zn^{2+}}}{x^{\circ}_{Zn^{2+}}}$$

We can now relate the molality of a solution to its mole fraction as follows. By definition, the molality, m_i, of a dissolved substance is equal to the number of moles of i dissolved in 1000 g of solvent. Thus for water solutions the molality is equal to the number of moles of solute per 1000/18.0 = 55.5 moles of water. The mole fraction is then related to the molality by the expression

$$x_i = \frac{m_i}{55.5 + m_i}$$

For dilute solutions it is often possible to neglect m_i in relation to 55.5, so the relation between x_i and m_i reduces to

$$x_i = \frac{m_i}{55.5}$$

We now substitute the expression for x_i for dilute solutions into the activity expression to give

$$a_{Zn^{2+}} = \frac{m_{Zn^{2+}}}{m^{\circ}_{Zn^{2+}}}$$

$$a_{Cu^{2+}} = \frac{m_{Cu^{2+}}}{m^{\circ}_{Cu^{2+}}}$$

But the standard-state solutions were chosen as 0.1 m. Therefore,

$$(a_{Zn^{2+}})_{out} = 10.0(m_{Zn^{2+}})_{out}$$

$$(a_{Cu^{2+}})_{in} = 10.0(m_{Cu^{2+}})_{in}$$

We now substitute the expressions for $(a_{Zn^{2+}})_{out}$ and $(a_{Cu^{2+}})_{in}$ into the expression for J_a:

$$J_a = \frac{10.0(m_{Zn^{2+}})_{out}}{10.0(m_{Cu^{2+}})_{in}} = \frac{(m_{Zn^{2+}})_{out}}{(m_{Cu^{2+}})_{in}}$$

Substitution of the expression for J_a into

$$E = E^{\circ} - \frac{RT}{\mathcal{N}F} \ln J_a$$

yields

$$E = E^{\circ} - \frac{RT}{\mathcal{N}F} \ln \frac{(m_{Zn^{2+}})_{out}}{(m_{Cu^{2+}})_{in}}$$

where $R = 1.987$ cal/g-mol °K

$T = 25°C = 298°K$

$\mathscr{N} = 2$ equiv/g-mol

$F = 23{,}068$ cal/volt-equiv

$E° = 1.10$ V

or

$$E = E° - \frac{(1.987)(298)\ (\text{cal °K/g-mol °K})}{(2.0)(23{,}068)\ (\text{cal/volt-equiv})\ (\text{equiv/g-mol})} \ln \frac{(m_{Zn^{2+}})_{out}}{(m_{Cu^{2+}})_{in}}$$

Therefore,

$$E = E° - 0.0128 \ln \frac{(m_{Zn^{2+}})_{out}}{(m_{Cu^{2+}})_{in}}$$

or, converting to base 10 logs,

$$E = 1.10 - 0.0295 \log \frac{(m_{Zn^{2+}})_{out}}{(m_{Cu^{2+}})_{in}}$$

(a) At the point where 90 per cent of the Cu^{2+} has been deposited, the molalities of Cu^{2+} and Zu^{2+} at their respective electrodes are

$$(m_{Cu^{2+}})_{in} = 0.1 - (0.90)(0.1) = 0.01$$

$$(m_{Zn^{2+}})_{out} = 0.1 + (0.90)(0.1) = 0.19$$

and the cell voltage is

$$E = 1.100 - 0.0295 \log \frac{0.19}{0.01}$$

or

$$E = 1.062 \text{ V}$$

(b) At the point where the cell voltage has dropped to zero, $E = 0$. Thus we may express the ratio of $(m_{Cu^{2+}})_{in}$ to $(m_{Zn^{2+}})_{out}$ from the relation

$$E = 0 = 1.100 - 0.0295 \log \frac{(m_{Zn^{2+}})_{out}}{(m_{Cu^{2+}})_{in}}$$

Thus

$$\log \frac{(m_{Zn^{2+}})_{out}}{(m_{Cu^{2+}})_{in}} = 37.4$$

We can now relate the molality of Zn^{2+} and Cu^{2+} to the fraction of Cu^{2+} deposited as

$$(m_{Cu^{2+}})_{in} = 0.1(1 - \underline{X})$$

$$(m_{Zn^{2+}})_{out} = 0.1(1 + \underline{X})$$

where $\underline{X}$ = fraction of Cu^{2+} deposited. Therefore,

$$\log \frac{0.1(1 + \underline{X})}{0.1(1 - \underline{X})} = 37.4$$

or

$$\frac{1 + \underline{X}}{1 - \underline{X}} = 10^{37.4}$$

A quick examination of this last relation indicates that $\underline{X}$ must be quite close to 1.0. Therefore, $1 + \underline{X} \longrightarrow 2.0$ and $(1 - \underline{X})$ is given by

$$1 - \underline{X} = \frac{2.0}{10^{37.4}} = 10^{-37.1} = 7.9 \times 10^{-38}$$

Thus, when the cell voltage is zero, the molalities of the copper and zinc ions at their respective electrodes are given by

$$(m_{Cu^{2+}})_{in} = 7.9 \times 10^{-38}$$

$$(m_{Zn^{2+}})_{out} = 0.20$$

As we shall show later, the cell concentration that yields a cell voltage of zero corresponds to the equilibrium conditions for the reaction

$$Cu^{2+} + Zn = Zn^{2+} + Cu$$

That is, if we placed metallic zinc and copper in a mixture of 0.2 m Zn^{2+} and 7.9×10^{-38} m Cu^{2+} at 25°C, we would find that no changes in the composition of either the Cu^{2+} or Zn^{2+} would occur. That is, the mixture is in *chemical equilibrium*!

13.2 Variation of Cell Voltage with Temperature

In Chapter 11 it was shown that $\Delta\mathscr{G}^\circ$ for most reactions was temperature dependent. Since electrochemical reactions obey the same rules as other reactions, $\Delta\mathscr{G}^\circ$, and hence E°, will be functions of temperature. We may express the variation of $\Delta\mathscr{G}^\circ$ for an electrochemical process in exactly the same fashion as for a nonelectrochemical process. That is,

$$\frac{\partial(\Delta\mathscr{G}^\circ/T)_P}{\partial T} = -\frac{\Delta\mathscr{H}^\circ}{T^2} \tag{13-17}$$

or

$$\frac{\partial(\Delta\mathscr{G}^\circ/T)}{\partial(1/T)} = \Delta\mathscr{H}^\circ \tag{13-17a}$$

Since E° is related to $\Delta\mathscr{G}^\circ$ by equation (13-14),

$$\Delta\mathscr{G}^\circ = -\mathscr{N}FE^\circ \tag{13-14}$$

We may express the variation of E° with T as

$$\frac{\partial(E^\circ/T)}{\partial(1/T)} = -\frac{\Delta\mathscr{H}^\circ}{\mathscr{N}F} \tag{13-18}$$

or

$$\frac{\partial(E^\circ/T)}{\partial(T)} = \frac{\Delta\mathscr{H}^\circ}{\mathscr{N}FT^2} \tag{13-18a}$$

As indicated in Chapter 11, integration of equation (13-18) or (13-18a) to determine the variation of $\Delta\mathscr{G}^\circ$ (or E°) with temperature requires a knowledge of $\Delta\mathscr{H}^\circ$ as a function of temperature. This function was shown to be

$$\Delta\mathscr{H}^\circ = \Delta\mathscr{H}^\circ_{T^*} + \int_{T^*}^{T} \Delta\mathscr{C}^\circ_P \, dT \tag{13-19}$$

where $\Delta\mathscr{H}^\circ_T$ = standard state $\Delta\mathscr{H}$ at any temperature T

$\Delta\mathscr{H}^\circ_{T^*}$ = standard state $\Delta\mathscr{H}$ at a temperature T^*

$\Delta\mathscr{C}^\circ_P = (\mathscr{C}^\circ_P)_{prod} - (\mathscr{C}^\circ_P)_{react}$

Thus knowledge of $\Delta\mathscr{H}^\circ$ at any one temperature, T^*, and the specific heats of the products and reactants is sufficient to allow calculation of $\Delta\mathscr{H}^\circ$ as a function of T. Once $\Delta\mathscr{H}^\circ$ is known as a function of T, we may integrate equation (13-18) or (13-18a) to determine the variation of E° with temperature.

When calculations of the type suggested above are performed for the commonly used galvanic cells it is found that E° increases with decreasing temperature. However, experience tells us that batteries always seem to perform better in warm weather than in cold weather. (Compare the number of dead automobile batteries encountered in the winter versus the number encountered in the summer.) This observation can be reconciled with the statement that the cell voltage increases with decreasing temperature by recalling the phenomenon of polarization. The standard cell voltage E° is measured when only an infinitesimal current is being drawn. However, when a finite current is withdrawn from the cell, polarization occurs and the cell voltage decreases. During low-temperature operation the ionic mobilities of the reacting species decrease significantly and polarization becomes more severe. Thus it is found that even though E° increases with decreasing temperature, the effect of polarization eliminates this slight increase and in general will cause a decrease in the usable cell voltage at a finite current drain.

SAMPLE PROBLEM 13-2. A fuel cell that combines pure gaseous hydrogen and oxygen over a catalytic electrode to produce steam and electrical energy is planned for use in our space program. Because of complete reaction between the H_2 and O_2, the product steam is essentially pure. Initial studies indicate that the cell will operate satisfactorily at a temperature of 1000°F and a pressure of 100 psia. Determine the reversible voltage of this cell.

For the reaction

$$H_2 + \tfrac{1}{2}O_2 = H_2O$$

at 25°C,

$$\Delta\mathscr{G}^\circ = -54{,}600 \frac{\text{cal}}{\text{g-mol}}$$

$$\Delta\mathscr{H}^\circ = -57{,}800 \frac{\text{cal}}{\text{g-mol}}$$

where the standard states are pure components at 1 atm pressure. Over the temperature range of interest the following average heat capacities apply:

$$(C_P)_{H_2} = 7.15 \text{ Btu/lb-mol °F}$$

$$(C_P)_{O_2} = 8.00 \text{ Btu/lb-mol °F}$$

$$(C_P)_{H_2O} = 9.25 \text{ Btu/lb-mol °F}$$

All gases may be assumed to behave ideally.

Solution: Determination of standard voltage, E°, for the fuel-cell reaction at 1000°F will allow determination of the actual cell voltage, E, from the Nernst equation:

$$E = E^\circ - \frac{RT}{\mathscr{N}F} \ln J_a$$

where

$$J_a = \frac{(a_{H_2O})_{out}}{[a_{H_2}(a_{O_2})^{1/2}]_{in}} = \frac{(\bar{f}_{H_2O}/f^\circ_{H_2O})_{out}}{[(\bar{f}_{H_2}/f^\circ_{H_2})(\bar{f}_{O_2}/f^\circ_{O_2})^{1/2}]_{in}}$$

Since the standard states are pure gases at 1 atm, $f^\circ_{H_2O} = f^\circ_{H_2} = f^\circ_{O_2} = 1$ atm. The partial fugacities of the H_2, O_2, and H_2O are then given by

$$\bar{f}_i = \gamma_i^\circ x_i f_i^\circ$$

But all components enter or leave the cell in their pure form (the standard state). Thus $\gamma_i^\circ x_i = 1.0$ and the partial fugacities reduce to

$$\bar{f}_i = f_i^\circ = f_i$$

Under the assumption of ideal gases, $f_i = P_T$, the total pressure. The activity ratio, J_a, thus reduces to

$$J_a = \frac{100/14.7}{(100/14.7)(100/14.7)^{1/2}} = 0.383$$

At the reaction conditions $T = 1000°\text{F} = 812°\text{K}$,

$$R = 1.987 \text{ cal/g-mol °K}$$

$$\mathscr{N} = 2 \text{ equiv/mol}$$

$$F = 23{,}068 \text{ cal/volt-equiv}$$

Thus the equation for the cell voltage reduces to

$$E = E^\circ - \frac{(1.987)(812)}{(2)(23{,}068)} \frac{\text{cal °K g-mol volt equiv}}{\text{g-mol °K equiv cal}} (\ln 0.383)$$

or

$$E = E^\circ + 3.35 \times 10^{-2} \text{ V}$$

We may calculate E° at 1000°F from E° at 25°C using the modified form of the Gibbs–Helmholtz equation:

$$\frac{\partial(E^\circ/T)}{\partial T} = \frac{\Delta\mathscr{H}^\circ}{\mathscr{N}FT^2}$$

$$E^\circ \text{ at } 25°\text{C} = -\frac{(\Delta\mathscr{G}^\circ)_{25°\text{C}}}{\mathscr{N}F} = \frac{54{,}600 \text{ cal/g-mol V}}{(2)(23{,}068) \text{ cal/g-mol}}$$

or

$$E^\circ_{25°\text{C}} = 1.185 \text{ V}$$

$$T = 25°\text{C} = 298°\text{K}$$

$$\left(\frac{E^\circ}{T}\right)_{298°\text{K}} = 3.98 \times 10^{-3} \text{ V/°K}$$

The Gibbs–Helmholtz equation is integrated to give

$$\left(\frac{E^\circ}{T}\right)_{T_2} - \left(\frac{E^\circ}{T}\right)_{298°\text{K}} = \int_{298°\text{K}}^{T_2} \frac{\Delta\mathscr{H}^\circ_T}{\mathscr{N}FT^2}\, dT$$

$T_2 = 812°\text{K}$, and $\Delta\mathscr{H}^\circ_T$ can be determined from the relation

$$\Delta\mathscr{H}^\circ_T = \Delta\mathscr{H}^\circ_{T^*} + \int_{T^*}^{T} \Delta(\mathscr{C}^\circ_P)\, dT$$

Substituting $\Delta\mathscr{H}^\circ$ at 298°K and the value of $\Delta(\mathscr{C}_P^\circ)$ gives

$$\Delta\mathscr{H}_T^\circ = -57{,}800 \text{ cal/g-mol} + \int_{298}^{T} \{9.25 - [7.15 + (\tfrac{1}{2})(8.00)]\}\, dT \text{ cal/g-mol}$$

or

$$\Delta\mathscr{H}_T^\circ = (-58{,}400 - 1.90T) \text{ cal/g-mol}$$

Substitution of the expression for $\Delta\mathscr{H}_T^\circ$ into the equation for $(E^\circ/T)_{T_2}$ gives

$$\left(\frac{E^\circ}{T}\right)_{812^\circ\text{K}} = 3.98 \times 10^{-3} \text{ V/°K} + \int_{298}^{812} \left[\frac{-58{,}400 - 1.90T}{(2)(23{,}068)(T^2)} dT\right] \text{V/°K}$$

or

$$\frac{E^\circ}{812^\circ\text{K}} = 3.98 \times 10^{-3} \text{ V/°K} + \int_{298}^{812} \left[\left(\frac{-1.26}{T^2} - \frac{4.14 \times 10^{-5}}{T}\right) dT\right] \text{V/°K}$$

which reduces to

$$E^\circ = 3.23 \text{ V} + 812^\circ\text{K}\left[126\left(\frac{1}{812} - \frac{1}{298}\right) - 4.14 \times 10^{-5} \ln \frac{812}{298}\right] \text{V/°K}$$

Solving for E° then yields

$$E^\circ = 1.09 \text{ V}$$

Thus the actual cell voltage is given by

$$E = (1.09 + 0.03) \text{ V} = 1.12 \text{ V}$$

13.3 Electrode Reactions and Standard Oxidation Potentials

In preceding sections we have shown how to predict the voltage that can be expected from any cell provided only that we know the cell voltage at some standard conditions of temperature and composition. Thus measurement and tabulation of E° at one temperature for all possible cell reactions would enable us to determine the reversible voltage for any cell at any condition of temperature and composition. Since it is possible, however, to conceive of many thousands of cell reactions (corresponding to a pair of electrode reactions), it is obviously not practical to tabulate E° at even one temperature for each of these reactions. This difficulty could be overcome if we were able to determine the potential difference due to each of the known electrode reactions when the electrode reactants and products were in their standard states. The E° for a given cell would then simply be the sum of the individual electrode potentials. Thus, for example, tabulation of 100 standard electrode potentials would allow calculation of E° for $(100 \times 99)/2 = 4950$ distinct cell reactions. Unfortunately, it is not possible to measure the absolute electrode potential for a single electrode. To avoid this problem the electrode potentials are measured against a reference electrode whose potential is arbitrarily set equal to zero so that the cell voltage is attributed completely to the test electrode. By convention, the hydrogen–hydrogen ion electrode (or simply the *hydrogen electrode*) is used as the reference electrode. When the reacting species in both the test and hydrogen electrodes are in their standard states, the electrode potentials are referred to as the *standard oxidation potentials*. The standard oxidation potentials have been measured and tabulated for a number of electrode reactions. A summary of some of the more useful standard electrode potentials is given in Table 13-1. The standard states are (1) pure solids, (2) pure gases at 1 atm, and (3) solutions of electrolytes at infinite dilution. That is, $f_i^\circ = 1\ m \times (\partial \bar{f}_i/\partial m_i)_{m_i \to 0}$

TABLE 13-1† Standard Oxidation Potentials at 25°C

Reaction	Standard Oxidation Potential, V
$Li = Li^+ + e^-$	3.045
$K = K^+ + e^-$	2.924
$Ba = Ba^{2+} + 2e^-$	2.90
$Ca = Ca^{2+} + 2e^-$	2.76
$Na = Na^+ + e^-$	2.711
$Mg = Mg^{2+} + 2e^-$	2.375
$Al = Al^{3+} + 3e^-$	1.706
$H_2 + 2OH^- = 2H_2O + 2e^-$	0.828
$Zn = Zn^{2+} + 2e^-$	0.763
$Fe = Fe^{2+} + 2e^-$	0.409
$Pb + SO_4^{2-} = PbSO_4 + 2e^-$	0.351
$Ni = Ni^{2+} + 2e^-$	0.23
$Sn = Sn^{2+} + 2e^-$	0.136
$Pb = Pb^{2+} + 2e^-$	0.126
$H_2 = 2H^+ + 2e^-$	0.000
$Cu^+ = Cu^{2+} + e^-$	−0.158
$Ag + Cl^- = AgCl + e^-$	−0.222
$2Hg + 2Cl^- = Hg_2Cl_2 + 2e^-$	−0.268
$Cu = Cu^{2+} + 2e^-$	−0.340
$4OH^- = O_2 + 2H_2O + 4e^-$	−0.401
$2I^- = I_2 + 2e^-$	−0.535
$Fe^{2+} = Fe^{3+} + e^-$	−0.770
$Ag = Ag^+ + e^-$	−0.800
$2Br^- = Br_2(l) + 2e^-$	−1.065
$2Br^- = Br_2(aq) + 2e$	−1.087
$2H_2O = O_2 + 4H^+ + 4e^-$	−1.229
$2Cl^- = Cl_2(g) + 2e^-$	−1.358
$Au = Au^{3+} + 3e^-$	−1.42
$Cl^- + 3H_2O = ClO_3^- + 6H^+ + 6e^-$	−1.45
$Cl^- + H_2O = HClO^- + H^+ + e^-$	−1.49
$PbSO_4 + 2H_2O = PbO_2 + SO_4^{2-} + 4H^+ + 2e^-$	−1.685
$2H_2O = H_2O_2 + 2H^+ + 2e^-$	−1.776
$2F^- = F_2 + 3e^-$	−2.87

† Selected from *Handbook of Chemistry and Physics*, Chemical Rubber Publishing Company, Cleveland, 50th ed., 1969.

By convention, an electrode reaction

$$M = M^{\mathscr{N}+} + \mathscr{N}e^- \tag{13-20}$$

has a positive electrode potential with respect to the hydrogen electrode when $\Delta\mathscr{G}^\circ$ for the cell reaction

$$M + \mathscr{N}H^+ = M^{\mathscr{N}+} \frac{\mathscr{N}}{2} H_2 \tag{13-21}$$

is negative, that is, when the reaction indicated in equation (13-21) will proceed spontaneously in the direction written, if the products and reactants are in their standard states.

SAMPLE PROBLEM 13-3. The following reactions have been suggested as possible energy sources (either as fuel cells or batteries) to power an experimental electric automobile:

$$Zn + CuSO_4 = ZnSO_4 + Cu$$
$$Pb + PbO + 2SO_4^{2-} + 4H^+ = 2PbSO_4 + 2H_2O$$
$$Mg + Cl_2 = MgCl_2$$
$$H_2 + \tfrac{1}{2}O_2 = H_2O$$

Assuming that these reactions are to be performed in aqueous solution, determine E° at 25°C for each cell. State clearly the standard states for all components.

Solution: The standard states are chosen as (1) pure gases at 1 atm, (2) pure solid phases, and (3) ionic solutions at the infinite-dilution standard state. With these standard states we may use Table 13-1 to obtain the standard electrode potentials for each of the electrodes in the reactions under consideration. The standard-state cell voltages are

Table S13-3a

Cell Reaction	Anode Reaction	Cathode Reaction
$Zn + CuSO_4 = Cu + ZnSO_4$	$Zn = Zn^{2+} + 2e^-$	$Cu^{2+} + 2e^- = Cu$
$Pb + PbO + 2SO_4^{2-} + 4H^+ = 2PbSO_4 + 2H_2O$	$Pb + SO_4^{2-} = PbSO_4 + 2e^-$	$PbO_2 + SO_4^{2-} + 4H^+ + 2e^- = PbSO_4 + 2H_2O$
$Mg + Cl_2 = MgCl_2$	$Mg = Mg^{2+} + 2e^-$	$Cl + 2e^- = 2Cl^-$
$H_2 + \frac{1}{2}O_2 = H_2O$	$H_2 = 2H^+ + 2e^-$	$2H^+ + \frac{1}{2}O_2 + 2e^- = H_2O$

Table S13-3b

Cell Reaction	Anode Potential, V	Cathode Potential, V	Cell Voltage (E°), V
$Zn + CuSO_4 = ZnSO_4 + Cu$	0.763	0.340	1.103
$Pb + PbO + 2SO_4^{2-} = 2PbSO_4 + 2H_2O$	0.351	1.685	2.036
$Mg + Cl_2 = MgCl_2$	2.375	1.358	3.733
$\frac{1}{2}O_2 + H_2 = H_2O$	0.0	1.229	1.229

then simply the sum of the electrode potentials. We may summarize the electrode reactions of the various cells as shown in Table S13-3a. The electrode potentials and cell voltages are then as shown in Table S13-3b.

13.4 Equilibrium in Electrochemical Reactions

In equation (13-16) the reversible voltage, E, of a cell was related to its standard voltage, E°, and the activities of the reacting species:

$$E = E^\circ - \frac{RT}{\mathscr{N}F} \ln J_a \tag{13-16a}$$

where $J_a = \prod_{\text{prod}} (a_i^{\alpha_i})_{\text{out}} / \prod_{\text{react}} (a_i^{-\alpha_i})_{\text{in}}$, the activity ratio. If the compositions within the cell are such that the cell voltage is zero, the activity ratio has the value

$$(RT \ln J_a)_{E=0} = -\Delta\mathscr{G}^\circ = \mathscr{N}FE^\circ \tag{13-22}$$

However, we have shown in earlier chapters that the equilibrium constant (or equilibrium activity ratio) is also given by

$$RT \ln K_a = -\Delta\mathscr{G}^\circ \tag{13-23}$$

Thus, when the reversible cell voltage is zero, the activity ratio equals the equilibrium constant, and the cell reaction must be at its equilibrium state.

Thus we see it is possible to use electrochemical methods to determine equilibrium constants for reactions that involve electron transfers. On the other hand, it is also possible to use equilibrium measurements to determine electrochemical properties for these reactions.

SAMPLE PROBLEM 13-4. It is proposed to reduce alumina (Al_2O_3) to pure aluminum with coke as the reducing agent according to the reaction

$$Al_2O_3(s) + 3C(s) = 2Al(l) + 3CO$$

(a) What is the maximum (or minimum) pressure at which this reaction will proceed spontaneously at 1000°C?

(b) Is this pressure a maximum or minimum?

(c) What is the reversible voltage required for the production of aluminum by electrolysis of alumina dissolved in fused cryolite at 1000°C and 1 atm? Graphite carbon electrodes are used, and CO rather than O_2 is the gaseous product. Assume that liquid aluminum is insoluble in the cryolite solution and that the activity of the alumina in the cryolite solution is close to unity.

(d) If the graphite electrodes of (c) are replaced with metallic ones so that oxygen is the gaseous product, determine the reversible voltage required for the electrolysis of the alumina solution.

At 1000°C, $E^\circ = 1.90$ V and $\mathscr{N} = 4$ for the reaction (O_2 is taken as specie 1)

$$O_2 + \tfrac{4}{3}Al(l) = \tfrac{2}{3}Al_2O_3(s) \tag{1}$$

while $\log_{10} K_a = -22.2$ for the reaction

$$2CO = O_2 + 2C \tag{2}$$

Standard states are pure components in their natural state at 1 atm and 1000°C. You may assume ideal gas behavior for the gaseous products.

Solution: (a) and (b) To determine the maximum (minimum) pressure at which the reaction

$$Al_2O_3(s) + 3C = 2Al(l) + 3CO$$

will proceed spontaneously, we need to determine the equilibrium constant. From the equilibrium constant the reaction pressure will be determined. For the first reaction,

$$O_2 + \tfrac{4}{3}Al(l) = \tfrac{2}{3}Al_2O_3(s) \tag{1}$$

$$E^\circ = 1.90 \text{ V} \qquad \mathscr{N} = 4$$

For the second reaction,

$$2CO = O_2 + 2C \tag{2}$$

$$\log K_a = -22.2$$

$\Delta\mathscr{G}^\circ$ for the above reactions may then be calculated as follows.

Reaction 1:

$$(\Delta\mathscr{G}^\circ)^1 = -\mathscr{N}FE^\circ = -[(4)(23{,}068)(\text{cal/g-mol V})]\, 1.90 \text{ V}$$

or

$$(\Delta\mathscr{G}^\circ)^1 = -175{,}000 \text{ (cal/g-mol)}$$

Reaction 2:

$$(\Delta\mathscr{G}^\circ)^2 = -RT \ln K_a = -[(1.987)(1273) \text{ (cal/g-mol)}]\,[-(2.303)(22.2)]$$

or

$$(\Delta\mathscr{G}^\circ)^2 = 129{,}000 \text{ (cal/g-mol)}$$

$\Delta\mathscr{G}^\circ$ for the overall reaction is then calculated from

$$\begin{aligned}(\Delta\mathscr{G}^\circ)_{\text{overall}} &= -\tfrac{3}{2}(\Delta\mathscr{G}^\circ)^1 - \tfrac{3}{2}(\Delta\mathscr{G}^\circ)^2 \\ &= -(\tfrac{3}{2})(-175{,}000 + 129{,}000) \text{ (cal/g-mol)}\end{aligned}$$

Therefore,

$$(\Delta\mathscr{G}^\circ)_{\text{overall}} = 69{,}000 \text{ cal/g-mol}$$

The equilibrium constant, K_2, for the overall reaction is then given by

$$\ln K_a = -\frac{(\Delta\mathscr{G}^\circ)_{\text{overall}}}{RT} = -\frac{(69{,}000)(\text{cal/g-mol})}{(1.987)(1273)(\text{cal/g-mol})}$$

or

$$K_a = 1.6 \times 10^{-12}$$

For the overall reaction to proceed spontaneously, the outlet activity ratio,

$$(J_a)_{\text{out}} = \left[\frac{(a_{\text{Al}})^2(a_{\text{CO}})^3}{a_{\text{Al}_2\text{O}_3}(a_{\text{C}})^3}\right]_{\text{out}}$$

must be less than K_a. Since the aluminum, aluminum oxide, and carbon are all in their standard states (pure solid Al_2O_3 and C, and pure liquid Al), their activities are unity. Thus the outlet activity ratio reduces to

$$(J_a)_{\text{out}} = (a_{\text{CO}})^3_{\text{out}}$$

The condition that the overall reaction occurs spontaneously becomes

$$(a_{CO})^3_{out} < 1.6 \times 10^{-12}$$

or

$$(a_{CO})_{out} < 1.17 \times 10^{-4}$$

The activity of CO is given by

$$a_{CO} = \frac{\bar{f}_{CO}}{f^\circ_{CO}} = \frac{y_{CO}P_T}{1 \text{ atm}}$$

Since CO is the only gaseous component, $y_{CO} = 1$ and a_{CO} is given by

$$(a_{CO})_{out} = P_T \text{ atm}^{-1}$$

Thus the pressure at which the reaction will proceed spontaneously is given by

$$P_T \text{ atm}^{-1} < 1.17 \times 10^{-4}$$

or

$$P_T < 1.17 \times 10^{-4} \text{ atm}$$

That is, $P = 1.17 \times 10^{-4}$ atm is the *maximum* pressure at which the reaction will proceed spontaneously.

(c) The reversible voltage for any reaction can be related to the standard-state voltage, and the activity ratio by the Nernst equation:

$$E = E^\circ - \frac{RT}{\mathscr{N}F} \ln J_a$$

If the electrolysis is performed at 1 atm, all components will appear in their standard states, so $J_a = 1.0$ and the Nernst equation reduces to

$$E = E^\circ$$

The standard-state voltage is calculated from the overall $\Delta\mathscr{G}^\circ$ as follows:

$$E^\circ = -\frac{(\Delta\mathscr{G}^\circ)_{overall}}{\mathscr{N}F}$$

where $\mathscr{N} = 4$ and $F = 23{,}068$ (cal/V)/g-equiv. $(\Delta\mathscr{G}^\circ)_{overall}$ as calculated in part (a) is substituted into the expression for E° to obtain

$$E^\circ = -\frac{69{,}000 \text{ cal/g-mol}}{(4)(23{,}068)} \text{ (cal/V)/g-mol}$$

or

$$E^\circ = -0.75 \text{ V}$$

Thus the reversible electrolysis voltage is -0.75 V, with the minus sign indicating that electrical energy must be supplied to the cell for the reduction to occur.

(d) If the graphite electrodes of part (c) are replaced with metallic electrodes, the reduction reaction becomes

$$\tfrac{2}{3}Al_2O_3 \rightleftharpoons \tfrac{4}{3}Al + O_2$$

Since this is the reverse of reaction (a),

$$E^\circ = -1.90 \text{ V}$$

for the reduction reaction. At 1 atm, all components are at their standard states, so the cell voltage equals E°. Thus

$$E = -1.90 \text{ V}$$

The minus sign again indicates that electrical energy must be supplied to the electrolysis cell.

Problems

13-1 For the reaction

$$CoCl_2(l) = Co(l) + Cl_2$$
$$\Delta\mathscr{G}^\circ = 36.3 \text{ kcal/g-mol at } 1500°C.$$

(a) Calculate the minimum electromotive force (emf) required for the electrolysis at 1500°C of pure liquid $CoCl_2$ to form pure liquid cobalt and chlorine gas at 1 atm pressure.

(b) Determine the reversible emf in the above electrolysis if chlorine gas is produced at a pressure of 380 mm Hg. Assume that chlorine is an ideal gas.

13-2 Caustic soda, hydrogen, and chlorine are obtained from the electrolysis of brine (NaCl solution) according to the reaction

$$2NaCl + 2H_2O = 2NaOH + H_2 + Cl_2$$

The electrolysis is performed at 25°C.

(a) Write the anode and cathode reactions.

(b) Determine the standard-state voltage for this electrolysis. What are the standard states?

(c) The cell is to operate in a continuous fashion—that is, brine will be fed continuously, while caustic, hydrogen, and chlorine will be continuously removed. The inlet brine is 3 per cent NaCl, and the electrolysis is to recover 95 per cent of the available chlorine. If the hydrogen and chlorine are collected at 15 atm, what is the minimum voltage required for the electrolysis? If the gases are collected at 1 atm, what is the minimum voltage? You may assume ideal gas behavior and that the activity coefficients of both NaCl and NaOH are independent of composition in the concentration range of interest.

13-3 Hydrogen and oxygen can be made by the electrolysis of water in a conventional U-shaped electrolytic cell. Determine the reversible voltage for the electrolysis if the cell operates at a pressure of 2 atm and 212°F. The H_2 gas and O_2 gas are both withdrawn at 2 atm pressure.

For the reaction

$$H_2 + \tfrac{1}{2}O_2 = H_2O$$

at 2000°F, $\log_{10} K = 6.50$, where the standard states are pure gases at 1 atm pressure. Over the temperature ranges in this problem, the following average specific heats apply (vapor-phase C_P):

$$(C_P)_{H_2} = 7.15 \text{ Btu/lb-mol °F}$$
$$(C_P)_{O_2} = 8.00 \text{ Btu/lb-mol °F}$$
$$(C_P)_{H_2O} = 9.25 \text{ Btu/lb-mol °F}$$
$$(\Delta\mathscr{H}_f^\circ)_{H_2O(g)} = -57.80 \text{ kcal/g-mol at 70°F}$$

All gases may be assumed to behave ideally.

13-4 An electrochemical cell consists of two electrodes dipping into a 0.1 N HCl solution. One electrode is lead coated with insoluble $PbCl_2$ and the other electrode is silver coated with insoluble AgCl. The overall reaction is

$$Pb + 2AgCl = PbCl_2 + 2Ag$$

The measured voltage for this reaction is 0.4900 V at 25°C and 0.4835 V at 60°C.

(a) Write the reactions that take place at each electrode.
(b) What is the free-energy change for the reaction at 25°C?
(c) What would be the measured voltage at 25°C if 0.2 *N* HCl was used?

13-5 A vapor-phase methanol-powered fuel cell is being considered as the power plant for a golf cart. The methanol will be oxidized by contact with air. The fuel cell will be operated at 1 atm pressure and 700°F to insure an adequate rate of reaction. If oxygen is supplied at the stoichiometric rate and the methane and oxygen are allowed to react completely, what is the maximum electrical energy that can be obtained from the fuel cell per lb_m of methanol feed? The only reaction is

$$CH_3OH + \tfrac{3}{2}O_2 = CO_2 + 2H_2O$$

The heat capacities at 1 atm for various species (gaseous) are given below.

methanol: $C_P^\circ = 4.394 + 24.274 \times 10^{-3}T(°K) - 6.855 \times 10^{-6}T^2$ cal/g-mol °K

oxygen: $C_P^\circ = 6.148 + 3.102 \times 10^{-3}T(°K) - 0.923 \times 10^{-6}T^2$ cal/g-mol °K

nitrogen: $C_P^\circ = 6.524 + 1.250 \times 10^{-3}T(°K)$ cal/g-mol °K

water: $C_P^\circ = 7.256 + 2.298 \times 10^{-3}T(°K) + 0.283 \times 10^{-6}T^2$ cal/g-mol °K

carbon dioxide: $C_P^\circ = 6.214 + 10.396 \times 10^{-3}T(°K) - 3.545 \times 10^{-6}T^2$ cal/g-mol °K

If 6 moles of electrons are transferred for each mole of methanol reacted, what is the reversible voltage for each fuel cell?

13-6 An electrochemical cell is constructed according to Fig. P13-6. Two columns of mercury of heights differing by 9 ft are in contact with an acidified mercurous nitrate solution through semipermeable parchment membranes. The membranes pass aqueous solutions of all ions freely but do not pass liquid mercury. When the electrodes are shorted, current passes but no change in the concentrations of the ions in solution takes

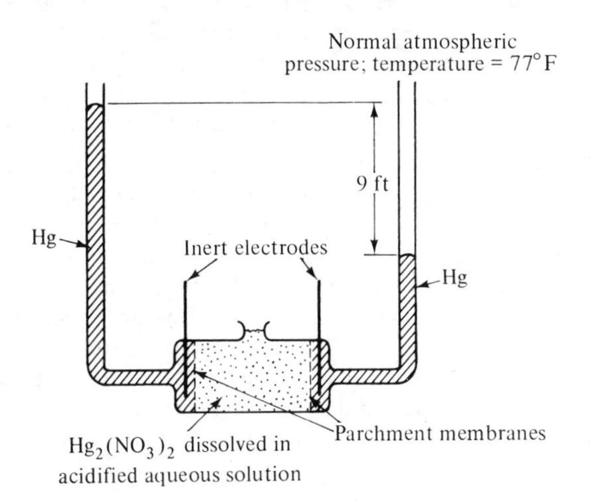

FIGURE P13-6

place. What reactions are taking place at each membrane? What is the overall reaction? For the conditions given, what zero-current voltage would be measured between the electrodes? The atomic weight of mercury is 200.6 and its density is 13.6 times that of water.

13-7 A moderately high temperature electrolytic cell may be constructed according to Fig. P13-7, and the potential between the gold–cadmium bar and the liquid cad-

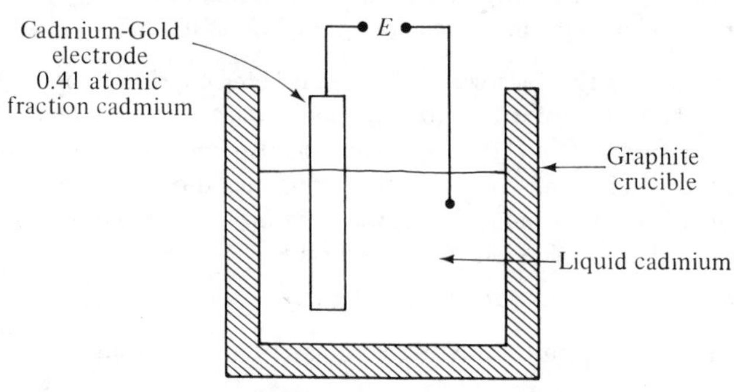

FIGURE P13-7

mium measured at different temperatures. Such measurements have been made above the melting point of cadmium (321°C). These are well represented by the empirical equation

$$E(\text{mV}) = 153.1 - 0.065\,[T(^\circ\text{K}) - 700]$$

In a separate experiment the heat of fusion of cadmium at 321°C has been measured as 12.9 cal/g. The atomic weights of gold and cadmium are 197.0 and 112.4, respectively. From these data determine the activity of the cadmium in the alloy bar at 321°C. Also calculate the entropy change at 321°C for the reaction Cd(s) $\longrightarrow$ Cd (0.41 atom-fraction in Au).

13-8 An electrochemical cell is constructed in a centrifuge according to Fig. P13-8. The columns of mercury differ in length by 12 cm. The semipermeable membranes pass all ions in the aqueous, acidified, mercurous nitrate solution but do not pass liquid mercury. During operation the electrodes are connected to a potentiometer on the outside by means of the slip rings on the centrifuge shaft. If the centrifuge is rotating at 2000 rpm, what zero-current voltage would be measured across the electrodes?

13-9 It is proposed to manufacture hydrogen and oxygen from slightly acidified water using a steady-flow process. The hydrogen and oxygen will both be used at another location and will be transported in cylinders at 2200 psia. Although the conventional method of manufacutring these gases is to electrolyze isothermally at about 25°C and 1 atm absolute, it has been suggested that the process might be carried out at 2200 psia and 25°C, thus saving the work of isothermally compressing the gases at 25°C. Starting with water at 1 atm and 25°C, compare the reversible heat and work loads in each continuous flow process as well as the reversible voltage. Use units of cal/g-mol H_2O for the work loads. It is agreed that this calculation may be made assuming ideal behavior of the gas phase. For the reaction

$$H_2 + \tfrac{1}{2}O_2 = H_2O$$

$$(\Delta\mathscr{G}^\circ)_{298} = -54{,}635 \text{ cal/g-mol}$$

when pure gases at 1 atm are chosen for the standard states.

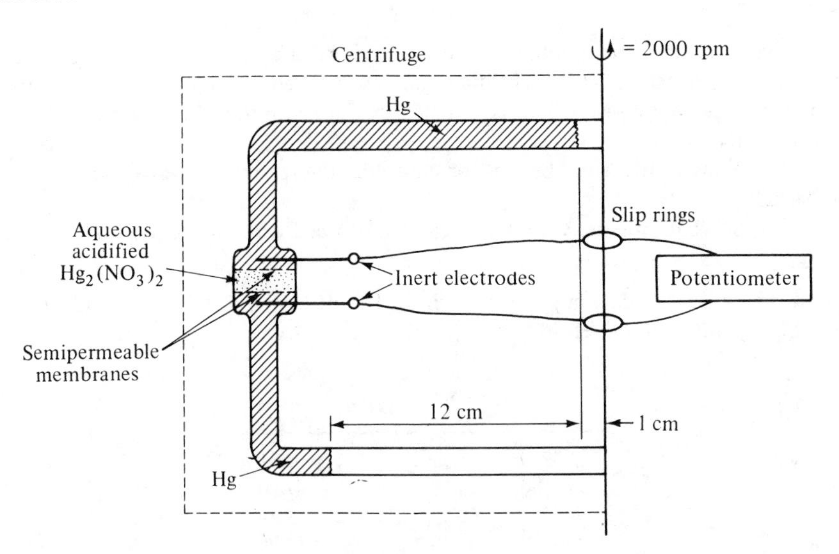

FIGURE P13-8

13-10 Water is to be electrolyzed in a Hoffman apparatus at 1 atm and 25°C as shown in Fig. P13-10. The anode reaction is

$$2OH^- = H_2O + \tfrac{1}{2}O_2 + 2\bar{e}$$

The cathode reaction is

$$2H^+ + 2\bar{e} = H_2$$

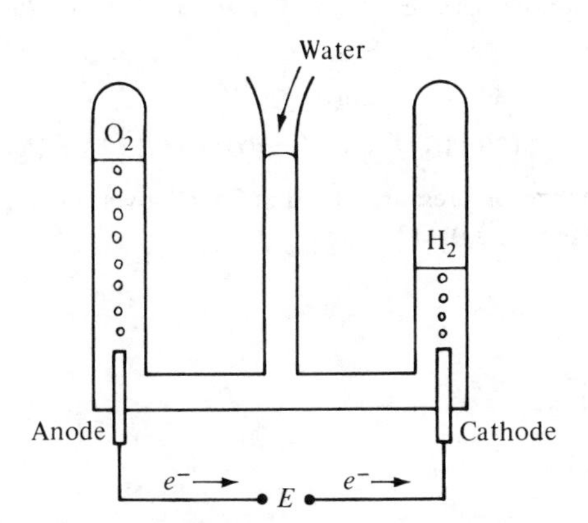

FIGURE P13-10

If the water has been made slightly acid (pH = 4.0) to increase its electrical conductivity, determine the reversible voltage needed for the electrolysis. You may assume that the side reaction

$$H_2O = H^+ + OH^-$$

is always at equilibrium. For this side reaction $K_a = 10^{-14}$ when the standard states are (1) pure water at 1 atm (liquid) and (2) ionic species at infinite dilution based on molality. That is,

$$f_i^\circ = (\partial \bar{f}_i / \partial m_i)_{m_i \to 0}$$

13-11 Magnesium metal is obtained by electrolyzing fused magnesium chloride, $MgCl_2$, in an electrolytic cell at 700°C. The magnesium so formed is a liquid, and is essentially insoluble in the molten $MgCl_2$. It is reported that at 1 atm the reversible cell voltage is approximately 5.5 V.

(a) What is the equilibrium constant for the reaction $MgCl_2 = Mg + Cl_2$ at 700°C?

(b) At what pressure would this reaction proceed without the need for electrical energy?

13-12 The following data apply to the cell Cd(l)| $CdCl_2$ in LiCl | $\underline{Cd}$ (in liquid solution with Sb) operated at 753°K:

x_{Cd}	E, mV
0.8960	3.31
0.8180	7.20
0.7497	11.56
0.6760	18.96
0.5880	29.69
0.5590	33.85
0.4340	52.00
0.4006	56.51
0.3745	59.90
0.3444	63.95

(a) Calculate the activity of cadmium in each solution.

(b) Find the vapor pressure of pure Cd at 753°K from the following data. For the vaporization reaction

$$\text{Cd(l)} = \text{Cd(g)}$$

$$\Delta\mathcal{G}^\circ = (26{,}110 + 4.97\,T\log T - 40.15\,T)\ \text{cal/g-mol}$$

(c) What is the vapor pressure of Cd at 753°K over a Cd–Sb alloy in which the mole fraction of Cd is 0.75? 0.40?

Irreversible Thermodynamics 14

Introduction

Traditionally the subjects of thermodynamics and rate processes have been developed in two rather isolated sequences in the engineering curriculum. The thermodynamicist has been content to manipulate the concepts of energy and entropy within the framework of equilibrium processes with the equilibrium state as his primary concern. He conceives of processes as originating and terminating in such states and accepts as his challenge the definition of those potentials that may produce change and the energy requirements accompanying such changes. The rate of change has been of minimal interest to him.

Likewise the kineticist whose major interest is the rate at which processes occur has focused his interest on reaction or transfer mechanism and the resistances or barriers that retard those changes. Models and mechanisms have been postulated, empiricism employed, and true driving forces (as proposed by thermodynamics) frequently obscured by the use of simplified parameters such as concentration instead of activity, fugacity, or chemical potential.

The development of irreversible thermodynamics utilizes the entropy producing character of spontaneously occurring processes to provide a link between classical thermodynamics and rate processes. The development which follows has as its primary purpose the bridging of this gap for the interested student. Of secondary importance is the demonstration of coupling phenomena between certain of the transport processes.

The application of thermodynamic formalism to nonequilibrium systems is hypothesized to be valid for systems not too far displaced from equilibrium. This condition limits the eventual application of the results but does not prove overly restrictive as many systems and phenomena can be shown to occur within this framework. Thus equations for property changes between equilibrium states, such as the Gibbs property relationship, are of assistance in formulating and interpreting the expressions for entropy production in nonequilibrium systems in terms of variables that are used in the studies of rate processes.

A general relationship between fluxes and forces is hypothesized which reduces to the transport expressions of Fourier, Fick, and Ohm for heat, mass, and electrical transport respectively under certain conditions. It is at this point where courses in rate processes ordinarily begin. However, a better understanding of the nature of the potentials or driving forces that produce certain fluxes is certain to improve one's ability to correlate and generalize transport phenomena.

In Chapter 5 entropy generation was discussed in relationship to the concept of lost work. Although this approach related entropy generation to a reduction in energy available for transformation to work, it failed to provide a useful means of calculating the generation term for all but a few simplified types of irreversible behavior, such as the frictional losses in the piston-cylinder example. In the following development entropy generation will be expressed as the product of thermodynamic forces (or potentials) such as T, P, or μ, and the fluxes (or flows) such as thermal, electrical, or mass which occur as a result of these forces. Thermodynamic relationships developed in earlier chapters for equilibrium systems will be used to justify such a relationship as well as to define precisely the thermodynamic forces and fluxes which must be used. Such an approach is facilitated by considering a lumped system such as that discussed in Sample Problem 5-1, and extending our analysis to include other than thermal energy flows. Finally we shall apply the results to systems in which continuous gradients in the thermodynamic potentials exist.

14.1 Entropy Generation in a Lumped System

Consider the following system: two subsystems I and II connected by a conducting but rigid, impermeable partition. The entire system, as shown in Fig. 14-1, is isolated from the surroundings such that the following conditions are imposed on any processes proceeding within:

$$
\begin{aligned}
&1.\ dS_{sys} = dS_{gen} = dS^{I} + dS^{II} \geq 0 \\
&2.\ dU_{sys} = 0; \qquad dU^{I} = -dU^{II} \\
&3.\ dV_{sys} = dV^{I} = dV^{II} = 0 \\
&4.\ dN_{sys} = dN^{I} = dN^{II} = 0
\end{aligned}
\tag{14-1}
$$

The two subsystems are assumed to be uniform throughout with any gradients confined to the diathermal (thermally conducting) partition separating them. Thus all classical thermodynamic relationships apply to each of the

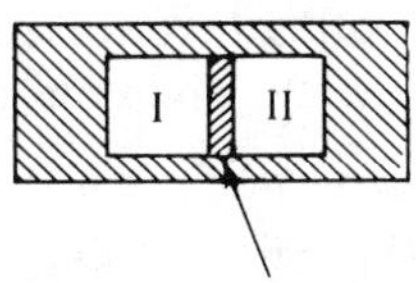

FIG. 14-1. Conducting partition.

subsystems. Using the property relationship developed in Chapter 6, we can write for each subsystem the expression for the differential change of entropy

$$dS_{\mathrm{I}} = \frac{1}{T^{\mathrm{I}}}\,dU^{\mathrm{I}} + \frac{P^{\mathrm{I}}}{T^{\mathrm{I}}}\,dV^{\mathrm{I}} - \frac{\mu^{\mathrm{I}}}{T^{\mathrm{I}}}\,dN^{\mathrm{I}} \tag{14-2}$$

$$dS^{\mathrm{II}} = \frac{1}{T^{\mathrm{II}}}\,dU^{\mathrm{II}} + \frac{P^{\mathrm{II}}}{T^{\mathrm{II}}}\,dV^{\mathrm{II}} - \frac{\mu^{\mathrm{II}}}{T^{\mathrm{II}}}\,dN^{\mathrm{II}} \tag{14-3}$$

Conditions 3 and 4 from equation (14-1) eliminate the last two terms in equations (14-2) and (14-3). From conditions 1 and 2 we conclude that

$$\begin{aligned} dS_{\mathrm{gen}} &= dS^{\mathrm{I}} + dS^{\mathrm{II}} \\ &= \frac{1}{T^{\mathrm{I}}}\,dU^{\mathrm{I}} + \frac{1}{T^{\mathrm{II}}}\,dU^{\mathrm{II}} \\ &= \left(\frac{1}{T^{\mathrm{I}}} - \frac{1}{T^{\mathrm{II}}}\right) dU^{\mathrm{I}} \end{aligned} \tag{14-4}$$

At equilibrium, dS_{gen} must go to zero since entropy will be maximized for given values of U, V, and N. Therefore at equilibrium

$$dS_{\mathrm{gen}} = 0 = \left(\frac{1}{T^{\mathrm{I}}} - \frac{1}{T^{\mathrm{II}}}\right) dU^{\mathrm{I}} \tag{14-5}$$

Since dU^{I} is independent and can take on nonzero values, $T^{\mathrm{I}} = T^{\mathrm{II}}$ at equilibrium as deduced earlier. For conditions where $T^{\mathrm{I}} \neq T^{\mathrm{II}}$, the entropy change of the system, dS_{sys}, will be observed to always be positive. When $T^{\mathrm{I}} > T^{\mathrm{II}}$, $dU^{\mathrm{I}} < 0$ and a negative potential, $[\Delta(1/T)]$, multiplied by a negative energy change, dU^{I}, yields $dS_{\mathrm{sys}} > 0$. For $T^{\mathrm{II}} > T^{\mathrm{I}}$ the signs on both potential and flow terms reverse; therefore as before $dS > 0$. The energy flow between subsystems is a thermal flow and can be represented by δQ^{I} or δQ^{II} by applying the energy balance separately to the two subsystems.

$$dU^{\mathrm{I}} = \delta Q^{\mathrm{I}} \qquad \text{and} \qquad dU^{\mathrm{II}} = \delta Q^{\mathrm{II}} \tag{14-6}$$

Thus we see for the total system (I + II) that the entropy generated by this process, dS_{gen}, at any instant in time can be expressed as the product of a potential difference (force), $\Delta(1/T)$, and a flow of thermal energy, δQ^{I}, by combining equations (14-5) and (14-6).

$$\begin{aligned} dS_{\mathrm{gen}} = \underbrace{\delta Q^{\mathrm{I}}}_{\text{thermal energy flow}} \underbrace{\left(\frac{1}{T^{\mathrm{I}}} - \frac{1}{T^{\mathrm{II}}}\right)}_{\Delta\text{ (potential)}} &> 0 \quad \text{for} \quad T^{\mathrm{I}} \neq T^{\mathrm{II}} \\ &= 0 \quad \text{for} \quad T^{\mathrm{I}} = T^{\mathrm{II}} \end{aligned} \tag{14-7}$$

Similarly for the transfer of matter, the equation can be written in such a way that μ/T is shown to be the appropriate potential. Consider the system as the two subsystems I and II with a permeable (but rigid) membrane separating the two subsystems. The following conditions are then imposed on the system:

$$\begin{aligned} &1.\ dS_{sys} = dS_{gen} = dS^{I} + dS^{II} \\ &2.\ dU_{sys} = 0; \qquad dU^{I} = -dU^{II} \\ &3.\ dV_{sys} = 0 = dV^{I} = dV^{II} \\ &4.\ dN_{i_{sys}} = 0; \qquad dN_i^{I} = -dN_i^{II} \end{aligned} \tag{14-8}$$

If the membrane is thermally conducting, then

$$dS_{gen} = \left(\frac{1}{T^{I}} - \frac{1}{T^{II}}\right) dU^{I} - \sum_i \left(\frac{\mu_i^{I}}{T^{I}} - \frac{\mu_i^{II}}{T^{II}}\right) dN_i^{I} \tag{14-9}$$

where the $\sum_i$ indicates the presence of several components. Notice that dN_i^{I} corresponds to the change in component i of subsystem I. A simple mass balance around subsystem I will show that this is also the amount of component i that flows into subsystem I during the process, $(\delta N_i^{I})_{in}$. Thus we observe that the dN_i term, although originally included because of the *change* in i content of the subsystem, also describes the *flow* of component i into the system. At equilibrium the change in dS is again zero and one concludes $T^{I} = T^{II}$ and hence $\mu_i^{I} = \mu_i^{II}$. If one also imposes the condition that the membrane be an insulator, then the equation reduces to

$$dS_{gen} = -\sum_i \left(\frac{\mu_i^{I}}{T^{I}} - \frac{\mu_i^{II}}{T^{II}}\right) dN_i^{I} = 0 \tag{14-10}$$

which, however, does not permit one to conclude that $\mu_i^{I} = \mu_i^{II}$ unless $T^{I} = T^{II}$, which need not be the case.

For a system in which the values of μ/T are not equal in two subsystems that are allowed to communicate, a mass flow will result. The entropy generation is given by the expression

$$dS_{gen} = -\sum_i \underbrace{\left(\frac{\mu_i^{I}}{T^{I}} - \frac{\mu_i^{II}}{T^{II}}\right)}_{\Delta\text{ (potential for mass transfer)}} \underbrace{dN_i^{I}}_{\text{component flow}} \tag{14-11}$$

Further observations that will be of importance in our later deliberations should also be made at this time. The transport of matter, dN_i, can also be accomplished with forces or potentials other than (μ_i/T). Charged species will be subjected to forces if an electrical potential difference exists across a system. These forces can cause migration even in the absence of a difference in (μ/T). Gravitational effects also may produce component transfer. It is thus necessary to provide for other possible forces in the potential term for component transfer. Let us now insert a generalized force (potential) Y_i so that the coefficient of dN_i now appears as follows:

$$dS_{gen} = -\sum_i \left[\left(\frac{\mu_i}{T} + Y_i\right)^{I} - \left(\frac{\mu_i}{T} + Y_i\right)^{II}\right] dN_i^{I} \tag{14-12}$$

At equilibrium

$$\left(\frac{\mu_i}{T} + Y_i\right)^{I} = \left(\frac{\mu_i}{T} + Y_i\right)^{II} \quad \text{for each component} \tag{14-13}$$

Thus the presence of an electrical field can produce a concentration (or chemical potential) difference in a system containing *charged* species. The interdependence of these potentials should not be confused with the coupling phenomena to be introduced later.

The number of moles of a component, n_i, in a system can change due to transfer into or out of the system, or due to a reaction within the system. Consider a closed system containing C components. Since mass is conserved but moles in general are not, there are some advantages to expressing component changes in mass units.

$$dm_i = \alpha_i M_i dX \tag{14-14}$$

where α_i = stoichiometric coefficient of component i in balanced reaction
+ if product
− if reactant

M_i = molecular weight

X = extent of reaction

For reaction:

$$N_2 + 3H_2 = 2NH_3$$

$$dX = \frac{dm_{N_2}}{(-1)M_{N_2}} = \frac{dm_{H_2}}{-3M_{H_2}} = \frac{dm_{NH_3}}{2M_{NH_3}} = \frac{dm_i}{\alpha_i M_i} \tag{14-15}$$

From the principle of conservation of mass

$$\sum_i dm_i = dm = 0 = \sum_i (M_i \alpha_i)\, dX \tag{14-16}$$

$$\frac{dm_i}{M_i} = dn_i = \alpha_i\, dX \tag{14-17}$$

Application of equation (14-12) to a closed reacting system that is uniform throughout yields

$$dS_{\text{gen}} = -\sum_i \left(\frac{\mu_i}{T} + Y_i\right) dn_i \tag{14-18}$$

The mole change, dn_i, due to transport is zero for a closed system, but can take on finite values if a reaction occurs within the system. Y_i is not considered a part of the potential for chemical reaction. Therefore, rewriting equation (14-18) for a reacting system using (14-17) to express dn_i yields

$$dS_{\text{gen}} = -\sum_i \frac{\mu_i}{T} \alpha_i\, dX \tag{14-19}$$

We choose to define a quantity A, which we term the chemical affinity, to characterize a particular chemical reaction as follows:

$$-A = \sum_i \mu_i\, \alpha_i \tag{14-20}$$

Thus for a closed system completely isolated from its surroundings, the entropy change resulting from chemical reaction is

$$dS_{\text{gen}} = \frac{A\, dX}{T} \geq 0 \tag{14-21}$$

For equilibrium the reaction potential must reduce to zero. Thus

$$\frac{A}{T} = -\frac{1}{T}\sum_i \mu_i \alpha_i = 0 \tag{14-22}$$

Again we observe that entropy generation occurring in an isolated system due to a chemical reaction can be expressed as a product of a potential and a flow. In this instance A/T represents the potential for chemical reaction and X the "reaction flow".

Taking the time derivative of the entropy generation expression for all these phenomena yields

$$\frac{dS_{\text{gen}}}{d\theta} = \dot{S}_{\text{gen}} = +\frac{\delta Q}{d\theta}\left[\Delta\left(\frac{1}{T}\right)\right] - \sum_i \frac{dn_i}{d\theta}\left[\Delta\left(\frac{\mu_i}{T} + Y_i\right)\right] + \sum_j \frac{A_j}{T}\frac{dX_j}{d\theta} \tag{14-23}$$

where the superscripts representing the subsystems have been deleted and the subscript j in the last term represents each independent reaction occurring in the system.

The rate of entropy production, $\dot{S}_{\text{gen}}$, is positive unless all potentials are identically zero in which case equilibrium exists within the system and $\dot{S}_{\text{gen}} = 0$. Examination of equation (14-23) shows that each term on the right side consists of a rate term times a potential change.

$\frac{\delta Q}{d\theta} = \dot{Q}$ rate of energy flow, for example, energy/time

$\Delta\left(\frac{1}{T}\right)$ represents the energy potential

$\frac{dn_i}{d\theta} = \dot{n}_i$ rate of component change, for example, moles/time

$\Delta\left(\frac{\mu_i}{T} + Y_i\right)$ component transfer potential

$\frac{dX_j}{d\theta} = \dot{X}_j$ reaction velocity of jth reaction

$\frac{A_j}{T}$ affinity or reaction potential of jth reaction

We generalize the above observation by writing

$$\dot{S}_{\text{gen}} = \sum_\ell \dot{W}_\ell Z_\ell \geq 0 \tag{14-24}$$

$\dot{W}_\ell$ generalized flow rate

Z_ℓ generalized potential

14.2 Entropy Generation in Systems with Property Gradients

Consideration to this point has been with equilibrium systems or subsystems between which changes were taking place. Such a prerequisite was necessary if the formalism of thermodynamics was to be used since the fundamental relation, $S = S(U, V, N, \text{etc.})$, loses its significance for systems not at equilibrium.

Thus we must face the rather perplexing question of what significance does a fundamental property like entropy have in a nonequilibrium system. We pointed out in Chapter 3 that it behaved as a state property only in the equilibrium state at which it assumed a unique (maximum) value subject to the constraints imposed on the system. Is it meaningful to talk about entropy, temperature, or indeed any of our thermodynamic properties for nonequilibrium conditions? It is not possible to completely answer this question without delving much further into the microscopic behavior of matter than it is appropriate to do in this text. Rather we shall offer a qualitative justification for the postulate of local equilibrium upon which the subsequent treatment of irreversible thermodynamics is based.

If one chooses to apply the analysis of Section 14.1 to differential systems within which property variations are of infinitesimal magnitudes but which are still sufficiently large to give meaning to macroscopic parameters, it seems reasonable to assume that the behavior of such elements can be adequately characterized by the local thermodynamic property values. On the microscopic scale, molecular distances are so small, the interactions so frequent, and the distribution of energies sufficiently broad that the behavior of any single molecule would be relatively unaffected by the presence of a macroscopic gradient unless that gradient were unusually steep. Experimental evidence can also be cited to justify the reasonableness of this conclusion.

Consider the rate of entropy generation occurring within a steady-state system across which a temperature gradient exists, as shown in Fig. 14-2. If the

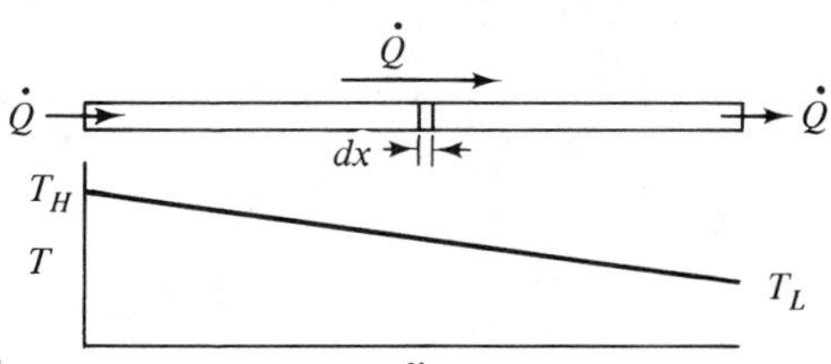

FIG. 14-2. System with a thermal gradient.

two ends are maintained at different temperatures, T_H and T_L, by supplying and removing thermal energy at a rate $\dot{Q}$ at each end, the rate of entropy generation can be calculated by applying equation (14-23) to successive differential elements within the system and then integrating over the entire system. For thermal energy flows alone equation (14-23) reduces to

$$\dot{S}_{\text{gen}} = \dot{Q}\Delta\left(\frac{1}{T}\right) \tag{14-25}$$

$\dot{Q}$ is simply the rate of energy flow and is constant along the system from element to element. For differential elements, dx in length,

$$\Delta\left(\frac{1}{T}\right) = \left(\frac{1}{T} + \frac{d(1/T)}{dx}\,dx\right) - \frac{1}{T} = \frac{d(1/T)}{dx}\,dx \tag{14-26}$$

Thus for a differential element, the rate of entropy production per unit length can be expressed as

$$\frac{d\dot{S}_{\text{gen}}}{dx} = \dot{Q}\frac{d(1/T)}{dx} \tag{14-27}$$

or

$$\frac{d\dot{S}_{\text{gen}}}{dx} = -\frac{\dot{Q}}{T^2}\left(\frac{dT}{dx}\right) \tag{14-27a}$$

If we divide both sides of (14-27) by $dy\,dz$, the cross sectional area normal to flow, then the following expression results:

$$\frac{d\dot{S}_{\text{gen}}}{dx\,dy\,dz} = \frac{\dot{Q}}{dy\,dz}\frac{d(1/T)}{dx} \tag{14-28}$$

If

$$\sigma = \frac{d\dot{S}_{\text{gen}}}{dx\,dy\,dz} \quad \text{rate of entropy generation per unit volume}$$

and

$$\dot{\boldsymbol{q}} = \frac{\dot{Q}}{dy\,dz} \quad \text{rate of heat flow per unit area (the heat flux),}$$

then (14-28) can be written as

$$\sigma = \underset{\text{flux}}{\dot{\boldsymbol{q}}}\underset{\text{force}}{\frac{d(1/T)}{dx}} \tag{14-29}$$

A similar treatment for one-dimensional component transfer using equation (14-23) would yield

$$\sigma = -\sum_i \dot{\boldsymbol{n}}_i \frac{d(\mu_i/T + Y_i)}{dx} = -\sum_i \frac{\boldsymbol{n}_i}{T}\left(\frac{d\mu_i}{dx} - \frac{\mu_i}{T}\frac{dT}{dx} + T\frac{dY_i}{dx}\right) \tag{14-30}$$

where

$$\dot{\boldsymbol{n}}_i = \frac{\text{moles } i}{\text{area-time}}$$

Entropy generation resulting from chemical reaction within a differential system can be handled by simply dividing the appropriate term in the differential form of equation (14-23) by the differential volume $dx\,dy\,dz$.

$$\frac{d\dot{S}_{\text{gen}}}{dx\,dy\,dz} = \sigma = \sum_j \frac{A_j}{T}\frac{d\dot{X}_j}{dx\,dy\,dz} = \sum_j \frac{A_j}{T}\dot{X}_j \tag{14-31}$$

where

$$\dot{X}_j = \frac{\text{velocity for } j\text{th reaction}}{\text{unit volume}}$$

Again for continuous systems we observe that σ, the entropy production per unit volume, is the sum of terms consisting of a flux times a force (potential gradient). We notice that all of the terms have one thing in common, namely that a nonzero potential gradient causes entropy generation within the system.

Although this analysis has not carried through a term representing the entropy production arising from electrical effects, it could have been included initially in the expression for entropy generation as

$$d\dot{S}_{\text{gen}} = \dot{I}\, d\left(\frac{\phi}{T}\right) \tag{14-32}$$

where ϕ represents the electrical potential and $\dot{I}$ the flow rate of electrical charge. Dividing by element volume, $dx\,dy\,dz$ yields

$$\sigma_{\text{elec}} = -\dot{\boldsymbol{i}}\frac{d(\phi/T)}{dx} \tag{14-33}$$

where

$$\dot{\boldsymbol{i}} = \frac{\text{charge}}{\text{area-time}} \qquad \text{or electrical current}$$

Thus for a system in which one dimensional gradients are permitted the rate of entropy generation (for thermal, component, electrical, and chemical reaction effects) is given by

$$\sigma = \dot{\boldsymbol{q}}\frac{d(1/T)}{dx} - \sum_i \dot{\boldsymbol{n}}_i \frac{d[(\mu_i/T) + Y_i]}{dx} - \dot{\boldsymbol{i}}\frac{d(\phi/T)}{dx} + \sum_j \frac{A_j}{T}\dot{\mathbf{X}}_j \tag{14-34}$$

As will be noted in equation (14-34), all fluxes have been designated with boldface type.

14.3 Flux-Force Interrelationship

The relationship of fluxes to potentials has been the major concern of kineticists and others active in the area of rate processes. The hypothesis of a linear variation between certain fluxes and their respective potentials was made and verified many years ago. The laws of Fourier, Fick, and Ohm represent formalized statements of this behavior for heat, mass, and electrical transfer processes, respectively. Phenomenological verification exists for systems near equilibrium. As gradients become steeper the linear postulates become less and less reliable.

One may choose to generalize to an even greater extent by representing each flux as consisting of contributions arising from all forces or potentials in the system. Again a linear relationship is postulated so that we will write:

$$\begin{aligned} J_1 &= L_{11}Z_1 + L_{12}Z_2 + \cdots + L_{1k}Z_k + \cdots + L_{1n}Z_n \\ J_j &= L_{j1}Z_1 + L_{j2}Z_2 + \cdots + L_{jk}Z_k + \cdots + L_{jn}Z_n \\ J_n &= L_{n1}Z_1 + L_{n2}Z_2 + \cdots + L_{nk}Z_k + \cdots + L_{nn}Z_n \end{aligned} \tag{14-35}$$

where J_j represents the jth flux, Z_j represents the jth potential or force, and $L_{11}, L_{22}, \ldots, L_{nn}$ are conjugate coefficients such as electrical or thermal conductivities and represent the dependence of the flux on the principal force. L_{12}, L_{21}, L_{jk}, etc. are coupling coefficients and relate fluxes to the other system forces. For example, the thermal diffusion coefficient is used to couple mass fluxes to the thermal driving force.

These equations are general expressions for different fluxes, such as heat, mass, and electrical curent. In most cases the off-diagonal contributions are negligible, and one obtains simple expressions such as Fourier's law for heat transfer. Experimental evidence shows that in reality the transport equations ordinarily used are special cases of these more general relationships in which the off-diagonal contributions to the flux are neglected. As we shall show, these off-diagonal terms cannot be neglected in certain instances.

A generalized expression for entropy generation, patterned after equation (14-24) but using fluxes J_ℓ rather than flow rates $\dot{W}_\ell$, can be written as

$$\sigma = \sum_{\ell=1}^{2} J_\ell Z_\ell \tag{14-36}$$

Additional information regarding the coefficients can be obtained by replacing the fluxes in the generalized entropy generation expression, equation (14-36), with equation (14-35). For a process with two driving forces we get for the entropy generation term

$$\begin{aligned}\sigma &= L_{11}Z_1^2 + L_{12}Z_2Z_1 + L_{21}Z_1Z_2 + L_{22}Z_2^2 > 0\\ &= L_{11}Z_1^2 + (L_{12} + L_{21})Z_1Z_2 + L_{22}Z_2 > 0\end{aligned} \tag{14-37}$$

The generation term must be positive; therefore, equation (14-87) is written as an inequality. Since Z_1 and Z_2 can take on positive or negative values, but Z_1^2 and Z_2^2 must be positive, one can conclude that L_{11} and L_{22} must be positive if the inequality is to hold for all possible Z's. However, L_{12} and L_{21} can be either positive or negative. If the polynomial is to have a value greater than zero for all possible Z's then equation (14-38) must hold.

$$(L_{12} + L_{21})^2 < 4L_{11}L_{22} \tag{14-38}$$

Equation (14-38) places an upper limit on the magnitude of the coupling coefficients relative to the primary coefficients.

Although there is a certain amount of arbitrariness in the selection of fluxes and forces to be used in equation (14-36), the rate of entropy production is independent of the choice of forces and fluxes. For example, the simple chain reaction

$$(1)\ A = B \qquad \text{and} \qquad (2)\ B = C \tag{14-39}$$

might also be represented as

$$(1')\ A = C \qquad \text{and} \qquad (2')\ B = C \tag{14-39a}$$

with comparable results. In the former case the chemical affinities are given by

$$\begin{aligned}A_1 &= \mu_A - \mu_B\\ A_2 &= \mu_B - \mu_C\end{aligned} \tag{14-40}$$

and the component changes in terms of the reaction velocities as

$$r_A = -\dot{\chi}_1; \qquad r_B = \dot{\chi}_1 - \dot{\chi}_2; \qquad r_C = \dot{\chi}_2 \tag{14-41}$$

For this choice equation (14-34) can be expressed as

$$T\sigma = A_1\dot{\chi}_1 + A_2\dot{\chi}_2 > 0 \tag{14-42}$$

In the second case, the affinities are

$$\begin{aligned}A_1' &= \mu_A - \mu_C = A_1 + A_2\\ A_2' &= \mu_B - \mu_C = A_2\end{aligned} \tag{14-43}$$

and the component changes are given in terms of the reaction velocities as

$$r_A = -\dot{\chi}_{1'}; \qquad r_B = -\dot{\chi}_{2'}; \qquad r_C = \dot{\chi}_{1'} + \dot{\chi}_{2'} \tag{14-44}$$

Equating component changes from equations (14-41) and (14-44) for each component yields

$$\dot{\chi}_1 = \dot{\chi}_{1'} \qquad \text{and} \qquad \dot{\chi}_2 = \dot{\chi}_{1'} + \dot{\chi}_{2'} = \dot{\chi}_1 + \dot{\chi}_{2'} \tag{14-45}$$

Entropy generation for the reactions in (14-39a) can be written as

$$T\sigma = A_{1'}\dot{\chi}_{1'} + A_{2'}\dot{\chi}_{2'} > 0 \tag{14-46}$$

or using (14-43) and (14-45)

$$T\sigma = (A_1 + A_2)\dot{\chi}_1 + A_2(\dot{\chi}_2 - \dot{\chi}_1) \tag{14-47}$$

Simplification of (14-47) yields

$$T\sigma = A_1\dot{\chi}_1 + A_2\dot{\chi}_2$$

which is identical to equation (14-42), the result obtained for the first choice of reaction forces and fluxes.

In this case the entropy production could be represented by either of the two linear combinations of fluxes and forces and identical results would be obtained. It is important to note that the fluxes and forces cannot be varied independently if the product is to represent entropy production.

14.4 Onsager's Reciprocity Relationships

The coupling of certain phenomena has been experimentally verified by many investigators. Onsager[1] employing the concept of microscopic reversibility and fluctuation theory first succeeded in showing that the linear coefficients conform to the following identity:

$$L_{jk} = L_{kj} \tag{14-48}$$

The jth flux is related to the kth force, Z_k, by the same proportionality constant that relates J_k, the kth flux, to the jth force, Z_j. Consideration of Curie's theorem at a later stage will show that in certain instances the coefficients must equal zero.

Two exceptions exist to the above statement of Onsager's conclusions. In magnetic fields where $L_{jk}(H) = L_{kj}(-H)$ and in systems where Coriolis forces are significant (that is, large rotational velocities) the equality does not hold. DeGroot[2] provides a more comprehensive discussion of this aspect of the subject. It is sufficient for our purposes to assume the validity of Onsager's relationship and employ it in subsequent operations as needed.

The coupling phenomenon and the Onsager equality may be illustrated by considering heat conduction in an anisotropic solid: the fluxes along the

[1] Onsager, L. *Phys. Rev. 37* (1931) and *38* (1931) 2265,

[2] DeGroot, S. R., *Thermodynamics of Irreversible Processes*, North-Holland Publishing Co., Amsterdam, 1958.

coordinate axes are symbolized by J_x and J_y and the temperature gradients along these axes are the forces, Z_x and Z_y. The effective conductivities along the x and y axes are determined from the known conductivities, k_m and k_n, along axes rotated α degrees with respect to the x-y coordinate system. Thus expressions for J_x and J_y in terms of linear forces can be treated in a similar manner. The x and y fluxes of heat are expressed in terms of the x and y temperature gradients as shown in equation (14-49).

$$J_x = k_{xx}Z_x + k_{xy}Z_y$$
$$J_y = k_{yx}Z_x + k_{yy}Z_y \tag{14-49}$$

As will be shown in Sample Problem 14-1, $k_{xy} = k_{yx}$, which is in agreement with Onsager's theory. Such an example, where all of the manipulations are understood, will hopefully make the reader more receptive to Onsager's generalization. A rigorous proof of the general reciprocity relationships is beyond the scope of this treatment.

SAMPLE PROBLEM 14-1. If the heat fluxes along the x and y axes of an anisotropic solid are expressed as follows:

$$J_x = k_{xx}Z_x + k_{xy}Z_y = \dot{q}_x \quad \text{heat flux along } x \text{ axis}$$

and

$$J_y = k_{yx}Z_x + k_{yy}Z_y = \dot{q}_y \quad \text{heat flux along } y \text{ axis}$$

show that $k_{yx} = k_{xy}$. Assume that a heat source exists at the origin and that $k_m > k_n$.

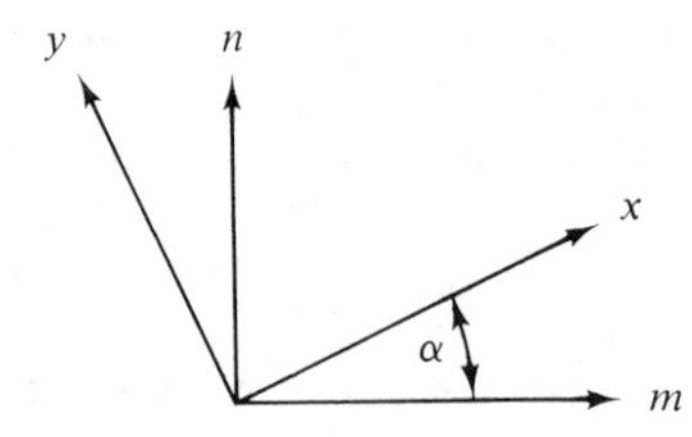

FIGURE S14-1

Solution:

$$J_x = \dot{q}_x = \dot{q}_m \cos\alpha + \dot{q}_n \sin\alpha$$

and

$$J_y = \dot{q}_y = -\dot{q}_m \sin\alpha + \dot{q}_n \cos\alpha$$

$$\dot{q}_m = -k_m \frac{\partial T}{\partial m} \quad \text{and} \quad \dot{q}_n = -k_n \frac{\partial T}{\partial n}$$

Substitution of the expressions for $\dot{q}_m$ and $\dot{q}_n$ into expression for J_x and J_y yields

$$J_x = -k_m(\cos\alpha)\frac{\partial T}{\partial m} - k_n(\sin\alpha)\frac{\partial T}{\partial n}$$

$$J_y = +k_m(\sin\alpha)\frac{\partial T}{\partial m} - k_n(\cos\alpha)\frac{\partial T}{\partial n}$$

But the following relationships among temperature derivatives are true:

$$\frac{\partial T}{\partial m} = \frac{\partial T}{\partial x}\frac{\partial x}{\partial m} + \frac{\partial T}{\partial y}\frac{\partial y}{\partial m}$$

$$\frac{\partial T}{\partial n} = \frac{\partial T}{\partial x}\frac{\partial x}{\partial n} + \frac{\partial T}{\partial y}\frac{\partial y}{\partial n}$$

Also

$$\frac{\partial y}{\partial m} = -\sin\alpha$$

$$\frac{\partial y}{\partial n} = \cos\alpha$$

$$\frac{\partial x}{\partial m} = \cos\alpha$$

$$\frac{\partial x}{\partial n} = \sin\alpha$$

Rewriting the expressions for J_x and J_y

$$J_x = -k_m \cos\alpha\left[\frac{\partial T}{\partial x}\cos\alpha + \frac{\partial T}{\partial y}(-\sin\alpha)\right] - k_n \sin\alpha\left(\frac{\partial T}{\partial x}\sin\alpha + \frac{\partial T}{\partial y}\cos\alpha\right)$$

$$J_y = \quad k_m \sin\alpha\left[\frac{\partial T}{\partial x}\cos\alpha + \frac{\partial T}{\partial y}(-\sin\alpha)\right] - k_n \cos\alpha\left(\frac{\partial T}{\partial x}\sin\alpha + \frac{\partial T}{\partial y}\cos\alpha\right)$$

Expressing J_x and J_y in terms of $\partial T/\partial x$ and $\partial T/\partial y$, the x and y forces, produces

$$J_x = -(k_m \cos^2\alpha + k_n \sin^2\alpha)\left(\frac{\partial T}{\partial x}\right) + (k_m - k_n)\sin\alpha\cos\alpha\left(\frac{\partial T}{\partial y}\right)$$

$$J_y = (k_m - k_n)\sin\alpha\cos\alpha\frac{\partial T}{\partial x} - (k_m \sin^2\alpha + k_n\cos^2\alpha)\frac{\partial T}{\partial y}$$

Relating $\partial T/\partial x$ to Z_x and $\partial T/\partial y$ to Z_y in our original expressions results in k_{xx}, k_{xy}, k_{yx}, and k_{yy} taking on the following values:

$$k_{xx} = -(k_m\cos^2\alpha + k_n\sin^2\alpha)$$

$$k_{xy} = (k_m - k_n)\cos\alpha\sin\alpha$$

$$k_{yx} = (k_m - k_n)\cos\alpha\sin\alpha$$

$$k_{yy} = -(k_m\sin^2\alpha + k_n\cos^2\alpha)$$

From the above it can be observed that $k_{xy} = k_{yx}$ when expressed in terms of the known conductivities k_m and k_n. The above equivalency between corresponding off-diagonal elements of the thermal conductivity tensor had been observed repeatedly long before Onsager utilized fluctuation theory and microscopic reversibility to develop the reciprocity relationships.

Reexamination of equation (14-38) in view of equation (14-48) provides an upper limit on the magnitudes of the off-diagonal coefficients relative to the primary coefficients. Since $L_{jk} = L_{kj}$, $(L_{jk} + L_{kj})^2 = (2L_{jk})^2 = 4L_{jk}^2$. Thus $4L_{jk}^2 < 4L_{jj}L_{kk}$ or

$$L_{jk} < \sqrt{L_{jj}L_{kk}} \tag{14-50}$$

Curie's theorem requires that certain coefficients be zero (that is, it prohibits coupling between certain fluxes and forces). To see why this occurs, rewrite the general expression for entropy generation in terms of linear coefficients and forces for a system with two forces yielding,

$$\begin{aligned}\sigma &= J_1 Z_1 + J_2 Z_2 \\ &= (L_{11} Z_1 + L_{12} Z_2) Z_1 + (L_{21} Z_1 + L_{22} Z_2) Z_2\end{aligned} \tag{14-51}$$

If Z_1 and Z_2 are both vectors, their product produces a scalar. Since all the coefficients must be scalars if they are to be properties of an isotropic medium, all terms in equation (14-51) are scalars including σ, the rate of entropy production per unit volume.

When the fluxes and forces present are of different tensorial rank, we must reexamine the problem. Consider for example the possibility of coupling between energy flow and chemical reaction. The flux and force for energy transfer are both vectors (tensors of rank 1) whereas those for a chemical reaction are both scalars (tensors of rank 0). The product of a scalar force with a tensor of rank 1 yields a vector component of the entropy production which is inconsistent with the scalar character of σ. Unless L_{12} is a first rank tensor, the product $L_{12} Z_1 Z_2$ is not a scalar and cannot contribute to σ. However, L_{12}, which is a property of an isotropic medium, cannot exhibit directional dependence and must therefore be zero.

For tensor forces with ranks differing by an even number, the coefficient would have to be a tensor of even rank. This type of coefficient is not inconsistent with our assumption of isotropy and thus coupling between phenomena with forces which differ by an even rank *is not prohibited.* Curie's theorem forbids coupling only in instances where the interacting forces and fluxes are tensors differing by an odd rank, for example, energy transfer and chemical reaction or energy transfer and viscous dissipation (a tensor of rank 2). It would not prohibit coupling between electrical and thermal transport processes nor between chemical reactions and viscous dissipation. The latter involves terms which differ by rank 2 and is therefore acceptable. Curie's theorem does not guarantee coupling; it merely restricts the processes which might be expected to interact. The basic premise is that the linear coefficients which relate nonconjugate fluxes and forces are independent of direction in isotropic media.

14.5 Coupled Processes

The generalized flux-force relationship, equation (14-35), provides for contributions to any of the fluxes from other than the principle or conjugate force subject to the restrictions imposed by Curie's theorem. This implies that

$$J_H = L_{HH} Z_H + L_{HE} Z_E + L_{H1} Z_1 + L_{H2} Z_2 + \cdots \tag{14-52}$$

where J, L, and Z have there usual meaning and

$$H = \text{heat}$$
$$E = \text{electrical}$$

$$1 = \text{component } 1$$
$$2 = \text{component } 2$$

Thus the first term $L_{HH}Z_H$ is Fourier's expression where L_{HH} relates to the thermal conductivity and Z_H to the temperature force. Equation (14-52) further predicts that heat (energy) transport will occur in smaller amounts if an electrical force or a chemical potential force is superimposed on the system. Similar expressions for electical and component fluxes could be written yielding for a binary system the following matrix:

$$J_H = \underbrace{L_{HH}Z_H}_{\text{Fourier}} + \underbrace{L_{HE}Z_E}_{\text{Peltier}} + \underbrace{L_{H1}Z_1}_{\text{Dufour}} + \underbrace{L_{H2}Z_2}_{\text{Dufour}} \tag{14-53a}$$

$$J_E = \underbrace{L_{EH}Z_H}_{\text{Seebeck}} + \underbrace{L_{EE}Z_E}_{\text{Ohm}} + L_{E1}Z_1 + L_{E2}Z_2 \tag{14-53b}$$

$$J_1 = \underbrace{L_{1H}Z_H}_{\text{Soret}} + L_{1E}Z_E + \underbrace{L_{11}Z_1}_{\text{Fick}} + L_{12}Z_2 \tag{14-53c}$$

$$J_2 = \underbrace{L_{2H}Z_H}_{\text{Soret}} + L_{2E}Z_E + L_{21}Z_1 + \underbrace{L_{22}Z_2}_{\text{Fick}} \tag{14-53d}$$

In each equation the main contribution to the flux arises from the underlined or conjugate term. Since all L_{jk} are nonzero for the above choice of fluxes and forces, contributions to each of the J's should result for all terms where $Z_j \neq 0$.

Though not large, these flows have proved quite useful in a number of instances. Principal interest has centered around thermal diffusional and electrical terms, all of which have the same tensorial rank.

14.6 Coupled Processes—Verification and Application

Isothermal Diffusion

The basic equation relating entropy production to diffusional processes alone can be obtained from equation (14-34).

$$\sigma = -\sum_{i=1}^{C} \dot{\boldsymbol{n}}_i \frac{d(\mu_i/T + Y_i)}{dx} \tag{14-54}$$

For constant temperature and $Y_i = 0$ the above equation applied to a multicomponent system in which there is *no bulk flow* (center of volume remains fixed) yields

$$T\sigma = -\sum_{i=1}^{C} \dot{\boldsymbol{n}}_i \frac{\partial \mu_i}{\partial x} \tag{14-55}$$

However, since the Onsager relationships apply only to independent forces and fluxes, it is essential that we eliminate any dependent forces or fluxes from equation (14-55). We know, for example, that the C diffusional fluxes are not independent, but must be related through the conservation of mass principle (or the continuity expression of fluid mechanics). For a system whose center of volume remains fixed in space the conservation of mass requires that

$$\sum_{i=1}^{C} \dot{\boldsymbol{n}}_i \bar{V}_i = 0 \tag{14-56}$$

Equation (14-56) then tells us that only $(C - 1)$ of the mass fluxes are independent. (From this we observe that coupling effects cannot occur in binary diffusion—a minimum of three diffusing species are required before direct coupling of the mass fluxes and chemical potential driving forces can occur.)

Likewise the chemical potential driving forces are not all independent; they are related through the Gibbs–Duhem equation

$$\sum_{i=1}^{C} n_i \frac{\partial \mu_i}{\partial x} = 0 \tag{14-57}$$

As in the case of the mass fluxes, we find only $(C - 1)$ independent forces.

Equation (14-57) can be put in a form that is somewhat more useful if we divide by V, the total volume of the system, and replace n_i/V by C_i, the concentration of specie i expressed as moles of specie i per unit volume.

$$\sum_{i=1}^{C} C_i \frac{d\mu_i}{dx} = 0 \tag{14-57a}$$

Since we wish to keep the algebraic complexity to a minimum in the following discussion, let us now restrict ourselves to consider only ternary diffusion. In this case we have only three diffusing species, and any one force and flux can be expressed in terms of the others from equations (14-56) and (14-57a). Thus

$$\dot{\boldsymbol{n}}_3 = -\frac{(\dot{\boldsymbol{n}}_1 \bar{V}_1 + \dot{\boldsymbol{n}}_2 \bar{V}_2)}{\bar{V}_3} \tag{14-58}$$

and

$$\frac{\partial \mu_3}{\partial x} = -\frac{[C_1(\partial \mu_1/\partial x) + C_2(\partial \mu_2/\partial x)]}{1 - C_1 - C_2} \tag{14-59}$$

Insertion of (14-58) and (14-59) in equation (14-55) yields

$$T\sigma = -\dot{\boldsymbol{n}}_1 \frac{\partial \mu_1}{\partial x} - \dot{\boldsymbol{n}}_2 \frac{\partial \mu_2}{\partial x} - \left(\dot{\boldsymbol{n}}_1 \frac{\bar{V}_1}{\bar{V}_3} + \dot{\boldsymbol{n}}_2 \frac{\bar{V}_2}{\bar{V}_3}\right)\left(\frac{C_1}{C_3}\frac{\partial \mu_1}{\partial x} + \frac{C_2}{C_3}\frac{\partial \mu_2}{\partial x}\right) \tag{14-60}$$

Rearrangement of equation (14-60) then gives

$$T\sigma = -\dot{\boldsymbol{n}}_1 \overbrace{\left[\frac{\partial \mu_1}{\partial x}\left(1 + \frac{C_1}{C_3}\frac{\bar{V}_1}{\bar{V}_3}\right) + \frac{\partial \mu_2}{\partial x}\frac{C_2}{C_3}\frac{\bar{V}_1}{\bar{V}_3}\right]}^{-Z'_1} - \dot{\boldsymbol{n}}_2 \overbrace{\left[\frac{\partial \mu_2}{\partial x}\left(1 + \frac{C_2}{C_3}\frac{\bar{V}_2}{\bar{V}_3}\right) + \frac{\partial \mu_1}{\partial x}\frac{C_1}{C_3}\frac{\bar{V}_2}{\bar{V}_3}\right]}^{-Z'_2} \tag{14-61}$$

or

$$T\sigma = \dot{n}_1 Z'_1 + \dot{n}_2 Z'_2 \tag{14-62}$$

where the modified forces Z'_1 and Z'_2 are defined as shown in equation (14-61). For the fluxes and forces defined by equation (14-61) the linear force-flux relationships [equations (14-53c) and (14-53d)] can be written as

$$\begin{aligned} \dot{n}_1 &= L_{11}Z'_1 + L_{12}Z'_2 \\ \dot{n}_2 &= L_{21}Z'_1 + L_{22}Z'_2 \end{aligned} \tag{14-63}$$

For independent fluxes and forces properly defined by equation (14-34) or transformations thereof, the L_{12} and L_{21} in equation (14-63) should be equal. Since most diffusion data is reported in terms of Fick's law diffusion coefficients, D_{ij}, it is necessary to relate D_{ij} to L_{ij} before the Onsager relationship can be applied. The fluxes may be expressed in terms of the diffusivities as

$$\begin{aligned} \dot{n}_1 &= -D_{11}\frac{\partial C_1}{\partial x} - D_{12}\frac{\partial C_2}{\partial x} \\ \dot{n}_2 &= -D_{21}\frac{\partial C_1}{\partial x} - D_{22}\frac{\partial C_2}{\partial x} \end{aligned} \tag{14-64}$$

where D_{ii} is equal to the ordinary diffusion coefficient in liter/cm-sec. The gradients in chemical potential can be expressed in terms of concentration gradients via the chain rule

$$\frac{\partial \mu_i}{\partial x} = \sum_{k=1}^{C-1} \frac{\partial \mu_i}{\partial C_k}\frac{\partial C_k}{\partial x} \tag{14-65}$$

Thus for the ternary system under consideration,

$$\frac{\partial \mu_1}{\partial x} = \frac{\partial \mu_1}{\partial C_1}\frac{\partial C_1}{\partial x} + \frac{\partial \mu_1}{\partial C_2}\frac{\partial C_2}{\partial x} \tag{14-66}$$

Substituting equation (14-66) for $\partial \mu_i/\partial x$ in the expressions for Z'_1 and Z'_2 in equation (14-61), we get

$$\dot{n}_1 = L_{11}Z'_1 + L_{12}Z'_2 = -D_{11}\frac{\partial C_1}{\partial x} - D_{12}\frac{\partial C_2}{\partial x} \tag{14-67}$$

$$\begin{aligned} \dot{n}_1 = &-L_{11}\left[\left(1 + \frac{C_1}{C_3}\frac{\bar{V}_1}{\bar{V}_3}\right)\left(\frac{\partial \mu_1}{\partial C_1}\frac{\partial C_1}{\partial x} + \frac{\partial \mu_1}{\partial C_2}\frac{\partial C_2}{\partial x}\right) + \frac{C_2}{C_3}\frac{\bar{V}_1}{\bar{V}_3}\left(\frac{\partial \mu_2}{\partial C_1}\frac{\partial C_1}{\partial x} + \frac{\partial \mu_2}{\partial C_2}\frac{\partial C_2}{\partial x}\right)\right] \\ &-L_{12}\left[\left(1 + \frac{C_2}{C_3}\frac{\bar{V}_2}{\bar{V}_3}\right)\left(\frac{\partial \mu_2}{\partial C_2}\frac{\partial C_2}{\partial x} + \frac{\partial \mu_2}{\partial C_1}\frac{\partial C_1}{\partial x}\right) + \frac{C_1}{C_3}\frac{\bar{V}_2}{\bar{V}_3}\left(\frac{\partial \mu_1}{\partial C_2}\frac{\partial C_2}{\partial x} + \frac{\partial \mu_1}{\partial C_1}\frac{\partial C_1}{\partial x}\right)\right] \\ = &-D_{11}\frac{\partial C_1}{\partial x} - D_{12}\frac{\partial C_2}{\partial x} \end{aligned} \tag{14-68}$$

A similar expression can be written for $\dot{n}_2$ such that L_{ij} can be related to D_{ij} as follows:

$$\begin{aligned} L_{11} &= \frac{dD_{11} - bD_{12}}{ad - bc}; \qquad L_{12} = \frac{aD_{12} - cD_{11}}{ad - bc} \\ L_{21} &= \frac{dD_{21} - bD_{22}}{ad - bc}; \qquad L_{22} = \frac{aD_{22} - cD_{21}}{ad - bc} \end{aligned} \tag{14-69}$$

where

$$a = \left[\left(1 + \frac{C_1\bar{V}_1}{C_3\bar{V}_3}\right)\frac{\partial \mu_1}{\partial C_1} + \frac{C_2\bar{V}_1}{C_3\bar{V}_3}\frac{\partial \mu_2}{\partial C_1}\right]$$

$$b = \left[\frac{C_1\bar{V}_2}{C_3\bar{V}_3}\frac{\partial \mu_1}{\partial C_1} + \left(1 + \frac{C_2\bar{V}_2}{C_3\bar{V}_3}\right)\frac{\partial \mu_2}{\partial C_1}\right]$$

and c and d are the same respectively as a and b except $\partial/\partial C_1$ is replaced with $\partial/\partial C_2$. Thus if the cross coefficients L_{12} and L_{21} are to be equal, then $aD_{12} + bD_{22}$ must equal $cD_{11} + dD_{21}$ (for $ad - bc \neq 0$). Table 14-1 shows the results for ten systems for which data is at least partially available. Eight of the ten systems exhibit excellent agreement with the Onsager theory.

TABLE 14-1†

System	L_{12}/L_{21}
LiCl-KCl-H_2O	1.03
LiCl-NaCl-H_2O	1.14
KCl-NaCl-H_2O	1.05
KCl-NaCl-H_2O	1.02
KCl-NaCl-H_2O	1.04
KCl-NaCl-H_2O	1.03
KCl-NaCl-H_2O	1.06
Raffinose-KCl-H_2O	0.42
Raffinose-KCl-H_2O	1.00
Raffinose-urea-H_2O	0.58

† D. G. Miller, *Chem. Rev.*, 60, 15 (1960).

Thermal Diffusion

In 1856 Ludwig observed concentration differences in samples of sulphate solution taken from parts of a system at different temperatures. Thirty years later, Soret, whose name has become associated with the thermal diffusion effect, explored the phenomenon more fully in salt solutions. Theoretical attempts to explain the phenomenon failed until the early part of the 20th century when Chapman and Enskog, working independently in the development of the kinetic theory of gases, predicted that the imposition of a temperature gradient on a gas should cause diffusion of the components. Chapman and co-workers were able to experimentally verify this prediction in gaseous systems.

The thermal diffusion, or Soret, effect appears in the two equations for mass flux, equations (14-53c) and (14-53d). The terms $L_{1H}Z_H$ and $L_{2H}Z_H$ provide for a diffusive flow of component 1 and 2, respectively, in a system with three or more components. For the linear Onsager force-flux relationships of equation (14-53) the mass flux is directly proportional to the magnitude of the temperature gradient imposed.

The concentration gradient produced by the temperature gradient immediately establishes a backflow of the diffusing components in accord with

Fick's law as indicated in equation (14-53). Thus knowledge of the normal diffusion coefficient is required for measurement of the Soret effect. If the normal diffusion coefficient is known, L_{1H} or L_{2H} can be obtained from the steady-state concentration profile produced by subjecting a closed system to a fixed temperature gradient. The experiment is, however, a difficult one, and extreme care must be taken to avoid convective flows arising from density differences or other miscellaneous forces within the systems.

The utilization of the Soret effect to separate components in a system is limited practically to systems of low thermal conductivity because the thermal flux across a system is directly proportional to the thermal conductivity. Gases and liquids are thus more logical media in which to study or utilize the Soret effect than are highly conducting metals.

Because of the highly irreversible (and hence basically inefficient) nature of the thermal flow associated with temperature gradients, practical applications of the Soret phenomenon are limited to more exotic materials that do not readily lend themselves to more conventional processes. Isotope separation is one application in which thermal diffusion has proved quite valuable. Typically the larger of the species tends to congregate in the cooler region and the lighter component in the hot region. Separation of uranium isotopes for the atomic energy program in massive thermal diffusion columns containing many stages is the largest scale example of an application of the Soret effect.

In treating thermal diffusion, it is important to define properly the forces and fluxes used in equations (14-34) and (14-53). Consider a *binary system* across which a temperature gradient is established. From the set of equations (14-53), two are applicable to the analysis of this problem, (14-53a) and (14-53c) (since J_2 is not an independent flux for a binary system). Thus we can write

$$\begin{aligned} J_H &= L_{HH}Z_H + L_{H1}Z_1 \\ J_1 &= L_{1H}Z_H + L_{11}Z_1 \end{aligned} \tag{14-70}$$

The energy flux across the system includes not only a conduction term but also an enthalpy contribution arising from the mass diffusion. The fluxes J_H and J_1 are thus defined as follows:

$$\begin{aligned} J_H &= \dot{\boldsymbol{q}}' = \dot{\boldsymbol{q}} + \sum_{i=1}^{2} \dot{\boldsymbol{n}}_i \bar{H}_i \\ J_1 &= \dot{\boldsymbol{n}}_1 \end{aligned} \tag{14-71}$$

and the corresponding forces

$$\begin{aligned} Z_H &= \frac{d(1/T)}{dx} = -\frac{dT}{T^2\,dx} \\ Z_1 &= -\frac{1}{T}\frac{d\mu_1}{dx} \qquad \text{assuming } Y_1 = 0 \end{aligned} \tag{14-72}$$

Let us assume for purposes of this development that $\dot{\boldsymbol{n}}_i$ is measured relative to the center of mass of the system. Thus

$$\dot{\boldsymbol{n}}_i = \rho_1(u_1 - u_m) \tag{14-73}$$

where ρ_1 = density of species 1
u_1 = velocity of species 1
u_m = velocity of center of mass

Then by continuity requirements

$$\sum_{i=1}^{2} \dot{\boldsymbol{n}}_i = \dot{\boldsymbol{n}}_1 + \dot{\boldsymbol{n}}_2 = 0 \tag{14-74}$$

or

$$\dot{\boldsymbol{n}}_1 = -\dot{\boldsymbol{n}}_2 \tag{14-74a}$$

Substituting equation (14-74a) into (14-34) yields

$$\sigma = -\frac{1}{T^2}\dot{q}'\frac{dT}{dx} - \dot{\boldsymbol{n}}_1\frac{d(\mu_1 - \mu_2)_T}{T\,dx} \tag{14-75}$$

The Gibbs-Duhem equation

$$\sum_{i=1}^{2} C_i\left(\frac{d\mu_i}{dx}\right)_{P,T} = 0 \tag{14-76}$$

may be used to eliminate μ_2 from equation (14-75) to give

$$\sigma = -\frac{1}{T^2}\dot{q}'\frac{dT}{dx} - \frac{\dot{\boldsymbol{n}}_1}{T}\left[\frac{d\mu_1}{dx}\left(1 + \frac{C_1}{C_2}\right)\right] \tag{14-77}$$

The chain rule, $(d\mu_1/dx) = (d\mu_1/dC_1)(dC_1/dx)$, can be used to express the chemical potential gradient in terms of concentration, if that is desired.

From equation (14-77) we see that the linear relationships between heat and mass transfer can be written as

$$\begin{aligned}\dot{q}' &= -\frac{L_{HH}}{T^2}\frac{dT}{dx} - \frac{L_{H1}}{T}\left[\left(1 + \frac{C_1}{C_2}\right)\frac{d\mu_1}{dx}\right]\\ \dot{\boldsymbol{n}}_1 &= -\frac{L_{1H}}{T^2}\frac{dT}{dx} - \frac{L_{11}}{T}\left[\left(1 + \frac{C_1}{C_2}\right)\frac{d\mu_1}{dx}\right]\end{aligned} \tag{14-78}$$

or in terms of concentrations

$$\begin{aligned}\dot{q}' &= -\frac{L_{HH}}{T^2}\frac{dT}{dx} - \frac{L_{H1}}{C_2 T}\rho\left(\frac{\partial\mu_1}{\partial C_1}\right)_{P,T}\frac{dC_1}{dx}\\ \dot{\boldsymbol{n}}_1 &= \frac{L_{1H}}{T^2}\frac{dT}{dx} - \frac{L_{11}}{C_2 T}\rho\left(\frac{\partial\mu_1}{\partial C_1}\right)_{P,T}\frac{dC_1}{dx}\end{aligned} \tag{14-79}$$

where $L_{H1} = L_{1H}$, according to Onsager's relationship.

If we define

$$\frac{L_{HH}}{T^2} = \text{thermal conductivity, } \lambda$$

$$\frac{\rho L_{H1}}{C_1 C_2 T^2} = \text{Dufour coefficient, } D''$$

$$\frac{L_{1H}}{C_1 T^2} = \text{thermal diffusion coefficient, } D'$$

$$\frac{L_{11}}{\rho C_2 T}\frac{\partial\mu_1}{\partial C_1} = \text{diffusivity, } D_1$$

We can rewrite equation (14-79) as

$$\dot{q}' = -\lambda\frac{dT}{dx} - C_1 TD''\frac{\partial\mu_1}{\partial C_1}\frac{dC_1}{dx}$$
$$\dot{n}_1 = -\quad C_1\ D'\frac{dT}{dx} - \rho D_1\frac{dC_1}{dx} \tag{14-80}$$

If a closed system is subjected to a temperature gradient, a stationary state will be reached for which $\dot{n}_1 = 0$. Thus

$$\frac{D'}{D_1} = -\frac{1}{C_1}\frac{dC_1}{dT} \tag{14-81}$$

The ratio D'/D_1 is termed the Soret coefficient. The thermal diffusion factor α is defined as

$$\alpha = \frac{D'}{D_1}T \tag{14-82}$$

The Soret coefficient for liquids and gases is typically of the order $10^{-3} - 10^{-5}$ reciprocal degrees. Assuming values of D_1 of 10^{-5} cm²/sec and 10^{-1} cm²/sec for liquids and gases, respectively, one obtains thermal diffusion coefficients of the order of $10^{-8} - 10^{-10}$ for liquids and $10^{-4} - 10^{-6}$ for gases. Dufour coefficients are measurable in gases where temperature differences of as much as 1°C are achievable with reasonable imposed concentration gradients.

Thermoelectric Effects

Thermoelectric phenomena were among the first coupling processes to be observed experimentally. Lord Kelvin attempted to explain these phenomena during the 1850s from a thermodynamic point of view. However, it was not until Onsager postulated his reciprocity relationships some 80 years later that this interrelationship between thermal and electrical flows was adequately explained on a theoretical basis.

Equation (14-34), rewritten to account for entropy production resulting solely from thermal and electrical flows, becomes

$$\sigma = \dot{q}\frac{d(1/T)}{dx} - \dot{i}\frac{d(\phi/T)}{dx} \tag{14-83}$$

Equations (14-53a and b) express generalized thermal and electrical flows (or fluxes) in terms of generalized forces and linear coefficients. The fluxes, J_H and J_E, and forces, Z_H and Z_E, must be consistent with those defined by equation (14-83), if the Onsager reciprocity relationships are to remain valid. Thus one could rewrite equations (14-53a and b) using the fluxes and forces in equation (14-83) as

$$\dot{q} = L_{HH}\frac{d(1/T)}{dx} + L_{HE}\frac{d(\phi/T)}{dx} \tag{14-84}$$

and

$$\dot{i} = L_{EH}\frac{d(1/T)}{dx} + L_{EE}\frac{d(\phi/T)}{dx} \tag{14-85}$$

The coupling coefficients L_{EH} and L_{HE} must, therefore, be equal under these conditions, a conclusion which has been corroborated experimentally. However, as was pointed out in Section 14.4, these are not the only fluxes and forces for which the Onsager postulate is valid. For example, equation (14-83) can be rewritten as

$$\sigma = -\frac{\dot{q}}{T^2}\frac{dT}{dx} - \frac{\dot{i}}{T}\left[\frac{d\phi}{dx} - \frac{\phi}{T}\frac{dT}{dx}\right] \tag{14-86}$$

Regrouping terms we have

$$\sigma = -(\dot{q} - \dot{i}\phi)\frac{1}{T^2}\frac{dT}{dx} - \dot{i}\frac{1}{T}\frac{d\phi}{dx}$$

where $(\dot{q} - \dot{i}\phi)$ and $\dot{i}$ are energy and electrical fluxes respectively and $-(1/T^2)(dT/dx)$ and $-(1/T)(d\phi/dx)$ are the corresponding forces. Thus the transformed fluxes become

$$\begin{aligned} J_H &= \dot{q}'' = \dot{q} - \dot{i}\phi \\ J_E &= \dot{i} \end{aligned} \tag{14-87}$$

and the transformed forces become

$$\begin{aligned} Z_H &= -\frac{1}{T^2}\frac{dT}{dx} \\ Z_E &= -\frac{1}{T}\frac{d\phi}{dx} \end{aligned} \tag{14-88}$$

This transformation is similar to that developed in equations (14-71) and (14-72) for thermal and mass flows. Just as equation (14-71) led to a modified energy flow $\dot{q}'$ which included both thermal and enthalpy transport contributions as a result of the thermal force, $-(1/T^2)(dT/dx)$, such is also the case here with the modified energy flux, $\dot{q}''$, which includes both thermal and electronic contributions. The advantage of this transformation is that it utilizes forces that consist of a temperature gradient, dT/dx, and an electrical potential gradient, $d\phi/dx$.

We can now rewrite equations (14-84) and (14-85) using these transformed fluxes and forces.

$$J_H = \dot{q}'' = L'_{HH}\left(-\frac{1}{T^2}\frac{dT}{dx}\right) + L'_{HE}\left(-\frac{1}{T}\frac{d\phi}{dx}\right) \tag{14-89}$$

and

$$J_E = \dot{i} = L'_{EH}\left(-\frac{1}{T^2}\frac{dT}{dx}\right) + L'_{EE}\left(-\frac{1}{T}\frac{d\phi}{dx}\right) \tag{14-90}$$

Since the transformations are consistent with equation (14-83), the coupling coefficients must conform to the Onsager relationships. Thus

$$L'_{HE} = L'_{EH} \tag{14-91}$$

The second term in equation (14-90), which relates the flux of electrical energy to the electrical potential gradient $d\phi/dx$ is simply Ohm's law. Thus

$$\dot{i} = -\frac{L'_{EE}}{T}\int_{x_1}^{x_2}\frac{d\phi}{dx}\,dx = \frac{\epsilon}{\mathscr{R}} \tag{14-92}$$

where ϵ = electromotive force and $\mathscr{R}$ = electrical resistance. Thus we see the coupling coefficient L'_{EE} is related to the electrical resistance $\mathscr{R}$ by the relationship

$$L'_{EE} = -\frac{T}{\mathscr{R}} \tag{14-93}$$

(Note since $1/\mathscr{R}$ = electrical conductance, we see that the conjugate coefficient L'_{EE} is merely the product of temperature and electrical conductance).

The coupling phenomena arising in equations (14-89) and (14-90) give rise to the Seebeck and Peltier effects, both of which have been put to practical use. The Seebeck effect, more commonly known as the thermocouple effect, results in an emf being generated as the result of a temperature difference. To illustrate the phenomenon consider a system such as shown in Fig. 14-3. Here

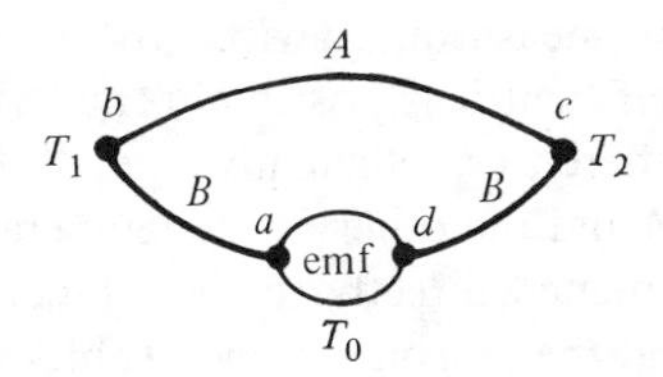

FIG. 14-3. The thermocouple circuit.

the junctions of two dissimilar metals, A and B (for example, chromel-alumel or copper-constantan), are maintained at different temperatures T_1 and T_2. An emf can be measured across the terminals which varies directly with the temperature difference $T_2 - T_1$. This emf is termed the Seebeck emf and will be observed to be a function of the materials A and B as well as the temperature difference between the two junctions.

Consider equation (14-90) applied to the thermocouple circuit shown in Fig. 14-3. If no current is permitted to flow, we can set $J_E = \dot{i} = 0$ and solve for the relationship between emf and temperature.

$$0 = -\frac{L'_{EE}}{T}\frac{d\phi}{dx} - \frac{L'_{EH}}{T^2}\frac{dT}{dx} \tag{14–94}$$

Solving for $d\phi/dT$ we get

$$\frac{d\phi}{dT} = -\frac{1}{T}\frac{L'_{EH}}{L'_{EE}} \equiv S^* \tag{14-94a}$$

where S^* is termed the entropy transport and L'_{EH} and L'_{EE} are assumed to be constant for a given material. Integration of equation (14-94a) around the circuit from a to d produces:

$$\epsilon_{a\to b} = \int_{\phi_a}^{\phi_b} d\phi = -\int_{T_0}^{T_1} \frac{1}{T}\left(\frac{L'_{EH}}{L'_{EE}}\right)_B dT = \alpha_B \ln\frac{T_0}{T_1} \tag{14-95a}$$

$$\epsilon_{b\to c} = \int_{\phi_b}^{\phi_c} d\phi = -\int_{T_1}^{T_2} \frac{1}{T}\left(\frac{L'_{EH}}{L'_{EE}}\right)_A dT = \alpha_A \ln\frac{T_1}{T_2} \tag{14-95b}$$

$$\epsilon_{c\to d} = \int_{\phi_c}^{\phi_d} d\phi = -\int_{T_2}^{T_0} \frac{1}{T}\left(\frac{L'_{EH}}{L'_{EE}}\right)_B dT = \alpha_B \ln\frac{T_2}{T_0} \tag{14-95c}$$

Summing these three equations yields

$$\epsilon_{a\to d} = \epsilon_{a\to b} + \epsilon_{b\to c} + \epsilon_{c\to d}$$
$$= \alpha_B \ln \frac{T_0}{T_1} + \alpha_A \ln \frac{T_1}{T_2} + \alpha_B \ln \frac{T_2}{T_0}$$

$$\epsilon_{a\to d} = (\alpha_A - \alpha_B) \ln \frac{T_1}{T_2} \tag{14-96}$$

Thus it can be seen that the emf for the circuit $\epsilon_{a\to d}$, is independent of T_0 and depends only on T_1 and T_2 and coefficients α_A and α_B for the two materials. The change in emf with respect to temperature for a given couple, $(d\epsilon_{AB}/dT)$, is termed the Seebeck coefficient. Differentiation of equation (14-96) produces

$$\frac{d\epsilon_{AB}}{dT} = \frac{\alpha_A - \alpha_B}{T} \tag{14-97}$$

The utility of this phenomenon is appreciated by anyone who has used thermocouples as a temperature measuring device. By selecting materials which exhibit a large Seebeck coefficient it is possible to make highly accurate measurements of temperature differences. Normally one of the junctions is maintained at the ice point (32°F) and the other at the temperature to be measured. The resulting emf is then related directly to the temperature difference between the reference junction and the sensing couple. Tables of emf versus T have been compiled for various materials which permit a direct conversion of the emf observed to the unknown temperature.

Examination of equations (14-89) and (14-90) suggests that if a current flows through a system, including a junction of two dissimilar materials maintained at a uniform temperature, a flow of thermal energy should occur to or from the junction. For a constant temperature system,

$$J_H = \dot{q}'' = -\frac{L'_{HE}}{T}\frac{d\phi}{dx} \tag{14-98}$$

$$J_E = \dot{i} = -\frac{L'_{EE}}{T}\frac{d\phi}{dx} \tag{14-99}$$

Dividing (14-98) by (14-99) yields

$$\left(\frac{\dot{q}''}{\dot{i}}\right)_{dt=0} = \frac{L'_{HE}}{L'_{EE}} \tag{14-100}$$

Equation (14-100) can be written for each material, A and B, in a circuit such as that shown in Fig. 14-4. Writing the energy balance about a junction as shown in Fig. 14-5 produces

$$\dot{q}''_A = \dot{i}\left(\frac{L'_{HE}}{L'_{EE}}\right)_A = \dot{i}\alpha'_A \tag{14-101}$$

$$\dot{q}''_B = \dot{i}\left(\frac{L'_{HE}}{L'_{EE}}\right)_B = \dot{i}\alpha'_B \tag{14-102}$$

$$\dot{q}''_A - \dot{q}''_B = \dot{i}(\alpha'_A - \alpha'_B) = \dot{q}_{AB} \tag{14-103}$$

where $\alpha' = L'_{HE}/L'_{EE}$ and $\dot{q}_{AB}$ is the *Peltier heat*. If $\alpha'_A > \alpha'_B$, $\dot{q}_{AB}$ is positive and energy must flow from the junction if its temperature is to remain constant. However, at the second junction where the current flows from B to A the signs on all terms in equation (14-103) are reversed and the Peltier heat flows to the

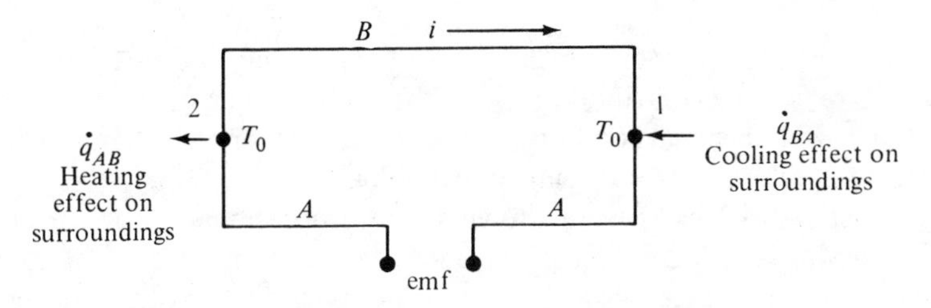

FIG. 14-4. Thermoelectric circuit.

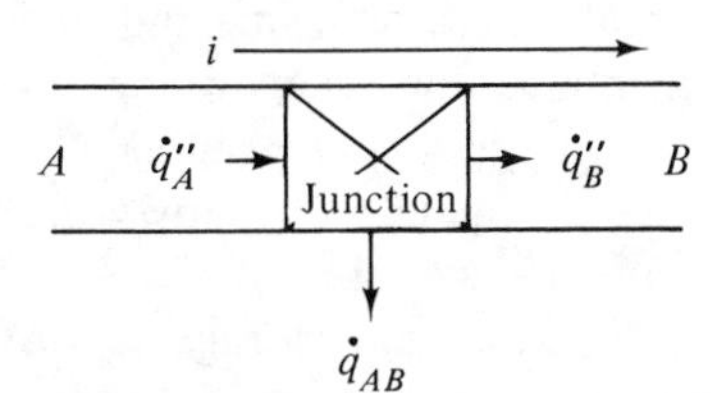

FIG. 14-5. Isothermal thermoelectric junction.

junction from the surroundings. Thus in a circuit as shown in Fig. 14-4 energy can be picked up at point 1 and released at point 2, producing a refrigeration effect in a system with no moving parts. This effect has been used commercially for refrigeration where the reliability associated with no moving parts outweighs the lower energy costs of conventional refrigeration units.

Rearrangement of equation (14-103) yields the *Peltier coefficient* for the system shown in Fig. 14-4

$$\left(\frac{\dot{q}_{AB}}{i}\right)_{\Delta T=0} = \alpha'_A - \alpha'_B = \Pi_{AB} \tag{14-104}$$

where Π is the Peltier coefficient which relates the amount of thermal energy transport at the junction due to the current flow. The Seebeck coefficient as given by equation (14-97) for the same system is

$$\frac{d\epsilon_{AB}}{dT} = \frac{\alpha_A - \alpha_B}{T} \tag{14-97}$$

Combination of (14-97) and (14-104) provides a test of the Onsager reciprocity relationship

$$\frac{d\epsilon_{AB}/dT}{\Pi/T_{AB}} = \frac{\alpha_A - \alpha_B}{\alpha'_A - \alpha'_B} \tag{14-105}$$

If the Onsager postulate is true, $L'_{EH} = L'_{HE}$ and $\alpha = \alpha'$ and measurement of the two coefficients should lead to a value of unity for the ratio in equation (14-105). Such behavior has been confirmed experimentally.

14.7 Application of Thermodynamics to Rate Processes

The preoccupation of the thermodynamicist with the equilibrium state has not been without its rewards to the engineer. Early emphasis was placed on energy and its transformation and provided a sound theoretical basis for the heat and power technology upon which much of our highly industrialized society is based. Likewise the contributions of Gibbs and others created a much refined understanding of chemical equilibrium and solution theory which has facilitated engineering developments in chemical technology.

In Section 14.3 rates were shown to be proportional to driving forces. The true driving forces are themselves state properties that provide a quantitative measure of the degree to which equilibrium states differ. Although engineers have designed chemical reactors and separations processes based on experience, experimentation, empiricism, and expediency, not until the thermodynamicist clearly identified the true driving forces for component transfer, the partial Gibbs free energy, could phenomena be completely explained and correlated, and predictions made. For many years it was common practice to use enthalpy rather than free energy as the criterion for reaction equilibrium, and concentrations, rather than partial fugacities or activities, as the criterion for diffusional equilibrium.

Similarly the transport properties of systems, diffusivities, conductivities, and so on, are state variables. Although progress has been made in relating these parameters to the thermodynamic state variables, much yet remains to be done in developing equations of state which include the transport properties of a medium. Nevertheless it is apparent that the better our understanding of the equilibrium state of matter, the more refined our procedures for predicting transport properties will become. Regardless of the increasingly sophisticated techniques for modeling and optimizing systems that mathematical analysis and the computer have produced, the engineer ultimately must have reliable property information to design a system.

Whereas the contributions of classical thermodynamics to rate processes have related mainly to a more precise definition of the true driving forces, it seems reasonable to expect irreversible thermodynamics, along with statistical mechanics, to yield increased insight into the transport properties themselves. The coupling phenomena, first observed experimentally, have been conveniently characterized and generalized within the framework of irreversible thermodynamics. Although much of the insight gained to date is qualitative in nature, the concept of entropy production unifies the diverse phenomena and provides a frame of reference within which improved methods for measuring, correlating, and deriving these coefficients can be formulated.

The Onsager relationships discussed in Section 14.4 aid immensely in obtaining quantitative results by reducing the number of independent measurements necessary to evaluate transport coefficients for coupled processes. The practical significance of this contribution cannot be fully evaluated until the technological value of such processes is determined. Nevertheless just the awareness of the coupling phenomena has already proved of value in explaining second-order effects arising in conventional rate processes.

Appendices

The Mollier Diagram

A

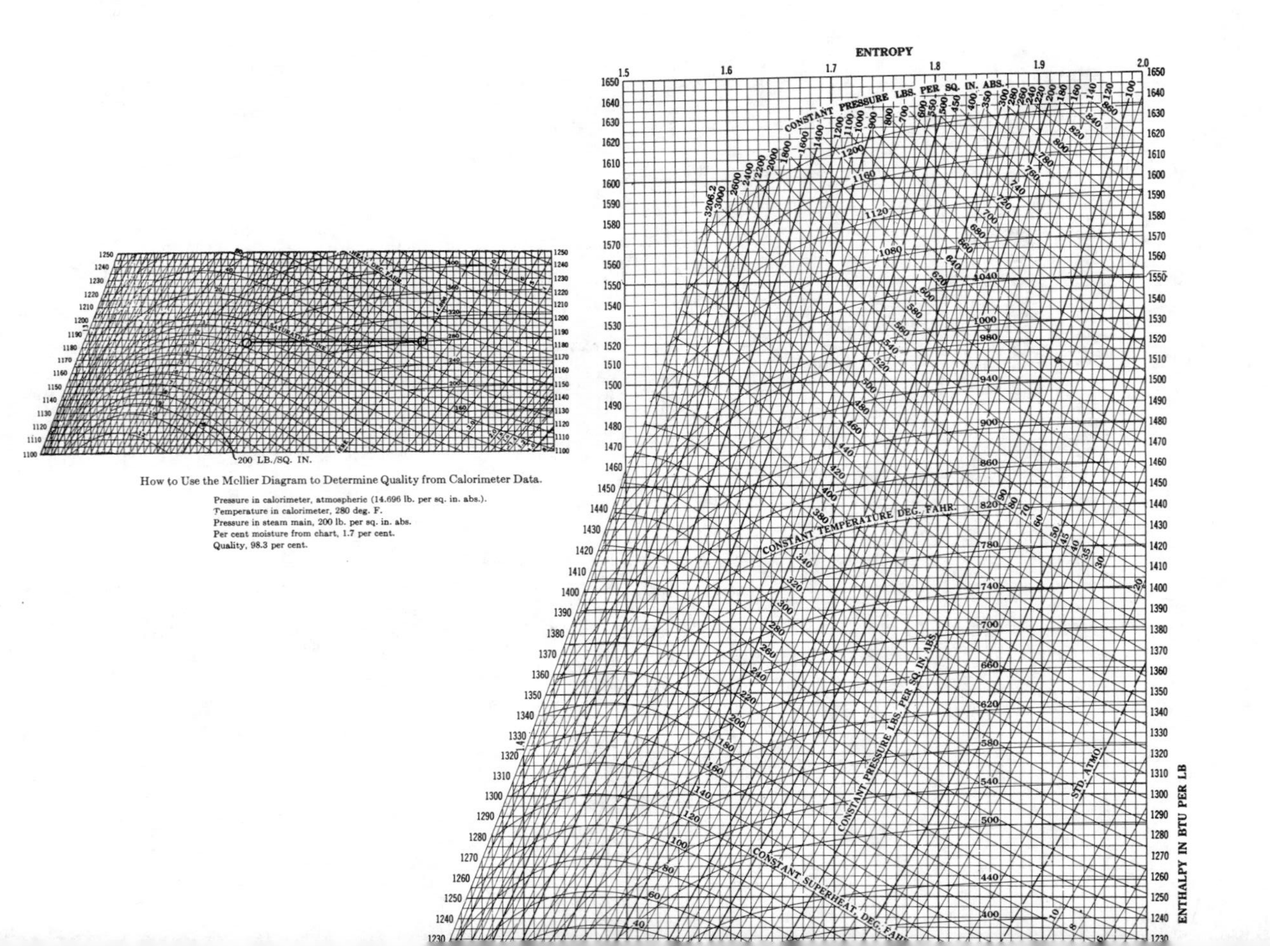

How to Use the Mollier Diagram to Determine Quality from Calorimeter Data.

Pressure in calorimeter, atmospheric (14.696 lb. per sq. in. abs.).
Temperature in calorimeter, 280 deg. F.
Pressure in steam main, 200 lb. per sq. in. abs.
Per cent moisture from chart, 1.7 per cent.
Quality, 98.3 per cent.

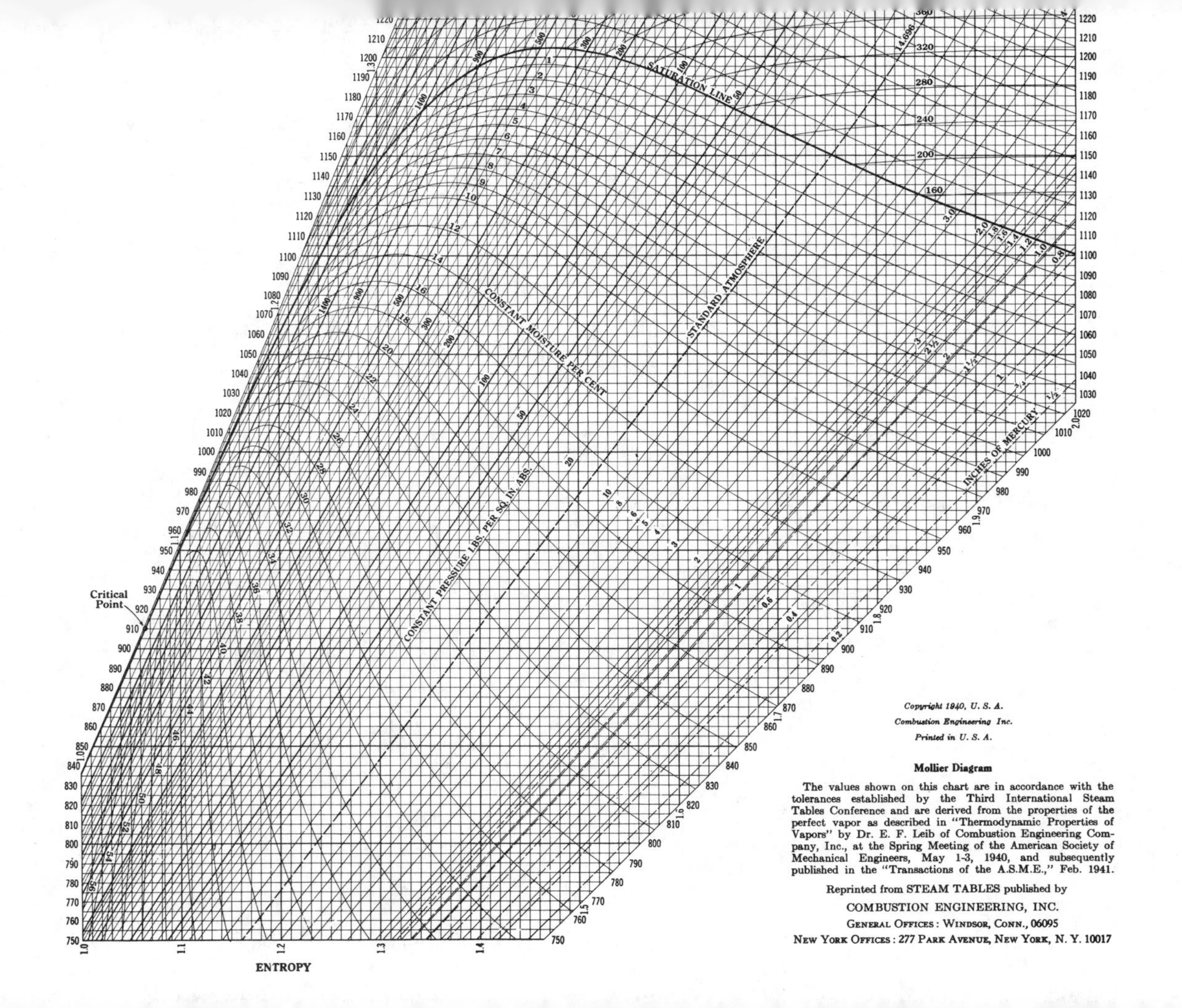
ENTROPY
SATURATION LINE
CONSTANT MOISTURE PER CENT
CONSTANT PRESSURE LBS. PER SQ. IN. ABS.
STANDARD ATMOSPHERE
INCHES OF MERCURY
Critical Point
Copyright 1940, U. S. A.
Combustion Engineering Inc.
Printed in U. S. A.
Mollier Diagram
The values shown on this chart are in accordance with the tolerances established by the Third International Steam Tables Conference and are derived from the properties of the perfect vapor as described in "Thermodynamic Properties of Vapors" by Dr. E. F. Leib of Combustion Engineering Company, Inc., at the Spring Meeting of the American Society of Mechanical Engineers, May 1-3, 1940, and subsequently published in the "Transactions of the A.S.M.E.," Feb. 1941.
Reprinted from STEAM TABLES published by
COMBUSTION ENGINEERING, INC.
GENERAL OFFICES: WINDSOR, CONN., 06095
NEW YORK OFFICES: 277 PARK AVENUE, NEW YORK, N. Y. 10017

Steam Tables B

Table 1. Saturated Steam: Temperature Table

Temp Fahr t	Abs Press. Lb per Sq In. p	Specific Volume: Sat. Liquid v_f	Specific Volume: Evap v_{fg}	Specific Volume: Sat. Vapor v_g	Enthalpy: Sat. Liquid h_f	Enthalpy: Evap h_{fg}	Enthalpy: Sat. Vapor h_g	Entropy: Sat. Liquid s_f	Entropy: Evap s_{fg}	Entropy: Sat. Vapor s_g	Temp Fahr t
32.0	0.08859	0.016022	3304.7	3304.7	0.0179	1075.5	1075.5	0.0000	2.1873	2.1873	**32.0**
34.0	0.09600	0.016021	3061.9	3061.9	1.996	1074.4	1076.4	0.0041	2.1762	2.1802	**34.0**
36.0	0.10395	0.016020	2839.0	2839.0	4.008	1073.2	1077.2	0.0081	2.1651	2.1732	**36.0**
38.0	0.11249	0.016019	2634.1	2634.2	6.018	1072.1	1078.1	0.0122	2.1541	2.1663	**38.0**
40.0	1.12163	0.016019	2445.8	2445.8	8.027	1071.0	1079.0	0.0162	2.1432	2.1594	**40.0**
42.0	0.13143	0.016019	2272.4	2272.4	10.035	1069.8	1079.9	0.0202	2.1325	2.1527	**42.0**
44.0	0.14192	0.016019	2112.8	2112.8	12.041	1068.7	1080.7	0.0242	2.1217	2.1459	**44.0**
46.0	0.15314	0.016020	1965.7	1965.7	14.047	1067.6	1081.6	0.0282	2.1111	2.1393	**46.0**
48.0	0.16514	0.016021	1830.0	1830.0	16.051	1066.4	1082.5	0.0321	2.1006	2.1327	**48.0**
50.0	0.17796	0.016023	1704.8	1704.8	18.054	1065.3	1083.4	0.0361	2.0901	2.1262	**50.0**
52.0	0.19165	0.016024	1589.2	1589.2	20.057	1064.2	1084.2	0.0400	2.0798	2.1197	**52.0**
54.0	0.20625	0.016026	1482.4	1482.4	22.058	1063.1	1085.1	0.0439	2.0695	2.1134	**54.0**
56.0	0.22183	0.016028	1383.6	1383.6	24.059	1061.9	1086.0	0.0478	2.0593	2.1070	**56.0**
58.0	0.23843	0.016031	1292.2	1292.2	26.060	1060.8	1086.9	0.0516	2.0491	2.1008	**58.0**
60.0	0.25611	0.016033	1207.6	1207.6	28.060	1059.7	1087.7	0.0555	2.0391	2.0946	**60.0**
62.0	0.27494	0.016036	1129.2	1129.2	30.059	1058.5	1088.6	0.0593	2.0291	2.0885	**62.0**
64.0	0.29497	0.016039	1056.5	1056.5	32.058	1057.4	1089.5	0.0632	2.0192	2.0824	**64.0**
66.0	0.31626	0.016043	989.0	989.1	34.056	1056.3	1090.4	0.0670	2.0094	2.0764	**66.0**
68.0	0.33889	0.016046	926.5	926.5	36.054	1055.2	1091.2	0.0708	1.9996	2.0704	**68.0**
70.0	0.36292	0.016050	868.3	868.4	38.052	1054.0	1092.1	0.0745	1.9900	2.0645	**70.0**
72.0	0.38844	0.016054	814.3	814.3	40.049	1052.9	1093.0	0.0783	1.9804	2.0587	**72.0**
74.0	0.41550	0.016058	764.1	764.1	42.046	1051.8	1093.8	0.0821	1.9708	2.0529	**74.0**
76.0	0.44420	0.016063	717.4	717.4	44.043	1050.7	1094.7	0.0858	1.9614	2.0472	**76.0**
78.0	0.47461	0.016067	673.8	673.9	46.040	1049.5	1095.6	0.0895	1.9520	2.0415	**78.0**
80.0	0.50683	0.016072	633.3	633.3	48.037	1048.4	1096.4	0.0932	1.9426	2.0359	**80.0**
82.0	0.54093	0.016077	595.5	595.5	50.033	1047.3	1097.3	0.0969	1.9334	2.0303	**82.0**
84.0	0.57702	0.016082	560.3	560.3	52.029	1046.1	1098.2	0.1006	1.9242	2.0248	**84.0**
86.0	0.61518	0.016087	227.5	527.5	54.026	1045.0	1099.0	0.1043	1.9151	2.0193	**86.0**
88.0	0.65551	0.016093	496.8	496.8	56.022	1043.9	1099.9	0.1079	1.9060	2.0139	**88.0**
90.0	0.69813	0.016099	468.1	468.1	58.018	1042.7	1100.8	0.1115	1.8970	2.0086	**90.0**
92.0	0.74313	0.016105	441.3	441.3	60.014	1041.6	1101.6	0.1152	1.8881	2.0033	**92.0**
94.0	0.79062	0.016111	416.3	416.3	62.010	1040.5	1102.5	0.1188	1.8792	1.9980	**94.0**
96.0	0.84072	0.016117	392.8	392.9	64.006	1039.3	1103.3	0.1224	1.8704	1.9928	**96.0**
98.0	0.89356	0.016123	370.9	370.9	66.003	1038.2	1104.2	0.1260	1.8617	1.9876	**98.0**
100.0	0.94924	0.016130	350.4	350.4	67.999	1037.1	1105.1	0.1295	1.8530	1.9825	**100.0**
102.0	1.00789	0.016137	331.1	331.1	69.995	1035.9	1105.9	0.1331	1.8444	1.9775	**102.0**
104.0	1.06965	0.016144	313.1	313.1	71.992	1034.8	1106.8	0.1366	1.8358	1.9725	**104.0**
106.0	1.1347	0.016151	296.16	296.18	73.99	1033.6	1107.6	0.1402	1.8273	1.9675	**106.0**
108.0	1.2030	0.016158	280.28	280.30	75.98	1032.5	1108.5	0.1437	1.8188	1.9626	**108.0**
110.0	1.2750	0.016165	265.37	265.39	77.98	1031.4	1109.3	0.1472	1.8105	1.9577	**110.0**
112.0	1.3505	0.016173	251.37	251.38	79.98	1030.2	1110.2	0.1507	1.8021	1.9528	**112.0**
114.0	1.4299	0.016180	238.21	238.22	81.97	1029.1	1111.0	0.1542	1.7938	1.9480	**114.0**
116.0	1.5133	0.016188	225.84	225.85	83.97	1027.9	1111.9	0.1577	1.7856	1.9433	**116.0**
118.0	1.6009	0.016196	214.20	214.21	85.97	1026.8	1112.7	0.1611	1.7774	1.9386	**118.0**
120.0	1.6927	0.016204	203.25	203.26	87.97	1025.6	1113.6	0.1646	1.7693	1.9339	**120.0**
122.0	1.7891	0.016213	192.94	192.95	89.96	1024.5	1114.4	0.1680	1.7613	1.9293	**122.0**
124.0	1.8901	0.016221	183.23	183.24	91.96	1023.3	1115.3	0.1715	1.7533	1.9247	**124.0**
126.0	1.9959	0.016229	174.08	174.09	93.96	1022.2	1116.1	0.1749	1.7453	1.9202	**126.0**
128.0	2.1068	0.016238	165.45	165.47	95.96	1021.0	1117.0	0.1783	1.7374	1.9157	**128.0**
130.0	2.2230	0.016247	157.32	157.33	97.96	1019.8	1117.8	0.1817	1.7295	1.9112	**130.0**
132.0	2.3445	0.016256	149.64	149.66	99.95	1018.7	1118.6	0.1851	1.7217	1.9068	**132.0**
134.0	2.4717	0.016265	142.40	142.41	101.95	1017.5	1119.5	0.1884	1.7140	1.9024	**134.0**
136.0	2.6047	0.016274	135.55	135.57	103.95	1016.4	1120.3	0.1918	1.7063	1.8980	**136.0**
138.0	2.7438	0.016284	129.09	129.11	105.95	1015.2	1121.1	0.1951	1.6986	1.8937	**138.0**
140.0	2.8892	0.016293	122.98	123.00	107.95	1014.0	1122.0	0.1985	1.6910	1.8895	**140.0**
142.0	3.0411	0.016303	117.21	117.22	109.95	1012.9	1122.8	0.2018	1.6534	1.8852	**142.0**
144.0	3.1997	0.016312	111.74	111.76	111.95	1011.7	1123.6	0.2051	1.6759	1.8810	**144.0**
146.0	3.3653	0.016322	106.58	106.59	113.95	1010.5	1124.5	0.2084	1.6684	1.8769	**146.0**
148.0	3.5381	0.016332	101.68	101.70	115.95	1009.3	1125.3	0.2117	1.6610	1.8727	**148.0**
150.0	3.7184	0.016343	97.05	97.07	117.95	1008.2	1126.1	0.2150	1.6536	1.8686	**150.0**
152.0	3.9065	0.016353	92.66	92.68	119.95	1007.0	1126.9	0.2183	1.6463	1.8646	**152.0**
154.0	4.1025	0.016363	88.50	88.52	121.95	1005.8	1127.7	0.2216	1.6390	1.8606	**154.0**
156.0	4.3068	0.016374	84.56	84.57	123.95	1004.6	1128.6	0.2248	1.6318	1.8566	**156.0**
158.0	4.5197	0.016384	80.82	80.83	125.96	1003.4	1129.4	0.2281	1.6245	1.8526	**158.0**
160.0	4.7414	0.016395	77.27	77.29	127.96	1002.2	1130.2	0.2313	1.6174	1.8487	**160.0**
162.0	4.9722	0.016406	73.90	73.92	129.96	1001.0	1131.0	0.2345	1.6103	1.8448	**162.0**
164.0	5.2124	0.016417	70.70	70.72	131.96	999.8	1131.8	0.2377	1.6032	1.8409	**164.0**
166.0	5.4623	0.016428	67.67	67.68	133.97	998.6	1132.6	0.2409	1.5961	1.8371	**166.0**
168.0	5.7223	0.016440	64.78	64.80	135.97	997.4	1133.4	0.2441	1.5892	1.8333	**168.0**
170.0	5.9926	0.016451	62.04	62.06	137.97	996.2	1134.2	0.2473	1.5822	1.8295	**170.0**
172.0	6.2736	0.016463	59.43	59.45	139.98	995.0	1135.0	0.2505	1.5753	1.8258	**172.0**
174.0	6.5656	0.016474	56.95	56.97	141.98	993.8	1135.8	0.2537	1.5684	1.8221	**174.0**
176.0	6.8690	0.016486	54.59	54.61	143.99	992.6	1136.6	0.2568	1.5616	1.8184	**176.0**
178.0	7.1840	0.016498	52.35	52.36	145.99	991.4	1137.4	0.2600	1.5548	1.8147	**178.0**

Table 1. Saturated Steam: Temperature Table—*Continued*

Temp Fahr t	Abs Press. Lb per Sq In. p	Specific Volume Sat. Liquid v_f	Evap v_{fg}	Sat. Vapor v_g	Enthalpy Sat. Liquid h_f	Evap h_{fg}	Sat. Vapor h_g	Entropy Sat. Liquid s_f	Evap s_{fg}	Sat. Vapor s_g	Temp Fahr t
180.0	7.5110	0.016510	50.21	50.22	148.00	990.2	1138.2	0.2631	1.5480	1.8111	**180.0**
182.0	7.850	0.016522	48.172	18.189	150.01	989.0	1139.0	0.2662	1.5413	1.8075	**182.0**
184.0	8.203	0.016534	46.232	46.249	152.01	987.8	1139.8	0.2694	1.5346	1.8040	**184.0**
186.0	8.568	0.016547	44.383	44.400	154.02	986.5	1140.5	0.2725	1.5279	1.8004	**186.0**
188.0	8.947	0.016559	42.621	42.638	156.03	985.3	1141.3	0.2756	1.5213	1.7969	**188.0**
190.0	9.340	0.016572	40.941	40.957	158.04	984.1	1142.1	0.2787	1.5148	1.7934	**190.0**
192.0	9.747	0.016585	39.337	39.354	160.05	982.8	1142.9	0.2818	1.5082	1.7900	**192.0**
194.0	10.168	0.016598	37.808	37.824	162.05	981.6	1143.7	0.2848	1.5017	1.7865	**194.0**
196.0	10.605	0.016611	36.348	36.364	164.06	980.4	1144.4	0.2879	1.4952	1.7831	**196.0**
198.0	11.058	0.016624	34.954	34.970	166.08	979.1	1145.2	0.2910	1.4888	1.7798	**198.0**
200.0	11.526	0.016637	33.622	33.639	168.09	977.9	1146.0	0.2940	1.4824	1.7764	**200.0**
204.0	12.512	0.016664	31.135	31.151	172.11	975.4	1147.5	0.3001	1.4697	1.7698	**204.0**
208.0	13.568	0.016691	28.862	28.878	176.14	972.8	1149.0	0.3061	1.4571	1.7632	**208.0**
212.0	14.696	0.016719	26.782	26.799	180.17	970.3	1150.5	0.3121	1.4447	1.7568	**212.0**
216.0	15.901	0.016747	24.878	24.894	184.20	967.8	1152.0	0.3181	1.4323	1.7505	**216.0**
220.0	17.186	0.016775	23.131	23.148	188.23	965.2	1153.4	0.3241	1.4201	1.7442	**220.0**
224.0	18.556	0.016805	21.529	21.545	192.27	962.6	1154.9	0.3300	1.4081	1.7380	**224.0**
228.0	20.015	0.016834	20.056	20.073	196.31	960.0	1156.3	0.3359	1.3961	1.7320	**228.0**
232.0	21.567	0.016864	18.701	18.718	200.35	957.4	1157.8	0.3417	1.3842	1.7260	**232.0**
236.0	23.216	0.016895	17.454	17.471	204.40	954.8	1159.2	0.3476	1.3725	1.7201	**236.0**
240.0	24.968	0.016926	16.304	16.321	208.45	952.1	1160.6	0.3533	1.3609	1.7142	**240.0**
244.0	26.826	0.016958	15.243	15.260	212.50	949.5	1162.0	0.3591	1.3494	1.7085	**244.0**
248.0	28.796	0.016990	14.264	14.281	216.56	946.8	1163.4	0.3649	1.3379	1.7028	**248.0**
252.0	30.883	0.017022	13.358	13.375	220.62	944.1	1164.7	0.3706	1.3266	1.6972	**252.0**
256.0	33.091	0.017055	12.520	12.538	224.69	941.4	1166.1	0.3763	1.3154	1.6917	**256.0**
260.0	35.427	0.017089	11.745	11.762	228.76	938.6	1167.4	0.3819	1.3043	1.6862	**260.0**
264.0	37.894	0.017123	11.025	11.042	232.83	935.9	1168.7	0.3876	1.2933	1.6808	**264.0**
268.0	40.500	0.017157	10.358	10.375	236.91	933.1	1170.0	0.3932	1.2823	1.6755	**268.0**
272.0	43.249	0.017193	9.738	9.755	240.99	930.3	1171.3	0.3987	1.2715	1.6702	**272.0**
276.0	46.147	0.017228	9.162	9.180	245.08	927.5	1172.5	0.4043	1.2607	1.6650	**276.0**
280.0	49.200	0.017264	8.627	8.644	249.17	924.6	1173.8	0.4098	1.2501	1.6599	**280.0**
284.0	52.414	0.01730	8.1280	8.1453	253.3	921.7	1175.0	0.4154	1.2395	1.6548	**284.0**
288.0	55.795	0.01734	7.6634	7.6807	257.4	918.8	1176.2	0.4208	1.2290	1.6498	**288.0**
292.0	59.350	0.01738	7.2301	7.2475	261.5	915.9	1177.4	0.4263	1.2186	1.6449	**292.0**
296.0	63.084	0.01741	6.8259	6.8433	265.6	913.0	1178.6	0.4317	1.2082	1.6400	**296.0**
300.0	67.005	0.01745	6.4483	6.4658	269.7	910.0	1179.7	0.4372	1.1979	1.6351	**300.0**
304.0	71.119	0.01749	6.0955	6.1130	273.8	907.0	1180.9	0.4426	1.1877	1.6303	**304.0**
308.0	75.433	0.01753	5.7655	5.7830	278.0	904.0	1182.0	0.4479	1.1776	1.6256	**308.0**
312.0	79.953	0.01757	5.4566	5.4742	282.1	901.0	1183.1	0.4533	1.1676	1.6209	**312.0**
316.0	84.688	0.01761	5.1673	5.1849	286.3	897.9	1184.1	0.4586	1.1576	1.6162	**316.0**
320.0	89.643	0.01766	4.8961	4.9138	290.4	894.8	1185.2	0.4640	1.1477	1.6116	**320.0**
324.0	94.826	0.01770	4.6418	4.6595	294.6	891.6	1186.2	0.4692	1.1378	1.6071	**324.0**
328.0	100.245	0.01774	4.4030	4.4208	298.7	888.5	1187.2	0.4745	1.1280	1.6025	**328.0**
332.0	105.907	0.01779	4.1788	4.1966	302.9	885.3	1188.2	0.4798	1.1183	1.5981	**332.0**
336.0	111.820	0.01783	3.9681	3.9859	307.1	882.1	1189.1	0.4850	1.1086	1.5936	**336.0**
340.0	117.992	0.01787	3.7699	3.7878	311.3	878.8	1190.1	0.4902	1.0990	1.5892	**340.0**
344.0	124.430	0.01792	3.5834	3.6013	315.5	875.5	1191.0	0.4954	1.0894	1.5849	**344.0**
348.0	131.142	0.01797	3.4078	3.4258	319.7	872.2	1191.1	0.5006	1.0799	1.5806	**348.0**
352.0	138.138	0.01801	3.2423	3.2603	323.9	868.9	1192.7	0.5058	1.0705	1.5763	**352.0**
356.0	145.424	0.01806	3.0863	3.1044	328.1	865.5	1193.6	0.5110	1.0611	1.5721	**356.0**
360.0	153.010	0.01811	2.9392	2.9573	332.3	862.1	1194.4	0.5161	1.0517	1.5678	**360.0**
364.0	160.903	0.01816	2.8002	2.8184	336.5	858.6	1195.2	0.5212	1.0424	1.5637	**364.0**
368.0	169.113	0.01821	2.6691	2.6873	340.8	855.1	1195.9	0.5263	1.0332	1.5595	**368.0**
372.0	177.648	0.01826	2.5451	2.5633	345.0	851.6	1196.7	0.5314	1.0240	1.5554	**372.0**
376.0	186.517	0.01831	2.4279	2.4462	349.3	848.1	1197.4	0.5365	1.0148	1.5513	**376.0**
380.0	195.729	0.01836	2.3170	2.3353	353.6	844.5	1198.0	0.5416	1.0057	1.5473	**380.0**
384.0	205.294	0.01842	2.2120	2.2304	357.9	840.8	1198.7	0.5466	0.9966	1.5432	**384.0**
388.0	215.220	0.01847	2.1126	2.1311	362.2	837.2	1199.3	0.5516	0.9876	1.5392	**388.0**
392.0	225.516	0.01853	2.0184	2.0369	366.5	833.4	1199.9	0.5567	0.9786	1.5352	**392.0**
396.0	236.193	0.01858	1.9291	1.9477	370.8	829.7	1200.4	0.5617	0.9696	1.5313	**396.0**
400.0	247.259	0.01864	1.8444	1.8630	375.1	825.9	1201.0	0.5667	0.9607	1.5274	**400.0**
404.0	258.725	0.01870	1.7640	1.7827	379.4	822.0	1201.5	0.5717	0.9518	1.5234	**404.0**
408.0	270.600	0.01875	1.6877	1.7064	383.8	818.2	1201.9	0.5766	0.9429	1.5195	**408.0**
412.0	282.894	0.01881	1.6152	1.6340	388.1	814.2	1202.4	0.5816	0.9341	1.5157	**412.0**
416.0	295.617	0.01887	1.5463	1.5651	392.5	810.2	1202.8	0.5866	0.9253	1.5118	**416.0**
420.0	308.780	0.01894	1.4808	1.4997	396.9	806.2	1203.1	0.5915	0.9165	1.5080	**420.0**
424.0	322.391	0.01900	1.4184	1.4374	401.3	802.2	1203.5	0.5964	0.9077	1.5042	**424.0**
428.0	336.463	0.01906	1.3591	1.3782	405.7	798.0	1203.7	0.6014	0.8990	1.5004	**428.0**
432.0	351.00	0.01913	1.30266	1.32179	410.1	793.9	1204.0	0.6063	0.8903	1.4966	**432.0**
436.0	366.03	0.01919	1.24887	1.26806	414.6	789.7	1204.2	0.6112	0.8816	1.4928	**436.0**
440.0	381.54	0.01926	1.19761	1.21687	419.0	785.4	1204.4	0.6161	0.8729	1.4890	**440.0**
444.0	397.56	0.01933	1.14874	1.16806	423.5	781.1	1204.6	0.6210	0.8643	1.4853	**444.0**
448.0	414.09	0.01940	1.10212	1.12152	428.0	776.7	1204.7	0.6259	0.8557	1.4815	**448.0**
452.0	431.14	0.01947	1.05764	1.07711	432.5	772.3	1204.8	0.6308	0.8471	1.4778	**452.0**
456.0	448.73	0.01954	1.01518	1.03472	437.0	767.8	1204.8	0.6356	0.8385	1.4741	**456.0**

Table 1. Saturated Steam: Temperature Table—*Continued*

Temp Fahr t	Abs Press. Lb per Sq In. p	Specific Volume Sat. Liquid v_f	Specific Volume Evap v_{fg}	Specific Volume Sat. Vapor v_g	Enthalpy Sat. Liquid h_f	Enthalpy Evap h_{fg}	Enthalpy Sat. Vapor h_g	Entropy Sat. Liquid s_f	Entropy Evap s_{fg}	Entropy Sat. Vapor s_g	Temp Fahr t
460.0	466.87	0.01961	0.97463	0.99424	441.5	763.2	1204.8	0.6405	0.8299	1.4704	**460.0**
464.0	485.56	0.01969	0.93588	0.95557	446.1	758.6	1204.7	0.6454	0.8213	1.4667	**464.0**
468.0	504.83	0.01976	0.89885	0.91862	450.7	754.0	1204.6	0.6502	0.8127	1.4629	**468.0**
472.0	524.67	0.01984	0.86345	0.88329	455.2	749.3	1204.5	0.6551	0.8042	1.4592	**472.0**
476.0	545.11	0.01992	0.82958	0.84950	459.9	744.5	1204.3	0.6599	0.7956	1.4555	**476.0**
480.0	566.15	0.02000	0.79716	0.81717	464.5	739.6	1204.1	0.6648	0.7871	1.4518	**480.0**
484.0	587.81	0.02009	0.76613	0.78622	469.1	734.7	1203.8	0.6696	0.7785	1.4481	**484.0**
488.0	610.10	0.02017	0.73641	0.75658	473.8	729.7	1203.5	0.6745	0.7700	1.4444	**488.0**
492.0	633.03	0.02026	0.70794	0.72820	478.5	724.6	1203.1	0.6793	0.7614	1.4407	**492.0**
496.0	656.61	0.02034	0.68065	0.70100	483.2	719.5	1202.7	0.6842	0.7528	1.4370	**496.0**
500.0	680.86	0.02043	0.65448	0.67492	487.9	714.3	1202.2	0.6890	0.7443	1.4333	**500.0**
504.0	705.78	0.02053	0.62938	0.64991	492.7	709.0	1201.7	0.6939	0.7357	1.4296	**504.0**
508.0	731.40	0.02062	0.60530	0.62592	497.5	703.7	1201.1	0.6987	0.7271	1.4258	**508.0**
512.0	757.72	0.02072	0.58218	0.60289	502.3	698.2	1200.5	0.7036	0.7185	1.4221	**512.0**
516.0	784.76	0.02081	0.55997	0.58079	507.1	692.7	1199.8	0.7085	0.7099	1.4183	**516.0**
520.0	812.53	0.02091	0.53864	0.55956	512.0	687.0	1199.0	0.7133	0.7013	1.4146	**520.0**
524.0	841.04	0.02102	0.51814	0.53916	516.9	681.3	1198.2	0.7182	0.6926	1.4108	**524.0**
528.0	870.31	0.02112	0.49843	0.51955	521.8	675.5	1197.3	0.7231	0.6839	1.4070	**528.0**
532.0	900.34	0.02123	0.47947	0.50070	526.8	669.6	1196.4	0.7280	0.6752	1.4032	**532.0**
536.0	931.17	0.02134	0.46123	0.48257	531.7	663.6	1195.4	0.7329	0.6665	1.3993	**536.0**
540.0	962.79	0.02146	0.44367	0.46513	536.8	657.5	1194.3	0.7378	0.6577	1.3954	**540.0**
544.0	995.22	0.02157	0.42677	0.44834	541.8	651.3	1193.1	0.7427	0.6489	1.3915	**544.0**
548.0	1028.49	0.02169	0.41048	0.43217	546.9	645.0	1191.9	0.7476	0.6400	1.3876	**548.0**
552.0	1062.59	0.02182	0.39479	0.41660	552.0	638.5	1190.6	0.7525	0.6311	1.3837	**552.0**
556.0	1097.55	0.02194	0.37966	0.40160	557.2	632.0	1189.2	0.7575	0.6222	1.3797	**556.0**
560.0	1133.38	0.02207	0.36507	0.38714	562.4	625.3	1187.7	0.7625	0.6132	1.3757	**560.0**
564.0	1170.10	0.02221	0.35099	0.37320	567.6	618.5	1186.1	0.7674	0.6041	1.3716	**564.0**
568.0	1207.72	0.02235	0.33741	0.35975	572.9	611.5	1184.5	0.7725	0.5950	1.3675	**568.0**
572.0	1246.26	0.02249	0.32429	0.34678	578.3	604.5	1182.7	0.7775	0.5859	1.3634	**572.0**
576.0	1285.74	0.02264	0.31162	0.33426	583.7	597.2	1180.9	0.7825	0.5766	1.3592	**576.0**
580.0	1326.17	0.02279	0.29937	0.32216	589.1	589.9	1179.0	0.7876	0.5673	1.3550	**580.0**
584.0	1367.7	0.02295	0.28753	0.31048	594.6	582.4	1176.9	0.7927	0.5580	1.3507	**584.0**
588.0	1410.0	0.02311	0.27608	0.29919	600.1	574.7	1174.8	0.7978	0.5485	1.3464	**588.0**
592.0	1453.3	0.02328	0.26499	0.28827	605.7	566.8	1172.6	0.8030	0.5390	1.3420	**592.0**
596.0	1497.8	0.02345	0.25425	0.27770	611.4	558.8	1170.2	0.8082	0.5293	1.3375	**596.0**
600.0	1543.2	0.02364	0.24384	0.26747	617.1	550.6	1167.7	0.8134	0.5196	1.3330	**600.0**
604.0	1589.7	0.02382	0.23374	0.25757	622.9	542.2	1165.1	0.8187	0.5097	1.3284	**604.0**
608.0	1637.3	0.02402	0.22394	0.24796	628.8	533.6	1162.4	0.8240	0.4997	1.3238	**608.0**
612.0	1686.1	0.02422	0.21442	0.23865	634.8	524.7	1159.5	0.8294	0.4896	1.3190	**612.0**
616.6	1735.9	0.02444	0.20516	0.22960	640.8	515.6	1156.4	0.8348	0.4794	1.3141	**616.0**
620.0	1786.9	0.02466	0.19615	0.22081	646.9	506.3	1153.2	0.8403	0.4689	1.3092	**620.0**
624.0	1839.0	0.02489	0.18737	0.21226	653.1	496.6	1149.8	0.8458	0.4583	1.3041	**624.0**
628.0	1892.4	0.02514	0.17880	0.20394	659.5	486.7	1146.1	0.8514	0.4474	1.2988	**628.0**
632.0	1947.0	0.02539	0.17044	0.19583	665.9	476.4	1142.2	0.8571	0.4364	1.2934	**632.0**
636.0	2002.8	0.02566	0.16226	0.18792	672.4	465.7	1138.1	0.8628	0.4251	1.2879	**636.0**
640.0	2059.9	0.02595	0.15427	0.18021	679.1	454.6	1133.7	0.8686	0.4134	1.2821	**640.0**
644.0	2118.3	0.02625	0.14644	0.17269	685.9	443.1	1129.0	0.8746	0.4015	1.2761	**644.0**
648.0	2178.1	0.02657	0.13876	0.16534	692.9	431.1	1124.0	0.8806	0.3893	1.2699	**648.0**
652.0	2239.2	0.02691	0.13124	0.15816	700.0	418.7	1118.7	0.8868	0.3767	1.2634	**652.0**
656.0	2301.7	0.02728	0.12387	0.15115	707.4	405.7	1113.1	0.8931	0.3637	1.2567	**656.0**
660.0	2365.7	0.02768	0.11663	0.14431	714.9	392.1	1107.0	0.8995	0.3502	1.2498	**660.0**
664.0	2431.1	0.02811	0.10947	0.13757	722.9	377.7	1100.6	0.9064	0.3361	1.2425	**664.0**
668.0	2498.1	0.02858	0.10229	0.13087	731.5	362.1	1093.5	0.9137	0.3210	1.2347	**668.0**
672.0	2566.6	0.02911	0.09514	0.12424	740.2	345.7	1085.9	0.9212	0.3054	1.2266	**672.0**
676.0	2636.8	0.02970	0.08799	0.11769	749.2	328.5	1077.6	0.9287	0.2892	1.2179	**676.0**
680.0	2708.6	0.03037	0.08080	0.11117	758.5	310.1	1068.5	0.9365	0.2720	1.2086	**680.0**
684.0	2782.1	0.03114	0.07349	0.10463	768.2	290.2	1058.4	0.9447	0.2537	1.1984	**684.0**
688.0	2857.4	0.03204	0.06595	0.09799	778.8	268.2	1047.0	0.9535	0.2337	1.1872	**688.0**
692.0	2934.5	0.03313	0.05797	0.09110	790.5	243.1	1033.6	0.9634	0.2110	1.1744	**692.0**
696.0	3013.4	0.03455	0.04916	0.08371	804.4	212.8	1017.2	0.9749	0.1841	1.1591	**696.0**
700.0	3094.3	0.03662	0.03857	0.07519	822.4	172.7	995.2	0.9901	0.1490	1.1390	**700.0**
702.0	3135.5	0.03824	0.03173	0.06997	835.0	144.7	979.7	1.0006	0.1246	1.1252	**702.0**
704.0	3177.2	0.04108	0.02192	0.06300	854.2	102.0	956.2	1.0169	0.0876	1.1046	**704.0**
705.0	3198.3	0.04427	0.01304	0.05730	873.0	61.4	934.4	1.0329	0.0527	1.0856	**705.0**
705.47*	3208.2	0.05078	0.00000	0.05078	906.0	0.0	906.0	1.0612	0.0000	1.0612	**705.47***

*Critical temperature

Table 2: Saturated Steam: Pressure Table *Critical pressure

s Press. b/Sq In. p	Temp Fahr t	Specific Volume Sat. Liquid v_f	Evap v_{fg}	Sat. Vapor v_g	Enthalpy Sat. Liquid h_f	Evap h_{fg}	Sat. Vapor h_g	Entropy Sat. Liquid s_f	Evap s_{fg}	Sat. Vapor s_g	Abs Press. Lb/Sq In. p
0.08865	32.018	0.016022	3302.4	3302.4	0.0003	1075.5	1075.5	0.0000	2.1872	2.1872	**0.08865**
0.25	59.323	0.016032	1235.5	1235.5	27.382	1060.1	1087.4	0.0542	2.0425	2.0967	**0.25**
0.50	79.586	0.016071	641.5	641.5	47.623	1048.6	1096.3	0.0925	1.9446	2.0370	**0.50**
1.0	101.74	0.016136	333.59	333.60	69.73	1036.1	1105.8	0.1326	1.8455	1.9781	**1.0**
5.0	162.24	0.016407	73.515	73.532	130.20	1000.9	1131.1	0.2349	1.6094	1.8443	**5.0**
10.0	193.21	0.016592	38.404	38.420	161.26	982.1	1143.3	0.2836	1.5043	1.7879	**10.0**
14.696	212.00	0.016719	26.782	26.799	180.17	970.3	1150.5	0.3121	1.4447	1.7568	**14.696**
15.0	213.03	0.016726	26.274	26.290	181.21	969.7	1150.9	0.3137	1.4415	1.7552	**15.0**
20.0	227.96	0.016834	20.070	20.087	196.27	960.1	1156.3	0.3358	1.3962	1.7320	**20.0**
30.0	250.34	0.017009	13.7266	13.7436	218.9	945.2	1164.1	0.3682	1.3313	1.6995	**30.0**
40.0	267.25	0.017151	10.4794	10.4965	236.1	933.6	1169.8	0.3921	1.2844	1.6765	**40.0**
50.0	281.02	0.017274	8.4967	8.5140	250.2	923.9	1174.1	0.4112	1.2474	1.6586	**50.0**
60.0	292.71	0.017383	7.1562	7.1736	262.2	915.4	1177.6	0.4273	1.2167	1.6440	**60.0**
70.0	302.93	0.017482	6.1875	6.2050	272.7	907.8	1180.6	0.4411	1.1905	1.6316	**70.0**
80.0	312.04	0.017573	5.4536	5.4711	282.1	900.9	1183.1	0.4534	1.1675	1.6208	**80.0**
90.0	320.28	0.017659	4.8779	4.8953	290.7	894.6	1185.3	0.4643	1.1470	1.6113	**90.0**
100.0	327.82	0.017740	4.4133	4.4310	298.5	888.6	1187.2	0.4743	1.1284	1.6027	**100.0**
110.0	334.79	0.01782	4.0306	4.0484	305.8	883.1	1188.9	0.4834	1.1115	1.5950	**110.0**
120.0	341.27	0.01789	3.7097	3.7275	312.6	877.8	1190.4	0.4919	1.0960	1.5879	**120.0**
130.0	347.33	0.01796	3.4364	3.4544	319.0	872.8	1191.7	0.4998	1.0815	1.5813	**130.0**
140.0	353.04	0.01803	3.2010	3.2190	325.0	868.0	1193.0	0.5071	1.0681	1.5752	**140.0**
150.0	358.43	0.01809	2.9958	3.0139	330.6	863.4	1194.1	0.5141	1.0554	1.5695	**150.0**
160.0	363.55	0.01815	2.8155	2.8336	336.1	859.0	1195.1	0.5206	1.0435	1.5641	**160.0**
170.0	368.42	0.01821	2.6556	2.6738	341.2	854.8	1196.0	0.5269	1.0322	1.5591	**170.0**
180.0	373.08	0.01827	2.5129	2.5312	346.2	850.7	1196.9	0.5328	1.0215	1.5543	**180.0**
190.0	377.53	0.01833	2.3847	2.4030	350.9	846.7	1197.6	0.5384	1.0113	1.5498	**190.0**
200.0	381.80	0.01839	2.2689	2.2873	355.5	842.8	1198.3	0.5438	1.0016	1.5454	**200.0**
210.0	385.91	0.01844	2.16373	2.18217	359.9	839.1	1199.0	0.5490	0.9923	1.5413	**210.0**
220.0	389.88	0.01850	2.06779	2.08629	364.2	835.4	1199.6	0.5540	0.9834	1.5374	**220.0**
230.0	393.70	0.01855	1.97991	1.99846	368.3	831.8	1200.1	0.5588	0.9748	1.5336	**230.0**
240.0	397.39	0.01860	1.89909	1.91769	372.3	828.4	1200.6	0.5634	0.9665	1.5299	**240.0**
250.0	400.97	0.01865	1.82452	1.84317	376.1	825.0	1201.1	0.5679	0.9585	1.5264	**250.0**
260.0	404.44	0.01870	1.75548	1.77418	379.9	821.6	1201.5	0.5722	0.9508	1.5230	**260.0**
270.0	407.80	0.01875	1.69137	1.71013	383.6	818.3	1201.9	0.5764	0.9433	1.5197	**270.0**
280.0	411.07	0.01880	1.63169	1.65049	387.1	815.1	1202.3	0.5805	0.9361	1.5166	**280.0**
290.0	414.25	0.01885	1.57597	1.59482	390.6	812.0	1202.6	0.5844	0.9291	1.5135	**290.0**
300.0	417.35	0.01889	1.52384	1.54274	394.0	808.9	1202.9	0.5882	0.9223	1.5105	**300.0**
350.0	431.73	0.01912	1.30642	1.32554	409.8	794.2	1204.0	0.6059	0.8909	1.4968	**350.0**
400.0	444.60	0.01934	1.14162	1.16095	424.2	780.4	1204.6	0.6217	0.8630	1.4847	**400.0**
450.0	456.28	0.01954	1.01224	1.03179	437.3	767.5	1204.8	0.6360	0.8378	1.4738	**450.0**
500.0	467.01	0.01975	0.90787	0.92762	449.5	755.1	1204.7	0.6490	0.8148	1.4639	**500.0**
550.0	476.94	0.01994	0.82183	0.84177	460.9	743.3	1204.3	0.6611	0.7936	1.4547	**550.0**
600.0	486.20	0.02013	0.74962	0.76975	471.7	732.0	1203.7	0.6723	0.7738	1.4461	**600.0**
650.0	494.89	0.02032	0.68811	0.70843	481.9	720.9	1202.8	0.6828	0.7552	1.4381	**650.0**
700.0	503.08	0.02050	0.63505	0.65556	491.6	710.2	1201.8	0.6928	0.7377	1.4304	**700.0**
750.0	510.84	0.02069	0.58880	0.60949	500.9	699.8	1200.7	0.7022	0.7210	1.4232	**750.0**
800.0	518.21	0.02087	0.54809	0.56896	509.8	689.6	1199.4	0.7111	0.7051	1.4163	**800.0**
850.0	525.24	0.02105	0.51197	0.53302	518.4	679.5	1198.0	0.7197	0.6899	1.4096	**850.0**
900.0	531.95	0.02123	0.47968	0.50091	526.7	669.7	1196.4	0.7279	0.6753	1.4032	**900.0**
950.0	538.39	0.02141	0.45064	0.47205	534.7	660.0	1194.7	0.7358	0.6612	1.3970	**950.0**
1000.0	544.58	0.02159	0.42436	0.44596	542.6	650.4	1192.9	0.7434	0.6476	1.3910	**1000.0**
1050.0	550.53	0.02177	0.40047	0.42224	550.1	640.9	1191.0	0.7507	0.6344	1.3851	**1050.0**
1100.0	556.28	0.02195	0.37863	0.40058	557.5	631.5	1189.1	0.7578	0.6216	1.3794	**1100.0**
1150.0	561.82	0.02214	0.35859	0.38073	564.8	622.2	1187.0	0.7647	0.6091	1.3738	**1150.0**
1200.0	567.19	0.02232	0.34013	0.36245	571.9	613.0	1184.8	0.7714	0.5969	1.3683	**1200.0**
1250.0	572.38	0.02250	0.32306	0.34556	578.8	603.8	1182.6	0.7780	0.5850	1.3630	**1250.0**
1300.0	577.42	0.02269	0.30722	0.32991	585.6	594.6	1180.2	0.7843	0.5733	1.3577	**1300.0**
1350.0	582.32	0.02288	0.29250	0.31537	592.3	585.4	1177.8	0.7906	0.5620	1.3525	**1350.0**
1400.0	587.07	0.02307	0.27871	0.30178	598.8	576.5	1175.3	0.7966	0.5507	1.3474	**1400.0**
1450.0	591.70	0.02327	0.26584	0.28911	605.3	567.4	1172.8	0.8026	0.5397	1.3423	**1450.0**
1500.0	596.20	0.02346	0.25372	0.27719	611.7	558.4	1170.1	0.8085	0.5288	1.3373	**1500.0**
1550.0	600.59	0.02366	0.24235	0.26601	618.0	549.4	1167.4	0.8142	0.5182	1.3324	**1550.0**
1600.0	604.87	0.02387	0.23159	0.25545	624.2	540.3	1164.5	0.8199	0.5076	1.3274	**1600.0**
1650.0	609.05	0.02407	0.22143	0.24551	630.4	531.3	1161.6	0.8254	0.4971	1.3225	**1650.0**
1700.0	613.13	0.02428	0.21178	0.23607	636.5	522.2	1158.6	0.8309	0.4867	1.3176	**1700.0**
1750.0	617.12	0.02450	0.20263	0.22713	642.5	513.1	1155.6	0.8363	0.4765	1.3128	**1750.0**
1800.0	621.02	0.02472	0.19390	0.21861	648.5	503.8	1152.3	0.8417	0.4662	1.3079	**1800.0**
1850.0	624.83	0.02495	0.18558	0.21052	654.5	494.6	1149.0	0.8470	0.4561	1.3030	**1850.0**
1900.0	628.56	0.02517	0.17761	0.20278	660.4	485.2	1145.6	0.8522	0.4459	1.2981	**1900.0**
1950.0	632.22	0.02541	0.16999	0.19540	666.3	475.8	1142.0	0.8574	0.4358	1.2931	**1950.0**
2000.0	635.80	0.02565	0.16266	0.18831	672.1	466.2	1138.3	0.8625	0.4256	1.2881	**2000.0**
2100.0	642.76	0.02615	0.14885	0.17501	683.8	446.7	1130.5	0.8727	0.4053	1.2780	**2100.0**
2200.0	649.45	0.02669	0.13603	0.16272	695.5	426.7	1122.2	0.8828	0.3848	1.2676	**2200.0**
2300.0	655.89	0.02727	0.12406	0.15133	707.2	406.0	1113.2	0.8929	0.3640	1.2569	**2300.0**
2400.0	662.11	0.02790	0.11287	0.14076	719.0	384.8	1103.7	0.9031	0.3430	1.2460	**2400.0**
2500.0	668.11	0.02859	0.10209	0.13068	731.7	361.6	1093.3	0.9139	0.3206	1.2345	**2500.0**
2600.0	673.91	0.02938	0.09172	0.12110	744.5	337.6	1082.0	0.9247	0.2977	1.2225	**2600.0**
2700.0	679.53	0.03029	0.08165	0.11194	757.3	312.3	1069.7	0.9356	0.2741	1.2097	**2700.0**
2800.0	684.96	0.03134	0.07171	0.10305	770.7	285.1	1055.8	0.9468	0.2491	1.1958	**2800.0**
2900.0	690.22	0.03262	0.06158	0.09420	785.1	254.7	1039.8	0.9588	0.2215	1.1803	**2900.0**
3000.0	695.33	0.03428	0.05073	0.08500	801.8	218.4	1020.3	0.9728	0.1891	1.1619	**3000.0**
3100.0	700.28	0.03681	0.03771	0.07452	824.0	169.3	993.3	0.9914	0.1460	1.1373	**3100.0**
3200.0	705.08	0.04472	0.01191	0.05663	875.5	56.1	931.6	1.0351	0.0482	1.0832	**3200.0**
3208.2*	705.47	0.05078	0.00000	0.05078	906.0	0.0	906.0	1.0612	0.0000	1.0612	**3208.2***

Table 3. Superheated Steam

Abs Press. Lb/Sq In. (Sat. Temp)		Sat. Water	Sat. Steam	200	250	300	350	400	450	500	600	700	800	900	1000	1100	1200
				Temperature — Degrees Fahrenheit													
1 (101.74)	Sh			98.26	148.26	198.26	248.26	298.26	348.26	398.26	498.26	598.26	698.26	798.26	898.26	998.26	1098.26
	v	0.01614	333.6	392.5	422.4	452.3	482.1	511.9	541.7	571.5	631.1	690.7	750.2	809.8	869.4	929.1	988.7
	h	69.73	1105.8	1150.2	1172.9	1195.7	1218.7	1241.8	1265.1	1288.6	1336.1	1384.5	1431.0	1480.8	1531.4	1583.0	1635.4
	s	0.1326	1.9781	2.0509	2.0841	2.1152	2.1445	2.1722	2.1985	2.2237	2.2708	2.3144	2.3512	2.3892	2.4251	2.4592	2.4918
5 (162.24)	Sh			37.76	87.76	137.76	187.76	237.76	287.76	337.76	437.76	537.76	637.76	737.76	837.76	937.76	1037.76
	v	0.01641	73.53	78.14	84.21	90.24	96.25	102.24	108.23	114.21	126.15	138.08	150.01	161.94	173.86	185.78	197.70
	h	130.20	1131.1	1148.6	1171.7	1194.8	1218.0	1241.3	1264.7	1288.2	1335.9	1384.3	1433.6	1483.7	1534.7	1586.7	1639.6
	s	0.2349	1.8443	1.8716	1.9054	1.9369	1.9664	1.9943	2.0208	2.0460	2.0932	2.1369	2.1776	2.2159	2.2521	2.2866	2.3194
10 (193.21)	Sh			6.79	56.79	106.79	156.79	206.79	256.79	306.79	406.79	506.79	606.79	706.79	806.79	906.79	1006.79
	v	0.01659	38.42	38.84	41.93	44.98	48.02	51.03	54.04	57.04	63.03	69.00	74.98	80.94	86.91	92.87	98.84
	h	161.26	1143.3	1146.6	1170.2	1193.7	1217.1	1240.6	1264.1	1287.8	1335.5	1384.0	1433.4	1483.5	1534.6	1586.6	1639.5
	s	0.2836	1.7879	1.7928	1.8273	1.8593	1.8892	1.9173	1.9439	1.9692	2.0166	2.0603	2.1011	2.1394	2.1757	2.2101	2.2430
14.696 * (212.00)	Sh				38.00	88.00	138.00	188.00	238.00	288.00	388.00	488.00	588.00	688.00	788.00	888.00	988.00
	v	0.0167	26.828		28.44	30.52	32.61	34.65	36.73	38.75	42.83	46.91	50.97	55.03	59.09	63.19	67.25
	h	180.07	1150.4		1169.2	1192.0	1215.4	1238.9	1262.1	1285.4	1333.0	1381.4	1430.5	1480.4	1531.1	1582.7	1635.1
	s	0.3120	1.7566		1.7838	1.8148	1.8446	1.8727	1.8989	1.9238	1.9709	2.0145	2.0551	2.0932	2.1292	2.1634	2.1960
15 (213.03)	Sh				36.97	86.97	136.97	186.97	236.97	286.97	386.97	486.97	586.97	686.97	786.97	886.97	986.97
	v	0.01673	26.290		27.837	29.899	31.939	33.963	35.977	37.985	41.986	45.978	49.964	53.946	57.926	61.905	65.882
	h	181.21	1150.9		1168.7	1192.5	1216.2	1239.9	1263.6	1287.3	1335.2	1383.8	1433.2	1483.4	1534.5	1586.5	1639.4
	s	0.3137	1.7552		1.7809	1.8134	1.8437	1.8720	1.8988	1.9242	1.9717	2.0155	2.0563	2.0946	2.1309	2.1653	2.1982
20 (227.96)	Sh				22.04	72.04	122.04	172.04	222.04	272.04	372.04	472.04	572.04	672.04	772.04	872.04	972.04
	v	0.01683	20.087		20.788	22.356	23.900	25.428	26.946	28.457	31.466	34.465	37.458	40.447	43.435	46.420	49.405
	h	196.27	1156.3		1167.1	1191.4	1215.4	1239.2	1263.0	1286.9	1334.9	1383.5	1432.9	1483.2	1534.3	1586.3	1639.3
	s	0.3358	1.7320		1.7475	1.7805	1.8111	1.8397	1.8666	1.8921	1.9397	1.9836	2.0244	2.0628	2.0991	2.1336	2.1665
25 (240.07)	Sh				9.93	59.93	109.93	159.93	209.93	259.93	359.93	459.93	559.93	659.93	759.93	859.93	959.93
	v	0.01693	16.301		16.558	17.829	19.076	20.307	21.527	22.740	25.153	27.557	29.954	32.348	34.740	37.130	39.518
	h	208.52	1160.6		1165.6	1190.2	1214.5	1238.5	1262.5	1286.4	1334.6	1383.3	1432.7	1483.0	1534.2	1586.2	1639.2
	s	0.3535	1.7141		1.7212	1.7547	1.7856	1.8145	1.8415	1.8672	1.9149	1.9588	1.9997	2.0381	2.0744	2.1089	2.1418
30 (250.34)	Sh					49.66	99.66	149.66	199.66	249.66	349.66	449.66	549.66	649.66	749.66	849.66	949.66
	v	0.01701	13.744			14.810	15.859	16.892	17.914	18.929	20.945	22.951	24.952	26.949	28.943	30.936	32.927
	h	218.93	1164.1			1189.0	1213.6	1237.8	1261.9	1286.0	1334.2	1383.0	1432.5	1482.8	1534.0	1586.1	1639.0
	s	0.3682	1.6995			1.7334	1.7647	1.7937	1.8210	1.8467	1.8946	1.9386	1.9795	2.0179	2.0543	2.0888	2.1217
35 (259.29)	Sh					40.71	90.71	140.71	190.71	240.71	340.71	440.71	540.71	640.71	740.71	840.71	940.71
	v	0.01708	11.896			12.654	13.562	14.453	15.334	16.207	17.939	19.662	21.379	23.092	24.803	26.512	28.220
	h	228.03	1167.1			1187.8	1212.7	1237.1	1261.3	1285.5	1333.9	1382.8	1432.3	1482.7	1533.9	1586.0	1638.9
	s	0.3809	1.6872			1.7152	1.7468	1.7761	1.8035	1.8294	1.8774	1.9214	1.9624	2.0009	2.0372	2.0717	2.1046
40 (267.25)	Sh					32.75	82.75	132.75	182.75	232.75	332.75	432.75	532.75	632.75	732.75	832.75	932.75
	v	0.01715	10.497			11.036	11.838	12.624	13.398	14.165	15.685	17.195	18.699	20.199	21.697	23.194	24.689
	h	236.14	1169.8			1186.6	1211.7	1236.4	1260.8	1285.0	1333.6	1382.5	1432.1	1482.5	1533.7	1585.8	1638.8
	s	0.3921	1.6765			1.6992	1.7312	1.7608	1.7883	1.8143	1.8624	1.9065	1.9476	1.9860	2.0224	2.0569	2.0899
45 ** (274.43)	Sh					25.57	75.57	125.57	175.57	225.57	325.57	425.57	525.57	625.57	725.57	825.57	925.57
	v	0.01722	9.403			9.782	10.503	11.206	11.897	12.584	13.939	15.284	16.623	17.959	19.292	20.623	21.954
	h	243.47	1172.1			1185.4	1210.8	1235.7	1260.2	1284.6	1333.3	1382.3	1432.0	1482.4	1533.6	1585.7	1638.8
	s	0.4021	1.6671			1.6849	1.7174	1.7472	1.7749	1.8010	1.8492	1.8934	1.9345	1.9730	2.0094	2.0439	2.0769
50 (281.02)	Sh					18.98	68.98	118.98	168.98	218.98	318.98	418.98	518.98	618.98	718.98	818.98	918.98
	v	0.1727	8.514			8.769	9.424	10.062	10.688	11.306	12.529	13.741	14.947	16.150	17.350	18.549	19.746
	h	250.21	1174.1			1184.1	1209.9	1234.9	1259.6	1284.1	1332.9	1382.0	1431.7	1482.2	1533.4	1585.6	1638.6
	s	0.4112	1.6586			1.6720	1.7048	1.7349	1.7628	1.7890	1.8374	1.8816	1.9227	1.9613	1.9977	2.0322	2.0652
55 ** (287.07)	Sh					12.93	62.93	112.93	162.93	212.93	312.93	412.93	512.93	612.93	712.93	812.93	912.93
	v	0.01733	7.787			7.947	8.550	9.134	9.706	10.270	11.385	12.489	13.587	14.682	15.775	16.865	17.954
	h	256.42	1176.0			1182.9	1208.9	1234.3	1259.1	1283.6	1132.6	1381.8	1431.6	1482.0	1533.3	1585.5	1638.5
	s	0.4196	1.6510			1.6602	1.6934	1.7238	1.7518	1.7781	1.8267	1.8710	1.9123	1.9507	1.9871	2.0217	2.0546
60 (292.71)	Sh					7.29	57.29	107.29	157.29	207.29	307.29	407.29	507.29	607.29	707.29	807.29	907.29
	v	0.1738	7.174			7.257	7.815	8.354	8.881	9.400	10.425	11.438	12.446	13.450	14.452	15.452	16.450
	h	262.21	1177.6			1181.6	1208.0	1233.5	1258.5	1283.2	1332.3	1381.5	1431.3	1481.8	1533.2	1585.3	1638.4
	s	0.4273	1.6440			1.6492	1.6934	1.7134	1.7417	1.7681	1.8168	1.8612	1.9024	1.9410	1.9774	2.0120	2.0450
65 (297.98)	Sh					2.02	52.02	102.02	152.02	202.02	302.02	402.02	502.02	602.02	702.02	802.02	902.02
	v	0.01743	6.653			6.675	7.195	7.697	8.186	8.667	9.615	10.552	11.484	12.412	13.337	14.261	15.183
	h	267.63	1179.1			1180.3	1207.0	1232.7	1257.9	1282.7	1331.9	1381.3	1431.1	1481.6	1533.0	1585.2	1638.3
	s	0.4344	1.6375			1.6390	1.6731	1.7040	1.7324	1.7590	1.8077	1.8522	1.8935	1.9321	1.9685	2.0031	2.0361
70 (302.93)	Sh						47.07	97.07	147.07	197.07	297.07	397.07	497.07	597.07	697.07	797.07	897.07
	v	0.01748	6.205				6.664	7.133	7.590	8.039	8.922	9.793	10.659	11.522	12.382	13.240	14.097
	h	272.74	1180.6				1206.0	1232.0	1257.3	1282.2	1331.6	1381.0	1430.9	1481.5	1532.9	1585.1	1638.2
	s	0.4411	1.6316				1.6640	1.6951	1.7237	1.7504	1.7993	1.8439	1.8852	1.9238	1.9603	1.9949	2.0279
75 (307.61)	Sh						42.39	92.39	142.39	192.39	292.39	392.39	492.39	592.39	692.39	792.39	892.39
	v	0.01753	5.814				6.204	6.645	7.074	7.494	8.320	9.135	9.945	10.750	11.553	12.355	13.155
	h	277.56	1181.9				1205.0	1231.2	1256.7	1281.7	1331.3	1380.7	1430.7	1481.3	1532.7	1585.0	1638.[illegible]
	s	0.4474	1.6260				1.6554	1.6868	1.7156	1.7424	1.7915	1.8361	1.8774	1.9161	1.9526	1.9872	2.0202

Sh = superheat, F
v = specific volume, cu ft per lb
h = enthalpy, Btu per lb
s = entropy, Btu per F per lb

*Values from STEAM TABLES, Properties of Saturated and Superheated Steam Published by COMBUSTION ENGINEERING, INC., Copyright 1940
**Values interpolated from ASME STEAM TABLES

Table 3. Superheated Steam – *Continued*

Press. /Sq In. . Temp)		Sat. Water	Sat. Steam	Temperature – Degrees Fahrenheit 350	400	450	500	550	600	700	800	900	1000	1100	1200	1300	1400
	Sh			37.96	87.96	137.96	187.96	237.96	287.96	387.96	487.96	587.96	687.96	787.96	887.96	987.96	1087.96
0	v	0.01757	5.471	5.801	6.218	6.622	7.018	7.408	7.794	8.560	9.319	10.075	10.829	11.581	12.331	13.081	13.829
2.04)	h	282.15	1183.1	1204.0	1230.5	1256.1	1281.3	1306.2	1330.9	1380.5	1430.5	1481.1	1532.6	1584.9	1638.0	1692.0	1746.8
	s	0.4534	1.6208	1.6473	1.6790	1.7080	1.7349	1.7602	1.7842	1.8289	1.8702	1.9089	1.9454	1.9800	2.0131	2.0446	2.0750
	Sh			33.74	83.74	133.74	183.74	233.74	283.74	383.74	483.74	583.74	683.74	783.74	883.74	983.74	1083.74
5	v	0.01762	5.167	5.445	5.840	6.223	6.597	6.966	7.330	8.052	8.768	9.480	10.190	10.898	11.604	12.310	13.014
5.26)	h	286.52	1184.2	1203.0	1229.7	1255.5	1280.8	1305.8	1330.6	1380.2	1430.3	1481.0	1532.4	1584.7	1637.9	1691.9	1746.8
	s	0.4590	1.6159	1.6396	1.6716	1.7008	1.7279	1.7532	1.7772	1.8220	1.8634	1.9021	1.9386	1.9733	2.0063	2.0379	2.0682
	Sh			29.72	79.72	129.72	179.72	229.72	279.72	379.72	479.72	579.72	679.72	779.72	879.72	979.72	1079.72
0	v	0.01766	4.895	5.128	5.505	5.869	6.223	6.572	6.917	7.600	8.277	8.950	9.621	10.290	10.958	11.625	12.290
0.28)	h	290.69	1185.3	1202.0	1228.9	1254.9	1280.3	1305.4	1330.2	1380.0	1430.1	1480.8	1532.3	1584.6	1637.8	1691.8	1746.7
	s	0.4643	1.6113	1.6323	1.6646	1.6940	1.7212	1.7467	1.7707	1.8156	1.8570	1.8957	1.9323	1.9669	2.0000	2.0316	2.0619
	Sh			25.87	75.87	125.87	175.87	225.87	275.87	375.87	475.87	575.87	675.87	775.87	875.87	975.87	1075.87
5	v	0.01770	4.651	4.845	5.205	5.551	5.889	6.221	6.548	7.196	7.838	8.477	9.113	9.747	10.380	11.012	11.643
4.13)	h	294.70	1186.2	1200.9	1228.1	1254.3	1279.8	1305.0	1329.9	1379.7	1429.9	1480.6	1532.1	1584.5	1637.7	1691.7	1746.6
	s	0.4694	1.6069	1.6253	1.6580	1.6876	1.7149	1.7404	1.7645	1.8094	1.8509	1.8897	1.9262	1.9609	1.9940	2.0256	2.0559
	Sh			22.18	72.18	122.18	172.18	222.18	272.18	372.18	472.18	572.18	672.18	772.18	872.18	972.18	1072.18
00	v	0.01774	4.431	4.590	4.935	5.266	5.588	5.904	6.216	6.833	7.443	8.050	8.655	9.258	9.860	10.460	11.060
7.82)	h	298.54	1187.2	1199.9	1227.4	1253.7	1279.3	1304.6	1329.6	1379.5	1429.7	1480.4	1532.0	1584.4	1637.6	1691.6	1746.5
	s	0.4743	1.6027	1.6187	1.6516	1.6814	1.7088	1.7344	1.7586	1.8036	1.8451	1.8839	1.9205	1.9552	1.9883	2.0199	2.0502
	Sh			18.63	68.63	118.63	168.63	218.63	268.63	368.63	468.63	568.63	668.63	768.63	868.63	968.63	1068.63
05	v	0.01778	4.231	4.359	4.690	5.007	5.315	5.617	5.915	6.504	7.086	7.665	8.241	8.816	9.389	9.961	10.532
1.37)	h	302.24	1188.0	1198.8	1226.6	1253.1	1278.8	1304.2	1329.2	1379.2	1429.4	1480.3	1531.8	1584.2	1637.5	1691.5	1746.4
	s	0.4790	1.5988	1.6122	1.6455	1.6755	1.7031	1.7288	1.7530	1.7981	1.8396	1.8785	1.9151	1.9498	1.9828	2.0145	2.0448
	Sh			15.21	65.21	115.21	165.21	215.21	265.21	365.21	465.21	565.21	665.21	765.21	865.21	965.21	1065.21
10	v	0.01782	4.048	4.149	4.468	4.772	5.068	5.357	5.642	6.205	6.761	7.314	7.865	8.413	8.961	9.507	10.053
4.79)	h	305.80	1188.9	1197.7	1225.8	1252.5	1278.3	1303.8	1328.9	1379.0	1429.2	1480.1	1531.7	1584.1	1637.4	1691.4	1746.4
	s	0.4834	1.5950	1.6061	1.6396	1.6698	1.6975	1.7233	1.7476	1.7928	1.8344	1.8732	1.9099	1.9446	1.9777	2.0093	2.0397
	Sh			11.92	61.92	111.92	161.92	211.92	261.92	361.92	461.92	561.92	661.92	761.92	861.92	961.92	1061.92
15	v	0.01785	3.881	3.957	4.265	4.558	4.841	5.119	5.392	5.932	6.465	6.994	7.521	8.046	8.570	9.093	9.615
8.08)	h	309.25	1189.6	1196.7	1225.0	1251.8	1277.9	1303.3	1328.6	1378.7	1429.0	1479.9	1531.6	1584.0	1637.2	1691.4	1746.3
	s	0.4877	1.5913	1.6001	1.6340	1.6644	1.6922	1.7181	1.7425	1.7877	1.8294	1.8682	1.9049	1.9396	1.9727	2.0044	2.0347
	Sh			8.73	58.73	108.73	158.73	208.73	258.73	358.73	458.73	558.73	658.73	758.73	858.73	958.73	1058.73
20	v	0.01789	3.7275	3.7815	4.0786	4.3610	4.6341	4.9009	5.1637	5.6813	6.1928	6.7006	7.2060	7.7096	8.2119	8.7130	9.2134
1.27)	h	312.58	1190.4	1195.6	1224.1	1251.2	1277.4	1302.9	1328.2	1378.4	1428.8	1479.8	1531.4	1583.9	1637.1	1691.3	1746.2
	s	0.4919	1.5879	1.5943	1.6286	1.6592	1.6872	1.7132	1.7376	1.7829	1.8246	1.8635	1.9001	1.9349	1.9680	1.9996	2.0300
	Sh			2.67	52.67	102.67	152.67	202.67	252.67	352.67	452.67	552.67	652.67	752.67	852.67	952.67	1052.67
30	v	0.01796	3.4544	3.4699	3.7489	4.0129	4.2672	4.5151	4.7589	5.2384	5.7118	6.1814	6.6486	7.1140	7.5781	8.0411	8.5033
7.33)	h	318.95	1191.7	1193.4	1222.5	1249.9	1276.4	1302.1	1327.5	1377.9	1428.4	1479.4	1531.1	1583.6	1636.9	1691.1	1746.1
	s	0.4998	1.5813	1.5833	1.6182	1.6493	1.6775	1.7037	1.7283	1.7737	1.8155	1.8545	1.8911	1.9259	1.9591	1.9907	2.0211
	Sh				46.96	96.96	146.96	196.96	246.96	346.96	446.96	546.96	646.96	746.96	846.96	946.96	1046.96
40	v	0.01803	3.2190		3.4661	3.7143	3.9526	4.1844	4.4119	4.8588	5.2995	5.7364	6.1709	6.6036	7.0349	7.4652	7.8946
3.04)	h	324.96	1193.0		1220.8	1248.7	1275.3	1301.3	1326.8	1377.4	1428.0	1479.1	1530.8	1583.4	1636.7	1690.9	1745.9
	s	0.5071	1.5752		1.6085	1.6400	1.6686	1.6949	1.7196	1.7652	1.8071	1.8461	1.8828	1.9176	1.9508	1.9825	2.0129
	Sh				41.57	91.57	141.57	191.57	241.57	341.57	441.57	541.57	641.57	741.57	841.57	941.57	1041.57
50	v	0.01809	3.0139		3.2208	3.4555	3.6799	3.8978	4.1112	4.5298	4.9421	5.3507	5.7568	6.1612	6.5642	6.9661	7.3671
8.43)	h	330.65	1194.1		1219.1	1247.4	1274.3	1300.5	1326.1	1376.9	1427.6	1478.7	1530.5	1583.1	1636.5	1690.7	1745.7
	s	0.5141	1.5695		1.5993	1.6313	1.6602	1.6867	1.7115	1.7573	1.7992	1.8383	1.8751	1.9099	1.9431	1.9748	2.0052
	Sh				36.45	86.45	136.45	186.45	236.45	336.45	436.45	536.45	636.45	736.45	836.45	936.45	1036.45
60	v	0.01815	2.8336		3.0060	3.2288	3.4413	3.6469	3.8480	4.2420	4.6295	5.0132	5.3945	5.7741	6.1522	6.5293	6.9055
3.55)	h	336.07	1195.1		1217.4	1246.0	1273.3	1299.6	1325.4	1376.4	1427.2	1478.4	1530.3	1582.9	1636.3	1690.5	1745.6
	s	0.5206	1.5641		1.5906	1.6231	1.6522	1.6790	1.7039	1.7499	1.7919	1.8310	1.8678	1.9027	1.9359	1.9676	1.9980
	Sh				31.58	81.58	131.58	181.58	231.58	331.58	431.58	531.58	631.58	731.58	831.58	931.58	1031.58
70	v	0.01821	2.6738		2.8162	3.0288	3.2306	3.4255	3.6158	3.9879	4.3536	4.7155	5.0749	5.4325	5.7888	6.1440	6.4983
8.42)	h	341.24	1196.0		1215.6	1244.7	1272.2	1298.8	1324.7	1375.8	1426.8	1478.0	1530.0	1582.6	1636.1	1690.4	1745.4
	s	0.5269	1.5591		1.5823	1.6152	1.6447	1.6717	1.6968	1.7428	1.7850	1.8241	1.8610	1.8959	1.9291	1.9608	1.9913
	Sh				26.92	76.92	126.92	176.92	226.92	326.92	426.92	526.92	626.92	726.92	826.92	926.92	1026.92
80	v	0.01827	2.5312		2.6474	2.8508	3.0433	3.2286	3.4093	3.7621	4.1084	4.4508	4.7907	5.1289	5.4657	5.8014	6.1363
3.08)	h	346.19	1196.9		1213.8	1243.4	1271.2	1297.9	1324.0	1375.3	1426.3	1477.7	1529.7	1582.4	1635.9	1690.2	1745.3
	s	0.5328	1.5543		1.5743	1.6078	1.6376	1.6647	1.6900	1.7362	1.7784	1.8176	1.8545	1.8894	1.9227	1.9545	1.9849
	Sh				22.47	72.47	122.47	172.47	222.47	322.47	422.47	522.47	622.47	722.47	822.47	922.47	1022.47
90	v	0.01833	2.4030		2.4961	2.6915	2.8756	3.0525	3.2246	3.5601	3.8889	4.2140	4.5365	4.8572	5.1766	5.4949	5.8124
7.53)	h	350.94	1197.6		1212.0	1242.0	1270.1	1297.1	1323.3	1374.8	1425.9	1477.4	1529.4	1582.1	1635.7	1690.0	1745.1
	s	0.5384	1.5498		1.5667	1.6006	1.6307	1.6581	1.6835	1.7299	1.7722	1.8115	1.8484	1.8834	1.9166	1.9484	1.9789
	Sh				18.20	68.20	118.20	168.20	218.20	318.20	418.20	518.20	618.20	718.20	818.20	918.20	1018.20
00	v	0.01839	2.2873		2.3598	2.5480	2.7247	2.8939	3.0583	3.3783	3.6915	4.0008	4.3077	4.6128	4.9165	5.2191	5.5209
1.80)	h	355.51	1198.3		1210.1	1240.6	1269.0	1296.2	1322.6	1374.3	1425.5	1477.0	1529.1	1581.9	1635.4	1689.8	1745.0
	s	0.5438	1.5454		1.5593	1.5938	1.6242	1.6518	1.6773	1.7239	1.7663	1.8057	1.8426	1.8776	1.9109	1.9427	1.9732

Sh = superheat, F
v = specific volume, cu ft per lb
h = enthalpy, Btu per lb
s = entropy, Btu per F per lb

Table 3. Superheated Steam—*Continued*

Abs Press. Lb/Sq In. (Sat. Temp)		Sat. Water	Sat. Steam	Temperature—Degrees Fahrenheit 400	450	500	550	600	700	800	900	1000	1100	1200	1300	1400	
210 (385.91)	Sh			14.09	64.09	114.09	164.09	214.09	314.09	414.09	514.09	614.09	714.09	814.09	914.09	1014.09	11
	v	0.01844	2.1822	2.2364	2.4181	2.5880	2.7504	2.9078	3.2137	3.5128	3.8080	4.1007	4.3915	4.6811	4.9695	5.2571	5
	h	359.91	1199.0	1208.02	1239.2	1268.0	1295.3	1321.9	1373.7	1425.1	1476.7	1528.8	1581.6	1635.2	1689.6	1744.8	1
	s	0.5490	1.5413	1.5522	1.5872	1.6180	1.6458	1.6715	1.7182	1.7607	1.8001	1.8371	1.8721	1.9054	1.9372	1.9677	1
220 (389.88)	Sh			10.12	60.12	110.12	160.12	210.12	310.12	410.12	510.12	610.12	710.12	810.12	910.12	1010.12	11
	v	0.01850	2.0863	2.1240	2.2999	2.4638	2.6199	2.7710	3.0642	3.3504	3.6327	3.9125	4.1905	4.4671	4.7426	5.0173	5
	h	364.17	1199.6	1206.3	1237.8	1266.9	1294.5	1321.2	1373.2	1424.7	1476.3	1528.5	1581.4	1635.0	1689.4	1744.7	1
	s	0.5540	1.5374	1.5453	1.5808	1.6120	1.6400	1.6658	1.7128	1.7553	1.7948	1.8318	1.8668	1.9002	1.9320	1.9625	1
230 (393.70)	Sh			6.30	56.30	106.30	156.30	206.30	306.30	406.30	506.30	606.30	706.30	806.30	906.30	1006.30	11
	v	0.01855	1.9985	2.0212	2.1919	2.3503	2.5008	2.6461	2.9276	3.2020	3.4726	3.7406	4.0068	4.2717	4.5355	4.7984	5.
	h	368.28	1200.1	1204.4	1236.3	1265.7	1293.6	1320.4	1372.7	1424.2	1476.0	1528.2	1581.1	1634.8	1689.3	1744.5	1
	s	0.5588	1.5336	1.5385	1.5747	1.6062	1.6344	1.6604	1.7075	1.7502	1.7897	1.8268	1.8618	1.8952	1.9270	1.9576	1
240 (397.39)	Sh			2.61	52.61	102.61	152.61	202.61	302.61	402.61	502.61	602.61	702.61	802.61	902.61	1002.61	11
	v	0.01860	1.9177	1.9268	2.0928	2.2462	2.3915	2.5316	2.8024	3.0661	3.3259	3.5831	3.8385	4.0926	4.3456	4.5977	4
	h	372.27	1200.6	1202.4	1234.9	1264.6	1292.7	1319.7	1372.1	1423.8	1475.6	1527.9	1580.9	1634.6	1689.1	1744.3	1
	s	0.5634	1.5299	1.5320	1.5687	1.6006	1.6291	1.6552	1.7025	1.7452	1.7848	1.8219	1.8570	1.8904	1.9223	1.9528	1
250 (400.97)	Sh				49.03	99.03	149.03	199.03	299.03	399.03	499.03	599.03	699.03	799.03	899.03	999.03	10
	v	0.01865	1.8432		2.0016	2.1504	2.2909	2.4262	2.6872	2.9410	3.1909	3.4382	3.6837	3.9278	4.1709	4.4131	4
	h	376.14	1201.1		1233.4	1263.5	1291.8	1319.0	1371.6	1423.4	1475.3	1527.6	1580.6	1634.4	1688.9	1744.2	1
	s	0.5679	1.5264		1.5629	1.5951	1.6239	1.6502	1.6976	1.7405	1.7801	1.8173	1.8524	1.8858	1.9177	1.9482	1
260 (404.44)	Sh				45.56	95.56	145.56	195.56	295.56	395.56	495.56	595.56	695.56	795.56	895.56	995.56	10
	v	0.01870	1.7742		1.9173	2.0619	2.1981	2.3289	2.5808	2.8256	3.0663	3.3044	3.5408	3.7758	4.0097	4.2427	4
	h	379.90	1201.5		1231.9	1262.4	1290.9	1318.2	1371.1	1423.0	1474.9	1527.3	1580.4	1634.2	1688.7	1744.0	1
	s	0.5722	1.5230		1.5573	1.5899	1.6189	1.6453	1.6930	1.7359	1.7756	1.8128	1.8480	1.8814	1.9133	1.9439	1
270 (407.80)	Sh				42.20	92.20	142.20	192.20	292.20	392.20	492.20	592.20	692.20	792.20	892.20	992.20	10
	v	0.01875	1.7101		1.8391	1.9799	2.1121	2.2388	2.4824	2.7186	2.9509	3.1806	3.4084	3.6349	3.8603	4.0849	4
	h	383.56	1201.9		1230.4	1261.2	1290.0	1317.5	1370.5	1422.6	1474.6	1527.1	1580.1	1634.0	1688.5	1743.9	1
	s	0.5764	1.5197		1.5518	1.5848	1.6140	1.6406	1.6885	1.7315	1.7713	1.8085	1.8437	1.8771	1.9090	1.9396	1
280 (411.07)	Sh				38.93	88.93	138.93	188.93	288.93	388.93	488.93	588.93	688.93	788.93	888.93	988.93	10
	v	0.01880	1.6505		1.7665	1.9037	2.0322	2.1551	2.3909	2.6194	2.8437	3.0655	3.2855	3.5042	3.7217	3.9384	4
	h	387.12	1202.3		1228.8	1260.0	1289.1	1316.8	1370.0	1422.1	1474.2	1526.8	1579.9	1633.8	1688.4	1743.7	1
	s	0.5805	1.5166		1.5464	1.5798	1.6093	1.6361	1.6841	1.7273	1.7671	1.8043	1.8395	1.8730	1.9050	1.9356	1
290 (414.25)	Sh				35.75	85.75	135.75	185.75	285.75	385.75	485.75	585.75	685.75	785.75	885.75	985.75	10
	v	0.01885	1.5948		1.6988	1.8327	1.9578	2.0772	2.3058	2.5269	2.7440	2.9585	3.1711	3.3824	3.5926	3.8019	4
	h	390.60	1202.6		1227.3	1258.9	1288.1	1316.0	1369.5	1421.7	1473.9	1526.5	1579.6	1633.5	1688.2	1743.6	1
	s	0.5844	1.5135		1.5412	1.5750	1.6048	1.6317	1.6799	1.7232	1.7630	1.8003	1.8356	1.8690	1.9010	1.9316	1
300 (417.35)	Sh				32.65	82.65	132.65	182.65	282.65	382.65	482.65	582.65	682.65	782.65	882.65	982.65	10
	v	0.01889	1.5427		1.6356	1.7665	1.8883	2.0044	2.2263	2.4407	2.6509	2.8585	3.0643	3.2688	3.4721	3.6746	3
	h	393.99	1202.9		1225.7	1257.7	1287.2	1315.2	1368.9	1421.3	1473.6	1526.2	1579.4	1633.3	1688.0	1743.4	1
	s	0.5882	1.5105		1.5361	1.5703	1.6003	1.6274	1.6758	1.7192	1.7591	1.7964	1.8317	1.8652	1.8972	1.9278	1
310 (420.36)	Sh				29.64	79.64	129.64	179.64	279.64	379.64	479.64	579.64	679.64	779.64	879.64	979.64	10
	v	0.01894	1.4939		1.5763	1.7044	1.8233	1.9363	2.1520	2.3600	2.5638	2.7650	2.9644	3.1625	3.3594	3.5555	3.
	h	397.30	1203.2		1224.1	1256.5	1286.3	1314.5	1368.4	1420.9	1473.2	1525.9	1579.2	1633.1	1687.8	1743.3	1
	s	0.5920	1.5076		1.5311	1.5657	1.5960	1.6233	1.6719	1.7153	1.7553	1.7927	1.8280	1.8615	1.8935	1.9241	1.
320 (423.31)	Sh				26.69	76.69	126.69	176.69	276.69	376.69	476.69	576.69	676.69	776.69	876.69	976.69	10
	v	0.01899	1.4480		1.5207	1.6462	1.7623	1.8725	2.0823	2.2843	2.4821	2.6774	2.8708	3.0628	3.2538	3.4438	3
	h	400.53	1203.4		1222.5	1255.2	1285.3	1313.7	1367.8	1420.5	1472.9	1525.6	1578.9	1632.9	1687.6	1743.1	1
	s	0.5956	1.5048		1.5261	1.5612	1.5918	1.6192	1.6680	1.7116	1.7516	1.7890	1.8243	1.8579	1.8899	1.9206	1
330 (426.18)	Sh				23.82	73.82	123.82	173.82	273.82	373.82	473.82	573.82	673.82	773.82	873.82	973.82	10
	v	0.01903	1.4048		1.4684	1.5915	1.7050	1.8125	2.0168	2.2132	2.4054	2.5950	2.7828	2.9692	3.1545	3.3389	3.
	h	403.70	1203.6		1220.9	1254.0	1284.4	1313.0	1367.3	1420.0	1472.5	1525.3	1578.7	1632.7	1687.5	1742.9	1
	s	0.5991	1.5021		1.5213	1.5568	1.5876	1.6153	1.6643	1.7079	1.7480	1.7855	1.8208	1.8544	1.8864	1.9171	1
340 (428.99)	Sh				21.01	71.01	121.01	171.01	271.01	371.01	471.01	571.01	671.01	771.01	871.01	971.01	10
	v	0.01908	1.3640		1.4191	1.5399	1.6511	1.7561	1.9552	2.1463	2.3333	2.5175	2.7000	2.8811	3.0611	3.2402	3.
	h	406.80	1203.8		1219.2	1252.8	1283.4	1312.2	1366.7	1419.6	1472.2	1525.0	1578.4	1632.5	1687.3	1742.8	1.
	s	0.6026	1.4994		1.5165	1.5525	1.5836	1.6114	1.6606	1.7044	1.7445	1.7820	1.8174	1.8510	1.8831	1.9138	1.
350 (431.73)	Sh				18.27	68.27	118.27	168.27	268.27	368.27	468.27	568.27	668.27	768.27	868.27	968.27	10
	v	0.01912	1.3255		1.3725	1.4913	1.6002	1.7028	1.8970	2.0832	2.2652	2.4445	2.6219	2.7980	2.9730	3.1471	3.
	h	409.83	1204.0		1217.5	1251.5	1282.4	1311.4	1366.2	1419.2	1471.8	1524.7	1578.2	1632.3	1687.1	1742.6	1
	s	0.6059	1.4968		1.5119	1.5483	1.5797	1.6077	1.6571	1.7009	1.7411	1.7787	1.8141	1.8477	1.8798	1.9105	1.
360 (434.41)	Sh				15.59	65.59	115.59	165.59	265.59	365.59	465.59	565.59	665.59	765.59	865.59	965.59	10
	v	0.01917	1.2891		1.3285	1.4454	1.5521	1.6525	1.8421	2.0237	2.2009	2.3755	2.5482	2.7196	2.8898	3.0592	3.
	h	412.81	1204.1		1215.8	1250.3	1281.5	1310.6	1365.6	1418.7	1471.5	1524.4	1577.9	1632.1	1686.9	1742.5	1
	s	0.6092	1.4943		1.5073	1.5441	1.5758	1.6040	1.6536	1.6976	1.7379	1.7754	1.8109	1.8445	1.8766	1.9073	1
380 (439.61)	Sh				10.39	60.39	110.39	160.39	260.39	360.39	460.39	560.39	660.39	760.39	860.39	960.39	10
	v	0.01925	1.2218		1.2472	1.3606	1.4635	1.5598	1.7410	1.9139	2.0825	2.2484	2.4124	2.5750	2.7366	2.8973	3.
	h	418.59	1204.4		1212.4	1247.7	1279.5	1309.0	1364.5	1417.9	1470.8	1523.8	1577.4	1631.6	1686.5	1742.2	1
	s	0.6156	1.4894		1.4982	1.5360	1.5683	1.5969	1.6470	1.6911	1.7315	1.7692	1.8047	1.8384	1.8705	1.9012	1

Sh = superheat, F
v = specific volume, cu ft per lb
h = enthalpy, Btu per lb
s = entropy, Btu per F per lb

Table 3. Superheated Steam—*Continued*

bs Press. b/Sq In. at. Temp)		Sat. Water	Sat. Steam	Temperature—Degrees Fahrenheit 450	500	550	600	650	700	800	900	1000	1100	1200	1300	1400	1500
	Sh			5.40	55.40	105.40	155.40	205.40	255.40	355.40	455.40	555.40	655.40	755.40	855.40	955.40	1055.40
400	v	0.01934	116.10	1.1738	1.2841	1.3836	1.4763	1.5646	1.6499	1.8151	1.9759	2.1339	2.2901	2.4450	2.5987	2.7515	2.9037
44.60)	h	424.17	1204.6	1208.8	1245.1	1277.5	1307.4	1335.9	1363.4	1417.0	1470.1	1523.3	1576.9	1631.2	1686.2	1741.9	1798.2
	s	0.6217	1.4847	1.4894	1.5282	1.5611	1.5901	1.6163	1.6406	1.6850	1.7255	1.7632	1.7988	1.8325	1.8647	1.8955	1.9250
	Sh			.60	50.60	100.60	150.60	200.60	250.60	350.60	450.60	550.60	650.60	750.60	850.60	950.60	1050.60
420	v	0.01942	1.1057	1.1071	1.2148	1.3113	1.4007	1.4856	1.5676	1.7258	1.8795	2.0304	2.1795	2.3273	2.4739	2.6196	2.7647
49.40)	h	429.56	1204.7	1205.2	1242.4	1275.4	1305.8	1334.5	1362.3	1416.2	1469.4	1522.7	1576.4	1630.8	1685.8	1741.6	1798.0
	s	0.6276	1.4802	1.4808	1.5206	1.5542	1.5835	1.6100	1.6345	1.6791	1.7197	1.7575	1.7932	1.8269	1.8591	1.8899	1.9195
	Sh				45.97	95.97	145.97	195.97	245.97	345.97	445.97	545.97	645.97	745.97	845.97	945.97	1045.97
440	v	0.01950	1.0554		1.1517	1.2454	1.3319	1.4138	1.4926	1.6445	1.7918	1.9363	2.0790	2.2203	2.3605	2.4998	2.6384
54.03)	h	434.77	1204.8		1239.7	1273.4	1304.2	1333.2	1361.1	1415.3	1468.7	1522.1	1575.9	1630.4	1685.5	1741.2	1797.7
	s	0.6332	1.4759		1.5132	1.5474	1.5772	1.6040	1.6286	1.6734	1.7142	1.7521	1.7878	1.8216	1.8538	1.8847	1.9143
	Sh				41.50	91.50	141.50	191.50	241.50	341.50	441.50	541.50	641.50	741.50	841.50	941.50	1041.50
460	v	0.01959	1.0092		1.0939	1.1852	1.2691	1.3482	1.4242	1.5703	1.7117	1.8504	1.9872	2.1226	2.2569	2.3903	2.5230
58.50)	h	439.83	1204.8		1236.9	1271.3	1302.5	1331.8	1360.0	1414.4	1468.0	1521.5	1575.4	1629.9	1685.1	1740.9	1797.4
	s	0.6387	1.4718		1.5060	1.5409	1.5711	1.5982	1.6230	1.6680	1.7089	1.7469	1.7826	1.8165	1.8488	1.8797	1.9093
	Sh				37.18	87.18	137.18	187.18	237.18	337.18	437.18	537.18	637.18	737.18	837.18	937.18	1037.18
480	v	0.01967	0.9668		1.0409	1.1300	1.2115	1.2881	1.3615	1.5023	1.6384	1.7716	1.9030	2.0330	2.1619	2.2900	2.4173
62.82)	h	444.75	1204.8		1234.1	1269.1	1300.8	1330.5	1358.8	1413.6	1467.3	1520.9	1574.9	1629.5	1684.7	1740.6	1797.2
	s	0.6439	1.4677		1.4990	1.5346	1.5652	1.5925	1.6176	1.6628	1.7038	1.7419	1.7777	1.8116	1.8439	1.8748	1.9045
	Sh				32.99	82.99	132.99	182.99	232.99	332.99	432.99	532.99	632.99	732.99	832.99	932.99	1032.99
500	v	0.01975	0.9276		0.9919	1.0791	1.1584	1.2327	1.3037	1.4397	1.5708	1.6992	1.8256	1.9507	2.0746	2.1977	2.3200
67.01)	h	449.52	1204.7		1231.2	1267.0	1299.1	1329.1	1357.7	1412.7	1466.6	1520.3	1574.4	1629.1	1684.4	1740.3	1796.9
	s	0.6490	1.4639		1.4921	1.5284	1.5595	1.5871	1.6123	1.6578	1.6990	1.7371	1.7730	1.8069	1.8393	1.8702	1.8998
	Sh				28.93	78.93	128.93	178.93	228.93	328.93	428.93	528.93	628.93	728.93	828.93	928.93	1028.93
520	v	0.01982	0.8914		0.9466	1.0321	1.1094	1.1816	1.2504	1.3819	1.5085	1.6323	1.7542	1.8746	1.9940	2.1125	2.2302
71.07)	h	454.18	1204.5		1228.3	1264.8	1297.4	1327.7	1356.5	1411.8	1465.9	1519.7	1573.9	1628.7	1684.0	1740.0	1796.7
	s	0.6540	1.4601		1.4853	1.5223	1.5539	1.5818	1.6072	1.6530	1.6943	1.7325	1.7684	1.8024	1.8348	1.8657	1.8954
	Sh				24.99	74.99	124.99	174.99	224.99	324.99	424.99	524.99	624.99	724.99	824.99	924.99	1024.99
540	v	0.01990	0.8577		0.9045	0.9884	1.0640	1.1342	1.2010	1.3284	1.4508	1.5704	1.6880	1.8042	1.9193	2.0336	2.1471
75.01)	h	458.71	1204.4		1225.3	1262.5	1295.7	1326.3	1355.3	1410.9	1465.1	1519.1	1573.4	1628.2	1683.6	1739.7	1796.4
	s	0.6587	1.4565		1.4786	1.5164	1.5485	1.5767	1.6023	1.6483	1.6897	1.7280	1.7640	1.7981	1.8305	1.8615	1.8911
	Sh				21.16	71.16	121.16	171.16	221.16	321.16	421.16	521.16	621.16	721.16	821.16	921.16	1021.16
560	v	0.01998	0.8264		0.8653	0.9479	1.0217	1.0902	1.1552	1.2787	1.3972	1.5129	1.6266	1.7388	1.8500	1.9603	2.0699
478.84)	h	463.14	1204.2		1222.2	1260.3	1293.9	1324.9	1354.2	1410.0	1464.4	1518.6	1572.9	1627.8	1683.3	1739.4	1796.1
	s	0.6634	1.4529		1.4720	1.5106	1.5431	1.5717	1.5975	1.6438	1.6853	1.7237	1.7598	1.7939	1.8263	1.8573	1.8870
	Sh				17.43	67.43	117.43	167.43	217.43	317.43	417.43	517.43	617.43	717.43	817.43	917.43	1017.43
580	v	0.02006	0.7971		0.8287	0.9100	0.9824	1.0492	1.1125	1.2324	1.3473	1.4593	1.5693	1.6780	1.7855	1.8921	1.9980
482.57)	h	467.47	1203.9		1219.1	1258.0	1292.1	1323.4	1353.0	1409.2	1463.7	1518.0	1572.4	1627.4	1682.9	1739.1	1795.9
	s	0.6679	1.4495		1.4654	1.5049	1.5380	1.5668	1.5929	1.6394	1.6811	1.7196	1.7556	1.7898	1.8223	1.8533	1.8831
	Sh				13.80	63.80	113.80	163.80	213.80	313.80	413.80	513.80	613.80	713.80	813.80	913.80	1013.80
600	v	0.02013	0.7697		0.7944	0.8746	0.9456	1.0109	1.0726	1.1892	1.3008	1.4093	1.5160	1.6211	1.7252	1.8284	1.9309
486.20)	h	471.70	1203.7		1215.9	1255.6	1290.3	1322.0	1351.8	1408.3	1463.0	1517.4	1571.9	1627.0	1682.6	1738.8	1795.6
	s	0.6723	1.4461		1.4590	1.4993	1.5329	1.5621	1.5884	1.6351	1.6769	1.7155	1.7517	1.7859	1.8184	1.8494	1.8792
	Sh				5.11	55.11	105.11	155.11	205.11	305.11	405.11	505.11	605.11	705.11	805.11	905.11	1005.11
650	v	0.02032	0.7084		0.7173	0.7954	0.8634	0.9254	0.9835	1.0929	1.1969	1.2979	1.3969	1.4944	1.5909	1.6864	1.7813
494.89)	h	481.89	1202.8		1207.6	1249.6	1285.7	1318.3	1348.7	1406.0	1461.2	1515.9	1570.7	1625.9	1681.6	1738.0	1794.9
	s	1.6828	1.4381		1.4430	1.4858	1.5207	1.5507	1.5775	1.6249	1.6671	1.7059	1.7422	1.7765	1.8092	1.8403	1.8701
	Sh					46.92	96.92	146.92	196.92	296.92	396.92	496.92	596.92	696.92	796.92	896.92	996.92
700	v	0.02050	0.6556			0.7271	0.7928	0.8520	0.9072	1.0102	1.1078	1.2023	1.2948	1.3858	1.4757	1.5647	1.6530
503.08)	h	491.60	1201.8			1243.4	1281.0	1314.6	1345.6	1403.7	1459.4	1514.4	1569.4	1624.8	1680.7	1737.2	1794.3
	s	0.6928	1.4304			1.4726	1.5090	1.5399	1.5673	1.6154	1.6580	1.6970	1.7335	1.7679	1.8006	1.8318	1.8617
	Sh					39.16	89.16	139.16	189.16	289.16	389.16	489.16	589.16	689.16	789.16	889.16	989.16
750	v	0.02069	0.6095			0.6676	0.7313	0.7882	0.8409	0.9386	1.0306	1.1195	1.2063	1.2916	1.3759	1.4592	1.5419
510.84)	h	500.89	1200.7			1236.9	1276.1	1310.7	1342.5	1401.5	1457.6	1512.9	1568.2	1623.8	1679.8	1736.4	1793.6
	s	0.7022	1.4232			1.4598	1.4977	1.5296	1.5577	1.6065	1.6494	1.6886	1.7252	1.7598	1.7926	1.8239	1.8538
	Sh					31.79	81.79	131.79	181.79	281.79	381.79	481.79	581.79	681.79	781.79	881.79	981.79
800	v	0.02087	0.5690			0.6151	0.6774	0.7323	0.7828	0.8759	0.9631	1.0470	1.1289	1.2093	1.2885	1.3669	1.4446
518.21)	h	509.81	1199.4			1230.1	1271.1	1306.8	1339.3	1399.1	1455.8	1511.4	1566.9	1622.7	1678.9	1735.7	1792.9
	s	0.7111	1.4163			1.4472	1.4869	1.5198	1.5484	1.5980	1.6413	1.6807	1.7175	1.7522	1.7851	1.8164	1.8464
	Sh					24.76	74.76	124.76	174.76	274.76	374.76	474.76	574.76	674.76	774.76	874.76	974.76
850	v	0.02105	0.5330			0.5683	0.6296	0.6829	0.7315	0.8205	0.9034	0.9830	1.0606	1.1366	1.2115	1.2855	1.3588
525.24)	h	518.40	1198.0			1223.0	1265.9	1302.8	1336.0	1396.8	1454.0	1510.0	1565.7	1621.6	1678.0	1734.9	1792.3
	s	0.7197	1.4096			1.4347	1.4763	1.5102	1.5396	1.5899	1.6336	1.6733	1.7102	1.7450	1.7780	1.8094	1.8395
	Sh					18.05	68.05	118.05	168.05	268.05	368.05	468.05	568.05	668.05	768.05	868.05	968.05
900	v	0.02123	0.5009			0.5263	0.5869	0.6388	0.6858	0.7713	0.8504	0.9262	0.9998	1.0720	1.1430	1.2131	1.2825
531.95)	h	526.70	1196.4			1215.5	1260.6	1298.6	1332.7	1394.4	1452.2	1508.5	1564.4	1620.6	1677.1	1734.1	1791.6
	s	0.7279	1.4032			1.4223	1.4659	1.5010	1.5311	1.5822	1.6263	1.6662	1.7033	1.7382	1.7713	1.8028	1.8329

Sh = superheat, F
v = specific volume, cu ft per lb
h = enthalpy, Btu per lb
s = entropy, Btu per F per lb

Table 3. Superheated Steam—*Continued*

Abs Press. Lb/Sq In. (Sat. Temp)		Sat. Water	Sat. Steam	Temperature—Degrees Fahrenheit 550	600	650	700	750	800	850	900	1000	1100	1200	1300	1400	150
950 (538.39)	Sh			11.61	61.61	111.61	161.61	211.61	261.61	311.61	361.61	461.61	561.61	661.61	761.61	861.61	961.
	v	0.02141	0.4721	0.4883	0.5485	0.5993	0.6449	0.6871	0.7272	0.7656	0.8030	0.8753	0.9455	1.0142	1.0817	1.1484	1.21
	h	534.74	1194.7	1207.6	1255.1	1294.4	1329.3	1361.5	1392.0	1421.5	1450.3	1507.0	1563.2	1619.5	1676.2	1733.3	1791
	s	0.7358	1.3970	1.4098	1.4557	1.4921	1.5228	1.5500	1.5748	1.5977	1.6193	1.6595	1.6967	1.7317	1.7649	1.7965	1.82
1000 (544.58)	Sh			5.42	55.42	105.42	155.42	205.42	255.42	305.42	355.42	455.42	555.42	655.42	755.42	855.42	955.
	v	0.02159	0.4460	0.4535	0.5137	0.5636	0.6080	0.6489	0.6875	0.7245	0.7603	0.8295	0.8966	0.9622	1.0266	1.0901	1.15
	h	542.55	1192.9	1199.3	1249.3	1290.1	1325.9	1358.7	1389.6	1419.4	1448.5	1505.4	1561.9	1618.4	1675.3	1732.5	1790
	s	0.7434	1.3910	1.3973	1.4457	1.4833	1.5149	1.5426	1.5677	1.5908	1.6126	1.6530	1.6905	1.7256	1.7589	1.7905	1.82
1050 (550.53)	Sh				49.47	99.47	149.47	199.47	249.47	299.47	349.47	449.47	549.47	649.47	749.47	849.47	949.
	v	0.02177	0.4222		0.4821	0.5312	0.5745	0.6142	0.6515	0.6872	0.7216	0.7881	0.8524	0.9151	0.9767	1.0373	1.09
	h	550.15	1191.0		1243.4	1285.7	1322.4	1355.8	1387.2	1417.3	1446.6	1503.9	1560.7	1617.4	1674.4	1731.8	1789
	s	0.7507	1.3851		1.4358	1.4748	1.5072	1.5354	1.5608	1.5842	1.6062	1.6469	1.6845	1.7197	1.7531	1.7848	1.81
1100 (556.28)	Sh				43.72	93.72	143.72	193.72	243.72	293.72	343.72	443.72	543.72	643.72	743.72	843.72	943.
	v	0.02195	0.4006		0.4531	0.5017	0.5440	0.5826	0.6188	0.6533	0.6865	0.7505	0.8121	0.8723	0.9313	0.9894	1.04
	h	557.55	1189.1		1237.3	1281.2	1318.8	1352.9	1384.7	1415.2	1444.7	1502.4	1559.4	1616.3	1673.5	1731.0	1789
	s	0.7578	1.3794		1.4259	1.4664	1.4996	1.5284	1.5542	1.5779	1.6000	1.6410	1.6787	1.7141	1.7475	1.7793	1.80
1150 (561.82)	Sh				39.18	89.18	139.18	189.18	239.18	289.18	339.18	439.18	539.18	639.18	739.18	839.18	939.
	v	0.02214	0.3807		0.4263	0.4746	0.5162	0.5538	0.5889	0.6223	0.6544	0.7161	0.7754	0.8332	0.8899	0.9456	1.00
	h	564.78	1187.0		1230.9	1276.6	1315.2	1349.9	1382.2	1413.0	1442.8	1500.9	1558.1	1615.2	1672.6	1730.2	1788
	s	0.7647	1.3738		1.4160	1.4582	1.4923	1.5216	1.5478	1.5717	1.5941	1.6353	1.6732	1.7087	1.7422	1.7741	1.80
1200 (567.19)	Sh				32.81	82.81	132.81	182.81	232.81	282.81	332.81	432.81	532.81	632.81	732.81	832.81	932.
	v	0.02232	0.3624		0.4016	0.4497	0.4905	0.5273	0.5615	0.5939	0.6250	0.6845	0.7418	0.7974	0.8519	0.9055	0.95
	h	571.85	1184.8		1224.2	1271.8	1311.5	1346.9	1379.7	1410.8	1440.9	1499.4	1556.9	1614.2	1671.6	1729.4	1787
	s	0.7714	1.3683		1.4061	1.4501	1.4851	1.5150	1.5415	1.5658	1.5883	1.6298	1.6679	1.7035	1.7371	1.7691	1.79
1300 (577.42)	Sh				22.58	72.58	122.58	172.58	222.58	272.58	322.58	422.58	522.58	622.58	722.58	822.58	922.
	v	0.02269	0.3299		0.3570	0.4052	0.4451	0.4804	0.5129	0.5436	0.5729	0.6287	0.6822	0.7341	0.7847	0.8345	0.88
	h	585.58	1180.2		1209.9	1261.9	1303.9	1340.8	1374.6	1406.4	1437.1	1496.3	1554.3	1612.0	1669.8	1727.9	1786
	s	0.7843	1.3577		1.3860	1.4340	1.4711	1.5022	1.5296	1.5544	1.5773	1.6194	1.6578	1.6937	1.7275	1.7596	1.79
1400 (587.07)	Sh				12.93	62.93	112.93	162.93	212.93	262.93	312.93	412.93	512.93	612.93	712.93	812.93	912.9
	v	0.02307	0.3018		0.3176	0.3667	0.4059	0.4400	0.4712	0.5004	0.5282	0.5809	0.6311	0.6798	0.7272	0.7737	0.819
	h	598.83	1175.3		1194.1	1251.4	1296.1	1334.5	1369.3	1402.0	1433.2	1493.2	1551.8	1609.9	1668.0	1726.3	1785
	s	0.7966	1.3474		1.3652	1.4181	1.4575	1.4900	1.5182	1.5436	1.5670	1.6096	1.6484	1.6845	1.7185	1.7508	1.781
1500 (596.20)	Sh				3.80	53.80	103.80	153.80	203.80	253.80	303.80	403.80	503.80	603.80	703.80	803.80	903.
	v	0.02346	0.2772		0.2820	0.3328	0.3717	0.4049	0.4350	0.4629	0.4894	0.5394	0.5869	0.6327	0.6773	0.7210	0.76
	h	611.68	1170.1		1176.3	1240.2	1287.9	1328.0	1364.0	1397.4	1429.2	1490.1	1549.2	1607.7	1666.2	1724.8	1783
	s	0.8085	1.3373		1.3431	1.4022	1.4443	1.4782	1.5073	1.5333	1.5572	1.6004	1.6395	1.6759	1.7101	1.7425	1.77
1600 (604.87)	Sh					45.13	95.13	145.13	195.13	245.13	295.13	395.13	495.13	595.13	695.13	795.13	895.1
	v	0.02387	0.2555			0.3026	0.3415	0.3741	0.4032	0.4301	0.4555	0.5031	0.5482	0.5915	0.6336	0.6748	0.71
	h	624.20	1164.5			1228.3	1279.4	1321.4	1358.5	1392.8	1425.2	1486.9	1546.6	1605.6	1664.3	1723.2	1782
	s	0.8199	1.3274			1.3861	1.4312	1.4667	1.4968	1.5235	1.5478	1.5916	1.6312	1.6678	1.7022	1.7347	1.76
1700 (613.13)	Sh					36.87	86.87	136.87	186.87	236.87	286.87	386.87	486.87	586.87	686.87	786.87	886.8
	v	0.02428	0.2361			0.2754	0.3147	0.3468	0.3751	0.4011	0.4255	0.4711	0.5140	0.5552	0.5951	0.6341	0.672
	h	636.45	1158.6			1215.3	1270.5	1314.5	1352.9	1388.1	1421.2	1483.8	1544.0	1603.4	1662.5	1721.7	1781.
	s	0.8309	1.3176			1.3697	1.4183	1.4555	1.4867	1.5140	1.5388	1.5833	1.6232	1.6601	1.6947	1.7274	1.758
1800 (621.02)	Sh					28.98	78.98	128.98	178.98	228.98	278.98	378.98	478.98	578.98	678.98	778.98	878.9
	v	0.02472	0.2186			0.2505	0.2906	0.3223	0.3500	0.3752	0.3988	0.4426	0.4836	0.5229	0.5609	0.5980	0.634
	h	648.49	1152.3			1201.2	1261.1	1307.4	1347.2	1383.3	1417.1	1480.6	1541.4	1601.2	1660.7	1720.1	1779.
	s	0.8417	1.3079			1.3526	1.4054	1.4446	1.4768	1.5049	1.5302	1.5753	1.6156	1.6528	1.6876	1.7204	1.751
1900 (628.56)	Sh					21.44	71.44	121.44	171.44	221.44	271.44	371.44	471.44	571.44	671.44	771.44	871.4
	v	0.02517	0.2028			0.2274	0.2687	0.3004	0.3275	0.3521	0.3749	0.4171	0.4565	0.4940	0.5303	0.5656	0.600
	h	660.36	1145.6			1185.7	1251.3	1300.2	1341.4	1378.4	1412.9	1477.4	1538.8	1599.1	1658.8	1718.6	1778.
	s	0.8522	1.2981			1.3346	1.3925	1.4338	1.4672	1.4960	1.5219	1.5677	1.6084	1.6458	1.6808	1.7138	1.745
2000 (635.80)	Sh					14.20	64.20	114.20	164.20	214.20	264.20	364.20	464.20	564.20	664.20	764.20	864.2
	v	0.02565	0.1883			0.2056	0.2488	0.2805	0.3072	0.3312	0.3534	0.3942	0.4320	0.4680	0.5027	0.5365	0.569
	h	672.11	1138.3			1168.3	1240.9	1292.6	1335.4	1373.5	1408.7	1474.1	1536.2	1596.9	1657.0	1717.0	1771.
	s	0.8625	1.2881			1.3154	1.3794	1.4231	1.4578	1.4874	1.5138	1.5603	1.6014	1.6391	1.6743	1.7075	1.738
2100 (642.76)	Sh					7.24	57.24	107.24	157.24	207.24	257.24	357.24	457.24	557.24	657.24	757.24	857.2
	v	0.02615	0.1750			0.1847	0.2304	0.2624	0.2888	0.3123	0.3339	0.3734	0.4099	0.4445	0.4778	0.5101	0.54
	h	683.79	1130.5			1148.5	1229.8	1284.9	1329.3	1368.4	1404.4	1470.9	1533.6	1594.7	1655.2	1715.4	1775.
	s	0.8727	1.2780			1.2942	1.3661	1.4125	1.4486	1.4790	1.5060	1.5532	1.5948	1.6327	1.6681	1.7014	1.733
2200 (649.45)	Sh					.55	50.55	100.55	150.55	200.55	250.55	350.55	450.55	550.55	650.55	750.55	850.5
	v	0.02669	0.1627			0.1636	0.2134	0.2458	0.2720	0.2950	0.3161	0.3545	0.3897	0.4231	0.4551	0.4862	0.516
	h	695.46	1122.2			1123.9	1218.0	1276.8	1323.1	1363.3	1400.0	1467.6	1530.9	1592.5	1653.3	1713.9	1774.
	s	0.8828	1.2676			1.2691	1.3523	1.4020	1.4395	1.4708	1.4984	1.5463	1.5883	1.6266	1.6622	1.6956	1.727
2300 (655.89)	Sh						44.11	94.11	144.11	194.11	244.11	344.11	444.11	544.11	644.11	744.11	844.1
	v	0.02727	0.1513				0.1975	0.2305	0.2566	0.2793	0.2999	0.3372	0.3714	0.4035	0.4344	0.4643	0.493
	h	707.18	1113.2				1205.3	1268.4	1316.7	1358.1	1395.7	1464.2	1528.3	1590.3	1651.5	1712.3	1773.
	s	0.8929	1.2569				1.3381	1.3914	1.4305	1.4628	1.4910	1.5397	1.5821	1.6207	1.6565	1.6901	1.72

Sh = superheat, F
v = specific volume, cu ft per lb
h = enthalpy, Btu per lb
s = entropy, Btu per F per lb

Table 3. Superheated Steam—*Continued*

Press. /Sq In. . Temp)		Sat. Water	Sat. Steam	700	750	800	850	900	950	1000	1050	1100	1150	1200	1300	1400	1500
	Sh			37.89	87.89	137.89	187.89	237.89	287.89	337.89	387.89	437.89	487.89	537.89	637.89	737.89	837.89
400	v	0.02790	0.1408	0.1824	0.2164	0.2424	0.2648	0.2850	0.3037	0.3214	0.3382	0.3545	0.3703	0.3856	0.4155	0.4443	0.4724
2.11)	h	718.95	1103.7	1191.6	1259.7	1310.1	1352.8	1391.2	1426.9	1460.9	1493.7	1525.6	1557.0	1588.1	1649.6	1710.8	1771.8
	s	0.9031	1.2460	1.3232	1.3808	1.4217	1.4549	1.4837	1.5095	1.5332	1.5553	1.5761	1.5959	1.6149	1.6509	1.6847	1.7167
	Sh			31.89	81.89	131.89	181.89	231.89	281.89	331.89	381.89	431.89	481.89	531.89	631.89	731.89	831.89
500	v	0.02859	0.1307	0.1681	0.2032	0.2293	0.2514	0.2712	0.2896	0.3068	0.3232	0.3390	0.3543	0.3692	0.3980	0.4259	0.4529
8.11)	h	731.71	1093.3	1176.7	1250.6	1303.4	1347.4	1386.7	1423.1	1457.5	1490.7	1522.9	1554.6	1585.9	1647.8	1709.2	1770.4
	s	0.9139	1.2345	1.3076	1.3701	1.4129	1.4472	1.4766	1.5029	1.5269	1.5492	1.5703	1.5903	1.6094	1.6456	1.6796	1.7116
	Sh			26.09	76.09	126.09	176.09	226.09	276.09	326.09	376.09	426.09	476.09	526.09	626.09	726.09	826.09
600	v	0.02938	0.1211	0.1544	0.1909	0.2171	0.2390	0.2585	0.2765	0.2933	0.3093	0.3247	0.3395	0.3540	0.3819	0.4088	0.4350
3.91)	h	744.47	1082.0	1160.2	1241.1	1296.5	1341.9	1382.1	1419.2	1454.1	1487.7	1520.2	1552.2	1583.7	1646.0	1707.7	1769.1
	s	0.9247	1.2225	1.2908	1.3592	1.4042	1.4395	1.4696	1.4964	1.5208	1.5434	1.5646	1.5848	1.6040	1.6405	1.6746	1.7068
	Sh			20.47	70.47	120.47	170.47	220.47	270.47	320.47	370.47	420.47	470.47	520.47	620.47	720.47	820.47
700	v	0.03029	0.1119	0.1411	0.1794	0.2058	0.2275	0.2468	0.2644	0.2809	0.2965	0.3114	0.3259	0.3399	0.3670	0.3931	0.4184
9.53)	h	757.34	1069.7	1142.0	1231.1	1289.5	1336.3	1377.5	1415.2	1450.7	1484.6	1517.5	1549.8	1581.5	1644.1	1706.1	1767.8
	s	0.9356	1.2097	1.2727	1.3481	1.3954	1.4319	1.4628	1.4900	1.5148	1.5376	1.5591	1.5794	1.5988	1.6355	1.6697	1.7021
	Sh			15.04	65.04	115.04	165.04	215.04	265.04	315.04	365.04	415.04	465.04	515.04	615.04	715.04	815.04
800	v	0.03134	0.1030	0.1278	0.1685	0.1952	0.2168	0.2358	0.2531	0.2693	0.2845	0.2991	0.3132	0.3268	0.3532	0.3785	0.4030
4.96)	h	770.69	1055.8	1121.2	1220.6	1282.2	1330.7	1372.8	1411.2	1447.2	1481.6	1514.8	1547.3	1579.3	1642.2	1704.5	1766.5
	s	0.9468	1.1958	1.2527	1.3368	1.3867	1.4245	1.4561	1.4838	1.5089	1.5321	1.5537	1.5742	1.5938	1.6306	1.6651	1.6975
	Sh			9.78	59.78	109.78	159.78	209.78	259.78	309.78	359.78	409.78	459.78	509.78	609.78	709.78	809.78
900	v	0.03262	0.0942	0.1138	0.1581	0.1853	0.2068	0.2256	0.2427	0.2585	0.2734	0.2877	0.3014	0.3147	0.3403	0.3649	0.3887
0.22)	h	785.13	1039.8	1095.3	1209.6	1274.7	1324.9	1368.0	1407.2	1443.7	1478.5	1512.1	1544.9	1577.0	1640.4	1703.0	1765.2
	s	0.9588	1.1803	1.2283	1.3251	1.3780	1.4171	1.4494	1.4777	1.5032	1.5266	1.5485	1.5692	1.5889	1.6259	1.6605	1.6931
	Sh			4.67	54.67	104.67	154.67	204.67	254.67	304.67	354.67	404.67	454.67	504.67	604.67	704.67	804.67
000	v	0.03428	0.0850	0.0982	0.1483	0.1759	0.1975	0.2161	0.2329	0.2484	0.2630	0.2770	0.2904	0.3033	0.3282	0.3522	0.3753
5.33)	h	801.84	1020.3	1060.5	1197.9	1267.0	1319.0	1363.2	1403.1	1440.2	1475.4	1509.4	1542.4	1574.8	1638.5	1701.4	1763.8
	s	0.9728	1.1619	1.1966	1.3131	1.3692	1.4097	1.4429	1.4717	1.4976	1.5213	1.5434	1.5642	1.5841	1.6214	1.6561	1.6888
	Sh				49.72	99.72	149.72	199.72	249.72	299.72	349.72	399.72	449.72	499.72	599.72	699.72	799.72
100	v	0.03681	0.0745		0.1389	0.1671	0.1887	0.2071	0.2237	0.2390	0.2533	0.2670	0.2800	0.2927	0.3170	0.3403	0.3628
0.28)	h	823.97	993.3		1185.4	1259.1	1313.0	1358.4	1399.0	1436.7	1472.3	1506.6	1539.9	1572.6	1636.7	1699.8	1762.5
	s	0.9914	1.1373		1.3007	1.3604	1.4024	1.4364	1.4658	1.4920	1.5161	1.5384	1.5594	1.5794	1.6169	1.6518	1.6847
	Sh				44.92	94.92	144.92	194.92	244.92	294.92	344.92	394.92	444.92	494.92	594.92	694.92	794.92
3200	v	0.04472	0.0566		0.1300	0.1588	0.1804	0.1987	0.2151	0.2301	0.2442	0.2576	0.2704	0.2827	0.3065	0.3291	0.3510
05.08)	h	875.54	931.6		1172.3	1250.9	1306.9	1353.4	1394.9	1433.1	1469.2	1503.8	1537.4	1570.3	1634.8	1698.3	1761.2
	s	1.0351	1.0832		1.2877	1.3515	1.3951	1.4300	1.4600	1.4866	1.5110	1.5335	1.5547	1.5749	1.6126	1.6477	1.6806
	Sh																
3300	v				0.1213	0.1510	0.1727	0.1908	0.2070	0.2218	0.2357	0.2488	0.2613	0.2734	0.2966	0.3187	0.3400
	h				1158.2	1242.5	1300.7	1348.4	1390.7	1429.5	1466.1	1501.0	1534.9	1568.1	1632.9	1696.7	1759.9
	s				1.2742	1.3425	1.3879	1.4237	1.4542	1.4813	1.5059	1.5287	1.5501	1.5704	1.6084	1.6436	1.6767
	Sh																
3400	v				0.1129	0.1435	0.1653	0.1834	0.1994	0.2140	0.2276	0.2405	0.2528	0.2646	0.2872	0.3088	0.3296
	h				1143.2	1233.7	1294.3	1343.4	1386.4	1425.9	1462.9	1498.3	1532.4	1565.8	1631.1	1695.1	1758.5
	s				1.2600	1.3334	1.3807	1.4174	1.4486	1.4761	1.5010	1.5240	1.5456	1.5660	1.6042	1.6396	1.6728
	Sh																
3500	v				0.1048	0.1364	0.1583	0.1764	0.1922	0.2066	0.2200	0.2326	0.2447	0.2563	0.2784	0.2995	0.3198
	h				1127.1	1224.6	1287.8	1338.2	1382.2	1422.2	1459.7	1495.5	1529.9	1563.6	1629.2	1693.6	1757.2
	s				1.2450	1.3242	1.3734	1.4112	1.4430	1.4709	1.4962	1.5194	1.5412	1.5618	1.6002	1.6358	1.6691
	Sh																
3600	v				0.0966	0.1296	0.1517	0.1697	0.1854	0.1996	0.2128	0.2252	0.2371	0.2485	0.2702	0.2908	0.3106
	h				1108.6	1215.3	1281.2	1333.0	1377.9	1418.6	1456.5	1492.6	1527.4	1561.3	1627.3	1692.0	1755.9
	s				1.2281	1.3148	1.3662	1.4050	1.4374	1.4658	1.4914	1.5149	1.5369	1.5576	1.5962	1.6320	1.6654
	Sh																
3800	v				0.0799	0.1169	0.1395	0.1574	0.1729	0.1868	0.1996	0.2116	0.2231	0.2340	0.2549	0.2746	0.2936
	h				1064.2	1195.5	1267.6	1322.4	1369.1	1411.2	1450.1	1487.0	1522.4	1556.8	1623.6	1688.9	1753.2
	s				1.1888	1.2955	1.3517	1.3928	1.4265	1.4558	1.4821	1.5061	1.5284	1.5495	1.5886	1.6247	1.6584
	Sh																
4000	v				0.0631	0.1052	0.1284	0.1463	0.1616	0.1752	0.1877	0.1994	0.2105	0.2210	0.2411	0.2601	0.2783
	h				1007.4	1174.3	1253.4	1311.6	1360.2	1403.6	1443.6	1481.3	1517.3	1552.2	1619.8	1685.7	1750.6
	s				1.1396	1.2754	1.3371	1.3807	1.4158	1.4461	1.4730	1.4976	1.5203	1.5417	1.5812	1.6177	1.6516
	Sh																
4200	v				0.0498	0.0945	0.1183	0.1362	0.1513	0.1647	0.1769	0.1883	0.1991	0.2093	0.2287	0.2470	0.2645
	h				950.1	1151.6	1238.6	1300.4	1351.2	1396.0	1437.1	1475.5	1512.2	1547.6	1616.1	1682.6	1748.0
	s				1.0905	1.2544	1.3223	1.3686	1.4053	1.4366	1.4642	1.4893	1.5124	1.5341	1.5742	1.6109	1.6452
	Sh																
4400	v				0.0421	0.0846	0.1090	0.1270	0.1420	0.1552	0.1671	0.1782	0.1887	0.1986	0.2174	0.2351	0.2519
	h				909.5	1127.3	1223.3	1289.0	1342.0	1388.3	1430.4	1469.7	1507.1	1543.0	1612.3	1679.4	1745.3
	s				1.0556	1.2325	1.3073	1.3566	1.3949	1.4272	1.4556	1.4812	1.5048	1.5268	1.5673	1.6044	1.6389

Temperature — Degrees Fahrenheit (columns 700–1500)

Sh = superheat, F
v = specific volume, cu ft per lb
h = enthalpy, Btu per lb
s = entropy, Btu per F per lb

Table 3. Superheated Steam—*Continued*

Abs Press. Lb/Sq In. (Sat. Temp)		Sat. Water	Sat. Steam	750	800	850	900	950	1000	1050	1100	1150	1200	1250	1300	1400	1500
4600	Sh																
	v			0.0380	0.0751	0.1005	0.1186	0.1335	0.1465	0.1582	0.1691	0.1792	0.1889	0.1982	0.2071	0.2242	0.240
	h			883.8	1100.0	1207.3	1277.2	1332.6	1380.5	1423.7	1463.9	1501.9	1538.4	1573.8	1608.5	1676.3	1742
	s			1.0331	1.2084	1.2922	1.3446	1.3847	1.4181	1.4472	1.4734	1.4974	1.5197	1.5407	1.5607	1.5982	1.633
4800	Sh																
	v			0.0355	0.0665	0.0927	0.1109	0.1257	0.1385	0.1500	0.1606	0.1706	0.1800	0.1890	0.1977	0.2142	0.229
	h			866.9	1071.2	1190.7	1265.2	1323.1	1372.6	1417.0	1458.0	1496.7	1533.8	1569.7	1604.7	1673.1	1740
	s			1.0180	1.1835	1.2768	1.3327	1.3745	1.4090	1.4390	1.4657	1.4901	1.5128	1.5341	1.5543	1.5921	1.627
5000	Sh																
	v			0.0338	0.0591	0.0855	0.1038	0.1185	0.1312	0.1425	0.1529	0.1626	0.1718	0.1806	0.1890	0.2050	0.220
	h			854.9	1042.9	1173.6	1252.9	1313.5	1364.6	1410.2	1452.1	1491.5	1529.1	1565.5	1600.9	1670.0	1737.
	s			1.0070	1.1593	1.2612	1.3207	1.3645	1.4001	1.4309	1.4582	1.4831	1.5061	1.5277	1.5481	1.5863	1.621
5200	Sh																
	v			0.0326	0.0531	0.0789	0.0973	0.1119	0.1244	0.1356	0.1458	0.1553	0.1642	0.1728	0.1810	0.1966	0.211
	h			845.8	1016.9	1156.0	1240.4	1303.7	1356.6	1403.4	1446.2	1486.3	1524.5	1561.3	1597.2	1666.8	1734.
	s			0.9985	1.1370	1.2455	1.3088	1.3545	1.3914	1.4229	1.4509	1.4762	1.4995	1.5214	1.5420	1.5806	1.616
5400	Sh																
	v			0.0317	0.0483	0.0728	0.0912	0.1058	0.1182	0.1292	0.1392	0.1485	0.1572	0.1656	0.1736	0.1888	0.203
	h			838.5	994.3	1138.1	1227.7	1293.7	1348.4	1396.5	1440.3	1481.1	1519.8	1557.1	1593.4	1663.7	1732.
	s			0.9915	1.1175	1.2296	1.2969	1.3446	1.3827	1.4151	1.4437	1.4694	1.4931	1.5153	1.5362	1.5750	1.610
5600	Sh																
	v			0.0309	0.0447	0.0672	0.0856	0.1001	0.1124	0.1232	0.1331	0.1422	0.1508	0.1589	0.1667	0.1815	0.195
	h			832.4	975.0	1119.9	1214.8	1283.7	1340.2	1389.6	1434.3	1475.9	1515.2	1552.9	1589.6	1660.5	1729.
	s			0.9855	1.1008	1.2137	1.2850	1.3348	1.3742	1.4075	1.4366	1.4628	1.4869	1.5093	1.5304	1.5697	1.605
5800	Sh																
	v			0.0303	0.0419	0.0622	0.0805	0.0949	0.1070	0.1177	0.1274	0.1363	0.1447	0.1527	0.1603	0.1747	0.188
	h			827.3	958.8	1101.8	1201.8	1273.6	1332.0	1382.6	1428.3	1470.6	1510.5	1548.7	1585.8	1657.4	1726.
	s			0.9803	1.0867	1.1981	1.2732	1.3250	1.3658	1.3999	1.4297	1.4564	1.4808	1.5035	1.5248	1.5644	1.600
6000	Sh																
	v			0.0298	0.0397	0.0579	0.0757	0.0900	0.1020	0.1126	0.1221	0.1309	0.1391	0.1469	0.1544	0.1684	0.181
	h			822.9	945.1	1084.6	1188.8	1263.4	1323.6	1375.7	1422.3	1465.4	1505.9	1544.6	1582.0	1654.2	1724.
	s			0.9758	1.0746	1.1833	1.2615	1.3154	1.3574	1.3925	1.4229	1.4500	1.4748	1.4978	1.5194	1.5593	1.596
6500	Sh																
	v			0.0287	0.0358	0.0495	0.0655	0.0793	0.0909	0.1012	0.1104	0.1188	0.1266	0.1340	0.1411	0.1544	0.166
	h			813.9	919.5	1046.7	1156.3	1237.8	1302.7	1358.1	1407.3	1452.2	1494.2	1534.1	1572.5	1646.4	1717.
	s			0.9661	1.0515	1.1506	1.2328	1.2917	1.3370	1.3743	1.4064	1.4347	1.4604	1.4841	1.5062	1.5471	1.5844
7000	Sh																
	v			0.0279	0.0334	0.0438	0.0573	0.0704	0.0816	0.0915	0.1004	0.1085	0.1160	0.1231	0.1298	0.1424	0.154
	h			806.9	901.8	1016.5	1124.9	1212.6	1281.7	1340.5	1392.2	1439.1	1482.6	1523.7	1563.1	1638.6	1711.1
	s			0.9582	1.0350	1.1243	1.2055	1.2689	1.3171	1.3567	1.3904	1.4200	1.4466	1.4710	1.4938	1.5355	1.573
7500	Sh																
	v			0.0272	0.0318	0.0399	0.0512	0.0631	0.0737	0.0833	0.0918	0.0996	0.1068	0.1136	0.1200	0.1321	0.143
	h			801.3	889.0	992.9	1097.7	1188.3	1261.0	1322.9	1377.2	1426.0	1471.0	1513.3	1553.7	1630.8	1704.
	s			0.9514	1.0224	1.1033	1.1818	1.2473	1.2980	1.3397	1.3751	1.4059	1.4335	1.4586	1.4819	1.5245	1.563
8000	Sh																
	v			0.0267	0.0306	0.0371	0.0465	0.0571	0.0671	0.0762	0.0845	0.0920	0.0989	0.1054	0.1115	0.1230	0.1338
	h			796.6	879.1	974.4	1074.3	1165.4	1241.0	1305.5	1362.2	1413.0	1459.6	1503.1	1544.5	1623.1	1698.1
	s			0.9455	1.0122	1.0864	1.1613	1.2271	1.2798	1.3233	1.3603	1.3924	1.4208	1.4467	1.4705	1.5140	1.5533
8500	Sh																
	v			0.0262	0.0296	0.0350	0.0429	0.0522	0.0615	0.0701	0.0780	0.0853	0.0919	0.0982	0.1041	0.1151	0.1254
	h			792.7	871.2	959.8	1054.5	1144.0	1221.9	1288.5	1347.5	1400.2	1448.2	1492.9	1535.3	1615.4	1691.7
	s			0.9402	1.0037	1.0727	1.1437	1.2084	1.2627	1.3076	1.3460	1.3793	1.4087	1.4352	1.4597	1.5040	1.5439
9000	Sh																
	v			0.0258	0.0288	0.0335	0.0402	0.0483	0.0568	0.0649	0.0724	0.0794	0.0858	0.0918	0.0975	0.1081	0.1179
	h			789.3	864.7	948.0	1037.6	1125.4	1204.1	1272.1	1333.0	1387.5	1437.1	1482.9	1526.3	1607.9	1685.3
	s			0.9354	0.9964	1.0613	1.1285	1.1918	1.2468	1.2926	1.3323	1.3667	1.3970	1.4243	1.4492	1.4944	1.534
9500	Sh																
	v			0.0254	0.0282	0.0322	0.0380	0.0451	0.0528	0.0603	0.0675	0.0742	0.0804	0.0862	0.0917	0.1019	0.111
	h			786.4	859.2	938.3	1023.4	1108.9	1187.7	1256.6	1318.9	1375.1	1426.1	1473.1	1517.3	1600.4	1679.0
	s			0.9310	0.9900	1.0516	1.1153	1.1771	1.2320	1.2785	1.3191	1.3546	1.3858	1.4137	1.4392	1.4851	1.5263
10000	Sh																
	v			0.0251	0.0276	0.0312	0.0362	0.0425	0.0495	0.0565	0.0633	0.0697	0.0757	0.0812	0.0865	0.0963	0.1054
	h			783.8	854.5	930.2	1011.3	1094.2	1172.6	1242.0	1305.3	1362.9	1415.3	1463.4	1508.6	1593.1	1672.8
	s			0.9270	0.9842	1.0432	1.1039	1.1638	1.2185	1.2652	1.3065	1.3429	1.3749	1.4035	1.4295	1.4763	1.5180
10500	Sh																
	v			0.0248	0.0271	0.0303	0.0347	0.0404	0.0467	0.0532	0.0595	0.0656	0.0714	0.0768	0.0818	0.0913	0.1001
	h			781.5	850.5	923.4	1001.0	1081.3	1158.9	1228.4	1292.4	1351.1	1404.7	1453.9	1500.0	1585.8	1666.7
	s			0.9232	0.9790	1.0358	1.0939	1.1519	1.2060	1.2529	1.2946	1.3371	1.3644	1.3937	1.4202	1.4677	1.5100

Temperature – Degrees Fahrenheit

Sh = superheat, F
v = specific volume, cu ft per lb
h = enthalpy, Btu per lb
s = entropy, Btu per F per lb

Table 3. Superheated Steam—*Continued*

Abs Press. Lb/Sq In. (Sat. Temp)		Sat. Water	Sat. Steam	Temperature—Degrees Fahrenheit 750	800	850	900	950	1000	1050	1100	1150	1200	1250	1300	1400	1500
11000	v			0.0245	0.0267	0.0296	0.0335	0.0386	0.0443	0.0503	0.0562	0.0620	0.0676	0.0727	0.0776	0.0868	0.0952
	h			779.5	846.9	917.5	992.1	1069.9	1146.3	1215.9	1280.2	1339.7	1394.4	1444.6	1491.5	1578.7	1660.6
	s			0.9196	0.9742	1.0292	1.0851	1.1412	1.1945	1.2414	1.2833	1.3209	1.3544	1.3842	1.4112	1.4595	1.5023
11500	v			0.0243	0.0263	0.0290	0.0325	0.0370	0.0423	0.0478	0.0534	0.0588	0.0641	0.0691	0.0739	0.0827	0.0909
	h			777.7	843.8	912.4	984.5	1059.8	1134.9	1204.3	1268.7	1328.8	1384.4	1435.5	1483.2	1571.8	1654.7
	s			0.9163	0.9698	1.0232	1.0772	1.1316	1.1840	1.2308	1.2727	1.3107	1.3446	1.3750	1.4025	1.4515	1.4949
12000	v			0.0241	0.0260	0.0284	0.0317	0.0357	0.0405	0.0456	0.0508	0.0560	0.0610	0.0659	0.0704	0.0790	0.0869
	h			776.1	841.0	907.9	977.8	1050.9	1124.5	1193.7	1258.0	1318.5	1374.7	1426.6	1475.1	1564.9	1648.8
	s			0.9131	0.9657	1.0177	1.0701	1.1229	1.1742	1.2209	1.2627	1.3010	1.3353	1.3662	1.3941	1.4438	1.4877
12500	v			0.0238	0.0256	0.0279	0.0309	0.0346	0.0390	0.0437	0.0486	0.0535	0.0583	0.0629	0.0673	0.0756	0.0832
	h			774.7	838.6	903.9	971.9	1043.1	1115.2	1184.1	1247.9	1308.8	1365.4	1418.0	1467.2	1558.2	1643.1
	s			0.9101	0.9618	1.0127	1.0637	1.1151	1.1653	1.2117	1.2534	1.2918	1.3264	1.3576	1.3860	1.4363	1.4808
13000	v			0.0236	0.0253	0.0275	0.0302	0.0336	0.0376	0.0420	0.0466	0.0512	0.0558	0.0602	0.0645	0.0725	0.0799
	h			773.5	836.3	900.4	966.8	1036.2	1106.7	1174.8	1238.5	1299.6	1356.5	1409.6	1459.4	1551.6	1637.4
	s			0.9073	0.9582	1.0080	1.0578	1.1079	1.1571	1.2030	1.2445	1.2831	1.3179	1.3494	1.3781	1.4291	1.4741
13500	v			0.0235	0.0251	0.0271	0.0297	0.0328	0.0364	0.0405	0.0448	0.0492	0.0535	0.0577	0.0619	0.0696	0.0768
	h			772.3	834.4	897.2	962.2	1030.0	1099.1	1166.3	1229.7	1291.0	1348.1	1401.5	1451.8	1545.2	1631.9
	s			0.9045	0.9548	1.0037	1.0524	1.1014	1.1495	1.1948	1.2361	1.2749	1.3098	1.3415	1.3705	1.4221	1.4675
14000	v			0.0233	0.0248	0.0267	0.0291	0.0320	0.0354	0.0392	0.0432	0.0474	0.0515	0.0555	0.0595	0.0670	0.0740
	h			771.3	832.6	894.3	958.0	1024.5	1092.3	1158.5	1221.4	1283.0	1340.2	1393.8	1444.4	1538.8	1626.5
	s			0.9019	0.9515	0.9996	1.0473	1.0953	1.1426	1.1872	1.2282	1.2671	1.3021	1.3339	1.3631	1.4153	1.4612
14500	v			0.0231	0.0246	0.0264	0.0287	0.0314	0.0345	0.0380	0.0418	0.0458	0.0496	0.0534	0.0573	0.0646	0.0714
	h			770.4	831.0	891.7	954.3	1019.6	1086.2	1151.4	1213.8	1275.4	1332.9	1386.4	1437.3	1532.6	1621.1
	s			0.8994	0.9484	0.9957	1.0426	1.0897	1.1362	1.1801	1.2208	1.2597	1.2949	1.3266	1.3560	1.4087	1.4551
15000	v			0.0230	0.0244	0.0261	0.0282	0.0308	0.0337	0.0369	0.0405	0.0443	0.0479	0.0516	0.0552	0.0624	0.0690
	h			769.6	829.5	889.3	950.9	1015.1	1080.6	1144.9	1206.8	1268.1	1326.0	1379.4	1430.3	1526.4	1615.9
	s			0.8970	0.9455	0.9920	1.0382	1.0846	1.1302	1.1735	1.2139	1.2525	1.2880	1.3197	1.3491	1.4022	1.4491
15500	v			0.0228	0.0242	0.0258	0.0278	0.0302	0.0329	0.0360	0.0393	0.0429	0.0464	0.0499	0.0534	0.0603	0.0668
	h			768.9	828.2	887.2	947.8	1011.1	1075.7	1139.0	1200.3	1261.1	1319.6	1372.8	1423.6	1520.4	1610.8
	s			0.8946	0.9427	0.9886	1.0340	1.0797	1.1247	1.1674	1.2073	1.2457	1.2815	1.3131	1.3424	1.3959	1.4433

Sh = superheat, F
v = specific volume, cu ft per lb
h = enthalpy, Btu per lb
s = entropy, Btu per F per lb

Reduced Property Correlation Charts

C

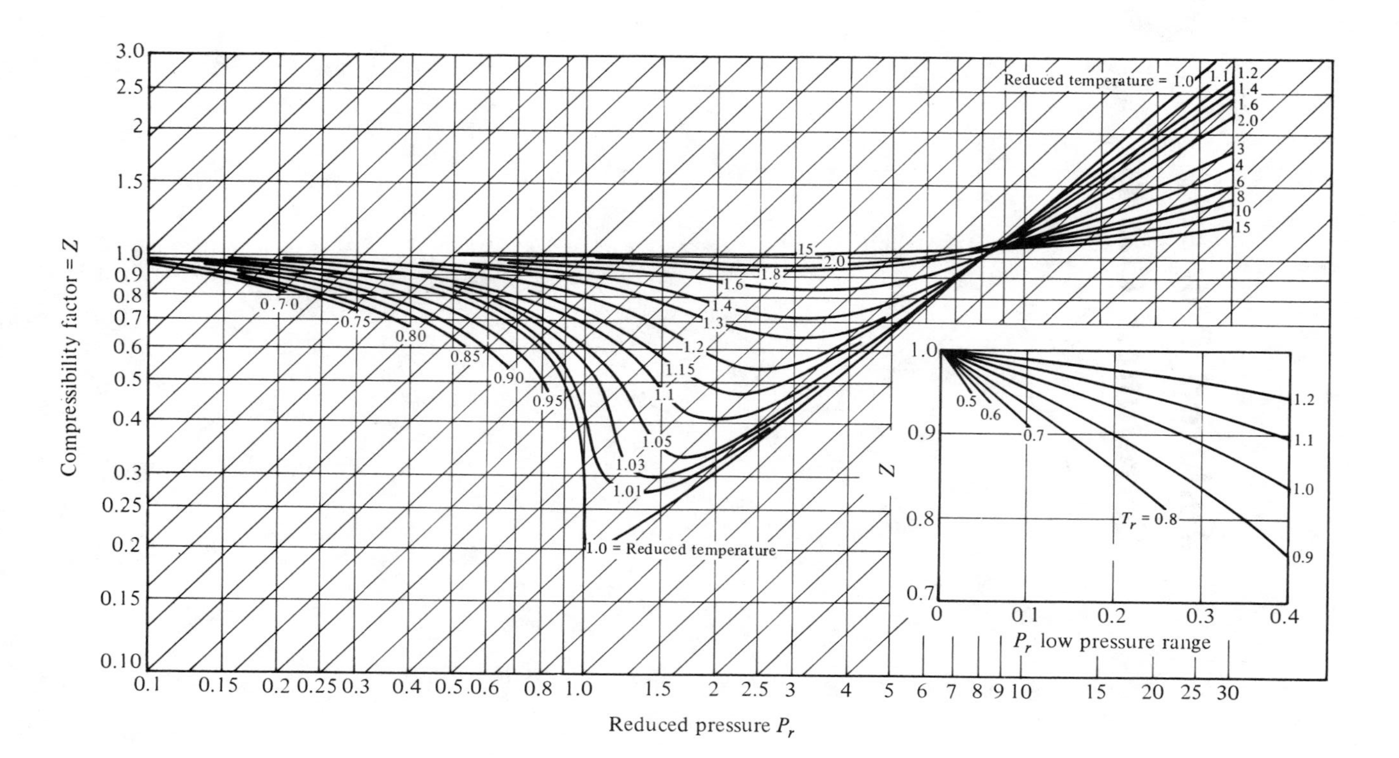

FIG. A-1. Compressibility factors of gases and vapors. (From O. A. Hougen and K. M. Watson, *Chemical Process Principles Charts*, 1st ed., John Wiley & Sons, New York, 1946.)

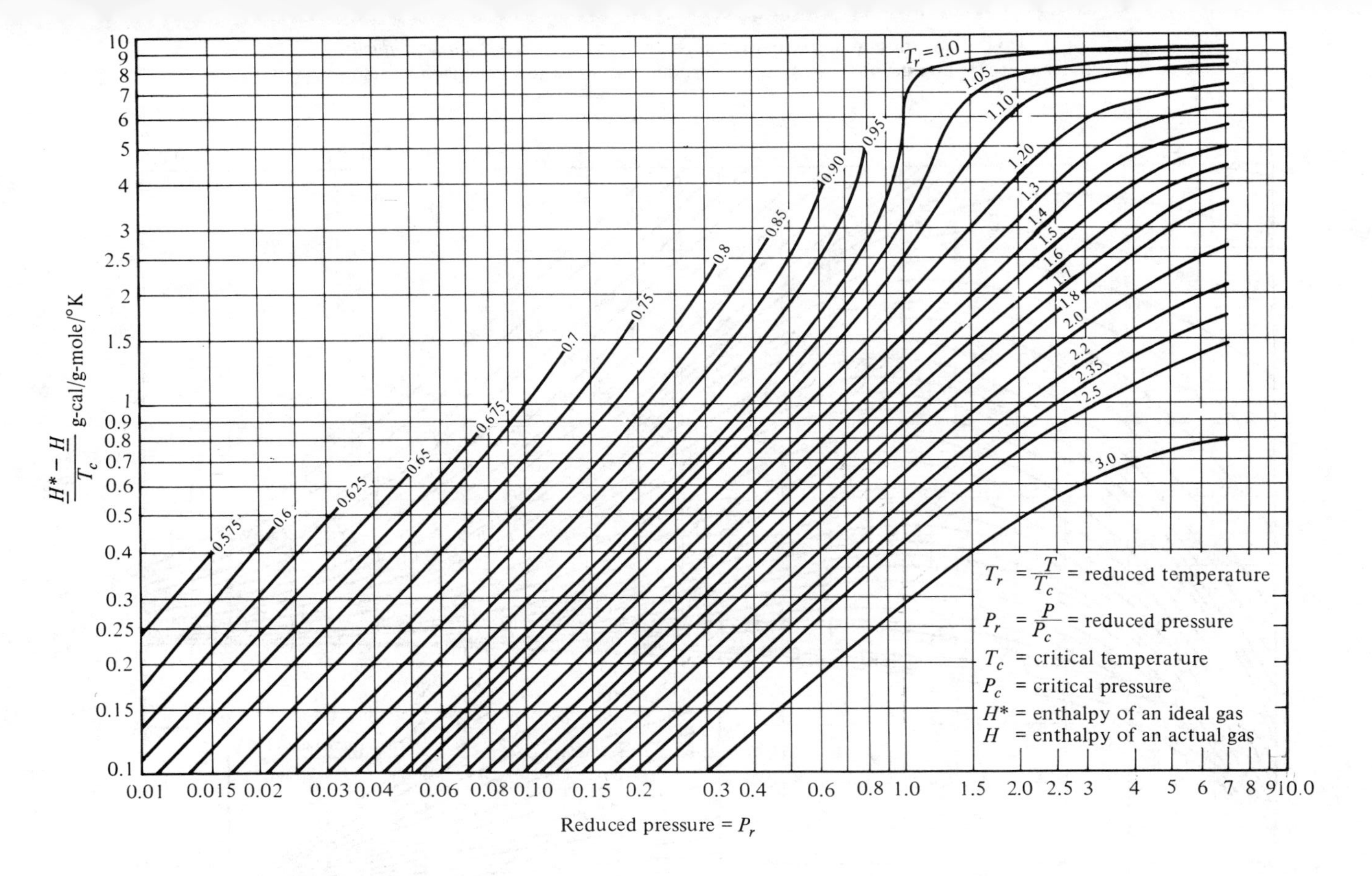

FIG. A-2. Enthalpy correction of gases and vapors due to lack of ideal behavior. (From O. A. Hougen and K. M. Watson, *Chemical Process Principles Charts*, 1st ed., John Wiley & Sons, New York, 1946.)

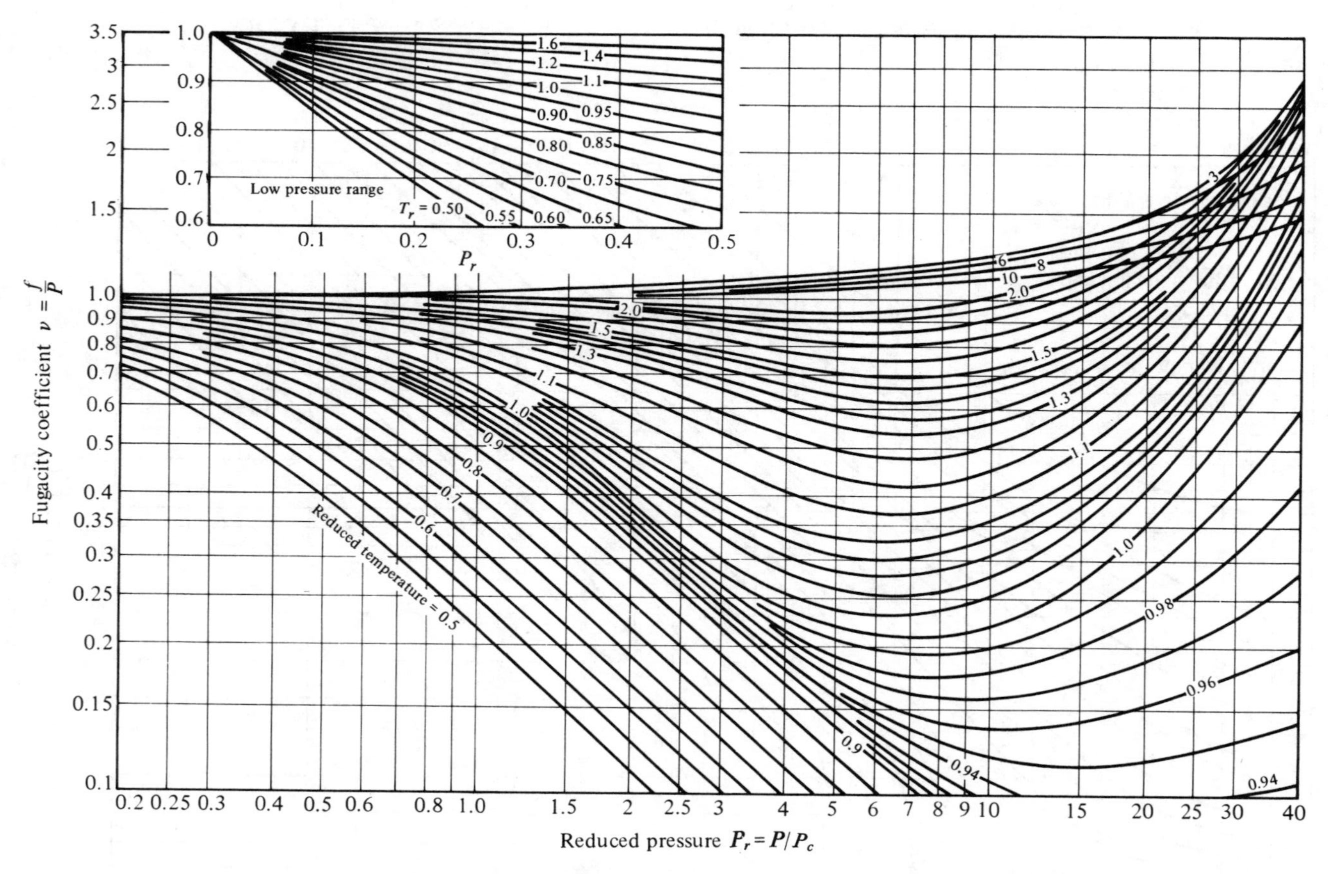

FIG. A-3. Fugacity coefficients of gases and vapors. (From O. A. Hougen and K. M. Watson, *Chemical Process Principles Charts*, 1st ed., John Wiley & Sons, New York, 1946.)

Property Charts for Selected Compounds

D

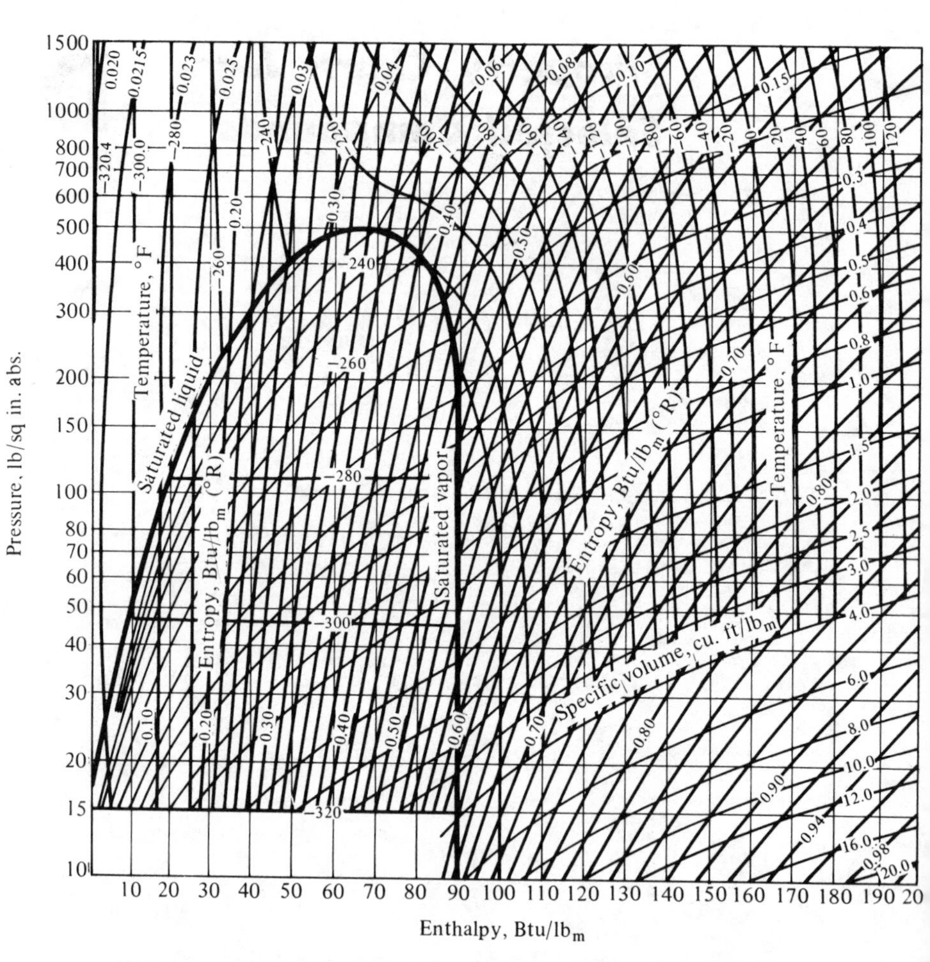

FIG. A-4. Mollier chart for nitrogen; $\underline{H} = 0$ and $\underline{S} = 0$ at -320.4°F and saturated liquid. [From G. S. Lin, "Molier Chart for Nitrogen," *C.E.P.*, *59*, 70 (1963).]

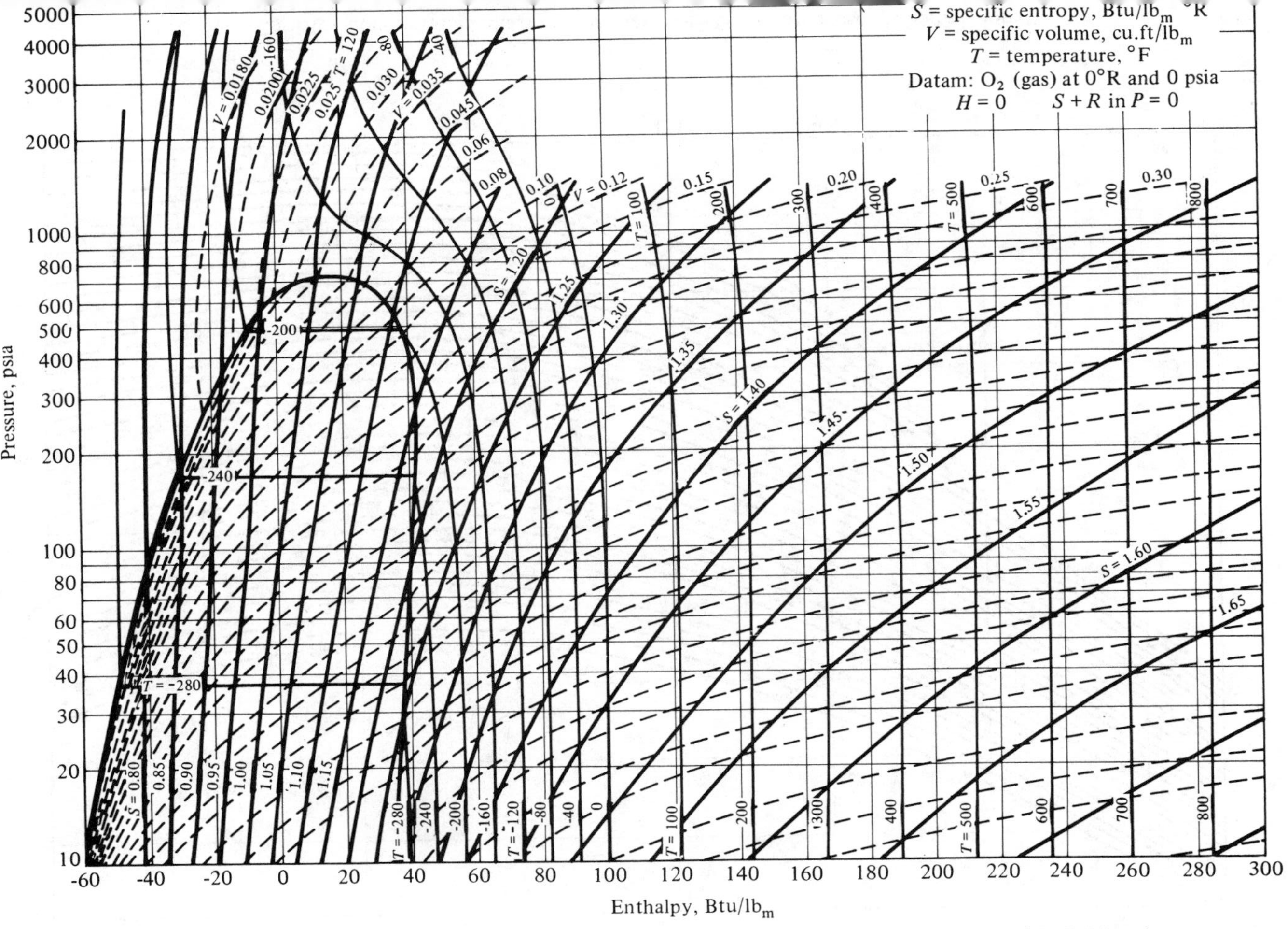

FIG. A-5. Oxygen pressure–enthalpy diagram. (Reproduced by permission from L. N. Canjar and F. S. Manning, *Thermodynamic Properties and Reduced Correlations for Gases*, copyright Gulf Publishing Co., Houston, 1967.)

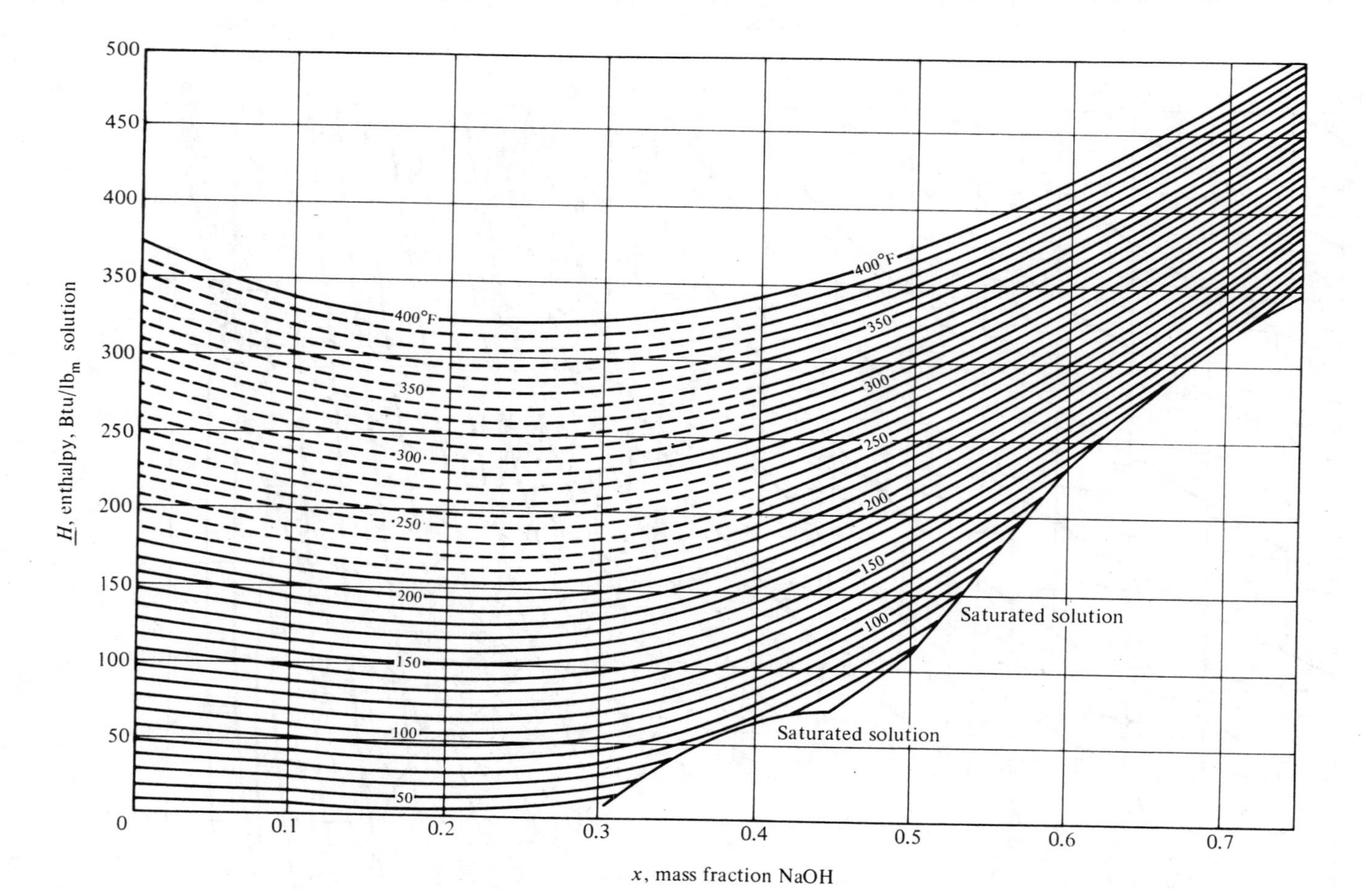

FIG. A-6. Enthalpy–concentration diagram for NaOH–H_2O. [Reproduced by permission.

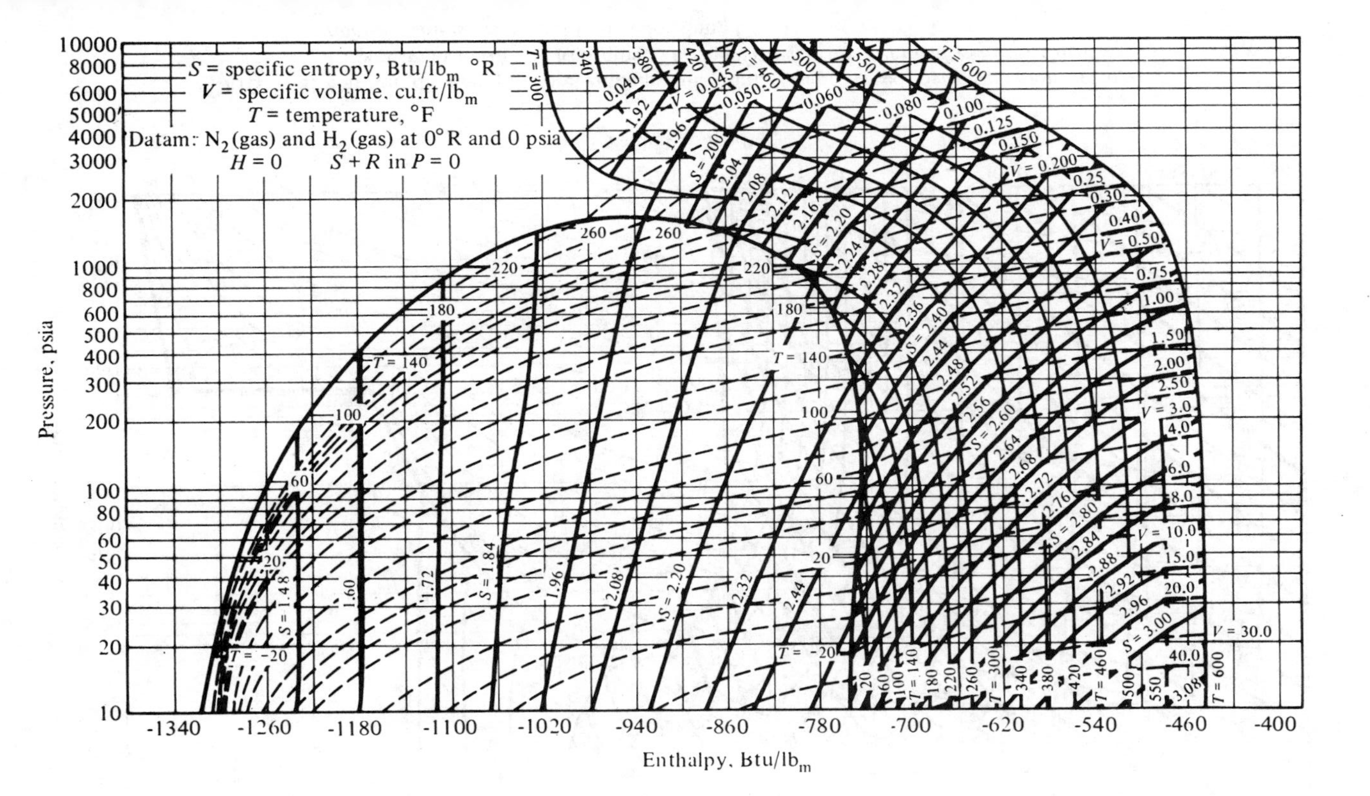

FIG. A-7. Ammonia pressure–enthalpy diagram. (Reproduced by permission from L. N. Canjar and F. S. Manning, *Thermodynamic Properties and Reduced Correlations for Gases*, copyright Gulf Publishing Co., Houston, 1967.)

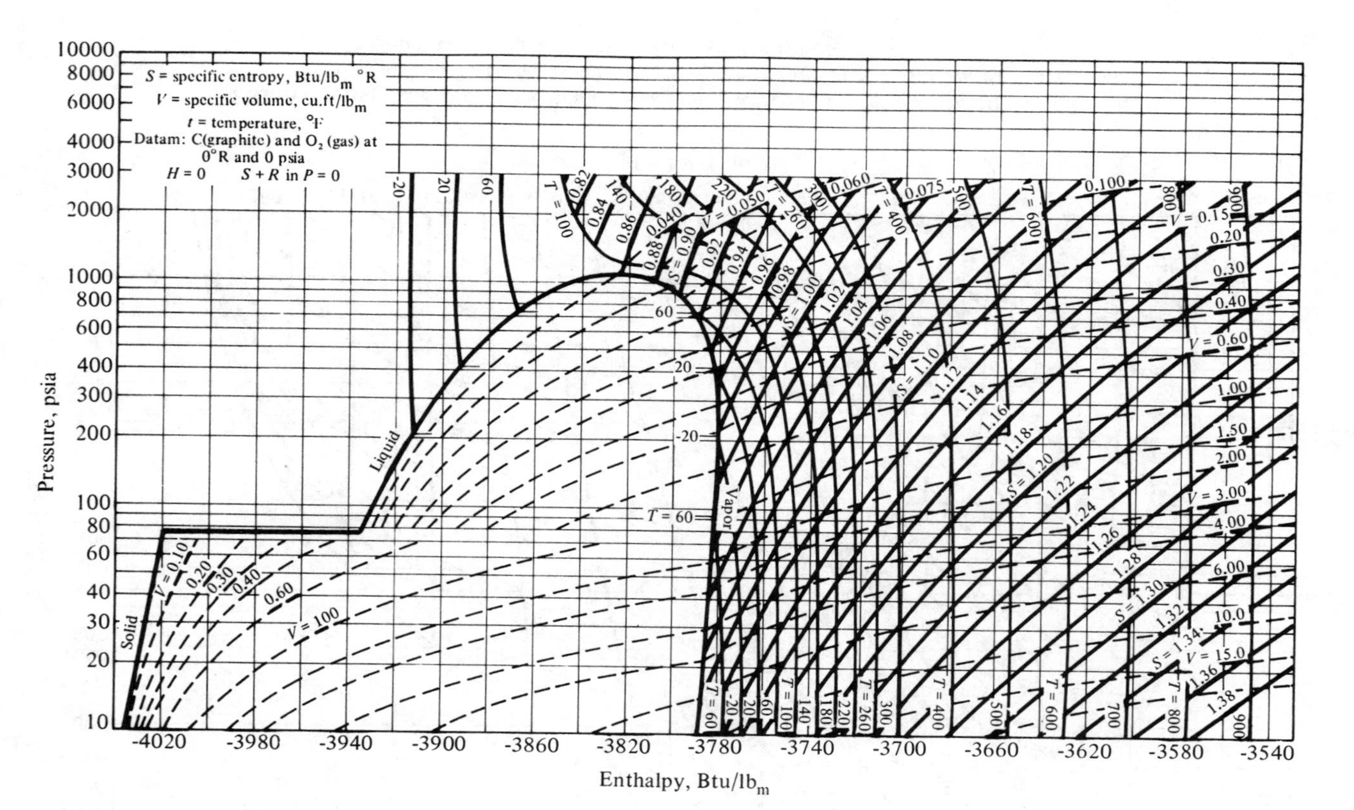

FIG. A-8. Carbon dioxide pressure–enthalpy diagram. (Reproduced by permission from L. N. Canjar and F. S. Manning, *Thermodynamic Properties and Reduced Correlations for Gases*, copyright Gulf Publishing Co., Houston, 1967.)

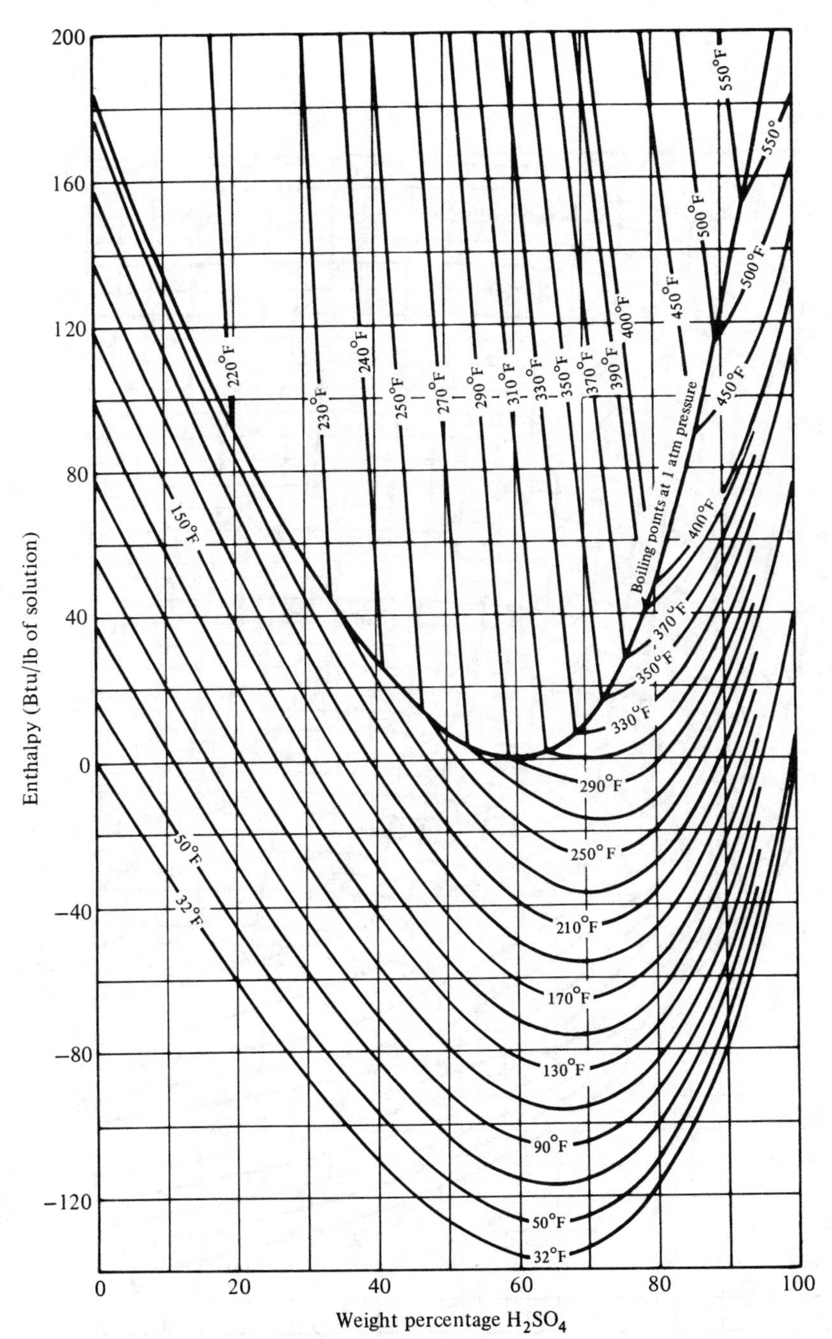

FIG. A-9. Enthalpy–concentration of sulfuric acid-water system. Reference states are pure components at 32°F and vapor pressures. (From O. A. Hougen and K. M. Watson, *Chemical Process Principles Charts*, 1st ed., John Wiley and Sons, New York, 1946.)

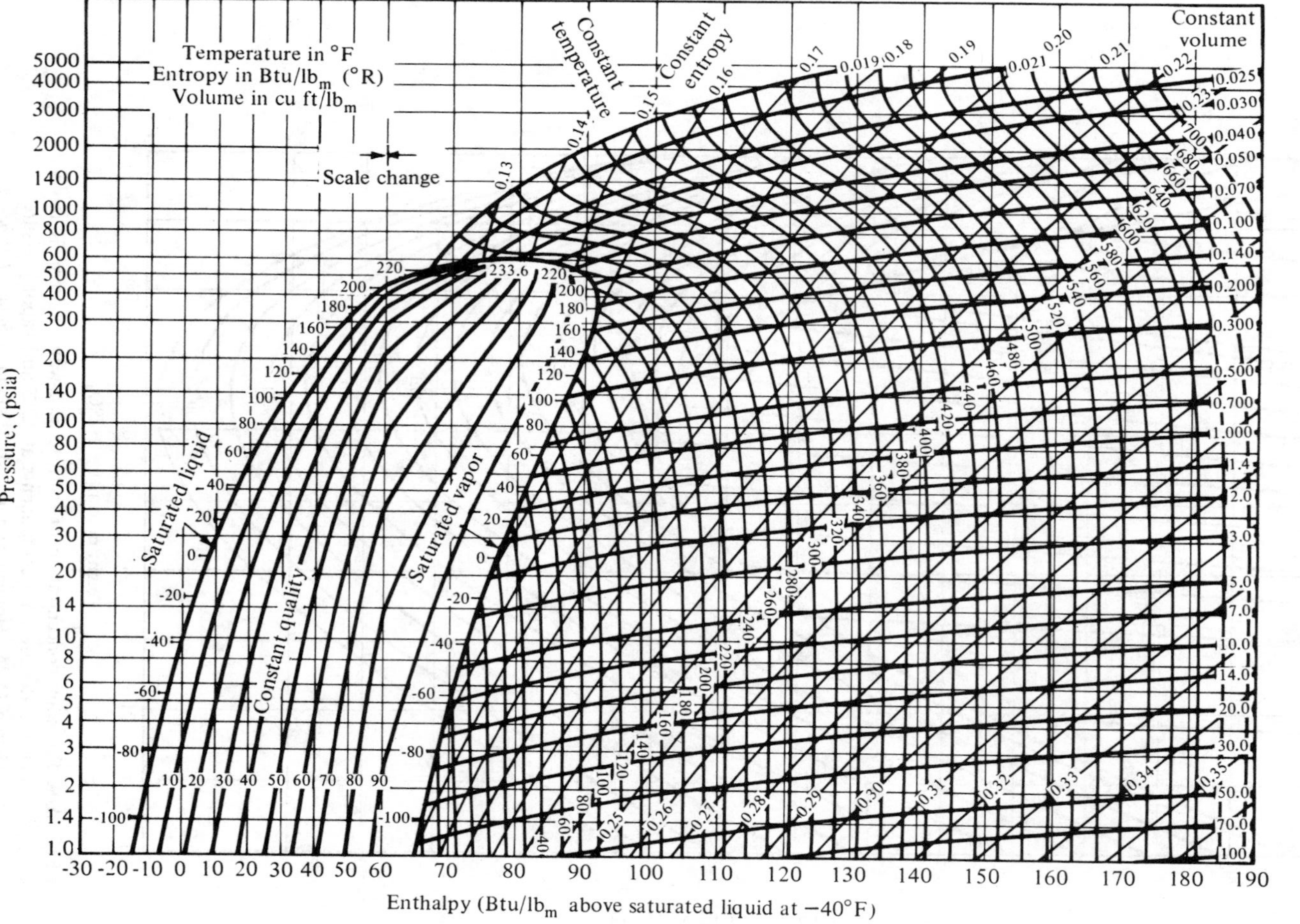

FIG. A-10. Pressure–enthalpy diagram for "Freon"-12 refrigerant. (Copyrighted by the "Freon" Products Division, E. I. du Pont de Nemours and Co., Inc., and re-

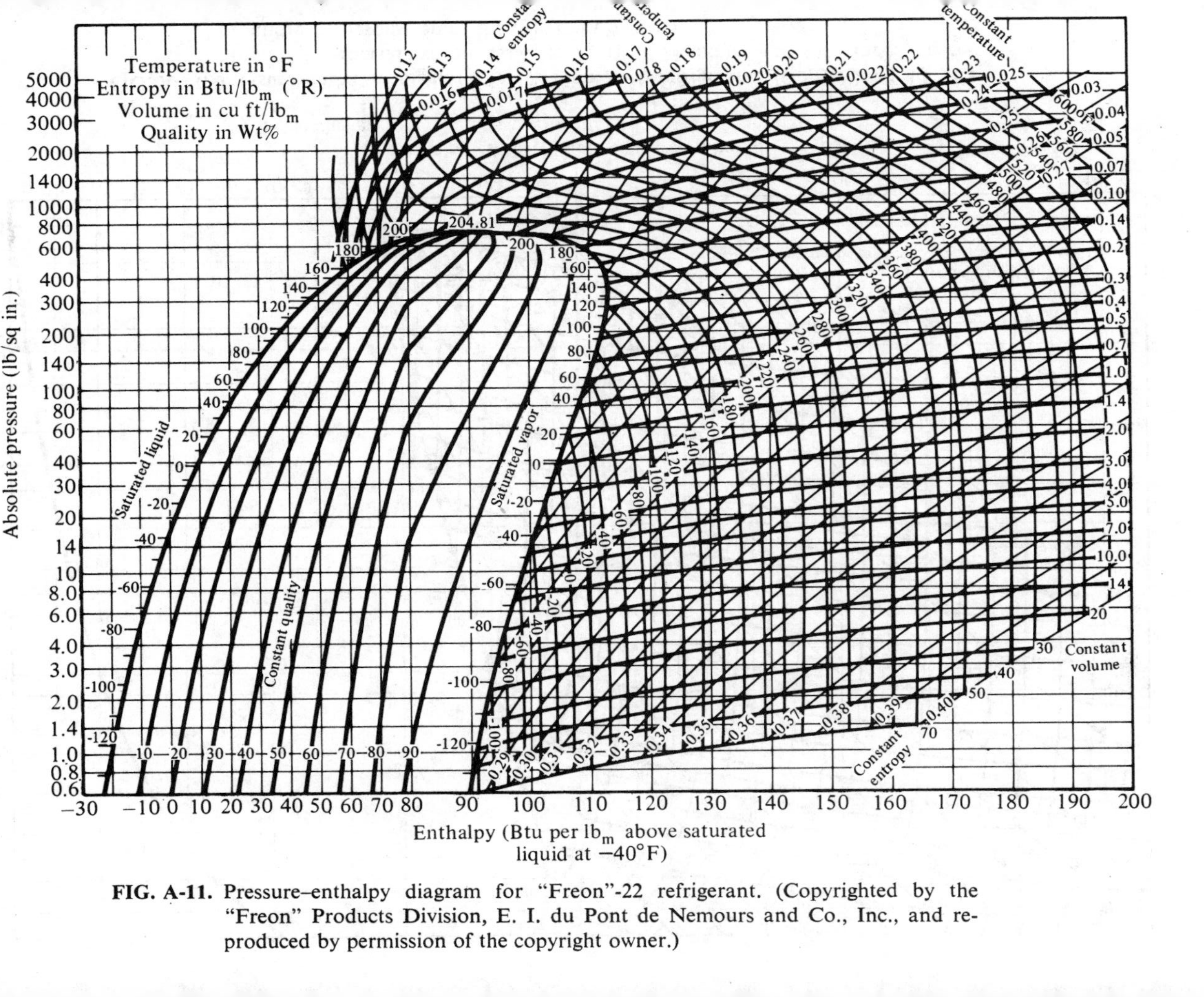

FIG. A-11. Pressure–enthalpy diagram for "Freon"-22 refrigerant. (Copyrighted by the "Freon" Products Division, E. I. du Pont de Nemours and Co., Inc., and reproduced by permission of the copyright owner.)

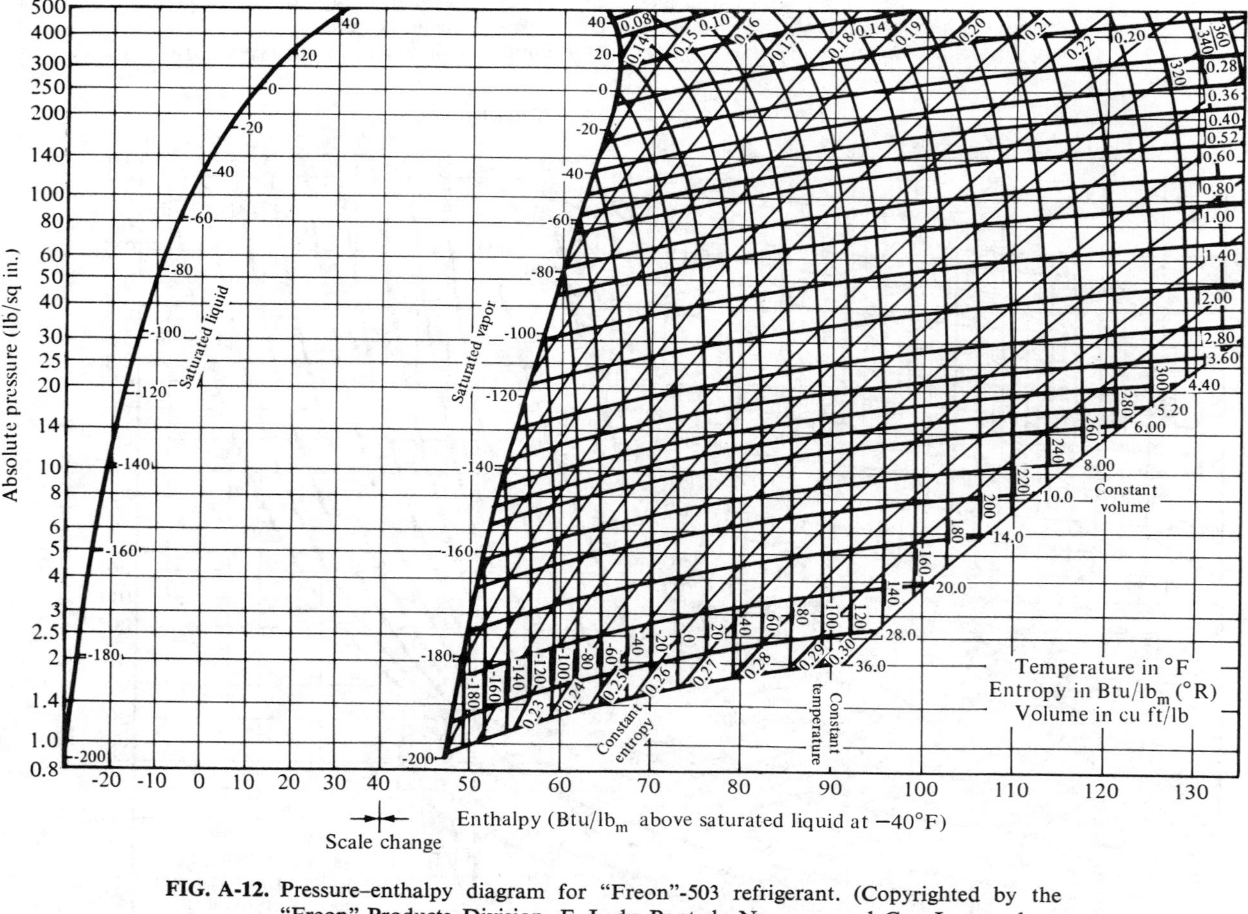

FIG. A-12. Pressure–enthalpy diagram for "Freon"-503 refrigerant. (Copyrighted by the "Freon" Products Division, E. I. du Pont de Nemours and Co., Inc., and reproduced by permission of the copyright owner.)

Tables of Thermodynamic Properties

E

TABLE A-1 Thermodynamic Data for 298.15° K†

Substance	$\Delta \underline{H}_f^\circ$ kcal/mole	$\Delta \underline{G}_f^\circ$ kcal/mole
Oxygen		
$O_{(g)}$	59.159	54.994
$O_{2(g)}$	0	0
Hydrogen		
$H_{(g)}$	52.089	48.575
$H_{2(g)}$	0	0
$H_2O_{(g)}$	−57.7979	−54.6357
$H_2O_{(l)}$	−68.3174	−56.6902
$H_2O_{2(g)}$	−31.83	
$H_2O_{2(l)}$	−44.84	
Fluorine		
$F_{(g)}$	18.3	14.2
$F_{2(g)}$	0	0
$HF_{(g)}$	−64.2	−64.7
Chlorine		
$Cl_{(g)}$	29.012	25.192
$Cl_{2(g)}$	0	0
$Cl_2 \cdot 8HO_{2(s)}$	−564.4	
$HCl_{(g)}$	−22.063	−22.769
Bromine		
$Br_{(g)}$	26.71	19.69
$Br_{2(g)}$	7.34	0.751
$Br_{2(l)}$	0	0
$HBr_{(g)}$	−8.66	−12.72
Iodine		
$I_{(g)}$	25.482	16.766
$I_{2(g)}$	14.876	4.63
$I_{2(s)}$	0	0
$HI_{(g)}$	6.20	0.31
Sulfur		
$S_{(g)}$	53.25	43.57
$S_{(s, rhombic)}$	0	0
$S_{(s, monoclinic)}$	0.071	0.023
$S_{8(g)}$	24.1	
$SO_{2(g)}$	−70.96	−71.79
$SO_{3(g)}$	−94.45	−88.52
$H_2S_{(g)}$	−4.815	−7.892
$H_2SO_{4(l)}$	−193.91	
Nitrogen		
$N_{(g)}$	85.565	81.471
$N_{2(g)}$	0	0
$NO_{(g)}$	21.600	20.719
$NO_{2(g)}$	8.091	12.390
$N_2O_{(g)}$	19.49	24.76
$N_2O_{4(g)}$	2.309	23.491
$NH_{3(g)}$	−11.04	−3.976
$N_2H_{4(l)}$	12.05	
$HNO_{3(l)}$	−41.404	−19.100
$NH_4NO_{3(s)}$	−87.27	
$NH_4Cl_{(s)}$	−75.38	−48.73
$(NH_4)_2SO_{4(s)}$	−281.86	−215.19
Phosphorus		
$P_{(g)}$	75.18	66.71
$P_{4(g)}$	13.12	5.82
$P_4O_{10(s)}$	−720.0	
$PH_{3(g)}$	2.21	4.36
$H_3PO_{4(s)}$	−306.2	
$PCl_{3(g)}$	−73.22	−68.42
$PCl_{5(g)}$	−95.35	−77.59
Arsenic		
$As_{(g)}$	60.64	50.74
$As_{(s)}$	0	0
$As_2O_{5(s)}$	−218.6	−184.6
$AsH_{3(g)}$	41.0	
$H_3AsO_{4(s)}$	−215.2	
Antimony		
$Sb_{(g)}$	60.8	51.1
$Sb_{(s)}$	0	0
$Sb_2O_{5(s)}$	−234.4	−200.5
$SbCl_{3(g)}$	−75.2	−72.3
$SbCl_{5(l)}$	−104.8	
Bismuth		
$Bi_{(g)}$	49.7	40.4
$Bi_{(s)}$	0	0
$Bi_2O_{3(s)}$	−137.9	−118.7
$Bi(OH)_{3(s)}$	−169.6	

† Table A-1 contains thermodynamic data for a number of selected elements and compounds. Most of the data are taken from *Selected Values of Chemical Thermodynamic Properties*, by F. D. Rossini and others, published by the National Bureau of Standards. The order followed for the elements (O, H, F, Cl, Br, I, S, N, etc.) is that employed in the Bureau publication. A compound is listed under that one of its elements which appears last in that order; for example, H_2O is listed under hydrogen, and HNO_3 is listed under nitrogen. The data include standard enthalpies of formation ($\Delta \underline{H}_f^\circ$) and standard chemical potentials or free energies of formation ($\Delta \underline{G}_f^\circ$) from the elements in their standard states at 1 atm and 25°C.

Substance	$\Delta \underline{H}_f^\circ$ kcal/mole	$\Delta \underline{G}_f^\circ$ kcal/mole
$BiCl_{3(s)}$	−90.61	−76.23
Carbon		
$C_{(g)}$	171.698	160.845
$C_{(s,\,diamond)}$	0.4532	0.6850
$C_{(s,\,graphite)}$	0	0
$CO_{(g)}$	−26.4157	−32.8079
$CO_{2(g)}$	−94.0518	−94.2598
$CH_{4(g)}$	−17.889	−12.140
$H_2CO_{(g)}$	−27.7	−26.2
$HCOOH_{(g)}$	−86.67	−80.24
$HCOOH_{(l)}$	−97.8	−82.7
$CH_3OH_{(g)}$	−48.08	−38.69
$CH_3OH_{(l)}$	−57.02	−39.73
$CCl_{4(g)}$	−25.5	−15.3
$CCl_{4(l)}$	−33.3	−16.4
$COCl_{2(g)}$	−53.30	−50.31
$CH_3Cl_{(g)}$	−19.6	−14.0
$CH_2Cl_{2(g)}$	−21.0	−14.0
$CHCl_{3(g)}$	−24.	−16.
$CHCl_{3(l)}$	−31.5	−17.1
$CS_{2(g)}$	27.55	15.55
$CS_{2(l)}$	21.0	15.2
$COS_{(g)}$	−32.80	−40.45
$CO(NH_2)_{2(s)}$	−79.634	−47.120
$C_2H_{2(g)}$	54.194	50.0
$C_2H_{4(g)}$	12.496	16.282
$C_2H_{6(g)}$	−20.236	−7.860
$(COOH)_{2(s)}$	−197.6	−166.8
$CH_3CHO_{(g)}$	−39.76	−31.96
$C_2H_4O_{(g)}$	−12.19	−2.79
$CH_3COOH_{(l)}$	−116.4	−93.8
$C_2H_5OH_{(g)}$	−56.24	−40.30
$C_2H_5OH_{(l)}$	−66.356	−41.77
$C_3H_{8(g)}$	−24.820	−5.614
$C_6G_{6(g)}$	19.820	30.989
Silicon		
$Si_{(g)}$	88.04	77.41
$Si_{(s)}$	0	0
$SiO_{2(s,\,quartz)}$	−205.4	−192.4
$H_2SiO_{3(s)}$	−270.7	
$H_4SiO_{4(s)}$	−340.6	
$SiF_{4(g)}$	−370.	−360.
$SiC_{(s)}$	−26.7	−26.1
Tin		
$Sn_{(g)}$	72.	64.
$Sn_{(s,\,gray)}$	−0.49	0.02
$Sn_{(s,\,white)}$	0	0
$SnO_{(s)}$	−68.4	−61.5
$SnO_{2(s)}$	−138.8	−124.2

Substance	$\Delta \underline{H}_f^\circ$ kcal/mole	$\Delta \underline{G}_f^\circ$ kcal/mole
$Sn(OH)_{2(s)}$	−138.3	−117.6
$Sn(OH)_{4(s)}$	−270.5	
$SnCl_{2(s)}$	−83.6	
$SnCl_{4(l)}$	−130.3	−113.3
Lead		
$Pb_{(g)}$	46.34	38.47
$Pb_{(s)}$	0	0
$PbO_{(s,\,yellow)}$	−52.07	−45.05
$PbO_{2(s)}$	−66.12	−52.34
$Pb_3O_{4(s)}$	−175.6	−147.6
$Pb(OH)_{2(s)}$	−123.0	−100.6
$PbCl_{2(s)}$	−85.85	−75.04
$PbS_{(s)}$	−22.54	−22.15
$PbSO_{4(s,\,II)}$	−219.50	−193.89
Zinc		
$Zn_{(g)}$	31.19	22.69
$Zn_{(s)}$	0	0
$ZnO_{(s)}$	−83.17	−76.05
$Zn(OH)_{2(s)}$	−153.5	
$ZnCl_{2(s)}$	−99.40	−88.255
$ZnS_{(s,\,II,\,sphalerite)}$	−48.5	−47.4
$ZnSO_{4(s)}$	−233.88	−208.31
Cadmium		
$Cd_{(g)}$	26.97	18.69
$Cd_{(s)}$	0	0
$CdO_{(s)}$	−60.86	−53.79
$Cd(OH)_{2(s)}$	−133.26	−112.46
$CdCl_{2(s)}$	−93.00	−81.88
$CdSO_{4(s)}$	−221.36	−195.99
Mercury		
$Hg_{(g)}$	14.54	7.59
$Hg_{(l)}$	0	0
$HgO_{(s,\,red)}$	−21.68	−13.990
$HgCl_{2(s)}$	−55.0	
$Hg_2Cl_{2(s)}$	−63.32	−50.350
$HgS_{(s,\,red)}$	−13.90	−11.67
$HgSO_{4(s)}$	−168.3	
$Hg_2SO_{4(s)}$	−177.34	−149.12
Copper		
$Cu_{(g)}$	81.52	72.04
$Cu_{(s)}$	0	0
$CuO_{(s)}$	−37.1	−30.4
$Cu_2O_{(s)}$	−39.84	−34.98
$CuCl_{(s)}$	−32.2	−28.4
$CuCl_{2(s)}$	−49.2	
$CuS_{(s)}$	−11.6	−11.7
$CuSO_{4(s)}$	−184.00	−158.2
$CuSO_4 \cdot H_2O_{(s)}$	−259.00	−219.2
$CuSO_4 \cdot 3H_2O_{(s)}$	−402.27	−334.6

Substance	$\Delta \underline{H}_f^\circ$ kcal/mole	$\Delta \underline{G}_f^\circ$ kcal/mole
$CuSO_4 \cdot 5H_2O_{(s)}$	−544.45	−449.3
Silver		
$Ag_{(g)}$	69.12	59.84
$Ag_{(s)}$	0	0
$Ag_2O_{(s)}$	−7.306	−2.586
$AgCl_{(s)}$	−30.362	−26.224
$Ag_2S_{(s)}$	−7.60	−9.62
$AgNO_{3(s)}$	−29.43	−7.69
Nickel		
$Ni_{(g)}$	101.61	90.77
$Ni_{(s)}$	0	0
$NiO_{(s)}$	−58.4	−51.7
$Ni(OH)_{2(s)}$	−128.6	−108.3
Cobalt		
$Co_{(g)}$	105.	94.
$Co_{(s)}$	0	0
$CoO_{(s)}$	−57.2	−51.0
Iron		
$Fe_{(g)}$	96.68	85.76
$Fe_{(s)}$	0	0
$Fe_2O_{3(s)}$	−196.5	−177.1
$Fe_3O_{4(s)}$	−267.0	−242.4
$Fe(OH)_{2(s)}$	−135.8	−115.57
$Fe(OH)_{3(s)}$	−197.0	
$Fe_3C_{(s)}$	5.0	3.5
Manganese		
$Mn_{(g)}$	68.34	58.23
$Mn_{(s)}$	0	0
$MnO_{(s)}$	−92.0	−86.8
$MnO_{2(s)}$	−124.5	−111.4
Chromium		
$Cr_{(g)}$	80.5	69.8
$Cr_{(s)}$	0	0
$CrO_{3(s)}$	−138.4	
$Cr_2O_{3(s)}$	−269.7	−250.2
Tungsten		
$W_{(g)}$	201.6	191.6
$W_{(s)}$	0	0
$WO_{3(s)}$	−200.84	−182.47
Titanium		
$Ti_{(g)}$	112.	101.
$Ti_{(s)}$	0	0
$TiO_{2(s, rutile)}$	−218.0	−203.8
Boron		
$B_{(g)}$	97.2	86.7
$B_{(s)}$	0	0
$B_2O_{3(s)}$	−302.0	−283.0
$H_3BO_{3(s)}$	−260.2	−230.2
Aluminum		
$Al_{(g)}$	75.0	65.3
$Al_{(s)}$	0	0
$Al_2O_{3(s)}$	−399.09	−376.77
Magnesium		
$Mg_{(g)}$	35.9	27.6
$Mg_{(s)}$	0	0
$MgO_{(s)}$	−143.84	−136.13
$Mg(OH)_{2(s)}$	−221.00	−199.27
$MgCl_{2(s)}$	−153.40	−141.57
$MgSO_{4(s)}$	−305.5	−280.5
$MgCO_{3(s)}$	−266.	−246.
Calcium		
$Ca_{(g)}$	46.04	37.98
$Ca_{(s)}$	0	0
$CaO_{(s)}$	−151.9	−144.4
$Ca(OH)_{2(s)}$	−235.80	−214.33
$CaF_{2(s)}$	−290.3	−277.7
$CaCl_{2(s)}$	−190.0	−179.3
$CaSO_{4(s, anhydrite)}$	−342.42	−315.56
$CaSO_4 \cdot 2H_2O_{(s)}$	−483.06	−429.19
$CaC_{2(s)}$	−15.0	−16.2
$CaCO_{3(s, calcite)}$	−288.45	−269.78
$CaCO_{3(s, aragonite)}$	−288.49	−269.53
Barium		
$Ba_{(g)}$	41.96	34.60
$Ba_{(s)}$	0	0
$BaO_{(s)}$	−133.4	−126.3
$Ba(OH)_{2(s)}$	−226.2	
$BaCl_{2(s)}$	−205.56	−193.8
$BaSO_{4(s)}$	−350.2	−323.4
$BaCO_{3(s)}$	−291.3	−272.2
Lithium		
$Li_{(g)}$	37.07	29.19
$Li_{(s)}$	0	0
$Li_2O_{(s)}$	−142.4	
$LiOH_{(s)}$	−116.45	−106.1
Sodium		
$Na_{(g)}$	25.98	18.67
$Na_{(s)}$	0	0
$Na_2O_{(s)}$	−99.4	−90.0
$NaOH_{(s)}$	−101.99	
$Na_2O_{2(s)}$	−120.6	
$NaF_{(s)}$	−136.0	−129.3
$NaCl_{(s)}$	−98.232	−91.785
$NaBr_{(s)}$	−86.030	
$NaI_{(s)}$	−68.84	
$Na_2SO_{3(s)}$	−260.6	−239.5
$Na_2SO_{4(s)}$	−330.90	−302.78
$NaNO_{3(s)}$	−111.54	−87.45

Substance	$\Delta \underline{H}_f^\circ$ kcal/mole	$\Delta \underline{G}_f^\circ$ kcal/mole	Substance	$\Delta \underline{H}_f^\circ$ kcal/mole	$\Delta \underline{G}_f^\circ$ kcal/mole
$Na_2CO_{3(s)}$	−270.3	−250.4	$KClO_{3(s)}$	−93.50	−69.29
$NaHCO_{3(s)}$	−226.5	−203.6	$KClO_{4(s)}$	−103.6	−72.7
Potassium			$KBr_{(s)}$	−93.73	−90.63
$K_{(g)}$	21.51	14.62	$KI_{(s)}$	−78.31	−77.03
$K_{(s)}$	0	0	$K_2SO_{4(s, II)}$	−342.66	−314.62
$K_2O_{(s)}$	−86.4		$KNO_{3(s)}$	−117.76	−93.96
$KOH_{(s)}$	−101.78		$KMnO_{4(s)}$	−194.4	−170.6
$KF_{(s)}$	−134.46	−127.42	$K_2CO_{3(s)}$	−273.93	
$KCl_{(s)}$	−104.175	−97.592			

TABLE A-2 Standard Heats of Formation at 25°C in Calories per Gram Mole†

Substance	Formula	State	ΔH^0_{f298}
Normal paraffins:			
n-Butane	C_4H_{10}	g	−30,150
n-Pentane	C_5H_{12}	g	−35,000
n-Hexane	C_6H_{14}	g	−39,960
Increment per C atom above C_6	....	g	−4,925
Normal monoolefins (1-alkenes):			
Propylene	C_3H_6	g	4,879
1-Butene	C_4H_8	g	−30
1-Pentene	C_5H_{10}	g	−5,000
1-Hexene	C_6H_{12}	g	−9,960
Increment per C atom above C_6	....	g	−4,925
Miscellaneous organic compounds:			
Benzene	C_6H_6	g	19,820
Benzene	C_6H_6	l	11,720
1,3-Butadiene	C_4H_6	g	26,330
Cyclohexane	C_6H_{12}	g	−29,430
Cyclohexane	C_6H_{12}	l	−37,340
Ethylbenzene	C_8H_{10}	g	7,120
Ethylene glycol	$C_2H_6O_2$	l	−108,580
Methylcyclohexane	C_7H_{14}	g	−36,990
Metylcyclohexane	C_7H_{14}	l	−45,450
Styrene	C_8H_8	g	35,220
Toluene	C_7H_8	g	11,950
Toluene	C_7H_8	l	2,870

† Selected from F. D. Rossini et al., "Selected Values of Physical and Thermodynamic Properties of Hydrocarbons and Related Compounds," A.P.I. Proj. 44, Carnegie Institute of Technology, Pittsburgh, 1953.

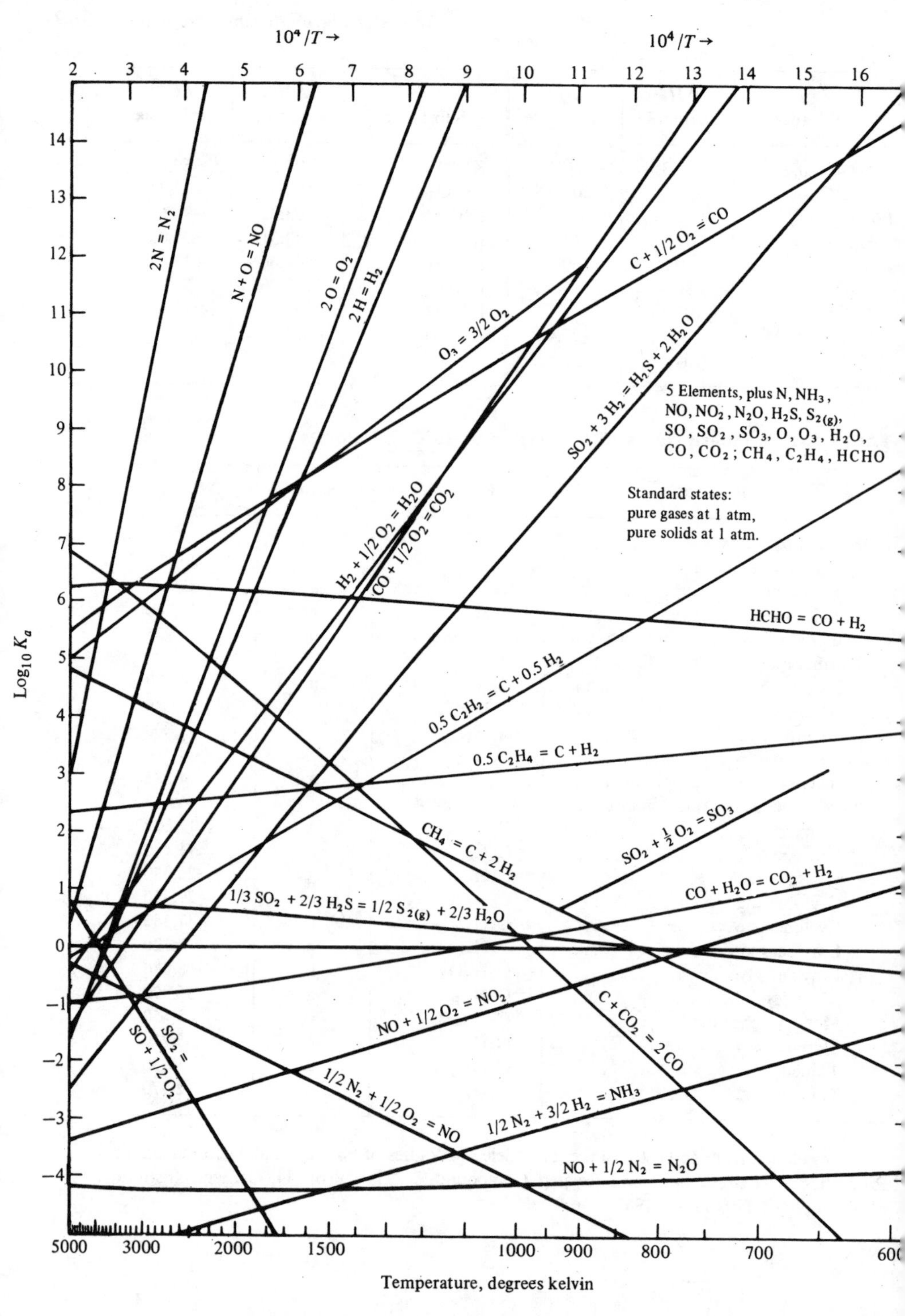

FIG. A-13. Combustion–related equilibrium constants. (Courtesy of Professor H. C. Hottel of the Massachusetts Institute of Technology.)

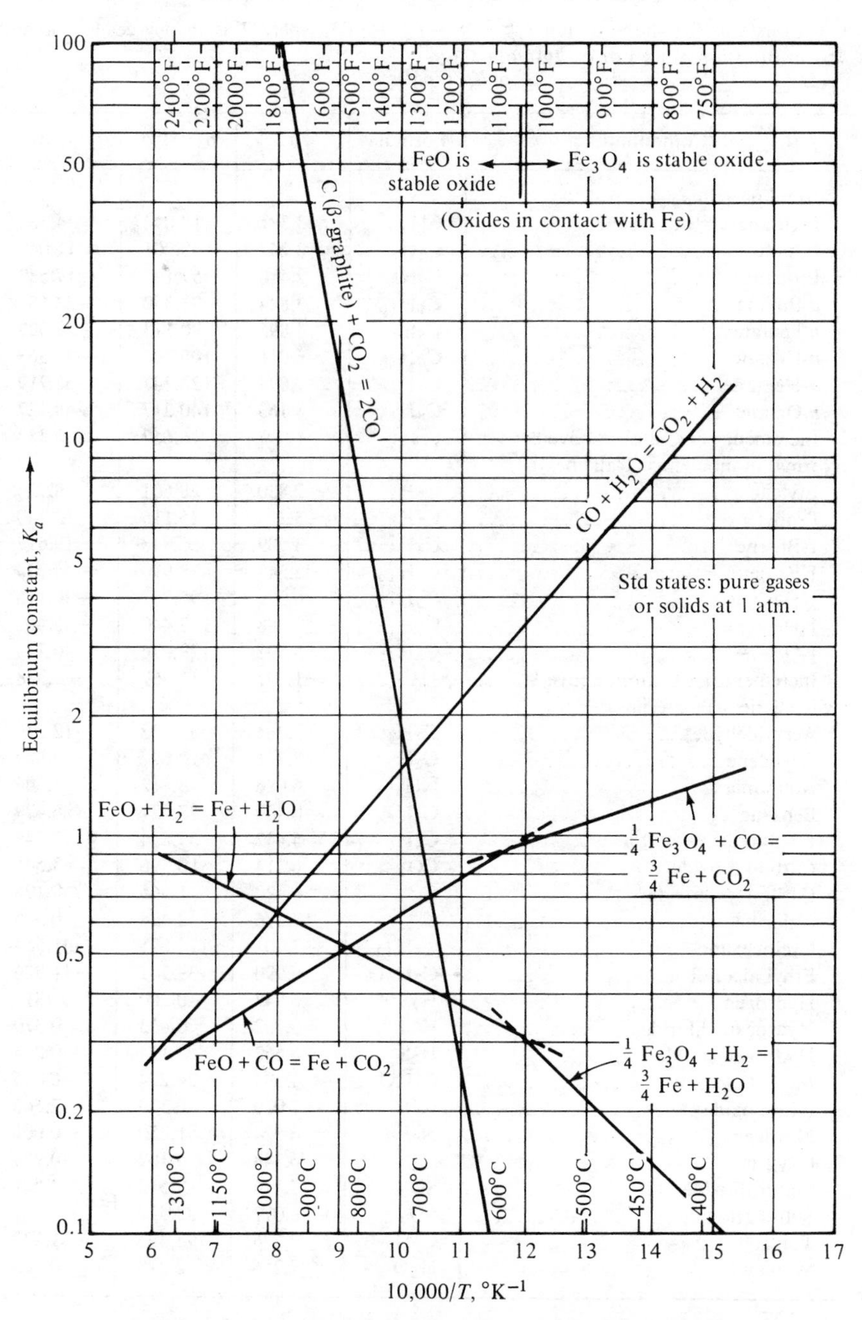

FIG. A-14. Equilibrium constants of the reduction of iron oxides and carbon dioxide. (From W. K. Lewis, A. H. Radasch, and H. C. Lewis, *Industrial Stoichiometry*, 2d ed., McGraw-Hill, Inc., New York, 1954.)

TABLE A-3 Molal Heat Capacities of Gases in the Ideal Gaseous State†

Constants for the equation $C_P^* = \alpha + \beta T + \gamma T^2$, where T is in degrees Kelvin and C_P^* is in Btu/(lb-mol)(°F) or cal/(g-mol)(°C).

298 to 1500°K

Compound	Formula	α	$\beta \times 10^3$	$\gamma \times 10^6$
Normal paraffins:				
Methane	CH_4	3.381	18.044	−4.300
Ethane	C_2H_6	2.247	38.201	−11.049
Propane	C_3H_8	2.410	57.195	−17.533
n-Butane	C_4H_{10}	3.844	73.350	−22.655
n-Pentane	C_5H_{12}	4.895	90.113	−28.039
n-Hexane	C_6H_{14}	6.011	106.746	−33.363
n-Heptane	C_7H_{16}	7.094	123.447	−38.719
n-Octane	C_8H_{18}	8.163	140.217	−44.127
Increment per C atom above 8	......	1.097	16.667	−5.338
Normal monoolefins (1-alkenes):				
Etylene	C_2H_4	2.830	28.601	−8.726
Propylene	C_3H_6	3.253	45.116	−13.740
1-Butene	C_4H_8	3.909	62.848	−19.617
1-Pentene	C_5H_{10}	5.347	78.990	−24.733
1-Hexene	C_6H_{12}	6.399	95.752	−30.116
1-Heptene	C_7H_{14}	7.488	112.440	−35.462
1-Octene	C_8H_{16}	8.592	129.076	−40.775
Increment per C atom above 8	......	1.097	16.667	−5.338
Miscellaneous materials:				
Acetaldehyde‡	C_2H_4O	3.364	35.722	−12.236
Acetylene	C_2H_2	7.331	12.622	−3.889
Ammonia	NH_3	6.086	8.812	−1.506
Benzene	C_6H_6	−0.409	77.621	−26.429
1,3-Butadiene	C_4H_6	5.432	53.224	−17.649
Carbon dioxide	CO_2	6.214	10.396	−3.545
Carbon monoxide	CO	6.420	1.665	−0.196
Chlorine	Cl_2	7.576	2.424	−0.965
Cyclohexane	C_6H_{12}	−7.701	125.675	−41.584
Ethyl alcohol	C_2H_6O	6.990	39.741	−11.926
Hydrogen	H_2	6.947	−0.200	0.481
Hydrogen chloride	HCl	6.732	0.433	0.370
Hydrogen sulfide	H_2S	6.662	5.134	−0.854
Methyl alcohol	CH_4O	4.394	24.274	−6.855
Nitric oxide	NO	7.020	−0.370	2.546
Nitrogen	N_2	6.524	1.250	−0.001
Oxygen	O_2	6.148	3.102	−0.923
Sulfur dioxide	SO_2	7.116	9.512	3.511
Sulfur trioxide	SO_3	6.077	23.537	−0.687
Toluene	C_7H_8	0.576	93.493	−31.227
Water	H_2O	7.256	2.298	0.283

† From J. M. Smith and H. C. Van Ness, *Introduction to Chemical Engineering Thermodynamics*, 2d ed., McGraw-Hill, Inc., New York, 1959.

‡ 298 to 1000°K.

TABLE A-4 Molal Heat Capacities of Solids†

Constants for the equation $C_P = a + bT + c/T^2$, where T is in degrees Kelvin and C_P is Btu/(lb-mol)(°F) or cal/(g-mol)(°C).

Solid	a	$b \times 10^3$	$c \times 10^{-5}$	Range, °K
CaO	11.67	1.08	−1.56	298–1800
$CaCO_3$	24.98	5.24	−6.20	298–1200
C (graphite)	4.10	1.02	−2.10	298–2300
Cu	5.41	1.50	......	298–1357
CuO	9.27	4.80	......	298–1250
Fe (α)	3.37	7.10	0.43	298–1033
FeO	12.38	1.62	−0.38	298–1200
Fe_2O_3	23.36	17.24	−3.08	298–1100
Fe_3O_4	39.92	18.86	−10.01	298–1100
FeS	15.20	....	......	298–412
NH_4Cl	11.80	32.00	......	298–458
Na	5.00	5.36	......	298–371
NaCl	10.98	3.90	......	298–1073
NaOH	19.2	....	......	298–595
S (rhombic)	3.58	6.24	......	298–469

† From J. M. Smith and H. C. Van Ness, *Introduction to Chemical Engineering Thermodynamics*, 2d ed., McGraw-Hill, Inc., New York, 1959.

Vapor Pressures of Selected Compounds

For vapor pressures between 10 and 1500 mm Hg Rossini[1] suggests the following form

$$\log_{10} P' = A - \frac{B}{T + C}$$

where P' = vapor pressure in mm Hg

T = temperature in °C

Values of the constants are given below:

Compound	A	B	C	Temp. range of applicability
Methane	6.61184	389.93	266.0	−183 to −152.5
Ethane	6.80266	656.40	256.0	−142 to − 75
Ethylene	6.74756	585.00	255.0	−153 to − 91
Propylene	6.81960	785.0	247.0	−112 to − 32
Propane	6.82973	813.20	248.0	−108 to − 25
1-Butene	6.84290	926.1	240.0	− 81 to + 12.5
n-Butane	6.83029	945.9	240.0	− 77 to + 19
n-Pentane	6.85221	1064.64	232.0	− 50 to 57
iso-Pentane	6.80380	1027.25	234.0	− 57 to 49
Benzene	6.89745	1206.35	220.237	+ 5.5 to 104
n-Hexane	6.87773	1171.53	224.366	− 25 to 92
Cyclohexane	6.84498	1203.526	222.863	6.6 to 105
Tolume	6.95334	1343.943	219.377	6 to 136
n-Heptane	6.90319	1268.586	216.954	− 2 to 123
n-Octane	6.92374	1355.126	209.517	19 to 152

Although these values may be used to predict vapor pressures greater than 1500 mm Hg, such useage is likely to lead to errors—especially for vapor pressures approaching the critical pressure. For such higher pressures where accurate predictions are needed, the following values of Thodos[2] should be used

$$\log_{10} P' = A + \frac{B}{T} + \frac{C}{T^2} + D\left(\frac{T}{T_d} - 1\right)^n$$

where P' is in mm Hg, T is in °K, and the last term is neglected for $T < T_d$, the various constants are given below:

Compound	T_d(°K)	A	B	C	D	n
Methane	118.83	6.18025	− 296.1	− 8,000	0.257	1.32
Ethane	204.74	6.73244	− 624.24	− 15,912	0.1842	1.963
Propane	261.20	6.80064	− 785.6	− 27,800	0.2102	2.236
n-Butane	312.30	6.78880	− 902.4	− 44,493	0.4008	2.40

[1] F. D. Rossini et al., A.P.I. Research Project 44, Nat. Bur. Stand. Circular C461 (1947).

[2] G. Thodos, *Ind. Eng. Chem. 42*, 8, p. 1514 (1950).

Compound	T_d(°K)	A	B	C	D	n
n-Pentane	357.79	6.77767	− 988.6	− 66,936	0.6550	2.46
n-Hexane	398.79	6.75933	−1054.9	− 92,720	0.9692	2.49
n-Heptane	436.34	6.74242	−1108.0	−121,489	1.3414	2.50
n-Octane	471.00	6.72908	−1151.6	−152,835	1.7706	2.50
n-Nonane	503.14	6.72015	−1188.2	−186,342	2.2438	2.50
n-Decane	533.13	6.71506	−1219.3	−221,726	2.7656	2.50
n-Dodecane	587.61	6.71471	−1269.7	−296,980	3.9302	2.50

Notation Index

Listed here are most of the symbols used in this text together with a brief descriptive name. Where applicable the dimensions are also given in terms of mass (M), length (L), time (t), temperature (T), and force (F). Because of the common use of thermal energy units such as Btu and cal, the symbolism TEU is substituted for the dimension $F \cdot L$ in many places. In general the dimension M can be construed as either mass or moles except where one or the other is specifically stated. Those symbols of particular importance or having unique use in this text are further identified with the number of the equation in which they are first introduced.

Latin Symbols

a	Acceleration (L/T^2)
a_i	Activity of species i in a mixture [(9-100), dimensionless)]
a_{ji}	Number of atoms of element j in chemical species i
A	Area (L^2)
A	Chemical affinity [(14-20), TEU/M)]
A	Helmholtz free energy (TEU)
$\underline{A}$	Specific Helmholtz free energy (TEU/M)
$\bar{A}_i$	Partial molar Helmholtz free energy (TEU/M)
A_i	ith chemical species in a chemical reaction
b_j	Total number of gram atoms of element j
C	Number of chemical species (components) in a mixture
C	Molar concentration (Moles/L^3)

C	Sonic Velocity (L/T)
C_P	Heat capacity at constant pressure (TEU/$M \cdot T$)
C_V	Heat capacity at constant volume (TEU/$M \cdot T$)
$\Delta \mathscr{C}_P^\circ$	Net heat capacity of stoichiometric reaction mixture [(11-78), TEU/$M \cdot T$)]
$(\mathscr{C}_P^\circ)_{exit}$	Total heat capacity of product stream [(11-88), TEU/$M \cdot T$]
COP	Coefficient of performance for refrigeration (7-37)
D	Diameter (L)
D	Diffusivity (L^2/t)
E	Total energy (TEU)
E	Electrical potential difference [(13-8), volts]
f	Degrees of freedom
f	Fanning friction factor [(8-18), dimensionless]
$\underline{f}$	Fugacity [(6-94), F/L^2]
$\bar{f}_i$	Partial fugacity [(9-48), F/L^2]
F	Force
F	Faraday constant [(13-20), coulombs/equivalent]
g	Acceleration of gravity (L/t^2)
g_c	Universal conversion factor (Table 2-1, $M \cdot L/F \cdot t^2$)
G	Mass flow rate per unit area [(S8-3), M/$L^2 \cdot t$]
G	Gibbs free energy (TEU)
$\underline{G}$	Specific Gibbs free energy (TEU/M)
$\bar{G}_i$	Partial molar Gibbs free energy (TEU/M)
$\Delta \underline{G}_f^\circ$	Standard state Gibbs free energy of formation (TEU/mole)
$\underline{\mathscr{G}}$	Gibbs free energy per mole of species l entering a reactor [(11-8), TEU/mole]
$\Delta \underline{\mathscr{G}}^\circ$	Standard state Gibbs free energy change of reaction [(11-16), TEU/mole]
h	Height (L)
H	Enthalpy (TEU)
$\underline{H}$	Specific enthalpy (TEU/M)
$\bar{H}_i$	Partial molar enthalpy (TEU/M)
$\Delta \underline{H}_f^\circ$	Standard-state enthalpy of formation (TEU/M)
$\underline{\mathscr{H}}^\circ$	Enthalpy per mole of species l entering a reactor [(11-84), TEU/M]
$\Delta \underline{\mathscr{H}}^\circ$	Standard state enthalpy change (heat) of reaction [(11-68), TEU/M]
$\dot{i}$	Electrical current [(14-33), charge/$L^2 \cdot t$]
J	Generalized flux (14-35)
J_a	Activity ratio [(11-21) and (11-31), dimensionless]
k	Boltzmann constant
k	Ratio of specific heats (6-83)
K_i	Equilibrium vaporization ratio (10-18)
K_a	Equilibrium constant (11-43); related equilibrium constants denoted $K_{\bar{f}}$, K_{f^0}, K_y, K_ν, K_P, and K_γ are identified at (11-53) and (11-58)

KE	Kinetic energy ($F \cdot L$ or TEU)
L	Length
LW	Total lost work ($F \cdot L$ or TEU)
$\underline{LW}$	Lost work per unit mass flowing (TEU/M)
lw	Local lost work per unit volume (TEU/L^3)
L_{ij}	Generalized coefficients relating generalized fluxes and potentials (14-35)
m	Molality (p. 587)
M	Molecular weight
M	Mass
$\dot{M}$	Mass flow rate (M/t)
M	Mach Number (8-87)
n	Number of moles
N	Number of competing chemical reactions (11-92)
$\mathscr{N}$	Number of moles of electrons liberated per mole of species l reacted (13-10)
p or P	Number of coexisting phases
P	Pressure (F/L^2)
P'	Vapor pressure (F/L^2)
P_i	Partial pressure (9-75, F/L^2)
$\mathscr{P}_i$	Pure component pressure [(9-121), F/L^2]
PE	Potential Energy ($F \cdot L$ or TEU)
q	Electrical charge transferred [(13-8), coulombs]
q_i	Molar volume of component in mixture [(9-151), L^3/mole]
q	Local heat transferred per unit area of boundary [(5-17), TEU/L^2]
Q	Thermal energy flux or heat (TEU)
$\underline{Q}$	Heat flow per unit of material flowing (TEU/M)
$\dot{Q}$	Rate of thermal energy flux (TEU/t)
$\mathscr{Q}$	Heat per mole of species l entering a reactor [(11-81), TEU/mole]
R	Ideal gas constant (Table 3-4)
S	Entropy (TEU/T)
$\underline{S}$	Specific entropy (TEU/$T \cdot M$)
$\overline{S}_i$	Partial molar entropy (TEU/$T \cdot M$)
T	Temperature
t	Time
u	Velocity (L/t)
U	Internal energy ($F \cdot L$ or TEU)
$\underline{U}$	Specific internal energy (TEU/M)
$\overline{U}_i$	Partial molar internal energy (TEU/M)
V	Volume (L^3)
$\underline{V}$	Specific volume (L^3/M)
$\overline{V}_i$	Partial molar volume (L^3/M)
W	Weight (F)

W Coefficient in regular solution theory (9-165)

W Mechanical energy flux or work transfer ($F \cdot L$ or TEU)

$\underline{W}$ Work transfer per unit of material flowing (TEU/M)

$\dot{W}$ Rate of mechanical energy flux (TEU/t)

$\mathscr{W}$ Work per mole of species l entering a reactor [(11-8), TEU/mole]

x_i Mole fraction of chemical specie i

X Measure of length or position (L)

X Extent of reaction [(11-26), moles]

$\underline{X}$ Fractional extent of reaction [(11-26), dimensionless]

y_i Mole fraction, particularly in gaseous phase

Y Generalized force or potential (14-12)

z_i Volumetric fraction of component in mixture (9-151)

Z Length or distance (L)

Z Generalized potential (14-35)

Z Compressibility factor (dimensionless)

$\mathscr{Z}_i$ Component compressibility factor (9-122)

Greek Symbols

α_i Stoichiometric coefficient [(11-2), dimensionless]

α_{ji} Stoichiometric coefficient for ith species in jth competing chemical reaction [(11-92), dimensionless]

γ_i Activity coefficient [(9-101), dimensionless]

ζ Euken coefficient [(6-58), $L^3 \cdot T/M$]

η Efficiency factor, actual quantity/ideal quantity

η Thermal efficiency, fraction of thermal energy supplied to a heat engine that is converted to work

$\underline{\eta}_i$ Portion of species i passing through a reactor unreacted (11-33)

θ Angle (radians)

θ Time (Chapter 14)

μ Joule–Thomson coefficient [(6-54), $L^2 \cdot T/F$]

μ Viscosity ($M/L \cdot t$ or $F \cdot t/L^2$)

μ_i Chemical potential [(9-29), TEU/M]

ν Fugacity coefficient [(6-96), dimensionless]

ν Kinematic viscosity [(S8-2), L^2/t]

ρ Density (M/L^3)

σ Volumetric rate of entropy generation [(14-29), TEU/$T \cdot L^3 \cdot t$]

Operators

δ Denotes infinitesimal change in path function

d Denotes infinitesimal change in a state property

Δ	Finite change in state property, final minus initial value
d	Total derivative operator
∂	Partial derivative operator
$\int$	Integral operator
$\oint$	Integral operator for closed path
ln	Natural logarithm operator (base e)
log	Common logarithm operator (base 10)
Π	Cumulative product operator
Σ	Cumulative summation operator

Subscripts

acc	Accumulation
c	Denotes value at critical state
gen	Generation
i, j, k, etc.	Component identification
in	Addition to system
H	High temperature reservior
L	Low temperature reservoir
m	With operator Δ denotes change upon mixing
m or mix	Mixture
out	Removal from system
prod	Product
r	Reduced property (3-19)
react	Reactant
res	Resistance of surroundings to change in system
rev	Reversible
soln	Solution or mixture
surr	Surroundings
syst	System
_(underbar)	Denotes specific quantity

Superscripts

E	With operator Δ denotes excess change in property upon mixing
L, V, G, I, II etc.	Phase identification
$^{-}$ (overbar)	Denotes partial molar quantity
$^{\cdot}$ (overdot)	Time rate of change
*	Denotes value of property in limit as $P \to 0$ ($\underline{V} \to \infty$)
$\circ$	Reference state

Index